Birkhäuser

Birkhäuser

Edward John Specht (1915–2011)
Harold Trainer Jones (1925–1995)
Keith G. Calkins
Donald H. Rhoads

Euclidean Geometry and its Subgeometries

Birkhäuser

Edward John Specht
Indiana University South Bend
South Bend, IN, USA

Harold Trainer Jones
Andrews University
Berrien Springs, MI, USA

Keith G. Calkins
Ferris State University
Big Rapids, MI, USA

Donald H. Rhoads
Andrews University
Berrien Springs, MI, USA

ISBN 978-3-319-23774-9 ISBN 978-3-319-23775-6 (eBook)
DOI 10.1007/978-3-319-23775-6

Library of Congress Control Number: 2015952314

Mathematics Subject Classification (2010): 51M05

Springer Cham Heidelberg New York Dordrecht London

Printed on acid-free paper

Springer International Publishing AG Switzerland is part of Springer Science+Business Media (www.springer.com)

Preface

"At last I said—Lincoln, you never can make a lawyer if you do not understand what demonstrate *means; and I left my situation in Springfield, went home to my fathers house, and stayed there till I could give any proposition in the six books of Euclid at sight. I then found out what* demonstrate *means, and went back to my law studies."*

–Abraham Lincoln, quoted by Henry Ketcham, in *The Life of Abraham Lincoln.*

For centuries, the study of Euclidean geometry has been considered an essential part of a literate person's education, both for the practical knowledge obtained and, more importantly, as an example of a deductive system in which non-obvious conclusions may be drawn from a collection of accepted statements. One of our esteemed colleagues has remarked to us that in view of the influence that Euclidean geometry has had on western civilization, "someone should do it right." This book is an attempt to do so.

Even though Hilbert's development of Euclidean geometry *Foundations of Geometry* (1899) [10] has been judged by some as only partially successful, it has set a standard for subsequent treatments, including this one. Our axioms are patterned largely after his, except that we base our treatment of congruence on reflections, rather than on congruence axioms.

Indeed, this book might well be regarded as a completion, updating, and expansion of the core of Hilbert's book. It began in the 1970s as a set of lecture notes for the teaching of an upper division undergraduate geometry course, and over the years has grown into a complete development of Euclidean geometry, emphasizing plane geometry. It is our intention to be completely rigorous, take no shortcuts, and "sweat the details."

We have used independent axioms, even though doing so means that getting to interesting theorems can be a daunting task. This is particularly evident in the last part of Chapter 4 and in the early part of Chapter 5, where extensive arguments are required to show that the behavior of lines, segments, rays, and their end points conforms to what we intuitively expect.

We present a total of 13 axioms in sequence, at each stage proving as many theorems as possible. The different geometric structures built up in this way are called *subgeometries*, and eventually we attain a full set of axioms for Euclidean geometry.

Chapter 1 introduces eight incidence axioms to form *incidence* geometry. From here, the development consists of two threads. The shorter of these threads (Chapters 2 and 3) is concerned with *incidence-parallel* or *affine* geometry. The second thread (*incidence-betweenness* geometry) consists of Chapters 4–10, with minor contributions from Chapter 3. Chapters 11–20 invoke the parallel Axiom PS and combine the two threads to produce *Euclidean* geometry.

Within the scheme just outlined, Chapter 5 (*Pasch* geometry) and Chapter 8 (*neutral* geometry) are pivotal. We want to develop as much theory as possible at each stage, so in Chapter 9 we create a rudimentary arithmetic on the set of *free segments*[1] using the machinery of neutral geometry alone. In this arithmetic, there is a natural definition for addition and for ordering, and for subtraction of a "smaller" free segment from a "larger" one. This arithmetic is eventually shown to coincide with the ordinary arithmetic of "positive" points on a line, which has been identified with the real numbers.

Indeed, a major goal is to identify the line with the real numbers, and to "coordinatize" the plane. To do this, we first define (in Chapter 14) the addition of points using *translations* and multiplication using *dilations*. (See the discussion of mappings on the next page.) This is completed in Chapters 17 and 18, where we construct isomorphisms between a subset of a line on the plane and the rational numbers, and also between a line and the real numbers, using the LUB axiom.

---

[1]A free segment is a congruence class of closed segments; two segments belong to the same free segment if they have the same "length."

Chapter 20 finishes the main development with two classical theorems, due to Menelaus and Ceva. In Chapter 21, we show that our axioms are consistent by showing that they are all true in ordinary coordinate space. It also shows (with two exceptions) that each of the axioms on our list is independent of the ones introduced earlier.

We give extensive study to bijections of a plane which preserve betweenness; we call these *belineations*. They are introduced in Chapter 7, and are a type of *collineation*. We list here the various types of belineation; their relationships are summarized in the last section of Chapter 19.

Isometry (a type of belineation) is defined in Chapter 8; the following types of belineations are isometries:

Reflection	Chs 8, 10, 12, 13;
Translation	Chs 3, 12, 14, 18;
Rotation	Chs 10, 12, 13, 18;
Glide reflection	Ch 12.

The following types of belineation are not isometries:

Dilation	Chs 3, 13, 14, 15, 17, 18;
Similarity (non-identity)	Ch 15.

The following type of belineation may or may not be an isometry:

Axial affinity	Chs 3, 12, 16, 19.

Two online collections of supplementary materials may be accessed from the home page of this book at www.springer.com. One of these contains solutions to starred exercises. The other includes an expanded treatment of coordinatization of the Euclidean plane; a development of complex numbers; an exploration of properties of polygons in the Pasch plane leading to a proof of the Jordan Curve Theorem; a development of arc length; a development of the circular functions (in a treatment originated by the first author, Specht); and a treatment of angle measure.

If used as a textbook, a one-semester course might consist of Chapters 1 through 5, emphasizing the detailed and rigorous proofs of Chapters 4 and 5 at the expense of omitting many standard results of geometry, which are contained in Chapter 8, neutral geometry. A different kind of course might summarize the last few theorems of Chapter 4 and the principal results of the early part of Chapter 5, omitting the detailed arguments, and move on to Chapter 8. This course would not be so rigorous as the first, but the student could see how to remedy the loss of rigor by revisiting the omitted or summarized sections.

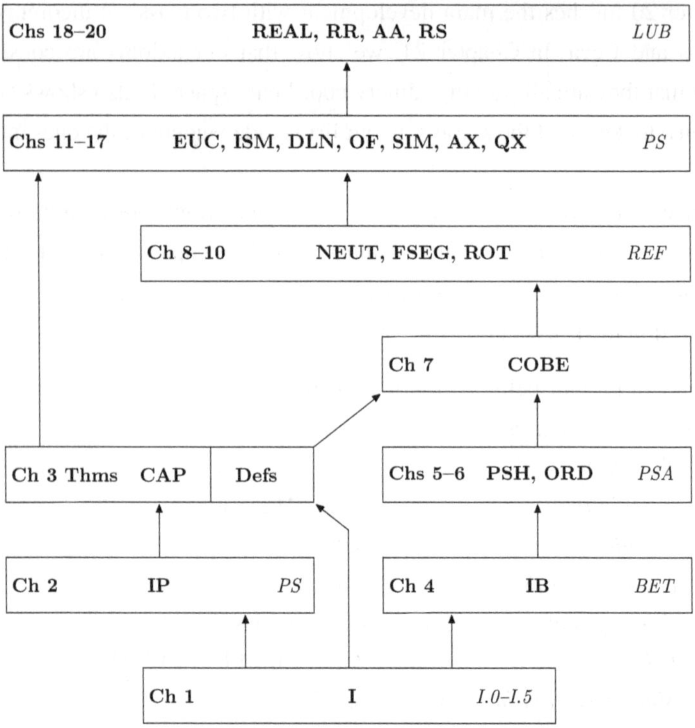

Fig. 1 Dependency chart for the main development

Notes on the dependency chart

(a) A rising arrow indicates that the upper depends on the lower.

(b) In each box, the first entry is a listing of chapters; the second lists acronyms used in these chapters; the third, in italics, names any axioms added in these chapters.

(c) The definitions in Chapter 3, as well as the first four theorems, might have been included in Chapter 1, and do not depend on Chapter 2. This division is shown by dividing the Chapter 3 box into a "Thms" section and "Defs" section; the former depends on Chapter 2, the latter on Chapter 1.

(d) Chapters 7–20 depend on the "Defs" part of Chapter 3; Chapters 11–20 depend on the "Thms" part; thus the placement of the arrows.

(e) Chapter 21 is not shown on the chart, as it is not part of the main development.

Edward J. Specht, the chief author of this work, was known to his friends for his devotion to mathematics and science. He received a B.S. degree from Walla Walla College, an M.S. from the University of Colorado, and, in 1949, a Ph.D. from the University of Minnesota. He chaired the Department of Mathematics at what is now Andrews University from 1947 to 1972, and was Professor of Mathematics at Indiana University South Bend from 1972 until his retirement in 1986. In 1984, Andrews University conferred upon him the honorary degree of Doctor of Science (D.Sc.). He began this project while at Indiana University and remained the force behind it until his death in 2011 at the age of 96.

Harold T. Jones, Ed's colleague for many years at Andrews University, assisted in this project during its initial years. Harold received an A.B. from what is now Washington Adventist University, an A.M. from Lehigh University, and in 1958 a Ph.D. from Brown University. He taught at Andrews University from 1952 until his retirement in 1991. His participation in this geometry project was, sadly, cut short by his death in 1995. Harold was doubtless responsible for many of the clarifying and sometimes lighthearted explanatory remarks in the early parts of the work.

Keith G. Calkins received his B.S., two M.S. degrees, and an M.A.T. degree from Andrews University; he received an M.S., as well as a Ph.D. in Physics (in 2005) from the University of Notre Dame. Keith was on the staff and faculty at Andrews for 32 years in several teaching and management capacities. Since 2011, he has taught a wide variety of courses in the Physical Sciences and Mathematics departments at Ferris State University. Keith keyboarded this entire work into LaTeX from Ed's hand-written manuscript, making corrections, establishing notational and editorial conventions, and frequently consulting with Ed. As a matter of interest, he took a geometry course at Andrews University in the 1980s in which an early draft of this book was used as a text.

Donald H. Rhoads received his B.A. from Andrews University, his M.A. from Rice University, and in 1968 a Ph.D. from the University of Michigan. Don taught mathematics at Andrews University from 1962 to 1964, from 1967 to 1972, and again from 1998 to 2006, during which time he served for six years as chair of the Department of Mathematics. He was drafted by Ed Specht to complete a chapter of this book, and subsequently has read the entire work, making corrections to proofs and references, and extensively revising and reorganizing several chapters.

As can be seen from these biographical sketches, all of the authors have been colleagues at one time or another at Andrews University—at least one of us (in many years two or three of us) worked there every year from 1947 to 2011, a span of 64 years.

We pay heartfelt tribute to both Edward Specht and Harold Jones; they were models of kindliness and dedication to their students, and were our beloved teachers and mentors.

At the end of a project so extensive as this one, thanks are due to many who have been of assistance and encouragement. We thank Indiana University South Bend for providing a grant-in-aid in support of research by the two initial authors into the Jordan Curve Theorem; a version of this material is part of the online supplementary material, which may be accessed from the home page of this book at www.springer.com.

We thank Rajiv Monsurate of Springer for his assistance with the technical aspects of LaTeX, including his welcome revision and polishing of our symbols for segments $\overline{AB}$, rays $\overrightarrow{AB}$, and lines $\overleftrightarrow{AB}$. We are greatly indebted to the several Birkhäuser reviewers who found errors and made many constructive suggestions.

We are especially indebted to Joel Weiner, Professor Emeritus of Mathematics, the University of Hawaii at Manoa; he read the first eleven chapters of the book with exceptional thoroughness, uncovering several major errors and generously supplying corrections. We are grateful to Professor Arlen Brown, who, as we have worked to finish this project, has been a constant source of encouragement.

We thank our wives, families, and friends for their patience with us at those times when we were "absent" while present in body. We are thankful for nurturing parents and teachers, and for the Divine Providence which over the years has enabled us in many ways to carry this project to completion.

The authors will be grateful to all who submit corrections or suggest improvements to the work.

Berrien Springs, MI, USA Keith G. Calkins
Bloomington, IN, USA Donald H. Rhoads

Contents

List of Figures

Chapter 1
Preliminaries and Incidence Geometry (I)

Acronym: *I*
Dependencies: *none*
New Axioms: *incidence Axioms I.0–I.5*
New Terms Defined (Section 1.6 and following): *universe, space, collinear, noncollinear, coplanar, noncoplanar, concurrent; the terms point, line, and plane are introduced, but not defined*

Abstract: This chapter contains a brief summary of several types of mathematical knowledge needed to read this book, including the elements of logic, set theory, mapping theory, and algebraic structures such as number systems and vector spaces. Definitions of basis, dimension, linear mappings, isomorphism, matrices and determinants are given; there is also discussion of the roles of axioms, theorems, and definitions in a mathematical theory. The main development of the book begins here with the statement of eight incidence axioms and proof of a few theorems including one from Desargues.

1.1 Introduction

Geometry began as a very practical subject. It was used, for example, to settle disputes regarding the sizes and shapes of parcels of land and to deal with other questions involving spatial relationships among concrete objects such as architectural elements and mechanical devices.

E.J. Specht et al., *Euclidean Geometry and its Subgeometries*,
DOI 10.1007/978-3-319-23775-6_1

As the Greek mathematicians carried out the task of organizing geometrical knowledge, they observed a structure in the interrelationships among the *statements* in geometry. To study this structure in detail, the geometers pursued a course of abstraction. For example, rectangular plots of land, rectangular pieces of wood, and rectangular pieces of cloth were replaced by the abstract concept of a rectangle, which was defined so as to embody those common characteristics of the concrete objects which were essential to a geometrical discussion about them. Hence, geometry became a collection of statements about relationships among such abstract concepts as points, lines, planes, triangles, rectangles, circles, cylinders, spheres, and polyhedra.

In a further step toward emphasizing structure, Euclid of Alexandria (active c. 300–265 BC), in his *Elements*, [6][1] identified certain statements which seemed to him to be "self-evident truths." He then undertook to show how each statement in geometry is a logical consequence of these "truths" together with other statements already shown to follow from them. Euclid's *Elements* set the tone for all subsequent development of geometry up to the present day, and is generally acknowledged to be one of the great triumphs of the human intellect.

Today, Euclid's "self-evident truths" are usually called *axioms*, and our attitude toward them has changed. Instead of regarding them as "self-evident truths," we treat them as mere starting points. We must start somewhere, and we agree that when we speak of "Euclidean geometry," we are starting with a set of statements similar to these. This book is a step in the task of determining how the various theorems in Euclidean geometry depend on these statements and on each other. We do not try to decide whether or not the axioms are "true," except in Chapter 21 where the whole point is to determine whether various axioms are true or false on models.

We employ what is sometimes referred to as a *synthetic* method of development. Our procedure is to begin with a few axioms which are in some sense more fundamental than the rest, proving as many theorems as we can from these. Then we add new axioms in sequence and at each stage see how many theorems we can prove. We continue this process until we come to the full set of axioms. At each stage, we call the structure we have developed up to that point a *geometry* or *subgeometry*. Thus we speak of "incidence geometry," "affine geometry," "Pasch geometry," and "neutral geometry," among others. The final result is, of course, Euclidean geometry. As Euclid said, there is no "royal road" to take us there—the difficult journey is the reward.

[1] Square bracketed numbers refer to entries in *References*, just before the *Index*.

The axioms we use for Euclidean geometry are not exactly those used by Euclid. As our understanding and standards of rigor have evolved over the centuries, gaps have been found in Euclid's structure. These have been remedied, and some of his concepts have been sharpened to conform to modern usage. Several such modernizations of Euclid's axioms have become widely known. Probably the three most significant are those set forth by David Hilbert (1862–1943) in his *Foundations of Geometry* (1899) [10]; George David Birkhoff (1824–1944) in 1932 (included in his *Basic Geometry* (1940))[3]; and Alfred Tarski (1901–1983) in 1959 [20].

The various sets of axioms have their associated advantages and disadvantages; Birkhoff's axioms are based on metric notions and angle measure; Tarski's axioms are encoded in the predicate calculus. We should also mention the modification of Birkhoff's axioms developed in the early 1960s by the School Mathematics Study Group (SMSG), one of the most successful aspects of the curricular reform known as the "*New Math*." The SMSG axioms are typically used in high school geometry textbooks, and are redundant, *i.e.* not independent, to facilitate rapidly proving significant results; they thus avoid the careful but sometimes tortuous development seen here.

Our incidence, betweenness, and plane separation axioms are close to those of Hilbert, but are stated somewhat differently. We access congruence using reflection mappings, yielding what we think is a more elegant and satisfying development than do Hilbert's axioms of congruence. As we stated in the *Preface*, this book could be regarded as a completion, updating, and expansion of the core of Hilbert's book.

The appeal of modern geometry, as we see it, lies in the fact that it reflects both aspects of its historical development. On the one hand, a person studying geometry gets ideas from drawing pictures, something not always possible when studying other mathematical subjects. On the other hand, the intuitive ideas gained from the pictures must then be subjected to the discipline of logic and proof. This interplay between imagination and intellectual discipline is not only a model for the way much of mathematical research proceeds, but also has long been a source of pleasure and fascination for mathematical intellects, from beginners to mature mathematicians.

Before we begin, we summarize a collection of facts about logic, sets, mappings and functions, algebraic structures, and the basic building blocks of axiomatic theory. We provide this material for reference; much of it may be familiar to the reader. At the end of this material we provide some discussion of the structure of this particular book, including the role of figures and exercises.

1.2 Elementary logic

Statements, propositions: In this discussion, a **statement** is a declarative statement (in the usual grammatical sense) which is either true or false—that is, it has **truth value**. In many treatments of logic, these are called **propositions**.

Logical operations: not, and, or: If p is a statement, the symbol $\neg p$ denotes the **negation** of p, the exact contrary of p, which is false whenever p is true and true whenever p is false.

If p and q are statements, the statement p **and** q (the **conjunction** of p and q) is true when both p and q are true, and false otherwise.

The statement p **or** q (the **disjunction** of p and q) is true when p is true, when q is true, and when both p and q are true; it is false when both p and q are false. This is the standard **inclusive** use of the word *or*.

To indicate the **exclusive or** (true in case one of p or q is true, but false when both or neither are true), we will often say **either** p **or** q. We never use the term *xor*. We may at times indicate the *exclusive or* by appending the words **but not both** to *either... or* or to *or*.

Logical operations: conditional, biconditional: The statement **if** p **then** q (the **conditional**) is true when both p and q are true, and is false when p is true and q is false. The conditional is always (vacuously) true when p is false. The statement p is called the **antecedent** or **hypothesis** of the conditional *if* p *then* q, and q is its **consequent** or **conclusion**. The **converse** of the conditional *if* p *then* q is the statement *if* q *then* p, its **inverse** is *if* $\neg p$ *then* $\neg q$, and its **contrapositive** is *if* $\neg q$ *then* $\neg p$. A conditional statement has the same truth value as its contrapositive, but there is no relation between the truth value of a conditional and either its converse or inverse. The converse and inverse of a conditional have the same truth value, since each is the contrapositive of the other.

The statement p **if and only if** q (the **biconditional**) means *(if p then q) and (if q then p)* and is often written p **iff** q.

Complex statements, logical equivalences: Many statements are constructed from other statements using the connectives *not, and, or*, and *if... then*; such statements are called **complex** statements. If p and q are complex statements constructed from a common set of simpler statements, p and q are said to be **logically equivalent** if they have the same truth value (either true or false) regardless of the contents of the simple statements that comprise them. In proofs, logically equivalent statements may be substituted freely for one another as needed to complete the argument.

Many of the most important logical equivalences involve negation; some of these are as follows:

$\neg(\neg p)$ is logically equivalent to p.

$\neg(p$ *and* $q)$ is logically equivalent to $(\neg p$ *or* $\neg q)$;

$\neg(p$ *or* $q)$ is logically equivalent to $(\neg p$ *and* $\neg q)$;

$(p$ *exclusive or* $q)$ is logically equivalent to $((p$ *or* $q)$ *and* $\neg(p$ *and* $q))$, that is, to $((p$ *or* $q)$ *and* $(\neg p$ *or* $\neg q))$;

$\neg(p$ *exclusive or* $q)$ is logically equivalent to $((\neg p$ *and* $\neg q)$ *or* $(p$ *and* $q))$; (this will be needed in Chapter 5);

$\neg($*if* p *then* $q)$ is logically equivalent to $(p$ *and* $\neg q)$;

Since the contrapositive of a conditional is equivalent to it, $\neg($*if* p *then* $q)$ is logically equivalent to $\neg$ (*if* $\neg q$ *then* $\neg p)$ which in turn is equivalent to $(\neg q$ *and* $\neg\neg p)$, or $(\neg q$ *and* $p)$.

Predicates, quantifiers: A **predicate** $p(x)$ is a statement that contains a **variable**, which can be thought of as a symbol for which various objects may be substituted. In the predicate $p(x)$ the symbol x denotes the variable, and different values of the variable yield different statements. The reader will need some acquaintance with what is called the "predicate calculus" involving the **quantifiers**, that is, **for all** and **for some** (including the customary variant of the latter involving the term **there exists**).

The negation of a quantified statement is obtained by interchanging the quantifiers "for all" and "for some" and negating the statement. Thus, the negation of *for all* $x, p(x)$ is the statement *for some* x, $\neg p(x)$.

Proofs: The reader will need to be familiar with the basic schemes for construction of proofs using rules of inference based on the above, such as **direct proof** and **indirect proof** (**proof by contraposition** or **proof by contradiction**).

Occasional use will be made of **proof by mathematical induction** in the following form: a predicate $p(n)$ defined on the natural numbers is true for all natural numbers, if it is proved that: 1) $p(1)$ is true, and 2) for every natural number m, if $p(m)$ is true then $p(m + 1)$ is true. The method of proof is to 1) show that the statement $p(1)$ is true (the **base case**); then 2) assume that $p(m)$ (the **induction hypothesis**) is true for an arbitrary m and infer from this the truth of $p(m + 1)$.

An equivalent form (sometimes called the "strong form") of mathematical induction is as follows: a predicate $p(n)$ on the natural numbers is true for all natural numbers if it is proved that: 1) $p(1)$ is true, and 2) for every natural number m, if $p(k)$ is true for all $k < m$, then $p(m)$ is true.

A common variant on the method of mathematical induction is to prove that a statement $p(m)$ is true for all $m \geq n_0$, where $n_0 > 1$, by simply beginning the process at $m = n_0$ rather than at $m = 1$. This is easily shown to be valid by rewriting the predicate as $p'(m) = p(m + n_0 - 1)$ so that $p'(1) = p(n_0)$.

Mathematical induction may also be used for definitions: a predicate $p(n)$ is defined on all the natural numbers if 1) $p(1)$ is defined, and 2) for every natural number m if $p(m)$ is defined then $p(m + 1)$ is defined.

Notation: The statement that "*if p then q* is true" will sometimes be symbolized by a double-lined arrow, as in $p \Rightarrow q$, and the statement that "*p if and only if q* is true" by "$p \Leftrightarrow q$." In definitions, "if" will have the definitional meaning of "if and only if" unless otherwise indicated, and the symbol $\square$ will be used to designate the end of a proof.

For more complete information about logic and its use in the construction of proofs, the reader may wish to consult a text on discrete mathematics such as Rosen, K., *Discrete Mathematics and Its Applications* (2003) [18].

1.3 Set theory

The Zermelo–Fraenkel set theory with the Axiom of Choice (ZFC) provides an adequate foundation for this work.

Membership and inclusion: The symbol $x \in \mathcal{A}$ says that the object x is a **member** of or **belongs to** the set $\mathcal{A}$. $x \notin \mathcal{A}$ says that the object x does not belong to set $\mathcal{A}$ (nonmembership), or, equivalently, $\neg(x \in \mathcal{A})$.

If $\mathcal{A}$ and $\mathcal{B}$ are mathematical objects, $\mathcal{A} = \mathcal{B}$ means they are the same object, or are **equal**. $\mathcal{A} \neq \mathcal{B}$ means that they are **not equal** or **unequal**. Two sets $\mathcal{A}$ and $\mathcal{B}$ are equal iff they contain the same members, and are not equal otherwise.

$\mathcal{A} \subseteq \mathcal{B}$ means that for all x, if $x \in \mathcal{A}$, then $x \in \mathcal{B}$. $\mathcal{A}$ is said to be a **subset** of $\mathcal{B}$. If $\mathcal{A}$ is a subset of $\mathcal{B}$ and there exists an $x \in \mathcal{B}$ such that $x \notin \mathcal{A}$, then $\mathcal{A}$ is a **proper** subset of $\mathcal{B}$. $\mathcal{A} = \mathcal{B}$ means that $\mathcal{A}$ and $\mathcal{B}$ contain exactly the same elements, that is, $\mathcal{A} \subseteq \mathcal{B}$ and $\mathcal{B} \subseteq A$. The symbolism $\mathcal{A} \subseteq \mathcal{B}$ indicates that it is possible for the two sets to be the same.

The symbol $\emptyset$ denotes the **null** or **empty set**, the set containing no elements. A **singleton** is a set having exactly one member; a **doubleton** is a set having exactly two members.

Truth sets of predicates: If $p(x)$ is a predicate, $\{x \mid p(x)\}$ is the set of all x for which the statement $p(x)$ is true. This set is commonly called the **truth set** of the predicate $p(x)$ and this type of construction is often referred to as the **set builder** notation.

Intersection, union, difference: $\mathcal{A} \cap \mathcal{B}$ is the set $\{x \mid x \in \mathcal{A}$ and $x \in \mathcal{B}\}$ and is called the **intersection** of $\mathcal{A}$ and $\mathcal{B}$.

$\mathcal{A} \cup \mathcal{B}$ is the set $\{x \mid x \in \mathcal{A}$ or $x \in \mathcal{B}\}$ and is called the **union** of $\mathcal{A}$ and $\mathcal{B}$.

$\mathcal{A} \setminus \mathcal{B}$ is the set $\{x \mid x \in \mathcal{A}$ and $x \notin \mathcal{B}\}$ and is called the **difference** of $\mathcal{A}$ and $\mathcal{B}$; informally, this operation is called **set subtraction**.

Listing elements and -tuples: When a set which has n elements is described by listing its elements, as in $\mathcal{A} = \{3, 5, 7, 9\}$, the order of the elements listed is irrelevant to the description; for example, $\{3, 5, 7, 9\} = \{5, 3, 7, 9\} = \{9, 5, 7, 3\}$. Frequently, however, we will need to list the elements of a set and at the same time specify which element is the first element, which one is second, and so on. The set whose elements are $a_1, a_2, \ldots, a_n$ where for each $k = 1, 2, \ldots, n$, a_k is specified as the "k-th" element is known as an **ordered n-tuple**, and is denoted by the symbol $(a_1, a_2, \ldots, a_n)$. If order is understood, we may refer to an ordered n-tuple simply as an **n-tuple**. Two (ordered) n-tuples $(a_1, a_2, \ldots, a_n)$ and $(b_1, b_2, \ldots, b_n)$ are equal iff $a_1 = b_1$, $a_2 = b_2, \ldots$, and $a_n = b_n$.

An *ordered 2-tuple* is called an **ordered pair** and is written (a, b), where a is the **first element** and b is the **second element** of the pair. Note that $(a, b) \neq (b, a)$ unless $a = b$. Occasionally we may speak of an "unordered pair" meaning simply a set with two elements.

An *ordered* 3-*tuple* is called an **ordered triple** and is written (a, b, c), where a is the **first element**, b is the **second element**, and c the **third element** of the triple.

The set of all n-tuples $(a_1, a_2, \ldots, a_n)$, where for all $i = 1, 2, \ldots, n$, $a_i \in \mathcal{A}_i$ is called the **Cartesian product**[2] of the sets $\mathcal{A}_1$,$\mathcal{A}_2, \ldots, \mathcal{A}_n$, and is denoted

$$\mathcal{A}_1 \times \mathcal{A}_2 \times \ldots \times \mathcal{A}_n.$$

Distinct and disjoint: When we say $\mathcal{A}_1$, $\mathcal{A}_2$, and $\mathcal{A}_3$ are **distinct sets**, we mean $\mathcal{A}_1 \neq \mathcal{A}_2$, $\mathcal{A}_1 \neq \mathcal{A}_3$, and $\mathcal{A}_2 \neq \mathcal{A}_3$. Expanding a bit on this theme, if $\{\mathcal{A}_1, \mathcal{A}_2, \mathcal{A}_3, \ldots, \mathcal{A}_n\}$ is any collection of mathematical objects (such as points, lines, planes, or what have you), saying that they are **distinct**, or **pairwise distinct** means that if $1 \leq k \neq l \leq n$, $\mathcal{A}_k \neq \mathcal{A}_l$. That is, no two objects in the collection are the same.

Two sets $\mathcal{A}$ and $\mathcal{B}$ are **disjoint** if $\mathcal{A} \cap \mathcal{B} = \emptyset$. If $\mathcal{A} \cap \mathcal{B} \neq \emptyset$, that is, $\mathcal{A} \cap \mathcal{B}$ contains some element x, $\mathcal{A}$ and $\mathcal{B}$ are said to **intersect**. A collection of sets $\{\mathcal{A}_1, \mathcal{A}_2, \mathcal{A}_3\}$ is said to be **pairwise disjoint** if $\mathcal{A}_1 \cap \mathcal{A}_2 = \emptyset$, $\mathcal{A}_1 \cap \mathcal{A}_3 = \emptyset$, and $\mathcal{A}_2 \cap \mathcal{A}_3 = \emptyset$, with obvious extensions to larger collections of sets. A collection of pairwise disjoint nonempty sets will clearly be distinct, but a collection of distinct sets is not necessarily pairwise disjoint.

1.4 Mappings, functions, cardinality, and relations

Mappings: A **mapping** (or **map**, for short) is a set of ordered pairs $(x, f(x))$ with the property that each first element x appears in exactly one pair $(x, f(x))$—that is, each x is "paired" with only one $f(x)$.[3]

The words **function** and **transformation** are synonyms for *mapping*. *Function* is commonly used in calculus where the emphasis is on mappings whose first and second elements are members of a number system, such as the set of real or complex numbers. *Transformation* is often used for mappings on vector spaces (to be defined later). In this book we will generally use the term *mapping*; here in the introduction we will use letters such as f, g, h to denote mappings even though they suggest the word *function*.

---

[2] After the French mathematician and philosopher René Descartes (1596–1650), inventor of the Cartesian coordinate system, which we will explore in the later chapters of this work. He has been called "the father of modern philosophy."

[3] You might prefer to think of a mapping as a *rule* which associates a single second element $f(x)$ with each first element x. Our definition is a bit more formal, as it does not depend on the undefined notion of a "rule."

Domain, range, and restriction: The set of all first elements of a mapping is its **domain**, and the set of all second elements is its **range.** We may state that "f is a mapping with domain $\mathcal{A}$ whose range is a subset of a set $\mathcal{B}$" by writing the symbol $f : \mathcal{A} \to \mathcal{B}$ or by saying that f **maps** $\mathcal{A}$ **to** (or **into**) $\mathcal{B}$.

If $\mathcal{E} \subseteq \mathcal{A}$, the **restriction** of f to $\mathcal{E}$ is the mapping (denoted by $f|\mathcal{E}$) consisting of the set of all ordered pairs of f whose first elements are members of $\mathcal{E}$; that is, the set $\{(x, f(x)) \mid x \in \mathcal{E}\}$.

Argument, value, image, and pre-image: If $f(x) = y$, we may say that f **maps** or **carries** x to y, or that y is the **image** of x under f, or that y is the **value** of f at x; in this case, x is sometimes said to be an **argument** for y. There may be more than one argument for a given value y. In some quarters it has become stylish, in this situation, to refer to x as an **input** for the mapping f, and to y as the **output** corresponding to x—a possible influence from the computer culture.

If f is a mapping with domain $\mathcal{A}$, and $\mathcal{E} \subseteq \mathcal{A}$, then the set $\{f(x) \mid x \in \mathcal{E}\}$ is the **image** of the set $\mathcal{E}$ under f, and is denoted by $f(\mathcal{E})$. If $\mathcal{F} \subseteq \mathcal{E}$, then $f(\mathcal{F}) \subseteq f(\mathcal{E})$. The range of a mapping f is the image $f(\mathcal{A})$ of its domain.

If $\mathcal{G}$ is any set, then $\{x \mid f(x) \in \mathcal{G}\}$ is the **pre-image** of $\mathcal{G}$ under f, and is denoted by $f^{-1}(\mathcal{G})$. (There are some subtleties to be observed here—see the paragraph below titled "Tricky notation.")

Composition of mappings: If $f : \mathcal{A} \to \mathcal{B}$ and $g : \mathcal{B} \to \mathcal{C}$ the **composition** $g \circ f$ is the mapping which maps $\mathcal{A}$ to $\mathcal{C}$ whose value for each $x \in \mathcal{A}$ is $g(f(x))$. In the case that $\mathcal{A} = \mathcal{B} = \mathcal{C}$, it is not necessarily true that $g \circ f = f \circ g$ (composition of mappings is not *commutative*). As an example, let f and g be defined as follows for each real number x: $f(x) = x+2$ and $g(x) = x^2$; then for each x, $(f \circ g)(x) = f(g(x)) = x^2+2$, whereas $(g \circ f)(x) = g(f(x)) = (x+2)^2 = x^2+4x+4$ which is not the same as x^2+2.

However, if $f : \mathcal{A} \to \mathcal{B}, g : \mathcal{B} \to \mathcal{C}$, and $h : \mathcal{C} \to \mathcal{D}$, it is always the case that $(h \circ g) \circ f = h \circ (g \circ f)$ (composition of mappings is *associative*). For if $x \in \mathcal{A}$ then

$$((h \circ g) \circ f)(x) = (h \circ g)(f(x)) = h(g(f(x))) = h((g \circ f)(x)) = h \circ (g \circ f)(x).$$

The notion of composition may easily be extended to any finite collection of mappings, provided that the domains and ranges match up properly. Thus, the composition of mappings $f_1, f_2, \ldots, f_n$ is the mapping whose value at each x is $f_1(f_2(\ldots(f_n(x))\ldots))$.

For definitions of the terms *commutative* and *associative* see the next section, elementary algebraic structures.

Onto, one-to-one, bijection: We say that a mapping f maps $\mathcal{A}$ **onto**, or is a **surjection onto** a set $\mathcal{B}$ iff for every $y \in \mathcal{B}$ there exists an $x \in \mathcal{A}$ such that $f(x) = y$. By definition, a mapping always maps its domain $\mathcal{A}$ onto its range $f(\mathcal{A})$, that is, onto the image of $\mathcal{A}$ under f. A mapping f is said to be **one-to-one** (sometimes written **1-1**), or an **injection**, in the case that for any two elements x and y belonging to its domain $\mathcal{A}$, if $x \neq y$ then $f(x) \neq f(y)$. If a mapping f with domain $\mathcal{A}$ is both one-to-one and onto the set $\mathcal{B}$, we say it is a **bijection** of $\mathcal{A}$ onto $\mathcal{B}$, or is a **one-to-one correspondence** between $\mathcal{A}$ and $\mathcal{B}$. If f is a bijection of $\mathcal{A}$ onto itself, we say f is a **bijection of** $\mathcal{A}$.

The composition of two mappings that are one-to-one is one-to-one, and the composition of two mappings of a set $\mathcal{A}$ onto itself is also a mapping of $\mathcal{A}$ onto itself. Thus if f and g are bijections of $\mathcal{A}$, $f \circ g$ is also a bijection of $\mathcal{A}$. To see this, let x and y be any distinct members of $\mathcal{A}$. Since g is a bijection of $\mathcal{A}$, it is one-to-one and $g(x) \neq g(y)$. Similarly, since f is a bijection of $\mathcal{A}$, $(f \circ g)(x) = f(g(x)) \neq f(g(y)) = (f \circ g)(y)$, so that $f \circ g$ is one-to-one. If $z \in \mathcal{A}$, since f is onto, there is an element $y \in \mathcal{A}$ such that $f(y) = z$; similarly, there exists an element $x \in \mathcal{A}$ such that $g(x) = y$ so that $f(g(x)) = z$ and $f \circ g$ is onto $\mathcal{A}$. Therefore $f \circ g$ is a bijection of $\mathcal{A}$.

Inverses and the identity: If f is one-to-one, the set $\{(f(x), x) \mid (x, f(x)) \in f\}$ is a mapping called its **inverse** f^{-1}. Its domain is the same as the range of f, and its range is the same as the domain of f. The inverse of a bijection is a bijection.

If $\mathcal{A}$ is any set, the **identity map** is the mapping ι with the property that for every $x \in A$, $\iota(x) = x$. The identity map is a bijection and is its own inverse. Note that if f is a one-to-one mapping with domain $\mathcal{A}$ and range $\mathcal{B}$, then $f \circ f^{-1}$ is the identity mapping on $\mathcal{B}$ and $f^{-1} \circ f$ is the identity mapping on $\mathcal{A}$.

Tricky notation: In an earlier paragraph we defined the pre-image of a set $\mathcal{G}$ under f as the set $\{x \mid f(x) \in \mathcal{G}\}$. But things get slightly tricky here: we use the notation $f^{-1}(\mathcal{G})$ to designate this set, *even if* f^{-1} *does not exist as a mapping*, that is, when f is not one-to-one—the pre-image of a set with only one element might contain many elements. Also, there is no necessity for $\mathcal{G}$ to be a subset of the range of f, or even to intersect it. In the latter case, $f^{-1}(\mathcal{G}) = \emptyset$.

Elementary mapping theory: There are some relations between set theory and mappings that will be of particular importance to us in Chapter 8 where we deal with reflections and isometries, both of which are bijections.

Let f be a mapping and let $\mathcal{A}$ and $\mathcal{B}$ be subsets of the domain of f. Then it is always true that $f(\mathcal{A} \cup \mathcal{B}) = f(\mathcal{A}) \cup f(\mathcal{B})$. But for intersection, all we can claim is that $f(\mathcal{A} \cap \mathcal{B}) \subseteq f(\mathcal{A}) \cap f(\mathcal{B})$. To illustrate this, let f be a mapping on the set of real numbers and define $f(1) = 1$ and $f(2) = 1$, and let $\mathcal{A} = \{1\}$ and $\mathcal{B} = \{2\}$. Then $f(\mathcal{A} \cap \mathcal{B}) = f(\emptyset) = \emptyset \neq \{1\} = f(\mathcal{A}) \cap f(\mathcal{B})$.

The situation is quite different for bijections, which are one-to-one and onto. Then it is always true that $f(\mathcal{A} \cap \mathcal{B}) = f(\mathcal{A}) \cap f(\mathcal{B})$. Thus, when we map a disjoint union of sets using a bijection, the image will be a disjoint union of the images of each of the individual sets in the union. We will sometimes also refer to these facts by the words **"elementary set theory."**

Finite and infinite sets: If there exists a one-to-one correspondence between two sets $\mathcal{A}$ and $\mathcal{B}$, then $\mathcal{A}$ and $\mathcal{B}$ are said to have the same **cardinal number** (the two sets **have the same number of elements**). A set $\mathcal{A}$ is said to be **finite** if it is empty (in which case it has zero elements), or if for some integer $n > 0$, there exists a one-to-one correspondence between $\mathcal{A}$ and the set $\{1, 2, 3, \ldots, n\}$ of natural numbers. We will normally express this fact by saying that the set **has** n **members**. Every subset of a finite set is finite. If $\mathcal{A} \subseteq \mathcal{B}$ are finite sets, $\mathcal{A}$ has m members, and $\mathcal{B}$ has n members, then $\mathcal{A}$ is a proper subset of $\mathcal{B}$ iff $m < n$. A set $\mathcal{A}$ is **infinite** if and only if it is not finite. A set $\mathcal{A}$ is infinite if it has an infinite subset.

Relations and their properties: A **relation** on a set $\mathcal{X}$ is a set of ordered pairs (x, y), where x and y are members of $\mathcal{X}$. If R is a relation on $\mathcal{X}$ we write $x R y$ to indicate that the pair $(x, y) \in R$. If a relation R satisfies the following three criteria, it is called an **equivalence relation**: 1) for every $x \in \mathcal{X}$, $x R x$ (R is **reflexive**); 2) if $x R y$ then $y R x$ (R is **symmetric**); and 3) if $x R y$ and $y R z$, then $x R z$ (R is **transitive**). If R is an equivalence relation on $\mathcal{X}$ and $x \in \mathcal{X}$, the **equivalence class** (denoted $[x]$) of x is $\{y \mid x R y\}$. The collection of all equivalence classes forms a **partition** of the underlying set $\mathcal{X}$—that is, every element of $\mathcal{X}$ belongs to exactly one of the equivalence classes.

1.5 Elementary algebraic structures

In modern mathematics, both *group* and *field* theory play a prominent role. In this work we will not be using these theories to any extent, but it will be convenient to have the following definitions on hand. We will use them principally for summarizing and organizing our knowledge.

Operations: An **operation**, or a **binary operation** on a set $\mathcal{E}$ is a mapping which maps $\mathcal{E} \times \mathcal{E}$ into $\mathcal{E}$; the image of (x, y) is indicated by writing the symbols x and y together, or with some symbol in between them, as in $x \cdot y$, $x + y$, or $x \circ y$.

Groups and their operations: A **group** is a nonempty set $\mathcal{G}$ together with an *operation* "·" such that conditions (G1) through (G4) are satisfied:

(G1) for every two elements x and y of $\mathcal{G}$, $x \cdot y \in \mathcal{G}$ (the group is **closed** under the operation ·);[4]

(G2) for any three elements x, y, and z of $\mathcal{G}$, $x \cdot (y \cdot z) = (x \cdot y) \cdot z$ (the operation · is **associative**);

(G3) there exists an element $e \in \mathcal{G}$ such that for every $x \in \mathcal{G}$, $e \cdot x = x \cdot e = x$ (e is the **identity** element for the operation ·); and

(G4) for any element $x \in \mathcal{G}$ there exists another element $x^{-1} \in \mathcal{G}$ such that $x \cdot x^{-1} = x^{-1} \cdot x = e$ (x^{-1} is the **inverse** of x under the operation ·).

If, in addition, the following condition (G5) is satisfied, $\mathcal{G}$ is said to be a **commutative** or **abelian**[5] group.

(G5) for any two elements x and y in $\mathcal{G}$ $x{\cdot}y = y{\cdot}x$ (the operation · is **commutative**).

A subset $\mathcal{H}$ of a group $\mathcal{G}$ is a **subgroup** of $\mathcal{G}$ if it forms a group under the operation of $\mathcal{G}$. To prove that a nonempty subset $\mathcal{H}$ is a subgroup of $\mathcal{G}$, it is sufficient to show that for every x and y in $\mathcal{H}$, both x^{-1} and $x \cdot y$ are members of $\mathcal{H}$.

A **semigroup** is a nonempty set $\mathcal{G}$ together with an *operation* "·" such that conditions (G1) and (G2) are satisfied.

Bijections forming a group: The following fact will come in handy in later chapters.

If $\mathcal{F}$ is any set of bijections on a set $\mathcal{A}$ such that a) the composition of any two bijections in $\mathcal{F}$ is again in $\mathcal{F}$ or is the identity ι, and b) the inverse of any member of $\mathcal{F}$ is in $\mathcal{F}$, then $\mathcal{F} \cup \{\iota\}$ is a group.

To see this, note that for any mapping f mapping $\mathcal{A}$ onto $\mathcal{A}$, $f \circ \iota = \iota \circ f = f$. G1 follows immediately from this and a); G2 follows from the associativity of functions under composition; G3 is true because $\iota \in \mathcal{F} \cup \{\iota\}$; and G4 follows from observation b) just above and the fact that the inverse of ι is ι.

---

[4]Technically, the condition (G1) is redundant here because the definition of an operation requires that the result be a member of the same set.

[5]After the Norwegian mathematician Niels Henrik Abel (1802–1829).

In various contexts, different symbols are used in place of the $\cdot$ we have used here; the real numbers form a group under the operation $+$ and the set of nonzero real numbers form a group under the operation of multiplication. Both of these are abelian groups. In situations where the composition of functions is possible, we use $\circ$ for this operation, and later in this book we will come upon several examples where sets of functions of particular types form groups under this operation. Very often, too, where the operation is well understood, we will not use any symbol at all, using juxtaposition instead; rather than writing $x \cdot y$ we will write simply xy.

Fields and their operations: A **field** is a nonempty set $\mathcal{F}$ together with two operations, which for convenience we will designate as $+$ (**addition**) and "$\cdot$" (**multiplication**), such that all the following conditions are satisfied:

(F1) $\mathcal{F}$ forms an abelian group under the operation $+$ (it is customary here to call $\mathcal{F}$ an **additive** group, to use the symbol 0 for the additive identity, and for each $x \in \mathcal{F}$ use the symbol $-x$ for its inverse);

(F2) $\mathcal{F} \setminus \{0\}$ forms an abelian group under the operation "$\cdot$" (that is, $\mathcal{F} \setminus \{0\}$ is a **multiplicative** group); we will generally use the symbol 1 for the multiplicative identity; and

(F3) for any three elements x, y, and z of $\mathcal{F}$, $x \cdot (y + z) = (x \cdot y) + (x \cdot z)$ (the **distributive law** of multiplication over addition holds).

If $\mathcal{F}$ is a field, then a subset $\mathcal{E}$ of $\mathcal{F}$ is a **subfield** of $\mathcal{F}$ if it is itself a field under the operations of $\mathcal{F}$. To prove that $\mathcal{E}$ is a subfield of $\mathcal{F}$, it is sufficient to show that both 0 and 1 are members of $\mathcal{E}$, and for every x and y in $\mathcal{E}$, $-x$, $x + y$, x^{-1} (where $x \neq 0$), and $x \cdot y$ are all members of $\mathcal{E}$.

Number systems: The reader should be familiar with the set $\mathbb{Z}$ of **integers** and its subset $\mathbb{N} = \{1, 2, 3, \ldots\}$, the **natural numbers**. If m and n are any integers such that $m < n$, the symbol $[m; n]$ will denote the set of all integers greater than or equal to m and less than or equal to n. Thus, for example, $[3; 7] = \{3, 4, 5, 6, 7\}$.

The reader will also need to be familiar with the field $\mathbb{R}$ of **real numbers** and its subfield $\mathbb{Q}$ of **rational numbers**. Between any two real numbers in $\mathbb{R}$ there is both a rational number and a nonrational (irrational) number. The numbers π, e (the base of natural logarithms), and $\sqrt{2}$ are examples of irrational numbers.

Another subfield of the real numbers which will be important in the last chapter of the book is the set of **real algebraic numbers,**[6] denoted by the symbol $\mathbb{A}$. This is the set of real numbers that are roots of polynomial equations having rational coefficients. The real number π is not an algebraic number, as it is not the solution of any polynomial equation with rational coefficients.

The sets $\mathbb{N}$, $\mathbb{Z}$, $\mathbb{Q}$, $\mathbb{A}$, and $\mathbb{R}$ are naturally **ordered** by defining $a < b$ iff there exists a number $c > 0$ such that $b = a + c$. For these number systems, there is also a natural concept of **betweenness**: b is said to be **between** the numbers a and c iff either $a < b < c$ or $c < b < a$. In symbols, this is written as either a–b–c or c–b–a.

Anticipating later developments, we point out that this order relation satisfies the properties of ordering as in Definition ORD.1 (Chapter 6), and this notion of betweenness satisfies Definition IB.1 (Chapter 4). Moreover, $\mathbb{R}$ has the LUB property, meaning that every set of real numbers which is bounded above has a least upper bound (the formal definition of which is given in Chapter 18). Every irrational number x is the least upper bound of the set of all rational numbers r such that $r < x$.

In these number systems the square of any (nonzero) number is positive. The square root of a number $a \geq 0$ is denoted by $\sqrt{a}$ and is the solution to the polynomial equation $x^2 = a$, and thus is an algebraic number if a is rational. Both the fields $\mathbb{R}$ of real numbers and $\mathbb{A}$ of real algebraic numbers contain the square roots of their non-negative members.

Vector spaces: A **vector space** (or **linear space**) over a field $\mathcal{F}$ consists of a set $\mathcal{V}$ of elements called **vectors** together with the field $\mathcal{F}$, whose members may be called either numbers or **scalars**; an operation "+" which denotes addition of vectors; and an operation "·" which multiplies a scalar times a vector, all satisfying the following conditions:

[6]More generally, an **algebraic number** is a **complex number** that is a root of a polynomial equation with rational coefficients. (A treatment of complex numbers is available in a supplement accessible from the home page of this book at www.springer.com.) The set of all algebraic numbers is a subfield of the complex numbers, as is the set of real numbers, and their intersection is the subfield of real algebraic numbers described here. The field of algebraic numbers is **algebraically closed**, meaning that any root of a polynomial with coefficients from the field is also a member of the field. The field of real algebraic numbers is not algebraically closed, as is readily seen from the fact that the polynomial equation $x^2 = -1$ has no real solution.

(V1) $\mathcal{V}$ forms an abelian group with respect to the operation $+$, with identity element O; the inverse of any vector A is $-A$.

(V2) The scalar product "·" obeys the following rules: for any scalars x and y belonging to $\mathcal{F}$ and any vectors A and B of $\mathcal{V}$, $x(yA) = (xy)A$, $1A = A$, $x(A+B) = xA + xB$, and $(x+y)A = xA + yA$.

It is customary to omit the dot symbol for scalar product, as we have done, and, where no confusion arises, to refer to the vector space by the name of its set of vectors.

If $\mathcal{V}$ is a vector space, then a subset $\mathcal{U}$ of $\mathcal{V}$ (equipped with the same field) is a **subspace** of $\mathcal{V}$ if it is itself a vector space under the operations of $\mathcal{V}$. A subset $\mathcal{U}$ is a subspace of $\mathcal{V}$ iff for all A and B in $\mathcal{U}$ and every $x \in \mathcal{F}$, both $A+B \in \mathcal{U}$ and $xA \in \mathcal{U}$. A subspace $\mathcal{U}$ is a **proper** subspace of $\mathcal{V}$ if there exists at least one point $A \in \mathcal{V}$ such that $A \notin \mathcal{U}$. $\mathcal{V}$ is a subspace of itself, and $\{O\}$ is the *trivial* subspace of $\mathcal{V}$.

Linear independence and dimension: A set $\{A_1, A_2, \ldots, A_n\}$ of vectors is **linearly dependent** iff there exists a set $\{x_1, x_2, \ldots, x_n\}$ of field elements, not all zero, such that $x_1A_1 + x_2A_2 + \ldots + x_nA_n = O$. A set that is not linearly dependent is **linearly independent**; that is, if $x_1A_1 + x_2A_2 + \ldots + x_nA_n = O$ then $x_i = 0$ for all $i \in \{1, 2, \ldots, n\}$. If for some numbers (field elements) $x_1, x_2, \ldots, x_n$, $X = x_1A_1 + x_2A_2 + \ldots + x_nA_n$, X is said to be a **linear combination** of the vectors $A_1, A_2, \ldots, A_n$. If every vector $X \in \mathcal{V}$ is a linear combination of the vectors in $\{A_1, A_2, \ldots, A_n\}$, we say that this set **spans** the space $\mathcal{V}$. A set of nonzero vectors which is both linearly independent and spans $\mathcal{V}$ is called a **basis** for $\mathcal{V}$, and every vector space has a basis.

Basis Theorem. *If $\{B_1, B_2, \ldots, B_m\}$ is a set of vectors that spans $\mathcal{V}$, and $\{A_1, A_2, \ldots, A_n\}$ is a set of linearly independent vectors in $\mathcal{V}$, then $n \le m$; if $n = m$, $\{A_1, A_2, \ldots, A_n\}$ spans $\mathcal{V}$.*

Proof. We outline the proof[7] of this result, which will be of basic importance to our development in Chapter 21.

Note first that for some numbers $x_1, x_2, \ldots, x_m$, $A_n = x_1B_1 + x_2B_2 + \ldots + x_mB_m$ and at least one of the x_is is nonzero. Dividing through by this x_i we see that B_i is a linear combination of the other Bs together with A_n. It follows that if, in the spanning set, this B_i is replaced by A_n, the resulting set $\{B_1, B_2, \ldots, B_{i-1}, A_n, B_{i+1} + \ldots, B_m\}$ spans $\mathcal{V}$.

[7]The proof is essentially that found in Halmos, *Finite Dimensional Vector Spaces* [9], pp. 9–14.

Then A_{n-1} is a linear combination of this new spanning set, and the coefficient of at least one of the Bs in the combination must be nonzero, for otherwise, the linear independence of the As would be contradicted. This B can be replaced by A_{n-1} in the spanning set, and the resulting set will span $\mathcal{V}$.

Repeat this process as many times as possible; if $n > m$, the Bs will be used up before the As, resulting in a spanning set that contains no Bs, and a list A_1, ..., A_j of As that have not yet been incorporated into the spanning set. In this case, the "un-substituted" As are linear combinations of the spanning set, which consists entirely of As; this contradicts our initial assumption that the As are linearly independent.

Therefore $n \leq m$, and the replacement process will stop when all the As have been used to replace Bs in the spanning set, leaving (possibly) some unreplaced Bs. It follows that in any vector space, the number of linearly independent vectors cannot exceed the number of vectors in a spanning set. Moreover, if $n = m$, $\{A_1, A_2, \ldots, A_n\}$ spans $\mathcal{V}$, since all the Bs have been displaced by As. □

Dimension: By the Basis Theorem, any two bases for a space have the same number of elements, because each basis is a linearly independent set and also spans the space. The number of elements in a basis is called the **dimension** of the space.

Dimension Criterion: If $\mathcal{U}$ is a subspace of $\mathcal{V}$, then $\mathcal{U} = \mathcal{V}$ iff the dimension of $\mathcal{U}$ is equal to the dimension of $\mathcal{V}$. Suppose the dimension of $\mathcal{U}$ is the same as that of $\mathcal{V}$; then a basis $\mathcal{B}$ for $\mathcal{U}$ is a linearly independent set in $\mathcal{V}$, having the same number of elements as a basis for $\mathcal{V}$, which is a spanning set for $\mathcal{V}$. By the Basis Theorem, $\mathcal{B}$ spans $\mathcal{V}$, so that every vector in $\mathcal{V}$ is a linear combination of the vectors of $\mathcal{B}$, and hence is a member of $\mathcal{U}$. Conversely, if $\mathcal{U}$ and $\mathcal{V}$ are the same space, they have the same dimensions.

In this work we will be mainly concerned with vector spaces of dimension 1, 2, or 3.

Since the vector space axioms are a subset of the field axioms, $\mathcal{F}$ is a vector space over itself, having dimension 1. If $A \neq O$ is a point of a vector space $\mathcal{V}$, $\{xA \mid x \in \mathcal{F}\}$ (that is, a "line" through the origin) is a vector subspace of $\mathcal{V}$ having dimension 1. Thus, the word *space* in *vector space* may at times mean "line"; it may also mean "plane," although not all lines (or planes) in a vector space are vector spaces.

The word *vector* in the term *vector space* does not include the notion of a "bound" vector as used in science, often visualized as an arrow whose tail can be located at any desired point. *Vector*, to us, means a point in a vector space, nothing more, nothing less; if we visualize it as an arrow, the tail of the arrow is *always* at the origin O and its head is at the point specified.

Linear mappings: A **linear mapping** (or **linear transformation** or **linear operator**) α on a vector space $\mathcal{V}$ is a mapping of $\mathcal{V}$ into $\mathcal{V}$ such that for all A and B in $\mathcal{V}$, and all field elements x and y, $\alpha(xA + yB) = x\alpha(A) + y\alpha(B)$. The mapping O is the mapping such that $O(A) = O$ for every $A \in \mathcal{V}$. The "negative" of the mapping α is the mapping $-\alpha$, which maps every $A \in \mathcal{V}$ to $-(\alpha(A))$. The **sum** of two linear mappings α and β on $\mathcal{V}$ is the mapping $\alpha + \beta$ such that for every $A \in \mathcal{V}$, $(\alpha + \beta)(A) = \alpha(A) + \beta(A)$. The **product** or **scalar product** of a field element x and a linear mapping α on $\mathcal{V}$ is the mapping $x\alpha$ such that for every $A \in \mathcal{V}$, $(x\alpha)(A) = x(\alpha(A))$. Any sum of linear mappings, and any scalar multiple of a linear mapping, is a linear mapping, as are the mapping O and the negative of any linear mapping. The set of all linear mappings on a vector space with the definitions of sum and scalar product as above is itself a vector space over the same field.

Vector spaces of n-tuples: Let $\mathcal{F}$ be a field and denote by $\mathcal{F}^n$ the set of all n-tuples $(a_1, a_2, \ldots, a_n)$ of elements of $\mathcal{F}$. Define the sum of two n-tuples $(a_1, a_2, \ldots, a_n)$ and $(b_1, b_2, \ldots, b_n)$ in $\mathcal{F}^n$ as

$$(a_1, a_2, \ldots, a_n) + (b_1, b_2, \ldots, b_n) = (a_1 + b_1, a_2 + b_2, \ldots, a_n + b_n),$$

and for any $t \in \mathcal{F}$ define the scalar product

$$t(a_1, a_2, \ldots, a_n) = (ta_1, ta_2, \ldots, ta_n).$$

With these definitions, $\mathcal{F}^n$ is a vector space, called **coordinate space over the field** $\mathcal{F}$.

If the additive identity of $\mathcal{F}$ is 0 and the multiplicative identity is 1, then $(0, 0, \ldots, 0)$ is the origin, or zero element of $\mathcal{F}^n$, and will often be denoted O. The set $\{\{(1, 0, 0, \ldots, 0), (0, 1, 0, \ldots, 0), (0, 0, 1, \ldots, 0), \ldots, (0, 0, 0, \ldots, 1)\}$ forms a basis for $\mathcal{F}^n$, so that $\mathcal{F}^n$ has dimension n.

Coordinate spaces will be explored further in Chapter 21, where they are used to show consistency and independence of our axioms.

More advanced vector space theory extends the notion of dimension to include spaces of *infinite* dimension. For example, the set of all real-valued functions defined on the unit interval [0,1] is a vector space having infinite dimension, where the sum $f + g$ of two functions f and g is defined by $(f + g)(x) = f(x) + g(x)$ for all

$x \in [0, 1]$, and for any real number t, $(tf)(x) = tf(x)$. This is only a hint at the extent and applicability of vector space theory; in this work, we will only scratch the surface.[8]

Isomorphisms: If $\mathcal{G}$ and $\mathcal{G}'$ are groups, an **isomorphism** (more elaborately, a **group isomorphism**) of $\mathcal{G}$ onto $\mathcal{G}'$ is a bijection Φ which preserves operations—that is, for every x and y in $\mathcal{G}$, $\Phi(x \cdot y) = \Phi(x) \odot \Phi(y)$, where "$\cdot$" is the operation of $\mathcal{G}$ and $\odot$ is the operation of $\mathcal{G}'$. If $\mathcal{F}$ and $\mathcal{F}'$ are fields, then an **isomorphism** (or **field isomorphism**) of $\mathcal{F}$ onto $\mathcal{F}'$ is a bijection Φ which preserves both field operations. If $\mathcal{V}$ and $\mathcal{U}$ are vector spaces, then an **isomorphism** (or **vector space isomorphism**) of $\mathcal{V}$ onto $\mathcal{U}$ is a bijection Φ which preserves both addition and scalar multiplication. We could also define isomorphisms between other types of algebraic systems.

If $\mathcal{A}$ and $\mathcal{B}$ are two isomorphic algebraic systems, we may say that $\mathcal{B}$ is an **isomorphic image** or **copy** of $\mathcal{A}$, and *vice versa*. Isomorphic systems are indistinguishable as to their algebraic structures. In particular, two vector spaces which are isomorphic have the same dimension.

It is well known that the isomorphic image of a group is a group, the isomorphic image of a field is a field, and the isomorphic image of a vector space is a vector space. Thus, if one can establish (as we do in later chapters) an isomorphism between a field $\mathcal{F}$ and another set $\mathcal{F}'$ which is equipped with two operations $+$ and "$\cdot$" , the set $\mathcal{F}'$ is automatically a field, and likewise for a vector space. This relieves us of the tedium of proving all the various field (or vector space) properties on the second set. All that is necessary is to show that the mapping (isomorphism) is a bijection onto the second set, and that the operations are preserved.

Matrices, determinants, and Cramer's rule: (1) A **matrix** is a rectangular array of numbers (members of $\mathbb{F}$) such as $\begin{pmatrix} a_{11} & a_{12} \\ a_{21} & a_{22} \end{pmatrix}$; it may also be denoted as $\begin{bmatrix} a_{11} & a_{12} \\ a_{21} & a_{22} \end{bmatrix}$. This one would be called a "square 2 by 2 matrix."

The array $\begin{pmatrix} a_{11} & a_{12}, & a_{13} \\ a_{21} & a_{22}, & a_{23} \\ a_{31} & a_{32}, & a_{33} \end{pmatrix}$ is a "square 3 by 3 matrix."

[8] There are a number of good books on vector space theory; *Finite Dimensional Vector Spaces* [9], by Paul Halmos (1916–2006), remains a classic. Originally published by Van Nostrand in 1958, it is still in print from Springer.

(2) The **determinant** $\begin{vmatrix} a_{11} & a_{12} \\ a_{21} & a_{22} \end{vmatrix}$ of a 2×2 matrix $\begin{pmatrix} a_{11} & a_{12} \\ a_{21} & a_{22} \end{pmatrix}$ is defined to be $a_{11}a_{22} - a_{12}a_{21}$.

The determinant $\begin{vmatrix} a_1 & b_1 & c_1 \\ a_2 & b_2 & c_2 \\ a_3 & b_3 & c_3 \end{vmatrix}$ of the 3×3 matrix $\begin{pmatrix} a_1 & b_1 & c_1 \\ a_2 & b_2 & c_2 \\ a_3 & b_3 & c_3 \end{pmatrix}$ is

$$a_1b_2c_3 - a_1b_3c_2 - b_1a_2c_3 + b_1a_3c_2 + c_1a_2b_3 - c_1a_3b_2.$$

Cramer's rule: If a_1, a_2, b_1, b_2, c_1, and c_2 are members of the field $\mathbb{F}$, and if the determinant $\begin{vmatrix} a_1 & b_1 \\ a_2 & b_2 \end{vmatrix} \neq 0$, then the simultaneous solution s, t to the equations $a_1s + b_1t = c_1$ and $a_2s + b_2t = c_2$ is

$$s = \frac{\begin{vmatrix} c_1 & b_1 \\ c_2 & b_2 \end{vmatrix}}{\begin{vmatrix} a_1 & b_1 \\ a_2 & b_2 \end{vmatrix}} \text{ and } t = \frac{\begin{vmatrix} a_1 & c_1 \\ a_2 & c_2 \end{vmatrix}}{\begin{vmatrix} a_1 & b_1 \\ a_2 & b_2 \end{vmatrix}}.$$

There is a version of Cramer's rule that solves three linear equations if the determinant of coefficients is nonzero, but we will not have occasion to use it.

The reader should be familiar with the addition and multiplication of matrices using the *row by column rule*, with scalar multiplication, as well as with the use of matrices to describe the behavior of linear mappings. We may occasionally use the method of calculating a determinant of a square matrix in which each *ij*th entry in an arbitrary row (or column) is multiplied by $(-1)^{i+j}$ times the determinant of the matrix obtained by deleting the *i*th row and *j*th column, and summing over all the entries in that row (or column).

1.6 The basic building blocks of axiomatic theory

It is impossible to assign meaning to every *term*[9] in a theory. To assign meaning to a term, we must use other terms, and the meanings of these in turn must be stated using yet other terms. This leads either to an infinite regression of terms and definitions or, more likely, to a circular "definition" (as in a dictionary).

[9] We use the words "*term*" and "*word*" interchangeably.

To avoid this, it is customary in mathematics to begin with certain **undefined terms** or **primitive notions**. By calling them "undefined" we mean that they are *initially* undefined; they acquire meaning when axioms are invoked, so in a sense, the axioms define them. Here, "**point**," "**line**," and "**plane**" are undefined terms. These are the primary building blocks of our theory, which is constructed using *definitions*, *axioms*, and *theorems*. Since it makes little sense to talk about nonexistent entities, we shall assume that *points*, *lines*, and *planes* exist, even though initially we do not know what they are.[10]

A **definition** assigns meaning to a word or symbol using *undefined* or *previously defined* words. Definitions do not add new content to our theory, but provide names and symbols which serve as shortcuts in our discourse, sparing us the trouble of writing out full descriptions which otherwise would quickly become very cumbersome. Be warned that in definitions we will often write "if" to mean "iff," and that some definitions are unacceptable. (See "*On 'good' and 'bad' definitions*" below.)

An **axiom** is a statement that gives meaning to *undefined* terms, states the relationship of such terms to other terms, or declares the existence of defined objects. Axioms are the starting points of our theory and are given without any justification or logical argument. Indeed, the set of axioms could be said to contain the entire theory.

The axioms must be **consistent**—meaning that the entire set of axioms does not give rise to any contradictions. Consistency of a set of axioms may be demonstrated by exhibiting an example (model) of a mathematical system in which all the axioms are true.

In this work we aim to make our axioms **independent**—meaning that no one of them can be derived from others that have been stated. Building a theory from independent axioms requires a lot more work (some of it tedious) than building

[10]"When we set out to construct a given discipline, we distinguish, first of all, a certain small group of expressions of this discipline that seem to us to be immediately understandable; the expressions in this group we call PRIMITIVE TERMS or UNDEFINED TERMS, and we employ them without explaining their meanings. At the same time we adopt the principle: not to employ any of the other expression[s] of the discipline under consideration, unless its meaning has first been determined with the help of primitive terms and of such expressions of the discipline whose meanings have been explained previously..." Alfred Tarski, *Introduction to Logic: and to the Methodology of Deductive Sciences*, 4th ed., page 118, Dover (1995) [21].

it from a carefully chosen set of axioms that are *not* independent. An extended discussion of *consistency* and *independence* of axioms will be found in Chapter 21.

In an axiomatic system, a statement is said to be "**true**" if and only if it can be deduced from the given axioms using the rules of logic. It is "**false**" if and only if its negation is true. *Axioms* are logical consequences of themselves, so are automatically considered true.

Within a given axiomatic system, it may be possible to construct a statement whose *truth value* cannot be determined. That is to say, the system of axioms may be *incomplete*.

A **theorem** is a statement about undefined or previously defined terms which has been **proved**, meaning that it has been shown to be a logical consequence of the axioms, set theory (which we have assumed), and previously proved theorems. Many theorems having special importance to the development will have names or descriptive labels.

A theorem that is a more or less immediate consequence of another theorem is called a **corollary** of the main theorem. A theorem that is used mainly for the proofs of other theorems is sometimes called a **lemma**. Items labeled **remark** are less formal in character and may contain easily proved theorems (and their proofs), which in turn may be cited in other proofs. An unproved statement that someone thinks is true is a **conjecture**.

What is possible to prove as a theorem in our axiomatic theory is entirely determined by the set of axioms we start with. This is why mathematicians fuss so much over the choice of axioms.

On "good" and "bad" definitions: Definitions must be succinct and concise. But it is inevitable that they will sometimes contain or imply statements. Any such statements must be *true* in order for the definition to be acceptable. If a statement is included that can't be proved, the definition is "bad" and must be discarded.

Suppose, for example, a definition specifies a name for "*the* plane which satisfies some property *p*." Implicit in this definition is the statement that there is only one such plane. If this can be proved to be true, the definition is "good." Otherwise, the definition is "bad."

1.7 Advice for the reader: labels, notation, figures, and exercises

Item labels and reference numbers: Theorems and their corollaries, lemmas, remarks, and definitions are usually (but not always) labeled with an acronym and a number, as in "**Theorem NEUT.15**." The acronym is intended to suggest the subject for the chapter (in this case, neutral geometry). In some cases, different parts of a chapter will have different acronyms. After the title of each chapter (except for Chapter 21) the acronyms used therein are listed in parentheses.

Numbers are assigned consecutively; occasionally, especially where it has been necessary to insert items late in the writing process, we have added decimal extensions, as in "**Remark PLGN.4.1**" and "**Definition PLGN.4.2**." Informal explanatory notes are often not given an acronym or number.

Numbers in square brackets [*n*] at the end of citations refer to the corresponding entry in "References."

Notational conventions: *Points* will be denoted by slanted capital Roman letters: $A, B, C, \ldots, X, Y, Z$. Both *lines* and *planes* will be denoted by calligraphic script capitals such as $\mathcal{E}$, $\mathcal{L}$, $\mathcal{M}$, $\mathcal{N}$, $\mathcal{P}$, or $\mathcal{Q}$, *etc. Space* will be denoted by letters in the form $\mathbb{U}$ or by calligraphic script, for instance $\mathcal{S}$. These symbols may also be used for other purposes–for instance, in later chapters $\mathcal{R}_{\mathcal{M}}$ routinely denotes a *reflection* over the line $\mathcal{M}$.

Figures: All four authors agree that the reader of a mathematical book should draw his or her own figures as an aid to understanding. But they have adhered to this ideal with varying degrees of rigor. There are some parts of the development in which it might be easy to construct figures, but doing so could be misleading; for instance, in Chapter 4 sides of a line are defined. At this point in the development, a line might could more than two sides, but any graphical portrayal would inevitably show a line having exactly two sides. This situation is not resolved until well into Chapter 5 after the Plane Separation Theorem is proved and its consequences explored. Thus there are no figures in Chapter 4 or the first part of Chapter 5.

In the final editing process we have removed some figures that now seem to us to deprive the reader of the proper pleasure of constructing his or her own, and we have inserted figures in other parts of the book where we think they might add clarity.

Exercises: Exercises in this book provide much more than routine practice of techniques learned in preceding sections. Each requires careful thought and possibly some ingenuity. The starred ones (*) (for which we provide solutions online at the home page for this book at www.springer.com) usually are an integral part of the development of the theory in the book. You cannot routinely skip them without missing much of what this book is about.

Even in those chapters (notably Chapters 5 on Pasch geometry, and Chapter 8 on neutral geometry) where there are a great many exercises, you will find it worthwhile to read each one, make a sketch and get in mind how to prove it, even if you don't actually put the details together. You can justify skipping an occasional exercise only if you are *quite sure* you could construct the proof if you had to, and feel it is a waste of your time to supply all the details. But beware that supplying all the details may look deceptively simple when you give a theorem a cursory glance.

It is often possible by exercising a certain amount of ingenuity to cut through a long and boring consideration of a list of cases. Indeed, much of what is beautiful and satisfying in mathematics has been motivated by a desire to avoid boring work. So, even if you can outline a "straightforward" proof of an exercise, and are therefore tempted to skip it, once in a while you might sharpen your mathematical insight by looking for a clever and more aesthetically satisfying proof. It is possible that you may, by such means, create new proofs of theorems and solutions for exercises that are more elegant than the ones we have given.

If that happens, please let us know. The authors are not geniuses, nor do we walk on water.[11]

1.8 Axioms for incidence geometry

Undefined terms: For our geometry, the words **point**, **line**, and **plane** are undefined terms. For now, the reader should try to avoid thinking of a *point* as a dot, a *line* as something long, straight, and thin, a *plane* as a flat expanse, or *space* as a solid. The familiar characteristics of lines and planes will emerge from the axioms to follow, and are entirely determined thereby.[12]

Definition I.0. **Space** $\mathbb{U}$ is the set of all points. We may think of space as the **universe** or the **universe of discourse**.

---

[11]Except for the third named author who, living in Michigan, routinely walks on (solidified) water.

[12] The following has been attributed to David Hilbert, as a way of saying that in proving geometric theorems we must use only the axioms, rather than any "real" interpretation of geometric objects: "One must be able to say at all times—instead of points, straight lines, and planes—tables, beer mugs, and chairs."

Axiom I.0. Lines and planes exist and are subsets of space $\mathbb{U}$.

Thus, $\mathbb{U}$ is *the set of "everything."* We may, as we have already done, employ the usual terminology of set theory which was introduced in Section 1.3, including the terms *member, belongs to, subset, union, intersection, disjoint*, and the like. It is quite correct to say things like "point A is a *member* of line $\mathcal{L}$" or "point A is a *member* of plane $\mathcal{P}$" when we mean $A \in \mathcal{L}$ or $A \in \mathcal{P}$. But this is geometry, so we may also say "point A *lies on* line $\mathcal{L}$ (or plane $\mathcal{P}$)" or sometimes "line $\mathcal{L}$ (plane $\mathcal{P}$) *goes through* (*contains*) point A." If $\mathcal{L} \subseteq \mathcal{P}$, we will often say "$\mathcal{L}$ is *contained in* $\mathcal{P}$" or "$\mathcal{P}$ *contains* $\mathcal{L}$."

Definition I.0.1. (A) "Points A, B, and C are **collinear**" means that there is a line $\mathcal{L}$ such that A, B, and C all lie on line $\mathcal{L}$. More generally, if $\mathcal{E}$ is any set of points, then $\mathcal{E}$ is **collinear** iff there exists a line $\mathcal{L}$ such that $\mathcal{E} \subseteq \mathcal{L}$. A set $\mathcal{E}$ is **noncollinear** iff there is no line containing all the points of $\mathcal{E}$.

(B) "Points A, B, C, and D are **coplanar**" means that there is a plane $\mathcal{P}$ such that A, B, C, and D all lie on $\mathcal{P}$. More generally, if $\mathcal{E}$ is any set of points, then $\mathcal{E}$ is **coplanar** iff there exists a plane $\mathcal{P}$ such that $\mathcal{E} \subseteq \mathcal{P}$. A set $\mathcal{E}$ is **noncoplanar** iff there is no plane containing all the points of $\mathcal{E}$.

(C) If $\mathbb{E}$ is a set of two or more lines, the lines in $\mathbb{E}$ are said to be **concurrent at a point** O if and only if the intersection of all members of $\mathbb{E}$ is $\{O\}$.

(D) A space on which the **incidence axioms** I.0 through I.5 are true is an **incidence space**, and a plane therein is an **incidence plane**. The geometry these axioms generate is **incidence geometry**.

Axiom I.1. There exists exactly one line through two distinct points.

Axiom I.2. There exists exactly one plane through three noncollinear points.

Axiom I.3. If two distinct points lie in a plane, then any line through the points is contained in the plane.

Axiom I.4. If two distinct planes have a nonempty intersection, then their intersection has at least two members.

Axiom I.5. **(A)** There exist at least two distinct points on every line.

(B) There exists at least one noncollinear set of three points on every plane.

(C) There exists at least one noncoplanar set of four points in space.

To visualize the meaning of these axioms you should feel free to draw pictures—this *is*, after all, *geometry*. In Chapter 21 we will exhibit a model in which all 13 axioms in this book are true. Thus, an incidence space and incidence planes actually exist, and incidence geometry is not vacuous.

What makes incidence geometry a new and interesting object is the fact that we must now get along without many of the familiar ideas of Euclid. For example, there is no concept of distance, so we must remember always to have the mental reservation that, even though two points in our picture seem to be farther apart than two other points, that has no meaning in the present context. Line segments do not have length either. In this strange world, we cannot tell whether two lines are perpendicular, whether two planes are perpendicular, whether a line is perpendicular to a plane, or even whether or not a point on a line is between two other points on the line. It may seem that there is very little we can do under such heavy restrictions, but we will find and prove a number of theorems.

Before we start, we have some comments about the axioms. Axiom I.1 really says two things: (1) if we have two points, there is a line containing both of them (**existence**), and (2) no other line contains these two points (**uniqueness**). Similarly, Axiom I.2 postulates both the existence and uniqueness of such a plane.

You might wonder why Axiom I.4 doesn't say: "If two planes intersect, their intersection is a line." To tell the truth, initially we had it that way. Later we saw, in light of Axiom I.1, it was enough to assume only the intersection contains two points. One of the properties of a good set of axioms is leanness, so we removed the unnecessary assumption, and will prove it as Theorem I.4.

1.9 A finite model for incidence geometry

Before we state any theorems about incidence geometry, it might be appropriate to give an example of a geometry which satisfies these axioms. It will help show just how seriously we take every word in the above definitions and axioms. According to Axiom I.0, three essential ingredients of a geometry are space, lines, and planes. It says space is a set of points. In this example, which is called a **model**, space is a set of eight points. Lacking originality, we will call them A, B, C, D, E, F, G, and H. Hence $\mathbb{U} = \{A, B, C, D, E, F, G, H\}$.

Now we will list which subsets of $\mathbb{U}$ are lines:

$$
\begin{array}{lllllll}
\{A,B\} & \{A,C\} & \{A,D\} & \{A,E\} & \{A,F\} & \{A,G\} & \{A,H\} \\
\{B,C\} & \{B,D\} & \{B,E\} & \{B,F\} & \{B,G\} & \{B,H\} & \\
\{C,D\} & \{C,E\} & \{C,F\} & \{C,G\} & \{C,H\} & & \\
\{D,E\} & \{D,F\} & \{D,G\} & \{D,H\} & & & \\
\{E,F\} & \{E,G\} & \{E,H\} & & & & \\
\{F,G\} & \{F,H\} & \text{and} & & & & \\
\{G,H\}. & & & & & &
\end{array}
$$

That is to say, every line contains exactly two points and every set containing exactly two points is a line in this geometry. Now we must specify which subsets of $\mathbb{U}$ are planes. They are:

$$
\begin{array}{llll}
\{A,B,C,D\} & \{A,B,E,F\} & \{A,B,G,H\} & \{A,C,E,G\} \\
\{A,C,F,H\} & \{A,D,E,H\} & \{E,F,G,H\} & \{C,D,G,H\} \\
\{C,D,E,F\} & \{B,D,F,H\} & \{B,D,E,G\} & \{B,C,F,G\} \\
\{A,D,F,G\} & \text{and} & \{B,C,E,H\}. &
\end{array}
$$

The sets which are planes do not lend themselves to as succinct a description as do those which are lines. However, the diagram in Figure 1.1 is a device for remembering which sets are lines and which sets are planes in this model.

If you think of this cube as a solid in Euclidean space—in other words, a cube like the ones you studied in high school geometry—then a set of points is a line in this new geometry iff the points lie on a line in Euclidean geometry. The planes you expect from Euclidean geometry are the first 12 listed. However, Axiom I.2 requires the last two which are otherwise unexpected.

If you don't find this memory device helpful or useful, forget it; nothing we say here depends on it. In fact, it has some inherent dangers. For example, you must avoid thinking there is anything on the line $\{A,B\}$ other than the two points A and B. The line segments joining A and B and other pairs of points in the picture are parts of the memory device, but are not part of our geometry. Finally, we must avoid thinking there are any other points on the plane $\{A,B,C,D\}$ other than the points A, B, C, and D, or that there is any concept here of congruence or perpendicularity—which might be inferred if one takes the display of a cube too seriously.

It is an illuminating exercise to verify that the incidence axioms are satisfied by this geometry. A quick check on Axiom I.1 might be to choose several pairs of

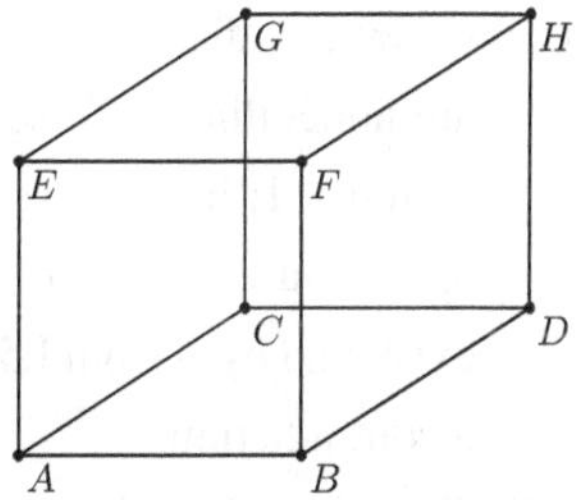

Fig. 1.1 Illustrating an 8 point model.

points, such as B and F, for example, and note that there is a line containing these two points and only one such line. A check for Axiom I.2 would be to choose sets of three points (which are noncollinear because no line contains three points) and verify that there is only one plane containing them. For instance, the only plane containing B, D, and H is $\{B, D, F, H\}$.

Checking all such possibilities is a tedious process that might best be left as an exercise; hopefully, we may find the memory device helpful in carrying it out. For example, if we choose the two points A and B, then there are three planes which contain these two points: $\{A, B, C, D\}$, $\{A, B, E, F\}$, and $\{A, B, G, H\}$. Of course, the line containing A and B, $\{A, B\}$, is a subset of each of these as required by Axiom I.3. To address Axiom I.4, we look for two planes with a nonempty intersection; $\{A, C, F, H\}$ and $\{B, D, F, H\}$ will do. Their intersection, $\{F, H\}$, contains at least two points as required by Axiom I.4. For Axiom I.5, it is not hard to check that every line has at least two different points and that each plane has as least three points which do not belong to the same line. Also, it is possible to find four points, A, B, C, and F, for example, which are not in the same plane.

As we state and prove some theorems in incidence geometry, we may want to look back at this model from time to time. It will often be an illuminating experience, but sometimes we will find it disappointing because the model is so simple.

1.10 Theorems for incidence geometry

Remark I.1 (Easy consequences of the axioms).

(A) Axiom I.5(C) says that space contains at least four points, therefore points exist, and space is nonempty. Axiom I.0 says that lines and planes exist and are subsets of space; Axiom I.5(A) and (B) says that each line contains at least two distinct points, and each plane contains at least three noncollinear points, so lines and planes are nonempty.

(B) Given any distinct points A and B there exists a point C such that A, B, and C are noncollinear. To see this, let $\mathcal{L}$ be the line through A and B guaranteed by Axiom I.1; if there were no point off $\mathcal{L}$ then all points of space would belong to $\mathcal{L}$, and every set of three points would be collinear. By Axiom I.0 planes exist, and by Axiom I.5(B) each one contains a set of three noncollinear points, a contradiction.

(C) There is a plane through any two points A and B of space. For by part (B) above there exists a point C not on the line $\mathcal{L}$ containing A and B. By Axiom I.2 there exists a plane $\mathcal{Q}$ containing A, B and C.

(D) (Criterion for noncollinear sets) If Axiom I.1 holds, to show that a set $\mathcal{E}$ is noncollinear it is sufficient to show that there is a line $\mathcal{L}$ containing two points of $\mathcal{E}$ which does not contain all the points of $\mathcal{E}$. For if $\mathcal{E}$ were collinear, there would be a line $\mathcal{M}$ containing $\mathcal{E}$; by Axiom I.1, $\mathcal{M} = \mathcal{L}$, so that $\mathcal{L}$ would contain all of $\mathcal{E}$, a contradiction.

(E) (Criterion for noncoplanar sets) If Axiom I.2 holds, then to show that a set $\mathcal{E}$ is noncoplanar, it is sufficient to show that there is a plane $\mathcal{P}$ containing three noncollinear points A, B, and C of $\mathcal{E}$ which does not contain all the points of $\mathcal{E}$. For if $\mathcal{E}$ were coplanar, there would be a plane $\mathcal{Q}$ containing $\mathcal{E}$, and by Axiom I.2 $\mathcal{Q} = \mathcal{P}$, so that $\mathcal{P}$ would contain all of $\mathcal{E}$, a contradiction.

Several of the exercises at the end of this chapter are similar to the statements just above and are about as easy to prove.

Definition I.2. (A) Let P and Q be distinct points. The line whose existence is asserted in Axiom I.1 is denoted by $\overleftrightarrow{PQ}$; this symbol is read "line PQ."

(B) Let P, Q, and R be noncollinear points. The plane whose existence is asserted in Axiom I.2 is denoted by $\overleftrightarrow{PQR}$.

Theorem I.3. *If $\mathcal{E}$ and $\mathcal{F}$ are distinct planes both of which contain line $\mathcal{L}$, then $\mathcal{E} \cap \mathcal{F} = \mathcal{L}$.*

Proof. Because both $\mathcal{E}$ and $\mathcal{F}$ contain $\mathcal{L}$, $\mathcal{L} \subseteq \mathcal{E} \cap \mathcal{F}$. Suppose there were some point A belonging to $\mathcal{E} \cap \mathcal{F}$ but not to $\mathcal{L}$. By Axiom I.5, there exist points B and C on $\mathcal{L}$. Then A, B, and C would all lie on $\mathcal{E}$, and they would all lie on $\mathcal{F}$. Hence by Axiom I.2 $\mathcal{E}$ would equal $\overleftrightarrow{ABC}$ and $\mathcal{F}$ would equal $\overleftrightarrow{ABC}$, from which it would follow that $\mathcal{E}$ would equal $\mathcal{F}$. But this contradicts the fact that $\mathcal{E}$ and $\mathcal{F}$ are distinct. This shows that there is no such point A, and we must conclude that $\mathcal{L} = \mathcal{E} \cap \mathcal{F}$. □

Theorem I.4. *If the intersection of two distinct planes is nonempty, then it is a line.*

Proof. Let the two distinct planes be $\mathcal{E}$ and $\mathcal{F}$. By Axiom I.4, there are two points, A and B, such that $\{A, B\} \subseteq \mathcal{E} \cap \mathcal{F}$. By Axiom I.1, there is one and only one line, $\overleftrightarrow{AB}$, containing A and B. By Axiom I.3, $\overleftrightarrow{AB} \subseteq \mathcal{E}$ and $\overleftrightarrow{AB} \subseteq \mathcal{F}$, so by Theorem I.3, $\mathcal{E} \cap \mathcal{F} = \overleftrightarrow{AB}$. □

Theorem I.5. *Given a plane $\mathcal{E}$ and a point A belonging to $\mathcal{E}$, there exists a line $\mathcal{L}$ such that $\mathcal{L} \subseteq \mathcal{E}$ and $A \notin \mathcal{L}$.*

Proof. By Axiom I.5 there exist three noncollinear points P, Q, and R belonging to $\mathcal{E}$. By Axiom I.1 there are three lines, $\overleftrightarrow{PQ}$, $\overleftrightarrow{QR}$, and $\overleftrightarrow{PR}$; by Axiom I.3, these lines are all contained in $\mathcal{E}$; and since P, Q, and R are noncollinear, the lines are distinct. The proof now splits into two cases.

(Case 1: The given point A happens to be one of the three points P, Q, or R, whose existence is assured by Axiom I.5.) In this case, the line determined by the other two points can be taken to be $\mathcal{L}$. This line does not contain A because if it did, P, Q, and R would be collinear.

(Case 2: The given point A does not coincide with any of the points P, Q, or R.) Then A cannot lie on more than one of the lines $\overleftrightarrow{PQ}$, $\overleftrightarrow{QR}$, and $\overleftrightarrow{PR}$. To see this, suppose for example that A belonged to both $\overleftrightarrow{PQ}$ and $\overleftrightarrow{QR}$. Then these two lines would have points A and Q in common, and hence by Axiom I.1 they would be the same line, contradicting the fact that they are distinct. Therefore $\mathcal{L}$ can be taken to be either of the two lines not containing A. □

Theorem I.6 (Two intersecting lines determine a plane). *Given lines $\mathcal{L}$ and $\mathcal{M}$ such that $\mathcal{L} \neq \mathcal{M}$ and $\mathcal{L} \cap \mathcal{M} \neq \emptyset$, there exists one and only one plane $\mathcal{E}$ such that $\mathcal{L} \subseteq \mathcal{E}$ and $\mathcal{M} \subseteq \mathcal{E}$.*

Proof. There are two things to be proved: (1) there is such a plane $\mathcal{E}$ (existence), and (2) there is not more than one such plane (uniqueness).

We first prove that there is such a plane $\mathcal{E}$. Since $\mathcal{L} \neq \mathcal{M}$ and $\mathcal{L} \cap \mathcal{M} \neq \emptyset$, by Exercise I.1 below, $\mathcal{L} \cap \mathcal{M}$ is a singleton $\{A\}$. By Axiom I.5 there exists a point B on $\mathcal{L}$ and distinct from A, and there exists a point C on $\mathcal{M}$ distinct from A. Since A, B, and C are noncollinear (if they were collinear, $\mathcal{L}$ and $\mathcal{M}$ would coincide, contrary to our assumption), by Axiom I.2 there is a plane $\mathcal{E}$ such that $\{A, B, C\} \subseteq \mathcal{E}$. By Axiom I.3, $\mathcal{L} \subseteq \mathcal{E}$ and $\mathcal{M} \subseteq \mathcal{E}$.

To prove that there is not more than one such plane, suppose on the contrary that there were a second plane $\mathcal{E}'$ containing both $\mathcal{L}$ and $\mathcal{M}$. Then all of the points A, B, and C defined above would belong to $\mathcal{E}'$. Hence by Axiom I.2, $\mathcal{E} = \mathcal{E}'$. □

Theorem I.7. *Let $\mathcal{E}$ be a plane. There exists a point P such that $P \notin \mathcal{E}$.*

Proof. By Axiom I.5 there exist points A, B, C, and D which are noncoplanar. If $\{A, B, C, D\}$ were a subset of $\mathcal{E}$, then A, B, C, and D would be coplanar. Hence at least one member of $\{A, B, C, D\}$ does not belong to $\mathcal{E}$, proving the theorem. □

Theorem I.8. *Let $\mathcal{S}$ and $\mathcal{T}$ be distinct planes whose intersection is the line $\mathcal{L}$, and let P be a member of $\mathcal{L}$; then there exist lines $\mathcal{M}$ and $\mathcal{N}$ such that $\mathcal{M} \subseteq \mathcal{S}$, $\mathcal{N} \subseteq \mathcal{T}$, $\mathcal{M} \neq \mathcal{L}$, $\mathcal{N} \neq \mathcal{L}$, and $\mathcal{M} \cap \mathcal{N} = \{P\}$. If $\mathcal{M}$ and $\mathcal{N}$ are any two lines satisfying these conditions, then there is exactly one plane $\mathcal{E}$ such that $\mathcal{M} \cup \mathcal{N} \subseteq \mathcal{E}$. Moreover, $\mathcal{E} \cap \mathcal{L} = \{P\}$.*

Proof. Since $\mathcal{S}$ and $\mathcal{T}$ are distinct, there is at least one point S and at least one point T such that $S \in \mathcal{S}$, $T \in \mathcal{T}$, $S \notin \mathcal{T}$, and $T \notin \mathcal{S}$. By Axiom I.1, there is one and only one line $\mathcal{M}$ containing P and S, and one and only one line $\mathcal{N}$ containing P and T. By Axiom I.3, $\mathcal{M} \subseteq \mathcal{S}$ and $\mathcal{N} \subseteq \mathcal{T}$. Since $S \neq T$, $\mathcal{M} \neq \mathcal{N}$. Therefore by Exercise I.1 below, $\mathcal{M} \cap \mathcal{N}$ contains one point. By the way $\mathcal{M}$ and $\mathcal{N}$ were defined, $P \in \mathcal{M}$ and $P \in \mathcal{N}$, so $\mathcal{M} \cap \mathcal{N} = \{P\}$. Moreover, since $S \notin \mathcal{T}$, and $S \in \mathcal{M}$, $\mathcal{M} \neq \mathcal{L}$; similarly, $\mathcal{N} \neq \mathcal{L}$.

Now let $\mathcal{M}$ and $\mathcal{N}$ be any two lines satisfying the conditions in the first part of the theorem. These lines are distinct because if $\mathcal{M}$ were equal to $\mathcal{N}$, then $\mathcal{M} \cap \mathcal{N}$ would be equal to $\mathcal{M}$, for example, and hence by Axiom I.5 would contain at least two points, which contradicts the fact that $\mathcal{M} \cap \mathcal{N} = \{P\}$.

By Theorem I.6, there is one and only one plane $\mathcal{E}$ such that $\mathcal{M} \cup \mathcal{N} \subseteq \mathcal{E}$. Now if $\mathcal{S}$ and $\mathcal{E}$ were equal then $\mathcal{N}$ would be a subset of $\mathcal{S}$ as well as a subset of $\mathcal{T}$. Therefore by Theorem I.3, $\mathcal{S} \cap \mathcal{T}$ would be $\mathcal{N}$. But $\mathcal{S} \cap \mathcal{T}$ is $\mathcal{L}$ by definition, and we have defined $\mathcal{N}$ so that $\mathcal{N} \neq \mathcal{L}$. This contradiction shows that $\mathcal{S}$ and $\mathcal{E}$ are distinct planes. Therefore by Theorem I.3, $\mathcal{S} \cap \mathcal{E} = \mathcal{M}$ because $\mathcal{M} \subseteq \mathcal{S}$ and $\mathcal{M} \subseteq \mathcal{E}$. By a similar argument, $\mathcal{T} \cap \mathcal{E} = \mathcal{N}$. Hence $\mathcal{L} \cap \mathcal{E} = (\mathcal{S} \cap \mathcal{T}) \cap \mathcal{E} = (\mathcal{S} \cap \mathcal{E}) \cap (\mathcal{T} \cap \mathcal{E}) = \mathcal{M} \cap \mathcal{N} = \{P\}$. □

Theorem I.9. *Let $\mathcal{P}_1$, $\mathcal{P}_2$, and $\mathcal{P}_3$ be distinct planes such that each of the sets $\mathcal{P}_1 \cap \mathcal{P}_2$, $\mathcal{P}_1 \cap \mathcal{P}_3$, $\mathcal{P}_2 \cap \mathcal{P}_3$ is nonempty. Then there exist lines $\mathcal{L}_1$, $\mathcal{L}_2$, and $\mathcal{L}_3$ such that $\mathcal{P}_2 \cap \mathcal{P}_3 = \mathcal{L}_1$, $\mathcal{P}_1 \cap \mathcal{P}_3 = \mathcal{L}_2$, and $\mathcal{P}_1 \cap \mathcal{P}_2 = \mathcal{L}_3$. Furthermore, one and only one of the following statements is true:*

(1) $\mathcal{L}_1 \cap \mathcal{L}_2 = \mathcal{L}_1 \cap \mathcal{L}_3 = \mathcal{L}_2 \cap \mathcal{L}_3 = \emptyset$,
(2) $\mathcal{L}_1 = \mathcal{L}_2 = \mathcal{L}_3$, *and*
(3) $\mathcal{L}_1 \cap \mathcal{L}_2 \cap \mathcal{L}_3$ *is a singleton.*

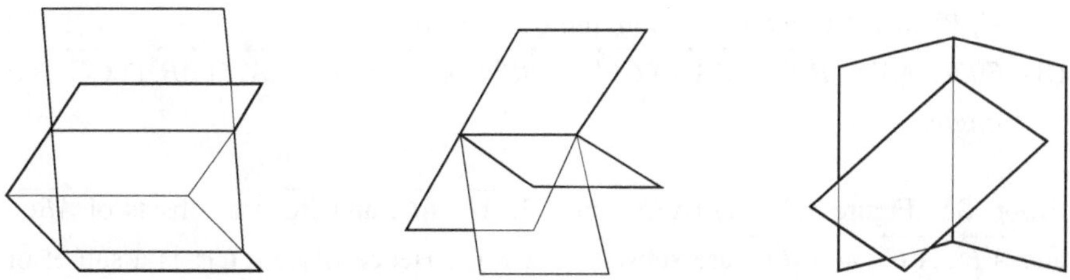

Fig. 1.2 For Theorem I.9, showing alternative (1) at left, alternative (2) in the middle, and alternative (3) on the right.

Proof. For a visualization, refer to Figure 1.2. By Theorem I.4, $\mathcal{P}_2 \cap \mathcal{P}_3$ is a line $\mathcal{L}_1$, $\mathcal{P}_1 \cap \mathcal{P}_3$ is a line $\mathcal{L}_2$, and $\mathcal{P}_1 \cap \mathcal{P}_2$ is a line $\mathcal{L}_3$. By set theory,

$$\mathcal{P}_1 \cap \mathcal{P}_2 \cap \mathcal{P}_3 = (\mathcal{P}_1 \cap \mathcal{P}_2) \cap (\mathcal{P}_2 \cap \mathcal{P}_3) \cap (\mathcal{P}_1 \cap \mathcal{P}_3) = \mathcal{L}_1 \cap \mathcal{L}_2 \cap \mathcal{L}_3.$$

Moreover, if X is any member of $\mathcal{L}_1 \cap \mathcal{L}_2$, then $X \in ((\mathcal{P}_2 \cap \mathcal{P}_3) \cap (\mathcal{P}_1 \cap \mathcal{P}_3)) = \mathcal{P}_1 \cap \mathcal{P}_2 \cap \mathcal{P}_3 \subseteq \mathcal{L}_3$. Hence $\mathcal{L}_1 \cap \mathcal{L}_2 \subseteq \mathcal{L}_3$.

Now there are two mutually exclusive possibilities: either

(A) $\mathcal{L}_1 \cap \mathcal{L}_2 = \mathcal{L}_1 \cap \mathcal{L}_3 = \mathcal{L}_2 \cap \mathcal{L}_3 = \emptyset$ (so that (1) is true), or
(B) $(\mathcal{L}_1 \cap \mathcal{L}_2) \cup (\mathcal{L}_1 \cap \mathcal{L}_3) \cup (\mathcal{L}_2 \cap \mathcal{L}_3) \neq \emptyset$.

In the latter case, at least one of the intersections is nonempty, and we can choose the notation so that $\mathcal{L}_1 \cap \mathcal{L}_2 \neq \emptyset$. Then either this set is a singleton or it is not. If it is a singleton, then from what we have said above, $\mathcal{L}_1 \cap \mathcal{L}_2 \cap \mathcal{L}_3 \subseteq \mathcal{L}_1 \cap \mathcal{L}_2 \subseteq \mathcal{L}_3$, and hence $\mathcal{L}_1 \cap \mathcal{L}_2 \cap \mathcal{L}_3$ is also a singleton, and (3) is true. If $\mathcal{L}_1 \cap \mathcal{L}_2$ is not a singleton, then by Exercise I.2, $\mathcal{L}_1 = \mathcal{L}_2$, and since $\mathcal{L}_1 \cap \mathcal{L}_2 \subseteq \mathcal{L}_3$, by Exercise I.3 $\mathcal{L}_1 = \mathcal{L}_2 = \mathcal{L}_3$, and (2) is true. □

Girard Desargues (1591–1661) lived in France and is especially known for his work with *projective geometry.*

Theorem I.10 (Proposition of Desargues, nonplanar version). *Let A, B, C, A', B', and C' be distinct points of the space $\mathbb{U}$ such that A, B, C are noncollinear, A', B', and C' are noncollinear, $\overleftrightarrow{ABC} \neq \overleftrightarrow{A'B'C'}$, $\overleftrightarrow{AB}$ and $\overleftrightarrow{A'B'}$ are concurrent at C_1* (i.e., *these two lines are distinct and have only the point C_1 in common), $\overleftrightarrow{AC}$ and $\overleftrightarrow{A'C'}$ are concurrent at B_1, and $\overleftrightarrow{BC}$ and $\overleftrightarrow{B'C'}$ are concurrent at A_1. Then:*

(A) *A_1, B_1, and C_1 are distinct and collinear. Moreover, none of the points A, B, C, A', B', or C' is on the line containing A_1, B_1, and C_1.*
(B) *Either $\overleftrightarrow{AA'} \cap \overleftrightarrow{BB'} = \overleftrightarrow{AA'} \cap \overleftrightarrow{CC'} = \overleftrightarrow{BB'} \cap \overleftrightarrow{CC'} = \emptyset$ or $\overleftrightarrow{AA'} \cap \overleftrightarrow{BB'} \cap \overleftrightarrow{CC'}$ is a singleton.*

Proof. See Figure 1.3. (A) By Axiom I.3, $\overleftrightarrow{AB}$, $\overleftrightarrow{AC}$, and $\overleftrightarrow{BC}$ are subsets of $\overleftrightarrow{ABC}$; and $\overleftrightarrow{A'B'}$, $\overleftrightarrow{A'C'}$, and $\overleftrightarrow{B'C'}$ are subsets of $\overleftrightarrow{A'B'C'}$. Hence $\{A_1, B_1, C_1\}$ is a subset of $\overleftrightarrow{ABC} \cap \overleftrightarrow{A'B'C'}$ (this follows from the properties of subsets and of the operation of intersection which one learns in elementary set theory). By Theorem I.4, $\overleftrightarrow{ABC} \cap \overleftrightarrow{A'B'C'}$ is a line $\mathcal{L}$. Thus A_1, B_1, and C_1 are collinear, and members of both $\overleftrightarrow{ABC}$ and $\overleftrightarrow{A'B'C'}$.

Claim 1: None of the points A, B, C, A', B', or C' belong to $\mathcal{L}$. We show first that $A \notin \mathcal{L}$; similar arguments show the other assertions.

If $A \in \mathcal{L} \subseteq \overleftrightarrow{A'B'C'}$, since $C_1 \in \overleftrightarrow{A'B'C'}$, by Axiom I.3 $B \in \overleftrightarrow{AC_1} \subseteq \overleftrightarrow{A'B'C'}$. Arguing similarly with A and B_1, we get that $C \in \overleftrightarrow{A'B'C'}$. Therefore the plane

Fig. 1.3 For Theorem I.10.

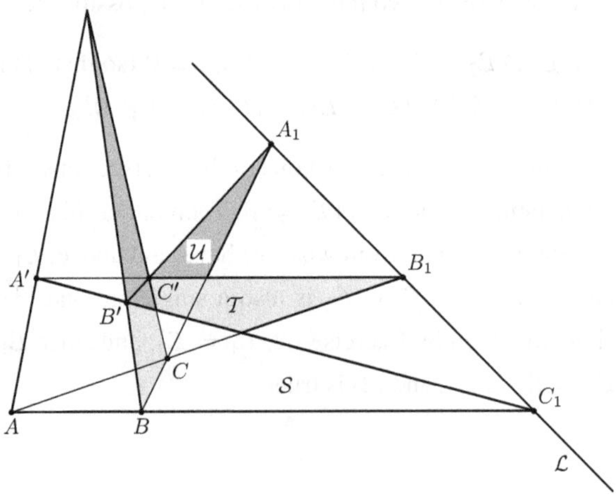

$\overleftrightarrow{A'B'C'}$ contains the three noncollinear points A, B, and C, so by Axiom I.2, $\overleftrightarrow{A'B'C'} = \overleftrightarrow{ABC}$, contradicting our hypothesis that $\overleftrightarrow{ABC} \neq \overleftrightarrow{A'B'C'}$. Therefore $A \notin \mathcal{L}$.

Claim 2: A_1, B_1, and C_1 are distinct. If $A_1 = B_1$, then $\overleftrightarrow{A_1B} = \overleftrightarrow{A_1C} = \overleftrightarrow{B_1C} = \overleftrightarrow{B_1A}$ so that A, B, and C are collinear. This contradicts our hypothesis that they are noncollinear. Similar arguments show that $A_1 \neq C_1$ and $B_1 \neq C_1$.

(B) We continue to use the notation of part (A) and we use its results.

Claim 3: The lines $\overleftrightarrow{AA'}$, $\overleftrightarrow{BB'}$, and $\overleftrightarrow{CC'}$ are pairwise distinct. If $\overleftrightarrow{AA'} = \overleftrightarrow{BB'}$, then A, B, A', and B' would be collinear, and hence $\overleftrightarrow{AB} = \overleftrightarrow{A'B'}$, contrary to the (given) fact that $\overleftrightarrow{AB}$ and $\overleftrightarrow{A'B'}$ are concurrent (only) at C_1. Hence $\overleftrightarrow{AA'} \neq \overleftrightarrow{BB'}$. By similar arguments, $\overleftrightarrow{AA'} \neq \overleftrightarrow{CC'}$, and $\overleftrightarrow{BB'} \neq \overleftrightarrow{CC'}$, proving the claim.

By Theorem I.6 there exist unique planes $\mathcal{S}$, $\mathcal{T}$, and $\mathcal{U}$ such that $\overleftrightarrow{AB} \cup \overleftrightarrow{A'B'} \subseteq \mathcal{S}$, $\overleftrightarrow{AC} \cup \overleftrightarrow{A'C'} \subseteq \mathcal{T}$, and $\overleftrightarrow{BC} \cup \overleftrightarrow{B'C'} \subseteq \mathcal{U}$.

Claim 4: $\mathcal{S}$, $\mathcal{T}$, and $\mathcal{U}$ are distinct. If $\mathcal{S} = \mathcal{T}$, then A, B, and C would lie in the same plane as A', B', and C', which would contradict the hypothesis that $\overleftrightarrow{ABC} \neq \overleftrightarrow{A'B'C'}$. Similar arguments show that $\mathcal{S} \neq \mathcal{U}$ and $\mathcal{T} \neq \mathcal{U}$.

Claim 5: $\mathcal{S} \cap \mathcal{T} = \overleftrightarrow{AA'}$, $\mathcal{S} \cap \mathcal{U} = \overleftrightarrow{BB'}$, and $\mathcal{T} \cap \mathcal{U} = \overleftrightarrow{CC'}$. By Theorem I.4 $\mathcal{S} \cap \mathcal{T}$ is a line, and this line contains both A and A'. Therefore $\mathcal{S} \cap \mathcal{T} \subseteq \overleftrightarrow{AA'}$ and by Exercise I.3 $\mathcal{S} \cap \mathcal{T} = \overleftrightarrow{AA'}$. The other two assertions follow by similar arguments.

Claim 6: $\overleftrightarrow{AB} \neq \mathcal{L}$, $\overleftrightarrow{BC} \neq \mathcal{L}$, $\overleftrightarrow{AC} \neq \mathcal{L}$, $\overleftrightarrow{A'B'} \neq \mathcal{L}$, $\overleftrightarrow{A'C'} \neq \mathcal{L}$, and $\overleftrightarrow{B'C'} \neq \mathcal{L}$. These assertions follow immediately from the observation (see Claim 1) that none of the points A, B, C, A', B', or C' belong to $\mathcal{L}$.

Claim 7: $\mathcal{S} \cap \mathcal{L} = \{C_1\}$, $\mathcal{T} \cap \mathcal{L} = \{B_1\}$, and $\mathcal{U} \cap \mathcal{L} = \{A_1\}$. Again, we argue only the first of these assertions, as the others follow similarly. Since $\mathcal{S}$ is the unique plane containing $\overleftrightarrow{AB} \cup \overleftrightarrow{A'B'}$, by the last assertion of Theorem I.8, $\mathcal{S} \cap \mathcal{L} = \{C_1\}$.

We can now complete the proof. By Claim 5, $\mathcal{S} \cap \mathcal{T} = \overleftrightarrow{AA'}$, $\mathcal{S} \cap \mathcal{U} = \overleftrightarrow{BB'}$, and $\mathcal{T} \cap \mathcal{U} = \overleftrightarrow{CC'}$. Thus we may apply Theorem I.9. Conclusion (2) of this theorem is ruled out by Claim 3, so that either

$$\overleftrightarrow{AA'} \cap \overleftrightarrow{BB'} = \overleftrightarrow{AA'} \cap \overleftrightarrow{CC'} = \overleftrightarrow{BB'} \cap \overleftrightarrow{CC'} = \emptyset$$

or

$$\overleftrightarrow{AA'} \cap \overleftrightarrow{BB'} \cap \overleftrightarrow{CC'} \text{ is a singleton.}$$ □

For a discussion of the significance of Desargues' Theorem, see David Hilbert, *The Foundations of Geometry*, Chapter V [10].

1.11 Exercises for incidence geometry

The following set of exercises consists of further theorems which can be proved from the incidence axioms alone. We strongly suggest that you review the item "Exercises" in Section 1.7 above, which explains the role of exercises in this book, which is different from their role in most textbooks. Answers to starred (*) exercises may be accessed from the home page for this book at www.springer.com.

Exercise I.1*. If $\mathcal{L}$ and $\mathcal{M}$ are distinct lines and if $\mathcal{L} \cap \mathcal{M} \neq \emptyset$, then $\mathcal{L} \cap \mathcal{M}$ is a singleton.

Exercise I.2*. (A) If A and B are distinct points, and if C and D are distinct points on $\overleftrightarrow{AB}$, then $\overleftrightarrow{CD} = \overleftrightarrow{AB}$.

(B) If A, B, and C are noncollinear points, and if D, E, and F are noncollinear points on $\overleftrightarrow{ABC}$, then $\overleftrightarrow{DEF} = \overleftrightarrow{ABC}$.

Exercise I.3*. If $\mathcal{L}$ and $\mathcal{M}$ are lines and $\mathcal{L} \subseteq \mathcal{M}$, then $\mathcal{L} = \mathcal{M}$.

Exercise I.4*. Let A and B be two distinct points, and let D, E, and F be three noncollinear points. If $\overleftrightarrow{AB}$ contains only one of the points D, E, and F, then each of the lines $\overleftrightarrow{DE}$, $\overleftrightarrow{EF}$, and $\overleftrightarrow{DF}$ intersects $\overleftrightarrow{AB}$ in at most one point.

Exercise I.5*. If $\mathcal{E}$ is a plane, $\mathcal{L}$ is a line such that $\mathcal{E} \cap \mathcal{L} \neq \emptyset$, and $\mathcal{L}$ is not contained in $\mathcal{E}$, then $\mathcal{E} \cap \mathcal{L}$ is a singleton.

Exercise I.6. Let $\mathcal{D}$ and $\mathcal{E}$ be distinct planes such that $\mathcal{D} \cap \mathcal{E} \neq \emptyset$, so that (by Theorem I.4) $\mathcal{D} \cap \mathcal{E}$ is a line $\mathcal{L}$; let P be a point on $\mathcal{D}$ but not on $\mathcal{L}$; and let Q be a point on $\mathcal{E}$ but not on $\mathcal{L}$. Then $\overleftrightarrow{PQ}$ and $\mathcal{L}$ are not coplanar.

Exercise I.7*. Given a line $\mathcal{L}$ and a point A not on $\mathcal{L}$, there exists one and only one plane $\mathcal{E}$ such that $A \in \mathcal{E}$ and $\mathcal{L} \subseteq \mathcal{E}$.

Exercise I.8*. Let A, B, C, and D be noncoplanar points. Then each of the triples $\{A, B, C\}$, $\{A, B, D\}$, $\{A, C, D\}$, and $\{B, C, D\}$ is noncollinear.

Exercise I.9. There exist four distinct planes such that no point is common to all of them.

Exercise I.10. Every plane $\mathcal{E}$ contains at least three lines $\mathcal{L}$, $\mathcal{M}$, and $\mathcal{N}$ such that $\mathcal{L} \cap \mathcal{M} \cap \mathcal{N} = \emptyset$.

Exercise I.11. Every plane contains (at least) three distinct lines.

Exercise I.12. Space contains (at least) six distinct lines.

Exercise I.13*. If $\mathcal{L}$ is a line contained in a plane $\mathcal{E}$, then there exists a point A belonging to $\mathcal{E}$ but not belonging to $\mathcal{L}$.

Exercise I.14. If P is a point in a plane $\mathcal{E}$, then there is a line $\mathcal{L}$ such that $P \in \mathcal{L}$ and $\mathcal{L} \subseteq \mathcal{E}$.

Exercise I.15. If a plane $\mathcal{E}$ has (exactly) three points, then each line contained in $\mathcal{E}$ has (exactly) two points.

Exercise I.16. If a plane $\mathcal{E}$ has exactly four points, and if all of the lines contained in $\mathcal{E}$ have the same number of points, then each line contained in $\mathcal{E}$ has (exactly) two points.

Exercise I.17. If each line in space has at least three points, then:

(1) Each point of a plane is a member of at least three lines of the plane;
(2) Each plane has at least seven points;
(3) Each plane contains at least seven lines.

Exercise I.18. In this exercise we will use the symbolism "$\mathcal{P} \parallel \mathcal{Q}$" to indicate that two planes $\mathcal{P}$ and $\mathcal{Q}$ do not intersect.

Consider what can happen if the restrictions of $\mathcal{P}_1 \cap \mathcal{P}_2$, $\mathcal{P}_1 \cap \mathcal{P}_3$, and $\mathcal{P}_2 \cap \mathcal{P}_3$ being nonempty are removed in Theorem I.9. Sketch at least four possibilities ($\mathcal{P}_1 \parallel \mathcal{P}_2 \parallel \mathcal{P}_3$ and $\mathcal{P}_1$, $\mathcal{P}_2$, and $\mathcal{P}_3$ are mutually disjoint, $\mathcal{P}_1 = \mathcal{P}_2 \parallel \mathcal{P}_3$, $\mathcal{P}_1 = \mathcal{P}_2 = \mathcal{P}_3$, $\mathcal{P}_1 \cap \mathcal{P}_2 = \emptyset$, but $\mathcal{P}_1 \cap \mathcal{P}_3 = \mathcal{L}_2$ and $\mathcal{P}_2 \cap \mathcal{P}_3 = \mathcal{L}_1$) and in each case determine what (if anything) similar to Theorem I.9 can be proved within incidence geometry.

Exercise I.19. Count the number of lines in the 8-point model. Compare this with $T_n = \frac{n(n+1)}{2}$, triangular numbers, for $n = 7$. Compare it also with ${}_nC_r = \frac{n!}{r!(n-r)!}$, the number of combinations of n items taken r at a time, where $n = 8$ and $r = 2$.

Exercise I.20. Count the number of planes in the 8-point model. Compare this with ${}_nC_r$ for $n = 8$ and $r = 3$. Note the reduction by a factor of four due to the fact that each plane has four points. Can you form a similar argument with $r = 4$?

Exercise I.21. Consider a 4-point model with the four points configured like the vertices of a tetrahedron. Label these points A, B, C, and D. Specify six lines and four planes and verify that this model satisfies the axioms and theorem of incidence geometry. Compare this with Exercises I.7, I.10, I.12, and I.13. How does Theorem I.9 apply in this geometry?

Chapter 2
Affine Geometry: Incidence with Parallelism (IP)

Acronym: *IP*
Dependencies: *Chapter 1*
New Axioms: *Axiom PS (strong parallel)*
New Terms Defined: *parallel, pencil, focal point, affine*

Abstract: This brief chapter introduces the notion of parallelism, discusses the two forms of the parallel axiom, defines affine geometry, and proves five elementary theorems relating to intersecting planes and parallel lines.

2.1 Parallelism and parallel axioms

The early part of the main development of this book (Chapters 4 through 10) does not invoke a parallel axiom, but does need the terminology of parallelism. For this reason (and to further our understanding of the history of geometry), we take this opportunity to digress for two chapters to discuss the parallel axiom and prove what can be proved at this stage.

There are two forms of the parallel axiom; one of them claims more than the other. We state both forms here, and whenever we use one, we will specify which one. Before we give these axioms, we need to define parallelism.

Definition IP.0. (A) Lines $\mathcal{L}$ and $\mathcal{M}$ are **parallel** (notation: $\mathcal{L} \parallel \mathcal{M}$) iff there is a plane that contains them both and $\mathcal{L} \cap \mathcal{M} = \emptyset$.

E.J. Specht et al., *Euclidean Geometry and its Subgeometries*,
DOI 10.1007/978-3-319-23775-6_2

(B) A line $\mathcal{L}$ and a plane $\mathcal{P}$ are **parallel** (notation: $\mathcal{L} \parallel \mathcal{P}$) iff $\mathcal{L} \cap \mathcal{P} = \emptyset$.

(C) Two planes $\mathcal{P}$ and $\mathcal{P}'$ are **parallel** (notation: $\mathcal{P} \parallel \mathcal{P}'$) iff $\mathcal{P} \cap \mathcal{P}' = \emptyset$.

(D) A set $\mathbb{E}$ of two or more distinct lines on a plane $\mathcal{P}$ is a **pencil** iff either (1) the members of $\mathbb{E}$ are concurrent at some point O, or (2) every member of $\mathbb{E}$ is parallel to every other member of $\mathbb{E}$. In case (1), the point O is the **focal point** of $\mathbb{E}$.

Notice that, by (A) alone, if two lines $\mathcal{L}$ and $\mathcal{M}$ are nonparallel (symbolically, $\mathcal{L} \nparallel \mathcal{M}$) and lie in the same plane, they are intersecting (this is Exercise I.1), and by (C), two planes are intersecting iff they are nonparallel. But beware that two nonparallel lines do not necessarily intersect—there may not be a plane that contains them both.

Axiom PS (Strong Form of the Parallel Axiom). Given a line $\mathcal{L}$ and a point P not belonging to $\mathcal{L}$, there exists *exactly* one line $\mathcal{M}$ such that $P \in \mathcal{M}$ and $\mathcal{L} \parallel \mathcal{M}$. (If such a line exists, it is denoted $\text{par}(P, \mathcal{L})$.)

Axiom PW (Weak Form of the Parallel Axiom). Given a line $\mathcal{L}$ and a point P not belonging to $\mathcal{L}$, there exists *at most* one line $\mathcal{M}$ such that $P \in \mathcal{M}$ and $\mathcal{L} \parallel \mathcal{M}$. Note that Axiom PW does not guarantee that such a line exists.

Axioms PS and PW have an interesting history. Euclid had an axiom in his *Elements* which is equivalent to PW, but it appears he wasn't sure it was as fully self-evident as his other "self-evident truths," since he postponed using it in his development as long as he could. His contemporaries and others who followed apparently felt the same way and tried to dispense with it as an axiom entirely by showing it to be a consequence of the other axioms. For the next 2,000 years, a favorite pastime was to try to prove it from the other axioms. It can be said that almost every great mathematician to live during those 20 centuries tried his hand at this proof. All these "proofs" turned out to be fallacious. Our Axiom PW was formulated in the fifth century by Proclus Lycaeus (412–485), but its equivalence to Euclid's original parallel postulate was most widely publicized by the Scotsman John Playfair (1748–1819) in the eighteenth century, so it is generally known as "Playfair's Axiom."

Saccheri (1667–1733) thought of dealing with the mystery by considering the collection of all the other axioms and the negation of Axiom PW, hoping to get a contradiction, which would prove PW to be a consequence of the other axioms. This new collection of axioms led to some really repugnant conclusions, but not to a contradiction. It isn't completely clear exactly how or when the truth of the matter dawned on the mathematical world, but it apparently happened during the first third of the nineteenth century, and it is quite clear that Gauss (1777–1855), Bolyai (1802–1860), and Lobachevsky (1792–1856) were all involved. For a very readable account of the whole matter, see *The History of Mathematics: An Introduction* by David M. Burton, 7th ed, McGraw Hill (2010) [4].

These three mathematicians (Gauss, Bolyai, and Lobachevsky) were apparently the first to recognize explicitly the idea upon which this development of geometry is founded: everyone is free to choose whatever axiom system pleases him/her, as long as the axioms in it are consistent (*i.e.*, don't contradict each other). You have already seen an example of this in incidence geometry. The geometry we get by choosing a particular system may seem weird to us, but that is not sufficient logical grounds for rejecting it.

Recall again the language of Axiom PS: given a line $\mathcal{L}$ and point $P \notin \mathcal{L}$, there is a line through P parallel to $\mathcal{L}$ *and* this is the only such line. A negation of Axiom PS, then, would say *either* 1) there is no such line through P parallel to $\mathcal{L}$ *or* 2) there is more than one such line. Either statement, separately, is a *denial* of Axiom PS. Adjoining denial 1) to the rest of Euclid's axioms yields *elliptic* geometry; adjoining denial 2) yields *hyperbolic (Lobachevskian)* geometry; both are *non-Euclidean* geometries.

This classification resulted from the way geometry developed in the nineteenth century. In it, Euclidean geometry is *parabolic.* Historically, elliptic geometry was also known as *Riemannian,* but modern usage tends to identify Riemannian geometry as a branch of *differential geometry.*

In Chapter 8, neutral geometry, we will prove that parallel lines exist, so that neutral geometry as we develop it there will be incompatible with elliptic geometry.

In hyperbolic geometry, there can be two coplanar intersecting lines $\mathcal{M}$ and $\mathcal{N}$ and a third line $\mathcal{L}$ lying in the same plane which intersects neither $\mathcal{M}$ nor $\mathcal{N}$, so that the intuitively appealing claim that any line must intersect one of two lines that intersect each other is false.

In this chapter we adopt Axiom PS rather than Axiom PW to explore a geometry involving only incidence and parallelism.

Definition IP.1. A plane $\mathcal{P}$ is an **affine plane** iff it is a subset of space where the incidence axioms[1] and Axiom PS hold. **Affine geometry** is the term used to describe the geometry of such a plane.

2.2 Theorems of affine geometry

Theorems IP.2 and IP.3 do not use either Axiom PS or Axiom PW, and could have been proved in incidence geometry. They do, however, use the terminology and notation just introduced in Definition IP.0.

Theorem IP.2. *Let $\mathcal{E}$ be a plane, and let $\mathcal{M}$ and $\mathcal{L}$ be parallel lines; if $\mathcal{L} \subseteq \mathcal{E}$ and $\mathcal{M} \nsubseteq \mathcal{E}$, then $\mathcal{M} \parallel \mathcal{E}$.*

Proof. By the definition of parallel lines (Definition IP.0(A)), there exists a plane $\mathcal{F}$ containing $\mathcal{L}$ and $\mathcal{M}$. Since $\mathcal{L} \subseteq \mathcal{E}$, $\mathcal{L} \subseteq \mathcal{F}$, and $\mathcal{E} \neq \mathcal{F}$ (because $\mathcal{M} \nsubseteq \mathcal{E}$), we have $\mathcal{L} = \mathcal{E} \cap \mathcal{F}$, by Theorem I.3 and Exercise I.3.

Now suppose there is a point P such that $P \in (\mathcal{M} \cap \mathcal{E})$. Since $P \in \mathcal{M} \subseteq \mathcal{F}$ and $P \in \mathcal{E}$, $P \in (\mathcal{E} \cap \mathcal{F}) = \mathcal{L}$, so $P \in (\mathcal{M} \cap \mathcal{L})$. This contradicts the fact that $\mathcal{M} \parallel \mathcal{L}$. This contradiction shows that there is no such point P, so $\mathcal{M} \parallel \mathcal{E}$. □

Theorem IP.3. *Let $\mathcal{L}$ be a line in a plane $\mathcal{F}$ and suppose $\mathcal{L}$ is parallel to a plane $\mathcal{E}$ that intersects $\mathcal{F}$. Then $\mathcal{E} \cap \mathcal{F}$ is a line $\mathcal{M}$ which is parallel to $\mathcal{L}$.*

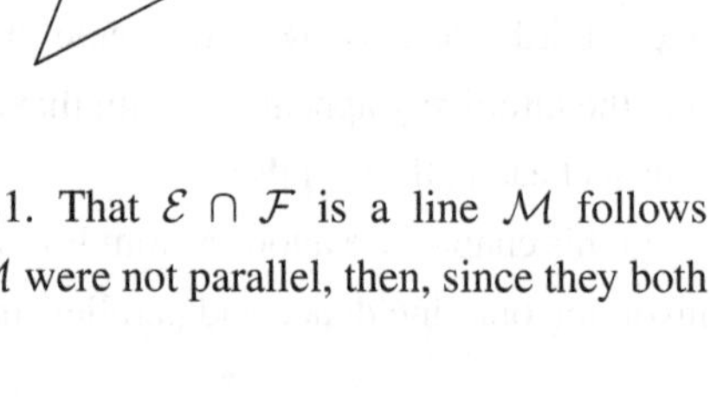

Fig. 2.1 For Theorem IP.3.

Proof. For a visualization, see Figure 2.1. That $\mathcal{E} \cap \mathcal{F}$ is a line $\mathcal{M}$ follows immediately from Theorem I.4. If $\mathcal{L}$ and $\mathcal{M}$ were not parallel, then, since they both

[1]Not all the incidence axioms apply to what goes on within a single plane; the ones that don't are Axiom I.2, Axiom I.4, and Axiom I.5(C); if later on we say that the incidence axioms hold for a plane we will mean that the *relevant* axioms hold.

are contained in $\mathcal{F}$, they would intersect in some point A, say. This A would belong to both $\mathcal{L}$ and $\mathcal{M}$ and since $\mathcal{M} = \mathcal{E} \cap \mathcal{F}$, this would contradict the fact that $\mathcal{L} \parallel \mathcal{E}$. Hence $\mathcal{L} \parallel \mathcal{M}$. □

Note that the proof of the next theorem depends on a parallel axiom.

Theorem IP.4. *If $\mathcal{F}$ is a plane containing two lines $\mathcal{L}$ and $\mathcal{M}$ which intersect at a point P, and if $\mathcal{E}$ is a plane which is parallel to both $\mathcal{L}$ and $\mathcal{M}$, then $\mathcal{E} \parallel \mathcal{F}$.*

Proof. Suppose $\mathcal{E}$ and $\mathcal{F}$ were intersecting. Then by Theorem I.4 their intersection would be a line $\mathcal{N}$. If $\mathcal{N}$ were parallel to both $\mathcal{L}$ and $\mathcal{M}$, we would have two lines through P which are both parallel to $\mathcal{N}$, which contradicts Axiom PS (and Axiom PW). Hence $\mathcal{N}$ is not parallel to both $\mathcal{L}$ and $\mathcal{M}$, and therefore must intersect at least one of them, since all three lines lie in $\mathcal{F}$. Suppose the notation is chosen so that $\mathcal{L}$ intersects $\mathcal{N}$. By Exercise I.1 their intersection is some point Q. Then $Q \in \mathcal{L}$ and $Q \in \mathcal{N} \subseteq \mathcal{E}$, so that $\mathcal{L} \cap \mathcal{E} \neq \emptyset$. This contradicts the hypothesis of the theorem that $\mathcal{L} \parallel \mathcal{E}$. Hence our assumption that $\mathcal{E}$ and $\mathcal{F}$ intersect is false, which means $\mathcal{E} \parallel \mathcal{F}$. □

Theorem IP.5. *Given a plane $\mathcal{E}$ and a point P not on $\mathcal{E}$, there exists exactly one plane $\mathcal{F}$ such that $P \in \mathcal{F}$ and $\mathcal{E} \parallel \mathcal{F}$.*

Proof. There are two things to prove: (I, existence) that there is such a plane $\mathcal{F}$, and (II, uniqueness) that there is not more than one such plane.

(I: existence) By Axiom I.5(B) there exist noncollinear points Q, R, and S belonging to $\mathcal{E}$. By Axiom I.1 there exist lines $\mathcal{L} = \overleftrightarrow{QR}$ and $\mathcal{M} = \overleftrightarrow{QS}$, and by Exercise I.1 $\mathcal{L} \cap \mathcal{M} = \{Q\}$. By Axiom I.3, $\mathcal{L}$ and $\mathcal{M}$ are contained in $\mathcal{E}$. By Exercise I.7 there exist planes $\mathcal{G}$ and $\mathcal{H}$ such that $P \in \mathcal{G}$, $\mathcal{L} \subseteq \mathcal{G}$, $P \in \mathcal{H}$, and $\mathcal{M} \subseteq \mathcal{H}$. By Theorem I.3, $\mathcal{G} \cap \mathcal{E} = \mathcal{L}$ and $\mathcal{H} \cap \mathcal{E} = \mathcal{M}$. By Axiom PS, there exists exactly one line $\mathcal{J}$ such that $P \in \mathcal{J}$ and $\mathcal{J} \parallel \mathcal{L}$, and there exists exactly one line $\mathcal{K}$ such that $P \in \mathcal{K}$ and $\mathcal{K} \parallel \mathcal{M}$. Now since $\mathcal{J} \parallel \mathcal{L}$, $\mathcal{J}$ and $\mathcal{L}$ must lie in the same plane (Definition IP.0), and by Exercise I.7, there is only one such plane, we have that $\mathcal{J} \subseteq \mathcal{G}$. Similarly $\mathcal{K} \subseteq \mathcal{H}$.

Now $\mathcal{J}$ and $\mathcal{K}$ are distinct lines because if they were the same line, then $\mathcal{L}$ and $\mathcal{M}$ would be two different lines through Q, both parallel to this line, contradicting Axiom PS. Moreover, $P \in \mathcal{J} \cap \mathcal{K}$. Therefore by Theorem I.6 there is a plane $\mathcal{F}$ such that $\mathcal{J} \subseteq \mathcal{F}$ and $\mathcal{K} \subseteq \mathcal{F}$; by Theorem IP.2 $\mathcal{L} \parallel \mathcal{F}$ and $\mathcal{M} \parallel \mathcal{F}$, and by Theorem IP.4, $\mathcal{F} \parallel \mathcal{E}$.

(II: uniqueness) We now show that there is not more than one plane satisfying the conditions in the theorem. To this end, suppose there were a plane $\mathcal{F}'$ such that $P \in \mathcal{F}'$, $\mathcal{E} \parallel \mathcal{F}'$, and $\mathcal{F}' \neq \mathcal{F}$. Then $\mathcal{F} \cap \mathcal{F}' \neq \emptyset$, so by Theorem I.4 $\mathcal{F} \cap \mathcal{F}'$ is a line. Call it $\mathcal{M}$. By Exercise I.13 there is a point Q on $\mathcal{F}$ such that $Q \notin \mathcal{M}$. By Axiom I.5(B) there exists a point R on $\mathcal{E}$. By Axiom I.2 there exists exactly one plane $\mathcal{G}$ containing P, Q, and R. By Theorem I.4, $\mathcal{G} \cap \mathcal{E}$, $\mathcal{G} \cap \mathcal{F}$, and $\mathcal{G} \cap \mathcal{F}'$ are lines we will call $\mathcal{N}$, $\mathcal{J}$, and $\mathcal{J}'$, respectively. By Exercise IP.7 below, $\mathcal{J} \parallel \mathcal{N}$ and $\mathcal{J}' \parallel \mathcal{N}$. Because of the way Q was chosen, $\mathcal{J} \neq \mathcal{J}'$. Since both $\mathcal{J}$ and $\mathcal{J}'$ contain P, this is a contradiction of Axiom PS, so our assumption that the plane $\mathcal{F}'$ exists is false, and the only plane through P parallel to $\mathcal{E}$ is $\mathcal{F}$. □

Theorem IP.6 (Transitivity of Parallelism). *If $\mathcal{L}$, $\mathcal{M}$, and $\mathcal{N}$ are distinct lines such that $\mathcal{L} \parallel \mathcal{M}$ and $\mathcal{M} \parallel \mathcal{N}$, then $\mathcal{L} \parallel \mathcal{N}$.*

Proof. If $\mathcal{L}$, $\mathcal{M}$, and $\mathcal{N}$ all lie in a plane, then this theorem is Exercise IP.2.

Suppose then that no plane contains all three lines. By Definition IP.0(A), there exist planes $\mathcal{F}$ and $\mathcal{G}$ such that $\mathcal{L} \subseteq \mathcal{F}$, $\mathcal{M} \subseteq \mathcal{F}$, $\mathcal{M} \subseteq \mathcal{G}$, and $\mathcal{N} \subseteq \mathcal{G}$. By hypothesis $\mathcal{L} \neq \mathcal{N}$ so there is a point P on $\mathcal{N}$ which does not belong to $\mathcal{L}$; then by Exercise I.7 there is a plane $\mathcal{H}$ such that $\mathcal{L} \subseteq \mathcal{H}$ and $P \in \mathcal{H}$.

We now show that $\mathcal{M}$ is not contained in $\mathcal{H}$. We know by hypothesis that $\mathcal{M} \subseteq \mathcal{G}$, so if $\mathcal{M} \subseteq \mathcal{H}$, it follows that $\mathcal{M} \subseteq (\mathcal{G} \cap \mathcal{H})$ which is a line by Theorem I.4, and by Exercise I.3, $\mathcal{G} \cap \mathcal{H} = \mathcal{M}$. Hence P would belong to $\mathcal{M}$, which would contradict the fact that $\mathcal{M} \parallel \mathcal{N}$. So $\mathcal{M}$ is not contained in $\mathcal{H}$, and we can apply Theorem IP.2 to get $\mathcal{M} \parallel \mathcal{H}$.

This permits us to apply Theorem IP.3 to get $\mathcal{G} \cap \mathcal{H}$ is a line $\mathcal{N}'$ and that $\mathcal{N}' \parallel \mathcal{M}$. Since $P \in \mathcal{H}$ by the way $\mathcal{H}$ was constructed, and $P \in \mathcal{N} \subseteq \mathcal{G}$ we have that $P \in \mathcal{G} \cap \mathcal{H} = \mathcal{N}'$. By Axiom PS there can be only one line through P which is parallel to $\mathcal{M}$, so $\mathcal{N} = \mathcal{N}'$, and $\mathcal{N} \subseteq \mathcal{H}$.

Now we have that $\mathcal{L} \subseteq \mathcal{H}$ and $\mathcal{N} \subseteq \mathcal{H}$. So by Definition IP.0 of parallel lines, if $\mathcal{L}$ and $\mathcal{N}$ were not parallel, they would intersect in some point A. But then there would be two different lines through A, both parallel to $\mathcal{M}$, which contradicts Axiom PS. So $\mathcal{L}$ and $\mathcal{N}$ must be parallel. □

2.3 Exercises for affine geometry

Answers to starred (*) exercises may be accessed from the home page for this book at www.springer.com.

Exercise IP.1*. If $\mathcal{L}$ and $\mathcal{M}$ are parallel lines, then there is exactly one plane containing both of them.

Exercise IP.2*. Let $\mathcal{L}$, $\mathcal{M}$, and $\mathcal{N}$ be distinct lines contained in a single plane.

(A) If $\mathcal{L} \parallel \mathcal{M}$ and $\mathcal{M} \parallel \mathcal{N}$, then $\mathcal{L} \parallel \mathcal{N}$.
(B) If $\mathcal{L}$ intersects $\mathcal{M}$, then $\mathcal{N}$ must intersect $\mathcal{L}$ or $\mathcal{M}$, possibly both.

Exercise IP.3*. Let $\mathbb{E}$ be a pencil of lines on the plane $\mathcal{P}$. If $\mathcal{L}$ and $\mathcal{M}$ are distinct members of $\mathbb{E}$ which intersect at the point O, then the members of $\mathbb{E}$ are concurrent at O.

Exercise IP.4*. Let $\mathcal{L}$, $\mathcal{M}$, and $\mathcal{N}$ be distinct lines in a plane $\mathcal{E}$ such that $\mathcal{L} \parallel \mathcal{M}$. Then if $\mathcal{L} \cap \mathcal{N} \neq \emptyset$, $\mathcal{M} \cap \mathcal{N} \neq \emptyset$.

Exercise IP.5*. Let $\mathcal{L}_1$, $\mathcal{L}_2$, $\mathcal{M}_1$, and $\mathcal{M}_2$ be lines on the plane $\mathcal{P}$ such that $\mathcal{L}_1$ and $\mathcal{L}_2$ intersect at a point, $\mathcal{L}_1 \parallel \mathcal{M}_1$, and $\mathcal{L}_2 \parallel \mathcal{M}_2$, then $\mathcal{M}_1$ and $\mathcal{M}_2$ intersect at a point.

Exercise IP.6*. Let $\mathcal{E}$ and $\mathcal{F}$ be planes such that $\mathcal{E} \parallel \mathcal{F}$, and let $\mathcal{L}$ be a line in $\mathcal{E}$. Then $\mathcal{L} \parallel \mathcal{F}$.

Exercise IP.7*. Let $\mathcal{E}$, $\mathcal{F}$, and $\mathcal{G}$ be planes such that $\mathcal{E} \parallel \mathcal{F}$, $\mathcal{E} \cap \mathcal{G} \neq \emptyset$, and $\mathcal{F} \cap \mathcal{G} \neq \emptyset$. Then $\mathcal{E} \cap \mathcal{G}$ is a line $\mathcal{L}$, $\mathcal{F} \cap \mathcal{G}$ is a line $\mathcal{M}$, and $\mathcal{L} \parallel \mathcal{M}$. See Figure 2.2.

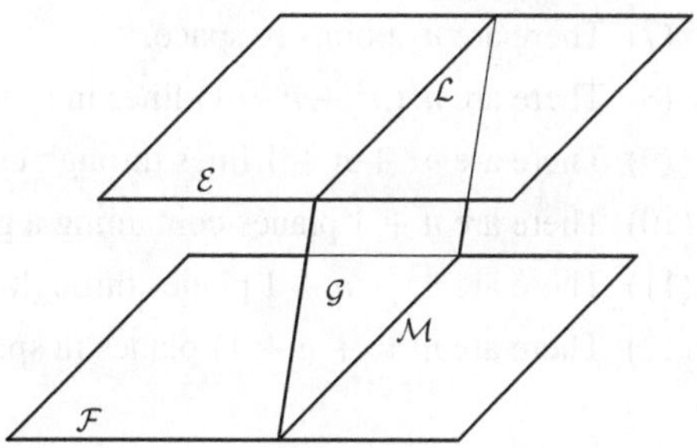

Fig. 2.2 For Exercise IP.7.

Exercise IP.8. If $\mathcal{E}$, $\mathcal{F}$, and $\mathcal{G}$ are distinct planes such that $\mathcal{E} \parallel \mathcal{F}$ and $\mathcal{F} \parallel \mathcal{G}$, then $\mathcal{E} \parallel \mathcal{G}$.

Exercise IP.9. If $\mathcal{E}$, $\mathcal{F}$, and $\mathcal{G}$ are distinct planes such that $\mathcal{E} \parallel \mathcal{F}$ and $\mathcal{E} \cap \mathcal{G} \neq \emptyset$, then $\mathcal{F} \cap \mathcal{G} \neq \emptyset$.

Exercise IP.10. If $\mathcal{L}$ and $\mathcal{M}$ are noncoplanar lines, then there exist planes $\mathcal{E}$ and $\mathcal{F}$ such that $\mathcal{E} \parallel \mathcal{F}$ and $\mathcal{L} \subseteq \mathcal{E}$, and $\mathcal{M} \subseteq \mathcal{F}$.

Exercise IP.11. Let $\mathcal{E}$ and $\mathcal{F}$ be parallel planes, and let $\mathcal{L}$ be a line which is parallel to $\mathcal{E}$ and which is not contained in $\mathcal{F}$. Then $\mathcal{L} \cap \mathcal{F} = \emptyset$.

Exercise IP.12. Let $\mathcal{E}$ and $\mathcal{F}$ be parallel planes, and let $\mathcal{L}$ be a line which is not parallel to $\mathcal{E}$ and which is not contained in $\mathcal{E}$. Then $\mathcal{L} \cap \mathcal{F} \neq \emptyset$.

Exercise IP.13. Given a plane $\mathcal{E}$ and a line $\mathcal{L}$ parallel to $\mathcal{E}$, there exists a plane $\mathcal{F}$ containing $\mathcal{L}$ and parallel to $\mathcal{E}$.

Exercise IP.14. Let n be a natural number greater than 1. If there exists a line which has exactly n points, then:

(1) Every line has exactly n points.
(2) For any point P and any plane $\mathcal{E}$ containing P, there are exactly $n + 1$ lines through P and contained in $\mathcal{E}$.
(3) For any line $\mathcal{L}$ and any plane $\mathcal{E}$ containing $\mathcal{L}$, there exist exactly $n - 1$ lines $\mathcal{L}_1, \ldots, \mathcal{L}_{n-1}$ such that $\mathcal{L}_k \parallel \mathcal{L}$ for each k in $[1; n - 1]$.
(4) Each plane contains $n(n + 1)$ lines.
(5) Each plane contains n^2 points.
(6) Given any plane $\mathcal{E}$, there exists exactly $n - 1$ planes $\mathcal{E}_1, \ldots, \mathcal{E}_{n-1}$ such that $\mathcal{E}_k \parallel \mathcal{E}$ for each k in $[1; n - 1]$.
(7) There are n^3 points in space.
(8) There are $n^2(n^2 + n + 1)$ lines in space.
(9) There are $n^2 + n + 1$ lines through each point.
(10) There are $n + 1$ planes containing a given line.
(11) There are $n^2 + n + 1$ planes through each point.
(12) There are $n(n^2 + n + 1)$ planes in space.

Chapter 3
Collineations of an Affine Plane (CAP)

Acronym: *CAP*
Dependencies: *Chapters 1 and 2*
New Axioms: *none*
New Terms Defined: *collineation, fixed point, fixed line, translation, parallel relation, dilation, axial affinity, stretch, shear*

Abstract: Collineations are bijections of a plane onto itself which map lines to lines; this chapter explores the elementary properties of collineations on an incidence plane on which the parallel axiom holds. Several types of collineations are studied, among them translations, dilations, and axial affinities.

This chapter consists of two parts which, in the dependency chart in the Preface, are called "Defs" and "Thms"; these parts are not labeled as such in this chapter, but are mixed together.

The part designated "Defs" includes Theorems CAP.1 through CAP.4, Definitions CAP.0 (*collineation*), CAP.6 (*translation*), CAP.10 (*parallel relation*), CAP.17 (*dilation*), and CAP.25 (*axial affinity*), together with Remark CAP.30, which anticipates the later definition of two subclasses of axial affinity, the *stretch* and the *shear*. These definitions and theorems are all valid on an incidence plane (that is, one on which the incidence axioms hold),[1] and thus depend only on Chapter 1.

[1] Again we remind the reader that this means the incidence axioms that apply to planes.

E.J. Specht et al., *Euclidean Geometry and its Subgeometries*,
DOI 10.1007/978-3-319-23775-6_3

The part designated "Thms" includes all other theorems and remarks in the chapter. The proofs of these theorems need a parallel axiom; thus, they are valid on an *affine plane*, and depend on Chapter 2. This need for a parallel axiom is why we included "affine plane" as part of the name of the chapter.

Later, Chapter 4 defines *betweenness*, and Chapter 7 introduces a type of collineation that preserves betweenness, called a *belineation.* The various types of mappings listed above, as well as *isometries*, are eventually shown to be belineations. These developments are briefly outlined in the Preface; Chapter 19 Section 19.2 contains a summary of the properties of these mappings and a chart comparing them.

We do not show the existence of collineations in this chapter. This will be done in later chapters, perhaps most importantly in Chapter 8 (neutral geometry), where the existence of isometries follows from Axiom REF.

In this chapter we are indebted to *Fundamentals of Mathematics*, Volume II, Behnke, *et al*, eds., published by MIT Press, Chapter 3, Affine and Projective Planes, by R. Lingenberg and A. Bauer [2].

3.1 Collineations of an incidence plane

In order to fully appreciate the following definition it might be well to review the terminology for images of sets mapped by functions, given briefly in Chapter 1, Section 1.4.

We use lowercase Greek letters, especially α, β, γ, δ, ϵ, ρ, τ, φ, and ψ for collineations. We have already used $\mathbb{E}$ for a set of lines and will also use $\mathbb{G}$ and $\mathbb{M}$ similarly.

Definition CAP.0. Let $\mathcal{P}$ be any plane containing points and lines, and let α be any mapping of $\mathcal{P}$ into itself.

(A) The mapping α is a **collineation** of $\mathcal{P}$ iff
  (i) α is a bijection of $\mathcal{P}$ onto itself and
  (ii) for every line $\mathcal{L}$ on $\mathcal{P}$, $\alpha(\mathcal{L})$ is a line on $\mathcal{P}$.

(B) Q is a **fixed point** of α iff $\alpha(Q) = Q$.

(C) $\mathcal{L}$ is a **fixed line** of α iff $\alpha(\mathcal{L}) = \mathcal{L}$.

Remark CAP.0.1. (A) The *identity* map ι is a collineation.

(B) A fixed line does not necessarily contain any fixed points, nor does a fixed point necessarily belong to some fixed line.

Theorem CAP.1. *Let $\mathcal{P}$ be a plane containing points and lines and let α, β, and γ be collineations of $\mathcal{P}$.*

(A) *If A and B are distinct points on $\mathcal{P}$, then $\alpha(\overleftrightarrow{AB}) = \overleftrightarrow{\alpha(A)\alpha(B)}$.*

(B) *If $\mathcal{L}$ and $\mathcal{M}$ are lines on $\mathcal{P}$, and if Q is a point on $\mathcal{P}$ such that $\mathcal{L} \cap \mathcal{M} = \{Q\}$, then $\alpha(\mathcal{L}) \cap \alpha(\mathcal{M}) = \{\alpha(Q)\}$.*

(C) *$\beta \circ \alpha$ is a collineation of $\mathcal{P}$. (The set of collineations is closed under the operation $\circ$.)*

(C') *$\gamma \circ (\beta \circ \alpha) = (\gamma \circ \beta) \circ \alpha$ (The operation $\circ$ is associative.)*

(D) *The identity ι is a collineation, and is an identity for the operation $\circ$.*

(D') *α^{-1} is a collineation of $\mathcal{P}$. (For each collineation α, there exists a collineation which is its inverse.)*

Proof. (A) Since α is one-to-one, $\alpha(A)$ and $\alpha(B)$ are distinct points. Since α is a collineation, $\alpha(\overleftrightarrow{AB})$ is a line. Since $\alpha(A)$ and $\alpha(B)$ belong to $\alpha(\overleftrightarrow{AB})$ and to $\overleftrightarrow{\alpha(A)\alpha(B)}$, by Axiom I.1, $\alpha(\overleftrightarrow{AB}) = \overleftrightarrow{\alpha(A)\alpha(B)}$.

(B) By elementary mapping theory, since α is a bijection,

$$\alpha(\mathcal{L}) \cap \alpha(\mathcal{M}) = \alpha(\mathcal{L} \cap \mathcal{M}) = \alpha(\{Q\}) = \{\alpha(Q)\}.$$

(C) By elementary mapping theory, $\beta \circ \alpha$ is a bijection of $\mathcal{P}$. If $\mathcal{L}$ is a line on $\mathcal{P}$, then so is $\alpha(\mathcal{L})$ because α is a collineation. Then since β is a collineation, $\beta(\alpha(\mathcal{L}))$ is a line on $\mathcal{P}$. By Definition CAP.0 $\beta \circ \alpha$ is a collineation of $\mathcal{P}$.

(C') Immediate from associativity of mappings (see Chapter 1, Section 1.4).

(D) Proof is trivial.

(D') Let $\mathcal{L}$ be any line on $\mathcal{P}$. By Axiom I.5, there are two distinct points, A and B say, on $\mathcal{L}$. By the elementary theory of mappings, α^{-1} is a one-to-one mapping of $\mathcal{P}$ onto itself because α is, so $\alpha^{-1}(A) \neq \alpha^{-1}(B)$. By part (A) above, $\alpha(\overleftrightarrow{\alpha^{-1}(A)\alpha^{-1}(B)}) = \overleftrightarrow{\alpha(\alpha^{-1}(A))\alpha(\alpha^{-1}(B))} = \overleftrightarrow{AB}$. On the other hand, $\alpha(\alpha^{-1}(\overleftrightarrow{AB})) = \overleftrightarrow{AB}$. Thus $\alpha(\overleftrightarrow{\alpha^{-1}(A)\alpha^{-1}(B)}) = \alpha(\alpha^{-1}(\overleftrightarrow{AB}))$, so that (since α is one-to-one) $\alpha^{-1}(\overleftrightarrow{AB}) = \alpha^{-1}(\mathcal{L}) = \overleftrightarrow{\alpha^{-1}(A)\alpha^{-1}(B)}$. We have shown that if $\mathcal{L}$ is any line on $\mathcal{P}$ then $\alpha^{-1}(\mathcal{L})$ is a line on $\mathcal{P}$. By Definition CAP.0, α^{-1} is a collineation on $\mathcal{P}$. □

Corollary CAP.2. *The set of collineations of a plane $\mathcal{P}$ onto itself forms a group under composition of mappings.*

Proof. This Corollary is an immediate consequence of Theorem CAP.1 and the definition of a group (see Chapter 1, Section 1.5). □

Theorem CAP.3. *Let $\mathcal{P}$ be a plane containing points and lines, and let α be a collineation on $\mathcal{P}$. If $\mathcal{L}$ and $\mathcal{M}$ are lines on $\mathcal{P}$ such that $\mathcal{L} \parallel \mathcal{M}$, then $\alpha(\mathcal{L}) \parallel \alpha(\mathcal{M})$.*

Proof. Since $\mathcal{L} \cap \mathcal{M} = \emptyset$, then by elementary mapping theory, $\alpha(\mathcal{L}) \cap \alpha(\mathcal{M}) = \alpha(\mathcal{L} \cap \mathcal{M}) = \alpha(\emptyset) = \emptyset$ so that $\alpha(\mathcal{L})$ and $\alpha(\mathcal{M})$ are disjoint, $\alpha(\mathcal{L}) \parallel \alpha(\mathcal{M})$. Here we have used Definition IP.0. □

Theorem CAP.4. *Let $\mathcal{P}$ be a plane containing points and lines, and let α be a collineation on $\mathcal{P}$.*

(A) *If A and B are fixed points of α, then $\overleftrightarrow{AB}$ is a fixed line of α.*
(B) *If $\mathcal{L}$ and $\mathcal{M}$ are fixed lines of α which intersect at Q, then Q is a fixed point of α.*

Proof. (A) By Definition CAP.0, $\alpha(A) = A$ and $\alpha(B) = B$. By Theorem CAP.1(A), $\alpha(\overleftrightarrow{AB}) = \overleftrightarrow{\alpha(A)\alpha(B)} = \overleftrightarrow{AB}$. Hence by Definition CAP.0 $\overleftrightarrow{AB}$ is a fixed line of α.
(B) By Theorem CAP.1(B) and Exercise I.1,
$$\{\alpha(Q)\} = \alpha(\mathcal{L}) \cap \alpha(\mathcal{M}) = \mathcal{L} \cap \mathcal{M} = \{Q\}$$
which completes the proof. □

Again, we emphasize that up to this point, we have not used a parallel axiom; the above development is valid in any space containing points, lines, and planes.

3.2 Collineations: mostly on translations

Theorem CAP.5. *Let α be a collineation on the affine plane $\mathcal{P}$. If $\mathcal{L}$ is a fixed line of α, and if Q is a fixed point of α such that $Q \notin \mathcal{L}$, then there exists a unique fixed line $\mathcal{M}$ of α containing Q such that $\mathcal{M} \parallel \mathcal{L}$.*

Proof. By Axiom PS, there exists a line $\mathcal{M}$ on $\mathcal{P}$ such that $Q \in \mathcal{M}$ and $\mathcal{M} \parallel \mathcal{L}$. By Theorem CAP.3, $\alpha(\mathcal{M}) \parallel \alpha(\mathcal{L}) = \mathcal{L}$. Since $\alpha(Q) = Q$, $Q \in \alpha(\mathcal{M})$. By Axiom PS, there is only one line through Q parallel to $\mathcal{L}$, so $\alpha(\mathcal{M}) = \mathcal{M}$. By Definition CAP.0, $\mathcal{M}$ is a fixed line of α. □

Definition CAP.6. Let $\mathcal{P}$ be a plane, and let α be a collineation of $\mathcal{P}$. α is a **translation** of $\mathcal{P}$ iff α has no fixed point, and for every line $\mathcal{L}$ on $\mathcal{P}$ either $\alpha(\mathcal{L}) \parallel \mathcal{L}$ or $\alpha(\mathcal{L}) = \mathcal{L}$. If $\mathcal{L}$ is a line, α is a translation, and $\alpha(\mathcal{L}) = \mathcal{L}$, then α is said to be a translation **along** $\mathcal{L}$. That is to say, α is a translation *along* any of its fixed lines.

Remark CAP.7. (1) If $\alpha(\mathcal{L}) = \mathcal{L}$, then $\mathcal{L}$ is a fixed line of α. A translation which has a fixed line is a translation along that line.

(2) The identity ι is not a translation, since it has fixed points.

(3) Definition CAP.6 is valid even in the absence of a parallel axiom.

Theorem CAP.8. *Let $\mathcal{P}$ be an affine plane, and let α be a translation of $\mathcal{P}$. Then:*

(A) *If $\mathcal{L}$ is a fixed line of α, then for every point Q on $\mathcal{L}$, $\mathcal{L} = \overleftrightarrow{Q\alpha(Q)}$.*

(B) *If Q is any point on $\mathcal{P}$, then $\overleftrightarrow{Q\alpha(Q)}$ is a fixed line of α.*

(C) *The set of all fixed lines of α is $\mathcal{H}_\alpha = \{\overleftrightarrow{X\alpha(X)} \mid X \in \mathcal{P}\}$; every member of $\mathcal{H}_\alpha$ is parallel to all other members.*

Proof. (A) Let Q be any member of $\mathcal{L}$. Since, by Definition CAP.6, α has no fixed points, Q is not a fixed point of α, and therefore $\alpha(Q) \neq Q$. Since $\mathcal{L}$ is a fixed line of α, $\alpha(Q) \in \mathcal{L}$. By Exercise I.2(A), $\mathcal{L} = \overleftrightarrow{Q\alpha(Q)}$.

(B) By Definition CAP.6, $\alpha(\overleftrightarrow{Q\alpha(Q)}) = \overleftrightarrow{Q\alpha(Q)}$ or $\alpha(\overleftrightarrow{Q\alpha(Q)}) \parallel \overleftrightarrow{Q\alpha(Q)}$. By Theorem CAP.1, $\alpha(\overleftrightarrow{Q\alpha(Q)}) = \overleftrightarrow{\alpha(Q)\alpha(\alpha(Q))}$. Hence $\overleftrightarrow{Q\alpha(Q)}$ and $\alpha(\overleftrightarrow{Q\alpha(Q)})$ are not parallel because they have the point $\alpha(Q)$ in common. Therefore $\alpha(\overleftrightarrow{Q\alpha(Q)}) = \overleftrightarrow{Q\alpha(Q)}$.

(C) Every fixed line for α is a member of $\mathcal{H}_\alpha$ by part (A), and every member of $\mathcal{H}_\alpha$ is a fixed line for α by part (B). Let A be any member of $\mathcal{P}$, and let B be a point off of $\overleftrightarrow{A\alpha(A)}$, so that $\overleftrightarrow{A\alpha(A)} \neq \overleftrightarrow{B\alpha(B)}$. If $\overleftrightarrow{A\alpha(A)}$ and $\overleftrightarrow{B\alpha(B)}$ were to intersect at a point Q, say, then by Theorem CAP.4(B), Q would be a fixed point of α, contrary to the fact that α has no fixed points. Hence $\overleftrightarrow{A\alpha(A)} \parallel \overleftrightarrow{B\alpha(B)}$. □

Theorem CAP.9. *If A and B are distinct points on the affine plane $\mathcal{P}$, there can exist no more than one translation α such that $\alpha(A) = B$.*

Proof. Suppose that α and β are translations of $\mathcal{P}$ such that $\alpha(A) = B$ and $\beta(A) = B$, and let X be any member of $\mathcal{P} \setminus \{A\}$.

(Case 1: $X \in \mathcal{P} \setminus \overleftrightarrow{AB}$.) By Theorem CAP.8, $\overleftrightarrow{X\alpha(X)} \parallel \overleftrightarrow{AB}$ and $\overleftrightarrow{X\beta(X)} \parallel \overleftrightarrow{AB}$. By Axiom PS, $\overleftrightarrow{X\alpha(X)} = \overleftrightarrow{X\beta(X)}$. By Theorem CAP.1, $\alpha(\overleftrightarrow{AX}) = \overleftrightarrow{\alpha(A)\alpha(X)} = \overleftrightarrow{B\alpha(X)}$ and $\beta(\overleftrightarrow{AX}) = \overleftrightarrow{\beta(A)\beta(X)} = \overleftrightarrow{B\beta(X)}$. By Definition IP.0, $\alpha(X) \notin \overleftrightarrow{AB}$ and $\beta(X) \notin \overleftrightarrow{AB}$.

If $\overleftrightarrow{AX} = \overleftrightarrow{B\alpha(X)}$, then $B = \alpha(A)$ (which is not equal to A) must belong to $\overleftrightarrow{AX}$ and by Exercise I.2 $\overleftrightarrow{AX} = \overleftrightarrow{AB}$; but this is impossible because $X \notin \overleftrightarrow{AB}$. Thus $\overleftrightarrow{AX} \neq \overleftrightarrow{B\alpha(X)}$ and similarly $\overleftrightarrow{AX} \neq \overleftrightarrow{B\beta(X)}$, so by Definition CAP.6, $\overleftrightarrow{AX} \parallel \overleftrightarrow{B\alpha(X)}$ and $\overleftrightarrow{AX} \parallel \overleftrightarrow{B\beta(X)}$.

By Axiom PS, $\overleftrightarrow{B\alpha(X)} = \overleftrightarrow{B\beta(X)}$. Since $\overleftrightarrow{X\alpha(X)} \cap \overleftrightarrow{B\alpha(X)} = \{\alpha(X)\}$, $\overleftrightarrow{X\beta(X)} \cap \overleftrightarrow{B\beta(X)} = \{\beta(X)\}$, $\overleftrightarrow{X\alpha(X)} = \overleftrightarrow{X\beta(X)}$ and $\overleftrightarrow{B\alpha(X)} = \overleftrightarrow{B\beta(X)}$, it follows from Exercise I.1 that $\alpha(X) = \beta(X)$.

(Case 2: $X \in \overleftrightarrow{AB} \setminus \{A\}$.) By Theorem CAP.8, $\overleftrightarrow{AB}$ is a fixed line of both α and β so $\alpha(X) \in \overleftrightarrow{AB}$ and $\beta(X) \in \overleftrightarrow{AB}$. Since $X \neq A$ and α and β are both one-to-one (injective), $\alpha(X) \neq \alpha(A) = B$ and $\beta(X) \neq \beta(A) = B$. By Exercise I.2, $\overleftrightarrow{AB} = \overleftrightarrow{B\alpha(X)} = \overleftrightarrow{B\beta(X)}$. Let Y be any member of $\mathcal{P} \setminus \overleftrightarrow{AB}$. Let $Z = \alpha(Y)$, which, by Case 1, is equal to $\beta(Y)$. By Theorem CAP.1, $\alpha(\overleftrightarrow{XY}) = \overleftrightarrow{\alpha(X)\alpha(Y)} = \overleftrightarrow{\alpha(X)Z}$ and $\beta(\overleftrightarrow{XY}) = \overleftrightarrow{\beta(X)\beta(Y)} = \overleftrightarrow{\beta(X)Z}$.

By essentially the argument used in Case 1, $\overleftrightarrow{\alpha(X)Z} \neq \overleftrightarrow{XY} \neq \overleftrightarrow{\beta(X)Z}$, so by Definition CAP.6, $\overleftrightarrow{XY} \parallel \overleftrightarrow{\alpha(X)Z}$ and $\overleftrightarrow{XY} \parallel \overleftrightarrow{\beta(X)Z}$. By Axiom PS, $\overleftrightarrow{\alpha(X)Z} = \overleftrightarrow{\beta(X)Z}$, so $\alpha(X) = \beta(X)$. □

Definition CAP.10. If $\mathcal{L}$ and $\mathcal{M}$ are lines on a plane $\mathcal{P}$, we will write $\mathcal{L} \mathbb{PE} \mathcal{M}$ iff $\mathcal{L} = \mathcal{M}$ or ($\mathcal{L} \neq \mathcal{M}$ and $\mathcal{L} \parallel \mathcal{M}$). That is, $\mathcal{L} \mathbb{PE} \mathcal{M}$ if and only if $\mathcal{L}$ and $\mathcal{M}$ are parallel or equal to each other. We will call $\mathbb{PE}$ the **parallel relation**. Note that this definition is valid even in the absence of a parallel axiom.

Remark CAP.11. Let $\mathcal{P}$ be an affine plane.

(A) The relation $\mathbb{PE}$ defined in Definition CAP.10 just above is an *equivalence relation* on the set of all lines on $\mathcal{P}$ and the set $\mathbb{M}_{\mathcal{M}} = \{\mathcal{L} \mid \mathcal{L} \subseteq \mathcal{P}$ and $\mathcal{L} \mathbb{PE} \mathcal{M}\}$ is the *equivalence class* of $\mathcal{M}$. To see this, note that reflexivity and symmetry of $\mathbb{PE}$ are quite obvious, and transitivity is Exercise CAP.1.

(B) If α is a collineation on $\mathcal{P}$, α is a translation of $\mathcal{P}$ iff α has no fixed point and for every $\mathcal{L}$, $\alpha(\mathcal{L}) \mathbb{PE} \mathcal{L}$. This is just a re-statement of Definition CAP.6.

Theorem CAP.12. *Let $\mathcal{P}$ be an affine plane.*

(A) *Under composition of mappings, the set of translations of $\mathcal{P}$, together with the identity mapping, is a group. (For definition of a group, see Chapter 1, Section 1.5)*

(B) *Let $\mathcal{M}$ be a line on $\mathcal{P}$ and let $\mathbb{M}_{\mathcal{M}} = \{\mathcal{L} \mid \mathcal{L} \subseteq \mathcal{P}$ and $\mathcal{L} \mathbin{\mathbb{PE}} M\}$, then $\mathbb{G}_{\mathcal{M}} = \{\alpha \mid \alpha$ is a translation whose set of fixed lines is $\mathbb{M}_{\mathcal{M}}$, or $\alpha = \iota\}$ is a group under composition of mappings; that is to say, for any given line, the set of all translations along that line, together with the identity ι, is a group.*

Proof. In Chapter 1, Section 1.4 we showed that the composition of mappings is associative, so that in particular, composition of translations is associative.

(A) First we show that if α is a translation, so is α^{-1}. By Theorem CAP.1 α^{-1} is a collineation on $\mathcal{P}$, and hence by Definition CAP.0 $\alpha^{-1}(\mathcal{L})$ is a line. Furthermore, $\alpha(\alpha^{-1}(\mathcal{L})) = \mathcal{L}$, and since α is a translation, by Remark CAP.11(B), $\alpha^{-1}(\mathcal{L}) \mathbin{\mathbb{PE}} \mathcal{L}$. Since α has no fixed points, α^{-1} has none, so is a translation on $\mathcal{P}$.

Now let $\mathcal{L}$ be any line on $\mathcal{P}$, and let α and β be translations on $\mathcal{P}$; then $(\beta \circ \alpha)(\mathcal{L}) = \beta(\alpha(\mathcal{L}))$. By Remark CAP.11(B), $\alpha(\mathcal{L}) \mathbin{\mathbb{PE}} \mathcal{L}$ and $\beta(\alpha(\mathcal{L})) \mathbin{\mathbb{PE}} \alpha(\mathcal{L})$. By Exercise CAP.1, $\beta(\alpha(\mathcal{L})) \mathbin{\mathbb{PE}} \mathcal{L}$.

Suppose that $\beta \circ \alpha$ has a fixed point X, and let $Y = \alpha(X)$. Then $\beta(\alpha(X)) = \beta(Y) = X$. We have just shown that α^{-1} is a translation and we know that $\alpha^{-1}(Y) = X$; by Theorem CAP.9 there can be only one translation mapping Y to X; it follows that $\beta = \alpha^{-1}$ and $\beta \circ \alpha = \iota$. Therefore if $\beta \circ \alpha \neq \iota$, $\beta \circ \alpha$ has no fixed point, and thus is a translation by Remark CAP.11(B).

The proof of (A) is complete once we observe that $\iota^{-1} = \iota$ and $\iota \circ \alpha = \alpha \circ \iota = \alpha$ for any translation α.

(B) To show that $\mathbb{G}_{\mathcal{M}}$ is a group, we need only show that if α and β are members of $\mathbb{G}_{\mathcal{M}}$, then so are $\beta \circ \alpha$ and α^{-1}. By hypothesis, $\iota \in \mathbb{G}_{\mathcal{M}}$, so if $\beta \circ \alpha = \iota$, there is nothing to prove.

For every member $\mathcal{L}$ of $\mathbb{M}_{\mathcal{M}}$, $\alpha^{-1}(\mathcal{L}) = \alpha^{-1}(\alpha(\mathcal{L})) = \iota(\mathcal{L}) = \mathcal{L}$ and $\beta(\alpha(\mathcal{L})) = \beta(\mathcal{L}) = \mathcal{L}$. This shows that every line in $\mathbb{M}_{\mathcal{M}}$ is fixed for these mappings.

Now let γ be any translation (either α^{-1} or $\beta \circ \alpha$) for which every line in $\mathbb{M}_{\mathcal{M}}$ is fixed. Suppose that $\mathcal{L}$ is a fixed line that does not belong to $\mathbb{M}_{\mathcal{M}}$. Then $\mathcal{L}$ is not parallel to $\mathcal{M}$, and by Definition IP.0, there exists a point P such that $\mathcal{L} \cap \mathcal{M} = \{P\}$. By Theorem CAP.4 P is a fixed point of γ, and by Definition CAP.6 this is impossible, since translations have no fixed points. Thus $\mathcal{L}$ cannot be a fixed line for γ, the set of fixed lines for γ is exactly $\mathbb{M}_{\mathcal{M}}$, and $\gamma \in \mathbb{G}_{\mathcal{M}}$. □

Theorem CAP.13. *Let $\mathcal{P}$ be an affine plane, $\mathcal{M}$ be a line on $\mathcal{P}$, and let $\mathbb{M}_{\mathcal{M}} = \{\mathcal{L} \mid \mathcal{L} \subseteq \mathcal{P}$ and $\mathcal{L}$ $\mathbb{PE}$ $M\}$. If α is a translation of $\mathcal{P}$ whose set of fixed lines is $\mathbb{M}_{\mathcal{M}}$, and β is a collineation on $\mathcal{P}$, then $\beta \circ \alpha \circ \beta^{-1}$ is a translation whose set of fixed lines is $\mathbb{M}_{\beta(\mathcal{M})}$.*

Proof. Let Q be any point on $\mathcal{P}$. If Q were a fixed point of $\beta \circ \alpha \circ \beta^{-1}$, then $\beta(\alpha(\beta^{-1}(Q)))$ would be equal to Q and $\alpha(\beta^{-1}(Q))$ would be equal to $\beta^{-1}(Q)$ and thus $\beta^{-1}(Q)$ would be a fixed point of α. This would contradict the fact that α has no fixed point. Hence $\beta \circ \alpha \circ \beta^{-1}$ has no fixed point.

Let $\mathcal{L}$ be any line on $\mathcal{P}$. Since α is a translation, $\alpha(\beta^{-1}(\mathcal{L}))$ $\mathbb{PE}$ $\beta^{-1}(\mathcal{L})$ by Remark CAP.11(B). By Theorem CAP.3, $\beta(\alpha(\beta^{-1}(\mathcal{L})))$ $\mathbb{PE}$ $\beta(\beta^{-1}(\mathcal{L})) = \mathcal{L}$. By Remark CAP.11, $\beta \circ \alpha \circ \beta^{-1}$ is a translation of $\mathcal{P}$. If $\mathcal{L}$ is a fixed line of α, then $\alpha(\mathcal{L}) = \mathcal{L}$, and so $(\beta \circ \alpha \circ \beta^{-1})(\beta(\mathcal{L})) = \beta(\alpha(\beta^{-1}(\beta(\mathcal{L})))) = \beta(\mathcal{L})$. Thus $\beta(\mathcal{L})$ is a fixed line of $\beta \circ \alpha \circ \beta^{-1}$. Conversely, if $\beta(\mathcal{L})$ is a fixed line of $\beta \circ \alpha \circ \beta^{-1}$, so that $(\beta \circ \alpha \circ \beta^{-1})\beta(\mathcal{L}) = \beta(\mathcal{L})$, then $\beta(\alpha(\mathcal{L})) = \beta(\alpha(\beta^{-1}(\beta(\mathcal{L})))) = \beta(\mathcal{L})$. Thus $\alpha(\mathcal{L}) = \mathcal{L}$ and $\mathcal{L}$ is a fixed line of α. Summarizing, $\mathcal{L}$ is a fixed line of α iff $\beta(\mathcal{L})$ is a fixed line of $\beta \circ \alpha \circ \beta^{-1}$. Hence the set of fixed lines of $\beta \circ \alpha \circ \beta^{-1}$ is $\mathbb{M}_{\beta(\mathcal{M})}$. □

Corollary CAP.14. *Let $\mathcal{P}$ be an affine plane and let α and β be translations of $\mathcal{P}$. If $\mathcal{L}$ is a fixed line of α, then $\mathcal{L}$ is also a fixed line of $\beta \circ \alpha \circ \beta^{-1}$, and α and $\beta \circ \alpha \circ \beta^{-1}$ have the same fixed lines.*

Proof. By Theorem CAP.13, $\beta(\mathcal{L})$ is a fixed line of $\beta \circ \alpha \circ \beta^{-1}$. By Theorem CAP.8, $\beta(\mathcal{L})$ $\mathbb{PE}$ $\mathcal{L}$, so that $\mathcal{L}$ is a fixed line of $\beta \circ \alpha \circ \beta^{-1}$. By the same argument, if $\mathcal{M}$ is a fixed line for $\beta \circ \alpha \circ \beta^{-1}$ it is also a fixed line for $\alpha = \beta^{-1} \circ \beta \circ \alpha \circ \beta^{-1} \circ \beta$. □

Theorem CAP.15. (A) *Let $\mathcal{P}$ be an affine plane and let α and β be translations of $\mathcal{P}$ having different fixed lines. Then the fixed lines of $\beta \circ \alpha$ are different from the fixed lines of both α and β.*

(B) *If there exist translations on an affine plane $\mathcal{P}$ with different (non-parallel) fixed lines, then for any translations α and β of $\mathcal{P}$, $\alpha \circ \beta = \beta \circ \alpha$.*

Proof. (A) Let $X \in \mathcal{P}$. By Theorem CAP.8(B), $\overleftrightarrow{X\alpha(X)}$ is a fixed line of α, $\overleftrightarrow{\alpha(X)\beta(\alpha(X))}$ is a fixed line for β, and $\overleftrightarrow{X\beta(\alpha(X))}$ is a fixed line for $\beta \circ \alpha$ which intersects both $\overleftrightarrow{X\alpha(X)}$ and $\overleftrightarrow{\alpha(X)\beta(\alpha(X))}$. By assumption $\overleftrightarrow{X\alpha(X)} \neq \overleftrightarrow{\alpha(X)\beta(\alpha(X))}$ and they are not parallel, so that by Exercise I.1 $\overleftrightarrow{X\alpha(X)} \cap \overleftrightarrow{\alpha(X)\beta(\alpha(X))} = \{\alpha(X)\}$. Then $\beta(\alpha(X)) \notin \overleftrightarrow{X\alpha(X)}$, so that $\overleftrightarrow{X\beta(\alpha(X))}$ is not equal to either $\overleftrightarrow{X\alpha(X)}$ or $\overleftrightarrow{\alpha(X)\beta(\alpha(X))}$ and is not parallel to either one.

(B) The proof divides into two cases:

(Case 1: α and β have different fixed lines which are not parallel.) Let X be any point of $\mathcal{P}$. We show that $\beta(\alpha(X)) = \alpha(\beta(X))$.

(i) By Theorem CAP.8(B), $\overleftrightarrow{\alpha(X))\beta(\alpha(X))}$ and $\overleftrightarrow{X\beta(X)}$ are both fixed lines for β. Hence these two lines are parallel or equal by Theorem CAP.8(C). But they cannot be equal, for then X, $\alpha(X)$, and $\beta(X)$ would be collinear, contradicting our hypothesis that α and β have different fixed lines.

By Theorem CAP.1 $\alpha(\overleftrightarrow{X\beta(X)}) = \overleftrightarrow{\alpha(X)\alpha(\beta(X))}$ which is parallel to $\overleftrightarrow{X\beta(X)}$ by Definition CAP.6. Since both $\overleftrightarrow{\alpha(X)\alpha(\beta(X))}$ and $\overleftrightarrow{\alpha(X)\beta(\alpha(X))}$ are parallel to the same line, by Theorem IP.6 they are parallel or equal to each other. They both contain the point $\alpha(X)$, so by Axiom PS they are the same line, which we will call $\mathcal{M}$. This is a fixed line for α.

(ii) By a similar argument both $\overleftrightarrow{\beta(X)\beta(\alpha(X))}$ and $\overleftrightarrow{\beta(X)\alpha(\beta(X))}$ are parallel or equal to $\overleftrightarrow{X\alpha(X)}$, and by Theorem IP.6 they are parallel or equal to each other. They both contain the point $\beta(X)$, so by Axiom PS they are the same line, which we will call $\mathcal{L}$. This is a fixed line for β.

The lines $\mathcal{L}$ and $\mathcal{M}$ are distinct and not parallel, so by Exercise I.1 their intersection is a single point. Both $\beta(\alpha(X))$ and $\alpha(\beta(X))$ belong to both $\mathcal{L}$ and to $\mathcal{M}$, therefore $\beta(\alpha(X)) = \alpha(\beta(X))$, which is what we wanted to prove.

(Case 2: α and β are translations with the same set $\mathbb{M}_{\mathcal{L}}$ of fixed lines.) There exists a translation γ having a fixed line $\mathcal{M}$ that is not in $\mathbb{M}_{\mathcal{L}}$. Otherwise, all translations would have the same fixed lines, contradicting our hypothesis that there exist translations with different (nonparallel) fixed lines.

Now $\mathcal{L}$ and $\mathcal{M}$ are distinct and nonparallel. By part (A) $\mathcal{L}$ is not a fixed line of either of the translations $\gamma \circ \beta$ or γ, so by case 1, $\gamma \circ (\beta \circ \alpha) = (\gamma \circ \beta) \circ \alpha = \alpha \circ (\gamma \circ \beta) = (\alpha \circ \gamma) \circ \beta = (\gamma \circ \alpha) \circ \beta = \gamma \circ (\alpha \circ \beta)$. Thus $\beta \circ \alpha = \alpha \circ \beta$. □

Theorem CAP.16. *Let $\mathcal{P}$ be an affine plane and let α be a collineation of $\mathcal{P}$. If α has no fixed point and if its set $\mathbb{E}$ of fixed lines is the pencil of all lines parallel to some given line $\mathcal{M}$, then α is a translation of $\mathcal{P}$.*

Proof. Let $\mathcal{L}$ be any line on $\mathcal{P}$ which is not a member of $\mathbb{E}$ and therefore intersects the lines of $\mathbb{E}$. Since α has no fixed point, for every member X of $\mathcal{L}$, $\alpha(X) \neq X$. Suppose there exists a point Q such that $\mathcal{L} \cap \alpha(\mathcal{L}) = \{Q\}$. If $Q \notin \mathcal{M}$, by Axiom PS, there exists a unique line $\mathcal{N}$ parallel to $\mathcal{M}$ containing Q; if $Q \in \mathcal{M}$ let $\mathcal{N} = \mathcal{M}$. By hypothesis, $\mathcal{N} \in \mathbb{E}$. Since $\mathcal{N}$ is a fixed line of α, $\alpha(Q) \in \mathcal{N}$. Since $\mathcal{L} \cap \alpha(\mathcal{L}) \cap \mathcal{N} = \{Q\}$, and $\alpha(Q)$ is in both $\alpha(\mathcal{L})$ and $\mathcal{N}$, $\alpha(Q) = Q$.

This contradicts our hypothesis that α has no fixed point, so the supposition that $\mathcal{L}$ and $\alpha(\mathcal{L})$ are not parallel is false and $\mathcal{L} \parallel \alpha(\mathcal{L})$. By Definition CAP.6, α is a translation of $\mathcal{P}$. □

3.3 Collineations: dilations

Definition CAP.17. Let $\mathcal{P}$ be a plane and let α be a collineation of $\mathcal{P}$. α is a **dilation** of $\mathcal{P}$ iff $\alpha \neq \iota$, α has a fixed point, and for every line $\mathcal{L}$ on $\mathcal{P}$, either $\alpha(\mathcal{L}) \parallel \mathcal{L}$, or $\alpha(\mathcal{L}) = \mathcal{L}$, i.e. $\alpha(\mathcal{L})$ $\mathbb{PE}$ $\mathcal{L}$.

Note (A) that this definition is valid even in the absence of a parallel axiom; and (B) this definition is identical to that for a *translation*, except that a translation has no fixed point.

Theorem CAP.18. *Let $\mathcal{P}$ be an affine plane and let α be a dilation of $\mathcal{P}$ such that O is a fixed point of α; then*

(A) *every line through O is a fixed line of α,*
(B) *α has no fixed point different from O,*
(C) *for every fixed line $\mathcal{L}$ of α, $O \in \mathcal{L}$, and*
(D) *if A is any point of $\mathcal{P} \setminus \{O\}$, then $\alpha(A)$ is collinear with O and A.*

Fig. 3.1 Showing action of a dilation; double-headed arrows show fixed lines.

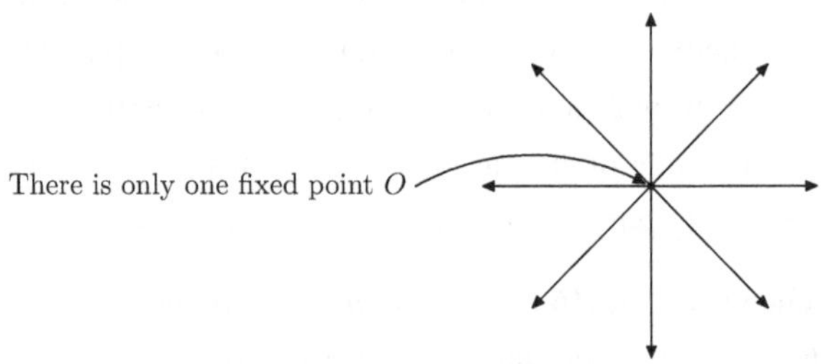

Proof. For a visualization, see Figure 3.1.

(A) Let $\mathcal{L}$ be any line through O. Since α is a collineation of $\mathcal{P}$, $\alpha(\mathcal{L})$ is a line on $\mathcal{P}$. Since O is a fixed point of α, $O \in \alpha(\mathcal{L})$, so that by Definition IP.0, $\mathcal{L}$ and $\alpha(\mathcal{L})$ are not parallel to each other. By Definition CAP.17, $\alpha(\mathcal{L}) = \mathcal{L}$.

(B) Suppose that α has a fixed point P distinct from O and let X be any point distinct from O and from P.

(Case 1: $X \notin \overleftrightarrow{OP}$.) By part (A), $\overleftrightarrow{OX}$ is a fixed line of α. Again, using part (A) but substituting P for O, $\overleftrightarrow{PX}$ is a fixed line of α. Thus by Theorem CAP.4(B), X is a fixed point of α.

(Case 2: $X \in \overleftrightarrow{OP}$.) By Exercise I.13 there exists a point Q not on $\overleftrightarrow{OP}$. By Case 1, Q is a fixed point for α. Then $X \notin \overleftrightarrow{QP}$ and substituting Q for P in Case 1, we have that X is a fixed point of α.

It follows that every X in $\mathcal{P}$ is a fixed point for α, and therefore α is the identity mapping ι, contradicting our assumption that $\alpha \neq \iota$. Therefore α has no fixed point P distinct from O.

(C) If $\mathcal{L}$ were a fixed line of α such that $O \notin \mathcal{L}$, then by part (A) for every point Q of $\mathcal{L}$, $\overleftrightarrow{OQ}$ would be a fixed line of α and $\{Q\} = \mathcal{L} \cap \overleftrightarrow{OQ}$. By Theorem CAP.4(B), Q would be a fixed point of α, contradicting the fact that O is the only fixed point of α.

(D) Let A be any point of $\mathcal{P}$ and $A \neq O$, and let $\mathcal{L} = \overleftrightarrow{AO}$. Since O is a fixed point, $O \in \alpha(\mathcal{L})$, so that $\alpha(\mathcal{L}) \not\parallel \mathcal{L}$; then, $\alpha(A) \in \alpha(\mathcal{L}) = \mathcal{L}$ by Definition CAP.17. □

Theorem CAP.19. *Let $\mathcal{P}$ be an affine plane and let α be a collineation of $\mathcal{P}$ such that for every line $\mathcal{L}$ on $\mathcal{P}$, $\alpha(\mathcal{L})$ $\mathbb{PE}$ $\mathcal{L}$. Then either $\alpha = \iota$, α is a translation of $\mathcal{P}$, or α is a dilation of $\mathcal{P}$.*

Proof. If $\alpha \neq \iota$, and if α has no fixed point, it is a translation by Definition CAP.6. If α has a fixed point, it is a dilation by Definition CAP.17. □

Theorem CAP.20. *Let δ be a dilation of an affine plane $\mathcal{P}$ with fixed point O.*

(A) *A line $\mathcal{L}$ is a fixed line for δ iff $O \in \mathcal{L}$.*

(B) *A line $\mathcal{L}$ is a fixed line for δ iff for some $Q \in \mathcal{P} \setminus \{O\}$, $\mathcal{L} = \overleftrightarrow{Q\delta(Q)}$.*

Proof. (A) This is Theorem CAP.18, parts (A) and (C), included for completeness.

(B) If $\mathcal{L}$ is a fixed line, by part (A), $O \in \mathcal{L}$; for any $Q \in \mathcal{L}$, $\delta(Q) \in \mathcal{L}$ so that by Exercise I.2 $\mathcal{L} = \overleftrightarrow{Q\delta(Q)}$. Conversely, let $Q \in \mathcal{P} \setminus \{O\}$, and suppose $\mathcal{L} = \overleftrightarrow{Q\delta(Q)}$ is not a fixed line, so that $\delta(\overleftrightarrow{Q\delta(Q)}) \neq \overleftrightarrow{Q\delta(Q)}$. By Theorem CAP.1(A), $\delta(\overleftrightarrow{Q\delta(Q)}) = \overleftrightarrow{\delta(Q)\delta(\delta(Q))}$, which is not parallel to $\overleftrightarrow{Q\delta(Q)}$. Therefore by Definition CAP.17, $\delta(\overleftrightarrow{Q\delta(Q)}) = \overleftrightarrow{Q\delta(Q)}$, and $\overleftrightarrow{Q\delta(Q)}$ is a fixed line for δ, contradicting our original assumption. □

Theorem CAP.21. *Let $\mathcal{P}$ be an affine plane. The set of dilations of $\mathcal{P}$ with fixed point O, together with ι form a group under the operation of composition of mappings. (For the definition of a group, see Chapter 1, Section 1.5.)*

Proof. In Chapter 1, Section 1.4 we showed that composition of mappings is associative, so that in particular the composition of dilations is associative. Let α and β be dilations of $\mathcal{P}$ with fixed point O and let $\mathcal{L}$ be any line on $\mathcal{P}$. O is a fixed point of α^{-1} and of $\beta \circ \alpha$. Since $\mathcal{L} = \alpha(\alpha^{-1}(\mathcal{L}))$ by Definition CAP.17, $\alpha^{-1}(\mathcal{L}) \mathbb{PE} \mathcal{L}$. By Definition CAP.17, α^{-1} is a dilation of $\mathcal{P}$. Since $\alpha(\mathcal{L}) \mathbb{PE} \mathcal{L}$ and $\beta(\alpha(\mathcal{L})) \mathbb{PE} \alpha(\mathcal{L})$, by Theorem IP.6, $\beta(\alpha(\mathcal{L})) \mathbb{PE} \mathcal{L}$. By Definition CAP.17, either $\alpha \circ \beta$ is a dilation of $\mathcal{P}$, or $\alpha \circ \beta = \iota$. □

Theorem CAP.22. *Let $\mathcal{P}$ be an affine plane and let α be a collineation of $\mathcal{P}$ such that α has one and only one fixed point O. If every line containing O is a fixed line for α, then*

(A) *every fixed line $\mathcal{M}$ for α contains O, and*
(B) *α is a dilation of $\mathcal{P}$.*

Proof. (A) We prove the contrapositive: suppose $\mathcal{M}$ is a fixed line of α not containing the point O. If Q is any point on $\mathcal{M}$, then by hypothesis $\overleftrightarrow{OQ}$ is a fixed line of α, and by Theorem CAP.4(B), Q is a fixed point of α. But $Q \neq O$ so is not a fixed point, a contradiction. Hence our supposition is false and $\mathcal{M}$ is not a fixed line of α.

(B) If $\mathcal{J}$ is any line on $\mathcal{P}$ such that $O \notin \mathcal{J}$, then by part (A), $\mathcal{J}$ is not a fixed line of α, so is distinct from $\alpha(\mathcal{J})$. Since $\alpha(O) = O$ and α is a bijection, O is the only point X such that $\alpha(X) = O$. Therefore, since $O \notin \mathcal{J}$, $O \notin \alpha(\mathcal{J})$. By part (A) $\alpha(\mathcal{J})$ is not a fixed line for α.

Assume $\mathcal{J}$ and $\alpha(\mathcal{J})$ are not parallel, so that they intersect at a point G. By hypothesis $\overleftrightarrow{OG}$ is a fixed line of α, and so is distinct from either $\mathcal{J}$ or $\alpha(\mathcal{J})$. Thus $\mathcal{J}$, $\alpha(\mathcal{J})$ and $\overleftrightarrow{OG}$ are distinct lines which are concurrent at the point G. Since $\overleftrightarrow{OG}$ is a fixed line, $\alpha(G)$ belongs to $\overleftrightarrow{OG}$; it also belongs to $\alpha(\mathcal{J})$, so by Exercise I.1 $\alpha(G) = G$, and G is a fixed point of α. This contradicts the hypothesis that there are no fixed points other than O. Hence our assumption is false, and $\mathcal{J} \parallel \alpha(\mathcal{J})$. By Definition CAP.17, α is a dilation of $\mathcal{P}$. □

Theorem CAP.23. *Let $\mathcal{P}$ be an affine plane, O be a point on $\mathcal{P}$, and δ a dilation of $\mathcal{P}$ with fixed point O.*

(A) *Let φ be a collineation of $\mathcal{P}$ such that $\varphi(O) = P$ (that is, $\varphi^{-1}(P) = O$), where P is some point of $\mathcal{P}$. Then $\varphi \circ \delta \circ \varphi^{-1}$ is a dilation with fixed point P.*

(B) *If φ is a collineation of $\mathcal{P}$ with fixed point O, then $\varphi \circ \delta \circ \varphi^{-1}$ is a dilation of $\mathcal{P}$ with fixed point O.*

(C) *If $P \neq O$ is a point of $\mathcal{P}$ and τ is a translation such that $\tau(P) = O$, then $\tau^{-1} \circ \delta \circ \tau$ is a dilation with fixed point P.*

Proof. (A) First we show that $\varphi^{-1}(P)$ is a fixed point of δ iff P is a fixed point for $\varphi \circ \delta \circ \varphi^{-1}$.

If $\varphi^{-1}(P)$ is a fixed point of δ, $\varphi \circ \delta \circ \varphi^{-1}(P) = \varphi \circ \varphi^{-1}(P) = P$, so that P is a fixed point for $\varphi \circ \delta \circ \varphi^{-1}$. If P is a fixed point for $\varphi \circ \delta \circ \varphi^{-1}$, $\varphi \circ \delta \circ \varphi^{-1}(P) = P$ so that $\delta \circ \varphi^{-1}(P) = \varphi^{-1} \circ \varphi \circ \delta \circ \varphi^{-1}(P) = \varphi^{-1}(P)$ and $\varphi^{-1}(P)$ is a fixed point for δ.

Since φ is one-to-one, if $\varphi \circ \delta \circ \varphi^{-1}$ were to have another fixed point, there would be another fixed point for δ, which is impossible since by assumption δ is a dilation (*cf* Theorem CAP.18(B)). Thus $\varphi \circ \delta \circ \varphi^{-1}$ has exactly one fixed point P.

Let $\mathcal{M}$ be any line containing P, and define $\mathcal{L} = \varphi^{-1}(\mathcal{M})$. Since φ is a collineation, so is φ^{-1}, and $\mathcal{L}$ is therefore a line. Since $P \in \mathcal{M}$, $O = \varphi^{-1}(P) \in \varphi^{-1}(\mathcal{M}) = \mathcal{L}$; by Theorem CAP.18(A), $\mathcal{L}$ is a fixed line for δ.

Then $\varphi \circ \delta \circ \varphi^{-1}(\mathcal{M}) = \varphi \circ \delta(\mathcal{L}) = \varphi(\mathcal{L}) = \mathcal{M}$ so that $\mathcal{M}$ is a fixed line for $\varphi \circ \delta \circ \varphi^{-1}$. By Theorem CAP.22, $\varphi \circ \delta \circ \varphi^{-1}$ is a dilation with fixed point P.

(B) To prove (B), let $P = O$ in part (A).

(C) To prove (C), let $\varphi = \tau^{-1}$ in part (A). Then $\varphi^{-1} = \tau$, and $\varphi^{-1}(P) = \tau(P) = O$. The translation τ is a collineation, so that φ is also a collineation. The result follows from part (A). □

Theorem CAP.24. *Let $\mathcal{P}$ be an affine plane, O be a point on $\mathcal{P}$, and let A and B be distinct points distinct from and collinear with O. Then there exists at most one dilation α of $\mathcal{P}$ with fixed point O such that $\alpha(A) = B$.*

Proof. If α and β are dilations of $\mathcal{P}$ each with fixed point O such that $\alpha(A) = B$ and $\beta(A) = B$, then A is a fixed point of $\beta^{-1} \circ \alpha$. By Theorem CAP.21 $\beta^{-1} \circ \alpha$ is a dilation; by Theorem CAP.18, if $\beta^{-1} \circ \alpha \neq \iota$ it has no fixed point $\neq O$. Since it has such a fixed point, it must be ι, so that $\beta = \alpha$. □

3.4 Collineations: axial affinities

Definition CAP.25. Let $\mathcal{P}$ be a plane and let α be a collineation of $\mathcal{P}$. Then α is an **axial affinity** of $\mathcal{P}$ iff $\alpha \neq \iota$ and there exists a line $\mathcal{M}$ (the **axis** of α) such that every point on $\mathcal{M}$ is a fixed point of α (and therefore $\mathcal{M}$ is a fixed line).

Note that this definition is valid even in the absence of a parallel axiom. Later (Remark CAP.30) we anticipate the definition in Chapter 16 of two subclasses of axial affinities, *stretches* and *shears*.

Theorem CAP.26. *Let $\mathcal{P}$ be an affine plane and let α be an axial affinity of $\mathcal{P}$ with axis $\mathcal{M}$.*

(A) *If $\mathcal{N}$ is a fixed line of α such that $\mathcal{N}$ and $\mathcal{M}$ are distinct and not parallel and if $\mathcal{L} \parallel \mathcal{N}$, then $\mathcal{L}$ is a fixed line of α.*
(B) *The set of fixed points of α is $\mathcal{M}$.*
(C) *For every point Q on $\mathcal{M}$, there exists at most one fixed line $\mathcal{L} \neq \mathcal{M}$ of α such that $Q \in \mathcal{L}$.*
(D) *If $\mathcal{L}$ and $\mathcal{N}$ are fixed lines of α such that $\mathcal{M}$, $\mathcal{L}$, and $\mathcal{N}$ are distinct, then $\mathcal{L} \parallel \mathcal{N}$.*

Proof. (A) By Theorem CAP.3, $\alpha(\mathcal{L}) \parallel \alpha(\mathcal{N})$. Since $\mathcal{N}$ is a fixed line of α, $\alpha(\mathcal{N}) = \mathcal{N}$. Thus $\mathcal{L} \parallel \mathcal{N} = \alpha(\mathcal{N}) \parallel \alpha(\mathcal{L})$, and by Exercise IP.2(A), $\mathcal{L} \mathbb{PE} \alpha(\mathcal{L})$.

Again by Exercise IP.2(A), $\mathcal{L}$ and $\mathcal{M}$ are not parallel; therefore there exists a point A such that $\mathcal{L} \cap \mathcal{M} = \{A\}$. By Definition CAP.25, $\alpha(A) = A$, and thus $A \in \alpha(\mathcal{L})$. By Axiom PS, $\alpha(\mathcal{L}) = \mathcal{L}$.

(B) If α had a fixed point off of $\mathcal{M}$, then by Exercise CAP.3, α would be the identity mapping ι. This contradicts Definition CAP.25. Hence α has no fixed point off of $\mathcal{M}$.

(C) Suppose there exist two distinct fixed lines $\mathcal{L}$ and $\mathcal{N}$ of α such that $Q \in \mathcal{L}$ and $Q \in \mathcal{N}$ and let X be any member of $\mathcal{P} \setminus \mathcal{M}$. By Axiom PS, there exist unique lines $\mathcal{G}$ and $\mathcal{H}$ on $\mathcal{P}$ such that $X \in \mathcal{G}$, $X \in \mathcal{H}$, $\mathcal{G} \parallel \mathcal{L}$, and $\mathcal{H} \parallel \mathcal{N}$. By part (A), each of the lines $\mathcal{G}$ or $\mathcal{H}$ is a fixed line of α. By Theorem CAP.4, X is a fixed point of α. Since X is any member of $\mathcal{P} \setminus \mathcal{M}$, and since every member of $\mathcal{M}$ is a fixed point of α, $\alpha = \iota$. This contradicts definition CAP.25. Hence our supposition that distinct fixed lines $\mathcal{L}$ and $\mathcal{M}$ exist is false.

(D) By part (C), $\mathcal{L}$ and $\mathcal{N}$ cannot intersect at a point on $\mathcal{M}$. If $\mathcal{L}$ and $\mathcal{N}$ were to intersect at a point off of $\mathcal{M}$, then by Theorem CAP.4, that point would be a fixed point of α, contrary to part (B). Hence $\mathcal{L} \parallel \mathcal{N}$. □

Theorem CAP.27. *Let $\mathcal{P}$ be an affine plane and let α be an axial affinity of $\mathcal{P}$ with axis $\mathcal{M}$.*

(A) *If $\mathcal{L}$ is a fixed line of α distinct from $\mathcal{M}$, then for every point $Q \in (\mathcal{L} \setminus \mathcal{M})$, $\mathcal{L} = \overleftrightarrow{Q\alpha(Q)}$.*

(B) *If there exists a fixed line $\mathcal{L}$ of α such that $\mathcal{L}$ and $\mathcal{M}$ are not parallel, then the set of fixed lines of α is $\{\mathcal{J} \mid \mathcal{J}$ is a line on $\mathcal{P}$ and $\mathcal{J} \mathbb{PE} \mathcal{L}\} \cup \{\mathcal{M}\}$.*

(C) *If there exists a fixed line $\mathcal{L}$ of α such that $\mathcal{L} \parallel \mathcal{M}$, then the set of fixed lines of α is $\{\mathcal{J} \mid \mathcal{J}$ is a line on $\mathcal{P}$ and $\mathcal{J} \mathbb{PE} \mathcal{M}\}$.*

Proof. (A) Since $\mathcal{L}$ is a fixed line of α, $\alpha(Q) \in \mathcal{L}$. By Axiom I.1, $\mathcal{L} = \overleftrightarrow{Q\alpha(Q)}$.

(B) If $\mathcal{J}$ is a line on $\mathcal{P}$ such that $\mathcal{J} \parallel \mathcal{L}$, then by Theorem CAP.26(A), $\mathcal{J}$ is a fixed line of α. If $\mathcal{J} \neq \mathcal{L}$ is a fixed line of α other than $\mathcal{M}$, then by Theorem CAP.26(D), $\mathcal{J} \parallel \mathcal{L}$.

(C) If $\mathcal{N}$ is a fixed line of α distinct from both $\mathcal{L}$ and $\mathcal{M}$, then by Theorem CAP.26(D) $\mathcal{N} \parallel \mathcal{L}$, and by Theorem IP.6 $\mathcal{N} \parallel \mathcal{M}$.

Conversely, if $\mathcal{N} \parallel \mathcal{M}$ and $\mathcal{N}$ is not a fixed line, $\alpha(\mathcal{N}) \neq \mathcal{N}$ so that for some point $Q \in \mathcal{N}$, $\alpha(Q) \notin \mathcal{N}$. If $\overleftrightarrow{Q\alpha(Q)}$ were parallel to $\mathcal{M}$, then both $\overleftrightarrow{Q\alpha(Q)}$ and $\mathcal{N}$ would be lines through Q parallel to $\mathcal{M}$, contrary to Axiom PS.

Therefore $\overleftrightarrow{Q\alpha(Q)}$ and $\mathcal{M}$ are not parallel, and there exists a point X such that $\{X\} = \mathcal{M} \cap \overleftrightarrow{Q\alpha(Q)}$. By Theorem CAP.1 $\alpha(\overleftrightarrow{Q\alpha(Q)}) = \overleftrightarrow{\alpha(Q)\alpha(\alpha(Q))}$ so that $\alpha(Q) \in \alpha(\overleftrightarrow{Q\alpha(Q)})$; also $\alpha(X) = X$ so that by Axiom I.1 $\alpha(\overleftrightarrow{Q\alpha(Q)}) = \overleftrightarrow{Q\alpha(Q)}$, which therefore is a fixed line.

But $\overleftrightarrow{Q\alpha(Q)}$ intersects $\mathcal{M}$ at the point X, contradicting the first part of part (C) of this proof, which says that every fixed line is parallel to $\mathcal{M}$. Therefore if $\mathcal{N} \parallel \mathcal{M}$, $\mathcal{N}$ is a fixed line. □

Theorem CAP.28. *Let $\mathcal{P}$ be an affine plane and $\mathcal{M}$ be a line on $\mathcal{P}$.*

(A) *Let $\mathcal{A}(\mathcal{M})$ be the set of axial affinities with axis $\mathcal{M}$, then under composition of mappings $\mathcal{A}(M) \cup \{\iota\}$ is a group. (For definition of a group, see Chapter 1, Section 1.5.)*

(B) *Let $\mathcal{L}$ be a line on $\mathcal{P}$ distinct from $\mathcal{M}$ and let $\mathcal{A}^*(\mathcal{M}, L)$ be the set of affinities of $\mathcal{P}$ with axis $\mathcal{M}$ such that $\mathcal{L}$ is a fixed line of every member α of $\mathcal{A}^*(\mathcal{M}, L)$, then $\mathcal{A}^*(\mathcal{M}, L) \cup \{\iota\}$ is a group under composition of mappings.*

Proof. In Chapter 1, Section 1.4 we showed that the composition of mappings is associative.

(A) Let α and β be axial affinities of $\mathcal{P}$ with axis $\mathcal{M}$. The set of fixed points of α^{-1} is $\mathcal{M}$, so α^{-1} is an affinity with axis $\mathcal{M}$. (cf Definition CAP.25) Furthermore, every member of $\mathcal{M}$ is a fixed point of $\beta \circ \alpha$. If $\beta \circ \alpha$ has a fixed point Q which is a member of $\mathcal{P} \setminus M$, then by Exercise CAP.3, $\beta \circ \alpha = \iota$.

Otherwise if $\beta \circ \alpha$ has no fixed point which is a member of $\mathcal{P} \setminus M$, then by Definition CAP.25 $\beta \circ \alpha$ is an axial affinity of $\mathcal{P}$ with axis $\mathcal{M}$. Hence $\mathcal{A}(M)$ is a group under composition of mappings.

(B) Let α and β be axial affinities of $\mathcal{P}$, each with axis $\mathcal{M}$ and with fixed line $\mathcal{L}$. By part (A) the set of fixed points for both α^{-1} and $\beta \circ \alpha$ is $\mathcal{M}$, and the set $\mathcal{A}(\mathcal{M})$ is a group. Thus to show that $\mathcal{A}^*(\mathcal{M}, L)$ is a subgroup, all we need to do is show that $\mathcal{L}$ is a fixed line for both $\beta \circ \alpha$ and α^{-1}.

$\mathcal{L}$ is a fixed line of α so that $\alpha(\mathcal{L}) = \mathcal{L}$, hence $\alpha^{-1}(\mathcal{L}) = \alpha^{-1}(\alpha(\mathcal{L})) = \mathcal{L}$, and $\mathcal{L}$ is a fixed line for α^{-1}. Since $\mathcal{L}$ is a fixed line of both α and β, $\beta(\alpha(\mathcal{L})) = \beta(\mathcal{L}) = \mathcal{L}$ so $\mathcal{L}$ is a fixed line of $\beta \circ \alpha$, as required. □

Theorem CAP.29. *Let $\mathcal{P}$ be an affine plane, $\mathcal{M}$ be a line on $\mathcal{P}$, and A and B be distinct members of $\mathcal{P} \setminus M$, then there exists at most one axial affinity α of $\mathcal{P}$ with axis $\mathcal{M}$ such that $\alpha(A) = B$.*

Proof. Let α and β be axial affinities of $\mathcal{P}$ with axis $\mathcal{M}$ such that $\alpha(A) = B$ and $\beta(A) = B$. Since $\mathcal{A}(M) \cup \{\iota\}$ is a group (cf Theorem CAP.28), $\beta^{-1} \circ \alpha$ is a member of $\mathcal{A}(M) \cup \{\iota\}$. Since $\beta^{-1}(\alpha(A)) = A$, A is a fixed point of $\beta^{-1} \circ \alpha$, *i.e.* $(\beta^{-1} \circ \alpha)(A) = A$. By Exercise CAP.3, $\beta^{-1} \circ \alpha = \iota$, i.e., $\alpha = \beta$. □

Remark CAP.30. In Chapter 16 (Axial affinities of a Euclidean plane) we will define two types of axial affinity with axis $\mathcal{M}$: φ will be a *stretch* if there exists a line $\mathcal{L}$ which is a fixed line for φ but is not parallel to $\mathcal{M}$, and the set of fixed lines of φ is $\{\mathcal{M}\} \cup \{\mathcal{J} \mid \mathcal{J} \mathbb{PE} \mathcal{L}\}$; φ will be a *shear* if the set of fixed lines of φ is $\{\mathcal{J} \mid \mathcal{J} \mathbb{PE} \mathcal{M}\}$. For a visualization see Figure 3.2.

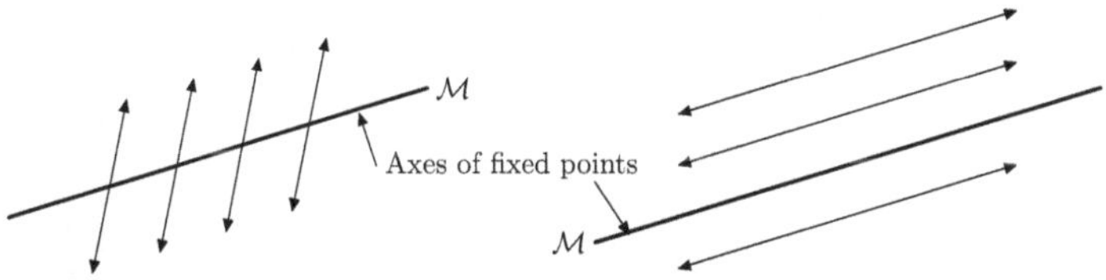

Fig. 3.2 Showing action of a stretch (left) and a shear (right); double-headed arrows show fixed lines.

In Chapter 8 (neutral geometry) we will meet a type of collineation called a *reflection*, which is a stretch. In Chapter 16 we will show the existence of stretches other than reflections.

3.5 Exercises for collineations

Answers to starred (*) exercises may be accessed from the home page for this book at www.springer.com.

Exercise CAP.1*. Let $\mathcal{P}$ be an affine plane and let $\mathcal{L}$, $\mathcal{M}$, and $\mathcal{N}$ be lines on $\mathcal{P}$. If $\mathcal{L}\ \mathbb{PE}\ M$ and $\mathcal{M}\ \mathbb{PE}\ N$, then $\mathcal{L}\ \mathbb{PE}\ N$.

Exercise CAP.2*. Let $\mathcal{P}$ be any plane where the incidence axioms hold, φ be a collineation of $\mathcal{P}$, and A, B, and C be points on $\mathcal{P}$.

(A) If A, B, and C are collinear, then $\varphi(A)$, $\varphi(B)$, and $\varphi(C)$ are collinear.
(B) If A, B, and C are noncollinear, then $\varphi(A)$, $\varphi(B)$, and $\varphi(C)$ are noncollinear.
(C) A, B, and C are collinear iff $\varphi(A)$, $\varphi(B)$, and $\varphi(C)$ are collinear.
(D) A, B, and C are noncollinear iff $\varphi(A)$, $\varphi(B)$, and $\varphi(C)$ are noncollinear.

Exercise CAP.3*. Let φ be a collineation of an affine plane $\mathcal{P}$, $\mathcal{M}$ a line on $\mathcal{P}$ such that every point on $\mathcal{M}$ is a fixed point of φ, and Q a fixed point of φ such that $Q \in (\mathcal{P} \setminus M)$. Then $\varphi = \iota$.

Exercise CAP.4*. Let $\mathcal{P}$ be an affine plane, $\mathcal{L}_1$, and $\mathcal{L}_2$ be parallel lines on $\mathcal{P}$, O_1 be a member of $\mathcal{L}_1$, O_2 be a member of $\mathcal{L}_2$, and τ be the translation (*cf* Theorem CAP.9) of $\mathcal{P}$ such that $\tau(O_1) = O_2$, then $\tau(\mathcal{L}_1) = \mathcal{L}_2$.

Exercise CAP.5*. Let $\mathcal{P}$ be an affine plane, φ be a dilation of $\mathcal{P}$ with fixed point O, and ψ be a stretch of $\mathcal{P}$ with axis $\mathcal{M}$ through O, then $\varphi \circ \psi = \psi \circ \varphi$. (We take Remark CAP.30 as a definition of a stretch.)

Chapter 4
Incidence and Betweenness (IB)

Acronym: *IB*
Dependencies: *Chapter 1*
New axiom: *Axiom BET (betweenness)*
New Terms Defined: *between, segment, ray, open, closed, endpoint, triangle, edge, opposite edge, convex, Q-side, halfplane, opposite side*

Abstract: This chapter defines a betweenness relation and uses it to define segments, rays, and triangles. A few theorems are proved in the resulting IB geometry. These are foundational for the rest of the development.

We now temporarily suspend the parallel Axioms PS and PW. This will launch us on a new "thread" of inquiry separate from that of Chapters 2 and 3, which will take us through Chapter 10, developing as much geometry as possible without invoking any parallel axiom. In Chapter 11 we will re-invoke Axiom PS, and combine the results of this new thread with that of Chapters 2 and 3 to get Euclidean geometry.

Meanwhile, in this chapter we will use only the axioms and results from Chapter 1, Theorems CAP.1 through CAP.4, the definitions from Chapter 3 that do not depend on a parallel axiom, and concepts we introduce here.

E.J. Specht et al., *Euclidean Geometry and its Subgeometries*,
DOI 10.1007/978-3-319-23775-6_4

4.1 Definition and properties of betweenness

One of the major defects in Euclid's treatment of geometry was his failure to deal with betweenness. As a result, if we take his axioms literally, it is possible to prove, for example, that all triangles are isosceles (having two edges of equal length).

To avoid such pitfalls, we construct a definition of betweenness to conform with our intuitive notion of that term. We do this by specifying a set called a betweenness relation. We do not describe this set by saying exactly what its members are; instead, we state various properties that describe how its members interact with points and lines, and with each other.

In this section, $\mathbb{U}$ is space as defined in Definition I.0, in which Axioms I.0 through I.5 hold. In this context, *collinearity* has meaning—that is, $\mathbb{U}$ contains subsets that are *lines*.

Definition IB.1. A **betweenness relation** on $\mathbb{U}$ (or one of its subsets) is a nonempty set $\mathbb{B}$ of ordered triples (A, B, C) of points having the following Properties B.0 through B.3. To indicate that a triple (A, B, C) is a member of $\mathbb{B}$ we will write A–B–C; this is read "B is **between** A and C."

B.0 (distinctness and collinearity): For any points A, B, and C, if A–B–C, then A, B, and C are distinct collinear points.

B.1 (symmetric property): For any points A, B, and C, if A–B–C, then C–B–A.

B.2 (trichotomy property): If A, B, and C are any distinct collinear points, exactly one of the following statements is true:

$$A\text{–}B\text{–}C, \quad B\text{–}A\text{–}C, \quad A\text{–}C\text{–}B.$$

B.3 (extension property): If A and B are any two distinct points, there exists a point C such that A–B–C.

You may wish to check the properties in this definition with a sketch or mental picture to assure yourself that this is indeed the betweenness you have known all your life.

Axiom BET. There exists a betweenness relation.

Definition IB.1.1. A space on which all the incidence axioms and Axiom BET are true is called **Incidence-Betweenness space** or simply **IB space**. A plane in this space will be called an **IB plane**, and the geometry of IB space **IB geometry**.

IB geometry is sufficiently rich to allow us to introduce several concepts that will be with us throughout the rest of the book, such as *triangle, convex set*, and *opposite sides of a line in a plane*. But we can prove only a few new theorems.

(A) Sets and ordered triples: Be sure to read this part carefully if you are unsure of your understanding of the term *ordered triple*. (*cf* Chapter 1 Section 1.3.)

There are six different ways to describe an (unordered) set containing exactly three points A, B, and C; we list them here: $\{A,B,C\} = \{B,A,C\} = \{A,C,B\} = \{C,B,A\} = \{C,A,B\} = \{B,C,A\}$.

An *ordered triple* is a set $\{A,B,C\}$ of points together with a one-to-one correspondence between this set and the set $\{1,2,3\}$. We denote such an ordered triple by listing the elements in the order specified by this correspondence, and enclosing the list in ordinary parentheses.

Thus the ordered triple (A,B,C) is the set $\{A,B,C\}$ where A corresponds to 1, B to 2, and C to 3; the ordered triple (C,A,B) is the same set where A corresponds to 2, B to 3, and C to 1. Any set $\{A,B,C\}$ can be "ordered" into six distinct (different) ordered triples, namely (A,B,C), (B,A,C), (A,C,B), (C,B,A), (C,A,B), and (B,C,A).

(B) Implications of Properties B.0–B.3: According to Definition IB.1, a *betweenness relation* $\mathbb{B}$ on space $\mathbb{U}$ is a collection of ordered triples of members of $\mathbb{U}$, which satisfies conditions B.0 through B.3. But not every set $\{A,B,C\}$ can be "ordered" by the process described in part (B) into an ordered triple that is a member of $\mathbb{B}$. Property B.0 says that this can be done only for sets $\{A,B,C\}$ consisting of distinct collinear points. Thus, our definition of betweenness is essentially a definition on lines.

In the coordinate plane it is not possible to get a member of $\mathbb{B}$ by ordering the points $(0,0,0)$, $(1,0,0)$, and $(0,1,0)$ because these points are not collinear. Also, it makes no sense to say that a point B is "between" A and A. One object cannot be "between" another (single) object.

On the other hand, Property B.2 says that if a set $\{A,B,C\}$ consists of distinct collinear points, these *can* be ordered in such a way that the resulting ordered triple (call it (D,E,F)) belongs to $\mathbb{B}$. In this case, Property B.1 says that the ordered triple (F,E,D) also belongs to $\mathbb{B}$. Thus, if $\{A,B,C\}$ can be ordered into a triple in $\mathbb{B}$, there are exactly two ways to do it.

(C) A betweenness relation is nonempty: We state in the first sentence of Definition IB.1 that a betweenness relation is nonempty, but this requirement is actually redundant. For by Axiom I.0, space $\mathbb{U}$ contains lines; by Axiom I.5(A) every line contains at least two points A and B; by property B.3 of Definition IB.1 there exists a point $C \in \overleftrightarrow{AB}$ such that A–B–C, so that the ordered triple (A, B, C) belongs to $\mathbb{B}$, which is therefore nonempty.

It is tempting to apply Property B.3 again to show that there exists another point $D \in \overleftrightarrow{AB}$ and eventually that there are infinitely many points in $\overleftrightarrow{AB}$. But the argument breaks down because it does not follow from A–B–C and A–C–D that A–B–D, as we will shortly discuss in Remark IB.4.2. Thus we may not conclude at this stage that $D \in \overleftrightarrow{AB}$. That there are infinitely many points on a line will be established later in Chapter 5 as Corollary PSH.22.2.

(D) Alterations to a betweenness relation: A certain amount of freedom is possible in defining a betweenness relation. Given a betweenness relation $\mathbb{B}$ containing (A, B, C) (and, by Property B.1, also containing (C, B, A)), we could define a new set $\mathbb{C}$ to be the same as $\mathbb{B}$ except that it contains (B, A, C) and (C, A, B) instead of (A, B, C) and (C, B, A). Then $\mathbb{C}$ would also be a betweenness relation.

For example, the standard betweenness relation for the integers includes the triples $(2, 3, 4)$ and $(4, 3, 2)$. If we let $\mathbb{C}$ contain the same ordered triples, with the exception that $(3, 2, 4)$ is substituted for $(2, 3, 4)$ and $(4, 2, 3)$ for $(4, 3, 2)$, it is quite easy to verify that all the Properties B.0 through B.3 hold for $\mathbb{C}$. Thus $\mathbb{C}$ is also a betweenness relation on the integers, even though it does not agree with our intuition. We will use this possibility in Chapter 21, Section 21.8, where, on the basis of models constructed there, we justify a number of the assertions made later in this chapter, particularly those in Remarks IB.3.1 and IB.4.2.

Definition IB.2. The symbol "A–B–C–D" means that A, B, C, and D are points such that A–B–C, A–B–D, A–C–D, and B–C–D.

Remark IB.2.1. By virtue of Property B.2, A–B–C–D is equivalent to the conjunction of $\neg(B$–A–$C)$, $\neg(A$–C–$B)$, $\neg(A$–D–$B)$, $\neg(B$–A–$D)$, $\neg(A$–D–$C)$, $\neg(C$–A–$D)$, $\neg(B$–D–$C)$, and $\neg(C$–B–$D)$.

By Property B.0, if A–B–C–D, then the points in each of the triples $\{A, B, C\}$, $\{A, B, D\}$, $\{A, C, D\}$, and $\{B, C, D\}$ are distinct and collinear. By Exercise I.2, A, B, C, and D are collinear.

Definition IB.3. A set $\mathcal{E}$ of points is a **segment** if there exist distinct points U and V such that one of the following holds:

$\mathcal{E} = \{X \mid U\text{–}X\text{–}V\}$, in which case $\mathcal{E}$ is called the **open segment** UV and is symbolized by $\overset{]\,[}{UV}$;

$\mathcal{E} = \{X \mid X = U \text{ or } X = V \text{ or } U\text{–}X\text{–}V\}$, in which case $\mathcal{E}$ is called the **closed segment** UV and is symbolized by $\overset{[\,]}{UV}$;

$\mathcal{E} = \{X \mid X = U \text{ or } U\text{–}X\text{–}V\}$, or $\mathcal{E} = \{X \mid X = V \text{ or } U\text{–}X\text{–}V\}$, in which cases $\mathcal{E}$ is denoted by $\overset{[\,[}{UV}$ or $\overset{]\,]}{UV}$, respectively. In these cases $\mathcal{E}$ is said to be a **half-open** segment (for the less optimistic, **half-closed**). There is no widely accepted verbiage to distinguish between $\overset{[\,[}{UV}$ and $\overset{]\,]}{UV}$.

The points U and V in this definition are called **endpoints** of $\mathcal{E}$.

Remark IB.3.1. Note that in this definition we did not say *the* endpoints—there's no guarantee here that a given segment does not have two different sets of endpoints.[1] Note also that there is nothing in Definition IB.3 that guarantees that an open segment is nonempty, or that there are any points between the endpoints of a closed segment. The proof that there are such points must wait until Chapter 5, Theorem PSH.22 (Denseness).

Definition IB.4. A set $\mathcal{E}$ of points is called a **ray** iff there exist distinct points A and B such that either $\mathcal{E} = \{X \mid X = A \text{ or } A\text{–}X\text{–}B \text{ or } X = B \text{ or } A\text{–}B\text{–}X\}$ or $\mathcal{E} = \{X \mid A\text{–}X\text{–}B \text{ or } X = B \text{ or } A\text{–}B\text{–}X\}$. In the first case, $\mathcal{E}$ is denoted by $\overset{[\rightarrow}{AB}$, which is read "the **closed ray** AB." In the second case, $\mathcal{E}$ is denoted by $\overset{]\rightarrow}{AB}$, which is read "the **open ray** AB." If $\mathcal{E}$ is a ray, then a point U is an **endpoint**, or **initial point**, of $\mathcal{E}$ iff there exists a point V such that $V \neq U$ and either $\mathcal{E} = \overset{]\rightarrow}{UV}$ or $\mathcal{E} = \overset{[\rightarrow}{UV}$.

Remark IB.4.1. If we make use of elementary logic and set theory, we can get the following simple relationships involving the above definitions:

(a) $\overset{]\,[}{AB} = \overset{[\,]}{AB} \setminus \{A, B\} = \overset{]\,]}{AB} \setminus \{B\} = \overset{[\,[}{AB} \setminus \{A\}$

(b) $\overset{[\,[}{AB} = \overset{[\,]}{AB} \setminus \{B\} = \overset{]\,[}{AB} \cup \{A\}$

(c) $\overset{]\,]}{AB} = \overset{[\,]}{AB} \setminus \{A\} = \overset{]\,[}{AB} \cup \{B\}$

(d) $\overset{[\,]}{AB} = \overset{]\,]}{AB} \cup \{A\} = \overset{[\,[}{AB} \cup \{B\} = \overset{]\,[}{AB} \cup \{A\} \cup \{B\}$

(e) $\overset{]\rightarrow}{AB} = \overset{[\rightarrow}{AB} \setminus \{A\}$

[1] In Chapter 21 we will construct Model DZIII for IB geometry and prove, in Theorem DZIII.4(A) that it is possible to have two segments $\overset{[\,]}{AB}$ and $\overset{[\,]}{CD}$ such that $\overset{[\,]}{AB} = \overset{[\,]}{CD}$ and yet $\{A, B\} \neq \{C, D\}$, i.e., two segments which are equal but have different endpoints.

(f) $\overset{[\to}{AB} = \overset{]\to}{AB} \cup \{A\}$

(g) $\overset{[\to}{AB} = \overset{[\,]}{AB} \cup \{X \mid A\text{–}B\text{–}X\} = \{A\} \cup \{B\} \cup \overset{]\,[}{AB} \cup \{X \mid A\text{–}B\text{–}X\}$

(h) $\overset{]\to}{AB} = \overset{]\,]}{AB} \cup \{X \mid A\text{–}B\text{–}X\} = \overset{]\,[}{AB} \cup \{B\} \cup \{X \mid A\text{–}B\text{–}X\}$

(i) $\overset{[\,]}{AB} \cap \{X \mid A\text{–}B\text{–}X\} = \emptyset$

(j) $\overset{]\,]}{AB} \cap \{X \mid A\text{–}B\text{–}X\} = \emptyset$

There may be a few others like this that we've missed, but hopefully we've listed more than enough of them to give the general idea. We'll be using these as well as the missing ones casually, without further reference, from now on.

Remark IB.4.2. Beware, however, of some relationships among these ideas that may seem just as appealing as those above but which are *not* consequences of the properties of betweenness we have stated so far. For example, it may seem obvious that a ray has a unique endpoint, and that if A–B–C, then $\overset{[\to}{BA} \cup \overset{[\to}{BC} = \overleftrightarrow{AC}$ and $\overset{[\,]}{AB} \cup \overset{[\,]}{BC} = \overset{[\,]}{AC}$. But it is not possible to prove any of these statements from what we have so far.[2]

These strange circumstances tell us that the properties of Definition IB.1 are inadequate to define a betweenness relation which conforms to our intuitive ideas of what betweenness means. One way to deal with this difficulty would be to add another property (B.4) to Definition IB.1, as follows:

Property B.4 (Not invoked). Let A, B, C, and D be collinear points.

(A) If A–B–C and A–C–D, then B–C–D.
(B) If A–B–C and B–C–D, then A–B–D.

Invoking this additional property as part of Definition IB.1 would provide a shortcut to the result of Theorem PSH.8 in Chapter 5, which is now a consequence of the Plane Separation Axiom (PSA). In our development, Theorem PSH.8 is fundamental to solving all the difficulties mentioned above. However, invoking Property B.4 would be only a partial measure that would not address the critical issues having to do with the "sides" of a line; we will meet these shortly. Moreover, Subsection 21.8.1 of Chapter 21 shows that Property B.4 would not imply "Denseness" (*cf* Theorem PSH.22).

---

[2]In Chapter 21 we will construct Model DZII for IB geometry and prove, in Theorem DZII.4, that all these statements are false.

While some treatments of geometry do invoke Property B.4, we do not take this path, preferring instead to invoke PSA, which, with some effort, will yield much more than this property would yield. So we put Property B.4 aside, and for now, content ourselves with proving a few theorems in IB geometry.

4.2 Theorems of Incidence-Betweenness geometry

In incidence geometry, lines are sets of points which interact with other sets in certain prescribed ways, but there is no language in that geometry to describe their internal structure. The introduction of betweenness gives them an internal structure, of which the next theorem gives us a first glimpse.

Theorem IB.5. *Let A and B be distinct points. Then*

$$\begin{aligned}\overleftrightarrow{AB} &= \{X\,|X\text{–}A\text{–}B\} \cup \{A\} \cup \{X\,|A\text{–}X\text{–}B\} \cup \{B\} \cup \{X\,|A\text{–}B\text{–}X\} \\ &= \{X\,|X\text{–}A\text{–}B\} \cup \overset{\sqsubset\sqsupset}{AB} \cup \{X\,|A\text{–}B\text{–}X\} \\ &= \{X\,|X\text{–}A\text{–}B\} \cup \overset{\sqsubset\!\rightarrow}{AB} \\ &= \{X\,|X\text{–}A\text{–}B\} \cup \{A\} \cup \overset{\sqsupset\!\rightarrow}{AB}\end{aligned}$$

and the sets in the unions are disjoint.

Proof. A point X belongs to $\overleftrightarrow{AB}$ iff A, B, and X are collinear. By Property B.2 of Definition IB.1, $X \in \overleftrightarrow{AB}$ iff exactly one of X–A–B or $X = A$ or A–X–B or $X = B$ or A–B–X holds. The first line in the formula given is an exact translation of this statement into set language, and the second, third, and fourth lines come from Definition IB.4.

The sets in the first line of the theorem are disjoint because A and B are distinct, and they do not belong to any of the sets $\{X\,|X\text{–}A\text{–}B\}$, $\{X\,|A\text{–}X\text{–}B\}$, $\{X\,|A\text{–}B\text{–}X\}$ by reason of Property B.0 of Definition IB.1; these latter are disjoint because, by Property B.2, no X can belong to more than one of them. It follows that the sets listed in each of the subsequent lines are disjoint. □

Corollary IB.5.1. *Let A, B, and C be distinct collinear points. Then*

(A) $A \notin \overset{\sqsubset\!\rightarrow}{BC}$ *iff A–B–C and*
(B) $A \notin \overset{\sqsupset\!\rightarrow}{BC}$ *iff A–B–C.*

Proof. First of all, since $\overset{\sqsubset\!\rightarrow}{BC} = \{B\} \cup \overset{\sqsupset\!\rightarrow}{BC}$ (see Remark IB.4.1 part(f)), $A \notin \overset{\sqsubset\!\rightarrow}{BC}$ is equivalent to $A \neq B$ and $A \notin \overset{\sqsupset\!\rightarrow}{BC}$. Since we are assuming to begin with that $A \neq B$, this conjunction is equivalent to $A \notin \overset{\sqsupset\!\rightarrow}{BC}$. So we need only prove the first statement in the corollary.

Since A, B, and C are collinear, $A \in \overleftrightarrow{BC}$. In the third line of the statement of Theorem IB.5 substitute A for X, B for A, and C for B. Since the sets in the theorem are disjoint, a point $A \notin \overset{\sqsubset\!\rightarrow}{BC}$ iff A–B–C. This is what we wished to prove. □

Corollary IB.5.2. *Let A and B be distinct points. Then* $\overset{\sqsubset\!\rightarrow}{AB}$ *and* $\overset{\sqsupset\!\rightarrow}{AB}$ *are both proper subsets of* $\overleftrightarrow{AB}$, $\overset{\sqsubset\!\sqsupset}{AB}$ *is a proper subset of* $\overset{\sqsubset\!\rightarrow}{AB}$, *and* $\overset{\sqsupset\!\sqsubset}{AB}$ *and* $\overset{\sqsupset\!\sqsupset}{AB}$ *are proper subsets of* $\overset{\sqsupset\!\rightarrow}{AB}$.

Proof. Exercise IB.6. □

Theorem IB.6. *For any two distinct points A and B,* $\overset{\sqsupset\!\rightarrow}{AB} \cup \overset{\sqsupset\!\rightarrow}{BA} = \overleftrightarrow{AB}$.

Proof. Applying Remark IB.4.1 (h), (b), (c), and (d) and line 2 of Theorem IB.5,

$$\begin{aligned}\overset{\sqsupset\!\rightarrow}{AB} \cup \overset{\sqsupset\!\rightarrow}{BA} &= \overset{\sqsupset\!\sqsupset}{AB} \cup \{X \mid A\text{–}B\text{–}X\} \cup \overset{\sqsupset\!\sqsupset}{BA} \cup \{X \mid B\text{–}A\text{–}X\}\\ &= \overset{\sqsupset\!\sqsubset}{AB} \cup \{A\} \cup \{B\} \cup \{X \mid A\text{–}B\text{–}X\} \cup \{X \mid B\text{–}A\text{–}X\}\\ &= \overset{\sqsubset\!\sqsupset}{AB} \cup \{X \mid X\text{–}A\text{–}B\} \cup \{X \mid A\text{–}B\text{–}X\}\\ &= \overleftrightarrow{AB}.\end{aligned}$$

□

Corollary IB.6.1. *For any two distinct points A and B,* $\overset{\sqsubset\!\rightarrow}{AB} \cup \overset{\sqsubset\!\rightarrow}{BA} = \overleftrightarrow{AB}$.

Proof. Exercise IB.7. □

With the concepts we have at our disposal at this time, we can define a triangle and prove one theorem about triangles.

Definition IB.7. A set $\mathcal{E}$ of points is a **triangle** iff there exist noncollinear points A, B, and C such that $\mathcal{E} = \overset{\sqsubset\!\sqsupset}{AB} \cup \overset{\sqsubset\!\sqsupset}{BC} \cup \overset{\sqsubset\!\sqsupset}{AC}$. This set is denoted by $\triangle ABC$, which is read "triangle ABC." A point U is a **corner** of $\mathcal{E}$ iff there exist points V and W such that U, V, and W are noncollinear and $\mathcal{E} = \triangle UVW$. A segment $\mathcal{J}$ is an **edge** of $\mathcal{E}$ iff there exist corners U and V of $\mathcal{E}$ such that $\mathcal{J} = \overset{\sqsubset\!\sqsupset}{UV}$. A corner U and an edge $\overset{\sqsubset\!\sqsupset}{VW}$ are **opposite** each other iff $U \notin \overleftrightarrow{VW}$.

Remark IB.7.1. Notice that a triangle is just the union of three segments; it does not include any points "inside" (whatever *that* means). Also, two edges of a triangle can intersect only at their common corner, for if these edges should intersect at some additional point, the lines they define would be the same by Axiom I.1, contradicting the noncollinearity of the corners. Here we are using the term *edge* in place of the more traditional "side," a term which we reserve for a "side" of a line, to be defined shortly.

Theorem IB.8. *Let A, B, and C be noncollinear points. Then*

$$\overleftrightarrow{AB} \cap \triangle ABC = \overset{\sqsubset\!\sqsupset}{AB}.$$

Proof. If A, B, and C are any noncollinear points, then by Definition IB.7, $\triangle ABC = \overset{\sqsubset\!\sqsupset}{AB} \cup \overset{\sqsubset\!\sqsupset}{BC} \cup \overset{\sqsubset\!\sqsupset}{AC}$. Hence

$$\begin{aligned}\overleftrightarrow{AB} \cap \triangle ABC &= \overleftrightarrow{AB} \cap (\overset{\sqsubset\!\sqsupset}{AB} \cup \overset{\sqsubset\!\sqsupset}{BC} \cup \overset{\sqsubset\!\sqsupset}{AC}) \\ &= (\overleftrightarrow{AB} \cap \overset{\sqsubset\!\sqsupset}{AB}) \cup (\overleftrightarrow{AB} \cap \overset{\sqsubset\!\sqsupset}{BC}) \cup (\overleftrightarrow{AB} \cap \overset{\sqsubset\!\sqsupset}{AC}). \qquad (*)\end{aligned}$$

Since A, B, and C are noncollinear, we have by Exercise I.1 that

$$\overleftrightarrow{AB} \cap \overleftrightarrow{BC} = \{B\} \text{ and } \overleftrightarrow{AB} \cap \overleftrightarrow{AC} = \{A\}.$$

From Theorem IB.5, $\overset{\sqsubset\!\sqsupset}{BC} \subseteq \overleftrightarrow{BC}$ and $\overset{\sqsubset\!\sqsupset}{AC} \subseteq \overleftrightarrow{AC}$, so that

$$\overleftrightarrow{AB} \cap \overset{\sqsubset\!\sqsupset}{BC} = \{B\} \text{ and } \overleftrightarrow{AB} \cap \overset{\sqsubset\!\sqsupset}{AC} = \{A\}.$$

Also, from Theorem IB.5, $\overleftrightarrow{AB} \cap \overset{\sqsubset\!\sqsupset}{AB} = \overset{\sqsubset\!\sqsupset}{AB}$. Rewriting (*) we have

$$\overleftrightarrow{AB} \cap \triangle ABC = \overset{\sqsubset\!\sqsupset}{AB} \cup \{B\} \cup \{A\} = \overset{\sqsubset\!\sqsupset}{AB}.$$ □

It would be nice to be able to prove more about triangles, but at this point it is impossible to prove much more. For example, it is impossible to prove the "obvious" fact that if $\triangle ABC = \triangle DEF$, then $\{A, B, C\} = \{D, E, F\}$. Indeed, in Chapter 21 Section 21.8 we will prove Theorem DZIII.4(B), which exhibits, in a model for IB geometry, two triangles which are equal to each other but have different corners.

In the face of such strange circumstances, it seems prudent to take evasive action and postpone any further theorems about triangles until we have stronger axioms to deal with them. We will discover this more congenial environment in Pasch geometry, to be introduced in the next chapter.

At this point we are in a position to introduce the important concept of convexity, but, as is the case with triangles, we can prove very little that is interesting about it. Since it is appropriate to the geometry we are discussing here, we now define it, and even though there are not many theorems or exercises about it, we leave you with the warm assurance that you will encounter these later.

Definition IB.9. Let $\mathcal{E}$ be a set of points. Then $\mathcal{E}$ is **convex** iff either (1) $\mathcal{E}$ is a singleton, or (2) $\mathcal{E}$ contains more than one point, and for every pair of points P and Q belonging to $\mathcal{E}$, $\overset{\dashv\vdash}{PQ} \subseteq \mathcal{E}$.

Theorem IB.10. *Every line is convex.*

Proof. Let $\mathcal{L}$ be a line, and let A and B be any points on $\mathcal{L}$. Then by Axiom I.1, $\mathcal{L} = \overleftrightarrow{AB}$; by Remark IB.4.1(d) and Theorem IB.5, $\overset{\dashv\vdash}{AB} \subseteq \overset{\sqsubset\!\sqsupset}{AB} \subseteq \overleftrightarrow{AB}$. □

The scenario for the next series of theorems—the last ones in IB geometry—is one which is familiar to anyone who has studied plane geometry: a plane with a line in it. The picture that comes to mind, of course, is a large (actually infinite) flat expanse with a line that separates the expanse into two "pieces," which we call the *sides* of $\mathcal{L}$; the segment joining two points lying on the same side does not intersect the line, whereas the segment joining a point on one side and a point on the other side does intersect the line. While the incidence and betweenness axioms impose certain restrictions on the behavior of points and lines in our geometry, they are not sufficient in themselves to force points and lines in a plane to behave according to this picture.

We deal now with an interesting question: Which of the features of the above "picture" are consequences of the axioms of IB geometry, and which features can be proved only after additional axioms have been introduced? As always, we start with the introduction of some terminology.

Definition IB.11. Let $\mathcal{P}$ be a plane, and let $\mathcal{L}$ be a line contained in $\mathcal{P}$. For any point Q in $\mathcal{P} \setminus \mathcal{L}$, the ***Q*-side** of $\mathcal{L}$ is the set

$$\{X \mid X = Q \text{ or } (X \in \mathcal{P} \setminus \{Q\} \text{ and } \overline{XQ} \cap \mathcal{L} = \emptyset)\}.$$

Note that this criterion is equivalent to

$$\{X \mid X = Q \text{ or } (X \in \mathcal{P} \setminus \{Q\} \text{ and } X \notin \mathcal{L} \text{ and } \overset{\circ}{\overline{XQ}} \cap \mathcal{L} = \emptyset)\}$$

because for $X \notin \mathcal{L}$ and $Q \notin \mathcal{L}$, $\overset{\circ}{\overline{XQ}} \cap \mathcal{L} = \emptyset$ iff $\overline{XQ} \cap \mathcal{L} = \emptyset$.

A subset $\mathcal{E}$ of $\mathcal{P}$ is a **side** of $\mathcal{L}$ iff there exists a point Q belonging to $\mathcal{P} \setminus \mathcal{L}$ such that $\mathcal{E}$ is the Q-side of $\mathcal{L}$. We say that $\mathcal{L}$ is an **edge** of $\mathcal{E}$ iff $\mathcal{E}$ is a side of $\mathcal{L}$. A subset $\mathcal{H}$ of a plane $\mathcal{P}$ is a **halfplane** of $\mathcal{P}$ iff there exists a line $\mathcal{L}$ in $\mathcal{P}$ such that $\mathcal{E}$ is a side of $\mathcal{L}$ and $\mathcal{H} = \mathcal{E} \cup \mathcal{L}$.

If $\mathcal{E}$ and $\mathcal{F}$ are sides of a line $\mathcal{L}$, they are **opposite sides** iff there exist points $P \in \mathcal{E}$, $Q \in \mathcal{L}$, and $R \in \mathcal{F}$ such that P–Q–R (that is to say, $\overset{\circ}{\overline{PR}} \cap \mathcal{L} = \{Q\}$).

If A, B, and Q are noncollinear points, then $\overrightarrow{ABQ}$ denotes the Q-side of $\overleftrightarrow{AB}$.

Remark IB.11.1. Every line is disjoint from any of its sides; for if X is a member of the Q-side of $\mathcal{L}$, $\overline{XQ} \cap \mathcal{L} = \emptyset$ so that in particular, $X \notin \mathcal{L}$.

Be careful not to be misled by the terminology introduced in Definition IB.11, which is heavily loaded toward the way things will *eventually* turn out. In Remark IB.12.1 below, we will point out several conclusions one might be tempted to draw, but which, at this point, are false.

Theorem IB.12. *Suppose P and Q are two distinct points not on $\mathcal{L}$.*

(A) *$P \in Q$-side of $\mathcal{L}$ iff $P = Q$ or ($P \neq Q$ and $\overline{PQ} \cap \mathcal{L} = \emptyset$) iff $Q \in P$-side of $\mathcal{L}$.*

(B) *P fails to be in the Q-side of $\mathcal{L}$ (and Q fails to be in the P-side of $\mathcal{L}$) iff ($P \neq Q$ and $\overline{PQ} \cap \mathcal{L} \neq \emptyset$).*

(C) *If P fails to be in the Q-side of $\mathcal{L}$ (and Q fails to be in the P-side of $\mathcal{L}$), then the Q-side of $\mathcal{L}$ and the P-side of $\mathcal{L}$ are opposite.*

Proof. Part (A) is an immediate consequence of Definition IB.11 and elementary logic, and part (B) is logically equivalent to (A).

(C) If P fails to be in the Q-side of $\mathcal{L}$, then by (B) $P \neq Q$ and $\overline{PQ} \cap \mathcal{L} \neq \emptyset$. Now $P \in P$-side of $\mathcal{L}$ and $Q \in Q$-side of $\mathcal{L}$, so by Definition IB.11 the Q-side and P-side of $\mathcal{L}$ are opposite. □

Remark IB.12.1. (I) Theorem IB.12 does not include a converse of part (C) because at this stage it can't be proved. We cannot show that if the Q-side of $\mathcal{L}$ and the P-side of $\mathcal{L}$ are opposite then P fails to be in the Q-side of $\mathcal{L}$. Indeed it may be that the Q-side of $\mathcal{L}$ and the P-side of $\mathcal{L}$ are opposite and at the same time $P \in Q$-side of $\mathcal{L}$! That is, $\overline{PQ} \cap \mathcal{L} = \emptyset$, but there exist points $A \in P$-side of $\mathcal{L}$ and $B \in Q$-side of $\mathcal{L}$ with $\overline{AB} \cap \mathcal{L} \neq \emptyset$.

(II) As an illustration for this and subsequent assertions, we anticipate Chapter 21. There, in Subsection 21.6.3, Definition DZI.1 defines Model DZI as the set $\mathbb{Z}^3$ of all ordered triples of integers, and Theorem DZI.5 proves this to be an IB space. The set $\mathcal{P}$ of all triples $(a, b, 0)$ (which we denote here as pairs (a, b)), where a and b are integers, is a plane in this model. We suggest making a simple sketch of the following to help keep things straight. Let $A = (-1, 0)$, $B = (0, 0)$, $C = (1, 0)$; $D = (-1, 1)$, $E = (0, 1)$, $F = (1, 1)$; $G = (-1, -1)$, $H = (0, -1)$, $I = (1, -1)$; then

the D-side of $\overleftrightarrow{AB}$ contains D, E, F, and H but not G or I;
the E-side of $\overleftrightarrow{AB}$ contains D, E, F, G and I but not H;
the G-side of $\overleftrightarrow{AB}$ contains G, H, I, and E but not D or F;
the H-side of $\overleftrightarrow{AB}$ contains G, H, I, D and F but not E.

(III) Returning to the assertion of part (I): in the illustration above, the D-side is opposite to the G-side because $\overline{GD} \cap \overleftrightarrow{AB} \neq \emptyset$; but both these sides contain the point H. Thus the H-side is opposite the D-side since it contains G, and also belongs to the D-side.

This also shows that given two sides $\mathcal{E}$ and $\mathcal{F}$ of $\mathcal{L}$, we cannot simply pick two arbitrary points, $P \in \mathcal{E}$ and $Q \in \mathcal{F}$, and determine whether or not $\mathcal{E}$ and $\mathcal{F}$ are opposite by checking whether or not $\overline{PQ} \cap \mathcal{L} \neq \emptyset$. To be sure, if we should find that $\overline{PQ} \cap \mathcal{L} \neq \emptyset$, that would show that $\mathcal{E}$ and $\mathcal{F}$ are opposite; but finding that $\overline{PQ} \cap \mathcal{L} = \emptyset$ would tell us nothing at all. Such a test must wait for the invocation of the Plane Separation Axiom in the next chapter.

(IV) It is also tempting to conclude that if a point Q belongs to the P-side of $\mathcal{L}$, then the Q-side of $\mathcal{L}$ is the same as the P-side of $\mathcal{L}$. But this is not so. Referring again to our illustration of part (II), the point H belongs to the D-side, but G, which also belongs to the H-side, does not belong to the D-side; hence, the D-side and the H-side are not the same. Again, once we have the Plane Separation Axiom at our disposal, this anomalous situation will be resolved.

(V) The term "opposite sides" might seem to imply that there are just two sides of a line (or at least that they come in pairs); but we can't prove that in IB geometry—at this point there is no way of telling how many sides a line in a plane has. In our illustration (II), $\overleftrightarrow{AB}$ has at least four sides, since none of the sides listed are the same.

Moreover, if $\mathcal{L}$ is a line in plane $\mathcal{P}$, and $\mathcal{E}$ is a side of $\mathcal{L}$, it is not correct to conclude that $\mathcal{P} \setminus (\mathcal{E} \cup \mathcal{L})$ is a side of $\mathcal{L}$. We can't even prove in IB geometry that opposite sides of a line are disjoint. Indeed, in our illustration from part (II), the D-side and the G-side of $\overleftrightarrow{AB}$ are opposite but not disjoint, as both contain the points E and H.

All the intuitively correct statements listed above, which we have shown to be false, *will* be provable after we introduce the Plane Separation Axiom. For now it is important not to assume they are true just because the terminology suggests that they might be.

Now that we've dwelt on some things we can *not* prove about a line in a plane, let's get to some of the things we *can* prove. First, it should be noted that by definition, a side of a line is nonempty.

Theorem IB.13. *Let $\mathcal{L}$ be a line in a plane $\mathcal{P}$. Then there is at least one side $\mathcal{E}$ of $\mathcal{L}$, and there is at least one side of $\mathcal{L}$ which is opposite to $\mathcal{E}$.*

Proof. By Exercise I.13, there is a point A in $\mathcal{P}$ which is not on $\mathcal{L}$. Then the set $\mathcal{E}$ defined by

$$\mathcal{E} = \{X \mid X = A \text{ or } (X \in \mathcal{P} \setminus \{A\} \text{ and } \overline{XA} \cap \mathcal{L} = \emptyset)\}$$

is nonempty and is a side of $\mathcal{L}$.

By Axiom I.5(A), there is a point B on $\mathcal{L}$, and by Property B.3 of Definition IB.1, there is a point C such that A–B–C. By Definition IB.11, the C-side of $\mathcal{L}$ is opposite $\mathcal{E}$. □

Theorem IB.14 and its corollaries, which we prove next, probably seem plausible to anyone who has studied a little geometry and who takes the time to look at the appropriate pictures; what may not seem plausible is why we chose these particular things to prove.

There are two reasons for our choices: (1) these particular theorems will come in very handy when we get around to proving the more "obvious" things (this becomes clear in retrospect, but don't be discouraged if you can't see the connection now), and (2) we tried to prove quite a few things, and these were the ones we succeeded in proving.

The following theorem is a fundamental building-block in the structure of theorems to follow. It will also make it possible to show later, as a consequence of Corollary PSH.22.2, that each side of a line in a plane contains a lot of points—in fact, an infinite number.

Theorem IB.14 (Side contains a ray). *Let $\mathcal{L}$ be a line in plane $\mathcal{P}$, and let P and Q be points such that $P \in \mathcal{L}$ and $Q \notin \mathcal{L}$. Then $\overrightarrow{PQ}$ is a subset of* the Q-side of $\mathcal{L}$.

Proof. Since $Q \notin \mathcal{L}$, $\overleftrightarrow{PQ} \cap \mathcal{L} = \{P\}$ by Axiom I.1. If $R \in \overrightarrow{PQ}$, by Definition IB.4 and Property B.2 of Definition IB.1 exactly one of P–R–Q, $R = Q$, or P–Q–R is true. If P–R–Q, $P \notin \overline{RQ}$ by Definition IB.3 and Property B.2. If P–Q–R, $P \notin \overline{RQ}$ for the same reasons. Thus $\overline{RQ} \cap \mathcal{L} = \emptyset$ and R belongs to the Q-side of $\mathcal{L}$. □

Corollary IB.14.1. *Let $\mathcal{P}$, $\mathcal{L}$, P, and Q be as in Theorem IB.14. Then $\overrightarrow{PQ} \cap \mathcal{L} = \emptyset$.*

Proof. Exercise IB.10. □

Corollary IB.14.2. *Let $\mathcal{P}$, $\mathcal{L}$, P, and Q be as in Theorem IB.14. Then $\overline{PQ}$ and $\overline{PQ}$ are subsets of* the Q-side of $\mathcal{L}$.

Proof. Exercise IB.11. □

Corollary IB.14.3. *For any triangle $\triangle ABC$, the edges $\overset{]\!-\![}{AB}$ and $\overset{]\!-\![}{AC}$ are subsets of $\overset{]\!\longrightarrow}{BCA}$, $\overset{]\!-\![}{AB}$ and $\overset{]\!-\![}{BC}$ are subsets of $\overset{]\!\longrightarrow}{ACB}$, and $\overset{]\!-\![}{AC}$ and $\overset{]\!-\![}{BC}$ are subsets of $\overset{]\!\longrightarrow}{ABC}$.*

Proof. Exercise IB.12. □

The following corollary will be useful in the next chapter.

Corollary IB.14.4. *Let $\mathcal{P}$ be an IB plane, $\mathcal{L}$ a line in $\mathcal{P}$, P a point of $\mathcal{P}$ not on $\mathcal{L}$, and let X and Y be distinct points belonging to* the P-side of $\mathcal{L}$. *Then there exists a point $Q \in$* the P-side of $\mathcal{L}$ *such that $\{X, Y, Q\}$ is noncollinear.*

Proof. If $\{X, Y, P\}$ is noncollinear, let $Q = P$. If $\{X, Y, P\}$ is collinear, we choose a point $Z \in \mathcal{L}$ as follows: if $\overleftrightarrow{XY} \cap \mathcal{L} = \emptyset$, let Z be any point of $\mathcal{L}$; if $\overleftrightarrow{XY}$ intersects $\mathcal{L}$, then since $\overleftrightarrow{XY} \neq \mathcal{L}$ by Exercise I.1, there exists one point W such that $\overleftrightarrow{XY} \cap \mathcal{L} = \{W\}$; choose Z to be any point of $\mathcal{L}$ such that $Z \neq W$.

In either case, by Theorem IB.14 the ray $\overset{]\!\rightarrow}{ZX} \subseteq$ the P-side of $\mathcal{L}$, and $\overset{]\!\rightarrow}{ZX} \cap \overleftrightarrow{XY} = \{X\}$. Let Q be any point of $\overset{]\!\rightarrow}{ZX}$ other than X. Then $\{X, Y, Q\}$ is noncollinear, and $Q \in$ the P-side of $\mathcal{L}$. □

4.3 Exercises for Incidence-Betweenness geometry

Answers to starred (*) exercises may be accessed from the home page for this book at www.springer.com.

Exercise IB.1. If A and B are distinct points, then there exist points E and F such that E–B–A and B–A–F.

Exercise IB.2*. Let A, B, C, and D be distinct collinear points, then A–B–C–D iff D–C–B–A.

Exercise IB.3. If A and B are any two distinct points, then $\overset{[\!-\!]}{AB} = \overset{[\!-\!]}{BA}$ and $\overset{]\!-\![}{AB} = \overset{]\!-\![}{BA}$.

Exercise IB.4*. If A and B are any two distinct points, then $\overset{]\!-\![}{AB} \subseteq \overset{]\!-\!]}{AB} \subseteq \overset{]\!\rightarrow}{AB} \subseteq \overleftrightarrow{AB}$, $\overset{]\!-\![}{AB} \subseteq \overset{[\!-\![}{AB} \subseteq \overset{[\!\rightarrow}{AB} \subseteq \overleftrightarrow{AB}$, and $\overset{[\!-\!]}{AB} \subseteq \overset{[\!\rightarrow}{AB} \subseteq \overleftrightarrow{AB}$.

Exercise IB.5. If $\overset{[\!\rightarrow}{AB} = \overset{[\!\rightarrow}{CD}$ or $\overset{[\!-\!]}{AB} = \overset{[\!-\!]}{CD}$, then $\overleftrightarrow{AB} = \overleftrightarrow{CD}$.

Exercise IB.6*. Prove Corollary IB.5.2. (See also Exercise IB.4.)

Exercise IB.7*. Prove Corollary IB.6.1.

Exercise IB.8*. If A and B are any two distinct points, then

(A) $\overset{[\longrightarrow}{AB} \cap \overset{[\longrightarrow}{BA} = \overset{[\;\;]}{AB}$,

(B) $\overset{]\longrightarrow}{AB} \cap \overset{]\longrightarrow}{BA} = \overset{]\;\;[}{AB}$,

(C) $\overset{[\longrightarrow}{AB} \cap \overset{]\longrightarrow}{BA} = \overset{[\;\;[}{AB}$, and

(D) $\overset{]\longrightarrow}{AB} \cap \overset{[\longrightarrow}{BA} = \overset{]\;\;]}{AB}$.

Exercise IB.9*. Let $\mathcal{L}$ be a line, and let A and B be distinct points such that $\mathcal{L} \neq \overleftrightarrow{AB}$. If $\overset{]\;\;[}{AB} \cap \mathcal{L} = \{R\}$, then $\overleftrightarrow{AB} \cap \mathcal{L} = \{R\}$.

Exercise IB.10*. Prove Corollary IB.14.1.

Exercise IB.11*. Prove Corollary IB.14.2.

Exercise IB.12. Prove Corollary IB.14.3.

Exercise IB.13. Space is convex.

Exercise IB.14. Every plane is convex.

Exercise IB.15*. If $\mathcal{G}$ is any collection of convex sets, and if the intersection of the members of $\mathcal{G}$ is nonempty, then the intersection is convex.

Exercise IB.16. Let $\mathcal{L}$ be a line and let $\mathcal{E}$ be a nonempty proper subset of $\mathcal{L}$ such that $\mathcal{E}$ is not a singleton. Then:

(1) $\mathcal{E}$ is not a segment iff for every pair of distinct points A and B on $\mathcal{L}$, there exists a point U such that A–U–B and $U \notin \mathcal{E}$, or there exists a point V such that A–B–V and $V \in \mathcal{E}$, or there exists a point W such that B–A–W and $W \in \mathcal{E}$.

(2) $\mathcal{E}$ is not a ray iff for every pair of distinct points A and B on $\mathcal{L}$, there exists a point U such that A–U–B and $U \notin \mathcal{E}$, or there exists a point V such that A–B–V and $V \notin \mathcal{E}$ and there exists a point W such that B–A–W and $W \notin \mathcal{E}$.

Exercise IB.17*. Let $\mathcal{P}$ be an IB plane, $\mathcal{L}$ and $\mathcal{M}$ be lines on $\mathcal{P}$, and O be a point such that $\mathcal{L} \cap \mathcal{M} = \{O\}$, then there exist points P and Q on $\mathcal{L}$ such that P and Q are on opposite sides of $\mathcal{M}$.

Exercise IB.18 (True or False?). Let $\mathcal{P}$ be an IB plane, and let $\mathcal{J}$, $\mathcal{K}$, and $\mathcal{L}$ be distinct lines on $\mathcal{P}$ such that $\mathcal{J} \cap \mathcal{L} \neq \emptyset$ and $\mathcal{K} \cap \mathcal{L} \neq \emptyset$. Then if U is a point on $\mathcal{J}$ but not on $\mathcal{L}$, there is a point V on $\mathcal{K}$ such that U and V are on opposite sides of $\mathcal{L}$.

Chapter 5
Pasch Geometry (PSH)

Acronym: *PSH*
Dependencies: *Chapters 1 and 4*
New Axioms: *Plane Separation Axiom PSA*
New Terms Defined: *Postulate of Pasch, Pasch plane, denseness, opposite rays, angle; quadrilateral, corner, edge, opposite edge, diagonal, rotund; trapezoid; inside, outside, enclosure, exclosure (of angle, triangle, and quadrilateral)*

Abstract: The first part of this chapter uses the Plane Separation Axiom to show that a line in a plane has two disjoint sides, and to prove the basic properties of segments, rays, and lines that are needed for a coherent geometry. The remainder of the chapter is a study of the basic interactions between lines, angles, triangles, and quadrilaterals, comprising Pasch geometry.

So far, we have not seen much in this book that most people would recognize as "real" geometry. To remedy this we need to surmount the difficulties we inherited from the previous chapter (IB geometry) where we could not prove several things that seem so natural to us—for instance, that a line has only two sides or that a triangle has only one set of corners (*cf* Remarks IB.4.2 and IB.12.1, and the note after Theorem IB.8). These anomalies arise in planes, and must be fixed in that context; indeed, most of the rest of the book is about the geometry of planes that are subsets of an IB space.

E.J. Specht et al., *Euclidean Geometry and its Subgeometries*,
DOI 10.1007/978-3-319-23775-6_5

We start by invoking the Plane Separation Axiom (PSA) on such planes; this will open the way to develop basic properties of angles, triangles, quadrilaterals and the like,[1] which most people think of as the "stuff" of geometry. But first a bit of history.

5.1 The Postulate of Pasch

In 1882 Moritz Pasch (1843–1930) published *Vorlesungen Über Neuere Geometrie*, (Lectures Over More Recent Geometry [17]) which embodied the beginnings of the modern axiomatization of geometry. Pasch documented many assumptions that were in Euclid and made clear many of the difficulties in his work that were due to intuition rather than sound mathematical reasoning from the stated assumptions. One of the cornerstones of his work is the following statement that has become known to geometers as the "Postulate of Pasch," or the "Pasch Postulate," or, in our proofs, simply as "Pasch."

The Postulate of Pasch *If A, B, and C are noncollinear points on an IB plane $\mathcal{P}$, and if $\mathcal{L}$ is a line on $\mathcal{P}$ such that $\mathcal{L} \neq \overleftrightarrow{AB}$, $\mathcal{L} \cap \overline{AB} \neq \emptyset$, and $C \notin \mathcal{L}$, then either $\mathcal{L} \cap \overline{AC} \neq \emptyset$ or $\mathcal{L} \cap \overline{BC} \neq \emptyset$ (but not both).*

Alternate form of the Postulate of Pasch Suppose A, B, and C are noncollinear points on plane $\mathcal{P}$ and $\mathcal{L}$ is a line on $\mathcal{P}$ containing none of these points. If $\overline{AC} \cap \mathcal{L} = \emptyset$ and $\overline{BC} \cap \mathcal{L} = \emptyset$, then $\overline{AB} \cap \mathcal{L} = \emptyset$.

The alternate form as stated above is not quite equivalent to Pasch, but is implied by it. To see this, let the overall hypothesis be A, B, and C are noncollinear points, $\mathcal{L}$ is a line not equal to $\overleftrightarrow{AB}$, and $C \notin \mathcal{L}$. Then Pasch says

$$\text{if } \mathcal{L} \cap \overline{AB} \neq \emptyset, \text{ then } (\mathcal{L} \cap \overline{AC} \neq \emptyset \text{ exclusive or } \mathcal{L} \cap \overline{BC} \neq \emptyset).$$

The contrapositive says (*cf* Chapter 1, Section 1.2)

$$\text{if } (\mathcal{L} \cap \overline{AC} = \emptyset \text{ and } \mathcal{L} \cap \overline{BC} = \emptyset) \text{ or (both } \mathcal{L} \cap \overline{AC} \neq \emptyset \text{ and } \mathcal{L} \cap \overline{BC} \neq \emptyset),$$
$$\text{then } \mathcal{L} \cap \overline{AB} = \emptyset.$$

Thus, any time we have Pasch, this alternate form will also be true.

[1]The process for doing this is somewhat complex and involves some subtleties. A reader desiring an overview of Pasch geometry (*cf* Definition PSH.7) without indulging in the details of a strict development may proceed as follows: first, peruse the statements below of the Postulate of Pasch and the Plane Separation Axiom; then accept Theorem PSH.12 (the Plane Separation Theorem) as an axiom, and go on from there. It must be noted, however, that Theorem PSH.8 is needed in the development following Theorem PSH.12.

The following proof has interest in its own right, and we will use it in the section titled "Pasch geometry," after the definition of the Pasch plane. We place it here because it is needed to facilitate the proof of Theorem PSH.6. It shows, among other things, that in a plane where the Pasch postulate holds, if two sides of a line have nonempty intersection, then they are the same side.

Theorem PSH.1. *Let $\mathcal{L}$ be a line in an IB plane $\mathcal{P}$ on which the Postulate of Pasch holds. Then if S is a point not on $\mathcal{L}$, and $Q \in$* the *S*-side of $\mathcal{L}$, the *Q*-side of $\mathcal{L} =$ the *S*-side of $\mathcal{L}$.

Proof. Let X be any point of the S-side of $\mathcal{L}$. By Definition IB.11 $\overline{XS} \cap \mathcal{L} = \emptyset$; since $Q \in$ the S-side of $\mathcal{L}$, $\overline{QS} \cap \mathcal{L} = \emptyset$.

(Case 1: X, Q, and S are noncollinear.) Then we may apply the alternate form of Pasch to $\triangle XQS$ to get $\overline{XQ} \cap \mathcal{L} = \emptyset$, and therefore $X \in$ the Q-side of $\mathcal{L}$.

(Case 2: X, Q, and S are collinear.) By Corollary IB.14.4, there exists a point $Y \in$ the S-side of $\mathcal{L}$ such that $Y \notin \overleftrightarrow{QS} = \overleftrightarrow{XS}$. Apply the alternate form of Pasch to $\triangle YSX$. Since $\overline{XS} \cap \mathcal{L} = \emptyset$ and $\overline{YS} \cap \mathcal{L} = \emptyset$, $\overline{XY} \cap \mathcal{L} = \emptyset$. Again apply the alternate form of Pasch to $\triangle YQS$. Since $\overline{QS} \cap \mathcal{L} = \emptyset$, $\overline{YQ} \cap \mathcal{L} = \emptyset$. Finally, apply the alternate form of Pasch to $\triangle YQX$, to get $\overline{XQ} \cap \mathcal{L} = \emptyset$. Therefore $X \in$ the Q-side of $\mathcal{L}$.

From these two cases, we see that the S-side of $\mathcal{L} \subseteq$ the Q-side of $\mathcal{L}$. By Theorem IB.12(A) $S \in$ the Q-side of $\mathcal{L}$. Reversing the roles of S and Q, the same proof shows that the Q-side of $\mathcal{L} \subseteq$ the S-side of $\mathcal{L}$, and hence the S-side of $\mathcal{L} =$ the Q-side of $\mathcal{L}$. □

5.2 The Plane Separation Axiom (PSA)

We do not base our development directly on the Pasch Postulate, but rather on an equivalent assumption called the "the Plane Separation Axiom" which appears (to us, at least) to deal with more fundamental ideas. The resulting geometry will be far richer than any we have seen so far. Many theorems that suggested themselves earlier but could not be proved, as well as a host of new ones, will now be within our reach. To celebrate this milestone in our development, we christen the new geometry this creates as *Pasch geometry*; without further hesitation, we add the Plane Separation Axiom (PSA) to the incidence (I.0–I.5) axioms and the betweenness Axiom BET.

Plane Separation Axiom (PSA). If $\mathcal{L}$ is any line, and if Q and R are points in opposite sides of $\mathcal{L}$, then $\overline{QR} \cap \mathcal{L} \neq \emptyset$.

The following statement is an almost trivial extension of Axiom PSA, and any citation of Axiom PSA should be understood to include it:

If $\mathcal{E}$ and $\mathcal{F}$ are opposite sides of $\mathcal{L}$, $Q \in \mathcal{E}$, and $R \in \mathcal{F}$, then by Definition IB.3 there exists a point $S \in \mathcal{L}$ with Q–S–R. By Exercise I.1 and elementary set theory $\overleftrightarrow{QR} \cap \mathcal{L} = \{S\}$, so that S is the single point of intersection of $\overleftrightarrow{QR}$ and $\mathcal{L}$. By Theorem IB.14, $\overrightarrow{SQ}$ is a subset of $\mathcal{E}$ and $\overrightarrow{SR}$ is a subset of $\mathcal{F}$.

Notice that Axiom PSA asserts something quite subtle: it says that if $\mathcal{E}$ and $\mathcal{F}$ are opposite sides of $\mathcal{L}$, that is, if *there exist* points $Q' \in \mathcal{E}$ and $R' \in \mathcal{F}$ such that $\overline{Q'R'} \cap \mathcal{L} \neq \emptyset$, then the same must be true *for all* points $Q \in \mathcal{E}$ and $R \in \mathcal{F}$.

In particular, this axiom neatly solves the quandary we were in the last chapter (*cf* Definition IB.11 and following discussion), where it was awkward to determine if two sides $\mathcal{E}$ and $\mathcal{F}$ of a line were opposite. Now, in the presence of Axiom PSA, all we need to do is pick arbitrary points in $\mathcal{E}$ and $\mathcal{F}$ and see if the segment connecting them intersects the line.

Thus, a line "separates" the plane; this is consistent with Hilbert's axiom system, in which the Postulate of Pasch is grouped with other axioms dealing with "betweenness."[2]

In Chapter 21 we will exhibit a model in which all our axioms are true (*cf* Subsection 21.5.8); this will show, among other things, that Pasch planes actually do exist, and Pasch geometry is not vacuous. Also, in Subsection 21.6.3 we will see a model for an IB plane on which PSA is false, showing that PSA is independent of the incidence and betweenness axioms. When axioms are chosen so they are independent of each other, an intricate logical development is usually required to reach key theorems. This is well illustrated by the rather complex path we must undertake to show Theorem PSH.12, which is fundamental to the rest of the book.

[2]It could be interesting to construct a theory in which space is divided into two half-spaces by a plane, in a manner analogous to the theory developed here which treats division of the plane into two half-planes by a line. We have not pursued this, but it is said to have been carried out by B. L. van der Waerden, in *De logische grondslagen der Euklidische meetkunde* (Dutch), Chr. Huygens 13, 65–84, 257–274 (1934) [22]. Axiomatizations of Pasch-like statements for hyperplanes (i. e., statements that hyperplanes divide the space into two half-spaces) have been presented by E. Sperner in *Die Ordnungsfunktionen einer Geometrie*, Math. Ann. 121, 107–130, (1949) [19]. For the significance of Sperner's work in ordered geometries, see H. Karzel, *Emanuel Sperner: Begründer einer neuen Ordnungstheorie*, Mitt. Math. Ges. Hamburg 25, 33–44 (2006) [12].

Theorem PSH.2 (Opposite sides of a line are disjoint). *Let $\mathcal{P}$ be an IB plane in which PSA holds, and let $\mathcal{E}$ and $\mathcal{F}$ be opposite sides of a line $\mathcal{L}$ in $\mathcal{P}$. Then $\mathcal{E} \cap \mathcal{F} = \emptyset$.*

Proof. By Definition IB.11, since $\mathcal{E}$ and $\mathcal{F}$ are opposite sides of $\mathcal{L}$, there exists a point P such that $\mathcal{E}$ = the P-side of $\mathcal{L}$. Suppose $\mathcal{E} \cap \mathcal{F} \neq \emptyset$, so that there exists a point $A \in \mathcal{E} \cap \mathcal{F}$. By Definition IB.11, $\overline{AP} \cap \mathcal{L} = \emptyset$. Since $A \in \mathcal{F}$, $P \in \mathcal{E}$ and $\mathcal{F}$ is opposite to $\mathcal{E}$, by PSA $\overline{AP} \cap \mathcal{L} \neq \emptyset$, a contradiction. □

Theorem PSH.3. *Let $\mathcal{P}$ be an IB plane in which PSA holds. Let A, B, and C be noncollinear points of $\mathcal{P}$, and let D, E, and F be points of $\mathcal{P}$ such that A–D–B, B–E–C, and A–F–C. Then D, E, and F are noncollinear.*

Proof. By Property B.0 of Definition IB.1, $\{A, D, B\}$, $\{B, E, C\}$, and $\{A, F, C\}$ are collinear. We show that D, E, and F are distinct points. If two of D, E, and F were the same, say $D = E$, then by Axiom I.1, $\overleftrightarrow{AB} = \overleftrightarrow{BC}$. This line contains A, B, and C, contradicting the hypothesis that these points are noncollinear.

Now assume that D, E, and F are collinear. Then by Property B.2 of Definition IB.1 one and only one of the following is true: D–E–F, or E–D–F, or D–F–E. We show that each of these possibilities leads to a contradiction.

If D–E–F, since $E \in \overleftrightarrow{BC}$, the D-side of $\overleftrightarrow{BC}$ is opposite the F-side. By Theorem IB.14 and Definition IB.3, $A \in \overrightarrow{BD} \subseteq$ (D-side of $\overleftrightarrow{BC}$), so A belongs to a side of $\overleftrightarrow{BC}$ which is opposite the F-side. Now recall that A–F–C, so by Definition IB.3 and Theorem IB.14, $A \in \overrightarrow{CF} \subseteq$ (F-side of $\overleftrightarrow{BC}$). We have shown that A belongs both to the F-side of $\overleftrightarrow{BC}$ and to a side opposite the F-side. This contradicts Theorem PSH.2, so the assumption that D–E–F is false.

In a similar way, we can show that both E–D–F and D–F–E lead to contradictions, so the proof is complete. □

Theorem PSH.4. *Let $\mathcal{P}$ be an IB plane in which PSA holds. Let A, B, and C be noncollinear points of $\mathcal{P}$ and let $\mathcal{L}$ be a line in $\mathcal{P}$ such that for some distinct points D and E, $\mathcal{L} \cap \overline{AB} = \{D\}$ and $\mathcal{L} \cap \overline{AC} = \{E\}$. Then $\mathcal{L} \cap \overline{BC} = \emptyset$.*

Proof. If $B \in \mathcal{L}$, then both the points B and D would belong to both $\overleftrightarrow{AB}$ and to $\mathcal{L}$, and by Axiom I.1 $\overleftrightarrow{AB} = \mathcal{L}$. This contradicts the assumption that $\mathcal{L} \cap \overline{AB}$ is a singleton, so that $B \notin \mathcal{L}$. Similarly $C \notin \mathcal{L}$. If X is any member of $\overline{BC}$, then by Theorem PSH.3, X, D, and E are noncollinear, so that $X \notin \mathcal{L}$. This shows that $\mathcal{L} \cap \overline{BC} = \emptyset$. □

At the beginning of the next section we will define a *Pasch plane*, and in Theorem PSH.11 we will prove that a line drawn in such a plane has only two sides. We haven't found a way to prove this in one easy step. The next theorem comes close to saying that the points off a line in a Pasch plane are the union of a pair of opposite sides of the line, but there is an interesting twist—a second line is involved.

Theorem PSH.5. *Let $\mathcal{P}$ be an IB plane in which PSA holds, $\mathcal{L}$ a line in $\mathcal{P}$, and let P, Q, and R be points on $\mathcal{P}$ such that $Q \in \mathcal{L}$, $\overleftrightarrow{PQ} \neq \mathcal{L}$, and P–Q–R. If X is any point not on $\mathcal{L}$ and not on $\overleftrightarrow{PQ}$, then X belongs to either the P-side or the R-side of $\mathcal{L}$. That is*

$$\mathcal{P} \setminus (\mathcal{L} \cup \overleftrightarrow{PQ}) = [(P\text{-side of } \mathcal{L}) \cup (R\text{-side of } \mathcal{L})] \setminus \overleftrightarrow{PQ}.$$

Moreover, $(P\text{-side of } \mathcal{L}) \cap (R\text{-side of } \mathcal{L}) = \emptyset$.

Proof. The last statement in the theorem is an obvious consequence of Theorem PSH.2 and is included for completeness.

Let X be any member of $\mathcal{P} \setminus (\mathcal{L} \cup \overleftrightarrow{PQ})$. We wish to show that either $X \in (P$-side of $\mathcal{L})$ or $X \in (R$-side of $\mathcal{L})$.

If $\overline{XP} \cap \mathcal{L} = \emptyset$, then since $X \neq P$, by Definition IB.11, X belongs to the P-side of $\mathcal{L}$.

Otherwise, if $\overline{XP} \cap \mathcal{L} \neq \emptyset$, observe first that the line $\overleftrightarrow{XP}$ is distinct from both $\mathcal{L}$ and from $\overleftrightarrow{PR}$, since X belongs to neither of these lines. Then by Exercise I.1, there exists a point S such that $\overline{XP} \cap \mathcal{L} = \{S\}$, and, since the only point of intersection of $\overleftrightarrow{XP}$ and $\overleftrightarrow{PR}$ is P, $S \neq Q$. By hypothesis we know that $Q \in \overline{PR}$, so we may apply Theorem PSH.4 to conclude that $\overline{XR} \cap \mathcal{L} = \emptyset$. By Definition IB.11, X belongs to the R-side of $\mathcal{L}$.

This shows that $\mathcal{P} \setminus (\mathcal{L} \cup \overleftrightarrow{PQ}) \subseteq [(P\text{-side of } \mathcal{L}) \cup (R\text{-side of } \mathcal{L})] \setminus \overleftrightarrow{PQ}$. The reverse inclusion is immediate from Definition IB.11, which defines a side of $\mathcal{L}$ to be disjoint from $\mathcal{L}$. □

The next theorem shows that if we add the Postulate of Pasch to the list of axioms (incidence and betweenness) for an IB plane, we get the same geometry as if we add the Plane Separation Axiom.

Theorem PSH.6 (Pasch is equivalent to PSA). *In the presence of Axioms I.0–I.5 and Axiom BET, the Postulate of Pasch is equivalent to the Plane Separation Axiom PSA.*

Proof. All points and lines in this proof will be in $\mathcal{P}$, an IB plane on which Axiom BET holds. In each case, the reader will find it helpful to sketch a figure.

(I: PSA $\Rightarrow$ Pasch) Let A, B, and C be noncollinear points in $\mathcal{P}$, and let $\mathcal{L}$ be a line such that $\mathcal{L} \neq \overleftrightarrow{AB}$, $\mathcal{L} \cap \overline{AB} \neq \emptyset$, and $C \notin \mathcal{L}$.

If $\mathcal{L}$ were to intersect $\overleftrightarrow{AB}$ in more than one point, then by Exercise I.2, $\mathcal{L}$ would be equal to $\overleftrightarrow{AB}$, which is false by hypothesis. Hence $\mathcal{L} \cap \overleftrightarrow{AB}$ is a singleton. By Definition IB.11, A and B are on opposite sides of $\mathcal{L}$. C belongs neither to $\mathcal{L}$ nor to $\overleftrightarrow{AB}$, so by Theorem PSH.5 C belongs either to the A-side or to the B-side of $\mathcal{L}$, but not to both.

If C belongs to the A-side of $\mathcal{L}$ but not to the B-side, by Definition IB.11, $\overline{AC} \cap \mathcal{L} = \emptyset$ and by PSA, $\overline{BC} \cap \mathcal{L} \neq \emptyset$. If C belongs to the B-side of $\mathcal{L}$ but not to the A-side, $\overline{BC} \cap \mathcal{L} = \emptyset$ and $\overline{AC} \cap \mathcal{L} \neq \emptyset$. This proves that the Pasch Postulate holds on $\mathcal{P}$.

(II: Pasch $\Rightarrow$ PSA) Let $\mathcal{E}$ = the U-side of $\mathcal{L}$ and $\mathcal{F}$ = the V-side of $\mathcal{L}$ be opposite sides of a line $\mathcal{L}$. By Definition IB.11, there exist points S in the U-side and T in the V-side of $\mathcal{L}$ such that $\overline{ST} \cap \mathcal{L} \neq \emptyset$. By Theorem PSH.1, S-side = U-side, and T-side = V-side of $\mathcal{L}$.

Let Q be any member of $\mathcal{E}$ = the U-side of $\mathcal{L}$, and R any member of $\mathcal{F}$ = the V-side of $\mathcal{L}$. Again, by Theorem PSH.1, Q-side = U-side = S-side, and R-side = V-side= T-side of $\mathcal{L}$.

We show, from $\overline{ST} \cap \mathcal{L} \neq \emptyset$, and repeated applications of the Proposition of Pasch, that $\overline{QR} \cap \mathcal{L} \neq \emptyset$.

Now Q and S are on the same side, and R and T are on the opposite side of $\mathcal{L}$. If the points Q, R, S, and T are not distinct, then either $Q = S$ or $R = T$ or both (in which case there is nothing to prove). The case where $Q = S$ but $R \neq T$ is covered in Cases 2 and 4 below. The case where $Q \neq S$ but $R = T$ is covered in Cases 3 and 5 below.

(Case 1: No three of the points Q, R, S, and T are collinear.) In this case, all points are distinct. Apply Pasch to $\triangle QST$; since $\overline{ST} \cap \mathcal{L} \neq \emptyset$ and $\overline{QS} \cap \mathcal{L} = \emptyset$, $\overline{QT} \cap \mathcal{L} \neq \emptyset$.

Now apply Pasch to $\triangle RQT$; since $\overline{QT} \cap \mathcal{L} \neq \emptyset$ and $\overline{RT} \cap \mathcal{L} = \emptyset$, then $\overline{QR} \cap \mathcal{L} \neq \emptyset$, which is the desired result.

(Case 2: Q, R, and S are collinear but T is not collinear with these points.) Apply Pasch to $\triangle SRT$; since $\overline{ST} \cap \mathcal{L} \neq \emptyset$ and $\overline{RT} \cap \mathcal{L} = \emptyset$, $\overline{SR} \cap \mathcal{L} \neq \emptyset$. If the points are not all distinct, $Q = S$, which completes the proof.

Otherwise, apply Pasch to $\triangle QST$; since $\overset{\dashv\vdash}{ST} \cap \mathcal{L} \neq \emptyset$ and $\overset{\dashv\vdash}{QS} \cap \mathcal{L} = \emptyset$, $\overset{\dashv\vdash}{QT} \cap \mathcal{L} \neq \emptyset$. Then apply Pasch to $\triangle QRT$; since $\overset{\dashv\vdash}{QT} \cap \mathcal{L} \neq \emptyset$ and $\overset{\dashv\vdash}{RT} \cap \mathcal{L} = \emptyset$, $\overset{\dashv\vdash}{QR} \cap \mathcal{L} \neq \emptyset$, the desired result.

(Case 3: Q, R, and T are collinear but S is not collinear with these points.) Interchange Q with R and interchange S with T, and the proof of this case is word-for-word as in Case 2.

(Case 4: Q, S, and T are collinear but R is not collinear with these points.) Apply Pasch to $\triangle SRT$; since $\overset{\dashv\vdash}{ST} \cap \mathcal{L} \neq \emptyset$ and $\overset{\dashv\vdash}{RT} \cap \mathcal{L} = \emptyset$, $\overset{\dashv\vdash}{RS} \cap \mathcal{L} \neq \emptyset$. If not all the points are distinct, then $Q = S$, which completes the proof.

Otherwise, apply Pasch to $\triangle QRS$; since $\overset{\dashv\vdash}{RS} \cap \mathcal{L} \neq \emptyset$ and $\overset{\dashv\vdash}{QS} \cap \mathcal{L} = \emptyset$, $\overset{\dashv\vdash}{QR} \cap \mathcal{L} \neq \emptyset$, which is the desired result.

(Case 5: R, S, and T are collinear but Q is not collinear with these points.) Interchange Q with R and interchange S with T, and the proof of this case is word-for-word as in Case 4.

(Case 6: Q, R, S and T are collinear and distinct.) By Corollary IB.14.4, there exists a point $X \in$ the S-side of $\mathcal{L} = \mathcal{E}$ such that $X \notin \overleftrightarrow{ST} = \overleftrightarrow{QR}$.

Apply the alternate form of Pasch to $\triangle QSX$. Since $\overset{\dashv\vdash}{SQ} \cap \mathcal{L} = \emptyset$ and $\overset{\dashv\vdash}{SX} \cap \mathcal{L} = \emptyset$, $\overset{\dashv\vdash}{QX} \cap \mathcal{L} = \emptyset$.

Apply Pasch to $\triangle STX$: $\overset{\dashv\vdash}{ST} \cap \mathcal{L} \neq \emptyset$ and $\overset{\dashv\vdash}{SX} \cap \mathcal{L} = \emptyset$ imply $\overset{\dashv\vdash}{TX} \cap \mathcal{L} \neq \emptyset$.

Apply Pasch to $\triangle RTX$: $\overset{\dashv\vdash}{TX} \cap \mathcal{L} \neq \emptyset$ and $\overset{\dashv\vdash}{RT} \cap \mathcal{L} = \emptyset$ imply $\overset{\dashv\vdash}{RX} \cap \mathcal{L} \neq \emptyset$.

Finally, apply Pasch to $\triangle QRX$; since $\overset{\dashv\vdash}{RX} \cap \mathcal{L} \neq \emptyset$ and $\overset{\dashv\vdash}{QX} \cap \mathcal{L} \neq \emptyset$, $\overset{\dashv\vdash}{QR} \cap \mathcal{L} \neq \emptyset$, which is the desired result. □

5.3 Pasch geometry

Definition PSH.7. An IB plane $\mathcal{P}$ for which the incidence, betweenness, and Plane Separation Axioms are true is called a **Pasch plane**. The geometry of such a plane is called **Pasch geometry**.

Theorem PSH.6 showed that the Postulate of Pasch holds in a Pasch plane! From now on all planes will be Pasch planes.

The following Theorem PSH.8 shows that the two statements labeled "Property B.4" in Chapter 4 (following Definition IB.1) are consequences of the other axioms adopted so far. This theorem, with its several corollaries, comprises a fundamental result about the behavior of line segments. This will be needed to prove the results leading to the Plane Separation Theorem (Theorem PSH.12), after which we will return to Theorem PSH.8 and use it in a more extensive exploration of the behavior of segments, rays, and lines.

Theorem PSH.8. *Let A, B, C, and D be distinct points in a Pasch plane $\mathcal{P}$.*

(A) *If A–B–C and A–C–D, then A, B, C, and D are collinear and both*

(1) *B–C–D and* (2) *A–B–D*

are true; that is, if A–B–C and A–C–D, then A–B–C–D.

(B) *If A–B–C and B–C–D, then A, B, C, and D are collinear and both*

(1) *A–B–D and* (2) *A–C–D*

are true; that is, if A–B–C and B–C–D, then A–B–C–D.[3]

Proof. If *A–B–C*, by Property B.0 of Definition IB.1, $B \in \overleftrightarrow{AC}$; similarly, if *A–C–D*, $D \in \overleftrightarrow{AC}$ so that A, B, C, and D are collinear. Thus if (A) is true, by Axiom I.5 there exists a point E on $\mathcal{P}$ not belonging to $\overleftrightarrow{AB}$. By Theorem IB.14, $\overrightarrow{CA} \subseteq$ (the A-side of $\overleftrightarrow{CE}$). Since *A–B–C*, by Definition IB.4 B belongs to $\overrightarrow{CA}$ and hence to the A-side of $\overleftrightarrow{CE}$. Since *A–C–D*, the A-side and the D-side are opposite sides of $\overleftrightarrow{CE}$. Therefore B and D are on opposite sides of $\overleftrightarrow{CE}$. By Axiom PSA, there exists a point Q such that $\overleftrightarrow{BD} \cap \overleftrightarrow{CE} = \{Q\}$ and *B–Q–D*. But $C \in \overleftrightarrow{BD}$ and $C \in \overleftrightarrow{CE}$, so by Exercise I.1 $Q = C$ and hence *B–C–D*. This proves (A)(1).

Before proving statement (A)(2) we turn to the proof of (B). The argument for collinearity is similar to that for (A). The argument for (B)(1), if *A–B–C* and *B–C–D* then *A–B–D*, is very much like but not exactly similar to the argument just above; the main point to note is that $\overleftrightarrow{BE}$ is used in place of $\overleftrightarrow{CE}$. We leave the details of this proof to the reader as Exercise PSH.1.

The conclusion *A–B–D* for (B)(1) is also part (A)(2), so part (A) is proved. To prove part (B)(2) apply this result where A and D are interchanged, and B and C are interchanged. The statement then becomes "if *D–C–B* and *C–B–A* then *D–C–A*," that is "if *B–C–D* and *A–B–C* then *A–C–D*." This is the result for part (B)(2). □

[3] Part (B) is sometimes called "Pasch's Theorem."

Corollary PSH.8.1. *Let A, B, C, and D be distinct coplanar points. If A–B–D and B–C–D, then A–B–C and A–C–D. In other words, if A–B–D and B–C–D, then A–B–C–D.*

Proof. From *A–B–D* and *B–C–D* it follows by Property B.1 of Definition IB.1 that *D–C–B* and *D–B–A*. Hence from Theorem PSH.8(A)(1), *C–B–A*. Another application of Property B.1 gives the first result. The proof of the second is similar, using Theorem PSH.8(A)(2). □

Corollary PSH.8.2. *Let A, B, C, and D be distinct coplanar points. If A–B–C and A–B–D, then exactly one of the following two statements is true:*

(1) *A–D–C and B–D–C;* (2) *A–C–D and B–C–D.*

In other words, if A–B–C and A–B–D, then either A–B–D–C or A–B–C–D.

Proof. By Property B.3 of Definition IB.1, either *A–D–C*, or *A–C–D*, or *C–A–D*. But if *C–A–D* and *A–B–D*, then by Corollary 1 *C–A–B*, which by Property B.3 contradicts *A–B–C*. Hence the assumption *C–A–D* is untenable, and we have either *A–D–C* or *A–C–D*, but not both, showing that alternatives (1) and (2) are mutually exclusive. The proof now splits into two parts.

(i) Suppose *A–D–C*. Together with *A–B–D*, this implies *B–D–C* by Theorem PSH.8(A)(1).
(ii) Suppose *A–C–D*. Together with *A–B–C*, this implies *B–C–D* by Theorem PSH.8(A)(1). □

Corollary PSH.8.3. *Let A, B, C, and D be distinct coplanar points. If A–C–D and B–C–D, then exactly one of the following two statements is true:*

(1) *A–B–C and A–B–D;* (2) *B–A–C and B–A–D.*

In other words, if A–C–D and B–C–D, then either A–B–C–D or B–A–C–D.

Proof. Exercise PSH.2(A). □

Corollary PSH.8.4. *Let A, B, C, and D be distinct coplanar points. If A–B–D and A–C–D, then exactly one of the following two statements is true:*

(1) *A–B–C and B–C–D;* (2) *A–C–B and C–B–D.*

In other words, if A–B–D and A–C–D, then either A–B–C–D or A–C–B–D.

Proof. Exercise PSH.2(B). □

Corollary PSH.8.5. *Let X, Y, and Z be distinct coplanar points. If X–Z–Y, then* (A) $\overset{\dashv\vdash}{XZ} \subseteq \overset{\dashv\vdash}{XY}$; (B) $\overset{\dashv\vdash}{ZY} \subseteq \overset{\dashv\vdash}{XY}$; (C) $\overset{\sqsubset\sqsupset}{XZ} \subseteq \overset{\sqsubset\sqsupset}{XY}$; *and* (D) $\overset{\sqsubset\sqsupset}{ZY} \subseteq \overset{\sqsubset\sqsupset}{XY}$.

Proof. (A) Let $W \in \overset{\dashv\vdash}{XZ}$, so that X–W–Z; by Theorem PSH.8(A) X–W–Y and $W \in \overset{\dashv\vdash}{XY}$.

(B) Follows immediately from (A) by interchanging the roles of X and Y.

(C) If $W \in \overset{\sqsubset\sqsupset}{XZ}$ either $W \in \overset{\dashv\vdash}{XZ} \subseteq \overset{\dashv\vdash}{XY} \subseteq \overset{\sqsubset\sqsupset}{XY}$ or $W = X$ or $W = Z$; both X and Z belong to $\overset{\sqsubset\sqsupset}{XY}$ so the result follows.

(D) The argument is similar to that for (C). □

Theorem PSH.9. *If $\mathcal{L}$ is a line in the plane $\mathcal{P}$, then each side of $\mathcal{L}$ is convex.*

Proof. Let P be a point on $\mathcal{P}$ off of $\mathcal{L}$, and let X and Y be distinct members of the P-side of $\mathcal{L}$. By Theorem PSH.1 the X-side of $\mathcal{L}$ = the P-side of $\mathcal{L}$ = the Y-side of $\mathcal{L}$ Thus $X \in$ the Y-side of $\mathcal{L}$. By Definition IB.11, $\mathcal{L} \cap \overset{\sqsubset\sqsupset}{XY} = \emptyset$. This proves that the P-side of $\mathcal{L}$ is convex. □

The next theorem fills a large gap in our picture. It finally assures us that any two opposite sides of a line in a Pasch plane constitute all of the points on the plane which are not on the line.

Theorem PSH.10. *If $\mathcal{L}$ is a line on plane $\mathcal{P}$ and if $\mathcal{D}$ and $\mathcal{E}$ are opposite sides of $\mathcal{L}$, then $\mathcal{D} \cap \mathcal{E} = \emptyset$, and $\mathcal{D} \cup \mathcal{E} = \mathcal{P} \setminus \mathcal{L}$.*

Proof. The fact that $\mathcal{D} \cap \mathcal{E} = \emptyset$ is just a restatement of Theorem PSH.2 and is included for completeness.

Since $\mathcal{D}$ and $\mathcal{E}$ are opposite sides of $\mathcal{L}$, by Definition IB.11 there exist points P, Q, and R such that $Q \in \mathcal{L}$, P–Q–R, $\mathcal{D}$ is the P-side of $\mathcal{L}$, and $\mathcal{E}$ is the R-side of $\mathcal{L}$. By Theorem PSH.5, we know that any point not on $\mathcal{L}$ and not on $\overleftrightarrow{PQ}$ belongs either to $\mathcal{D}$ or to $\mathcal{E}$. Hence we need only prove that every point not on $\mathcal{L}$ and on $\overleftrightarrow{PQ}$ belongs either to $\mathcal{D}$ or to $\mathcal{E}$.

Let S be any point on $\mathcal{L}$ different from Q. There exists a point Y such that P–S–Y. Since $Y \notin \mathcal{L}$ and $Y \notin \overleftrightarrow{PQ}$, Y is a member of either $\mathcal{D}$ or $\mathcal{E}$. By Definition IB.11, Y is on a side opposite $\mathcal{D}$ and hence is in $\mathcal{E}$, which is therefore the Y-side of $\mathcal{L}$.

Applying Theorem PSH.5 again, every point on $\mathcal{L}$ and not on $\overleftrightarrow{PY}$ is a member of either $\mathcal{D}$ or $\mathcal{E}$. Every point on $\overleftrightarrow{PR}$ other than P and Q fails to be on $\overleftrightarrow{PY}$ and therefore belongs to either $\mathcal{D}$ or $\mathcal{E}$. We know already that $P \in \mathcal{D}$, so every point of $\overleftrightarrow{PR}$ not on $\mathcal{L}$ is a member of either $\mathcal{D}$ or $\mathcal{E}$. □

Corollary PSH.10.1. *With the same hypotheses as in Theorem PSH.10,* $\mathcal{L} \cup \mathcal{E} = \mathcal{P} \setminus \mathcal{D}$, $\mathcal{L} \cup \mathcal{D} = \mathcal{P} \setminus \mathcal{E}$, $\mathcal{E} = \mathcal{P} \setminus (\mathcal{L} \cup \mathcal{D})$ *and* $\mathcal{D} = \mathcal{P} \setminus (\mathcal{L} \cup \mathcal{E})$.

This corollary (which is easily proved using elementary set theory) says that a side of a line $\mathcal{L}$ completely determines the side opposite to it.

Theorem PSH.11. *There can be only one pair* $(\mathcal{H}_1, \mathcal{H}_2)$ *of sides for a line* $\mathcal{L}$ *in a Pasch plane.*

Proof. Let $\mathcal{H}_1$ be a side of $\mathcal{L}$, and $\mathcal{H}_2$ a side opposite to $\mathcal{H}_1$. Suppose $\mathcal{K}_1$ is a side of $\mathcal{L}$ and $\mathcal{K}_2$ is an opposite side. Let A be a point of $\mathcal{K}_1$; by Theorem PSH.1 $\mathcal{K}_1$ = the A-side of $\mathcal{L}$. By Theorem PSH.10, since $A \notin \mathcal{L}$, either $A \in \mathcal{H}_1$ or $A \in \mathcal{H}_2$. Choose the notation so that $A \in \mathcal{H}_1$. Then, again by Theorem PSH.1, $\mathcal{K}_1$ = the A-side of $\mathcal{L} = \mathcal{H}_1$. By Theorem PSH.10, $\mathcal{K}_2 = \mathcal{P} \setminus (\mathcal{H}_1 \cup \mathcal{L}) = \mathcal{H}_2$. □

We may now speak of *the two* sides of a line $\mathcal{L}$ in a plane $\mathcal{P}$. This legitimizes the use of the word "same" in connection with sides of a line. If two sides have a point in common, they are the same set (by Theorem PSH.1) and if two sides are both opposite to the same side, they are the same set. Thus we may speak of "the" opposite side of a side.

We might define a relation $\equiv$ on $\mathcal{P} \setminus \mathcal{L}$ as follows: for any two points P and Q in $\mathcal{P} \setminus \mathcal{L}$, $P \equiv Q$ iff P and Q belong to the same side of $\mathcal{L}$. By Definition IB.11, $\equiv$ is reflexive; by Theorem PSH.1, $\equiv$ is symmetric; and by the observation just above, $\equiv$ is transitive. Therefore $\equiv$ is an equivalence relation. But it isn't a very interesting one, since it has only two equivalence classes, which are the two sides of $\mathcal{L}$.

The next theorem summarizes the results we have obtained so far in completing the picture we described at the beginning of this chapter. We mark it as a major milestone here because it is sometimes called the Plane Separation Axiom and used as an axiom by geometers who are not concerned—as we are here—with the detailed logical relationships between its various provisions and the axioms of incidence and betweenness.

Theorem PSH.12 (Plane Separation Theorem). *If* $\mathcal{L}$ *is a line on a Pasch plane* $\mathcal{P}$, *then there exists a unique pair* $(\mathcal{H}_1, \mathcal{H}_2)$ *of convex subsets of* $\mathcal{P}$ *such that*

(I) $\mathcal{P} = \mathcal{H}_1 \cup \mathcal{H}_2 \cup \mathcal{L}$;
(II) *the sets* $\mathcal{H}_1$, $\mathcal{H}_2$, *and* $\mathcal{L}$ *are pairwise disjoint;*
(III) $\mathcal{H}_1$ *and* $\mathcal{H}_2$ *are the opposite sides of* $\mathcal{L}$;

(IV) *If* P_1 *and* P_2 *are any points in* $\mathcal{P} \setminus \mathcal{L}$, *then*

(A) $\overset{\rceil\!\!-\!\!\lceil}{P_1P_2} \cap \mathcal{L} = \emptyset$ *iff* P_1 *and* P_2 *belong to the same side (either* $\mathcal{H}_1$ *or* $\mathcal{H}_2$*) of* $\mathcal{L}$*; or equivalently*

(B) $\overset{\rceil\!\!-\!\!\lceil}{P_1P_2} \cap \mathcal{L} \neq \emptyset$ *iff one of the points* P_1, P_2 *belongs to* $\mathcal{H}_1$ *and the other to* $\mathcal{H}_2$.

Proof. By Axiom I.5, there exists a point A belonging to $\mathcal{P} \setminus \mathcal{L}$. Let $\mathcal{H}_1$ = the A-side of $\mathcal{L}$. Let B be any point of $\mathcal{L}$. Then by Property B.3 of Definition IB.1 there exists a point C such that A–B–C. Let $\mathcal{H}_2$ = the C-side of $\mathcal{L}$. Then by Definition IB.11, $\mathcal{H}_1$ and $\mathcal{H}_2$ are opposite sides of $\mathcal{L}$, showing (III). (I) and (II) are true by Theorem PSH.10.

The uniqueness of the two sides is Theorem PSH.11, and this theorem also shows that $\mathcal{H}_2$ is the only side opposite $\mathcal{H}_1$ (and, *mutatis mutandis*, $\mathcal{H}_1$ is the only side opposite $\mathcal{H}_2$). The convexity of the sides is Theorem PSH.9.

Finally, Definition IB.11 (combined with Theorem PSH.1) says that $\overset{[\!-\!]}{P_1P_2} \cap \mathcal{L} = \emptyset$ iff P_1 and P_2 belong to the same side of $\mathcal{L}$. Since these two points do not belong to $\mathcal{L}$, this is equivalent to saying that $\overset{\rceil\!\!-\!\!\lceil}{P_1P_2} \cap \mathcal{L} = \emptyset$. This shows (IV)(A); (IV)(B) is the contrapositive of part (IV)(A). □

Remark PSH.12.1. We conclude this section by observing that if $\overleftrightarrow{AB}$ and $\overleftrightarrow{CD}$ are parallel lines in a Pasch plane $\mathcal{P}$, each of them is a subset of a single side of the other; in particular, the points A and B are on the same side of $\overleftrightarrow{CD}$, and C and D are on the same side of $\overleftrightarrow{AB}$.

For if $\overleftrightarrow{AB}$, say, does not lie entirely in one side of $\overleftrightarrow{CD}$, by Theorem PSH.12(I) it must contain a point of $\overleftrightarrow{CD}$, or a point on the other side, so that, by Theorem PSH.12(IV)(B) it contains a point of $\overleftrightarrow{CD}$. In either case the lines are not parallel (*cf* Exercise PSH.14).

5.4 Segments, rays, lines, and their properties

We now enter into a detailed consideration of the behavior of segments, rays, and lines. This development depends largely on Theorem PSH.8 and its corollaries.

Theorem PSH.13. *Let A, B, and C be points such that A–B–C. Then*

(A) $\{X\,|A\text{–}B\text{–}X\} = \overrightarrow{BC}$,

(B) $\overleftrightarrow{AB} = \overset{\sqsubset\sqsupset}{AB} \cup \overrightarrow{BC} = \{A\} \cup \{B\} \cup \overset{\sqsupset\sqsubset}{AB} \cup \overrightarrow{BC}$,

(C) $\overset{\sqsubset\sqsupset}{AB} \cap \overrightarrow{BC} = \emptyset$, *and*

(D) $\overrightarrow{AB} = \{B\} \cup \overset{\sqsupset\sqsubset}{AB} \cup \overrightarrow{BC} = \overset{\sqsupset\sqsubset}{AB} \cup \overleftrightarrow{BC}$.

Proof. By Corollary PSH.8.2, *A–B–X*, in the presence of *A–B–C*, implies one and only one of the following statements is true: (1) *B–X–C*; (2) $X = C$; (3) *B–C–X*. By Definition IB.4, $\overrightarrow{BC} = \{X\,|B\text{–}X\text{–}C$, or $X = C$, or $B\text{–}C\text{–}X\}$, so we have shown $\{X\,|A\text{–}B\text{–}X\} \subseteq \overrightarrow{BC}$.

To show the reverse inclusion we start with the overarching assumption *A–B–C*. Let $X \in \overrightarrow{BC}$ and use Definition IB.4 as stated above. If *B–X–C* by Corollary PSH.8.1, *A–B–X*. If $X = C$, then *A–B–X*. If *B–C–X* by Theorem 8(B) *A–B–X*. This completes the proof of $\{X\,|A\text{–}B\text{–}X\} = \overrightarrow{BC}$.

By part (A) and Remark IB.4.1(g), $\overleftrightarrow{AB} = \overset{\sqsubset\sqsupset}{AB} \cup \overrightarrow{BC}$; by part (A) and Remark IB.4.1(i), $\overset{\sqsubset\sqsupset}{AB} \cap \overrightarrow{BC} = \emptyset$; and by part (A) and Remark IB.4.1(h)

$$\overrightarrow{AB} = \overset{\sqsupset\sqsubset}{AB} \cup \{B\} \cup \{X\,|A\text{–}B\text{–}X\} = \overset{\sqsupset\sqsubset}{AB} \cup \{B\} \cup \overrightarrow{BC}. \qquad \square$$

Corollary PSH.13.1. *Let A and B be distinct points and let C and D be points such that A–B–C and A–B–D. Then* $\overrightarrow{BC} = \overrightarrow{BD}$ *and* $\overleftrightarrow{BC} = \overleftrightarrow{BD}$.

Proof. By Theorem PSH.13(B), $\overleftrightarrow{AB} = \overset{\sqsubset\sqsupset}{AB} \cup \overrightarrow{BC} = \overset{\sqsubset\sqsupset}{AB} \cup \overrightarrow{BD}$. From Definition IB.3 and Property B.2 of Definition IB.1, $\overset{\sqsubset\sqsupset}{AB} \cap \overrightarrow{BC} = \overset{\sqsubset\sqsupset}{AB} \cap \overrightarrow{BD} = \emptyset$, so from elementary set theory $\overrightarrow{BC} = \overrightarrow{BD}$. By Remark IB.4.1(e) we get $\overleftrightarrow{BC} = \overleftrightarrow{BD}$. $\square$

Theorem PSH.14. *Let A, B, C, and D be distinct collinear points. Then exactly one of the following twelve statements is true:*

A–B–C–D	*A–D–B–C*	*B–C–A–D*
A–B–D–C	*A–D–C–B*	*B–D–A–C*
A–C–B–D	*B–A–C–D*	*C–A–B–D*
A–C–D–B	*B–A–D–C*	*C–B–A–D.*

Proof. By Definition IB.2 and Property B.2 of Definition IB.1, not more than one of the above statements is true. By Property B.2 one and only one of the following three statements is true:

A–B–C   *A–C–B*   *B–A–C*;

and one and only one for the following statements is true:

$$A\text{–}B\text{–}D \qquad A\text{–}D\text{–}B \qquad B\text{–}A\text{–}D.$$

This gives rise to nine mutually exclusive and exhaustive possibilities:

(1) *A–B–C* and *A–B–D*. By Corollary PSH.8.2, either *A–B–C–D* or *A–B–D–C*.
(2) *A–B–C* and *A–D–B*. By Theorem PSH.8(A), *A–D–B–C*.
(3) *A–B–C* and *B–A–D*. By Property B.1 of Definition IB.1, *D–A–B* and *A–B–C*, so by Theorem PSH.8(B), *D–A–B–C* and by Exercise IB.2, *C–B–A–D*.
(4) *A–C–B* and *A–B–D*. By Theorem PSH.8(A), *A–C–B–D*.
(5) *A–C–B* and *A–D–B*. By Corollary PSH.8.4, either *A–C–D–B* or *A–D–C–B*.
(6) *A–C–B* and *B–A–D*. By Property B.1 of Definition IB.1, *B–C–A* and *B–A–D*, so by Theorem PSH.8(A), *B–C–A–D*.
(7) *B–A–C* and *A–B–D*. By Property B.1, *D–B–A* and *B–A–C*, so by Theorem PSH.8 (B) *D–B–A–C*, and by Exercise IB.2, *C–A–B–D*.
(8) *B–A–C* and *A–D–B*. By Property B.1, *B–D–A* and *B–A–C*, so by Theorem PSH.8(A), *B–D–A–C*.
(9) *B–A–C* and *B–A–D*. By Corollary PSH.8.2, either *B–A–C–D* or *B–A–D–C*.

A perusal of this list shows at least one of the twelve possibilities in the theorem *must* occur. Putting this together with the statement at the beginning that at most one can occur, we have the statement in the theorem. □

The proof of the preceding theorem involved a lot of tedious checking and can hardly be called elegant. But the theorem is worth the work. One reason is that, until now, our concept of line has had serious limitations because it has given us no insight into the "internal structure" of a line. Lines have been special sets of points that satisfied certain axioms which described how these sets relate to each other and to other special kinds of sets, such as planes. It can be quite difficult to tell whether or not a particular set of points is a line unless we have a complete list of all the sets that are lines in the geometry, as in our model of incidence geometry. As we will see in Theorem PSH.15 (a consequence of Theorem PSH.14) we can now identify a line if we know something about its internal structure as described by the betweenness relation.

Theorem PSH.15. *If A, B, and C are points such that A–B–C, then*

(A) $\overleftrightarrow{AB} = \overleftrightarrow{BC} = \overleftrightarrow{AC}$.
(B) $\overleftrightarrow{AB}$ *is the union of the disjoint sets* $\overrightarrow{BA}$, $\{B\}$, *and* $\overrightarrow{BC}$.

(C) *$\overline{AC}$ is the union of the disjoint sets $\{A, B, C\}$, $\overset{\vdash\dashv}{AB}$, and $\overset{\vdash\dashv}{BC}$.*
(D) *$\overset{\vdash\dashv}{AC}$ is the union of the disjoint sets $\{B\}$, $\overset{\vdash\dashv}{AB}$, and $\overset{\vdash\dashv}{BC}$.*

Proof. (A) Follows from Property B.0 of Definition IB.1 and Exercise I.2.

(B) By Theorem IB.5 line 4, $\overleftrightarrow{AB} = \overleftrightarrow{BC} = \{X \mid X\text{–}B\text{–}C\} \cup \{B\} \cup \overrightarrow{BC}$, and the sets in the union are disjoint. By Definition IB.4, $\overrightarrow{BA} = \{X \mid B\text{–}X\text{–}A$ or $X = A$ or $B\text{–}A\text{–}X\}$.

If $X\text{–}B\text{–}C$ and $X \neq A$, by Corollary PSH.8.3 either $X\text{–}A\text{–}B$ or $A\text{–}X\text{–}B$, so that $\{X \mid X\text{–}B\text{–}C\} \subseteq \overrightarrow{BA}$.

Conversely, suppose $X \in \overrightarrow{BA}$; we know that $A\text{–}B\text{–}C$. If $B\text{–}X\text{–}A$, that is $A\text{–}X\text{–}B$, by Theorem PSH.8(A)(1) $X\text{–}B\text{–}C$. If $B\text{–}A\text{–}X$, that is, $X\text{–}A\text{–}B$, by Theorem PSH.8(B)(2), $X\text{–}B\text{–}C$. If $X = A$, then $X\text{–}B\text{–}C$. Therefore $\overrightarrow{BA} \subseteq \{X \mid X\text{–}B\text{–}C\}$, so that $\overleftrightarrow{AB} = \overrightarrow{BA} \cup \{B\} \cup \overrightarrow{BC}$.

(C) First note that by Definition IB.3, $X \in \overline{AC}$ iff $X = A$ or $A\text{–}X\text{–}C$ or $X = C$. If $A\text{–}X\text{–}C$ (together with the assumption $A\text{–}B\text{–}C$), by Corollary PSH.8.4, either $A\text{–}X\text{–}B$, $B\text{–}X\text{–}C$, or $X = B$; then if $X \in \overline{AC}$, exactly one of $X = A$, $A\text{–}X\text{–}B$, $X = B$, $B\text{–}X\text{–}C$ or $X = C$ is true, so that $\overline{AC} \subseteq \{A, B, C\} \cup \overset{\vdash\dashv}{AB} \cup \overset{\vdash\dashv}{BC}$.

Conversely, if $A\text{–}X\text{–}B$, then by Theorem PSH.8(A)(2), $A\text{–}X\text{–}C$. If $B\text{–}X\text{–}C$ by Corollary PSH.8.1, $A\text{–}X\text{–}C$. Since $\{A, B, C\} \subseteq \overline{AC}$, it follows that $\{A, B, C\} \cup \overset{\vdash\dashv}{AB} \cup \overset{\vdash\dashv}{BC} \subseteq \overline{AC}$ so that (C) is proved. The sets are disjoint by Property B.2 of Definition IB.1.

(D) By Remark IB.4.1(d), $\overset{\vdash\dashv}{AC} = \overline{AC} \setminus \{A, C\}$, so by part (C) $\overset{\vdash\dashv}{AC}$ is the union of the disjoint sets $\{B\}$, $\overset{\vdash\dashv}{AB}$, and $\overset{\vdash\dashv}{BC}$. □

Theorem PSH.16. *Let A, B, and C be points such that $C \in \overrightarrow{AB}$. Then $\overrightarrow{AB} = \overrightarrow{AC}$ and $\overset{\sqsubset\!\rightarrow}{AB} = \overset{\sqsubset\!\rightarrow}{AC}$.*

Proof. Since $C \in \overrightarrow{AB}$, by Definition IB.4, either $C = B$, $A\text{–}C\text{–}B$, or $A\text{–}B\text{–}C$. If $B = C$, then clearly $\overrightarrow{AB} = \overrightarrow{AC}$ and $\overset{\sqsubset\!\rightarrow}{AB} = \overset{\sqsubset\!\rightarrow}{AC}$.

Suppose that $A\text{–}C\text{–}B$. By Properties B.3 and B.1 of Definition IB.1 there is a point P such that $P\text{–}A\text{–}C$. By Theorem PSH.8(B)(1), $P\text{–}A\text{–}B$. By the Corollary PSH.13.1, $P\text{–}A\text{–}C$ and $P\text{–}A\text{–}B$ together imply $\overrightarrow{AB} = \overrightarrow{AC}$ and $\overset{\sqsubset\!\rightarrow}{AB} = \overset{\sqsubset\!\rightarrow}{AC}$, the desired conclusion.

If $A\text{–}B\text{–}C$, interchanging the roles of B and C in the above argument yields the conclusion $\overrightarrow{AC} = \overrightarrow{AB}$ and $\overset{\sqsubset\!\rightarrow}{AC} = \overset{\sqsubset\!\rightarrow}{AB}$, again the desired conclusion. □

Theorem PSH.17. *Let A, B, C, and D be points on a Pasch plane $\mathcal{P}$ such that $A \neq B$ and $C \neq D$. Then*

$$\overset{\sqsubset\!\rightarrow}{CD} \subseteq \overset{\sqsubset\!\rightarrow}{AB} \textit{ iff } \big((C = A \text{ and } D \in \overrightarrow{AB}) \text{ or } (C \in \overrightarrow{AB} \textit{ and } A\text{–}C\text{–}D)\big).$$

Proof. I: $\overrightarrow{CD} \subseteq \overrightarrow{AB} \Rightarrow \big((C = A \text{ and } D \in \overrightarrow{AB}) \text{ or } (C \in \overrightarrow{AB} \text{ and } A\text{–}C\text{–}D)\big)$.

By Definition IB.4, C and D belong to $\overrightarrow{CD} \subseteq \overrightarrow{AB}$. Either $C = A$ or $C \neq A$.

(I.1) If $C = A$, $D \neq A$ so that by Remark IB.4.1(f) $D \in \overrightarrow{AB}$, which is the first half of the disjunction we wish to prove.

(I.2) Now suppose $C \neq A$. Then by Theorem IB.4.1(f) $C \in \overrightarrow{AB}$, and by Theorem PSH.16, $\overrightarrow{AB} = \overrightarrow{AC}$ and $\overrightarrow{AB} = \overrightarrow{AC}$.

Claim: $D \neq A$. For suppose $D = A$. Then since $C \neq D$, by Property B.3 of Definition IB.1 there exists a point X with C–D–X; by Definition IB.4 $X \in \overrightarrow{CD} \subseteq \overrightarrow{AB}$ which is equal to $\overrightarrow{AC}$ as we have seen. Since $D = A$, C–A–X. But $X \in \overrightarrow{AC}$ which by Definition IB.4 means that either $X = A$ or $X = C$ (both of which we know to be false by the definition of X) or A–X–C or A–C–X, which are in contradiction to C–A–X by Property B.2 of Definition IB.1. This proves the claim.

From Theorem IB.4.1(f) it follows that $D \in \overrightarrow{AB}$ so that by Theorem PSH.16, $\overrightarrow{AD} = \overrightarrow{AB} = \overrightarrow{AC}$, and $\overrightarrow{AD} = \overrightarrow{AB} = \overrightarrow{AC}$.

Since $C \neq A$, $C \neq D$, and $D \neq A$, by Property B.2 of Definition IB.1 one and only one of A–C–D, C–A–D, or C–D–A is true.

C–A–D is not true because by Definition IB.4 and Property B.2 of Definition IB.1, it implies $C \notin \overrightarrow{AD} = \overrightarrow{AB}$, contradicting our hypothesis $\overrightarrow{CD} \subseteq \overrightarrow{AB}$.

C–D–A is not true either. To see this note first that by Property B.3 of Definition IB.1 there is a point P such that D–A–P. Now on the one hand, C–D–A and D–A–P imply by Theorem PSH.8(B)(1) C–D–P, which means $P \in \overrightarrow{CD} \subseteq \overrightarrow{AB}$. But on the other hand, D–A–P implies by Definition IB.4 and Property B.2 $P \notin \overrightarrow{AD} = \overrightarrow{AB}$. This contradiction shows C–D–A is false.

We are left with only the first alternative which we wished to prove.

II. $\overrightarrow{CD} \subseteq \overrightarrow{AB} \Leftarrow \big((C = A \text{ and } D \in \overrightarrow{AB}) \text{ or } (C \in \overrightarrow{AB} \text{ and } A\text{–}C\text{–}D)\big)$.

(II.1) If $C = A$ and $D \in \overrightarrow{AB}$, then by Theorem PSH.16, $\overrightarrow{CD} = \overrightarrow{AB}$ and $\overrightarrow{CD} = \overrightarrow{AB}$.

(II.2) If $C \in \overrightarrow{AB}$, then by Theorem PSH.16, $\overrightarrow{AC} = \overrightarrow{AB}$. By Theorem PSH.13, if A–C–D, then $\{X \mid A\text{–}C\text{–}X\} = \overrightarrow{CD}$ and $\overrightarrow{AC} = \overline{AC} \cup \overrightarrow{CD}$, so $\overrightarrow{CD} \subseteq \overrightarrow{AB}$. □

The proof of Theorem PSH.19 becomes much easier once we know that rays are convex sets. Therefore we will prove the following theorem at this point.

Theorem PSH.18 (Convexity of segments and rays). *Let A and B be distinct points on a Pasch plane $\mathcal{P}$. Then each of the following sets is convex:* (1) $\overset{]\!\rightarrow}{AB}$, (2) $\overset{[\!\rightarrow}{AB}$, (3) $\overset{]-[}{AB}$, (4) $\overset{[-[}{AB}$, (5) $\overset{]-]}{AB}$, *and* (6) $\overset{[-]}{AB}$.

Proof. (1) Let U and V be distinct members of $\overset{]\!\rightarrow}{AB}$. By Definition IB.9 we must show $\overset{]-[}{UV} \subseteq \overset{]\!\rightarrow}{AB}$. To do this we let X be any member of $\overset{]-[}{UV}$ and show $X \in \overset{]\!\rightarrow}{AB}$. The points A, B, U, and V are collinear, and by Property B.2 of Definition IB.1, exactly one of A–U–V, U–A–V, or A–V–U is true.

If U–A–V, by Theorem PSH.15(B) $\overleftrightarrow{AB} = \overset{]\!\rightarrow}{AU} \cup \{A\} \cup \overset{]\!\rightarrow}{AV}$ and these sets are disjoint; therefore $B \in \overset{]\!\rightarrow}{AU}$ or $B \in \overset{]\!\rightarrow}{AV}$ but not both. If $B \in \overset{]\!\rightarrow}{AU}$, by Theorem PSH.16 $\overset{]\!\rightarrow}{AU} = \overset{]\!\rightarrow}{AB}$ so that $V \in \overset{]\!\rightarrow}{AU}$ and either A–U–V or A–V–U, in contradiction to our assumption. (A similar proof holds if $B \in \overset{]\!\rightarrow}{AV}$.)

Therefore by Property B.2 of Definition IB.1 either A–U–V or A–V–U. Choose the notation so that A–U–V. Since $V \in \overset{]\!\rightarrow}{AB}$, by Theorem PSH.16, $\overset{]\!\rightarrow}{AV} = \overset{]\!\rightarrow}{AB}$. If X is any point of $\overset{]-[}{UV}$, by Definition IB.3, U–X–V. By Corollary PSH.8.1, A–U–V and U–X–V imply A–X–V. By Definition IB.4, $X \in \overset{]\!\rightarrow}{AV} = \overset{]\!\rightarrow}{AB}$. This shows $\overset{]-[}{UV} \subseteq \overset{]\!\rightarrow}{AB}$ and completes the proof of this part.

(2) Let U and V be distinct members of $\overset{[\!\rightarrow}{AB}$. As above, we must show $\overset{]-[}{UV} \subseteq \overset{[\!\rightarrow}{AB}$. If U and V both belong to $\overset{]\!\rightarrow}{AB}$, then, since $\overset{[\!\rightarrow}{AB}$ is convex by part (1) above, $\overset{]-[}{UV} \subseteq \overset{]\!\rightarrow}{AB} \subseteq \overset{[\!\rightarrow}{AB}$, and we are done.

If either $U = A$ or $V = A$, choose the notation so $U = A$. Since U and V are distinct, $V \in \overset{]\!\rightarrow}{AB}$ and by Theorem PSH.16 $\overset{[\!\rightarrow}{AV} = \overset{[\!\rightarrow}{AB}$. By Remark IB.4.1(g), $\overset{[-]}{AV} \subseteq \overset{[\!\rightarrow}{AV}$. Since $U = A$ we have shown $\overset{[-]}{UV} \subseteq \overset{[\!\rightarrow}{AB}$, which is what we wished to prove in this part.

(3) Let U and V be distinct members of $\overset{]-[}{AB}$. We show $\overset{[-]}{UV} \subseteq \overset{]-[}{AB}$. By Property B.2 of Definition IB.1 exactly one of A–U–V, U–A–V, or A–V–U is true.

We show that U–A–V is impossible. We know that both U and V are members of $\overset{]-[}{AB}$ so that by Definition IB.3 A–U–B and A–V–B. If U–A–V by Theorem PSH.8(B) B–U–V and also U–V–B, which are incompatible by Property B.2 of Definition IB.1.

Therefore either A–U–V or A–V–U is true. Choose the notation so A–U–V. Let X be any member of $\overset{]-[}{UV}$, so U–X–V. By Corollary PSH.8.1 A–X–V. Since $V \in \overset{]-[}{AB}$ we have A–V–B. Putting this with A–X–V and using Theorem PSH.8(A)(2), we get A–X–B, which is to say $X \in \overset{]-[}{AB}$. We have shown $\overset{]-[}{UV} \subseteq \overset{]-[}{AB}$, and since we already know that $U \in \overset{]-[}{AB}$ and $V \in \overset{]-[}{AB}$, the desired result is proved.

The proof of parts (4–6) is left to the reader as Exercise PSH.48. □

Theorem PSH.19. *Let A, B, C, and D be points such that $A \neq B$ and $C \neq D$. Then $\overrightarrow{AB} \cap \overrightarrow{CD}$ is a ray iff ($\overrightarrow{AB} \subseteq \overrightarrow{CD}$ or $\overrightarrow{CD} \subseteq \overrightarrow{AB}$).*

Proof. (I) $\overrightarrow{AB} \cap \overrightarrow{CD}$ is a ray $\Rightarrow$ ($\overrightarrow{AB} \subseteq \overrightarrow{CD}$ or $\overrightarrow{CD} \subseteq \overrightarrow{AB}$).

We prove the equivalent statement: ($\overrightarrow{AB} \not\subseteq \overrightarrow{CD}$ and $\overrightarrow{CD} \not\subseteq \overrightarrow{AB}$) $\Rightarrow \overrightarrow{AB} \cap \overrightarrow{CD}$ is not a ray (the contrapositive).

Since $\overrightarrow{AB} \not\subseteq \overrightarrow{CD}$ and $\overrightarrow{CD} \not\subseteq \overrightarrow{AB}$,

there exists a point $S \in \overrightarrow{AB}$ such that $S \notin \overrightarrow{CD}$ (1)

and there exists a point $T \in \overrightarrow{CD}$ such that $T \notin \overrightarrow{AB}$. (2)

By Property B.2 of Definition IB.1 either S–C–T, C–S–T, or C–T–S. We will show that the first is true by showing that the second and third are false. If C–S–T, since C and T are members of $\overrightarrow{CD}$, $S \in \overrightarrow{CD}$ by Theorem PSH.18 (convexity of rays), contradicting (1). Now suppose C–T–S; both $C \in \overrightarrow{CD}$ and $T \in \overrightarrow{CD}$ so $T \in \overset{\circ}{\overrightarrow{CD}}$. By Theorem PSH.16 $\overrightarrow{CD} = \overrightarrow{CT}$. It follows from Definition IB.4 that $S \in \overrightarrow{CD}$, again contradicting (1), so that S–C–T.

Similarly, either S–A–T, A–S–T, or A–T–S. If A–T–S, since A and S are members of $\overrightarrow{AB}$, $T \in \overrightarrow{AB}$ by convexity of rays, contradicting (2). Now suppose A–S–T; both $A \in \overrightarrow{AB}$ and $S \in \overrightarrow{AB}$; so $S \in \overset{\circ}{\overrightarrow{AB}}$ and by Theorem PSH.16 $\overrightarrow{AB} = \overrightarrow{AS}$. It follows from Definition IB.4 that $T \in \overrightarrow{AB}$, again contradicting (2), so that S–A–T; therefore both end points of the rays belong to $\overline{ST}$.

There is no point X of $\overrightarrow{CD}$ such that T–S–X, for if there were, $S \in \overrightarrow{CD}$ because the ray is convex (*cf* Theorem PSH.18), and this contradicts (1). Neither is there a point $Y \in \overrightarrow{AB}$ such that S–T–Y by a similar argument. Therefore there is no point $Z \in \overrightarrow{AB} \cap \overrightarrow{CD}$ with T–S–Z or S–T–Z; neither S nor T is in $\overrightarrow{AB} \cap \overrightarrow{CD}$, so if $Z \in \overrightarrow{AB} \cap \overrightarrow{CD}$, S–Z–T. Therefore by Exercise PSH.4(A), $\overrightarrow{AB} \cap \overrightarrow{CD}$ is not a ray.

(II) ($\overrightarrow{AB} \subseteq \overrightarrow{CD}$ or $\overrightarrow{CD} \subseteq \overrightarrow{AB}$) $\Rightarrow \overrightarrow{AB} \cap \overrightarrow{CD}$ is a ray.

By set theory, if $\overrightarrow{AB} \subseteq \overrightarrow{CD}$, then $\overrightarrow{AB} \cap \overrightarrow{CD} = \overrightarrow{AB}$, and if $\overrightarrow{CD} \subseteq \overrightarrow{AB}$, then $\overrightarrow{AB} \cap \overrightarrow{CD} = \overrightarrow{CD}$. □

Theorem PSH.20. *Let A, B, P, and Q be points on a Pasch plane such that $A \neq B$, $P \neq Q$, and $\overrightarrow{AB} \cap \overrightarrow{PQ}$ is a ray. If $R \in \overset{\circ}{\overrightarrow{QP}}$, then $\overrightarrow{AB} \cap \overrightarrow{QR}$ is not a ray.*

Proof. Since $\overrightarrow{AB} \cap \overrightarrow{PQ}$ is a ray, by Theorem PSH.19, $\overrightarrow{AB} \subseteq \overrightarrow{PQ}$ or $\overrightarrow{PQ} \subseteq \overrightarrow{AB}$.

(I) If $\overrightarrow{AB} \subseteq \overrightarrow{PQ}$ and $R \in \overset{\circ}{\overrightarrow{QP}}$, then $\overrightarrow{AB} \cap \overrightarrow{QR}$ is not a ray.

By Theorem PSH.17, if $\overrightarrow{AB} \subseteq \overrightarrow{PQ}$, then ($A = P$ and $B \in \overrightarrow{PQ}$) or ($A \in \overset{\circ}{\overrightarrow{PQ}}$ and P–A–B).

(I.1) If $A = P$, $B \in \overset{]\rightarrow}{PQ}$, and $R \in \overset{]\rightarrow}{QP}$, then $\overset{[\rightarrow}{AB} \cap \overset{[\rightarrow}{QR}$ is not a ray.

By Theorem PSH.16, if $A = P$ and $B \in \overset{]\rightarrow}{PQ}$, then $\overset{[\rightarrow}{AB} = \overset{[\rightarrow}{PQ}$, and if $R \in \overset{]\rightarrow}{QP}$, then $\overset{[\rightarrow}{QR} = \overset{[\rightarrow}{QP}$. By Exercise IB.8, $\overset{[\rightarrow}{PQ} \cap \overset{[\rightarrow}{QP} = \overset{[-]}{PQ}$. Hence $\overset{[\rightarrow}{AB} \cap \overset{[\rightarrow}{QR} = \overset{[-]}{PQ}$, which by Definition IB.4 is not a ray.

(I.2) If $A \in \overset{]\rightarrow}{PQ}$, P–A–B, and $R \in \overset{]\rightarrow}{QP}$, then $\overset{[\rightarrow}{AB} \cap \overset{[\rightarrow}{QR}$ is not a ray.

By Theorem PSH.16, if $A \in \overset{]\rightarrow}{PQ}$, then $\overset{[\rightarrow}{PA} = \overset{[\rightarrow}{PQ}$, and if $R \in \overset{]\rightarrow}{QP}$, then $\overset{[\rightarrow}{QR} = \overset{[\rightarrow}{QP}$. By Theorem PSH.13, if P–A–B, then $\overset{[\rightarrow}{AB} \subseteq \overset{]\rightarrow}{PA} = \overset{]\rightarrow}{PQ}$. By set theory, $\overset{[\rightarrow}{AB} \subseteq \overset{[\rightarrow}{PQ}$ implies $\overset{[\rightarrow}{AB} \cap \overset{[\rightarrow}{QR} \subseteq \overset{[\rightarrow}{PQ} \cap \overset{[\rightarrow}{QR} = \overset{[\rightarrow}{PQ} \cap \overset{[\rightarrow}{QP}$. By Exercise IB.8 $\overset{[\rightarrow}{PQ} \cap \overset{[\rightarrow}{QP} = \overset{[-]}{PQ}$. Since $\overset{[\rightarrow}{AB} \cap \overset{[\rightarrow}{QR}$ is a subset of $\overset{[-]}{PQ}$, by the Exercise PSH.4(B), it is not a ray.

(II) If $\overset{[\rightarrow}{PQ} \subseteq \overset{[\rightarrow}{AB}$ and $R \in \overset{]\rightarrow}{QP}$, then $\overset{[\rightarrow}{AB} \cap \overset{[\rightarrow}{QR}$ is not a ray.

By Theorem PSH.17, if $\overset{[\rightarrow}{PQ} \subseteq \overset{[\rightarrow}{AB}$, then ($P = A$ and $Q \in \overset{]\rightarrow}{AB}$) or ($P \in \overset{]\rightarrow}{AB}$ and A–P–Q).

(II.1) If $P = A$, $Q \in \overset{]\rightarrow}{AB}$, and $R \in \overset{]\rightarrow}{QP}$, then $\overset{[\rightarrow}{AB} \cap \overset{[\rightarrow}{QR}$ is not a ray.

The proof of this statement is similar to that of (I.1) above.

(II.2) If $P \in \overset{]\rightarrow}{AB}$, A–P–Q, and $R \in \overset{]\rightarrow}{QP}$, then $\overset{[\rightarrow}{AB} \cap \overset{[\rightarrow}{PR}$ is not a ray.

By Theorem PSH.16, $P \in \overset{]\rightarrow}{AB}$ implies $\overset{[\rightarrow}{AP} = \overset{[\rightarrow}{AB}$ and $R \in \overset{]\rightarrow}{QP}$ implies $\overset{[\rightarrow}{QR} = \overset{[\rightarrow}{QP}$. By Definition IB.4, A–P–Q implies $A \in \overset{]\rightarrow}{QP}$ and $Q \in \overset{]\rightarrow}{AP}$. Hence by Theorem PSH.16 $\overset{[\rightarrow}{QA} = \overset{[\rightarrow}{QP}$ and $\overset{[\rightarrow}{AQ} = \overset{[\rightarrow}{AP}$. By Exercise IB.8 $\overset{[\rightarrow}{AQ} \cap \overset{[\rightarrow}{QR} = \overset{[-]}{AQ}$. Thus summarizing the above statements, $\overset{[\rightarrow}{AB} \cap \overset{[\rightarrow}{QR} = \overset{[\rightarrow}{AQ} \cap \overset{[\rightarrow}{QA} = \overset{[-]}{AQ}$. By Exercise PSH.4(B) $\overset{[\rightarrow}{AB} \cap \overset{[\rightarrow}{QR}$ is not a ray. □

Corollary PSH.20.1. *Let A, B, P, and Q be points on a Pasch plane $\mathcal{P}$ such that $A \neq B$, $P \neq Q$, and $\overset{[\rightarrow}{AB} \cap \overset{[\rightarrow}{PQ}$ is a ray. If R is a point distinct from Q such that $\overset{[\rightarrow}{AB} \cap \overset{[\rightarrow}{QR}$ is a ray, then P–Q–R.*

We undertake a detailed proof of this corollary as an illustration of the way the contrapositive may be used in proofs.

Proof. We rewrite both statements in such a way that each is the contrapositive of the other.

Theorem PSH.20 says that $\overset{[\rightarrow}{AB} \cap \overset{[\rightarrow}{PQ}$ is a ray, so by Theorem PSH.19, $\overset{[\rightarrow}{AB} \subseteq \overset{[\rightarrow}{PQ}$ or $\overset{[\rightarrow}{PQ} \subseteq \overset{[\rightarrow}{AB}$; hence A, B, P, and Q are collinear. It also says $R \in \overset{]\rightarrow}{QP}$ so that R is collinear with A, B, P, and Q and $R \neq Q$. Thus it does not change the meaning of Theorem PSH.20 to re-write it as follows:

Let A, B, P, Q, and R be collinear points on a Pasch plane such that $A \neq B$, $P \neq Q$, $R \neq Q$ and $\overset{[\rightarrow}{AB} \cap \overset{[\rightarrow}{PQ}$ is a ray. If $R \in \overset{]\rightarrow}{QP}$, then $\overset{[\rightarrow}{AB} \cap \overset{[\rightarrow}{QR}$ is not a ray.

Now the corollary says (as does the Theorem) that $\overrightarrow{AB} \cap \overrightarrow{PQ}$ is a ray, so again A, B, P, and Q are collinear. It also says $\overrightarrow{AB} \cap \overrightarrow{QR}$ is a ray so R is collinear with A, B, Q, and P.

For such an R, P–Q–R means that $R \notin \overrightarrow{QP}$. For $R \in \overrightarrow{QP}$ by Definition IB.4 means that either $R = P$, Q–R–P, or Q–P–R. If P–Q–R, by Property B.2 of Definition IB.1 none of these can be true so $R \notin \overrightarrow{QP}$. Conversely, since $R \neq Q$ by Property B.2 exactly one of Q–R–P, Q–P–R, or P–Q–R is true; if $R \notin \overrightarrow{QP}$ both Q–R–P and Q–P–R are false, so P–Q–R is true.

The corollary now reads *Let A, B, P, Q and R be collinear points on a Pasch plane $\mathcal{P}$ such that $A \neq B$, $P \neq Q$, $R \neq Q$, and $\overrightarrow{AB} \cap \overrightarrow{PQ}$ is a ray. If $\overrightarrow{AB} \cap \overrightarrow{QR}$ is a ray, then $R \notin \overrightarrow{QP}$.*

The first sentence in Theorem PSH.20 is identical to that in the corollary; the second sentence of each is the contrapositive of the other, and thus the theorem and its corollary are logically equivalent. □

Theorem PSH.21. (A) *Let A, B, and C be points such that A–B–C. Then $\overset{\vdash\dashv}{AB} \subseteq \overset{\vdash\dashv}{AC}$.*

(B) *Let A and B be distinct points. If $C \in \overrightarrow{AB}$, then $\overset{\vdash\dashv}{AB} \cap \overset{\vdash\dashv}{AC}$ is an open segment.*

(C) *Let A, B, and C be points such that B–A–C, and let D be any member of $\overleftrightarrow{BC} \setminus \{A\}$. Then $\overset{\vdash\dashv}{AD} \cap \overset{\vdash\dashv}{BC}$ is an open segment.*

We remind the reader that according to Definition IB.3 an open segment is always nonempty. Thus part (C) of the above theorem says that if one of two open segments has an end point belonging to the other, then the intersection of the two segments is nonempty, and is an open segment.

Proof. (A) This is Corollary PSH.8.5, and is included here for completeness.

(B) If $B = C$, then $\overset{\vdash\dashv}{AB} \cap \overset{\vdash\dashv}{AC} = \overset{\vdash\dashv}{AB}$, which is an open segment. If $B \neq C$, then by Definition IB.4, either A–B–C, or A–C–B. If A–B–C, then by part (I), $\overset{\vdash\dashv}{AB} \subseteq \overset{\vdash\dashv}{AC}$, and hence $\overset{\vdash\dashv}{AB} \cap \overset{\vdash\dashv}{AC} = \overset{\vdash\dashv}{AB}$, an open segment. If A–C–B, then by part (A), $\overset{\vdash\dashv}{AB} \cap \overset{\vdash\dashv}{AC} = \overset{\vdash\dashv}{AC}$, again, an open segment.

(C) By Theorem PSH.15(D), elementary set theory, and the fact that $\overset{\vdash\dashv}{AD} \cap \{A\} = \emptyset$,

$$\overset{\vdash\dashv}{AD} \cap \overset{\vdash\dashv}{BC} = \overset{\vdash\dashv}{AD} \cap (\overset{\vdash\dashv}{AB} \cup \overset{\vdash\dashv}{AC} \cup \{A\}) = (\overset{\vdash\dashv}{AD} \cap \overset{\vdash\dashv}{AB}) \cup (\overset{\vdash\dashv}{AD} \cap \overset{\vdash\dashv}{AC}) \quad (1)$$

By Theorem PSH.15(B), $\overrightarrow{AB} \cup \overrightarrow{AC} = \overleftrightarrow{BC} \setminus \{A\}$ and $\overrightarrow{AB} \cap \overrightarrow{AC} = \emptyset$. Since $D \in \overleftrightarrow{BC} \setminus \{A\}$, either $D \in \overrightarrow{AB}$ or $D \in \overrightarrow{AC}$.

Suppose $D \in \overrightarrow{AB}$. By Theorem PSH.16, $\overrightarrow{AD} = \overrightarrow{AB}$ and by Remark IB.4.1(h) $\overset{\vdash\dashv}{AD} \subseteq \overrightarrow{AD} = \overrightarrow{AB}$ which is disjoint from $\overrightarrow{AC}$ and thus from $\overset{\vdash\dashv}{AC}$. Hence

$\overset{\dashv\vdash}{AD} \cap \overset{\dashv\vdash}{AC} = \emptyset$, and (1) becomes $\overset{\dashv\vdash}{AD} \cap \overset{\dashv\vdash}{BC} = \overset{\dashv\vdash}{AD} \cap \overset{\dashv\vdash}{AB}$, which is an open segment by part (B), since $D \in \overset{\dashv\to}{AB}$.

The case where $D \in \overset{\dashv\to}{AC}$ can be treated similarly. □

Theorem PSH.22 (Denseness property). *If A and B are distinct points, then there exists a point C such that A–C–B. That is, $\overset{\dashv\vdash}{AB} \neq \emptyset$.*

Proof. By Axiom I.5(B) there exists a point D not belonging to $\overleftrightarrow{AB}$. By Property B.3 of Definition IB.1 there exists a point E such that A–D–E and a point F such that E–B–F. Note that F cannot be on the line $\overleftrightarrow{AE}$ because then $A = B$; the intersections of the lines $\overleftrightarrow{DF}$ and $\overleftrightarrow{EF}$ with $\overleftrightarrow{AE}$ are not the same, and so they are distinct lines. Hence by Axiom I.1 they have only one point (F) in common.

Also, by Property B.2 it is false that E–F–B, since it is true that E–B–F. Therefore by Definition IB.3 $\overset{\dashv\vdash}{EB}$ has no point in common with $\overleftrightarrow{DF}$, and $F \neq B$ so that $B \notin \overleftrightarrow{DF}$.

Now $\overleftrightarrow{DF} \cap \overset{\dashv\vdash}{AE} = \{D\} \neq \emptyset$ and it follows from the Postulate of Pasch (which applies by Theorem PSH.6) that the line $\overleftrightarrow{DF}$ must intersect $\overset{\dashv\vdash}{AB}$ at some point C, and by Definition IB.3 A–C–B. □

Corollary PSH.22.1. *Let $\overset{\dashv\vdash}{AB}$ be any open segment. For every natural number n, there exists a subset $\mathcal{B} \subseteq \overset{\dashv\vdash}{AB}$ containing $2^n - 1$ points, that is, there exists a bijection $f : \{1, 2, 3, \ldots, 2^n - 1\} \to \mathcal{B}$.*

Proof. We give a proof by induction. By Theorem PSH.21 there exists a point belonging to $\overset{\dashv\vdash}{AB}$, so the assertion is true for $n = 1$. Suppose now it is true for $n = k$, i.e. $\overset{\dashv\vdash}{AB}$ contains a subset with $2^k - 1$ points. We call these points $P_1, P_2, \ldots, P_m$, where $m = 2^k - 1$, and we will suppose their names have been chosen so that A–P_1–P_2–…–P_m–B. Now by Theorem PSH.21, there exist points $Q_1, \ldots, Q_{m+1}$ such that A–Q_1–P_1–Q_2–P_2–Q_3–…–Q_m–P_m–Q_{m+1}–B. That is to say, $\overset{\dashv\vdash}{AB}$ contains at least $2m + 1 = 2(2^k - 1) + 1 = 2^{k+1} - 2 + 1 = 2^{k+1} - 1$ points. We have shown that if the statement in the theorem is true for $n = k$, then it is true for $n = k + 1$, which completes the induction. □

Corollary PSH.22.2. *Every open segment $\overset{\dashv\vdash}{AB}$ contains an infinite number of points.*

Proof. Suppose $\overset{\dashv\vdash}{AB}$ is a finite set having n elements; by two applications of Theorem PSH.22 there exist points M and M' of $\overset{\dashv\vdash}{AB}$ such that A–M–M'–B. Thus if there are n elements of $\overset{\dashv\vdash}{AB}$, $n \geq 2$. By Corollary PSH.22.1, there exists a subset $\mathcal{B}$

of $\overset{]\!-\!-\![}{AB}$ having $2^n - 1$ elements; if $\mathcal{B}$ is a proper subset of $\overset{]\!-\!-\![}{AB}$, $2^n - 1 < n$; if $\mathcal{B} = \overset{]\!-\!-\![}{AB}$, then $2^n - 1 = n$; in either case, $2^n - 1 \leq n$ which is false. Therefore $\overset{]\!-\!-\![}{AB}$ is an infinite set. □

Corollary PSH.22.3. *Every convex set which is not a singleton is infinite.*

Proof. This is immediate from Corollary PSH.22.2 above and Definition IB.9. □

5.5 Uniqueness of endpoints and edges

We said in IB geometry that it is possible to have a single ray with more than one endpoint or a single segment with two different sets of endpoints. The next set of theorems proves these situations can't happen in Pasch geometry.

The thoughtful reader might ask why it is necessary to prove that the endpoint of a ray is unique, or the set of endpoints of a segment is unique. After all, these matters seem "intuitively obvious," being well specified in Definitions IB.3 and IB.4. The answer lies in the fact that we may define rays and segments either geometrically, as we do in their definitions, or as sets of points without reference to endpoints.

If in a proof, two segments are shown to have exactly the same points it may become important to know that the endpoints of one are the same as the endpoints of the other. The particular circumstances of a given proof may or may not make it easy to show this; but it would be better to have settled the issue once and for all.

The method we use to do this is to first identify those points designated as endpoints in the definitions of rays and segments; then prove (in Theorem PSH.23) that these endpoints have certain properties, and finally (Theorems PSH.24 and PSH.25) prove that no points other than the original endpoints have those properties.

In an analogous manner, Theorem PSH.32 proves certain properties of corners (of angles, triangles, or quadrilaterals) and Theorems PSH.33, PSH.34, and PSH.35 show that no other points have these properties.

Theorem PSH.23. *Let $\mathcal{P}$ be a Pasch plane, let $\mathcal{E}$ be a ray or a segment on $\mathcal{P}$, let $\mathcal{L}$ be the line containing $\mathcal{E}$, and let U be an endpoint of $\mathcal{E}$. Then there exist points V and W such that $W{-}U{-}V$, $\overset{]\!-\!-\![}{UV} \subseteq \mathcal{E}$, and $\overset{]\!-\!-\![}{WU} \subseteq \mathcal{L} \setminus \mathcal{E}$.*

Proof. If $\mathcal{E}$ is a ray, let V be any point of the ray other than U. If $\mathcal{E}$ is a segment, let V be the other endpoint of $\mathcal{E}$. Definitions IB.3, IB.4, and Theorem PSH.16 show that exactly one of the following statements is true:

(1) $\mathcal{E} = \overset{]\!\longrightarrow}{UV}$, (2) $\mathcal{E} = \overset{[\!\longrightarrow}{UV}$, (3) $\mathcal{E} = \overset{]\!-\!-\![}{UV}$, (4) $\mathcal{E} = \overset{[\!-\!-\![}{UV}$, (5) $\mathcal{E} = \overset{]\!-\!-\!]}{UV}$, (6) $\mathcal{E} = \overset{[\!-\!-\!]}{UV}$.

(I) By Remark IB.4.1, $\overline{UV}$ is a subset of each of the above six sets.

(II) By Properties B.1 and B.3 of Definition IB.1 there exists a point W such that W–U–V. Let Y be any member of $\overline{WU}$, so that by Definition IB.3 W–Y–U. By Theorem PSH.8(A), W–Y–U and W–U–V imply Y–U–V. By Property B.0 $Y \in \mathcal{L}$, but referring to Definitions IB.3 and IB.4, we see Y doesn't belong to any of the six sets enumerated above. Hence $Y \in (\mathcal{L} \setminus \mathcal{E})$. That is to say, $\overline{WU} \subseteq \mathcal{L} \setminus \mathcal{E}$. □

Theorem PSH.24. *Let A and B be distinct points on a Pasch plane $\mathcal{P}$.*

(A) *A is the unique endpoint of $\overrightarrow{AB}$ and of $\overrightarrow{AB}$; and*

(B) *if A, B, C, and D are points on $\mathcal{P}$ such that $A \neq B$, $C \neq D$, and $\overrightarrow{AB} = \overrightarrow{CD}$, then $A = C$ and $B \in \overrightarrow{CD}$.*

Proof. (A) Let $\mathcal{E} = \overrightarrow{AB}$ or $\mathcal{E} = \overrightarrow{AB}$. We note first that if U is an endpoint of $\mathcal{E}$, then $U \in \overleftrightarrow{AB}$. For by Exercise PSH.47(A), if $U \notin \overleftrightarrow{AB}$, then for every point V in $\mathcal{P}$, $\overrightarrow{UV}$ is not a subset of $\overleftrightarrow{AB}$. By Theorem PSH.15, $\mathcal{E} \subseteq \overleftrightarrow{AB}$, so by Definition IB.4, U is not an endpoint of $\mathcal{E}$. Hence our task is to prove: if U is any member of $\overleftrightarrow{AB} \setminus \{A\}$, then U is not an endpoint of $\mathcal{E}$.

By Properties B.1 and B.3 of Definition IB.1 there exists a point C such that C–A–B. By Theorem PSH.15 $\overrightarrow{AB} \cap \overrightarrow{AC} = \emptyset$. Assume that U is an endpoint for $\overrightarrow{AB}$ or $\overrightarrow{AB}$, and $U \neq A$; by Theorem PSH.15 either $U \in \overrightarrow{AB}$ or $U \in \overrightarrow{AC}$.

(Case 1: $U \in \overrightarrow{AB}$.) By Theorem PSH.22 there exists a point P such that A–P–U, and by Property B.3 there exists a point Q such that A–U–Q. By Theorem PSH.8(A) A–P–U–Q. By Definition IB.4 both $P \in \overrightarrow{AU}$ and $Q \in \overrightarrow{AU}$, so by Theorem PSH.18 (convexity) $\overline{PQ} \subseteq \overrightarrow{AU}$. Also by Theorem PSH.16 $\overrightarrow{AU} = \overrightarrow{AB}$.

Since U is an endpoint of $\mathcal{E}$, then by Theorems PSH.23 and PSH.15, there is a point W such that $\overline{WU} \subseteq \overrightarrow{AC}$.

Since P–U–Q and $W \neq U$, we may apply Theorem PSH.21 to conclude that $\overline{UW} \cap \overline{PQ}$ is a (nonempty) open segment. And since $\overline{PQ} \subseteq \overrightarrow{AB}$, $\overline{UW} \cap \overline{PQ} \subseteq \overrightarrow{AB}$, which is disjoint from $\overrightarrow{AC}$, and hence from $\overline{UW}$ so we have a contradiction, showing that U is not an endpoint for $\mathcal{E}$.

(Case 2: $U \in \overrightarrow{AC}$.) Since U is an endpoint of $\mathcal{E}$, by Theorem PSH.23 there is a point V such that $\overline{UV} \subseteq \mathcal{E} \subseteq \overrightarrow{AB}$.

By Property B.3 we may let X be a point such that X–U–A. By Theorem PSH.21 $\overline{UV} \cap \overline{AX}$ is a nonempty open segment. Since $X \in \overrightarrow{AC}$, $\overrightarrow{AX} = \overrightarrow{AC}$

and hence by Remark IB.4.1(h) $\overset{]-[}{AX} \subseteq \overset{]\to}{AX} = \overset{]\to}{AC}$ and $\overset{]-[}{UV} \cap \overset{]-[}{AX} \subseteq \overset{]\to}{AC}$, which is disjoint from $\overset{]\to}{AB}$, hence from $\overset{]-[}{UV}$. This is a contradiction, so again, U is not an endpoint for $\mathcal{E}$.

Therefore no point other than A can be an endpoint for $\overset{]\to}{AB}$ or $\overset{[\to}{AB}$.

(B) Asserting $A = C$ is simply a restatement of part (A). Since $B \neq A$ and $A = C$, $B \neq C$. But then by Definition IB.4, $B \in \overset{]\to}{CD}$. □

We may henceforth refer to *the* endpoint of a ray.

Theorem PSH.25. *If A and B are distinct points on a Pasch plane $\mathcal{P}$, then $\{A, B\}$ is the set of endpoints of each of the segments $\overset{]-[}{AB}$, $\overset{]-]}{AB}$, $\overset{[-[}{AB}$, and $\overset{[-]}{AB}$. This can also be stated as follows: If A, B, C, and D are points on a Pasch plane $\mathcal{P}$ such that $A \neq B$, $C \neq D$, and $\overset{[-]}{AB} = \overset{[-]}{CD}$, then $\{A, B\} = \{C, D\}$.*

Proof. By Exercise PSH.47(B), if a point U does not belong to the line $\overleftrightarrow{AB}$, then it is not an endpoint of any of the segments in the theorem. Therefore, yet another way of stating the theorem is: *If U is any member of $\overleftrightarrow{AB} \setminus \{A, B\}$, then U is not an endpoint of any of the segments $\overset{]-[}{AB}$, $\overset{]-]}{AB}$, $\overset{[-[}{AB}$, or $\overset{[-]}{AB}$.*

Let $\mathcal{E} = \overset{]-[}{AB}$, or $\mathcal{E} = \overset{]-]}{AB}$, or $\mathcal{E} = \overset{[-[}{AB}$, or $\mathcal{E} = \overset{[-]}{AB}$, and assume U is an endpoint of $\mathcal{E}$ different from A or B. By Property B.3 of Definition IB.1 there exist points C and D such that C–A–B (or B–A–C) and A–B–D. By Theorem PSH.15(B), $\overleftrightarrow{AB} = \overset{]\to}{BA} \cup \overset{]\to}{BD} \cup \{B\}$, and the sets in this union are disjoint. Since B–A–C, by Theorem PSH.13(D) and Exercise IB.3, $\overset{]\to}{BA} = \{A\} \cup \overset{]-[}{AB} \cup \overset{]\to}{AC}$, and these sets are disjoint by Theorem PSH.15(B). Putting this into the preceding equality gives $\overleftrightarrow{AB} = \{A, B\} \cup \overset{]\to}{AC} \cup \overset{]-[}{AB} \cup \overset{]\to}{BD}$ and the sets in this union are disjoint.

Since U is an endpoint of $\mathcal{E}$ and $U \notin \{A, B\}$; there are three cases: $U \in \overset{]-[}{AB}$, $U \in \overset{]\to}{AC}$, and $U \in \overset{]\to}{BD}$. We show that these are all impossible.

(Case I: $U \in \overset{]-[}{AB}$.) By Theorem PSH.23 there exists a point W such that $\overset{]-[}{WU} \subseteq \overleftrightarrow{AB} \setminus \mathcal{E} \subseteq \overleftrightarrow{AB} \setminus \overset{]-[}{AB}$, that is, $\overset{]-[}{WU} \cap \overset{]-[}{AB} = \emptyset$. On the other hand, since A–U–B, and $W \neq U$, we may apply Theorem PSH.21(C) to conclude that $\overset{]-[}{WU} \cap \overset{]-[}{AB}$ is an open segment, and thus nonempty. This contradicts the previous sentence; therefore U cannot be an endpoint for $\mathcal{E}$.

(Case II: $U \in \overset{]\to}{AC}$.) By Theorem PSH.16 $\overset{]\to}{AC} = \overset{]\to}{AU}$. By Property B.3 of Definition IB.1 there exists a point P such that P–U–A; then $P \in \overset{]\to}{AU}$ so $\overset{]\to}{AP} = \overset{]\to}{AU} = \overset{]\to}{AC}$.

By Theorem PSH.23 there exists a point $V \neq U$ such that $\overset{\circ}{\overline{UV}} \subseteq \mathcal{E} \subseteq \overline{AB}$. By Remark IB.4.1(d) and (h) $\overline{AB} \subseteq \overrightarrow{AB}$ and by Theorem PSH.15(B) $\overrightarrow{AB} \cap \overset{\circ}{\overrightarrow{AC}} = \emptyset$, so $\overset{\circ}{\overline{UV}}$ and $\overset{\circ}{\overrightarrow{AC}}$ are disjoint.

On the other hand, since *P*–*U*–*A*, and $V \neq U$, we may apply Theorem PSH.21(C) to conclude that $\overset{\circ}{\overline{UV}} \cap \overset{\circ}{\overline{AP}}$ is an open segment, and thus nonempty. But by Remark IB.4.1(h) $\overset{\circ}{\overline{AP}} \subseteq \overset{\circ}{\overrightarrow{AP}} = \overset{\circ}{\overrightarrow{AC}}$ so that $\overset{\circ}{\overline{UV}} \cap \overset{\circ}{\overrightarrow{AC}}$ is nonempty. This contradicts the fact that $\overset{\circ}{\overline{UV}}$ and $\overset{\circ}{\overrightarrow{AC}}$ are disjoint, and therefore *U* cannot be an endpoint for $\mathcal{E}$.

(Case III: $U \in \overset{\circ}{\overrightarrow{BD}}$.) The proof in this case is just like that for Case II and can be obtained from it by replacing every "*A*" by "*B*" and every "*C*" by "*D*." □

We may henceforth refer to *the* endpoints of a segment.

The next theorem goes back to the scenario involving a line $\mathcal{L}$ in a Pasch plane $\mathcal{E}$ and the two sides of $\mathcal{L}$. It confirms some concepts we ordinarily carry in our intuitive picture of this situation. In particular it assures us we may refer to *the* edge of a given halfplane.

Theorem PSH.26. *If $\mathcal{L}$ is a line in plane $\mathcal{P}$, and if $\mathcal{H}$ is a side of $\mathcal{L}$, then:*

(A) *$\mathcal{H}$ contains at least three noncollinear points; and*
(B) *$\mathcal{L}$ is uniquely determined by $\mathcal{H}$, i.e., there exists no line distinct from $\mathcal{L}$ which is an edge of $\mathcal{H}$.*

Proof. (A) By Axiom I.5 there exist distinct points *P* and *Q* on $\mathcal{L}$. By Definition IB.11 there exists a point *R* belonging to $\mathcal{H}$. By Theorem IB.14 $\overset{\circ}{\overrightarrow{PR}}$ and $\overset{\circ}{\overrightarrow{QR}}$ are each contained in $\mathcal{H}$. By Theorem PSH.22 (denseness) there exist points *S* and *T* such that *P*–*S*–*R* and *Q*–*T*–*R*. By Definition IB.4 $S \in \overset{\circ}{\overrightarrow{PR}}$ and $T \in \overset{\circ}{\overrightarrow{QR}}$. Hence $S \in \mathcal{H}$ and $T \in \mathcal{H}$ by Theorem IB.14.

By Theorem IB.5 $S \in \overleftrightarrow{PR}$. Now if *R*, *S*, and *T* were collinear we would have $T \in \overleftrightarrow{PR}$. Since $T \in \overleftrightarrow{QR}$ by definition we would then have $T \in \overleftrightarrow{PR} \cap \overleftrightarrow{QR}$. But by Exercise I.1, $\overleftrightarrow{PR} \cap \overleftrightarrow{QR} = \{R\}$. This contradicts the definition of *T* so our assumption that *R*, *S*, and *T* are collinear is false, hence they are noncollinear.

(B) Let $\mathcal{H}$ be a side of a line $\mathcal{L}$, and suppose $\mathcal{L}' \neq \mathcal{L}$ is another edge of $\mathcal{H}$. Since $\mathcal{L} \neq \mathcal{L}'$, there is at least one point *P* on $\mathcal{L}'$ which is not on $\mathcal{L}$. By Theorem PSH.12 $\mathcal{L}$ has another side $\mathcal{H}^*$. We have assumed that $\mathcal{L}'$ is an edge of $\mathcal{H}$, so by Definition IB.11, $P \notin \mathcal{H}$. By Theorem PSH.12, $P \in \mathcal{H}^*$. Also by Theorem PSH.12 there exists a point *Q* in $\mathcal{H}$ such that for some point *R*, $\overset{\circ}{\overline{PQ}} \cap \mathcal{L} = \{R\}$, since *P* and *Q* are on different sides of $\mathcal{L}$.

By Theorem IB.14, $\overrightarrow{PQ} \subseteq$ the Q-side of $\mathcal{L}'$ and since $R \in \overline{PQ} \subseteq \overrightarrow{PQ}$, R belongs to the Q-side of $\mathcal{L}' = \mathcal{H}$. Here we have used Theorem PSH.13(D). Since $Q \in \mathcal{H}$ (the Q-side of $\mathcal{L}'$), $R \in \mathcal{H}$ by Theorem PSH.12. But since $R \in \mathcal{L}$, and $\mathcal{L}$ is an edge of $\mathcal{H}$, $R \notin \mathcal{H}$, a contradiction.

Hence our assumption $\mathcal{L} \neq \mathcal{L}'$ is false. □

5.6 Uniqueness of corners of angles, *etc*

Definition PSH.27. Let A, B, C, and D be points such that $A \neq B$ and $C \neq D$. The rays $\overrightarrow{AB}$ and $\overrightarrow{CD}$ are **opposite** iff $A = C$ and B–A–D.

Theorem PSH.28. *Let A, B, and C be distinct points such that* $\overrightarrow{AB} \neq \overrightarrow{AC}$. *Then* $\overrightarrow{AB}$ *and* $\overrightarrow{AC}$ *are opposite iff A, B, and C are collinear. This is equivalent to saying that* $\overrightarrow{AB}$ *and* $\overrightarrow{AC}$ *are nonopposite iff A, B, and C are noncollinear.*

Proof. If $\overrightarrow{AB}$ and $\overrightarrow{AC}$ are opposite, then B–A–C by Definition PSH.27, and therefore A, B, and C are collinear. Conversely, if A, B, and C are collinear and $\overrightarrow{AB} \neq \overrightarrow{AC}$, by Theorem PSH.16 $B \notin \overrightarrow{AC}$ and $C \notin \overrightarrow{AB}$. Therefore neither A–B–C nor$A$–C–B so that by the Trichotomy Property B.2 of Definition IB.1, B–A–C. □

Definition PSH.29. An **angle** is the union of two distinct nonopposite rays having the same endpoint. The common endpoint of the rays is the **corner** of the angle.

By Theorem PSH.28, $\mathcal{E}$ is an angle iff there exist noncollinear points A, B, and C such that $\mathcal{E} = \overrightarrow{AB} \cup \overrightarrow{AC}$. In this case, $\mathcal{E}$ is denoted by $\angle BAC$.

Notice how our definition excludes both what are sometimes called a "straight angle" and a "zero angle" in high-school geometry. What is usually called a "reflex angle" could be considered as a kind of "complement" to what we call an angle. The *inside* of a reflex angle would correspond to the *outside* of the "complementary" angle. Here we are using the terminology to be introduced in Definition PSH.36.

Theorem PSH.30. *If A, B, and C are noncollinear points and if D and E are points such that* $D \in \overrightarrow{AB}$ *and* $E \in \overrightarrow{AC}$, *then* $\angle DAE = \angle BAC$.

Proof. This theorem is an immediate consequence of Theorem PSH.16 and Definition PSH.29. □

Definition PSH.31. A subset $\mathcal{E}$ of a Pasch plane $\mathcal{P}$ is a **quadrilateral** iff there exist points A, B, C, and D on $\mathcal{P}$ such that all three of the following are true:

(1) each of the triples $\{A, B, C\}$, $\{B, C, D\}$, $\{C, D, A\}$, and $\{D, A, B\}$ is a set of noncollinear points;
(2) $\overline{AB} \cap \overline{CD} = \overline{BC} \cap \overline{DA} = \emptyset$;
(3) $\mathcal{E} = \overline{AB} \cup \overline{BC} \cup \overline{CD} \cup \overline{DA}$.

If $\mathcal{E}$ is a quadrilateral, then it is denoted by $\square ABCD$ or $\langle A, B, C, D\rangle$. A point U is a **corner** of $\mathcal{E}$ iff there exist points V, W, and X such that $\mathcal{E} = \square UVWX$.

A segment $\mathcal{J}$ is an **edge** of $\mathcal{E}$ iff there exist corners U and V in $\mathcal{E} = \square ABCD$ such that $\mathcal{J} = \overline{UV} \subseteq \mathcal{E}$. Edges $\mathcal{J}$ and $\mathcal{K}$ in a given representation of $\mathcal{E}$ are **opposite** iff $\mathcal{J} \cap \mathcal{K} = \emptyset$.

Corners U and W are **opposite** iff $\overset{\circ}{UW} \cap \mathcal{E} = \emptyset$. A segment $\mathcal{J}$ is a **diagonal** of $\mathcal{E}$ iff there exist corners U and W of $\mathcal{E}$ such that U and W are opposite and $\mathcal{J} = \overline{UW}$.

$\mathcal{E}$ is **rotund** iff for every line $\mathcal{L}$ containing an edge of $\mathcal{E}$, the corners of $\mathcal{E}$ not on $\mathcal{L}$ are on the same side of $\mathcal{L}$.

A quadrilateral $\mathcal{E}$ is a **trapezoid** iff there exist opposite edges $\mathcal{E}_1$ and $\mathcal{E}_2$ of $\mathcal{E}$ such that if $\mathcal{L}_1$ is the line containing $\mathcal{E}_1$ and $\mathcal{L}_2$ is the line containing $\mathcal{E}_2$, then $\mathcal{L}_1 \parallel \mathcal{L}_2$.

Theorem PSH.32. *Let $\mathcal{E}$ be an angle, a triangle, or a quadrilateral in a Pasch plane $\mathcal{P}$, and let U be a corner of $\mathcal{E}$. Then there exist points V and W such that U, V, and W are noncollinear, and $\overline{UV} \cup \overline{UW} \subseteq \mathcal{E}$. Furthermore, there exist points V' and W' such that V'–U–V, W'–U–W, and $\overset{\circ}{UV'} \cup \overset{\circ}{UW'} \subseteq \mathcal{P} \setminus \mathcal{E}$. That is, $(\overset{\circ}{UV'} \cup \overset{\circ}{UW'}) \cap \mathcal{E} = \emptyset$.*

Proof. (Case 1: $\mathcal{E}$ is an angle and U is a corner of $\mathcal{E}$.) By the remark following Definition PSH.29 there exist points V and W such that U, V, and W are noncollinear and $\mathcal{E} = \angle VUW = \overrightarrow{UV} \cup \overrightarrow{UW}$. By Corollary IB.5.2 $\overline{UV} \cup \overline{UW} \subseteq \overrightarrow{UV} \cup \overrightarrow{UW} = \mathcal{E}$, which is half of what we wished to prove in this case.

By Property B.3 of Definition IB.1 there exist points V' and W' such that V–U–V' and W–U–W'. By Theorem PSH.15 and Remark IB.4.1(f) $\overleftrightarrow{UV} \setminus \overrightarrow{UV} = \overset{\circ}{UV'}$ and $\overleftrightarrow{UW} \setminus \overrightarrow{UW} = \overset{\circ}{UW'}$. Hence

$$\overset{\circ}{UV'} \cup \overset{\circ}{UW'} = (\overleftrightarrow{UV} \cup \overleftrightarrow{UW}) \setminus (\overrightarrow{UV} \cup \overrightarrow{UW}) = \overleftrightarrow{UV} \cup \overleftrightarrow{UW} \setminus \mathcal{E} \subseteq \mathcal{P} \setminus \mathcal{E};$$

here we have used a number of facts about unions, relative complements, and subsets from elementary set theory, as well as the fact that $\overleftrightarrow{UV} \cap \overleftrightarrow{UW} = \{U\}$. By Corollary IB.5.2, $\overset{\circ}{UV'} \cup \overset{\circ}{UW'} \subseteq \mathcal{P} \setminus \mathcal{E}$.

(Case 2: $\mathcal{E}$ is a triangle and U is a corner of $\mathcal{E}$.) By Definition IB.7 there exist points V and W such that U, V, and W are noncollinear and $\mathcal{E} = \overline{UV} \cup \overline{UW} \cup \overline{VW}$. Hence $\overline{UV} \cup \overline{UW} \subseteq \mathcal{E}$, which is the first statement we wished to prove in this case.

By Property B.3 of Definition IB.1 there exist points V' and W' such that V–U–V' and W–U–W'. By Theorem PSH.15 and Remark IB.4.1(f), $\overleftrightarrow{UV} \setminus \overrightarrow{UV} = \mathring{\overrightarrow{UV'}}$ and $\overleftrightarrow{UW} \setminus \overrightarrow{UW} = \mathring{\overrightarrow{UW'}}$. By Corollary IB.5.2 $\overline{UV} \cup \overline{UW} \subseteq \overrightarrow{UV} \cup \overrightarrow{UW}$ and $\mathring{\overline{UV'}} \cup \mathring{\overline{UW'}} \subseteq \mathring{\overrightarrow{UV'}} \cup \mathring{\overrightarrow{UW'}}$. Using an argument similar to Case 1 above we get $\mathring{\overline{UV'}} \cup \mathring{\overline{UW'}} \subseteq \mathcal{P} \setminus \mathcal{E}$.

(Case 3: $\mathcal{E}$ is a quadrilateral and U is a corner of $\mathcal{E}$.) By Definition PSH.31 there exist points U, V, W, and X such that U, V, and W are noncollinear; V, W, and X are noncollinear; W, X, and U are noncollinear; $\overline{UV} \cap \overline{XW} = \overline{UX} \cap \overline{VW} = \emptyset$; and $\mathcal{E} = \overline{UV} \cup \overline{VW} \cup \overline{WX} \cup \overline{XU}$. Hence $\overline{UV} \cup \overline{UX} \subseteq \mathcal{E}$, which completes the first part of the proof in this case.

We now show that there exists a point V' such that V–U–V' and $\mathring{\overline{UV'}} \subseteq (\mathcal{P} \setminus \mathcal{E})$. First note by the noncollinearity conditions above and Exercise I.1 $\overleftrightarrow{UV} \cap \overleftrightarrow{UX} = \{U\}$ and $\overleftrightarrow{UV} \cap \overleftrightarrow{VW} = \{V\}$. By Property B.3 of Definition IB.1 there exists a point T such that V–U–T. If W and X are on the same side of $\overleftrightarrow{UV}$, then by Theorem PSH.12 $\overline{WX}$ is a subset of a side of $\overleftrightarrow{UV}$, so $\overline{WX} \cap \overleftrightarrow{UV} = \emptyset$. If we put these statements together we get $\mathcal{E} \cap \overleftrightarrow{UV} = \overline{UV}$. By Theorem PSH.15(C) $\overline{UV} \cap \mathring{\overline{UT}} = \emptyset$. Hence if we take $V' = T$, we have $\mathring{\overline{UV'}} \subseteq (\mathcal{P} \setminus \mathcal{E})$.

If W and X are on opposite sides of $\overleftrightarrow{UV}$, then by Theorem PSH.12 there exists a point S such that $\overleftrightarrow{UV} \cap \mathring{\overline{WX}} = \{S\}$. Then by Property B.2 of Definition IB.1 S–U–V, or U–S–V, or U–V–S. But the second of these possibilities is ruled out by the definition of a quadrilateral (PSH.31). The same definition tells us $\mathcal{E} \cap \overleftrightarrow{UV} = \overline{UV} \cup \{S\}$. Now if U–V–S we take V' to be the point T defined above, so that by Theorem PSH.8(B) V'–U–V–S and by virtue of what has just been said, $\mathring{\overline{UV'}} \subseteq \mathcal{P} \setminus \mathcal{E}$, and we are done. In case S–U–V we know by Theorem PSH.22 (denseness) there is a point V' such that S–V'–U. Then by Theorem PSH.8(A) S–V'–U–V, and by reasoning similar to that above, we again get $\mathring{\overline{UV'}} \subseteq \mathcal{P} \setminus \mathcal{E}$. □

Theorem PSH.33 (The corner of an angle is unique). *Every angle has exactly one corner.*

Proof. By Definition PSH.29, an angle is the union of two distinct nonopposite rays having the same endpoint. The corner of the angle is defined to be this point of intersection.

Each of these rays defines a line; the two lines so defined are distinct because the rays defining them are distinct and nonopposite; they must intersect because the rays intersect. A corner of the angle must be a point of intersection of the rays, and therefore a point of intersection of the two lines. By Exercise I.1 there is only one such point. □

Theorem PSH.34. *Let A, B, and C be noncollinear points on a Pasch plane $\mathcal{P}$; then the set of corners of $\triangle ABC$ is $\{A, B, C\}$.*

Proof. By Definition IB.7, a point U is a corner of a triangle $\triangle ABC$ iff there exist points V and W such that U, V, and W are noncollinear and $\triangle UVW = \triangle ABC$. This shows immediately that A, B, and C are corners of $\triangle ABC$.

If U is any corner of $\triangle ABC$, U, V, and W are noncollinear and $\triangle UVW = \triangle ABC$, then by definition the edge $\overline{UV}$ is a subset of $\triangle ABC$. By two successive applications of Theorem PSH.22 we see that $\overline{UV}$ contains at least four points, so that at least two of them belong to the same line $\overleftrightarrow{AB}$, $\overleftrightarrow{AC}$, or $\overleftrightarrow{BC}$. Therefore by Exercise I.2, $\overline{UV}$ is a subset of one of these lines. Similarly, $\overline{UW}$ is a subset of one of these lines. Since U, V, and W are noncollinear, these segments are subsets of different lines.

It follows that any corner of $\triangle ABC$ belongs to two of the lines $\overleftrightarrow{AB}$, $\overleftrightarrow{AC}$, or $\overleftrightarrow{BC}$. The only points which satisfy this criterion are A, B, and C. Therefore these are the only possible corners for the triangle. □

It is possible to construct a different proof of Theorem PSH.34 using Theorem PSH.32. See Exercise PSH.58.

Theorem PSH.35. *Let A, B, C, and D be points on a Pasch plane $\mathcal{P}$ such that A, B, and C are noncollinear, B, C, D are noncollinear, C, D, and A are noncollinear, D, A, B are noncollinear, and that $\overline{AB} \cap \overline{CD} = \overline{AD} \cap \overline{BC} = \emptyset$. Then the set of corners of $\square ABCD$ is $\{A, B, C, D\}$.*

Proof. All the points A, B, C, and D are corners of $\square ABCD$ by Definition PSH.31, and every corner of $\square ABCD$ is a point of $\square ABCD$. Therefore all that is needed to prove the theorem is to show that no member of $\mathring{AB} \cup \mathring{BC} \cup \mathring{CD} \cup \mathring{DA}$ is a corner of $\square ABCD$.

Choose the notation so that $U \in \mathring{AB}$, that is, A–U–B, and assume U is a corner of $\square ABCD$. From this point on the proof is identical to that of Theorem PSH.34. □

5.7 Mostly about angles

Definition PSH.36. Let A, B, and C be noncollinear points in a Pasch plane $\mathcal{P} = \overleftrightarrow{ABC}$.

(A) The **inside of** $\angle BAC$ (notation: ins $\angle BAC$) is $\overrightarrow{ABC} \cap \overrightarrow{ACB}$.
The **outside of** $\angle BAC$ (out $\angle BAC$) is $\overleftrightarrow{ABC} \setminus (\angle BAC \cup \text{ins}\,\angle BAC)$.
The **enclosure of** $\angle BAC$ (enc $\angle BAC$) is $\angle BAC \cup \text{ins}\,\angle BAC$.
The **exclosure of** $\angle BAC$ (exc $\angle BAC$) is $\angle BAC \cup \text{out}\,\angle BAC$.

(B) The **inside of** $\triangle ABC$ (ins $\triangle ABC$) is $\overrightarrow{ABC} \cap \overrightarrow{ACB} \cap \overrightarrow{BCA}$.
The **outside of** $\triangle ABC$ (out $\triangle ABC$) is $\overleftrightarrow{ABC} \setminus (\triangle ABC \cup \text{ins}\,\triangle ABC)$.
The **enclosure of** $\triangle ABC$ (enc $\triangle ABC$) is $\triangle ABC \cup \text{ins}\,\triangle ABC$.
The **exclosure of** $\triangle ABC$ (exc $\triangle ABC$) is $\triangle ABC \cup \text{out}\,\triangle ABC$.

A consequence of this definition, which we will state as part (B) of Theorem PSH.41, is that $\angle BAC$, ins $\angle BAC$, and out $\angle BAC$ are disjoint sets whose union is $\mathcal{P}$. Similarly, the consequence that $\triangle ABC$, ins $\triangle ABC$, and out $\triangle ABC$ are disjoint sets whose union is $\mathcal{P}$ is part (B) of Theorem PSH.46. The next theorem shows that a segment connecting a point on each ray of an angle must intersect its inside.

Theorem PSH.37. *Let A, B, and C be noncollinear points, let $P \in \overrightarrow{AB}$, and $Q \in \overrightarrow{AC}$. Then $\overline{PQ} \subseteq$ ins $\angle BAC$.*

Proof. By Theorem IB.14, $Q \in \overrightarrow{AC} \subseteq$ the C-side of $\overleftrightarrow{AB} = \overrightarrow{ABC}$ and likewise $P \in \overrightarrow{AB} \subseteq$ the B-side of $\overleftrightarrow{AC} = \overrightarrow{ACB}$. By the same theorem $\overrightarrow{PQ} \subseteq \overrightarrow{ABC}$ and $\overrightarrow{QP} \subseteq \overrightarrow{ACB}$. By Exercise IB.8, Definition PSH.36, and elementary set theory, $\overline{PQ} = \overrightarrow{PQ} \cap \overrightarrow{QP} \subseteq \overrightarrow{ABC} \cap \overrightarrow{ACB} =$ ins $\angle BAC$. See Figure 5.1. □

Corollary IB.37.1. *Let A, B, and C be noncollinear points. Then* ins $\angle BAC \neq \emptyset$.

Proof. Since P and Q are distinct points, by Theorem PSH.22 there is a point R such that P–R–Q and $R \in \overline{PQ} \subseteq$ ins $\angle BAC$. □

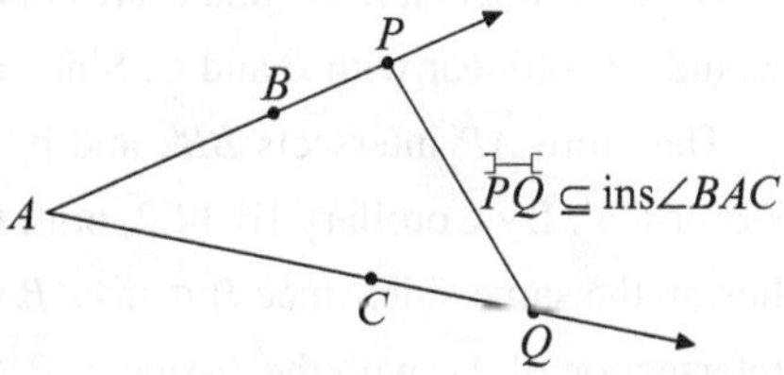

Fig. 5.1 For Theorem PSH.37.

Theorem PSH.38. *Let A, B, and C be noncollinear points.*

(A) $\overleftrightarrow{AC} \cap \overrightarrow{ABC} = \overrightarrow{AC}$; *that is, if P is on a side of a line,* $\overrightarrow{AP}$ *is the intersection of* $\overleftrightarrow{AP}$ *with that side.*

(B) *If* $P \in \text{ins}\,\angle BAC$, *then* $\overrightarrow{AP} = \overleftrightarrow{AP} \cap \text{ins}\,\angle BAC = \overleftrightarrow{AP} \cap \overrightarrow{ABC} \cap \overrightarrow{ACB}$.

(C) *If P is on the C-side of* $\overleftrightarrow{AB}$, *and if B and C are on opposite sides of* $\overleftrightarrow{AP}$, *then* $P \in \text{ins}\,\angle BAC$.

Proof. (A) By Property B.2 of Definition IB.1 there exists a point C' such that $C\text{–}A\text{–}C'$. By Theorem IB.14, $\overrightarrow{AC} \subseteq \overrightarrow{ABC}$ and $\overrightarrow{AC'} \subseteq \overrightarrow{ABC'}$. By Definition IB.11 the C-side and the C'-side are opposite sides of $\overleftrightarrow{AB}$ and therefore by Theorem PSH.12 these two sides are disjoint. Hence $\overrightarrow{AC'} \cap \overrightarrow{ABC} = \emptyset$. Also $A \in \overleftrightarrow{AB}$ and again by PSH.12 $\overleftrightarrow{AB} \cap \overrightarrow{ABC} = \emptyset$ so $A \notin \overrightarrow{ABC}$. By Theorem PSH.15 and the distributive laws for union and intersection,

$$\begin{aligned}\overleftrightarrow{AC} \cap \overrightarrow{ABC} &= (\overrightarrow{AC} \cup \{A\} \cup \overrightarrow{AC'}) \cap \overrightarrow{ABC}\\ &= (\overrightarrow{AC} \cap \overrightarrow{ABC}) \cup (\{A\} \cap \overrightarrow{ABC}) \cup (\overrightarrow{AC'} \cap \overrightarrow{ABC})\\ &= \overrightarrow{AC} \cup (\{A\} \cap \overrightarrow{ABC}) \cup (\overrightarrow{AC'} \cap \overrightarrow{ABC})\\ &= \overrightarrow{AC} \cup \emptyset \cup \emptyset = \overrightarrow{AC}.\end{aligned}$$

(B) By Definition PSH.36 $P \in \text{ins}\,\angle CAB = \overrightarrow{ABC} \cap \overrightarrow{ACB}$. By part (A), $\overrightarrow{AP} = \overleftrightarrow{AP} \cap \overrightarrow{ABC}$ and $\overrightarrow{AP} = \overleftrightarrow{AP} \cap \overrightarrow{ACB}$ so by elementary set theory, $\overrightarrow{AP} = \overleftrightarrow{AP} \cap \overrightarrow{ABC} \cap \overrightarrow{ACB}$.

(C) Since B and C are on opposite sides of $\overleftrightarrow{AP}$ by Axiom PSA there exists a point Q such that $\{Q\} = \overleftrightarrow{AP} \cap \overline{BC}$. By Theorem IB.14, $\overline{BC} \subseteq C$-side of $\overleftrightarrow{AB}$, so that Q, C, and P are on the same side of $\overleftrightarrow{AB}$. By part (A) $\overrightarrow{AP} = \overleftrightarrow{AP} \cap \overrightarrow{ABC}$ so that $Q \in \overrightarrow{AP}$. By Theorem PSH.37, $\overline{BC} \subseteq \text{ins}\,\angle BAC$, hence $Q \in \angle BAC$ and by part (B) $P \in \text{ins}\,\angle BAC$.

□

Theorem PSH.39 (Crossbar theorem). *Let A, B, and C be noncollinear points. If P is any member of* ins $\angle BAC$, $\overrightarrow{AP} \cap \overline{BC}$ *is nonempty and therefore a singleton* $\{Q\}$.

Proof. For a visualization see Figure 5.2. By Property B.3 of Definition IB.1 there exists a point B' such that $B\text{–}A\text{–}B'$. By Axiom I.1, $\overleftrightarrow{AB} = \overleftrightarrow{B'B}$.

Observe first that B, B', and C are noncollinear, for if they were collinear, $A \in \overline{BB'}$ would be collinear with B and C. Since $P \notin \overleftrightarrow{AB}$, $P \notin \overleftrightarrow{BB'}$; and $\overleftrightarrow{AP} \cap \overleftrightarrow{BB'} = \{A\}$.

Therefore $\overleftrightarrow{AP}$ intersects $\overline{BB'}$, and by Theorem PSH.6, $\overleftrightarrow{AP}$ must intersect either $\overline{CB}$ or $\overline{CB'}$. By Corollary IB.14.2, both these sets lie on the C-side of $\overleftrightarrow{BB'}$. Also, P lies on the same side, since $P \in \text{ins}\,\angle BAC$, and by Theorem IB.14, so does $\overrightarrow{AP}$, the intersection of $\overleftrightarrow{AP}$ with the C-side of $\overleftrightarrow{BB'}$. Thus $\overrightarrow{AP}$ must intersect either $\overline{CB}$ or $\overline{CB'}$.

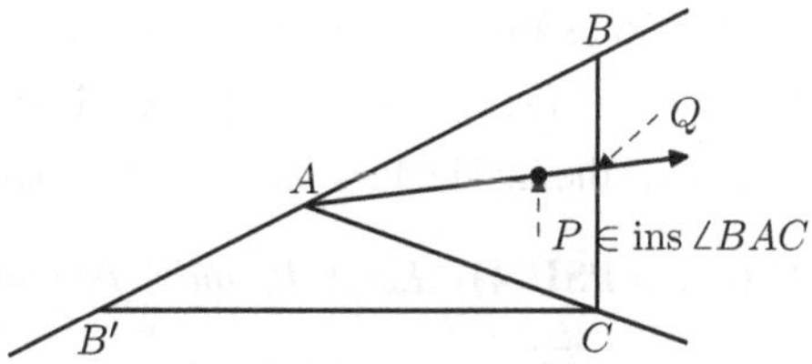

Fig. 5.2 For Theorem PSH.39.

By Definition IB.11, B and B' are on opposite sides of $\overleftrightarrow{AC}$, and by Theorem PSH.10 the B-side and B'-side are disjoint. By Corollary IB.14.2, $\overline{CB'}$ lies on the the B'-side of $\overleftrightarrow{AC}$, whereas $\overline{CB}$ lies on the B-side. Since P also lies on the B-side, $\overrightarrow{AP}$ is also on that side and hence intersects $\overline{CB}$ at some point Q.

If there were a second such point Q', both would belong to $\overleftrightarrow{AP} \cap \overleftrightarrow{BC}$ and by Exercise I.1 $\overleftrightarrow{AP} = \overleftrightarrow{BC}$ which is impossible since A, B, and C are not collinear. □

Corollary PSH.39.1. *If, in Theorem PSH.39, Q is the point such that $\overrightarrow{AP} \cap \overrightarrow{BC} = \{Q\}$, then $\overleftrightarrow{AP} \cap \overline{BC} = \{Q\}$ and $Q \in$ ins $\angle BAC$.*

Proof. By Corollary IB.5.2, $\overrightarrow{AP}$ is a subset of $\overleftrightarrow{AP}$ and $\overline{BC}$ is a subset of $\overleftrightarrow{BC}$ so that $Q \in \overrightarrow{AP} \cap \overline{BC} \subseteq \overleftrightarrow{AP} \cap \overleftrightarrow{BC}$. By Exercise I.1 $\overleftrightarrow{AP} \cap \overleftrightarrow{BC} = \{Q\}$, since $\overleftrightarrow{AP} \neq \overleftrightarrow{BC}$. $Q \in$ ins $\angle BAC$ by Theorem PSH.37. □

Corollary PSH.39.2. *If A, B, and C are noncollinear points and if $P \in$ ins $\angle BAC$, then B and C are on opposite sides of the line $\overleftrightarrow{AP}$.*

Proof. By Theorem PSH.39, $\overrightarrow{AP}$ intersects $\overline{BC}$ at some point Q. By Definition IB.11, B and C are on opposite sides of $\overleftrightarrow{AP}$. □

Theorem PSH.40. *Let A, B, and C be noncollinear points, let $P \in \overrightarrow{AB}$ and $Q \in \overrightarrow{AC}$. Then $\overleftrightarrow{PQ} \cap$ ins $\angle BAC = \overline{PQ}$.*

Proof. By Theorem PSH.37 $\overline{PQ} \subseteq$ ins $\angle BAC$, and since $\overline{PQ} \subseteq \overleftrightarrow{PQ}$,

$$\overline{PQ} \subseteq \overleftrightarrow{PQ} \cap \text{ins } \angle BAC.$$

Conversely, if $X \in \overleftrightarrow{PQ} \cap$ ins $\angle BAC$, by the Crossbar theorem PSH.39 there is a single point Y such that $\{Y\} = \overrightarrow{AX} \cap \overline{PQ} \subseteq \overleftrightarrow{AX} \cap \overleftrightarrow{PQ}$. Since $X \in \overleftrightarrow{AX}$ and $X \in \overleftrightarrow{PQ}$, $X \in \overleftrightarrow{AX} \cap \overleftrightarrow{PQ}$ so that by Exercise I.1, $X = Y \in \overline{PQ}$. □

The following theorem says that insides and outsides of angles behave "as they should." It says (A) the lines containing the edges of an angle do not intersect its inside (but they do intersect its outside); (B) an angle is disjoint from its inside and outside, and the inside and outside are disjoint. (C) The inside of an angle is defined in PSH.36 as the intersection of two half-planes; here we show that its outside is the union of the opposite halfplanes. Part (D) states that when we make an angle "smaller," the inside does indeed get "smaller" and the outside gets "bigger."

Theorem PSH.41. *Let A, B, and C be noncollinear points. Then*

(A) $(\overleftrightarrow{AB} \cup \overleftrightarrow{AC}) \cap \text{ins } \angle BAC = \emptyset$;

(B) $\text{ins } \angle BAC \cup \angle BAC \cup \text{out } \angle BAC = \mathcal{P}$, *and the sets in this union are pairwise disjoint; that is to say, an angle, its inside and its outside are mutually disjoint;*

(C) $\text{out } \angle BAC = \mathcal{E} \cup \mathcal{F}$, *where* $\mathcal{E}$ *is the side of* $\overleftrightarrow{AB}$ *opposite C and* $\mathcal{F}$ *is the side of* $\overleftrightarrow{AC}$ *opposite B; and*

(D) *if* $D \in \overrightarrow{AB} \cup \text{ins } \angle BAC$ *and* $E \in \overrightarrow{AC} \cup \text{ins } \angle BAC$, *then* $\text{ins } \angle DAE \subseteq \text{ins } \angle BAC$ *and* $\text{out } \angle BAC \subseteq \text{out } \angle DAE$.

Proof. (A) By Definition PSH.36, $\text{ins } \angle BAC = \overrightarrow{ABC} \cap \overrightarrow{ACB}$. By Theorem PSH.12, $\overleftrightarrow{AB} \cap \overrightarrow{ABC} = \emptyset$ and $\overleftrightarrow{AC} \cap \overrightarrow{ACB} = \emptyset$. Hence a point on either of these lines fails to belong to at least one of the sides listed, and thus fails to belong to their intersection $\text{ins } \angle BAC$.

(B) That the union of the sets equals $\mathcal{P}$ follows immediately from Definition PSH.36 (A). We examine each pair of sets to see that each pair is disjoint:

(1) $\angle BAC \cap \text{ins } \angle BAC = \emptyset$ since $\angle BAC \subseteq \overleftrightarrow{AB} \cup \overleftrightarrow{AC}$ and by part (A), $(\overleftrightarrow{AB} \cup \overleftrightarrow{AC}) \cap \text{ins } \angle BAC = \emptyset$;

(2) by Definition PSH.36, $\text{out } \angle BAC \cap (\angle BAC \cup \text{ins } \angle BAC) = \emptyset$ and therefore $\text{out } \angle BAC \cap \text{ins } \angle BAC = \emptyset$ and $\text{out } \angle BAC \cap \text{ins } \angle BAC = \emptyset$.

(C) To prove this part we consider two cases:

(Case 1: $X \notin \overleftrightarrow{AB} \cup \overleftrightarrow{AC}$.) Then $X \notin \angle BAC$ and since by definition

$$\text{out } \angle BAC = \mathcal{P} \setminus (\angle BAC \cup \text{ins } \angle BAC),$$

$$X \in \text{out } \angle BAC \Leftrightarrow X \notin \text{ins } \angle BAC \Leftrightarrow X \notin \overrightarrow{ABC} \cap \overrightarrow{ACB}$$
$$\Leftrightarrow X \in \mathcal{E} \text{ or } X \in \mathcal{F} \Leftrightarrow X \in \mathcal{E} \cup \mathcal{F}.$$

(Case 2: $X \in \overleftrightarrow{AB} \cup \overleftrightarrow{AC}$.) Then

$$X \in \text{out } \angle BAC \Leftrightarrow X \notin \angle BAC \Leftrightarrow X\text{–}A\text{–}B \text{ or } X\text{–}A\text{–}C,$$

which follows from part (A) above and Definition PSH.29.

Now by Definition IB.11, X–A–B means that X is on the side of $\overleftrightarrow{AC}$ opposite B, that is, $X \in \mathcal{F}$. By a similar argument, X–A–C means that $X \in \mathcal{E}$. Thus $X \in \text{out } \angle BAC \Leftrightarrow X \in \mathcal{E} \cup \mathcal{F}$.

(D) First we will show that $\text{ins } \angle DAE \subseteq \text{ins } \angle BAC$.

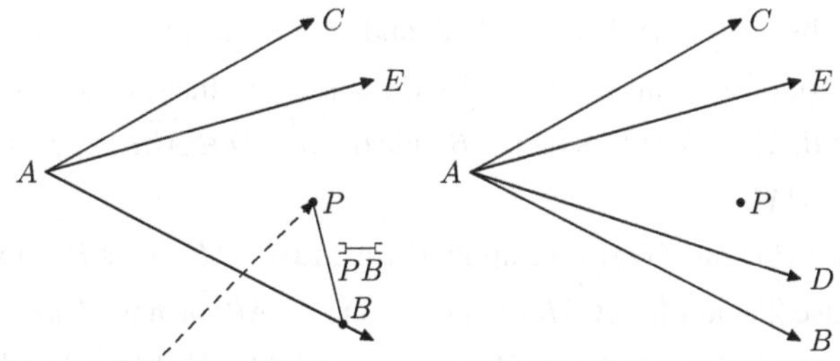

Case 2: If $P \in \text{ins} \angle BAE$ then $P \in \text{ins} \angle BAC$ Case 3: If $P \in \text{ins} \angle DAE$ then $P \in \text{ins} \angle BAC$

Fig. 5.3 For Theorem PSH.41(D).

(Case 1: Both points D and E belong to $\angle BAC$.) (Note that by hypothesis D and E cannot belong to the same ray in $\angle BAC$). In this case the proof is trivial.

(Case 2: Either $D \in \overrightarrow{AB}$ and $E \in \text{ins} \angle BAC$, or $E \in \overrightarrow{AC}$ and $D \in \text{ins} \angle BAC$.) Since the two alternatives are symmetric it should be sufficient entertainment to prove only the first one. For a visualization see Figure 5.3.

$E \in \text{ins} \angle BAC \subseteq \overrightarrow{ABC}$ so by Theorem PSH.12 $\overrightarrow{ABE} = \overrightarrow{ABC}$. That is, the E-side and the C-side of $\overleftrightarrow{AB}$ are the same.

Let $P \in \text{ins} \angle BAE = \overrightarrow{ABE} \cap \overrightarrow{AEB} = \overrightarrow{ABC} \cap \overrightarrow{AEB}$. Then P belongs to the B-side of $\overleftrightarrow{AE}$.

P is also on the $E = C$-side of $\overleftrightarrow{AB}$. If $\overline{PB}$ were to intersect $\overleftrightarrow{AC}$, it would necessarily intersect $\overrightarrow{AC}$ because the intersection would lie on the C-side of $\overleftrightarrow{AB}$ and $\overleftrightarrow{AC} \cap \overrightarrow{ABC} = \overrightarrow{AC}$ by Theorem PSH.38. But $\overrightarrow{AC}$ is on the side of $\overleftrightarrow{AE}$ opposite B because C and B are on opposite sides of $\overleftrightarrow{AE}$ by Corollary PSH.39.2. Therefore $\overline{PB} \cap \overleftrightarrow{AC} = \emptyset$.

It follows from Definition IB.11 that P and B are on the same side of $\overleftrightarrow{AC}$, so $P \in \overrightarrow{ACB}$ and hence $P \in \text{ins} \angle BAC$.

(Case 3: Both D and E belong to ins $\angle BAC$.) We may choose the notation so that $E \in C$-side of $\overleftrightarrow{AD}$. By Theorem PSH.12

both D and E belong to $\overrightarrow{ABC}$ so that $\overrightarrow{ABC} = \overrightarrow{ABD} = \overrightarrow{ABE}$,

and

both D and E belong to $\overrightarrow{ACB}$ so that $\overrightarrow{ACB} = \overrightarrow{ACD} = \overrightarrow{ACE}$.

Since $E \in \overrightarrow{ACD}$ and by our choice of notation $E \in \overrightarrow{ADC}$, $E \in \text{ins} \angle DAC$.

Also, by Corollary PSH.39.2, C and D are on opposite sides of $\overleftrightarrow{AE}$, and by assumption $E \in$ ins $\angle BAC$ so by the same corollary, C is on the side of $\overleftrightarrow{AE}$ opposite B. Therefore D is on the B-side of $\overleftrightarrow{AE}$, ($D \in \overrightarrow{AEB}$) and since $D \in \overrightarrow{ABE}$, $D \in$ ins $\angle BAE$.

Then if $P \in$ ins $\angle DAE$, we apply Case 2 to $\angle DAC$ to get $P \in$ ins $\angle DAC$. We apply Case 2 again to $\angle CAB$ to get $P \in$ ins $\angle CAB$, proving Case 3.

Finally, we show that out $\angle BAC \subseteq$ out $\angle DAE$. We have already shown that ins $\angle DAE \subseteq$ ins $\angle BAC$. By definition of D and E, we know that $\angle DAE \subseteq \angle BAC \cup$ ins $\angle BAC$. Taking unions, we have

$$\angle DAE \cup \text{ins } \angle DAE \subseteq \angle BAC \cup \text{ins } \angle BAC,$$

and taking complements, we have by Definition PSH.36

$$\text{out } \angle BAC \subseteq \text{out } \angle DAE,$$

which is the desired result. □

Corollary PSH.41.1. *Every angle is nonconvex.*

Proof. Let $\angle BAC$ be any angle, and let $P \in \overrightarrow{AB}$, and $Q \in \overrightarrow{AC}$; then by Theorem PSH.37 and Theorem PSH.22 there is a point $R \in \overline{PQ} \subseteq$ ins $\angle BAC$, and by Theorem PSH.41(B) ins $\angle BAC$ and $\angle BAC$ are disjoint so that $R \notin \angle BAC$. Then $\overline{PQ}$ is not a subset of $\angle BAC$ and by Definition IB.9, $\angle BAC$ is not convex. □

Theorem PSH.42. *Let P and Q be distinct points, and let $\mathcal{H}$ be a side of $\overleftrightarrow{PQ}$. Let A and B be members of $\mathcal{H} \cup \overleftrightarrow{PQ}$ such that A, B, and P are noncollinear. Then* ins $\angle APB \subseteq \mathcal{H}$.

Proof. Exercise PSH.30. See Figure 5.4. □

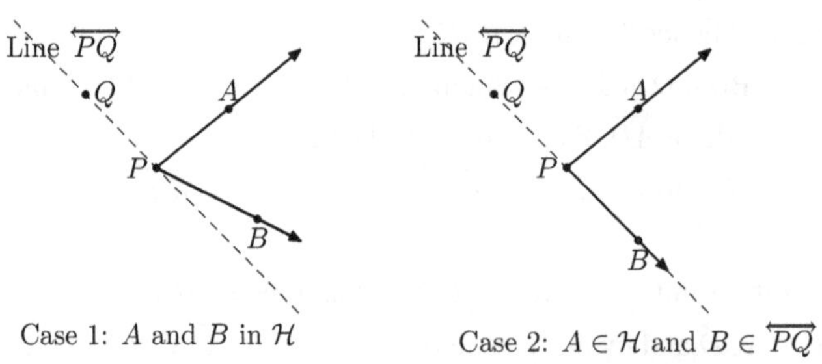

Fig. 5.4 For Theorem PSH.42.

The following theorem summarizes the possible ways that a line can intersect an angle.

Theorem PSH.43. *On a Pasch plane, let $\mathcal{L}$ be a line and $\angle BAC$ be an angle; if $\mathcal{L} \cap \angle BAC \neq \emptyset$, then $\mathcal{L} \cap \angle BAC$ is exactly one of the following alternatives:*

(1) *the single point A where $\mathcal{L} \cap \operatorname{ins} \angle BAC = \emptyset$; in this case, the sets $\angle BAC \setminus \{A\}, \overset{\mapsto}{AB}, \overset{\mapsto}{AC}$, and $\operatorname{ins} \angle BAC$ all are subsets of the B-side $(= C\text{-side})$ of $\mathcal{L}$;*

(2) *a single point P where $\mathcal{L} \cap \operatorname{ins} \angle BAC \neq \emptyset$; in this case, P may be any point of $\angle BAC$; let $Q \neq P$ and $Q \in \mathcal{L} \cap \operatorname{ins} \angle BAC$; then*

   (a) $\overset{\mapsto}{PQ} = \mathcal{L} \cap \operatorname{ins} \angle BAC$,
   (b) $\{X \mid X\text{–}P\text{–}Q\} = \mathcal{L} \cap \operatorname{out} \angle BAC$, *and*
   (c) $\mathcal{L} = \overleftrightarrow{PQ} = \{X \mid X\text{–}P\text{–}Q\} \cup \{P\} \cup \overset{\mapsto}{PQ}$;

(3) *exactly two points P and Q, in which case*

   (a) *no ray of $\angle BAC$ contains both P and Q,*
   (b) $P \neq A \neq Q$,
   (c) $\overset{\vdash\!\dashv}{PQ} = \mathcal{L} \cap \operatorname{ins} \angle BAC$,
   (d) $\{X \mid X\text{–}P\text{–}Q\} \cup \{X \mid P\text{–}Q\text{–}X\} = \mathcal{L} \cap \operatorname{out} \angle BAC$, *and*
   (e) $\mathcal{L} = \overleftrightarrow{PQ} = \{X \mid X\text{–}P\text{–}Q\} \cup \{P\} \cup \overset{\vdash\!\dashv}{PQ} \cup \{Q\} \cup \{X \mid P\text{–}Q\text{–}X\}$;

(4) *more than two points, in which case*

   (a) *$\mathcal{L}$ contains an entire ray of $\angle BAC$, and*
   (b) $\mathcal{L} \cap \operatorname{ins} \angle BAC = \emptyset$.

Proof. It is clear that exactly one of the alternatives (1), (2), (3), or (4) holds. For a visualization see Figure 5.5.

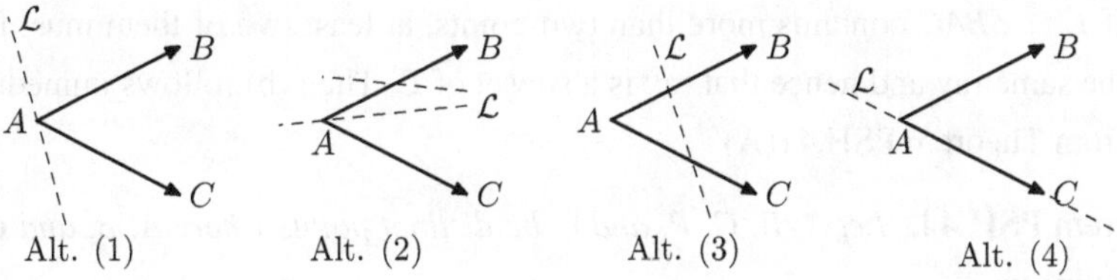

Fig. 5.5 For Theorem PSH.43.

(1) Suppose the intersection is the single point A and $\mathcal{L} \cap \text{ins}\,\angle BAC = \emptyset$. If there are points P and $Q \in \angle BAC$ that are on different rays of the angle and are on opposite sides of $\mathcal{L}$, then by Axiom PSA there exists a point $R \in \overline{PQ} \cap \mathcal{L}$. By Theorem PSH.40 $R \in \text{ins}\,\angle BAC$ which is impossible by hypothesis.

If there are points P and $Q \in \angle BAC$ that are both on one ray of the angle, and are on opposite sides of $\mathcal{L}$, then by Theorem PSH.12(IV)(B) there is a point $R \in \mathcal{L}$ belonging to that ray, and $R \neq A$. But by hypothesis, there is only one point of intersection between $\angle BAC$ and $\mathcal{L}$, so this is impossible.

Therefore all the points of $\overrightarrow{AB}$ and $\overrightarrow{AC}$ are on the same side of $\mathcal{L}$, and by Theorem PSH.42, ins $\angle BAC$ is on that same side.

(2) Since $Q \in \text{ins}\,\angle BAC$, by Theorem PSH.41(A) $Q \notin \angle BAC$; Q is a member of both $\overrightarrow{ABC}$ and $\overrightarrow{ACB}$ by Definition PSH.36, so by Theorem IB.13, $\overrightarrow{PQ} \subseteq \overrightarrow{ABC}$ and $\overrightarrow{PQ} \subseteq \overrightarrow{ACB}$; thus $\overrightarrow{PQ} \subseteq \text{ins}\,\angle BAC$ (Definition PSH.36) and $\overrightarrow{PQ} \subseteq \mathcal{L} \cap \text{ins}\,\angle BAC$. Similarly, if $X–P–Q$, X belongs to at least one of the sides opposite to $\overrightarrow{ABC}$ or $\overrightarrow{ACB}$ and hence (PSH.41(C)) to out $\angle BAC$, so that $\{X \mid X–P–Q\} \subseteq \mathcal{L} \cap \text{out}\,\angle BAC$. Equality (c), and consequently (a) and (b) follow from Theorem IB.5.

(3) If $\mathcal{L} \cap \angle BAC$ contains exactly two points P and Q, not both of them can be in the same ray because then $\mathcal{L}$ would contain that ray by Axiom I.1 and their intersection would contain more than two points. Hence neither P nor Q is equal to A, establishing (a) and (b).

Assume for convenience that $P \in \overrightarrow{AB}$ and $Q \in \overrightarrow{AC}$. If $X–P–Q$, then X is on the side of $\overleftrightarrow{AB}$ opposite Q, that is, opposite C. Likewise, if $P–Q–X$, then X is on the side of $\overleftrightarrow{AC}$ opposite P, that is, opposite B. It follows that $\{X \mid X–P–Q\} \cup \{X \mid P–Q–X\} \subseteq \mathcal{L} \cap \text{out}\,\angle BAC$, by Theorem PSH.41(C). Also, by Theorem PSH.40, $\overline{PQ} = \mathcal{L} \cap \text{ins}\,\angle BAC$. The equalities (d) and (e) follow immediately from Theorem IB.5.

(4) If $\mathcal{L} \cap \angle BAC$ contains more than two points, at least two of them must lie on the same ray and hence that ray is a subset of $\mathcal{L}$. Then (b) follows immediately from Theorem PSH.41(A). □

Theorem PSH.44. *Let A, B, C, P, and Q be distinct points where A, B, and C are noncollinear.*

(A) *If* $P \in \text{ins}\,\angle BAC$ *and* $Q \in \text{out}\,\angle BAC$, *then* $\overline{PQ} \cap \angle BAC$ *is a singleton.*
(B) *If* $A \notin \overline{PQ}$, *and if* $\overline{PQ} \cap \angle BAC = \{R\}$ *for some point R, then* $P \in \text{ins}\,\angle BAC$ *if and only if* $Q \in \text{out}\,\angle BAC$.

See Figure 5.6. Notice that there is nothing in Theorem PSH.44 that guarantees that every line intersecting ins $\angle BAC$ must intersect $\angle BAC$. This, however, is true for a Pasch plane where Axiom PS holds, as will be proved in Theorem EUC.2.

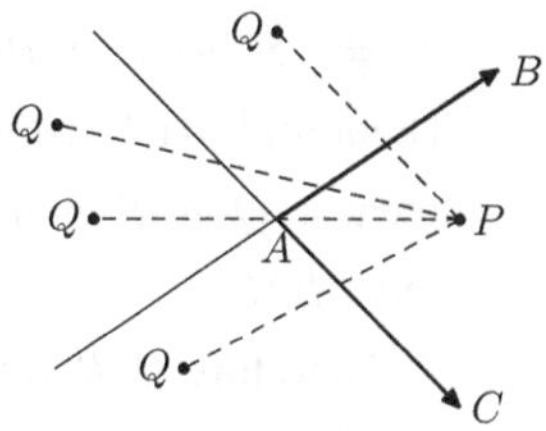

Fig. 5.6 Showing four possible locations for a segment.

Proof. (A) The line $\overleftrightarrow{PQ}$ contains a point of ins $\angle BAC$ which rules out alternatives (1) and (4) of Theorem PSH.43.

If alternative (2) holds, $\overleftrightarrow{PQ} \cap \angle BAC = \{R\}$ for some point R; $P \in$ ins $\angle BAC$ and $Q \in$ out $\angle BAC$ so by Theorem PSH.43(2)(b) Q–R–P so $R \in \overline{PQ}$, and there is only one point of intersection of $\angle BAC$ with $\overline{PQ}$ because there is only one with $\overleftrightarrow{PQ}$.

If (3) holds, $\overleftrightarrow{PQ}$ contains exactly 2 points R and S of $\angle BAC$, which must belong to two different rays, say $R \in \overrightarrow{AB}$ and $S \in \overrightarrow{AC}$. Then by Theorem PSH.43(3)(c) R–P–S, since $P \in$ ins $\angle BAC$. By part (3)(d) either Q–R–P–S or R–P–S–Q, because $Q \in$ out $\angle BAC$. In the first case, $\overline{PQ} \cap \angle BAC = \{R\}$ and in the second case $\overline{PQ} \cap \angle BAC = \{S\}$ and in either case the intersection is a singleton.

(B) We look first at $\overleftrightarrow{PQ}$. Note first that $A \notin \overleftrightarrow{PQ}$, for otherwise $\overleftrightarrow{PQ}$ would contain a ray of $\angle BAC$ and $\overline{PQ}$ would then contain many elements of $\angle BAC$, which contradicts the hypothesis that $\overline{PQ} \cap \angle BAC = \{R\}$. Therefore alternative (1) of Theorem PSH.43 is ruled out. If $\overleftrightarrow{PQ}$ contains more than two points of $\angle BAC$, then it would contain one of the rays and hence A, so alternative (4) is also ruled out; thus either alternative (2) or (3) holds.

If alternative (2) of Theorem PSH.43 holds, $\overleftrightarrow{PQ} \cap \angle BAC = \{R\}$. Now Q–R–P so by (2)(b), if $P \in$ ins $\angle BAC, Q \in$ out $\angle BAC$. Conversely, if $Q \in$ out $\angle BAC$, then there exists some $P' \in$ ins $\angle BAC$ such that Q–R–P' (in alternative (2), the line intersects the inside of the angle). By Corollary PSH.8.2 we have either Q–R–P–P' or Q–R–P'–P. In either case, $P \in \overrightarrow{RP'} = \overleftrightarrow{RP'} \cap$ ins $\angle BAC$ by Theorem PSH.43(a).

If alternative (3) holds, there is a point $S \neq R$ such that $\overleftrightarrow{PQ} \cap \angle BAC = \{R, S\}$. If $P \in$ ins $\angle BAC$, by (c) $P \in \overline{RS}$, then R–P–S. We know that Q–R–P so Q–R–P–S and Q–R–S and hence, by (3)(d), $Q \in$ out $\angle BAC$.

Conversely, suppose $Q \in$ out $\angle BAC$. By hypothesis, $P \neq R$ and $P \neq S$ because $\overline{PQ} \cap \angle BAC = \{R\} \neq \{S\}$. Since Q is outside, by Theorem PSH.43(3)(d) we have either Q–R–S or R–S–Q. The last is impossible because Q–R–P yields P–R–S–Q which would force two intersections of $\overline{PQ}$ and $\angle BAC$.

Therefore Q–R–S and again using Q–R–P either Q–R–P–S or Q–R–S–P, by Corollary PSH.8.2. Again, Q–R–S–P is impossible, because it would force two intersections of $\overline{PQ}$ and $\angle BAC$. Therefore we have Q–R–P–S which shows that $P \in$ ins $\angle BAC$, by Theorem PSH.43(3)(c). □

Theorem PSH.45. *Let A, B, and C be noncollinear points; let D be a point such that A–C–D, thus extending the edge* $\overline{AC}$ *of* $\triangle ABC$. *Choose a point* $F \in \overline{BC}$ *(so that B–F–C) and let G be a point such that A–F–G. Then* $G \in$ ins $\angle BCD$.

Later, in Definition NEUT.79, $\angle BCD$ will be designated as an *outside angle* of $\triangle ABC$.)

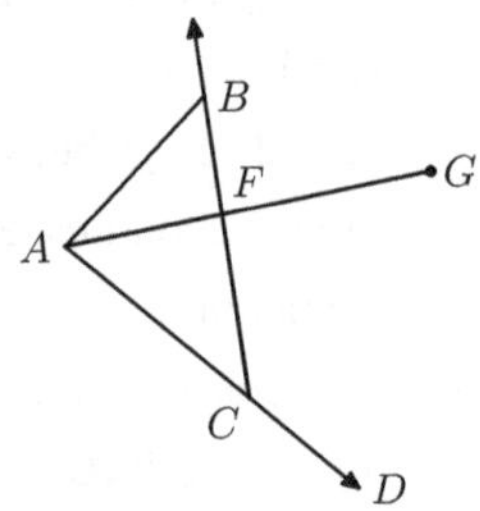

Fig. 5.7 Showing G on the inside of an *outside angle* of $\triangle ABC$.

Proof. By Definition IB.11 A and G are on opposite sides of $\overleftrightarrow{BC}$ and A and D are on opposite sides of $\overleftrightarrow{BC}$. By Theorem PSH.12 G and D are on the same side of $\overleftrightarrow{BC}$, i.e., $G \in \overrightarrow{BCD}$. By Definition IB.4 $B \in \overrightarrow{CF}$ and $G \in \overrightarrow{AF}$. By Theorem IB.14 $B \in \overrightarrow{ADF}$ so that the F-side and the B-side of $\overleftrightarrow{AD}$ are the same. By the same theorem, $G \in \overrightarrow{ADF} = \overrightarrow{ADB}$. By Definition PSH.36, $G \in$ ins $\angle BCD$, proving the theorem. See Figure 5.7. □

5.8 Mostly about triangles

Theorem PSH.46 (Analogous to Theorem PSH.41 for angles). *Let A, B, and C be noncollinear points. Then*

(A) $(\overleftrightarrow{AB} \cup \overleftrightarrow{BC} \cup \overleftrightarrow{CA}) \cap \text{ins}\,\triangle ABC = \emptyset$;

(B) $\text{ins}\,\triangle ABC \cup \triangle ABC \cup \text{out}\,\triangle ABC = \mathcal{P}$ *and the sets in this union are pairwise disjoint;*

(C) $\text{ins}\,\triangle ABC = \text{ins}\,\angle BAC \cap \text{ins}\,\angle ABC = \text{ins}\,\angle BAC \cap \overrightarrow{BCA}$;

(D) $\text{out}\,\triangle ABC = \mathcal{E} \cup \mathcal{F} \cup \mathcal{G}$, *where*

$\mathcal{E}$ = *the side of* $\overleftrightarrow{AB}$ *opposite C,*

$\mathcal{F}$ = *the side of* $\overleftrightarrow{BC}$ *opposite A, and*

$\mathcal{G}$ = *the side of* $\overleftrightarrow{CA}$ *opposite B.*

Proof. (A) By Definition PSH.36 (B), $\text{ins}\,\triangle ABC = \overrightarrow{ABC} \cap \overrightarrow{BCA} \cap \overrightarrow{CAB}$. By Theorem PSH.12, $\overleftrightarrow{AB} \cap \overrightarrow{ABC} = \emptyset$, $\overleftrightarrow{BC} \cap \overrightarrow{BCA} = \emptyset$, and $\overleftrightarrow{CA} \cap \overrightarrow{CAB} = \emptyset$. Hence a point on any of these lines fails to belong to at least one of the sides listed, and thus fails to belong to their intersection $\text{ins}\,\triangle ABC$.

(B) Exercise PSH.49.

(C) Exercise PSH.50.

(D) To prove this part we consider two classes of points in $\mathcal{P}$:

If $X \notin \overleftrightarrow{AB} \cup \overleftrightarrow{BC} \cup \overleftrightarrow{CA}$, then $X \notin \triangle ABC$ and since by definition

$$\text{out}\,\angle BAC = \mathcal{P} \setminus (\triangle ABC \cup \text{ins}\,\triangle ABC),$$

it follows that

$$\begin{aligned} X \in \text{out}\,\triangle ABC &\Leftrightarrow X \notin \text{ins}\,\triangle ABC \\ &\Leftrightarrow X \notin \overrightarrow{ABC} \cap \overrightarrow{BCA} \cap \overrightarrow{CAB} \\ &\Leftrightarrow X \in \mathcal{E} \text{ or } X \in \mathcal{F} \text{ or } X \in \mathcal{G} \\ &\Leftrightarrow X \in \mathcal{E} \cup \mathcal{F} \cup \mathcal{G}. \end{aligned}$$

If $X \in \overleftrightarrow{AB} \cup \overleftrightarrow{BC} \cup \overleftrightarrow{CA}$, then

$$\begin{aligned} X \in \text{out}\,\triangle ABC &\Leftrightarrow X \notin \triangle ABC \\ &\Leftrightarrow X \notin \overline{AB} \cup \overline{BC} \cup \overline{CA} \\ &\Leftrightarrow \text{one of the following holds:} \end{aligned}$$

X–A–B, A–B–X, X–B–C, B–C–X, X–C–A, or C–A–X. This fact follows from Definition IB.7 and Theorem IB.5. By Definition IB.11,

X–B–C or C–A–X iff X belongs to the side of $\overleftrightarrow{AB}$ opposite C or $X \in \mathcal{E}$;

A–B–X or X–C–A iff X belongs to the side of $\overleftrightarrow{BC}$ opposite A or $X \in \mathcal{F}$; and

X–A–B or B–C–X iff X belongs to the side of $\overleftrightarrow{AC}$ opposite B or $X \in \mathcal{G}$.

Thus $X \in \text{out}\,\triangle ABC$ if and only if $X \in \mathcal{E} \cup \mathcal{F} \cup \mathcal{G}$. □

The next three theorems embark on an analysis of the possible ways a line can intersect a triangle. Theorem PSH.47 says that a line that intersects a triangle in exactly two points also intersects its inside; when this happens, the intersection of the line with the inside is the open segment connecting the two points of intersection.

The next two theorems provide a converse for PSH.47. Theorem PSH.48 shows that if a line contains a corner of a triangle, and also intersects its inside, then it intersects the triangle in exactly two points. PSH.49 shows that if a line does not contain any corner of a triangle, the same result follows. Finally, Theorem PSH.50 combines all these results, adds detail, and summarizes the possibilities for the intersection of a line and a triangle.

Theorem PSH.47 (Analogous to PSH.37, but for triangles). *Let A, B, and C be noncollinear points, and let P and Q belong to $\triangle ABC$. If no edge of $\triangle ABC$ contains both P and Q, then* $\overset{\dashv\vdash}{PQ} \subseteq \text{ins}\,\triangle ABC$.

Proof. Exercise PSH.24. □

Corollary PSH.47.1. *For any triangle $\triangle ABC$, $\text{ins}\,\triangle ABC \neq \emptyset$, and $\triangle ABC$ is nonconvex.*

Proof. Using Theorem PSH.22 let P and Q be members (not endpoints) of two different edges of $\triangle ABC$. By the same theorem there is a point R such that P–R–Q and $R \in \overset{\dashv\vdash}{PQ} \subseteq \text{ins}\,\triangle ABC$. This shows that $\text{ins}\,\triangle ABC \neq \emptyset$. By Theorem PSH.46(B), $\text{ins}\,\triangle ABC \cap \triangle ABC = \emptyset$, so that $R \notin \triangle ABC$ and $\overset{\dashv\vdash}{PQ}$ is not a subset of $\triangle ABC$. By Definition IB.9, $\triangle ABC$ is not convex. □

Theorem PSH.48. *Let A, B, and C be noncollinear points. If P is any member of* $\text{ins}\,\triangle ABC$, *there exists a point* $Q \in \overset{\dashv\vdash}{BC}$ *such that*

(1) $\overset{\dashv\rightarrow}{AP} \cap \overset{\dashv\vdash}{BC} = \{Q\}$,
(2) $\overleftrightarrow{AP} \cap \triangle ABC = \{A, Q\}$,
(3) $\overset{\dashv\vdash}{AQ} \subseteq \text{ins}\,\triangle ABC$, *and*
(4) $\overset{\dashv\rightarrow}{AQ} \setminus \overset{\sqsubset\sqsupset}{AQ} \subseteq \text{out}\,\triangle ABC$.

Note: In the summary Theorem PSH.50 we will show that the intersection of $\overleftrightarrow{AQ}$ and $\text{ins}\,\triangle ABC$ is exactly $\overset{\dashv\vdash}{AQ}$, which shows that $P \in \overset{\dashv\vdash}{AQ}$ and A–P–Q.

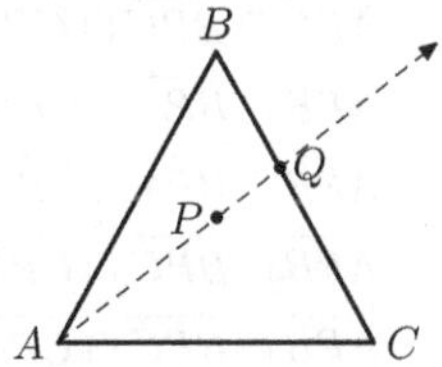

Fig. 5.8 For Theorem PSH.48.

Proof. See Figure 5.8. (1) By Theorem PSH.46(C), $P \in \text{ins}\,\angle BAC$ so that by the Crossbar theorem PSH.39, there exists $Q \in \overline{BC}$ such that $\overrightarrow{AP} \cap \overline{BC} = \{Q\}$.

(2) If there were more than two points in the intersection, two of them would lie on one edge of $\triangle ABC$ and that edge would be a subset of $\overleftrightarrow{AP}$ and P would belong to one of the lines containing an edge which is ruled out by Theorem PSH.46(A).

(3)(4) Proofs are Exercise PSH.19. □

Theorem PSH.49. *Let A, B, and C be noncollinear points and let $\mathcal{L}$ be a line such that $\mathcal{L} \cap \text{ins}\,\triangle ABC \neq \emptyset$ and $\mathcal{L} \cap \{A, B, C\} = \emptyset$. If $P \in \mathcal{L} \cap \text{ins}\,\triangle ABC$ and $Q \neq P$ is any point of $\mathcal{L}$, then*

(1) *$\overrightarrow{PQ}$ intersects exactly one of the segments $\overline{AC}, \overline{BC}$ or $\overline{AB}$ in exactly one point,*

(2) *$\mathcal{L} = \overleftrightarrow{PQ}$ intersects exactly two of the segments $\overline{AC}, \overline{BC}$ or $\overline{AB}$, and thus $\mathcal{L}$ intersects $\triangle ABC$ in exactly two points D and E, and*

(3) *$\overline{DE} \subseteq \text{ins}\,\triangle ABC$.*

In the summary Theorem PSH.50 we will show that the intersection of $\mathcal{L} = \overleftrightarrow{PQ}$ and $\text{ins}\,\triangle ABC$ is exactly $\overline{DE}$, which shows that $P \in \overline{DE}$ and $D\text{–}P\text{–}E$.

Proof. To prove part (1), note that $Q \notin \overleftrightarrow{AP}$, for otherwise $\overleftrightarrow{AP} = \mathcal{L}$ so that $A \in \mathcal{L}$, which is false by hypothesis. By similar arguments, Q is not a member of any of the lines $\overleftrightarrow{AP}$, $\overleftrightarrow{BP}$, and $\overleftrightarrow{CP}$, and by Theorem PSH.12, Q is a member of one side or the other of each of these lines. Thus, Q belongs to

either $\overrightarrow{APC}$ or $\overrightarrow{APB}$, and
either $\overrightarrow{BPA}$ or $\overrightarrow{BPC}$, and
either $\overrightarrow{CPB}$ or $\overrightarrow{CPA}$.

There are eight possible intersections of a side from each pair listed, as follows:

$\overrightarrow{APC} \cap \overrightarrow{BPA} \cap \overrightarrow{CPB}$
$\overrightarrow{APC} \cap \overrightarrow{BPA} \cap \overrightarrow{CPA} \subseteq \overrightarrow{APC} \cap \overrightarrow{CPA} = \text{ins}\,\angle APC$
$\overrightarrow{APC} \cap \overrightarrow{BPC} \cap \overrightarrow{CPB} \subseteq \overrightarrow{BPC} \cap \overrightarrow{CPB} = \text{ins}\,\angle BPC$

$$\overrightarrow{APC} \cap \overrightarrow{BPC} \cap \overrightarrow{CPA} \subseteq \overrightarrow{APC} \cap \overrightarrow{CPA} = \text{ins } \angle APC$$
$$\overrightarrow{APB} \cap \overrightarrow{BPA} \cap \overrightarrow{CPB} \subseteq \overrightarrow{APB} \cap \overrightarrow{BPA} = \text{ins } \angle APB$$
$$\overrightarrow{APB} \cap \overrightarrow{BPA} \cap \overrightarrow{CPA} \subseteq \overrightarrow{APB} \cap \overrightarrow{BPA} = \text{ins } \angle APB$$
$$\overrightarrow{APB} \cap \overrightarrow{BPC} \cap \overrightarrow{CPB} \subseteq \overrightarrow{BPC} \cap \overrightarrow{CPB} = \text{ins } \angle BPC$$
$$\overrightarrow{APB} \cap \overrightarrow{BPC} \cap \overrightarrow{CPA}$$

If we can show that $\overrightarrow{APC} \cap \overrightarrow{BPA} \cap \overrightarrow{CPB}$ and $\overrightarrow{APB} \cap \overrightarrow{BPC} \cap \overrightarrow{CPA}$ are empty, it will follow that Q will be a member of exactly one of the sets ins $\angle APC$, ins $\angle BPC$, and ins $\angle APB$. Then by the Crossbar theorem PSH.39, $\overrightarrow{PQ}$ intersects exactly one of the segments $\overline{AC}$, $\overline{BC}$ or $\overline{AB}$ in exactly one point.

Let A', B', and C' be the points such that $\overrightarrow{AP} \cap \overline{BC} = \{A'\}$, $\overrightarrow{BP} \cap \overline{AC} = \{B'\}$, and $\overrightarrow{CP} \cap \overline{AB} = \{C'\}$ as guaranteed by Theorem PSH.48(1). Since $B' \in \overrightarrow{APC}$ and also $B' \in \overleftrightarrow{BP}$, $\overrightarrow{APC} \cap \overrightarrow{BPA} = \overrightarrow{APB'} \cap \overrightarrow{B'PA} = \text{ins } \angle APB'$. Both A and $B' \in \overrightarrow{CPA}$ so that by Theorem PSH.42, ins $\angle APB' \subseteq \overrightarrow{CPA}$, which is disjoint from $\overrightarrow{CPB}$. Therefore $\overrightarrow{APC} \cap \overrightarrow{BPA} \cap \overrightarrow{CPB} = \emptyset$. A similar argument shows that $\overrightarrow{APB} \cap \overrightarrow{BPC} \cap \overrightarrow{CPA} = \emptyset$. This completes the proof of part (1).

The proof of (2) is Exercise PSH.25. By Theorem PSH.47, $\overline{DE} \subseteq$ ins $\triangle ABC$, thus proving part (3). □

Theorem PSH.50. *If the intersection of a line $\mathcal{L}$ with a triangle $\triangle ABC$ is nonempty, then*

(A) $\mathcal{L} \cap$ enc $\triangle ABC$ *is either a single point or a closed segment, and*
(B) $\mathcal{L} \cap \triangle ABC$ *is exactly one of the following alternatives:*

  (1) *a single point S, in which case*

   (a) $S \in \{A, B, C\}$, *the set of corners of* $\triangle ABC$,
   (b) $\mathcal{L} \cap$ ins $\triangle ABC = \emptyset$ *and* $\mathcal{L} \cap$ ins $\angle TSU = \emptyset$ *where T and U are the corners of $\triangle ABC$ other than S, and*
   (c) $\triangle ABC \setminus \{S\}$, $\overrightarrow{ST}$, $\overrightarrow{SU}$, ins $\triangle ABC$ *and* ins $\angle TSU$ *all are subsets of the T-side (= U-side) of $\mathcal{L}$;*

  (2) *exactly two points P and Q, in which case*

   (a) *no edge of $\triangle ABC$ contains both P and Q, and at least one of P and Q is not a corner,*
   (b) $\mathcal{L} \cap$ ins $\triangle ABC = \overline{PQ}$,

(c) $\mathcal{L} \cap \text{out}\,\triangle ABC = \{X\,|X\text{–}P\text{–}Q\} \cup \{X\,|P\text{–}Q\text{–}X\}$, *and*

(d) $\mathcal{L} = \overleftrightarrow{PQ} = \{X\,|X\text{–}P\text{–}Q\} \cup \{P\} \cup \overline{PQ} \cup \{Q\} \cup \{X\,|P\text{–}Q\text{–}X\}$;

(3) *more than two points, in which case*

(a) $\mathcal{L}$ *contains an edge* $\mathcal{E}$ *of* $\triangle ABC$, *and*

(b) $\mathcal{L} \cap \text{ins}\,\triangle ABC = \emptyset$.

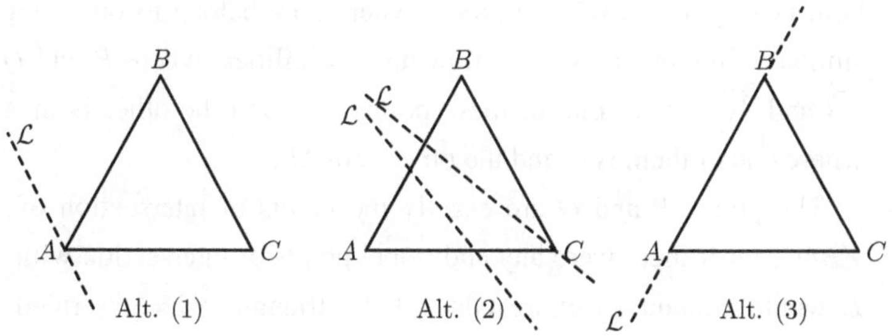

Fig. 5.9 For Theorem PSH.50(B).

Proof. (A) Follows immediately from (B).

(B) Clearly exactly one of the alternatives (1), (2), or (3) holds. For a visualization, see Figure 5.9.

(1) If the intersection is a single point S, suppose $S \notin \{A, B, C\}$. Then S is a member of either $\overline{AB}$, $\overline{BC}$, or $\overline{CA}$. Suppose for the moment $S \in \overline{AB}$. Then $C \notin \mathcal{L}$ since there is only one point of intersection, so Theorem PSH.6 applies and there is a second intersection, which is a contradiction. Similar proofs will show that S is not a member of $\overline{BC}$ or $\overline{CA}$. Hence $S = A$ or B or C.

Now T and U are defined to be the corners of $\triangle ABC$ other than S; if T and U are on different sides of $\mathcal{L}$, by Axiom PSA, $\mathcal{L} \cap \overline{TU} \neq \emptyset$, where $\overline{TU}$ is an edge of $\triangle ABC$ and this contradicts the assumption that there is only one point of intersection. Then by Theorem IB.13 both $\overrightarrow{ST}$ and $\overrightarrow{SU}$ are subsets of the T-side of $\mathcal{L}$, and by Theorem PSH.42, ins $\angle TSU$ is a subset of the T-side of $\mathcal{L}$ which is disjoint from $\mathcal{L}$.

Finally, $\overline{TU} \subseteq \text{ins}\,\angle TSU$ (by Theorem PSH.37) and $\text{ins}\,\triangle ABC \subseteq \text{ins}\,\angle TSU$ (Definition PSH.36) and both these sets are disjoint from $\mathcal{L}$. From these observations all the conclusions of alternative (1) follow.

(2) If $\mathcal{L} \cap \triangle ABC$ contains exactly two points P and Q, not both of them can be in the same edge, and not both can be corners, for in either case by Axiom I.1 $\overleftrightarrow{PQ} = \mathcal{L}$ would contain an edge and there would be more than two points of intersection. This proves (a). Result (d) follows immediately from Theorem IB.5.

The points P and Q both belong to the (open) rays of at least one of $\angle BAC$, $\angle ABC$, or $\angle ACB$. We shall prove part (b) in the case where they both belong to $\angle BAC$ (the proofs where they belong to other angles are similar). This includes the following possibilities: where P and Q are in $\overset{\vdash\dashv}{AB}$ and $\overset{\vdash\dashv}{AC}$, where one of these points is B and the other is in $\overset{\vdash\dashv}{AC}$, and where one of them is C and the other is in $\overset{\vdash\dashv}{AB}$.

The points P and Q are exactly the points of intersection of $\mathcal{L}$ with $\angle BAC$, for if there were any additional points of intersection with $\angle BAC$, $\mathcal{L}$ would contain an entire edge of the triangle, which is ruled out by hypothesis.

By Theorem PSH.46(C) ins $\triangle ABC$ = ins $\angle BAC \cap \overset{\vdash\rightarrow}{BCA} \subseteq$ ins $\angle BAC$, so that $\mathcal{L} \cap$ ins $\triangle ABC \subseteq \mathcal{L} \cap$ ins $\angle BAC = \overset{\vdash\dashv}{PQ}$ (Theorem PSH.40). Then, by Theorem PSH.47, $\overset{\vdash\dashv}{PQ} \subseteq$ ins $\triangle ABC$, so that $\mathcal{L} \cap$ ins $\angle BAC = \overset{\vdash\dashv}{PQ} = \mathcal{L} \cap$ ins $\triangle ABC$, proving part (b).

By Theorem PSH.46(B), $\mathcal{P}$ = ins $\triangle ABC \cup \triangle ABC \cup$ out $\triangle ABC$ and the sets in this union are pairwise disjoint. Intersecting this with $\mathcal{L}$, we get

$$\mathcal{L} = \mathcal{L} \cap \mathcal{P} = (\mathcal{L} \cap \text{ins } \triangle ABC) \cup (\mathcal{L} \cap \triangle ABC) \cup (\mathcal{L} \cap \text{out } \triangle ABC) \qquad (*)$$

and all sets in parentheses are disjoint. We know that $\mathcal{L} \cap \triangle ABC = \{P, Q\}$; using (b) and (d),

$$\mathcal{L} = \{X \mid X\text{–}P\text{–}Q\} \cup \{P\} \cup \overset{\vdash\dashv}{PQ} \cup \{Q\} \cup \{X \mid P\text{–}Q\text{–}X\}$$
$$= \{X \mid X\text{–}P\text{–}Q\} \cup (\mathcal{L} \cap \triangle ABC) \cup (\mathcal{L} \cap \text{ins } \triangle ABC) \cup \{X \mid P\text{–}Q\text{–}X\}.$$

From this equation and equation (*), and the fact that all the sets in parentheses (or braces) are disjoint, out $\triangle ABC = \{X \mid X\text{–}P\text{–}Q\} \cup \{X \mid P\text{–}Q\text{–}X\}$ proving part (c).

(3) If $\mathcal{L} \cap \triangle ABC$ contains more than two points, they are collinear and hence must all belong to one edge of $\triangle ABC$ so that edge is a subset of $\mathcal{L}$, and by Theorem PSH.46(A), $\mathcal{L}$ is disjoint from ins $\triangle ABC$. □

Lemma PSH.51. *Let A, B, and C be noncollinear points and let* $P \in$ ins $\triangle ABC$. *Let* B' *and* C' *be points such that* $P\text{–}B'\text{–}B$ *and* $P\text{–}C'\text{–}C$. *Then* $P \in \overset{\vdash\rightarrow}{B'C'A}$.

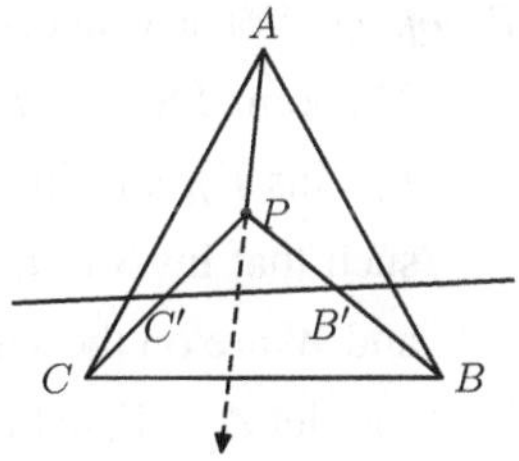

Fig. 5.10 For Lemma PSH.51.

Proof. See Figure 5.10. Since $P \in \text{ins}\,\triangle ABC$, $P \in \text{ins}\,\angle BAC$ by Definition PSH.36, so by the Crossbar theorem PSH.39, $\overrightarrow{AP}$ has a single point of intersection with $\overline{BC}$, which we will call Q. By Theorem PSH.37, $Q \in \overline{BC} \subseteq \text{ins}\,\angle BPC$ and again by the Crossbar theorem, there exists a unique point P' such that $\overrightarrow{PQ} \cap \overleftrightarrow{B'C'} = \{P'\}$.

Now $\overline{AP} \cup \overrightarrow{PQ} = \overrightarrow{AQ} \subseteq \overleftrightarrow{AQ}$ has only a single intersection with $\overleftrightarrow{B'C'}$ by Exercise I.1, and P' is that single intersection. Since $\overline{AP} \cap \overrightarrow{PQ} = \emptyset$ by Theorem IB.5, there can be no intersection of $\overleftrightarrow{B'C'}$ with $\overline{AP}$. By Definition IB.11, $P \in \overrightarrow{B'C'}A$. □

For those who have studied general topology, the following result may be of some interest. Together with the result of Exercise PSH.52(A), it is what is needed to show that the set of all insides of triangles in a Pasch plane forms a *base* for some topology on the plane, thus enabling continuity arguments. Actually it shows a bit more —all that is really needed is that $\text{ins}\,\triangle ABC \subseteq \text{ins}\,\mathcal{T} \cap \text{ins}\,\mathcal{S}$.

Theorem PSH.52. *Let $\mathcal{T}$ and $\mathcal{S}$ be triangles in a Pasch plane and let P be a point such that $P \in \mathcal{T} \cap \mathcal{S}$. Then there exists a triangle $\triangle ABC$ such that $P \in \text{enc}\,\triangle ABC \subseteq \text{ins}\,\mathcal{T} \cap \text{ins}\,\mathcal{S}$.*

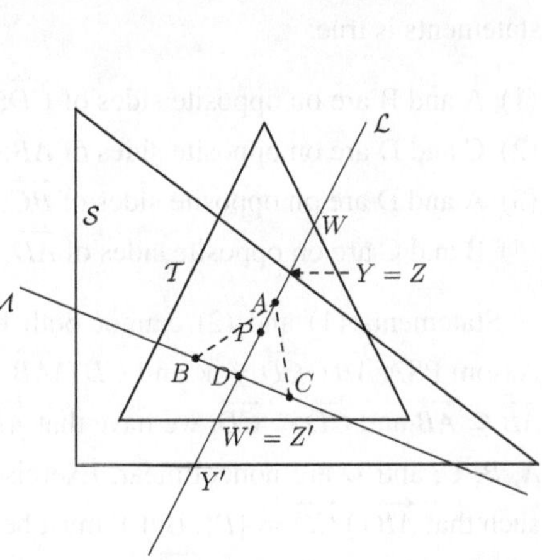

Fig. 5.11 For Theorem PSH.52.

Proof. (I) For a visualization see Figure 5.11. Let $\mathcal{L}$ be a line containing P. By Theorem PSH.50 $\mathcal{L}$ intersects $\mathcal{T}$ in exactly two points W and W' (because $P \in \operatorname{ins}\mathcal{T}$), and $\operatorname{ins}\mathcal{T} \cap \mathcal{L} = \overset{]\!-\!-\![}{WW'}$. Likewise, we can find points Y and Y' such that $\operatorname{ins}\mathcal{S} \cap \mathcal{L} = \overset{]\!-\!-\![}{YY'}$. For convenience we can label Y and Y' so that Y and W are on the same side of P, that is, either P–Y–W or P–W–Y; in the first case let $Z = Y$, in the second let $Z = W$. Likewise either P–Y'–W' or P–W'–Y'; in the first case let $Z' = Y'$, in the second let $Z' = W'$. Then $\overset{]\!-\!-\![}{ZZ'} = \overset{]\!-\!-\![}{WW'} \cap \overset{]\!-\!-\![}{YY'}$, and $P \in \overset{]\!-\!-\![}{ZZ'} \subseteq (\operatorname{ins}\mathcal{T} \cap \operatorname{ins}\mathcal{S})$. Let A and D be points on $\mathcal{L}$ with Z–A–P–D–Z'. Then $P \in \overset{[\!-\!-\!]}{DA} \subseteq (\operatorname{ins}\mathcal{T} \cap \operatorname{ins}\mathcal{S})$.

(II) Let $\mathcal{M} \neq \mathcal{L}$ be a line containing D. We may now apply the reasoning of part (I) to $\mathcal{M}$ (in place of $\mathcal{L}$) and point D (in place of P) to obtain points B and C on $\mathcal{M}$ such that B–D–C and $D \in \overset{[\!-\!-\!]}{BC} \subseteq \operatorname{ins}\mathcal{T} \cap \operatorname{ins}\mathcal{S}$.

Then A, B, and C are all members of $\operatorname{ins}\mathcal{T} \cap \operatorname{ins}\mathcal{S}$. By two applications of Exercise PSH.42, $\operatorname{enc} \triangle ABC \subseteq (\operatorname{ins}\mathcal{T} \cap \operatorname{ins}\mathcal{S})$. □

5.9 Mostly about quadrilaterals

Theorem PSH.53. *A quadrilateral is not rotund iff exactly one of its corners belongs to the inside of the triangle whose corners are the other three corners of the quadrilateral.*

Proof. By Definition PSH.31 $\square ABCD$ is not rotund iff at least one of the following statements is true:

(1) A and B are on opposite sides of $\overleftrightarrow{CD}$;
(2) C and D are on opposite sides of $\overleftrightarrow{AB}$;
(3) A and D are on opposite sides of $\overleftrightarrow{BC}$;
(4) B and C are on opposite sides of $\overleftrightarrow{AD}$.

Statements (1) and (2) cannot both be true. If they were we would have by Axiom PSA $\overset{]\!-\!-\![}{AB} \cap \overleftrightarrow{CD} \neq \emptyset$ and $\overset{]\!-\!-\![}{CD} \cap \overleftrightarrow{AB} \neq \emptyset$ and then, since, by Corollary IB.5.2, $\overset{]\!-\!-\![}{AB} \subseteq \overleftrightarrow{AB}$ and $\overset{]\!-\!-\![}{CD} \subseteq \overleftrightarrow{CD}$, we have that $\overleftrightarrow{AB} \cap \overleftrightarrow{CD} \neq \emptyset$. Since by Definition PSH.29 A, B, C, and D are noncollinear, Exercise I.1 implies there is exactly one point P such that $\overleftrightarrow{AB} \cap \overleftrightarrow{CD} = \{P\}$. But P must belong to $\overset{]\!-\!-\![}{AB}$ because if it did not, the above relationships would imply $\overleftrightarrow{AB}$ intersected $\overleftrightarrow{CD}$ in some other point besides P, so A, B, C, and D would be collinear, a contradiction. In a similar way we can infer $P \in \overset{]\!-\!-\![}{CD}$. Therefore $\overset{]\!-\!-\![}{AB}$ and $\overset{]\!-\!-\![}{CD}$ would intersect, contradicting Definition PSH.31(2).

Similar reasoning shows (3) and (4) cannot both be true.

If (1) is true, by Axiom PSA there exists a point P such that $\overline{AB} \cap \overleftrightarrow{CD} = \{P\}$. By Property B.2 of Definition IB.1, exactly one of C–D–P, D–C–P, or C–P–D holds. The last of these contradicts $\overline{AB} \cap \overline{CD} = \emptyset$, which we know to be true by Definition PSH.31(2); so this is ruled out.

Suppose that C–D–P. Since $A \notin \overleftrightarrow{CD} = \overleftrightarrow{CP}$, $\angle PAC = \angle BAC$ is defined. By Theorem PSH.40, since $C \in \overrightarrow{AC}$ and $P \in \overrightarrow{AB}$, $\overline{CP} \subseteq \text{ins}\,\angle BAC$, and since $D \in \overline{CP}$, $D \in \text{ins}\,\angle BAC$. By Theorem PSH.39, there exists a point Q such that $\overrightarrow{AD} \cap \overline{BC} = \{Q\}$, and thus by Definition IB.3, B–Q–C. By Theorem IB.5 $Q \in \overleftrightarrow{AD}$. By Definition IB.11 statement (4) is true. Moreover, since $P \in \overline{AB}$ by Corollary IB.14.2 $P \in \overrightarrow{BCA}$ and $D \in \overrightarrow{BCA}$ follows by the same corollary. Therefore by Theorem PSH.46(C) $D \in \text{ins}\,\triangle ABC$.

On the other hand if D–C–P, then by arguments similar to those just used, $C \in \text{ins}\,\triangle BAD$ and (3) is true.

Again, similar reasoning shows that if (2) is true, then either (3) is true and $B \in \text{ins}\,\triangle ACD$, or (4) is true and $A \in \text{ins}\,\triangle BCD$. Thus there are four mutually exclusive possibilities:

(i) Statement (1) and (4) are true, (2) and (3) are false, and $D \in \text{ins}\,\triangle ABC$.
(ii) Statement (1) and (3) are true, (2) and (4) are false, and $C \in \text{ins}\,\triangle ABD$.
(iii) Statement (2) and (3) are true, (1) and (4) are false, and $B \in \text{ins}\,\triangle ACD$.
(iv) Statement (2) and (4) are true, (1) and (3) are false, and $A \in \text{ins}\,\triangle BCD$.

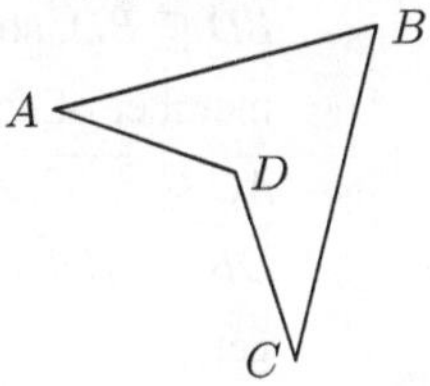

Fig. 5.12 For Theorem PSH.53 alternative (i).

Figure 5.12 illustrates alternative (i) above. The reader will find it quite easy to construct figures for the other alternatives (ii) through (iv). □

Theorem PSH.53.1. *If a quadrilateral is a trapezoid, then it is rotund.*

Proof. Let $\mathcal{T}$ be a trapezoid on a Pasch plane $\mathcal{P}$. By Definition PSH.31 there exist points A, B, C, and D such that $\mathcal{T} = \square ABCD$ and $\overleftrightarrow{AD} \parallel \overleftrightarrow{BC}$ or $\overleftrightarrow{AB} \parallel \overleftrightarrow{CD}$.

Suppose $\mathcal{T}$ were not rotund. By Theorem PSH.53, exactly one of its corners belongs to the inside of the triangle whose corners are the other three corners of $\mathcal{T}$. Without loss of generality assume that $C \in \text{ins}\,\triangle ABD$. By Theorem PSH.46(C) $\text{ins}\,\triangle ABD \subseteq \text{ins}\,\angle ABD$ and $\text{ins}\,\triangle ABD \subseteq \text{ins}\,\angle ADB$. Then by the Crossbar theorem PSH.39, $\overleftrightarrow{BC} \cap \overline{AD} \neq \emptyset$, contradicting $\overleftrightarrow{AD} \parallel \overleftrightarrow{BC}$, and $\overleftrightarrow{CD} \cap \overline{AB} \neq \emptyset$, contradicting $\overleftrightarrow{AB} \parallel \overleftrightarrow{CD}$. Therefore $\mathcal{T}$ is rotund. □

Theorem PSH.54. (A) *The diagonals of a quadrilateral intersect iff the quadrilateral is rotund.*

(B) *A quadrilateral is rotund iff every corner P is inside the "opposite" angle ∠XYZ, where Y is the corner opposite P and X and Z are the corners adjacent to Y.*

Proof. (A) Let A, B, C, and D be the corners of a quadrilateral.

(I: If □ABCD is rotund, then $\overline{AC}$ and $\overline{BD}$ intersect at a point.) By Definition PSH.31 $C \in \overrightarrow{ABD}$ and $C \in \overrightarrow{ADB}$. By Definition PSH.36 $C \in \text{ins}\,\angle BAD$. By Theorem PSH.39 $\overrightarrow{AC} \cap \overline{BD} = \{E\}$ for some point E. By a similar argument $\overrightarrow{CA} \cap \overline{BD} = \{E'\}$. By Exercise IB.4 $\overline{BD} \subseteq \overrightarrow{BD} \subseteq \overleftrightarrow{BD}$ and $\overline{CA} \subseteq \overrightarrow{CA} \subseteq \overleftrightarrow{CA}$. Using this and Exercise I.1 we get $\overleftrightarrow{BD} \cap \overleftrightarrow{CA} = \{E\} = \{E'\}$, whence $E = E'$ and $\overline{AC} \cap \overline{BD} = \{E\}$.

(II: If $\overline{AC}$ and $\overline{BD}$ intersect at a point, then $\square ABCD$ is rotund.) Suppose $\overline{AC} \cap \overline{BD} = \{P\}$ for some point P; by Theorem IB.5 $\overline{AC} \subseteq \overleftrightarrow{AC}$ and $\overline{BD} \subseteq \overleftrightarrow{BD}$, so that by Exercise I.1, $\overleftrightarrow{AC} \cap \overleftrightarrow{BD} = \{P\}$. By Exercise IB.4 P is a member of each of the open rays $\overrightarrow{AC}$, $\overrightarrow{CA}$, $\overrightarrow{DB}$, and $\overrightarrow{BD}$. By Corollary IB.14.2 $\overline{AC} \subseteq \overrightarrow{ABC}$, $\overline{AC} \subseteq \overrightarrow{ADC}$, $\overline{CA} \subseteq \overrightarrow{CBA}$, $\overline{CA} \subseteq \overrightarrow{CDA}$, $\overline{BD} \subseteq \overrightarrow{BCD}$, $\overline{BD} \subseteq \overrightarrow{BAD}$, $\overline{DB} \subseteq \overrightarrow{ADB}$, and $\overline{DB} \subseteq \overrightarrow{DCB}$. Since $P \in \overline{AC} = \overline{CA}$ and $P \in \overline{BD} = \overline{DB}$ we get

(1) P and C both belong to $\overrightarrow{ABC}$, and P and D both belong to $\overrightarrow{ABD}$;
(2) P and C both belong to $\overrightarrow{ADC}$, and P and B both belong to $\overrightarrow{ADB}$;
(3) P and B both belong to $\overrightarrow{CDB}$, and P and A both belong to $\overrightarrow{CDA}$;
(4) P and A both belong to $\overrightarrow{BCA}$, and P and D both belong to $\overrightarrow{BCD}$.

As was noted in the paragraph after the proof of Theorem PSH.11, if two sides of a line intersect they are the same side. Both the C-side of $\overleftrightarrow{AB}$ and the D-side of $\overleftrightarrow{AB}$ contain the point P, so these are the same, and C and D are on the same side of $\overleftrightarrow{AB}$; likewise, B and C are on the same side of $\overleftrightarrow{AD}$, A and B are on the same side of $\overleftrightarrow{CD}$, and A and D are on the same side of $\overleftrightarrow{BC}$. By Definition PSH.31 $\square ABCD$ is rotund.

(B) If a quadrilateral is rotund and P is one of its corners, let Y be the corner opposite P, and let X and Z be the corners adjacent to Y. Then by rotundity, X and P belong to the same side of $\overleftrightarrow{YZ}$; also P and Z belong to the same side of $\overleftrightarrow{YX}$; therefore $P \in \text{ins} \angle XYZ$.

Conversely, let $\overleftrightarrow{XY}$ be any line containing an edge $\overline{XY}$ of a quadrilateral; let P be the corner opposite X, and Q be the corner which is neither X, Y, nor P. Neither P nor Q is on $\overleftrightarrow{XY}$. Then by hypothesis,

$$P \in \text{ins} \angle QXY = \text{the } Q\text{-side of } \overleftrightarrow{XY} \cap \text{the } Y\text{-side of } \overleftrightarrow{QX};$$

hence P and Q are on the same side of $\overleftrightarrow{XY}$. Since $\overline{XY}$ can be chosen to be any edge, this shows that $\square ABCD$ is rotund, by Definition PSH.31. □

Definition PSH.55. Let A, B, C, and D be points on a Pasch plane $\mathcal{P}$ such that the sets $\{A, B, C\}$, $\{B, C, D\}$, $\{C, D, A\}$, $\{D, A, B\}$ are noncollinear, and $\overline{AB} \cap \overline{CD} = \overline{AD} \cap \overline{BC} = \emptyset$.

If $\square ABCD$ is rotund, then the **inside** of $\square ABCD$ (ins $\square ABCD$) is

$$\overrightarrow{ABC} \cap \overrightarrow{BCD} \cap \overrightarrow{BCA} \cap \overrightarrow{DAB}.$$

If $\square ABCD$ is nonrotund, then using Theorem PSH.53 we choose the notation so that $C \in \text{ins} \triangle ABD$.

The **inside** of nonrotund $\square ABCD$ (ins $\square ABCD$) is ins $\triangle ABD \setminus$ enc $\triangle BCD$.

The **enclosure** of $\square ABCD$ (enc $\square ABCD$) is $\square ABCD \cup$ ins $\square ABCD$.

The **outside** of $\square ABCD$ (out $\square ABCD$) is $\mathcal{P} \setminus$ enc $\square ABCD$.

The **exclosure** of $\square ABCD$ (exc $\square ABCD$) is $\mathcal{P} \setminus$ ins $\square ABCD$.

Theorem PSH.56 (Mapping Segments). *Let A, B, C, and D be points such that $A \neq B$ and $C \neq D$. Then there exists a one-to-one mapping Φ of $\overline{AB}$ onto $\overline{CD}$ having the following properties:*

(I) *$\Phi(A) = C$ and $\Phi(B) = D$;*

(II) *if R, S, and T are members of $\overline{AB}$ then R–S–T iff $\Phi(R)$–$\Phi(S)$–$\Phi(T)$;*

(III) *if U and V are distinct members of $\overline{AB}$ then $\Phi(\overline{UV}) = \overline{\Phi(U)\Phi(V)}$ and $\Phi(\mathring{\overline{UV}}) = \mathring{\overline{\Phi(U)\Phi(V)}}$.*

Proof. The proof we give will be something more than a sketch, leaving several sub-arguments to the reader. We consider several different cases reflecting the various possible relationships among the points A, B, C, and D, being careful to ensure these cases cover all the possibilities. First, either (1) $C \in \overleftrightarrow{AB}$ or (2) $C \notin \overleftrightarrow{AB}$. If (1) is true then there are two possible subcases: either (1a): $D \notin \overleftrightarrow{AB}$ or (1b): $D \in \overleftrightarrow{AB}$. The subcase (1a) can be further subdivided into two "sub-subcases": subcase ($1a_1$) where $A = C$, or ($1a_2$) where $A \neq C$. Case (2) breaks into two subcases: either (2c), $D \in \overleftrightarrow{BC}$ or (2d), $D \notin \overleftrightarrow{BC}$.

(Subcase ($1a_1$): $A = C$ and $D \notin \overleftrightarrow{AB}$.) Using Properties B.3 and B.1 of Definition IB.1 let Q be a point such that Q–D–B. A simple argument involving Theorem PSH.37, which is left to the reader, shows $\overline{AB} \subseteq \text{ins}\,\angle AQD$. Hence we may use Theorem PSH.39 (Crossbar) to construct a mapping Φ by letting $\Phi(A) = A$ and $\Phi(B) = D$, and by letting $\Phi(X)$ be the member of $\overline{AD}$ such that $\overrightarrow{QX} \cap \overline{AD} = \{\Phi(X)\}$ for each X in $\overline{AB}$. This shows property (I) for Φ in this subcase.

It is possible to interchange the roles of $\overline{AD}$ and $\overline{AB}$ in the above argument and then construct a mapping Ψ by letting $\Psi(A) = A$ and $\Psi(D) = B$, and by letting $\Psi(Y)$ be the member of $\overline{AB}$ such that $\overrightarrow{QY} \cap \overline{AB} = \{\Psi(Y)\}$ for each Y in $\overline{AD}$.

From these definitions of Φ and Ψ it follows that $\Psi(\Phi(X)) = X$ for every X in $\overline{AB}$, and $\Phi(\Psi(Y)) = Y$ for every Y in $\overline{AD}$. To prove the first of these statements let X be any member of $\overline{AB}$. Then by the definition of Φ above and Theorem PSH.16 $\Phi(X) \in \overrightarrow{QX} = \overrightarrow{Q\Phi(X)}$. Therefore by the definition of Ψ above $\overrightarrow{QX} \cap \overline{AB} = \overrightarrow{Q\Phi(X)} \cap \overline{AB} = \{\Psi(\Phi(X)\}$. We save this fact in our memory and notice that since $X \in \overline{AB}$ by the definition of X, and $X \in \overrightarrow{QX}$ by Definition IB.4, we have $\{X\} \subseteq \overrightarrow{QX} \cap \overline{AB}$. Elementary arguments left to the reader show A, B, and Q are noncollinear and therefore $\overrightarrow{QX} \cap \overline{AB}$ is a singleton. Hence $\{X\} = \overrightarrow{QX} \cap \overline{AB}$.

Putting this together with the statement we saved above gives $\{X\} = \{\Psi(\Phi(X))\}$, or $X = \Psi(\Phi(X))$, which is what we wished to prove. A similar proof shows $\Phi(\Psi(Y)) = Y$ for every Y in $\overline{AD}$. From these facts it follows by elementary mapping theory that Φ is a one-to-one mapping of $\overline{AB}$ onto $\overline{AD}$, that Ψ is a one-to-one mapping of $\overline{AD}$ onto $\overline{AB}$, and that Φ and Ψ are inverses of each other.

To prove property (II) for subcase ($1a_1$), let R, S, and T be any members of $\overline{AB}$ such that R–S–T, and let Q be as above. Then arguing as above we get $\overrightarrow{QR}$ and $\overrightarrow{QT}$ intersect $\overline{AD}$ in points $\Phi(R)$ and $\Phi(T)$, respectively. Since $S \in \overline{RT}$, $S \in \text{ins}\,\angle RQT$ follows by Theorem PSH.37. But $\angle RQT = \angle\Phi(R)Q\Phi(T)$ by Theorem PSH.30 so $S \in \text{ins}\,\angle\Phi(R)Q\Phi(T)$ and by Theorem PSH.39 $\overrightarrow{QS}$ intersects $\overline{\Phi(R)\Phi(T)}$ in a point $\Phi(S)$. Hence $\Phi(R)$–$\Phi(S)$–$\Phi(T)$.

Suppose now that R, S, and T belong to $\overline{AB}$, that $\Phi(R)$–$\Phi(S)$–$\Phi(T)$, and Q is as above. An argument similar to those above shows R–S–T, proving property (II).

Property (III) is an immediate consequence of (II) and Definition IB.3. This completes the argument for subcase (1a$_1$).

(Subcase (1a$_2$): $C \in \overleftrightarrow{AB}$ and $D \notin \overleftrightarrow{AB}$, and $A \neq C$.) Let E be a member of ins $\angle ACD$. By subcase (1a$_i$) there exist one-to-one mappings Φ_1, Φ_2, and Φ_3 such that $\Phi_1(\overline{AB}) = \overline{AE}$, $\Phi_1(A) = A$, $\Phi_1(B) = E$, $\Phi_2(\overline{AE}) = \overline{CE}$, $\Phi_2(A) = C$, $\Phi_2(E) = E$, $\Phi_3(\overline{CE}) = \overline{CD}$, $\Phi_3(C) = C$, and $\Phi_3(E) = D$. Furthermore, each of the mappings Φ_1, Φ_2, and Φ_3 can be constructed so that it has the properties (II) and (III). Let $\Phi = \Phi_3 \circ \Phi_2 \circ \Phi_1$. Then from the definition of composition and from its elementary properties it follows that Φ is a one-to-one mapping of $\overline{AB}$ onto $\overline{CD}$ such that $\Phi(A) = C$, $\Phi(B) = D$, and Φ has properties (II) and (III).

(Subcase (1b): $C \in \overleftrightarrow{AB}$ and $D \in \overleftrightarrow{AB}$.) Let E be any point off of $\overleftrightarrow{AB}$. By subcase (1a$_1$) there exist one-to-one mappings Φ_1, Φ_2, and Φ_3 such that $\Phi_1(\overline{AB}) = \overline{AE}$, $\Phi_1(A) = A$, $\Phi_1(B) = E$, $\Phi_2(\overline{AE}) = \overline{CE}$, $\Phi_2(A) = C$, $\Phi_2(E) = E$, $\Phi_3(\overline{CE}) = \overline{CD}$, $\Phi_3(C) = C$, and $\Phi_3(E) = D$. The remainder of the proof of this case is similar to that of the preceding case and the details are left to the reader.

(Subcase (2c): $C \notin \overleftrightarrow{AB}$ and $D \in \overleftrightarrow{BC}$.) By subcase (1a$_{ii}$) there exists a one-to-one mapping Φ_1 such that $\Phi_1(\overline{AB}) = \overline{BC}$, $\Phi_1(A) = C$, and $\Phi_1(B) = B$, and by subcase (1b) there exists a one-to-one mapping Φ_2 such that $\Phi_2(\overline{BC}) = \overline{CD}$, $\Phi_2(C) = C$, and $\Phi_2(B) = D$. The remainder of the proof is similar to those above and the details are left to the reader.

(Subcase (2d): $C \notin \overleftrightarrow{AB}$ and $D \notin \overleftrightarrow{BC}$.) It is left to the reader to introduce appropriate mappings Φ_1 and Φ_2 whose composition will provide the required mapping as in the above cases.

Theorem PSH.57. *Let $\mathcal{L}$ be a line on a Pasch plane $\mathcal{P}$ and let A, B, and C be distinct collinear points on $\mathcal{P}$, none of which is on $\mathcal{L}$. If $\mathcal{L}$ and $\overset{\vdash\dashv}{AB}$ intersect at the point P, then either $\mathcal{L} \cap \overset{\vdash\dashv}{AC} = \{P\}$ and $\mathcal{L} \cap \overset{\vdash\dashv}{BC} = \emptyset$, or $\mathcal{L} \cap \overset{\vdash\dashv}{AC} = \emptyset$ and $\mathcal{L} \cap \overset{\vdash\dashv}{BC} = \{P\}$.*

Proof. By Exercise I.1 $\mathcal{L} \cap \overleftrightarrow{AB} = \{P\}$. By Property B.2 of Definition IB.1 one and only one of the following statements is true: A–B–C, B–A–C, A–C–B. By Exercise I.2, $\overleftrightarrow{AB} = \overleftrightarrow{AC} = \overleftrightarrow{BC}$.

(A) If A–B–C, then by Theorem PSH.15 $\overset{\vdash\dashv}{AB} \subseteq \overset{\vdash\dashv}{AC}$ and $\overset{\vdash\dashv}{AB} \cap \overset{\vdash\dashv}{BC} = \emptyset$. By Theorem IB.5 $\overline{AC} \subseteq \overleftrightarrow{AC} = \overleftrightarrow{AB}$. If we put these statements together with the fact $\mathcal{L} \cap \overset{\vdash\dashv}{AB} = \{P\}$, we get by elementary set theory $\mathcal{L} \cap \overset{\vdash\dashv}{AC} = \{P\}$ and $\mathcal{L} \cap \overset{\vdash\dashv}{BC} = \emptyset$.

(B) If B–A–C, then an argument similar to (A) shows $\mathcal{L} \cap \overset{\vdash\dashv}{BC} = \{P\}$ and $\mathcal{L} \cap \overset{\vdash\dashv}{AC} = \emptyset$.

(C) If A–C–B, then by Theorem PSH.15 $\overset{\vdash\dashv}{AC}$ and $\overset{\vdash\dashv}{BC}$ are both subsets of $\overset{\vdash\dashv}{AB}$ and $\overset{\vdash\dashv}{AC} \cap \overset{\vdash\dashv}{BC} = \emptyset$. Hence by elementary set theory either $\mathcal{L} = \overset{\vdash\dashv}{AC} = \{P\}$ and $\mathcal{L} \cap \overset{\vdash\dashv}{BC} = \emptyset$ or $\mathcal{L} = \overset{\vdash\dashv}{AC} = \emptyset$ and $\mathcal{L} \cap \overset{\vdash\dashv}{BC} = \{P\}$. □

5.10 Exercises for Pasch geometry

Answers to starred (*) exercises may be accessed from the home page for this book at www.springer.com.

In the following exercises, all points and lines are in a Pasch plane.

Exercise PSH.0*. (A) Let $\mathcal{P}$ be a Pasch plane, $\mathcal{L}$ and $\mathcal{M}$ be lines, O be a point on $\mathcal{P}$ such that $\mathcal{L} \cap \mathcal{M} = \{O\}$. If $\mathcal{H}$ is a side of $\mathcal{L}$, then $\mathcal{M} \cap \mathcal{H} \neq \emptyset$.

(B) Let $\mathcal{P}$ be a Pasch plane and let $\mathcal{J}$, $\mathcal{K}$, and $\mathcal{L}$ be distinct lines on $\mathcal{P}$ such that $\mathcal{J} \cap \mathcal{L} \neq \emptyset$ and $\mathcal{K} \cap \mathcal{L} \neq \emptyset$. If U is a point on $\mathcal{J}$ but is not on $\mathcal{L}$, then there is a point V on $\mathcal{K}$ such that U and V are on opposite sides of $\mathcal{L}$.

Exercise PSH.1*. Complete the details of the proof of Theorem PSH.8, part (B)(1).

Exercise PSH.2*. (A) Prove Corollary PSH.8.3.

(B) Prove Corollary PSH.8.4.

Exercise PSH.3*. Let A, B, C, and D be points such that A–B–C–D and let P and Q be points such that P–A–B and C–D–Q. Then:

(A) $\overleftrightarrow{AB} = \overleftrightarrow{AC} = \overleftrightarrow{AD} = \overleftrightarrow{BC} = \overleftrightarrow{BD} = \overleftrightarrow{CD}$; the points A, B, C, and D are collinear;

(B) $\overleftrightarrow{BC}$ is the union of the disjoint sets $\{B, C\}$, $\overrightarrow{BA}$, $\overset{\vdash\dashv}{BC}$, and $\overrightarrow{CD}$;

(C) $\overleftrightarrow{BC}$ is the union of the disjoint sets $\{A, B, C, D\}$, $\overrightarrow{AP}$, $\overset{\vdash\dashv}{AB}$, $\overset{\vdash\dashv}{BC}$, $\overset{\vdash\dashv}{CD}$, and $\overrightarrow{DQ}$;

(D) $\overline{AD}$ is the union of the disjoint sets $\{A, B, C, D\}$, $\overset{\vdash\dashv}{AB}$, $\overset{\vdash\dashv}{BC}$, and $\overset{\vdash\dashv}{CD}$;

(E) $\overleftrightarrow{AD}$ is the union of the sets $\{X \mid X\text{–}A\text{–}D\}$, $\overline{AD}$, and $\{X \mid A\text{–}D\text{–}X\}$, which are all disjoint.

Exercise PSH.4*. (A) Let A and B be distinct points on a Pasch plane $\mathcal{P}$ and let $\mathcal{E}$ be a nonempty subset of $\overset{\vdash\dashv}{AB}$. Then $\mathcal{E}$ is not a ray.

(B) Let A and B be distinct points on a Pasch plane $\mathcal{P}$ and let $\mathcal{E}$ be a nonempty subset of $\overline{AB}$. Then $\mathcal{E}$ is not a ray.

Exercise PSH.5*. Let A, B, C, D, and E be points on plane $\mathcal{P}$ such that A, B, and C are noncollinear, A–B–D, and A–C–E. Then $D \in \text{ins}\,\angle BCE$.

Exercise PSH.6. Let A, B, C, D, and E be as in Exercise PSH.5. Then $\overleftrightarrow{AB} \cap (\text{ins}\,\angle BCE) = \overrightarrow{BD}$ and $\overleftrightarrow{AB} \cap (\text{out}\,\angle BCE) = \overrightarrow{BA}$.

Exercise PSH.7. Let A, B, C, D, and E be as in Exercise PSH.5. Then there exists a point F such that $\overset{\circ}{\overline{BE}} \cap \overset{\circ}{\overline{CD}} = \{F\}$.

Exercise PSH.8*. Let O, A, B, A', and B' be points on $\mathcal{P}$ such that O, A, and B are noncollinear, B–O–B', and A–O–A'. Let X be any member of ins $\angle AOB$, and let X' be any point such that X–O–X'. Then $\overleftrightarrow{OX} \cap \text{ins}\,\angle A'OB' = \overrightarrow{OX'}$.

Exercise PSH.9*. Let O, A, B, A', and B' be points on a Pasch plane $\mathcal{P}$ such that O, A, and B are noncollinear, B'–O–B, and A'–O–A, let X be any member of ins $\angle AOB$ and X' be any point such that X'–O–X.

(A) $\overleftrightarrow{OX} \cap \overset{\circ}{\overline{A'B'}} = \overrightarrow{OX'} \cap \overset{\circ}{\overline{A'B'}}$ is a singleton, *i.e.*, there exists a point Y such that $\overleftrightarrow{OX} \cap \overset{\circ}{\overline{A'B'}} = \{Y\}$.

(B) Let $X \in \overline{AB}$; if $X \in \overset{\circ}{\overline{AB}}$ define $\Omega(X) = Y$, where Y is as in part (A); if $X = A$ define $Y = A'$, and if $X = B$ define $Y = B'$. Then the mapping Ω maps $\overline{AB}$ onto $\overline{A'B'}$ and is one-to-one, hence is a bijection.

Exercise PSH.10. If A and B are distinct points, then $\{A, B\}$ is nonconvex.

Exercise PSH.11. Let $\mathcal{P}$ be a Pasch plane, $\mathcal{L}$ be a line on $\mathcal{P}$, and let $\mathcal{J}$ be a side of $\mathcal{L}$. If $P \in \mathcal{L}$ and $Q \in \mathcal{J}$, then $\overset{\circ}{\overline{PQ}} \subseteq \mathcal{J}$.

Exercise PSH.12*. Let A, B, and C be noncollinear points on a Pasch plane. If $D \in \text{ins}\,\angle BAC$, by Corollary PSH.39.2 B and C are on opposite sides of $\overleftrightarrow{AD}$. Prove that $\overrightarrow{AB} \subseteq \overrightarrow{ADB}$, $\overrightarrow{AC} \subseteq \overrightarrow{ADC}$, $B \in \text{out}\,\angle CAD$, and $C \in \text{out}\,\angle BAD$.

Exercise PSH.13*. Let A, B, and C be noncollinear points on a Pasch plane, and let P and Q be members of ins $\angle BAC$. Then if $P \in \text{ins}\,\angle BAQ$, $Q \in \text{ins}\,\angle CAP$.

Exercise PSH.14 (Key exercise)*.

(A) Let $\mathcal{E}$ be a convex subset of plane $\mathcal{P}$ and let $\mathcal{L}$ be a line on $\mathcal{P}$. If $\mathcal{E} \cap \mathcal{L} = \emptyset$, then $\mathcal{E}$ is a subset of a side of $\mathcal{L}$.

(B) If a line $\mathcal{M}$, or a segment or a ray does not intersect $\mathcal{L}$, then that line, segment, or ray lies entirely on one side of $\mathcal{L}$.

Exercise PSH.15*. Let A, B, and C be noncollinear points on a Pasch plane $\mathcal{P}$ and let $\mathcal{L}$ be a line on $\mathcal{P}$. If $\{A, B, C\} \cap \mathcal{L} = \emptyset$, then either $\mathcal{L} \cap \triangle ABC = \emptyset$ or $\mathcal{L}$ intersects two and only two edges of $\triangle ABC$. Moreover, $\mathcal{L} \cap \triangle ABC$ is a doubleton.

Exercise PSH.16*. The inside of every angle is convex and the inside of every triangle is convex.

Exercise PSH.17*. Let $\mathcal{P}$ be a Pasch plane and let A, B, and C be noncollinear points on $\mathcal{P}$.

(A) If $D \in \text{ins}\,\angle BAC$, then $\overrightarrow{AD} \subseteq \text{ins}\,\angle BAC$.
(B) $\text{ins}\,\angle BAC = \bigcup_{D \in \overline{BC}} \overrightarrow{AD}$

Exercise PSH.18 (Angle analog for Exercise PSH.32)*. Let A, B, and C be noncollinear points on a Pasch plane $\mathcal{P}$ and let D be a member of $\text{ins}\,\angle BAC$. Then $\text{ins}\,\angle BAC$ is the union of the disjoint sets $\overrightarrow{AD}$, $\text{ins}\,\angle BAD$ and $\text{ins}\,\angle DAC$.

Exercise PSH.19*. Prove parts (3) and (4) of Theorem PSH.48.

Exercise PSH.20*. The union of a line and one of its sides is convex (*i.e.*, a halfplane is convex).

Exercise PSH.21*. Let $\mathcal{A}$ be any subset of plane $\mathcal{P}$ having at least two members and let $\mathcal{B}$ be the union of all segments $\overline{PQ}$ such that $P \in \mathcal{A}$ and $Q \in \mathcal{A}$. Is $\mathcal{B}$ necessarily convex?

Exercise PSH.22*. If A, B, and C are noncollinear points, then both $\text{enc}\,\angle ABC$ and $\text{enc}\,\triangle ABC$ are convex sets.

Exercise PSH.23*. Construct a proof of part (A) of Theorem PSH.44 without referring to Theorem PSH.43; that is, using principally the definitions of inside, outside, and Theorem PSH.41(C).

Exercise PSH.24*. Prove Theorem PSH.47.

Exercise PSH.25*. Prove part (2) of Theorem PSH.49.

Exercise PSH.26*. Let A, B, and C be noncollinear points, let E be any member of $\overline{AC}$, and let F be any member of $\overline{AB}$. Then $\overline{BE}$ and $\overline{CF}$ intersect in a point O which belongs to $\text{ins}\,\triangle ABC$.

Exercise PSH.27*. Let A, B, and C be noncollinear points on plane $\mathcal{P}$, let Q be a member of ins $\angle ABC$, and R a member of ins $\angle ACB$. Then $\overrightarrow{BQ}$ and $\overrightarrow{CR}$ intersect at a point O which belongs to ins $\angle ABC$.

Exercise PSH.28*. Let A, B, and C be noncollinear points and suppose $P \in \overrightarrow{AB}$ and $Q \in$ ins $\angle BAC$. Then $\overline{PQ} \subseteq$ ins $\angle BAC$.

Exercise PSH.29*. Let A, B, and C be noncollinear points and suppose $P \in \triangle ABC$ and $Q \in$ ins $\triangle ABC$. Then $\overline{PQ} \subseteq$ ins $\triangle ABC$.

Exercise PSH.30*. Prove Theorem PSH.42.

Exercise PSH.31*. Let P and Q be distinct points on plane $\mathcal{P}$, let $\mathcal{H}$ be a side of $\overleftrightarrow{PQ}$ in $\mathcal{P}$, and let A and B be members of $\mathcal{H}$ such that A, B, and P are noncollinear. Then either $B \in$ ins $\angle APQ$ or $A \in$ ins $\angle BPQ$.

Exercise PSH.32 (Side analog for Exercise PSH.18). Let P, O, and Q be points such that P–O–Q, and let R be a point off of $\overleftrightarrow{OP}$. Then $\overrightarrow{OPR}$ is the union of the disjoint sets ins $\angle POR$, $\overrightarrow{OR}$, and ins $\angle QOR$.

Exercise PSH.33. Let A, B, and C be noncollinear points and let B' and C' be points such that B–A–B' and C–A–C'. Then out $\angle BAC$ is the union of the disjoint sets $\overrightarrow{AB'}$, $\overrightarrow{AC'}$, ins $\angle BAC'$, ins $\angle CAB'$, and ins $\angle B'AC'$.

Exercise PSH.34. Let A, B, and C be noncollinear points and let E be a member of out $\angle BAC$. Then $\overrightarrow{AE}$ is a subset of out $\angle BAC$.

Exercise PSH.35. Let A, B, and C be noncollinear points and let P and Q be members of (enc $\angle BAC \setminus \{A\}$) such that P, Q, and A are noncollinear. Then ins $\angle PAQ \subseteq$ ins $\angle BAC$. Note: try solving this before reading the proof of Theorem PSH.41(D).

Exercise PSH.36*. Let $\mathcal{L}$ be a line and let $\mathcal{H}$ be a side of $\mathcal{L}$. If A, B, and C are noncollinear members of $\mathcal{H}$, then enc $\triangle ABC \subseteq \mathcal{H}$.

Exercise PSH.37. Let A, B, C, R, and S be points such that A, B, and C are noncollinear, $R \in \overline{AB}$, and $S \in \overline{AC}$. Then $\overleftrightarrow{RS} \cap \overline{BC} = \emptyset$ and $\overleftrightarrow{BC} \cap \overline{RS} = \emptyset$.

Exercise PSH.38. Let $\mathcal{T}$ be a triangle, let P be a member of ins $\mathcal{T}$, and let Q be a point distinct from P. Then there exists a point R such that $\mathcal{T} \cap \overrightarrow{PQ} = \{R\}$, ins $\mathcal{T} \cap \overrightarrow{PQ} = \overline{PR}$, and out $\mathcal{T} \cap \overrightarrow{PQ} = \overrightarrow{PQ} \setminus \overline{PR}$.

Exercise PSH.39. Let A, B, and C be noncollinear points on plane $\mathcal{P}$, let P be a member of $\triangle ABC$, let Q be a member of ins $\triangle ABC$, and let R be a point such that Q–P–R. Then $R \in$ out $\triangle ABC$, $\overset{]\!\rightarrow}{QP} \cap \text{ins}\,\triangle ABC = \overset{[\,\,[}{QP}$, and $\overset{]\!\rightarrow}{QP} \cap \text{out}\,\triangle ABC = \overset{]\!\rightarrow}{QP} \setminus \overset{[\,\,]}{PQ}$.

Exercise PSH.40. Let $\mathcal{T}$ be a triangle, let P be a member of ins $\mathcal{T}$ and Q be a member of out $\mathcal{T}$. Then there exists a point R such that $\overleftrightarrow{PQ} \cap \mathcal{T} = \{R\}$, $\overset{[\,\,[}{PR} = \overset{[\,\,]}{PQ} \cap \text{ins}\,\mathcal{T}$, and $\overset{]\,\,]}{RQ} = \overset{[\,\,]}{PQ} \cap \text{out}\,\mathcal{T}$.

Exercise PSH.41. Let $\mathcal{T}$ be a triangle and let P, Q, and R be noncollinear members of enc $\mathcal{T}$. Then ins $\triangle PQR \subseteq$ ins $\mathcal{T}$.

Exercise PSH.42*. Let A, B, and C be noncollinear points and let P, Q, and R be noncollinear members of ins $\triangle ABC$. Then enc $\triangle PQR \subseteq$ ins $\triangle ABC$.

Exercise PSH.43. Let A, B, and C be noncollinear points on Pasch plane $\mathcal{P}$, let O be a member of ins $\triangle ABC$, let A' be any point between O and A, let B' be any point between O and B, and let C' be any point between O and C. Then $O \in$ ins $\triangle A'B'C'$, and enc $\triangle A'B'C' \subseteq$ ins $\triangle ABC$.

Exercise PSH.44. Let A, B, and C be noncollinear points. Then:

(a) There exist points P and Q such that A is between P and Q, $\angle BAC \cap \overset{[\,\,]}{PQ} = \{A\}$, and P and Q are both members of out $\angle BAC$.
(b) If P and Q are any points satisfying the conditions in (a) above, then B and C are on the same side of $\overleftrightarrow{PQ}$.

Exercise PSH.45*. Let $\mathcal{E}$ be a nonempty convex subset of the plane $\mathcal{P}$, and let A, B, and C be noncollinear members of $\mathcal{E}$. Then enc $\triangle ABC \subseteq \mathcal{E}$.

Exercise PSH.46. Let A, B, and C be noncollinear points and let O be a member of ins $\triangle ABC$. Then:

$$\text{ins}\,\triangle ABC = \overset{[\,\,[}{OA} \cup \overset{]\,\,[}{OB} \cup \overset{]\,\,[}{OC} \cup \text{ins}\,\triangle OAB \cup \text{ins}\,\triangle OAC \cup \text{ins}\,\triangle OBC.$$

Exercise PSH.47*. Let $\mathcal{P}$ be a Pasch plane and A, B, and U be noncollinear points. Then for every point V in $\mathcal{P}$,

(A) $\overset{]\!\rightarrow}{UV}$ is not a subset of $\overleftrightarrow{AB}$; and
(B) $\overset{]\,\,[}{UV}$ is not a subset of $\overleftrightarrow{AB}$.

Exercise PSH.48*. Prove parts 4–6 of Theorem PSH.18.

Exercise PSH.49*. Prove Theorem PSH.46(B).

Exercise PSH.50*. Prove Theorem PSH.46(C).

Exercise PSH.51*. Let $\mathcal{P}$ be a Pasch plane, O, B, and R be noncollinear points on $\mathcal{P}$, C be a member of ins $\angle ROB$ and B' be a point on $\overleftrightarrow{OB}$ such that B–O–B', then $R \in$ ins $\angle COB'$.

Exercise PSH.52*. (A) Let X be any point on a Pasch plane $\mathcal{P}$, then there exists a triangle $\mathcal{T}$ such that $X \in \text{ins}\,\mathcal{T}$.

(B) Let P and Q be distinct points on plane $\mathcal{P}$. Then there exist triangles $\mathcal{T}$ and $\mathcal{U}$ such that $P \in \text{ins}\,\mathcal{T}$, $Q \in \text{ins}\,\mathcal{U}$, $\text{enc}\,\mathcal{T} \subseteq \text{out}\,\mathcal{U}$, and $\text{enc}\,\mathcal{U} \subseteq \text{out}\,\mathcal{T}$.

Exercise PSH.53*. Let $\mathcal{P}$ be a Pasch plane, $\mathcal{L}$ and $\mathcal{L}'$ be distinct lines on $\mathcal{P}$, O be a member of $\mathcal{P} \setminus (\mathcal{L} \cup \mathcal{L}')$, A, B, and C be points on $\mathcal{L}$ such that A–B–C and A', B', and C' be points on $\mathcal{L}'$ such that A–O–A', B–O–B', and C–O–C', then A'–B'–C'.

Exercise PSH.54*. Let A, B, and C be points on a Pasch plane $\mathcal{P}$ such that A–B–C. Then $\overrightarrow{AB} \cap \overrightarrow{CB} = \overline{AC}$.

Exercise PSH.55 (Sets bounded by two parallel lines). Let $\mathcal{P}$ be the plane containing parallel lines $\mathcal{L}_1$ and $\mathcal{L}_2$, let P_1 and P_2 be points on $\mathcal{L}_1$ and $\mathcal{L}_2$, respectively, and let Q_1 and Q_2 be points on $\overleftrightarrow{P_1P_2}$ such that Q_1–P_1–P_2 and P_1–P_2–Q_2, $\mathcal{Q}_1$ be the Q_1-side of $\mathcal{L}_1$, let $\mathcal{Q}_1^*$ be the P_2-side of $\mathcal{L}_1$, let $\mathcal{Q}_2$ be the Q_2-side of $\mathcal{L}_2$, let $\mathcal{Q}_2^*$ be the P_1-side of $\mathcal{L}_2$, and let $\mathcal{Q} = \mathcal{Q}_1^* \cap \mathcal{Q}_2^*$. Then $\mathcal{Q}_1 \cap \mathcal{Q}_2 = \mathcal{Q}_1 \cap \mathcal{Q} = \mathcal{Q}_2 \cap \mathcal{Q} = \emptyset$; each of the sets $\mathcal{Q}_1$, $\mathcal{Q}_1^*$, $\mathcal{Q}_2$, $\mathcal{Q}_2^*$, and $\mathcal{Q}$ is convex; and $\mathcal{Q}_1 \cup \mathcal{Q}_2 \cup \mathcal{Q} = \mathcal{P} \setminus (\mathcal{L}_1 \cup \mathcal{L}_2)$.

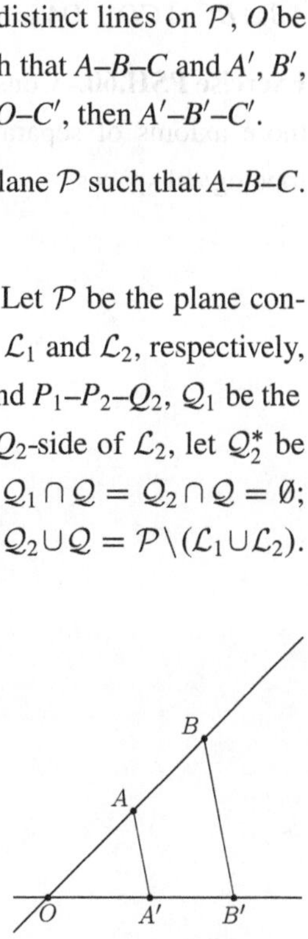

Fig. 5.13 For Exercise PSH.56(I).

Exercise PSH.56*. See Figure 5.13; note also that the symbol "$\|$" is defined in Chapter 2, Definition IP.1.

Let O, A, B, A', and B' be distinct points on a Pasch plane $\mathcal{P}$ such that $\overleftrightarrow{AB} \cap \overleftrightarrow{A'B'} = \{O\}$ and $\overleftrightarrow{AA'} \parallel \overleftrightarrow{BB'}$, then

(I) O–A–B iff O–A'–B',

(II) O–B–A iff O–B'–A', and

(III) A–O–B iff A'–O–B'.

Exercise PSH.57*. Let $\mathcal{L}$ and $\mathcal{M}$ be distinct lines in a Pasch plane, let A, B, and C be points of $\mathcal{L}$, and let D, E, and F be points of $\mathcal{M}$ such that $\overleftrightarrow{AD} \parallel \overleftrightarrow{BE} \parallel \overleftrightarrow{CF}$. Then A–B–C iff D–E–F.

Exercise PSH.58*. Prove Theorem PSH.34 using the result of Theorem PSH.32. That is, show that if A, B, and C are noncollinear points on a Pasch plane $\mathcal{P}$, then the set of corners of $\triangle ABC$ is $\{A, B, C\}$.

Exercise PSH.59*. Let A, B, C, and D be points on a Pasch plane $\mathcal{P}$ such that $\overline{AB} \cup \overline{BC} \cup \overline{CD} \cup \overline{DA}$ is a quadrilateral; then if $\overleftrightarrow{AB} \parallel \overleftrightarrow{CD}$, this quadrilateral is rotund.

Exercise PSH.60. Consult a book on projective geometry and compare/contrast those axioms of separation with those involving the open sets used to classify topological spaces.

Chapter 6
Ordering a Line in a Pasch Plane (ORD)

Acronym: *ORD*

Dependencies: *Chapters 1, 4, and 5*

New Axioms: *none*

New Terms Defined: *less than, greater than, less or equal, greater or equal, bounded above, bounded below, bounded, unbounded, maximum, minimum*

Abstract: This chapter defines order relations on lines and derives their properties, including transitivity and trichotomy. The concepts of maximum, minimum, upper bound, and lower bound of a subset of an ordered line are developed, as well as the connections between order, segments, and rays. Ordering will assume great importance in later chapters which develop the correspondence between a line and the set of all rational (or real) numbers.

Up to this point, we have never spoken of one point on a line being to the right or to the left of another point on the same line. It may seem silly to discuss this because one can usually settle the question for any two points on any nonvertical line (whatever *that* is) by just looking at a picture. But we want to continue to make our treatment of geometry independent of particular pictures, and this chapter is an attempt to deal with such questions in the same spirit of logical precision to which we have adhered thus far.

We do this by introducing a relation among the points of a line which is similar to the relation *is less than* (symbolized by "$<$") among the rational numbers and the real numbers. In fact, we will use the same symbol for this geometrical relation as

E.J. Specht et al., *Euclidean Geometry and its Subgeometries*,
DOI 10.1007/978-3-319-23775-6_6

the one for the numerical relation. Hence we will be writing such things as "$A < B$," where A and B are points on a line. When you read this symbol to yourself, you can say "A is less than B," or "A is to the left of B," or "B is to the right of A," or whatever other expression is most comfortable for you. Of course, if you choose "A is less than B," you must be careful to remember that this does not necessarily mean that the point A is less than the point B in some numerical sense. The reason that the above three verbalizations of the relation are appealing is that the relation we will define has two important properties that the relations *is less than*, *is to the left of*, *is to the right of* also have, namely, transitivity and trichotomy. The fact that our relation "<" has these two properties will be the subject of two early theorems in this chapter. We start with some definitions.

Definition ORD.1. Let $\mathcal{L}$ be a line, let $\mathcal{E}$ be a convex subset of $\mathcal{L}$ which is not a singleton, and let O and P be distinct members of $\mathcal{E}$.

(A) For any two distinct points X and Y of $\mathcal{E}$, X is **less than** Y (notation: $X < Y$) iff $\overrightarrow{XY} \cap \overrightarrow{OP}$ is a ray. When two such points O and P have been chosen and the relation "<" has been defined in this way, we describe this situation by saying that the points of $\mathcal{E}$ are **ordered** by the relation "<" so that $O < P$.

(B) Let the points of $\mathcal{E}$ be ordered by the relation "<" so that $O < P$. Then for any two distinct points X and Y of $\mathcal{E}$, X is **greater than** Y (notation: $X > Y$) iff $Y < X$; X is **less than or equal to** Y (notation: $X \leq Y$) iff $X < Y$ or $X = Y$; and X is **greater than or equal to** Y (notation: $X \geq Y$) iff $Y \leq X$.

Remark ORD.2. The initial choice of two points (here O and P) on which to base the definition of ordering is entirely arbitrary. There is nothing sacred about ordering $\mathcal{E}$ so that $O < P$. We could just as well reverse the roles of O and P in the above definition and define $P < O$, and then for any two distinct points X and Y, define X to be *less than* Y (notation: $X < Y$) iff $\overrightarrow{YX} \cap \overrightarrow{OP}$ is a ray. Then we could infer all the properties of "<" from the properties of ">" by the fact that $X < Y$ iff $Y > X$.

Remark ORD.3. It should be noted that the meaning of the symbol "<" for a particular geometry depends heavily on the particular betweenness relation in that geometry because the definition of a ray depends on betweenness. Hence, if, on a particular Pasch plane $\mathcal{P}$, a new betweenness relation is introduced, this could conceivably change which sets are rays and which are not, and that, in turn, could change which points are less than others.

6.1 Theorems for ordering

Theorem ORD.4 (Transitivity of the relation "<"). *Let $\mathcal{L}$ be a line, $\mathcal{E}$ be a convex subset of $\mathcal{L}$ which is not a singleton, O and P be distinct members of $\mathcal{E}$, and suppose that the members of $\mathcal{E}$ are ordered by "<" so that $O < P$. If X, Y, and Z are any three distinct members of $\mathcal{E}$ such that $X < Y$ and $Y < Z$, then $X < Z$.*

Proof. Suppose $X < Y$ and $Y < Z$, so that by Definition ORD.1, $\overrightarrow{OP} \cap \overrightarrow{XY}$ and $\overrightarrow{OP} \cap \overrightarrow{YZ}$ are rays. If $\overrightarrow{OP} \cap \overrightarrow{XY}$ is a ray, by the contrapositive of Theorem PSH.20 $Z \notin \overrightarrow{YX}$. Since $Z \in \overleftrightarrow{YX}$, by Theorem IB.5 and Property B.1 of Definition IB.1, X–Y–Z, and therefore $Z \in \overrightarrow{XY}$. By Theorem PSH.16, $\overrightarrow{XZ} = \overrightarrow{XY}$. Hence $\overrightarrow{OP} \cap \overrightarrow{XZ} = \overrightarrow{OP} \cap \overrightarrow{XY}$, and we already know that the latter intersection is a ray. Therefore by Definition ORD.1, $X < Z$. □

Theorem ORD.5 (Trichotomy for ordering "<"). *Let $\mathcal{L}$ be a line, $\mathcal{E}$ be a convex subset of $\mathcal{L}$ which is not a singleton, O and P be distinct members of $\mathcal{E}$, and suppose that the members of $\mathcal{E}$ are ordered by "<" so that $O < P$. If X and Y are any two members of $\mathcal{E}$, then one and only one of the following statements is true:*

$$X = Y, \qquad X < Y, \qquad Y < X.$$

Proof. If $X \neq Y$, then either $\overrightarrow{OP} \cap \overrightarrow{XY}$ is a ray or it is not. If it is, then by Definition ORD.1 $X < Y$. If it is not, we wish to show that $Y < X$, *i.e.*, that $\overrightarrow{OP} \cap \overrightarrow{YX}$ is a ray.

By the contrapositive of Theorem PSH.19, $\overrightarrow{XY} \not\subseteq \overrightarrow{OP}$ and $\overrightarrow{OP} \not\subseteq \overrightarrow{XY}$. This means that there exist points S and T such that $S \in \overrightarrow{XY}$, $S \notin \overrightarrow{OP}$, $T \in \overrightarrow{OP}$, and $T \notin \overrightarrow{XY}$. By Theorem PSH.16, $\overrightarrow{XS} = \overrightarrow{XY}$ and $\overrightarrow{OT} = \overrightarrow{OP}$. Since $S \notin \overrightarrow{OP} = \overrightarrow{OT}$ and $T \notin \overrightarrow{XY} = \overrightarrow{XS}$, but all of these points belong to $\mathcal{L}$, we have by Theorem IB.5 that S–O–T, T–X–S, and T–X–Y, whence T–O–S and Y–X–T. Then by Definition IB.4 $O \in \overrightarrow{TS}$, $X \in \overrightarrow{TS}$, and $T \in \overrightarrow{YX}$. Theorem PSH.16 then gives $\overrightarrow{TO} = \overrightarrow{TS}$, $\overrightarrow{TX} = \overrightarrow{TS}$, and $\overrightarrow{YX} = \overrightarrow{YT}$, so that $\overrightarrow{TO} = \overrightarrow{TX}$. We will use these assorted facts freely in what follows.

Since S–O–T, by Theorem PSH.13, $\overrightarrow{TO} = \overline{TO} \cup \{O\} \cup \overrightarrow{OS}$, and the sets in this union are mutually disjoint, so we may use this as a basis for splitting the proof into three cases at this point.

(Case 1: $Y = O$.) If $Y = O$, then T–X–O, and by Theorem PSH.16, $\overrightarrow{YX} = \overrightarrow{OX} = \overrightarrow{OT} = \overrightarrow{OP}$. Hence $\overrightarrow{YX} \cap \overrightarrow{OP} = \overrightarrow{OP}$, which is a ray, as we wished to prove.

(Case 2: $Y \in \overrightarrow{OS}$.) If $Y \in \overrightarrow{OS}$, then by Theorem PSH.16, $\overrightarrow{OY} = \overrightarrow{OS}$. Since T–O–S, by Theorem IB.5 $T \notin \overrightarrow{OS} = \overrightarrow{OY}$ and therefore T–O–Y. By Theorem PSH.13 $\overrightarrow{OT} \subseteq \overrightarrow{YT}$. Hence $\overrightarrow{YX} \cap \overrightarrow{OP} = \overrightarrow{YT} \cap \overrightarrow{OT} = \overrightarrow{OT}$, which is a ray, as we wished to prove.

(Case 3: $Y \in \overline{TO}$.) If $Y \in \overline{TO} = \overline{OT}$, then by (i) following Definition IB.4, $Y \in \overrightarrow{OT}$, so by Theorem PSH.16, $\overrightarrow{OT} = \overrightarrow{OY}$. Also from $Y \in \overline{OT}$ it follows that O–Y–T, whence by Theorem PSH.13, $\overrightarrow{YT} \subseteq \overrightarrow{OY}$.

From these we can say that $\overrightarrow{YX} \cap \overrightarrow{OP} = \overrightarrow{YT} \cap \overrightarrow{OT} = \overrightarrow{YT} \cap \overrightarrow{OY} = \overrightarrow{YT}$, which is a ray, as we wished to prove. □

Theorem ORD.6. *Let $\mathcal{L}$ be a line, $\mathcal{E}$ be a convex subset of $\mathcal{L}$ which is not a singleton, O and P be distinct members of $\mathcal{E}$, and suppose that the members of $\mathcal{E}$ are ordered by "<" so that $O < P$. Then for all points X, Y, and Z of $\mathcal{E}$, X–Y–Z iff $X < Y < Z$ or $Z < Y < X$.*

Proof. (I: If X–Y–Z, then $X < Y < Z$ or $Z < Y < X$.) By Property B.0 of Definition IB.1, X, Y, and Z are distinct. By Theorem ORD.5 either $X < Y$ or $Y < X$, so we may split the proof into two cases.

(Case I.A: $X < Y$.) By definition $\overrightarrow{XY} \cap \overrightarrow{OP}$ is a ray. Since X–Y–Z, $\overrightarrow{XZ} = \overrightarrow{XY}$ so $\overrightarrow{XZ} \cap \overrightarrow{OP}$ is a ray. Now $Y \in \overline{ZX}$ so by Theorem PSH.20, $\overrightarrow{OP} \cap \overrightarrow{ZY}$ is not a ray. Thus Z is not less than Y. Hence by Trichotomy (ORD.5), $Y < Z$.

(Case I.B: $Y < X$.) Suppose $Y < X$. Then $\overrightarrow{YX} \cap \overrightarrow{OP}$ is a ray. By Theorem PSH.19, either $\overrightarrow{YX} \subseteq \overrightarrow{OP}$ or $\overrightarrow{OP} \subseteq \overrightarrow{YX}$. Suppose $\overrightarrow{OP} \subseteq \overrightarrow{YX}$. Then $\overrightarrow{OP} \subseteq \overrightarrow{ZY}$ since $\overrightarrow{YX} \subseteq \overrightarrow{ZY}$ by Theorem PSH.13. Hence Theorem PSH.19 implies that $\overrightarrow{OP} \cap \overrightarrow{ZY}$ is a ray, so that $Z < Y$.

Now suppose $\overrightarrow{YX} \subseteq \overrightarrow{OP}$; if $Y = O$, then by Theorem PSH.16 $\overrightarrow{YX} = \overrightarrow{OP}$ and by Theorem PSH.13 since Z–Y–X, $\overrightarrow{ZY} \cap \overrightarrow{YX} = \overrightarrow{ZY} \cap \overrightarrow{OP}$ is a ray, hence $Z < Y$.

If $Y \neq O$, by Theorem PSH.17 $Y \in \overrightarrow{OP}$ and O–Y–X. By Property B.2 of Definition IB.1 and Corollary PSH.8.3 the following three possibilities exist.

(a) If $O = Z$, then $\overrightarrow{OP} = \overrightarrow{OY} = \overrightarrow{ZY}$ so $\overrightarrow{OP} \cap \overrightarrow{ZY} = \overrightarrow{ZY}$.
(b) If Z–O–Y–X, then $\overrightarrow{OP} \cap \overrightarrow{ZY} = \overrightarrow{OY} \cap \overrightarrow{ZY} = \overrightarrow{OY}$ by Theorem PSH.17.
(c) If O–Z–Y–X, then $\overrightarrow{OP} \cap \overrightarrow{ZY} = \overrightarrow{OY} \cap \overrightarrow{ZY} = \overrightarrow{ZY}$ by Theorem PSH.17.

In each of (a) through (c), $Z < Y$.

(II: If $X < Y < Z$ or $Z < Y < X$, then X–Y–Z.) We deal with each of the possibilities in the hypothesis as separate cases.

(Case II.A: $X < Y < Z$.) If $X < Y < Z$, then by Definition ORD.1, X and Y are distinct, Y and Z are distinct, and $\overset{[\rightarrow}{XY} \cap \overset{[\rightarrow}{OP}$ and $\overset{[\rightarrow}{YZ} \cap \overset{[\rightarrow}{OP}$ are both rays. By Corollary PSH.20.1, X–Y–Z.

(Case II.B: $Z < Y < X$.) If $Z < Y < X$, X and Y are distinct, Y and Z are distinct, and $\overset{[\rightarrow}{ZY} \cap \overset{[\rightarrow}{OP}$ and $\overset{[\rightarrow}{YX} \cap \overset{[\rightarrow}{OP}$ are both rays. By Corollary PSH.20.1, Z–Y–X, that is, X–Y–Z. □

Theorem ORD.7. *Let O and P be distinct points on a Pasch plane $\mathcal{P}$ and suppose the points of $\overleftrightarrow{OP}$ are ordered so that $O < P$.*

(I) *If A and B are points on $\overleftrightarrow{OP}$ such that $A < B$, then there exist points C, D, and E such that $D < A < C < B < E$.*

(II) *If A and B are points on $\overleftrightarrow{OP}$ such that $A < B$, then*

$$\begin{aligned}
\overset{]\,[}{AB} &= \{X|A < X < B\} = \{X|B > X > A\},\\
\overset{[\,[}{AB} &= \{X|A \le X < B\} = \{X|B > X \ge A\},\\
\overset{]\,]}{AB} &= \{X|A < X \le B\} = \{X|B \ge X > A\},\\
\overset{[\,]}{AB} &= \{X|A \le X \le B\} = \{X|B \ge X \ge A\},\\
\overset{]\rightarrow}{AB} &= \{X|A < X\} = \{X|X > A\},\\
\overset{[\rightarrow}{AB} &= \{X|A \le X\} = \{X|X \ge A\};
\end{aligned}$$

while if A and B are points on $\overleftrightarrow{OP}$ such that $B < A$, then

$$\begin{aligned}
\overset{]\,[}{AB} &= \{X|A > X > B\} = \{X|B < X < A\},\\
\overset{[\,[}{AB} &= \{X|A \ge X > B\} = \{X|B < X \le A\},\\
\overset{]\,]}{AB} &= \{X|A > X \ge B\} = \{X|B \le X < A\},\\
\overset{[\,]}{AB} &= \{X|A \ge X \ge B\} = \{X|B \le X \le A\},\\
\overset{]\rightarrow}{AB} &= \{X|A > X\} = \{X|X < A\},\\
\overset{[\rightarrow}{AB} &= \{X|A \ge X\} = \{X|X \le A\}.
\end{aligned}$$

(III) *Let C be any member of $\overleftrightarrow{OP}$.*

(A) *If $\mathcal{D} = \{X|X > C\}$ and $\mathcal{D}$ is nonempty, then there exists a member D of $\overleftrightarrow{OP}$ such that $D > C$ and $\mathcal{D} = \overset{]\rightarrow}{CD}$;*

(B) *If $\mathcal{E} = \{X|X < C\}$ and $\mathcal{E}$ is nonempty, then there exists a member E of $\overleftrightarrow{OP}$ such that $E < C$ and $\mathcal{E} = \overset{]\rightarrow}{CE}$.*

Proof. (I) Since A and B are distinct, by Theorem PSH.22 (Denseness) there exists a point C such that A–C–B. By Theorem ORD.6, either $A < C < B$ or $B < C < A$. But the alternative $B < C < A$ implies that $B < A$ by Theorem ORD.4, and by Theorem ORD.5 this contradicts our assumption that

$A < B$, so we must conclude that $A < C < B$. By Properties B.3 and B.1 of Definition IB.1, there exist points D and E such that D–A–C and C–B–E. If we put these relationships together with the assumptions that $A < C$ and $C < B$ and argue as above, we get $D < A < C$ and $C < B < E$. This conjunction can be written $D < A < C < B < E$.

(II) Exercise ORD.6.

(III) Let D and E be any members of $\mathcal{D}$ and $\mathcal{E}$, respectively. Then by part (II) above, $\mathcal{D} = \overset{\sqsupset\!\rightarrow}{CD}$ and $\mathcal{E} = \overset{\sqsupset\!\rightarrow}{CE}$. □

Definition ORD.8. Let O and P be distinct points on a Pasch plane $\mathcal{P}$, let the points on $\overleftrightarrow{OP}$ be ordered so that $O < P$, and let $\mathcal{E}$ be a nonempty subset of $\overleftrightarrow{OP}$. Then:

(A) $\mathcal{E}$ is **bounded above** iff there exists a member A of $\overleftrightarrow{OP}$ such that $X \leq A$ for all X in $\mathcal{E}$. Such a point A is an **upper bound** of $\mathcal{E}$.

(B) $\mathcal{E}$ is **bounded below** iff there exists a member A of $\overleftrightarrow{OP}$ such that $A \leq X$ for all X in $\mathcal{E}$. Such a point A is an **lower bound** of $\mathcal{E}$.

(C) $\mathcal{E}$ is **bounded** iff $\mathcal{E}$ is bounded above and bounded below. $\mathcal{E}$ is **unbounded** iff it is not bounded.

(D) $\mathcal{E}$ has a **maximum** (largest element) iff there exists a member A of $\mathcal{E}$ such that $X \leq A$ for all $X \in \mathcal{E}$. If $\mathcal{E}$ has a maximum, it is denoted by $\max \mathcal{E}$.

(E)) $\mathcal{E}$ has a **minimum** (smallest element) iff there exists a member A of $\mathcal{E}$ such that $X \geq A$ for all $X \in \mathcal{E}$. If $\mathcal{E}$ has a minimum, it is denoted by $\min \mathcal{E}$.

Remark ORD.9. Let $\mathcal{M}$ be a line in a Pasch plane which is ordered by the order relation $<$. Let $\mathcal{E}$ be a nonempty subset of $\mathcal{M}$.

(A) If D is an upper bound of $\mathcal{E}$, and $E > D$, then E is an upper bound of $\mathcal{E}$.

(B) If F is a lower bound of $\mathcal{E}$, and $G < F$, then G is a lower bound of $\mathcal{E}$.

(C) Every segment in $\mathcal{M}$ is bounded.

(D) Every ray which is a subset of $\mathcal{M}$ is unbounded, and is either bounded above or bounded below.

(E) Let A and B be distinct points on the line $\mathcal{M}$. If $A < B$, then $\max \overset{\sqsubset\!\sqsupset}{AB} = B$, $\min \overset{\sqsubset\!\sqsupset}{AB} = A$, $\min \overset{\sqsubset\!\rightarrow}{AB} = A$.

(F) Using Theorem ORD.7(I), the reader may easily confirm that if $A < B$, each of the sets $\overset{\sqsupset\!\sqsubset}{AB}$, $\overset{\sqsupset\!\sqsupset}{AB}$, $\overset{\sqsupset\!\rightarrow}{BA}$, and $\overset{\sqsubset\!\rightarrow}{BA}$ has no minimum, and each of the sets $\overset{\sqsupset\!\sqsubset}{AB}$, $\overset{\sqsubset\!\sqsubset}{AB}$, $\overset{\sqsubset\!\rightarrow}{AB}$, and $\overset{\sqsupset\!\rightarrow}{AB}$ has no maximum.

Theorem ORD.10 (Finite sets are bounded). *Let $\mathcal{L}$ be a line which is ordered according to Definition ORD.1. Then every nonempty finite subset of $\mathcal{L}$ has a maximum and a minimum.*

Proof. Let $\mathcal{E}$ be a nonempty finite subset of $\mathcal{L}$. In elementary set theory (see Chapter 1) a nonempty set $\mathcal{S}$ is defined to be *finite* iff there exists a natural number p such that the number of elements in $\mathcal{S}$ is p. It is also shown that every subset of a finite set is finite.

We propose to prove the following statement by mathematical induction. If $\mathcal{E}$ is any subset of $\mathcal{L}$ having n members, where n is any natural number, then $\mathcal{E}$ has a maximum and a minimum.

(I) If $n = 1$, so that there exists a point A such that $\mathcal{E} = \{A\}$, then by Definition ORD.8, $\max \mathcal{E} = \min \mathcal{E} = A$. Hence the above statement is true when $n = 1$.

(II) If $n = 2$, then there exist distinct points A and B such that $\mathcal{E} = \{A, B\}$. Since $A \neq B$, by Theorem ORD.5, either $A < B$ or $A > B$. If $A < B$, then by Definition ORD.8, $\min \mathcal{E} = A$ and $\max \mathcal{E} = B$. Similarly, if $B < A$, then $\min \mathcal{E} = B$ and $\max \mathcal{E} = A$. Hence the above statement is true for $n = 2$.

(III) To complete the proof by induction, we must show that *if* the above statement is true for any natural number k, say, *then* it must be true for $k + 1$. To do this, we suppose that it is true for some k, *i.e.*, we suppose that every subset $\mathcal{E}$ of $\mathcal{L}$ having k members has a maximum and a minimum, and we use this (the *induction assumption*) to prove that every subset $\mathcal{S}$ of $\mathcal{L}$ having $k+1$ members has a maximum and a minimum. Let A be any member of $\mathcal{S}$, which member we keep fixed for the remainder of this argument. Then the set $\mathcal{S} \setminus \{A\}$ contains k members, so the induction assumption says that $\mathcal{S} \setminus \{A\}$ has both a maximum U and a minimum V.

From (II) above, $\{A, U\}$ has a maximum M, and $\{A, V\}$ has a minimum N. We will have completed the induction argument when we have shown that M is a maximum of $\mathcal{S}$ and N is a minimum of $\mathcal{S}$. To this end, let X be any member of $\mathcal{S}$. Since $\mathcal{S} = (\mathcal{S} \setminus \{A\}) \cup \{A\}$, and the sets in this union are disjoint, either $X = A$ or $X \in (\mathcal{S} \setminus A)$. If $X = A$, then directly from the definition of M and N $X \leq M$ and $N \leq X$. On the other hand, if $X \in (\mathcal{S} \setminus A)$, then by the definitions of U, V, M, and N, $X \leq U \leq M$ and $N \leq V \leq X$, so by Theorem ORD.4, $X \leq M$ and $N \leq X$. We have shown that every member X of $\mathcal{S}$ satisfies the requirements in Definition ORD.8, so $M = \max \mathcal{S}$ and $N = \min \mathcal{S}$. □

Theorem ORD.11. *Let $\mathcal{L}$ be a line which is ordered according to Definition ORD.1. Let $\mathcal{E}$ be a nonempty subset of $\mathcal{L}$.*

(I) *If $\mathcal{E}$ is bounded above and $\mathcal{U}$ is the set of its upper bounds, and if $\mathcal{E}$ has a maximum, then $\mathcal{U}$ has a minimum and* $\min \mathcal{U} = \max \mathcal{E}$.

(II) *If $\mathcal{E}$ is bounded below and $\mathcal{L}$ is its set of lower bounds, and if $\mathcal{E}$ has a minimum, then $\mathcal{L}$ has a maximum and* $\max \mathcal{L} = \min \mathcal{E}$.

(III) *If $\mathcal{E}$ is bounded above, and if the set of upper bounds of $\mathcal{E}$ has a minimum M, and $M \in \mathcal{E}$, then $\mathcal{E}$ has a maximum and* $\max \mathcal{E} = M$.

(IV) *If $\mathcal{E}$ is bounded below, and if the set of lower bounds of $\mathcal{E}$ has a maximum N, and $N \in \mathcal{E}$, then $\mathcal{E}$ has a minimum and* $\min \mathcal{E} = N$.

Proof. The proof of item (I) is a direct application of Definition ORD.8. Suppose U is the maximum of $\mathcal{E}$; U is a member of $\mathcal{U}$, since it is an upper bound. If there were a member $V \in \mathcal{U}$ with $V < U$, V would not be an upper bound of $\mathcal{E}$ and hence not a member of $\mathcal{U}$. Hence U is the minimum of the set $\mathcal{U}$ of upper bounds of $\mathcal{E}$. Proofs of (II)–(IV) are also easy consequences of Definition ORD.8. □

Theorem ORD.12. *Let A and B be distinct points in a line $\mathcal{L}$ which is ordered according to Definition ORD.1. Then each of the following sets is infinite: $\overleftrightarrow{AB}$, $\overset{]\!\!\longrightarrow}{AB}$, $\overset{[\!\!\longrightarrow}{AB}$, $\overset{]\!-\!-\![}{AB}$, $\overset{[\!-\!-\![}{AB}$, $\overset{]\!-\!-\!]}{AB}$, and $\overset{[\!-\!-\!]}{AB}$.*

Proof. From elementary set theory we recall that a set is infinite if it is not finite, and every set having an infinite subset is infinite. Since $\overset{]\!-\!-\![}{AB}$ is a subset of all of the sets listed above, we need only show that $\overset{]\!-\!-\![}{AB}$ is infinite. This has already been proved in Corollary PSH.22.2. The alternative proof we give here uses ordering.

By Theorem PSH.22, $\overset{]\!-\!-\![}{AB}$ is nonempty. Order the points on $\mathcal{L}$ so that $A < B$. Assume $\overset{]\!-\!-\![}{AB}$ is finite. Then by Theorem ORD.10, $\overset{]\!-\!-\![}{AB}$ has a minimum Q. By the definition of a minimum, $Q \in \overset{]\!-\!-\![}{AB}$. By Theorem ORD.7(II), $A < Q < B$. By Theorem ORD.7(I), there exists a point R such that $A < R < Q < B$. Using Theorem ORD.7(II) again shows that $R \in \overset{]\!-\!-\![}{AB}$. Hence we have found an element of $\overset{]\!-\!-\![}{AB}$ which is smaller than Q, which is a minimum of $\overset{]\!-\!-\![}{AB}$. This is a contradiction of Definition ORD.8, so our assumption that $\overset{]\!-\!-\![}{AB}$ is finite is false. □

6.2 Exercises for ordering

Answers to starred (*) exercises may be accessed from the home page for this book at www.springer.com.

Exercise ORD.1*. Let A, B, C, and D be points such that A–B–C–D. If the points on $\overleftrightarrow{AD}$ are ordered so that $A < D$, then $A < B < C < D$.

Exercise ORD.2. Let O and P be distinct points, and let $\mathcal{E}$ be a nonempty finite subset of $\overleftrightarrow{OP}$ which has n elements. Then there exists a mapping θ of $[1; n]$ onto $\mathcal{E}$ such that for every member k of $[1; n-1]$, $\theta(k) < \theta(k+1)$, and every member of $\{\theta(j) | j \in [1; k]\}$ is less than every member of $\mathcal{E} \setminus \{\theta(j) | j \in [1; k]\}$.

Exercise ORD.3. Let $\mathbb{D}$ be the field of dyadic rational numbers,[1] let $\mathbb{D}'$ be equal to $\mathbb{D} \cap [0; 1]$, and let A and B be distinct points on a Pasch plane $\mathcal{P}$. Then there exists a mapping θ of $\mathbb{D}'$ into $\overset{\sqsubset\!\sqsupset}{AB}$ such that, for all members r and s of $\mathbb{D}'$, $r < s$ iff $\theta(r) < \theta(s)$.

Exercise ORD.4*. Let $\mathcal{E}$ be a convex subset of a line $\mathcal{M}$. If $\mathcal{E}$ is not a singleton, then $\mathcal{E}$ is infinite.

Exercise ORD.5. Let $\mathcal{E}$ be an infinite convex subset of a line $\mathcal{M}$. If A is a member of $\mathcal{E}$, B is a member of $\mathcal{M} \setminus E$, and C is a point such that A–B–C, then $\overrightarrow{BC}$ is a subset of $\mathcal{M} \setminus E$.

Exercise ORD.6*. Prove Theorem ORD.7 part (II).

Exercise ORD.7*. Let A and B be distinct points on a Pasch plane $\mathcal{P}$ and let C and D be distinct members of $\overset{\sqsubset\!\sqsupset}{AB}$, then $\overset{\sqsupset\!\sqsubset}{CD} \subseteq \overset{\sqsupset\!\sqsubset}{AB}$ and $\overset{\sqsubset\!\sqsupset}{CD} \subseteq \overset{\sqsubset\!\sqsupset}{AB}$.

Exercise ORD.8*. Let O, A, B, and C be collinear points on a Pasch plane $\mathcal{P}$ such that $O < A < B$ and $O < A < C$, then there exists a point D such that $D > \max\{B, C\}$.

Exercise ORD.9*. Let $\mathcal{P}$ be a Pasch plane, and let $\mathcal{L}$ and $\mathcal{L}'$ be distinct lines on $\mathcal{P}$, O be a member of $\mathcal{P} \setminus (\mathcal{L} \cup \mathcal{L}')$. Suppose further that a line through O intersects $\mathcal{L}$ iff it intersects $\mathcal{L}'$, and that each of the intersections of every such line with $\mathcal{L}$ or $\mathcal{L}'$ is a singleton.

[1] Dyadic rationals are numbers that can be written in the form $\frac{a}{2^b}$ where a is an integer and b is a natural number greater than 0.

Let A and B be distinct points on $\mathcal{L}$, A' be the point such that $\overleftrightarrow{OA} \cap \mathcal{L}' = \{A'\}$ and B' be the point such that $\overleftrightarrow{OB} \cap \mathcal{L}' = \{B'\}$. Order the points on $\mathcal{L}$ so that $A < B$, and order the points on $\mathcal{L}'$ so that $A' < B'$.

For every $X \in \mathcal{L}$ let $\varphi(X)$ be the point on $\mathcal{L}'$ such that $\overleftrightarrow{OX} \cap \mathcal{L}' = \{\varphi(X)\}$.

(A) φ is a bijection of $\mathcal{L}$ onto $\mathcal{L}'$.

Let X, Y, and Z be any distinct points on $\mathcal{L}$.

(B) X–Y–Z iff $\varphi(X)$–$\varphi(Y)$–$\varphi(Z)$.

(C) $\varphi(\overline{XY}) = \overline{\varphi(X)\varphi(Y)}$.

(D) $\varphi(\overrightarrow{XY}) = \overrightarrow{\varphi(X)\varphi(Y)}$.

(E) $\varphi(\mathcal{L}) = \varphi(\overleftrightarrow{XY}) = \overleftrightarrow{\varphi(X)\varphi(Y)} = \mathcal{L}'$.

(F) If $X < Y$, then $\varphi(X) < \varphi(Y)$.

Chapter 7
Collineations Preserving Betweenness (COBE)

Acronym: *COBE*

Dependencies: *Chapters 1, 3 (definitions and Theorems CAP.1– CAP.4), 4, 5, and 6*

New Axioms: *none*

New Terms Defined: *belineation*

Abstract: A belineation is a bijection of a plane that preserves betweenness. This chapter shows that every belineation on a Pasch plane is a collineation, and explores the interactions between belineations and segments, rays, lines, sides of a line, angles, and triangles.

The principal result of this chapter is Theorem COBE.5, which establishes that a belineation on a Pasch plane carries segments to segments, rays to rays, and angles to angles, *etc.* In Chapter 8 we define reflections and isometries, which are belineations, and restate Theorem COBE.5 in that context as Theorem NEUT.15; this result is essential to the development of neutral geometry. Many citations of Theorem COBE.5 are interchangeable with citations of Theorem NEUT.15.

In later chapters we show that dilations, symmetries, and axial affinities are belineations (*cf* Theorem DLN.8, Theorem SIM.2, and Theorem AX.4). Applications are found in Chapter 14, where the properties of a line as an ordered field are established (*cf* Theorem OF.10(C)), in Chapter 15 in the proof of Theorem SIM.9, and in Chapter 17 (Theorem QX.2) where multiplication on a line is defined.

It has been conjectured that every collineation on a Pasch plane where Axiom PS holds is a belineation; if this were true, it would be unnecessary to prove

E.J. Specht et al., *Euclidean Geometry and its Subgeometries*,
DOI 10.1007/978-3-319-23775-6_7

either Theorem DLN.8 (Chapter 13) or Theorem AX.4 (Chapter 16). We have not succeeded either in proving this conjecture or in constructing a counterexample, that is, a model of a Pasch plane with Axiom PS in which there is a collineation which is not a belineation. The authors would be grateful to anyone who could provide either.

Definition COBE.1. A bijection φ on a Pasch plane $\mathcal{P}$ is a **belineation**[1] if it *preserves betweenness*; that is, for any points A, B, and C on $\mathcal{P}$, if A–B–C, then $\varphi(A)$–$\varphi(B)$–$\varphi(C)$.

Theorem COBE.2. *Let $\mathcal{P}$ be a Pasch plane; then every belineation φ on $\mathcal{P}$ is a collineation. More specifically, for any distinct points A and B on $\mathcal{P}$, $\varphi(\overleftrightarrow{AB}) = \overleftrightarrow{\varphi(A)\varphi(B)}$.*

Proof. (I: $\varphi(\overleftrightarrow{AB}) \subseteq \overleftrightarrow{\varphi(A)\varphi(B)}$) Let $X \in \overleftrightarrow{AB}$. By Theorem IB.5, exactly one of X–A–B, $X = A$, A–X–B, $X = B$ or A–B–X is true. Since φ preserves betweenness, exactly one of $\varphi(X)$–$\varphi(A)$–$\varphi(B)$, $\varphi(X) = \varphi(A)$, $\varphi(A)$–$\varphi(X)$–$\varphi(B)$, $\varphi(X) = \varphi(B)$, or $\varphi(A)$–$\varphi(B)$–$\varphi(X)$ is true. Thus by Theorem IB.5, $\varphi(X) \in \overleftrightarrow{\varphi(A)\varphi(B)}$, so that $\varphi(\overleftrightarrow{AB}) \subseteq \overleftrightarrow{\varphi(A)\varphi(B)}$.

(II: $\overleftrightarrow{\varphi(A)\varphi(B)} \subseteq \varphi(\overleftrightarrow{AB})$) Suppose that $\overleftrightarrow{\varphi(A)\varphi(B)} \nsubseteq \varphi(\overleftrightarrow{AB})$; then there exists a point $D \in \overleftrightarrow{\varphi(A)\varphi(B)}$ such that $D \notin \varphi(\overleftrightarrow{AB})$; since φ is a bijection, there exists a point $C \in \mathcal{P}$ such that $\varphi(C) = D \in \overleftrightarrow{\varphi(A)\varphi(B)}$, and $C \notin \overleftrightarrow{AB}$ (for if $C \in \overleftrightarrow{AB}$, then $\varphi(C) = D \in \varphi(\overleftrightarrow{AB})$, which is false by assumption). By Exercise I.2

$$\overleftrightarrow{\varphi(A)\varphi(C)} = \overleftrightarrow{\varphi(A)\varphi(B)} = \overleftrightarrow{\varphi(B)\varphi(C)}.$$

By Part I,

$$\varphi(\overleftrightarrow{AC}) \subseteq \overleftrightarrow{\varphi(A)\varphi(C)} = \overleftrightarrow{\varphi(A)\varphi(B)} \text{ and}$$
$$\varphi(\overleftrightarrow{BC}) \subseteq \overleftrightarrow{\varphi(B)\varphi(C)} = \overleftrightarrow{\varphi(A)\varphi(B)}.$$

Now let X be any point of $\mathcal{P} \setminus (\overleftrightarrow{AB} \cup \overleftrightarrow{AC} \cup \overleftrightarrow{BC})$, and let $Y \in \overline{AB}$, so that $\varphi(Y) \in \varphi(\overleftrightarrow{AB}) \subseteq \overleftrightarrow{\varphi(A)\varphi(B)}$.

Then $\overleftrightarrow{XY}$ intersects $\overline{AB}$ but does not contain any of the points A, B, or C. By the Postulate of Pasch, $\overleftrightarrow{XY}$ must intersect either $\overline{AC}$ or $\overline{BC}$; let Z be this point of intersection. Since both $\varphi(\overleftrightarrow{AC})$ and $\varphi(\overleftrightarrow{BC})$ are subsets of $\overleftrightarrow{\varphi(A)\varphi(B)}$, and Z is a member of either $\overleftrightarrow{AC}$ or $\overleftrightarrow{BC}$, $\varphi(Z) \in \overleftrightarrow{\varphi(A)\varphi(B)}$. Since $X \in \overleftrightarrow{YZ}$, by Part I

[1] We hope the reader is not offended by this rather odd name; believe it or not, we bandied about some other names that were even stranger—such as *betweeneation.*

$\varphi(X) \in \varphi(\overleftrightarrow{YZ}) \subseteq \overleftrightarrow{\varphi(Y)\varphi(Z)}$ which by Exercise I.2 is equal to $\overleftrightarrow{\varphi(A)\varphi(B)}$, because both $\varphi(Y)$ and $\varphi(Z)$ are members of $\overleftrightarrow{\varphi(A)\varphi(B)}$.

We have shown that $\varphi(\overleftrightarrow{AC})$, $\varphi(\overleftrightarrow{BC})$, and $\varphi(\overleftrightarrow{AB})$ are all subsets of $\overleftrightarrow{\varphi(A)\varphi(B)}$, and that $\varphi(X) \in \overleftrightarrow{\varphi(A)\varphi(B)}$. Therefore $\varphi(\mathcal{P}) \subseteq \overleftrightarrow{\varphi(A)\varphi(B)}$. By assumption φ is a bijection, so that $\mathcal{P} = \varphi(\mathcal{P}) \subseteq \overleftrightarrow{\varphi(A)\varphi(B)}$. But Axiom I.5(B) says that $\mathcal{P}$ contains at least three noncollinear points. This is a contradiction; therefore, $\overleftrightarrow{\varphi(A)\varphi(B)} \subseteq \varphi(\overleftrightarrow{AB})$. □

Theorem COBE.3. *If φ is a belineation on a Pasch plane $\mathcal{P}$, so is φ^{-1}.*

Proof. Assume that φ is a belineation; by definition, it is a bijection, and by elementary mapping theory, its inverse is also a bijection.

Let A, B, and C be any points on $\mathcal{P}$; we show that if A–B–C, then $\varphi^{-1}(A)$–$\varphi^{-1}(B)$–$\varphi^{-1}(C)$.

By Theorem COBE.2 φ is a collineation, and by Theorem CAP.4(D') φ^{-1} is a collineation, so that $\varphi^{-1}(A)$, $\varphi^{-1}(B)$, and $\varphi^{-1}(C)$ are collinear, and we may apply the trichotomy Property B.2 of Definition IB.1.

If $\varphi^{-1}(A)$–$\varphi^{-1}(C)$–$\varphi^{-1}(B)$, then

$$\varphi(\varphi^{-1}(A))\text{–}\varphi(\varphi^{-1}(C))\text{–}\varphi(\varphi^{-1}(B))\,,$$

that is, A–C–B. This is false by the trichotomy Property B.2 of Definition IB.1. By a similar argument, $\varphi^{-1}(B)$–$\varphi^{-1}(A)$–$\varphi^{-1}(C)$ is false; hence by the trichotomy property, $\varphi^{-1}(A)$–$\varphi^{-1}(B)$–$\varphi^{-1}(C)$. □

Remark COBE.4. Summarizing Theorem COBE.2 and Theorem COBE.3, φ is a belineation iff its inverse is a belineation, in which case both it and its inverse are collineations. We have already seen in Theorem CAP.1(D') that a bijection is a collineation iff its inverse is a collineation.

Theorem COBE.5. *Let $\mathcal{P}$ be a Pasch plane, A, B, and C be noncollinear points on $\mathcal{P}$, and let φ be a belineation of $\mathcal{P}$; then:*

(1) $\varphi(\overleftrightarrow{AB}) = \overleftrightarrow{\varphi(A)\varphi(B)}$.
(2) $\varphi(\overset{\rbrack\!\!\longrightarrow}{AB}) = \overset{\rbrack\!\!\longrightarrow}{\varphi(A)\varphi(B)}$,
(3) $\varphi(\overset{\lbrack\!\!\longrightarrow}{AB}) = \overset{\lbrack\!\!\longrightarrow}{\varphi(A)\varphi(B)}$,
(4) $\varphi(\overset{\rbrack\!-\!\lbrack}{AB}) = \overset{\rbrack\!-\!\lbrack}{\varphi(A)\varphi(B)}$,
(5) $\varphi(\overset{\lbrack\!-\!\rbrack}{AB}) = \overset{\lbrack\!-\!\rbrack}{\varphi(A)\varphi(B)}$,
(6) $\varphi(\overset{\rbrack\!-\!\rbrack}{AB}) = \overset{\rbrack\!-\!\rbrack}{\varphi(A)\varphi(B)}$,
(7) $\varphi(\overset{\lbrack\!-\!\lbrack}{AB}) = \overset{\lbrack\!-\!\lbrack}{\varphi(A)\varphi(B)}$,

(8) $\varphi(\angle ABC) = \angle\varphi(A)\varphi(B)\varphi(C)$,

(9) φ(the C-side of $\overleftrightarrow{AB}$) = the $\varphi(C)$-side of $\overleftrightarrow{\varphi(A)\varphi(B)}$,

(10) $\varphi(\triangle ABC) = \triangle\varphi(A)\varphi(B)\varphi(C)$,

(11) $\varphi(\text{ins } \angle BAC) = \text{ins } \angle\varphi(B)\varphi(A)\varphi(C)$,

(12) $\varphi(\text{ins } \triangle ABC) = \text{ins } \triangle\varphi(A)\varphi(B)\varphi(C)$.

(13) *If A, B, C, and D are points on $\mathcal{P}$ and $\square ABCD$ is a quadrilateral, then $\varphi(\square ABCD)$ is a quadrilateral, and $\varphi(\square ABCD) = \square\varphi(A)\varphi(B)\varphi(C)\varphi(D)$.*

Proof. (1) This is Theorem COBE.2.

(2) We first prove that $\varphi(\overrightarrow{AB}\setminus\{B\}) = \overrightarrow{\varphi(A)\varphi(B)}\setminus\{\varphi(B)\}$. Let Y be any member of $\varphi(\overrightarrow{AB}\setminus\{B\})$; then there exists a member X of $\overrightarrow{AB}\setminus\{B\}$ such that $Y = \varphi(X)$. By Definition IB.4, either A–X–B or A–B–X. By Definition COBE.1, $\varphi(A)$–$\varphi(X)$–$\varphi(B)$ or $\varphi(A)$–$\varphi(B)$–$\varphi(X)$. By Definition IB.4, $Y \in \overrightarrow{\varphi(A)\varphi(B)}\setminus\{\varphi(B)\}$. Thus

$$\varphi(\overrightarrow{AB}\setminus\{B\}) \subseteq \overrightarrow{\varphi(A)\varphi(B)}\setminus\{\varphi(B)\}.$$

By this result, since φ^{-1} is a belineation,

$$\varphi^{-1}(\overrightarrow{\varphi(A)\varphi(B)}\setminus\{\varphi(B)\}) \subseteq \overrightarrow{\varphi^{-1}(\varphi(A))\varphi^{-1}(\varphi(B))}\setminus\{\varphi^{-1}(\varphi(B))\} = \overrightarrow{AB}\setminus\{B\}$$

Applying φ to both sides, we have

$$\overrightarrow{\varphi(A)\varphi(B)}\setminus\{\varphi(B)\} \subseteq \varphi(\overrightarrow{AB}\setminus\{B\}).$$

Combining these results, $\varphi(\overrightarrow{AB}\setminus\{B\}) = \overrightarrow{\varphi(A)\varphi(B)}\setminus\{\varphi(B)\}$. Now φ is one-to-one and $\overrightarrow{AB}\setminus\{B\}$ and $\{B\}$ are disjoint, so by elementary set theory,

$$\varphi(\overrightarrow{AB}) = \varphi(\overrightarrow{AB}\setminus\{B\} \cup \{B\}) = \varphi(\overrightarrow{AB}\setminus\{B\}) \cup \varphi(\{B\})$$
$$= \overrightarrow{\varphi(A)\varphi(B)}\setminus\{\varphi(B)\} \cup \{\varphi(B)\} = \overrightarrow{\varphi(A)\varphi(B)}.$$

(3) Using part (2) and elementary set theory,

$$\varphi(\overrightarrow{AB}) = \varphi(\{A\} \cup \overrightarrow{AB}) = \{\varphi(A)\} \cup \varphi(\overrightarrow{AB})$$
$$= \varphi(A) \cup \overrightarrow{\varphi(A)\varphi(B)} = \overrightarrow{\varphi(A)\varphi(B)}.$$

(4) Let Y be any member of $\varphi(\overline{AB})$; then there exists a member X of $\overline{AB}$ such that $Y = \varphi(X)$. By Definition IB.3, A–X–B. By Definition COBE.1, $\varphi(A)$–$\varphi(X)$–$\varphi(B)$. By Definition IB.3, $Y \in \overline{\varphi(A)\varphi(B)}$. Thus $\varphi(\overline{AB}) \subseteq \overline{\varphi(A)\varphi(B)}$. Since φ^{-1} is a belineation,

$$\varphi^{-1}(\overline{\varphi(A)\varphi(B)}) \subseteq \overline{\varphi^{-1}(\varphi(A))\varphi^{-1}(\varphi(B))} = \overline{AB},$$

and applying φ to both sides,

$$\overline{\varphi(A)\varphi(B)} = \varphi(\varphi^{-1}(\overline{\varphi(A)\varphi(B)})) \subseteq \varphi(\overline{AB}).$$

Therefore $\varphi(\overline{AB}) = \overline{\varphi(A)\varphi(B)}$.

(5) Since $\varphi(A)$ and $\varphi(B)$ both belong to $\varphi(\overline{AB})$ and to $\overline{\varphi(A)\varphi(B)}$, using part (5) we get $\varphi(\overline{AB}) = \overline{\varphi(A)\varphi(B)}$.

(6) By part (4) and elementary set theory,

$$\varphi(\overline{AB}) = \varphi(\overline{AB} \cup \{B\}) = \varphi(\overline{AB}) \cup \{\varphi(B)\}$$
$$= \overline{\varphi(A)\varphi(B)} \cup \{\varphi(B)\} = \overline{\varphi(A)\varphi(B)}.$$

(7) The argument is similar to that for part (6).

(8) By Definition PSH.29 and elementary set theory,

$$\varphi(\angle BAC) = \varphi(\overrightarrow{AB} \cup \overrightarrow{AC}) = \varphi(\overrightarrow{AB}) \cup \varphi(\overrightarrow{AC})$$
$$= \overrightarrow{\varphi(A)\varphi(B)} \cup \overrightarrow{\varphi(A)\varphi(C)} = \angle\varphi(B)\varphi(A)\varphi(C).$$

(9) Let Y be any member of $\varphi(C\text{-side of } \overleftrightarrow{AB} \setminus \{C\})$; then there exists a member X of (C-side of $\overleftrightarrow{AB} \setminus \{C\}$) such that $Y = \varphi(X)$. By Definition IB.11, $\overline{CX} \cap \overleftrightarrow{AB} = \emptyset$. By elementary set theory and part (5), $\emptyset = \varphi(\overline{CX} \cap \overleftrightarrow{AB}) = \overline{\varphi(C)\varphi(X)} \cap \overleftrightarrow{\varphi(A)\varphi(B)}$. By Definition IB.11, Y belongs to $\varphi(C)$-side of $\overleftrightarrow{\varphi(A)\varphi(B)} \setminus \{\varphi(C)\}$. Thus

$$\varphi(C\text{-side of } \overleftrightarrow{AB} \setminus \{C\}) \subseteq (\varphi(C)\text{-side of } \overleftrightarrow{\varphi(A)\varphi(B)}) \setminus \{\varphi(C)\}.$$

Now φ^{-1} is a belineation, so in the above we may substitute φ^{-1} for φ, $\varphi(A)$ for A, $\varphi(B)$ for B, and $\varphi(C)$ for C. Then

$$\varphi^{-1}(\varphi(C)\text{-side of } \overleftrightarrow{\varphi(A)\varphi(B)} \setminus \{\varphi(C)\})$$
$$\subseteq (\varphi^{-1}(\varphi(C))\text{-side of } \overleftrightarrow{\varphi^{-1}(\varphi(A))\varphi^{-1}(\varphi(B))}) \setminus \{\varphi^{-1}(\varphi(C))\}$$
$$= C\text{-side of } \overleftrightarrow{AB} \setminus \{C\}.$$

Applying φ to both sides we have

$$\varphi(C)\text{-side of } \overleftrightarrow{\varphi(A)\varphi(B)} \setminus \{\varphi(C)\} \subseteq \varphi(C\text{-side of } \overleftrightarrow{AB}) \setminus \{C\}.$$

Therefore,

$$\varphi(C\text{-side of } \overleftrightarrow{AB}) \setminus \{C\} = (\varphi(C)\text{-side of } \overleftrightarrow{\varphi(A)\varphi(B)} \setminus \{\varphi(C)\}).$$

By elementary set theory, $\varphi(C\text{-side of } \overleftrightarrow{AB}) = \varphi(C)\text{-side of } \overleftrightarrow{\varphi(A)\varphi(B)}$.

(10) By Definition IB.7, part (5) above, and elementary set theory,

$$\varphi(\triangle ABC) = \varphi(\overline{AB} \cup \overline{BC} \cup \overline{AC})$$
$$= \varphi(\overline{AB}) \cup \varphi(\overline{BC}) \cup \varphi(\overline{AC})$$
$$= \overline{\varphi(A)\varphi(B)} \cup \overline{\varphi(B)\varphi(C)} \cup \overline{\varphi(A)\varphi(C)}$$
$$= \triangle\varphi(A)\varphi(B)\varphi(C).$$

(11) By Definition PSH.36(A) and elementary set theory,

$$\varphi(\text{ins } \angle BAC) = \varphi(\text{the } B\text{-side of } \overleftrightarrow{AC} \cap \text{the } C\text{-side of } \overleftrightarrow{AB})$$
$$= \text{the } \varphi(B)\text{-side of } \overleftrightarrow{\varphi(A)\varphi(C)} \cap \text{the } \varphi(C)\text{-side of } \overleftrightarrow{\varphi(A)\varphi(B)}$$
$$= \text{ins}(\triangle\varphi(B)\varphi(A)\varphi(C)).$$

(12) By parts (9), (11), Theorem PSH.46(C) and elementary set theory,

$$\begin{aligned}\varphi(\text{ins } \triangle ABC) &= \varphi((\text{ins } \angle BAC) \cap (\text{the } A\text{-side of } \overleftrightarrow{BC})) \\ &= \text{ins } \angle\varphi(B)\varphi(A)\varphi(C) \cap (\text{the } \varphi(A)\text{-side of } \overleftrightarrow{\varphi(B)\varphi(C)}), \\ &= \text{ins } \triangle\varphi(B)\varphi(A)\varphi(C).\end{aligned}$$

(13) By Definition PSH.31, since $\square ABCD$ is a quadrilateral, $\square ABCD = \overline{AB} \cup \overline{BC} \cup \overline{CD} \cup \overline{DA}$, all the triples $\{A, B, C\}$, $\{B, C, D\}$, $\{C, D, A\}$, and $\{D, A, B\}$ are noncollinear, and $\overline{AB} \cap \overline{CD} = \overline{BC} \cap \overline{DA} = \emptyset$. By part (5) above and elementary set theory,

$$\begin{aligned}\varphi(\square ABCD) &= \varphi(\overline{AB} \cup \overline{BC} \cup \overline{CD} \cup \overline{DA}) \\ &= \varphi(\overline{AB}) \cup \varphi(\overline{BC}) \cup \varphi(\overline{CD}) \cup \varphi(\overline{DA}) \\ &= \overline{\varphi(A)\varphi(B)} \cup \overline{\varphi(B)\varphi(C)} \cup \overline{\varphi(C)\varphi(D)} \cup \overline{\varphi(D)\varphi(A)} \\ &= \square\varphi(A)\varphi(B)\varphi(C)\varphi(D).\end{aligned}$$

By Exercise CAP.2, all the triples $\{\varphi(A), \varphi(B), \varphi(C)\}$, $\{\varphi(B), \varphi(C), \varphi(D)\}$, $\{\varphi(C), \varphi(D), \varphi(A)\}$, and $\{\varphi(D), \varphi(A), \varphi(B)\}$ are noncollinear. Since φ is a bijection,

$$\overline{\varphi(A)\varphi(B)} \cap \overline{\varphi(C)\varphi(D)} = \overline{\varphi(B)\varphi(C)} \cap \overline{\varphi(D)\varphi(A)} = \emptyset.$$

By Definition PSH.31, $\varphi(\square ABCD)$ is a quadrilateral. □

Theorem COBE.6. *Let $\mathcal{P}$ be a Pasch plane, φ a belineation of $\mathcal{P}$, and let A and B be distinct points on $\mathcal{P}$. Let the points on $\overleftrightarrow{AB}$ be ordered so that $A < B$ (cf Definition ORD.1), and let the points on $\overleftrightarrow{\varphi(A)\varphi(B)}$ be ordered so that $\varphi(A) < \varphi(B)$. Then for any points X and Y of $\overleftrightarrow{AB}$, $X < Y$ iff $\varphi(X) < \varphi(Y)$.*

Proof. If $X < Y$, by Definition ORD.1 (and the fact that $A < B$) $\overrightarrow{XY} \cap \overrightarrow{AB}$ is a ray. By Theorem PSH.19 $\overrightarrow{XY} \subseteq \overrightarrow{AB}$ or $\overrightarrow{AB} \subseteq \overrightarrow{XY}$.

By the elementary properties of mappings, $\varphi(\overrightarrow{XY}) \subseteq \varphi(\overrightarrow{AB})$ or $\varphi(\overrightarrow{AB}) \subseteq \varphi(\overrightarrow{XY})$. By Theorem COBE.5(3) $\overrightarrow{\varphi(X)\varphi(Y)} \subseteq \overrightarrow{\varphi(A)\varphi(B)}$ or $\overrightarrow{\varphi(A)\varphi(B)} \subseteq \overrightarrow{\varphi(X)\varphi(Y)}$.

By Theorem PSH.19 $\overrightarrow{\varphi(X)\varphi(Y)}) \cap \overrightarrow{\varphi(A)\varphi(B)}$ is a ray. By Definition ORD.1 (and the fact that $\varphi(A) < \varphi(B)$) $\varphi(X) < \varphi(Y)$.

Conversely, if $\varphi(X) < \varphi(Y)$, applying φ^{-1} to both sides, the argument above yields $X < Y$. □

There are no exercises for this chapter.

Chapter 8
Neutral Geometry (NEUT)

Acronym: *NEUT*
Dependencies: *Chapters 1, 3 (definitions and Theorems CAP.1–CAP.4), 4, 5, and 6*
New axiom: *Axiom REF*
New Terms Defined: *mirror mapping over a line, reflection set (of mirror mappings), reflection over a line, angle reflection, fixed segment, line of symmetry, isometry, congruent, midpoint, neutral plane, complementary mapping of a reflection, perpendicular, vertical angles, supplementary angles, bisecting ray, right angle, perpendicular bisecting line, kite; smaller, larger, smaller or congruent, larger or congruent (angle ordering); outside angle of a triangle; right, obtuse, acute (angle); right, obtuse, acute (triangle); maximal angle, maximal edge (of a triangle); hypotenuse, leg (of a triangle); altitude, base (of a triangle); foot (of a line)*

Abstract: This chapter deals with neutral geometry, which is central to the entire book. It begins with definitions of mirror mappings and reflections over lines. Every line is an axis for some reflection. A line of symmetry for a set is a line whose reflection maps that set onto itself. Every angle has a line of symmetry, its angle bisector. Compositions of reflections are isometries, and isometric sets are congruent. These concepts provide access to the standard congruence theorems. Reflections are used to define perpendicularity, the perpendicular bisector and midpoint of a segment, and to prove the existence of a line (not necessarily unique) through a given point parallel to a given line. Ordering of angles is defined, leading to the notions of acute angle, obtuse angle, and maximal angle of a triangle.

E.J. Specht et al., *Euclidean Geometry and its Subgeometries*,
DOI 10.1007/978-3-319-23775-6_8

As can be seen by surveying either the abstract above or the list of new terms defined, this is a wide-ranging chapter; it is the place where many of the most familiar results of plane geometry are developed, and where we deal, finally, with *congruence*.

The geometry in this chapter is called *neutral* geometry[1] since no commitment is made regarding a parallel axiom. It builds on Pasch geometry, invoking Axioms I.0 through I.5 (the incidence axioms), Axiom BET (betweenness), and the Plane Separation Axiom.

In contrast with Hilbert's axioms in which congruence is an undefined term given meaning by axioms, we use *reflections* to define congruence, taking a two-step approach. First, in Definition NEUT.1 we define a *mirror mapping* as a type of bijection on a Pasch plane $\mathcal{P}$ which is both an axial affinity and a belineation, and is its own inverse. Its properties are based on our everyday observations about reflection in a physical mirror (*cf* Remark NEUT.1.0).

Unfortunately, we cannot base neutral geometry on mirror mappings—one reason being that, as shown in Exercise NEUT.0, there can be more than one mirror mapping over a line in a coordinate plane.[2] Our notion of perpendicularity arises from the behavior of reflections, and allowing multiple reflections over a line would lead to the existence of many perpendiculars to a line at a point; this would not be helpful. Nor do we have, for mirror mappings, any assurance about several other kinds of behavior that are essential to our development, involving characteristics not of just one mapping, but of the entire family of mirror mappings.

Therefore, in Definition NEUT.2 we specify a subset of the collection of all possible mirror mappings, called a *reflection set*, which is defined by the six Properties R.1 through R.6 of Definition NEUT.2. Members of the reflection set are called *reflections*. We wish this list of properties could be shorter, but we haven't succeeded in proving any of them from the others. Finally, we state a single reflection Axiom REF, which asserts simply that a reflection set exists.

Even though we take no stand in neutral geometry with regard to a parallel axiom, in Theorem NEUT.48(B) we will prove Property PE:

[1]Historically, geometry developed without regard to a parallel axiom was referred to as *absolute* geometry.

[2]A coordinate plane is a Pasch plane, satisfying the incidence, betweenness, and Plane Separation axioms; this will be shown in Chapter 21, Sections 21.5.1 through 21.5.4.

Property PE (Parallel Existence). *Given a neutral plane $\mathcal{P}$ and a line $\mathcal{L}$ on $\mathcal{P}$, for every point Q belonging to $\mathcal{P} \setminus \mathcal{L}$, there exists a line $\mathcal{M}$ through Q which is parallel to $\mathcal{L}$.*

There is no claim of uniqueness—Property PE falls short of Axiom PS. However, if PE is joined with PW, then we get PS; the reader may wish to compare this with the discussion of Axioms PS and PW following Definition IP.0 in Chapter 2.

Since the incidence, betweenness, plane separation axioms, and the reflection axiom to be introduced here imply Property PE, they are incompatible with elliptic geometry, in which there are no parallel lines; to make them so, we would have to weaken our definition of reflection. Once we have developed neutral geometry, we may accept the Parallel Axiom by accepting either Axiom PW or PS, yielding Euclidean geometry, as in Chapter 11. Or we may deny it by saying that given a line $\mathcal{L}$ and a point P not on that line, there may exist more than one line through P parallel to $\mathcal{L}$. This denies both Axioms PW and PS and yields hyperbolic geometry, which we do not pursue.

This chapter is loosely based on the development in *Fundamentals of Mathematics*, Volume II, Behnke, *et al*, eds., published by MIT Press; particularly Chapter 4 by J. Diller and J. Boczeck, and Chapter 5 by F. Bachmann, W. Pejas, H. Wolff, and A. Bauer [2].[3]

8.1 Mirror mappings and their elementary properties

Definition NEUT.1. Let $\mathcal{L}$ be any line contained in a Pasch plane $\mathcal{P}$. A **mirror mapping** over the line $\mathcal{L}$ is a mapping φ of $\mathcal{P}$ into $\mathcal{P}$ which satisfies

(A) if X is any point on $\mathcal{L}$, $\varphi(X) = X$; that is, every point of $\mathcal{L}$ is a fixed point for φ;
(B) for every member X of $\mathcal{P} \setminus \mathcal{L}$, X and $\varphi(X)$ are on opposite sides of $\mathcal{L}$;
(C) for every member X of $\mathcal{P} \setminus \mathcal{L}$, $\varphi(\varphi(X)) = X$; and
(D) for all points A, B, and C on $\mathcal{P}$ such that $A–B–C$, $\varphi(A)–\varphi(B)–\varphi(C)$; that is to say, *betweenness is preserved* by φ.

[3]Other authors introduce axioms on motions of an ordered plane to develop neutral geometry. See Doneddu, A., *Étude de géométries planes ordonnées*, Rend. Circ. Mat. Palermo, II. Ser. 19, 27–68 (1970) [5], and Lumiste, U., *Foundations of geometry*, Estonian Mathematical Society, Tartu, 2009 [13].

The line $\mathcal{L}$ is the **axis** of the mirror mapping φ.

Remark NEUT.1.0. We have based the properties of a mirror mapping on our observations of reflections in a physical mirror. Property (A) of Definition NEUT.1 says that the reflecting surface of a mirror appears stationary. Property (B) says the objects seen in a mirror appear to be on the opposite side of the reflecting surface from their actual positions. Property (C) hints that the original objects are themselves reflections of their images (a slightly mind-twisting thought!). Property (D) says that the images of objects lined up in order are in the same order as the objects themselves.

Items NEUT.1.1 through NEUT.1.8 use the terminology and notation of Definition NEUT.1, and describe the elementary properties of mirror mappings. They are stated and proved somewhat informally.

Remark NEUT.1.1. *A mirror mapping has only one axis of fixed points.* If φ is a mirror mapping over $\mathcal{L}$, by Property (A) of Definition NEUT.1, every point of $\mathcal{L}$ is a fixed point of φ. If $A \notin \mathcal{L}$, by Property (B) A cannot be a fixed point for φ. Therefore $\mathcal{L}$ is the set of all fixed points for φ. If there were a second axis $\mathcal{N}$, at least one of its points would not belong to $\mathcal{L}$ and hence would not be a fixed point, contradicting the assumption that $\mathcal{N}$ is an axis.

Remark NEUT.1.2. *The identity map ι is not a mirror mapping.* This follows from Property (B) of Definition NEUT.1.

Remark NEUT.1.3. *For any mirror mapping φ, $\varphi \circ \varphi = \iota$, and φ is a bijection of $\mathcal{P}$.* Let A be any point of $\mathcal{P}$. If $A \in \mathcal{L}$, by Property (A) of Definition NEUT.1, $\varphi(\varphi(A)) = \varphi(A) = A$; if $A \notin \mathcal{L}$, by Property (C) $\varphi(\varphi(A)) = A$. This shows that $\varphi \circ \varphi = \iota$, and also that φ maps $\mathcal{P}$ onto $\mathcal{P}$.

Note that $A \in \mathcal{L}$ iff $\varphi(A) \in \mathcal{L}$; for if $A \in \mathcal{L}$, by Property (A) $\varphi(A) = A \in \mathcal{L}$; if $A \notin \mathcal{L}$, by Property (B), $\varphi(A) \notin \mathcal{L}$. To show that φ is one-to-one, let $\varphi(A) = \varphi(B)$. If $A \in \mathcal{L}$, $\varphi(A) = \varphi(B) \in \mathcal{L}$ and hence $B \in \mathcal{L}$. Thus $A = \varphi(A) = \varphi(B) = B$. If $A \notin \mathcal{L}$, by Property (C) $A = \varphi(\varphi(A)) = \varphi(\varphi(B))$. Then $\varphi(\varphi(B)) \notin \mathcal{L}$ so that $\varphi(B) \notin \mathcal{L}$ and $B \notin \mathcal{L}$; by Property (C) $\varphi(\varphi(B)) = B$ so that $A = B$. Thus φ is a bijection of $\mathcal{P}$.

Remark NEUT.1.4. *A–B–C iff $\varphi(A)$–$\varphi(B)$–$\varphi(C)$.* Let A, B, and C be any points of $\mathcal{P}$; by Property (D), if A–B–C then $\varphi(A)$–$\varphi(B)$–$\varphi(C)$. If $\varphi(A)$–$\varphi(B)$–$\varphi(C)$, by Property (D) $\varphi(\varphi(A))$–$\varphi(\varphi(B))$–$\varphi(\varphi(C))$, and by Remark NEUT.1.3 above, this is A–B–C.

Remark NEUT.1.5. *A mirror mapping φ is both a belineation and a collineation.* This follows immediately from Remark NEUT.1.4 above and Theorem COBE.2.

Remark NEUT.1.6. *A mirror mapping is an axial affinity.* This follows immediately from Definition CAP.25 using Remark NEUT.1.2, Remark NEUT.1.5, and Property (A) of Definition NEUT.1.

Remark NEUT.1.7. It can be shown that if every point O of $\mathcal{L}$ is contained in some line $\overleftrightarrow{A\varphi(A)}$ $(A \notin \mathcal{L})$, where φ is a mapping obeying properties (B) through (D) of Definition NEUT.1, then Property (A) holds. This is Exercise NEUT.83.

Theorem NEUT.1.8. *A mirror mapping maps endpoints of segments and rays to end points.*

Proof. We show this for closed segments. Let A, B, C, and D be points of $\mathcal{P}$ where $A \neq B$ and $C \neq D$. If φ is a mirror mapping and $\varphi(\overline{AB}) = \overline{CD}$, then either $\varphi(A) = C$ and $\varphi(B) = D$ or $\varphi(A) = D$ and $\varphi(B) = C$. For if $\varphi(A) \in \overline{CD}$ and is not an endpoint, C–$\varphi(A)$–D; by Property (D) of Definition NEUT.1, $\varphi(C)$–$\varphi(\varphi(A))$–$\varphi(D)$, and by Remark NEUT.1.3, $\varphi(C)$–A–$\varphi(D)$. This is impossible since A is an endpoint of $\overline{AB}$, and both $\varphi(C)$ and $\varphi(D)$ are members of $\overline{AB}$ (again by Remark NEUT.1.3). □

8.2 Reflection sets and the reflection axiom

Definition NEUT.2. A set $\mathcal{E}$ of mirror mappings on a Pasch plane is said to be a **reflection set** if it satisfies Properties R.1 through R.6 listed below.

R.1 (Existence) For every line $\mathcal{L}$ in the plane $\mathcal{P}$, $\mathcal{E}$ contains a mirror mapping $\mathcal{R}_{\mathcal{L}}$ over $\mathcal{L}$.

R.2 (Uniqueness) $\mathcal{E}$ contains no more than one mirror mapping $\mathcal{R}_{\mathcal{L}}$ over a line $\mathcal{L}$ in $\mathcal{P}$.

R.3 (Closure) If φ is a mirror mapping over a line $\mathcal{L}$ and φ is the composition of two or more mirror mappings in $\mathcal{E}$, then $\varphi \in \mathcal{E}$.

R.4 (Linear scaling) If A, B, and C are distinct points on the plane such that $C \in \overrightarrow{AB}$, and for some composition α of mirror mappings in $\mathcal{E}$, $\alpha(\overline{AB}) = \overline{AC}$, then $B = C$.

R.5 ("Angle reflection") For every angle $\angle AOB$ in the plane $\mathcal{P}$, there exists a mirror mapping $\mathcal{R}_{\mathcal{L}} \in \mathcal{E}$ such that $\mathcal{R}_{\mathcal{L}}(\angle AOB) = \angle AOB$.

R.6 ("Existence of a midpoint") For any closed segment $\overline{AB} \in \mathcal{P}$ there exists a point $M \in \overline{AB}$ and a composition α of mirror mappings belonging to $\mathcal{E}$ such that $\alpha(\overline{AM}) = \overline{MB}$.

Since we have not yet officially defined either *angle reflection* or *midpoint* we put "Angle reflection" and "Existence of a midpoint" in quotation marks to emphasize that at this point they are only labels.

In Chapter 21 (Subsection 21.7.3) we will show the independence of various of the Properties R.1 through R.6 by exhibiting sets of mirror mappings that are not reflection sets.

Axiom REF. There exists a reflection set $\mathbb{REF}$.

Definition NEUT.3. (A) A member of a reflection set $\mathbb{REF}$ will be called a **line reflection**, or simply a **reflection** over $\mathcal{L}$, and will be denoted by the symbol $\mathcal{R}_{\mathcal{L}}$.

A mapping α of a Pasch plane $\mathcal{P}$ into $\mathcal{P}$ is an **isometry** of $\mathcal{P}$ iff either α is the identity mapping of $\mathcal{P}$, or is the composition of a finite number (at least one) of reflections over lines in $\mathcal{P}$; that is, for some natural number $n \geq 1$ and every $k = 1, 2, ..., n$, $\mathcal{R}_{\mathcal{L}_k}$ is a reflection over a line $\mathcal{L}_k$ in $\mathcal{P}$, and $\alpha = \mathcal{R}_{\mathcal{L}_1} \circ \cdots \circ \mathcal{R}_{\mathcal{L}_n}$.

(B) Let $\mathcal{S}$ and $\mathcal{T}$ be nonempty subsets of a Pasch plane $\mathcal{P}$. $\mathcal{S}$ is **congruent** to $\mathcal{T}$ (notation: $\mathcal{S} \cong \mathcal{T}$) iff there exists an isometry α of $\mathcal{P}$ such that $\alpha(\mathcal{S}) = \mathcal{T}$.

(C) Let A and B be distinct points on the plane $\mathcal{P}$. M is a **midpoint** of $\overline{AB}$ iff A–M–B and $\overline{AM} \cong \overline{BM}$.

(D) A line $\mathcal{L}$ in a Pasch plane $\mathcal{P}$ is a **line of symmetry** for a nonempty set $\mathcal{S}$ iff $\mathcal{R}_{\mathcal{L}}(\mathcal{S}) = \mathcal{S}$, where $\mathcal{R}_{\mathcal{L}}$ is a reflection over the line $\mathcal{L}$. The reflection $\mathcal{R}_{\mathcal{L}}$ may sometimes be referred to as the reflection **implementing** the symmetry. It should be noted that since all the points of the line $\mathcal{L}$ are fixed points, $\mathcal{L}$ is trivially a line of symmetry for any of its subsets.

If $\mathcal{S} = \angle AOB$, we will say that $\mathcal{R}_{\mathcal{L}}$ is the **angle reflection** for $\angle AOB$. An *angle* reflection is also a *line* reflection; the terminology merely reminds us that it is being applied in a certain way.

A ray $\overrightarrow{OD}$ is a **bisecting ray** of $\angle AOB$ iff $\overleftrightarrow{OD}$ is a line of symmetry for $\angle AOB$ and $D \in \text{ins}\,\angle AOB$.

Remark NEUT.4 (Partial restatement of Definition NEUT.2). Using Definition NEUT.3 we may state some of the properties of Definition NEUT.2 more succinctly. Let $\mathcal{P}$ be a Pasch plane on which mirror mappings and reflections are defined.

R.4 (Linear scaling) may be restated as: If A, B, and C are distinct points on $\mathcal{P}$ such that $C \in \overrightarrow{AB}$ and $\overline{AB} \cong \overline{AC}$, then $B = C$.

R.5 (Angle reflection) may be restated as: For any angle $\angle AOB$ in the plane $\mathcal{P}$, there exists an angle reflection $\mathcal{R}_{\mathcal{L}} \in \mathbb{REF}$ for $\angle AOB$.

R.6 (Existence of a midpoint) may be restated as: For any closed segment $\overline{AB} \subseteq \mathcal{P}$ there exists a point $M \in \overline{AB}$ such that $\overline{AM} \cong \overline{MB}$; that is, M is a midpoint of $\overline{AB}$.

Remark NEUT.5. (A) Property R.3 (closure) may appear a bit puzzling at first; if it were applied to a single mirror mapping, it would be vacuous. But it specifies that if the mirror mapping in question is a composition of *two or more* mirror mappings *that belong to* $\mathcal{E}$, then it must belong to $\mathcal{E}$. This imposes a requirement.

(B) Property R.4 (linear scaling) may also appear a bit mysterious. However, it is absolutely pivotal to the development, being as close as we can come to declaring that isometry preserves distance, without actually having a notion of distance.[4]

(C) Property R.6 of Definition NEUT.2 does not imply uniqueness of a midpoint of a segment; for now we will assume that it can have more than one. The proof that it can have only one midpoint is Theorem NEUT.50.

(D) Throughout the remainder of the book, any citation of one of Properties R.1 through R.6 of Definition NEUT.2 will be understood to include a reference to Axiom REF, which establishes that the cited property is in force.

Remark NEUT.6 (On angle reflections).

(A) Property R.5 of Definition NEUT.2 by itself does not imply uniqueness of lines of symmetry or angle reflections; for now we will assume that there could be two reflections $\mathcal{R}_{\mathcal{L}}$ and $\mathcal{R}_{\mathcal{L}'}$ such that

$$\mathcal{R}_{\mathcal{L}}(\angle AOB) = \angle AOB = \mathcal{R}_{\mathcal{L}'}(\angle AOB).$$

In Theorem NEUT.26 we will show that there is only one angle reflection for an angle, and hence, by Remark NEUT.1.1, there can be only one line of symmetry.

(B) If O, A, and B are noncollinear points and $\mathcal{R}_{\mathcal{L}}(\overrightarrow{OA}) = \overrightarrow{OB}$, then by Remark NEUT.1.3 $\mathcal{R}_{\mathcal{L}}(\overrightarrow{OB}) = \mathcal{R}_{\mathcal{L}}(\mathcal{R}_{\mathcal{L}}(\overrightarrow{OA})) = \overrightarrow{OA}$. By Definition PSH.29

---

[4]In a conversation many years ago a theologian friend cracked "you mathematicians have humor without humor." Here we have "distance without distance."

$\angle AOB = \overrightarrow{OA} \cup \overrightarrow{OB}$. Thus $\mathcal{R}_{\mathcal{L}}$ maps $\angle AOB$ onto itself, and therefore by Definition NEUT.3(D) $\mathcal{R}_{\mathcal{L}}$ is an angle reflection and $\mathcal{L}$ a line of symmetry for $\angle AOB$. However at this stage there is no guarantee that $\mathcal{R}_{\mathcal{L}}(O) = O$ (*cf* part (C) below).

(C) Property R.5 of Definition NEUT.2 says only that $\mathcal{R}_{\mathcal{L}}(\angle AOB) = \angle AOB$ in the sense of set equality; at this stage, there is no guarantee that $\mathcal{R}_{\mathcal{L}}(\overrightarrow{OA}) = \overrightarrow{OB}$ or $\mathcal{R}_{\mathcal{L}}(\overrightarrow{OB}) = \overrightarrow{OA}$, or, for that matter, that $O \in \mathcal{L}$. These things will be shown in Theorem NEUT.20, and once they have been established, any citation of Property R.5 will be understood to include these facts.

Remark NEUT.7. (A) Since the identity mapping ι of $\mathcal{P}$ is an isometry, every nonempty set $\mathcal{T}$ of $\mathcal{P}$ is congruent to itself, that is, $\mathcal{T} \cong \mathcal{T}$.

(B) Even though we know by assumption that there is only one reflection over a given line, the possibility will still exist that two sets $\mathcal{S}$ and $\mathcal{T}$ might be congruent to each other by means of two different isometries. That is, we might have $\alpha(\mathcal{S}) = \mathcal{T}$ and $\beta(\mathcal{S}) = \mathcal{T}$ where $\alpha \neq \beta$. Thus in some situations (for instance in the congruence Theorems NEUT.62, NEUT.64, and NEUT.65), to achieve complete clarity it will be necessary to specify the isometry by which the congruence is achieved. See also Remark NEUT.61.

Definition NEUT.8. $\mathcal{P}$ is a **neutral plane** if it is a Pasch plane in which Axiom REF holds. The geometry resulting from applying Axiom REF to a Pasch plane is **neutral geometry**.

Remark NEUT.9. In Chapter 21 (Theorem LC.33) we will show that neutral geometry is not vacuous, by showing that in a coordinate plane (which is a Pasch plane) it is possible to construct a set of reflections that satisfies all the properties of both Definitions NEUT.1 and NEUT.2.

Throughout the remainder of this chapter (except for Section 8.6, Constructed mirror mappings) our universe of discourse is a neutral plane $\mathcal{P}$; all lines, rays, and segments will be subsets of this plane. It should be noted, however, that when we invoke a neutral plane as the universe in a theorem, there is no presumption that all the properties of Definition NEUT.2 will be used.

The only property of Definition NEUT.2 that we invoke immediately is Property R.1, existence of a reflection over a given line; the first uses of Property R.2 (uniqueness) and of Property R.3 (closure) occur in the proof of Theorem NEUT.30. The first use of Property R.4 (linear scaling) will be in the proof of Theorem NEUT.23; of Property R.5 (angle reflection), in the proof of Theorem NEUT.35; and of Property R.6 (existence of midpoint), in Theorem NEUT.50.

8.3 Congruence, isometries, and lines of symmetry

While we state the theorems of this section in terms of reflections, isometry, and other notions defined in Definition NEUT.3, their proofs do not depend on Properties R.2 through R.6 of Definition NEUT.2, and remain valid if "mirror mapping' is substituted for "reflection," "composition of mirror mappings" is substituted for "isometry," and the definitions of congruence and midpoint are altered accordingly.

Theorem NEUT.10. *Let $\mathcal{E}$ be a nonempty subset of the neutral plane $\mathcal{P}$, $\mathcal{M}$ a line on $\mathcal{P}$, and let $\mathcal{R}_{\mathcal{M}}$ be a reflection over $\mathcal{M}$. If either $\mathcal{R}_{\mathcal{M}}(\mathcal{E}) \subseteq \mathcal{E}$ or $\mathcal{E} \subseteq \mathcal{R}_{\mathcal{M}}(\mathcal{E})$, then $\mathcal{R}_{\mathcal{M}}(\mathcal{E}) = \mathcal{E}$, so that $\mathcal{M}$ is a line of symmetry for $\mathcal{E}$.*

Proof. By Remark NEUT.1.3, $\mathcal{R}_{\mathcal{M}}(\mathcal{R}_{\mathcal{M}}(\mathcal{E})) = \mathcal{E}$.

If $\mathcal{R}_{\mathcal{M}}(\mathcal{E}) \subseteq \mathcal{E}$, we may apply $\mathcal{R}_{\mathcal{M}}$ to both sides to get $\mathcal{E} = \mathcal{R}_{\mathcal{M}}(\mathcal{R}_{\mathcal{M}}(\mathcal{E})) \subseteq \mathcal{R}_{\mathcal{M}}(\mathcal{E})$, and $\mathcal{E} = \mathcal{R}_{\mathcal{M}}(\mathcal{E})$. Likewise, if $\mathcal{E} \subseteq \mathcal{R}_{\mathcal{M}}(\mathcal{E})$, $\mathcal{R}_{\mathcal{M}}(\mathcal{E}) \subseteq \mathcal{R}_{\mathcal{M}}(\mathcal{R}_{\mathcal{M}}(\mathcal{E})) = \mathcal{E}$, and again $\mathcal{R}_{\mathcal{M}}(\mathcal{E}) = \mathcal{E}$. In either case, $\mathcal{M}$ is a line of symmetry for $\mathcal{E}$. □

Theorem NEUT.11. *Let α be an isometry of the neutral plane $\mathcal{P}$; then α is a bijection of $\mathcal{P}$ onto itself, and α^{-1} is an isometry of $\mathcal{P}$. Furthermore, any finite composition of isometries of $\mathcal{P}$ is an isometry of $\mathcal{P}$.*

Proof. By Definition NEUT.3(A), if α is an isometry other than the identity ι, there exists a natural number $n \geq 1$ such that for every $k \in [1; n]$, $\mathcal{M}_k$ is a line on $\mathcal{P}$ and $\alpha = \mathcal{R}_{\mathcal{M}_1} \circ \cdots \circ \mathcal{R}_{\mathcal{M}_n}$. Then α^{-1} exists and equals $\mathcal{R}_{\mathcal{M}_n} \circ \cdots \circ \mathcal{R}_{\mathcal{M}_1}$. Hence α^{-1} is an isometry of $\mathcal{P}$. By Remark NEUT.1.3, each of the mappings $\mathcal{R}_{\mathcal{M}}$ is a bijection, and every composition of bijections is a bijection.

A finite composition of finite compositions of reflections is a finite composition of reflections, so a finite composition of isometries is an isometry. □

Corollary NEUT.12. *The set of isometries of a neutral plane is a group under composition of mappings.*

Proof. This follows directly from Theorem NEUT.11 and the note *Bijections forming a group* in Chapter 1, Section 1.5. □

Theorem NEUT.13. *Let $\mathcal{S}$ and $\mathcal{T}$ be any nonempty subsets of the neutral plane $\mathcal{P}$ such that $\mathcal{S} \cong \mathcal{T}$ and let α be any isometry of $\mathcal{P}$. Then $\alpha(\mathcal{S}) \cong \alpha(\mathcal{T})$.*

Proof. By Definition NEUT.3(B) there exists an isometry β of $\mathcal{P}$ such that $\mathcal{T} = \beta(\mathcal{S})$, so that $\alpha(\mathcal{T}) = \alpha(\beta(\mathcal{S})) = (\alpha \circ \beta)(\mathcal{S}) = (\alpha \circ \beta \circ \alpha^{-1})(\alpha(\mathcal{S}))$. By Theorem NEUT.11, $\alpha \circ \beta \circ \alpha^{-1}$ is an isometry of $\mathcal{P}$, so $\alpha(\mathcal{S}) \cong \alpha(\mathcal{T})$. □

Theorem NEUT.14 (Congruence is an equivalence relation). *Let $\mathcal{S}$, $\mathcal{T}$, and $\mathcal{U}$ be nonempty subsets of the neutral plane $\mathcal{P}$. Then*

(1) *$\mathcal{S} \cong \mathcal{S}$ (congruence is **reflexive**);*
(2) *If $\mathcal{S} \cong \mathcal{T}$, then $\mathcal{T} \cong \mathcal{S}$ (congruence is **symmetric**);*
(3) *If $\mathcal{S} \cong \mathcal{T}$ and $\mathcal{T} \cong \mathcal{U}$, then $\mathcal{S} \cong \mathcal{U}$ (congruence is **transitive**).*

Proof. (1) Since the identity mapping ι is an isometry and $\mathcal{S} = \iota(\mathcal{S})$, $\mathcal{S} \cong \mathcal{S}$.

(2) If $\mathcal{S} \cong \mathcal{T}$, then by Definition NEUT.3(B) there exists an isometry α of $\mathcal{P}$ such that $\alpha(\mathcal{S}) = \mathcal{T}$, but then $\mathcal{S} = \alpha^{-1}(\mathcal{T})$. Since by Theorem NEUT.11, α^{-1} is an isometry of $\mathcal{P}$, $\mathcal{T} \cong \mathcal{S}$.

(3) If $\mathcal{S} \cong \mathcal{T}$ and $\mathcal{T} \cong \mathcal{U}$, then by Definition NEUT.3(B) there exist isometries α and β of $\mathcal{P}$ such that $\mathcal{T} = \alpha(\mathcal{S})$ and $\mathcal{U} = \beta(\mathcal{T})$. But then $\mathcal{U} = \beta(\alpha(\mathcal{S})) = (\beta \circ \alpha)(\mathcal{S})$. By Theorem NEUT.11 $\beta \circ \alpha$ is an isometry of $\mathcal{P}$. Hence $\mathcal{S} \cong \mathcal{U}$. □

Since isometries play such a major role in neutral geometry, we now restate Theorem COBE.5 (from Chapter 7) explicitly for them.

Theorem NEUT.15 (Properties of isometry). *Let A, B, and C be noncollinear points on the neutral plane $\mathcal{P}$. If φ is an isometry, or, for that matter, a mirror mapping, or any finite composition of mirror mappings of $\mathcal{P}$, φ is a belineation, and hence a collineation. Moreover, the following properties hold.*

(1) $\varphi(\overleftrightarrow{AB}) = \overleftrightarrow{\varphi(A)\varphi(B)}$.
(2) $\varphi(\overset{]\!\rightarrow}{AB}) = \overset{]\!\longrightarrow}{\varphi(A)\varphi(B)}$,
(3) $\varphi(\overset{[\!\rightarrow}{AB}) = \overset{[\!\longrightarrow}{\varphi(A)\varphi(B)}$,
(4) $\varphi(\overset{]\!-\!\![}{AB}) = \overset{]\!\longrightarrow\!\![}{\varphi(A)\varphi(B)}$,
(5) $\varphi(\overset{[\!-\!\!]}{AB}) = \overset{[\!\longrightarrow\!\!]}{\varphi(A)\varphi(B)}$,
(6) $\varphi(\overset{]\!-\!\!]}{AB}) = \overset{]\!\longrightarrow\!\!]}{\varphi(A)\varphi(B)}$,
(7) $\varphi(\overset{[\!-\!\![}{AB}) = \overset{[\!\longrightarrow\!\![}{\varphi(A)\varphi(B)}$,

(8) $\varphi(\angle ABC) = \angle\varphi(A)\varphi(B)\varphi(C)$,

(9) φ(the C-side of $\overleftrightarrow{AB}$) = the $\varphi(C)$-side of $\overleftrightarrow{\varphi(A)\varphi(B)}$,

(10) $\varphi(\triangle ABC) = \triangle\varphi(A)\varphi(B)\varphi(C)$,

(11) $\varphi(\text{ins}\,\angle BAC) = \text{ins}\,\angle\varphi(B)\varphi(A)\varphi(C)$,

(12) $\varphi(\text{ins}\,\triangle ABC) = \text{ins}\,\triangle\varphi(A)\varphi(B)\varphi(C)$.

(13) *If A, B, C, and D are points on $\mathcal{P}$ and $\square ABCD$ is a quadrilateral, then $\varphi(\square ABCD)$ is a quadrilateral, and* $\varphi(\square ABCD) = \square\varphi(A)\varphi(B)\varphi(C)\varphi(D)$.

Proof. By Remarks NEUT.1.3 and NEUT.1.4, every mirror mapping is a belineation, and hence, by Theorem COBE.2, a collineation. By a simple induction argument, every composition of a finite number of belineations is a belineation. Each of the above properties, then, follows directly from the corresponding property listed in Theorem COBE.5. □

Remark NEUT.16. This extends Theorem NEUT.1.8 to isometries. Let A and B be distinct points of $\mathcal{P}$. If φ is an isometry of $\mathcal{P}$, then by Theorem NEUT.15(5) $\varphi(\overset{\sqsubset\sqsupset}{AB}) = \overset{\sqsubset\!\!-\!\!-\!\!\sqsupset}{\varphi(A)\varphi(B)}$. Thus the end points of the image of the closed segment $\overset{\sqsubset\sqsupset}{AB}$ under the isometry φ are the images of the end points A and B. In other words, an isometry maps end points of a closed segment to end points of the image segment.

Also, if X is a point interior to $\overset{\sqsubset\sqsupset}{AB}$, an isometry φ maps X to a point interior to the image segment. For if A–X–B then $\varphi(A)$–$\varphi(X)$–$\varphi(B)$, since by Theorem NEUT.15, φ is a belineation.

Remark NEUT.17. A closed segment $\overset{\sqsubset\sqsupset}{AB}$ cannot be congruent to an open segment $\overset{\sqsupset\sqsubset}{CD}$; neither an open or a closed segment can be congruent to a half-open-half-closed segment such as $\overset{\sqsubset\sqsubset}{EF}$ or $\overset{\sqsupset\sqsupset}{EF}$. This follows easily from Theorem NEUT.15(4) through (7). It can also be proved independently of these results, using only the fact that an isometry preserves betweenness. This is Exercise NEUT.81.

Definition NEUT.18. A segment $\overset{\sqsubset\sqsupset}{PQ}$ is a **fixed segment** for a mapping α iff $\alpha(\overset{\sqsubset\sqsupset}{PQ}) = \overset{\sqsubset\sqsupset}{PQ}$; in particular, by Theorem NEUT.15(4), if $\alpha(P) = Q$ and $\alpha(Q) = P$, then $\overset{\sqsubset\sqsupset}{PQ}$ is a fixed segment for α.

Remark NEUT.19. Let $\mathcal{P}$ be a neutral plane, $\mathcal{L}$ and $\mathcal{M}$ be lines on $\mathcal{P}$, and φ be an isometry of $\mathcal{P}$, which by Theorem NEUT.15 is a collineation and a belineation.

(A) From Theorem CAP.1(B), if $\mathcal{L}$ and $\mathcal{M}$ intersect at the point O, then $\varphi(\mathcal{L})$ and $\varphi(\mathcal{M})$ intersect at $\varphi(O)$.

(B) From Theorem CAP.3, if $\mathcal{L} \parallel \mathcal{M}$, then $\varphi(\mathcal{L}) \parallel \varphi(\mathcal{M})$.

(C) From Theorem CAP.4(A), if A and B are fixed points of φ, $\overleftrightarrow{AB}$ is a fixed line of φ.

(D) From Theorem CAP.4(B), if $\mathcal{L}$ and $\mathcal{M}$ are fixed lines of φ which intersect at the point Q, then Q is a fixed point of φ.

8.4 Lines of symmetry and fixed lines

Theorems NEUT.20 and NEUT.22 establish important properties of angle reflections and line reflections, respectively; however, as in the previous section (Section 8.3) none of the proofs in this section call upon Properties R.2 through R.6 of Definition NEUT.2. Thus these theorems and proofs remain valid if "mirror mapping" is substituted for "reflection," "composition of mirror mappings" is substituted for "isometry," and the definitions of congruence and midpoint are altered accordingly.

Theorem NEUT.20 (Angle reflection properties). *Let A, B, and C be non-collinear points on the neutral plane $\mathcal{P}$. Then $\mathcal{M}$ is a line of symmetry and $\mathcal{R}_{\mathcal{M}}$ an angle reflection for $\angle BAC$ iff $\mathcal{R}_{\mathcal{M}}(A) = A$, $\mathcal{R}_{\mathcal{M}}(B) \in \overrightarrow{AC}$, and $\mathcal{R}_{\mathcal{M}}(C) \in \overrightarrow{AB}$. In this case the following are all true:*

(A) $A \in \mathcal{M}$, $\mathcal{R}_{\mathcal{M}}(\overrightarrow{AB}) = \overrightarrow{AC}$ and $\mathcal{R}_{\mathcal{M}}(\overrightarrow{AC}) = \overrightarrow{AB}$,

(B) $\mathcal{R}_{\mathcal{M}}(\overleftrightarrow{AB}) = \overleftrightarrow{A\mathcal{R}_{\mathcal{M}}(B)} = \overleftrightarrow{AC}$*and*$\mathcal{R}_{\mathcal{M}}(\overleftrightarrow{AC}) = \overleftrightarrow{A\mathcal{R}_{\mathcal{M}}(C)} = \overleftrightarrow{AB}$;

(C) $\mathcal{R}_{\mathcal{M}}(\overline{AB}) = \overline{A\mathcal{R}_{\mathcal{M}}(B)}$ *and* $\mathcal{R}_{\mathcal{M}}(\overline{AC}) = \overline{A\mathcal{R}_{\mathcal{M}}(C)}$;

(D) *$\mathcal{R}_{\mathcal{M}}$ maps only points of $\overleftrightarrow{AB}$ to $\overleftrightarrow{AC}$, and only points of $\overleftrightarrow{AC}$ to $\overleftrightarrow{AB}$;*

(E) *there exists a point $D \in \mathcal{M}$ such that*

  (1) $\mathcal{M} \cap \overline{B\mathcal{R}_{\mathcal{M}}(B)} = \{D\}$, *so that B–D–$\mathcal{R}_{\mathcal{M}}(B)$,*

  (2) $\mathcal{R}_{\mathcal{M}}(\overline{BD}) = \overline{\mathcal{R}_{\mathcal{M}}(B)D}$, *so that* $\overline{BD} \cong \overline{\mathcal{R}_{\mathcal{M}}(B)D}$, that is, D is a midpoint of $\overline{B\mathcal{R}_{\mathcal{M}}(B)}$,

  (3) $\overrightarrow{AD} \subseteq \text{ins}\,\angle BAC$, *and*

  (4) $\mathcal{R}_{\mathcal{M}}(\angle DAB) = \angle DAC$, so that $\angle DAB \cong \angle DAC$.

Proof. By Definition NEUT.3(D) $\mathcal{M}$ is a line of symmetry for $\angle BAC$ iff there exists a line reflection $\mathcal{R}_{\mathcal{M}}$ over $\mathcal{M}$ such that $\mathcal{R}_{\mathcal{M}}(\angle BAC) = \angle BAC$. If this is true, by Theorem NEUT.15(8)

$$\angle BAC = \mathcal{R}_{\mathcal{M}}(\angle BAC) = \angle \mathcal{R}_{\mathcal{M}}(B)\mathcal{R}_{\mathcal{M}}(A)\mathcal{R}_{\mathcal{M}}(C),$$

so that $\mathcal{R}_{\mathcal{M}}(A)$ and A are both the corner of this angle, which by Theorem PSH.33 is unique, and $\mathcal{R}_{\mathcal{M}}(A) = A$. Since $\mathcal{M}$ is the set of all fixed points for $\mathcal{R}_{\mathcal{M}}, A \in \mathcal{M}$. Also by Theorem NEUT.15(2),

$$\mathcal{R}_{\mathcal{M}}(\overrightarrow{AB}) = \overrightarrow{\mathcal{R}_{\mathcal{M}}(A)\mathcal{R}_{\mathcal{M}}(B)} = \overrightarrow{A\mathcal{R}_{\mathcal{M}}(B)}, \text{ and}$$
$$\mathcal{R}_{\mathcal{M}}(\overrightarrow{AC}) = \overrightarrow{\mathcal{R}_{\mathcal{M}}(A)\mathcal{R}_{\mathcal{M}}(C)} = \overrightarrow{A\mathcal{R}_{\mathcal{M}}(C)}.$$

By Exercise NEUT.4, neither $\overleftrightarrow{AB}$ nor $\overleftrightarrow{AC}$ is a line of symmetry for $\angle BAC$, hence each can intersect the line of symmetry $\mathcal{M}$ in only the one point A, so neither B nor C belongs to $\mathcal{M}$. By Definition NEUT.1(B), B and $\mathcal{R}_{\mathcal{M}}(B)$ are on opposite sides of $\mathcal{M}$.

If B and C were on the same side $\mathcal{E}$ of $\mathcal{M}$, then by Theorem IB.14 $\overrightarrow{AB}$ and $\overrightarrow{AC}$ would be subsets of $\mathcal{E}$, and their images under $\mathcal{R}_{\mathcal{M}}$ would be on the other side of $\mathcal{M}$, so that $\mathcal{R}_{\mathcal{M}}$ could not map $\angle BAC$ into itself, contradicting our hypothesis that $\mathcal{R}_{\mathcal{M}}$ is an angle reflection for $\angle BAC$. Thus B and C are on opposite sides of $\mathcal{M}$. Since B and $\mathcal{R}_{\mathcal{M}}(B)$ are on opposite sides of $\mathcal{M}$, by Theorem PSH.12 (plane separation), C and $\mathcal{R}_{\mathcal{M}}(B)$ are on the same side.

Now $\mathcal{R}_{\mathcal{M}}(B)$ is in $\angle BAC$ but not in $\overrightarrow{AB}$, so $\mathcal{R}_{\mathcal{M}}(B) \in \overrightarrow{AC}$, and by Theorem PSH.16 $\mathcal{R}_{\mathcal{M}}(\overrightarrow{AB}) = \overrightarrow{A\mathcal{R}_{\mathcal{M}}(B)} = \overrightarrow{AC}$. By similar reasoning, interchanging the roles of B and C, $\mathcal{R}_{\mathcal{M}}(C) \in \overrightarrow{AB}$, and $\mathcal{R}_{\mathcal{M}}(\overrightarrow{AC}) = \overrightarrow{AB}$. This proves half of the main assertion of the theorem, and also proves part (A).

Conversely, suppose $\mathcal{R}_{\mathcal{M}}(A) = A$, $\mathcal{R}_{\mathcal{M}}(B) \in \overrightarrow{AC}$, and $\mathcal{R}_{\mathcal{M}}(C) \in \overrightarrow{AB}$. Then by Theorem NEUT.15(3) $\mathcal{R}_{\mathcal{M}}(\overrightarrow{AB}) = \overrightarrow{\mathcal{R}_{\mathcal{M}}(A)\mathcal{R}_{\mathcal{M}}(B)} = \overrightarrow{AC}$ and $\mathcal{R}_{\mathcal{M}}(\overrightarrow{AC}) = \overrightarrow{\mathcal{R}_{\mathcal{M}}(A)\mathcal{R}_{\mathcal{M}}(C)} = \overrightarrow{AB}$, so that $\mathcal{R}_{\mathcal{M}}(\angle BAC) = \angle BAC$. Thus $\mathcal{R}_{\mathcal{M}}$ is an angle reflection for $\angle BAC$.

Parts (B) and (C) follow directly from Theorem NEUT.15 and part (A).

If Y is any point of the plane, since $\mathcal{R}_{\mathcal{M}}$ maps onto the plane, there exists a point X such that $Y = \mathcal{R}_{\mathcal{M}}(X)$. Then if $Y \in \overleftrightarrow{AB}$, $\mathcal{R}_{\mathcal{M}}(Y) \in \overleftrightarrow{AC}$. But $\mathcal{R}_{\mathcal{M}}(Y) = \mathcal{R}_{\mathcal{M}}(\mathcal{R}_{\mathcal{M}}(X)) = X$ so $X \in \overleftrightarrow{AC}$. Therefore the only points that map to $\overleftrightarrow{AB}$ are those of $\overleftrightarrow{AC}$; a similar argument shows that the only points that map to $\overleftrightarrow{AC}$ are those of $\overleftrightarrow{AB}$, proving assertion (D).

Since $\mathcal{R}_{\mathcal{M}}(B)$ and B are on opposite sides of $\mathcal{M}$, by Theorem PSH.12 (plane separation) there exists a point D such that $\overline{B\mathcal{R}_{\mathcal{M}}(B)} \cap \mathcal{M} = \{D\}$. Since $\mathcal{R}_{\mathcal{M}}(D) = D$, $\mathcal{R}_{\mathcal{M}}(\overline{BD}) = \overline{\mathcal{R}_{\mathcal{M}}(B)\mathcal{R}_{\mathcal{M}}(D)} = \overline{\mathcal{R}_{\mathcal{M}}(B)D}$. This, together with Definition NEUT.3(C), shows parts (1) and (2) of (E).

By Theorem PSH.37 $\overline{B\mathcal{R}_{\mathcal{M}}(B)} \subseteq \text{ins}\,\angle BA(\mathcal{R}_{\mathcal{M}}(B)) = \text{ins}\,\angle BAC$ so that $D \in \text{ins}\,\angle BAC$. By Theorem PSH.38(B), $\overrightarrow{AD} \subseteq \text{ins}\,\angle BAC$, showing (E)(3). Finally, since $\mathcal{R}_{\mathcal{M}}(\overrightarrow{AB}) = \overrightarrow{AC}$ and $\mathcal{R}_{\mathcal{M}}(\overrightarrow{AC}) = \overrightarrow{AB}$, $\mathcal{R}_{\mathcal{M}}(\angle DAB) = \angle DAC$, proving part (E)(4). □

Remark NEUT.20.1. (A) In part (E) above, we were careful to speak of D as "*a*" midpoint of $\overline{B\mathcal{R}_{\mathcal{M}}(B)}$; since we have not yet proved that midpoints are unique (which we will do in Theorem NEUT.50), we cannot speak of *the* midpoint of a segment. Also, we have not yet proved (we will do so in Theorem NEUT.26) that lines of symmetry for angles and angle reflections are unique, so we have taken care to speak of $\mathcal{M}$ as "*a*" line of symmetry and $\mathcal{R}_{\mathcal{M}}$ as '*an*" angle reflection for $\angle BAC$.

(B) (Important convention!) From this point forward, the reader should assume that whenever Property R.5 of Definition NEUT.2 is invoked, Theorem NEUT.20 is also invoked without reference. Thus, whenever we state that there exists a line $\mathcal{M}$ of symmetry or angle reflection $\mathcal{R}_{\mathcal{M}}$ for an angle $\angle BAC$, it will be understood that $\mathcal{R}_{\mathcal{M}}(A) = A$, $\mathcal{R}_{\mathcal{M}}(B) \in \overrightarrow{AC}$ and $\mathcal{R}_{\mathcal{M}}(C) \in \overrightarrow{AB}$.

(C) In the following, we will frequently use results (1) through (13) from Theorem NEUT.15, and while these will often be referenced, there will be a tendency to do so less and less as we assume the reader's habits are established. This will be especially true of the lower-numbered results involving lines, rays, and segments. Be warned!

Theorem NEUT.21. *Let A, B, and C be noncollinear points on the neutral plane $\mathcal{P}$ and let $\mathcal{M}$ be a line of symmetry of $\angle BAC$. Then $\mathcal{M}$ is a line of symmetry of* ins $\angle BAC$.

Proof. By Theorem NEUT.20 $\mathcal{R}_{\mathcal{M}}(B) \in \overrightarrow{AC}$ and $\mathcal{R}_{\mathcal{M}}(C) \in \overrightarrow{AB}$, so that $\angle BAC = \angle BA\mathcal{R}_{\mathcal{M}}(B)$ and $\angle BAC = \angle\mathcal{R}_{\mathcal{M}}(C)AC$. Let X be any member of ins $\angle BAC$; then by Definition PSH.36(A), $X \in \mathcal{R}_{\mathcal{M}}(B)$-side of $\overleftrightarrow{AB}$. By Theorem NEUT.15(9), and the fact that $\mathcal{R}_{\mathcal{M}}(A) = A$,

$$\mathcal{R}_{\mathcal{M}}(X) \in \mathcal{R}_{\mathcal{M}}(\mathcal{R}_{\mathcal{M}}(B))\text{-side of } \overleftrightarrow{\mathcal{R}_{\mathcal{M}}(A)\mathcal{R}_{\mathcal{M}}(B)} = B\text{-side of } \overleftrightarrow{AC}$$

Also $X \in \mathcal{R}_{\mathcal{M}}(C)$-side of $\overleftrightarrow{AC}$, and by the same theorem

$$\mathcal{R}_{\mathcal{M}}(X) \in \mathcal{R}_{\mathcal{M}}(\mathcal{R}_{\mathcal{M}}(C))\text{-side of } \overleftrightarrow{\mathcal{R}_{\mathcal{M}}(A)\mathcal{R}_{\mathcal{M}}(C)} = C\text{-side of } \overleftrightarrow{AB}.$$

Thus $\mathcal{R}_{\mathcal{M}}(X) \in \text{ins}\,\angle BAC$; by Theorem NEUT.10, $\mathcal{M}$ is a line of symmetry for ins $\angle BAC$. □

Theorem NEUT.22 (Behavior of line reflections). *Let $\mathcal{M}$ be a line on the neutral plane and let $\mathcal{R}_{\mathcal{M}}$ be a reflection over $\mathcal{M}$.*

(A) *If $A \notin \mathcal{M}$, $\overleftrightarrow{A\mathcal{R}_{\mathcal{M}}(A)}$ is a fixed line for $\mathcal{R}_{\mathcal{M}}$, $\overline{A\mathcal{R}_{\mathcal{M}}(A)}$ is a fixed segment, and $\overleftrightarrow{A(\mathcal{R}_{\mathcal{M}}(A))} \neq \mathcal{M}$.*
(B) *If $\mathcal{L}$ is a fixed line and $\mathcal{L} \neq \mathcal{M}$, then for some $A \notin \mathcal{M}$, $\mathcal{L} = \overleftrightarrow{A\mathcal{R}_{\mathcal{M}}(A)}$.*
(C) *For any line $\mathcal{L}$ such that $\mathcal{L} \neq \mathcal{M}$, the following statements are equivalent; the equivalence of (1) and (2) summarizes (A) and (B):*

(1) *$\mathcal{L}$ is a fixed line for $\mathcal{R}_{\mathcal{M}}$;*
(2) *for some $A \notin \mathcal{M}$, $\mathcal{L} = \overleftrightarrow{A\mathcal{R}_{\mathcal{M}}(A)}$;*
(3) *for some $A \in \mathcal{L}$ such that $A \notin \mathcal{M}$, $\mathcal{R}_{\mathcal{M}}(A) \in \mathcal{L}$;*
(4) *for every $A \in \mathcal{L}$ such that $A \notin \mathcal{M}$, $\mathcal{L} = \overleftrightarrow{A\mathcal{R}_{\mathcal{M}}(A)}$.*
(5) *$\mathcal{M}$ is a line of symmetry for $\mathcal{L}$.*

(D) *There is at most one fixed line $\mathcal{L} \neq \mathcal{M}$ for $\mathcal{R}_{\mathcal{M}}$ through a point A.*
(E) *Let $\overleftrightarrow{A\mathcal{R}_{\mathcal{M}}(A)} \neq \mathcal{M}$ and $\overleftrightarrow{B\mathcal{R}_{\mathcal{M}}(B)} \neq \mathcal{M}$ be distinct fixed lines for $\mathcal{R}_{\mathcal{M}}$. Then $\overleftrightarrow{A\mathcal{R}_{\mathcal{M}}(A)} \parallel \overleftrightarrow{B\mathcal{R}_{\mathcal{M}}(B)}$.*
(F) *Every fixed line $\mathcal{L} = \overleftrightarrow{A\mathcal{R}_{\mathcal{M}}(A)}$ for $\mathcal{R}_{\mathcal{M}}$ intersects $\mathcal{M}$ at exactly one point D; moreover, A–D–$\mathcal{R}_{\mathcal{M}}(A)$, and D is a midpoint of the fixed segment $\overline{A\mathcal{R}_{\mathcal{M}}(A)}$.*

Notice that there is no claim here that every line which is parallel to a fixed line for $\mathcal{R}_{\mathcal{M}}$ is a fixed line. This will be proved (on a Euclidean plane) as Corollary EUC.3.1.

Proof. (A) By Theorem NEUT.15(1) and Definition NEUT.1(C),
$$\mathcal{R}_{\mathcal{M}}(\overleftrightarrow{A(\mathcal{R}_{\mathcal{M}}(A))}) = \overleftrightarrow{(\mathcal{R}_{\mathcal{M}}(A))(\mathcal{R}_{\mathcal{M}}(\mathcal{R}_{\mathcal{M}}(A)))} = \overleftrightarrow{A(\mathcal{R}_{\mathcal{M}}(A))};$$
by Definition CAP.0(C), $\overleftrightarrow{A(\mathcal{R}_{\mathcal{M}}(A))}$ is a fixed line for $\mathcal{R}_{\mathcal{M}}$. By Definition NEUT.18, $\overline{A\mathcal{R}_{\mathcal{M}}(A)}$ is a fixed segment for $\mathcal{R}_{\mathcal{M}}$.

(B) Pick $A \in \mathcal{L}$ such that $A \notin \mathcal{M}$. Since $\mathcal{L}$ is a fixed line, $\mathcal{R}_{\mathcal{M}}(A) \in \mathcal{L}$. Since both A and $\mathcal{R}_{\mathcal{M}}(A)$ belong to $\mathcal{L}$, by Exercise I.2, $\mathcal{L} = \overleftrightarrow{A(\mathcal{R}_{\mathcal{M}}(A))}$.

(C) Part (A) is "(2) implies (1)"; part (B) is "(1) implies (2)." (3) is equivalent to (2): if $A \in \mathcal{L}$ and $\mathcal{R}_{\mathcal{M}}(A) \in \mathcal{L}$, by Exercise I.2, $\mathcal{L} = \overleftrightarrow{A\mathcal{R}_{\mathcal{M}}(A)}$; conversely, if $\mathcal{L} = \overleftrightarrow{A\mathcal{R}_{\mathcal{M}}(A)}$, then $\mathcal{R}_{\mathcal{M}}(A) \in \mathcal{L}$.

That (4) implies (2) is trivial; conversely, if (1) is true, $\mathcal{L}$ is a fixed line; then for every $A \in \mathcal{L}$, $\mathcal{R}_{\mathcal{M}}(A) \in \mathcal{L}$.

(5) is equivalent to (1): since $\mathcal{L} \neq \mathcal{M}$, by Definition CAP.0(C), $\mathcal{L}$ is a fixed line for $\mathcal{R}_{\mathcal{M}}$ iff $\mathcal{R}_{\mathcal{M}}(\mathcal{L}) = \mathcal{L}$, by Definition NEUT.4, this is true iff $\mathcal{M}$ is a line of symmetry for $\mathcal{L}$.

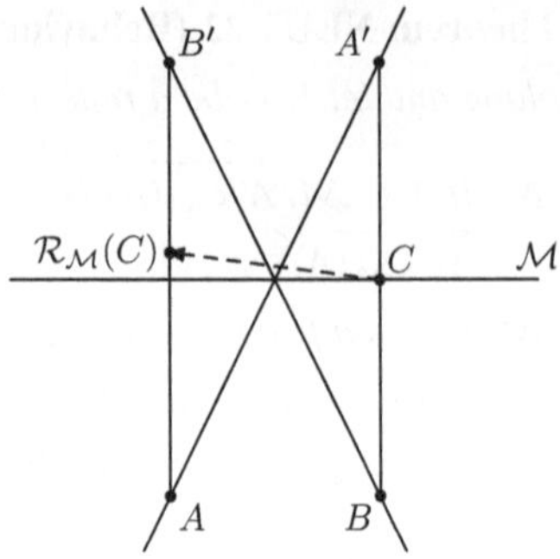

Fig. 8.1 For the construction in Theorem NEUT.22(D).

(D) If $\overleftrightarrow{A\mathcal{R}_{\mathcal{M}}(A)}$ and $\overleftrightarrow{B\mathcal{R}_{\mathcal{M}}(B)}$ are distinct fixed lines for $\mathcal{R}_{\mathcal{M}}$ which intersect at some point Q, then by Theorem CAP.4(B) Q is a fixed point for $\mathcal{R}_{\mathcal{M}}$, so that $Q \in \mathcal{M}$.

We may choose the notation so that A and B are on the same side of $\mathcal{M}$. (See Figure 8.1 for a visualization.) Let $A' = \mathcal{R}_{\mathcal{M}}(A)$ and $B' = \mathcal{R}_{\mathcal{M}}(B)$; then A' and B' are on the same side of $\mathcal{M}$. Since A and A' are on opposite sides of $\mathcal{M}$, and B and B' are on opposite sides of $\mathcal{M}$, by Theorem PSH.12 (PSA) A–Q–A' and B–Q–B'.

B is on the opposite side of $\overleftrightarrow{AA'}$ from B', and $\overset{\sqsupset\!-\!\sqsubset}{BA'} \subseteq \overset{\sqsupset\!\rightarrow}{A'B}$ which is a subset of the B-side of $\overleftrightarrow{AA'}$. Likewise, B' is on the opposite side of $\overleftrightarrow{AA'}$ from B, and $\overset{\sqsupset\!-\!\sqsubset}{B'A} \subseteq \overset{\sqsupset\!\rightarrow}{AB'}$ which is a subset of the side of $\overleftrightarrow{AA'}$ opposite B.

Then by Theorem NEUT.15(4) $\mathcal{R}_{\mathcal{M}}(\overset{\sqsupset\!-\!\sqsubset}{BA'}) = \overset{\sqsupset\!-\!\sqsubset}{B'\mathcal{R}_{\mathcal{M}}(A')} = \overset{\sqsupset\!-\!\sqsubset}{B'A}$ so that $\mathcal{R}_{\mathcal{M}}(\overset{\sqsupset\!-\!\sqsubset}{BA'})$ and $\overset{\sqsupset\!-\!\sqsubset}{BA'}$ are on opposite sides of $\overleftrightarrow{AA'}$. Since B and A' are on opposite sides of $\mathcal{M}$, by Theorem PSH.12 there is a point C such that $\{C\} = \overset{\sqsupset\!-\!\sqsubset}{BA'} \cap \mathcal{M}$. Then $\mathcal{R}_{\mathcal{M}}(C)$ is on the opposite side of $\overleftrightarrow{AA'}$ from C, a contradiction to the fact that $C \in \mathcal{M}$ is a fixed point of $\mathcal{R}_{\mathcal{M}}$.

(E) If $\overleftrightarrow{A\mathcal{R}_{\mathcal{M}}(A)}$ and $\overleftrightarrow{B\mathcal{R}_{\mathcal{M}}(B)}$ are not parallel, then they intersect, and this is impossible by part (D).

(F) If $\mathcal{L} \neq \mathcal{M}$ is a fixed line, by part (A)(3) there is a point $A \notin \mathcal{M}$ such that $\mathcal{L} = \overleftrightarrow{A\mathcal{R}_{\mathcal{M}}(A)}$. By Definition NEUT.1(B) A and $\mathcal{R}_{\mathcal{M}}(A)$ are on opposite sides of $\mathcal{M}$; hence by Theorem PSH.12 there exists a point D such that $\overset{\sqsupset\!-\!\sqsubset}{A\mathcal{R}_{\mathcal{M}}(A)} \cap \mathcal{M} = \{D\}$. By Exercise I.1, D is the only such point of intersection. By Theorem NEUT.15(5) $\mathcal{R}_{\mathcal{M}}(\overline{AD}) = \overline{\mathcal{R}_{\mathcal{M}}(A)\mathcal{R}_{\mathcal{M}}(D)} = \overline{\mathcal{R}_{\mathcal{M}}(A)D}$ so that $\overline{\mathcal{R}_{\mathcal{M}}(A)D} \cong \overline{AD}$; therefore D is a midpoint of $\overline{A\mathcal{R}_{\mathcal{M}}(A)}$. □

In part (F) above, we were careful to speak of D as "*a*" midpoint of $\overline{A\mathcal{R}_{\mathcal{M}}(A)}$; since we have not yet proved that midpoints are unique (which we will do in Theorem NEUT.50), we cannot speak of *the* midpoint of a segment.

8.5 Uniqueness of angle reflections

The proof of the next theorem is our first use of Property R.4 of Definition NEUT.2 (linear scaling). This theorem strengthens Theorem CAP.4; not only is $\overleftrightarrow{AB}$ a fixed line, but every point on it is a fixed point.

Theorem NEUT.23. *Let φ be an isometry of the neutral plane $\mathcal{P}$. If A and B are distinct fixed points of φ, then every point on $\overleftrightarrow{AB}$ is a fixed point of φ.*

Proof. (I) Let X be any member of $\overrightarrow{AB}\setminus\{B\}$. Since $\varphi(\overrightarrow{AB}) = \overrightarrow{\varphi(A)\varphi(B)} = \overrightarrow{AB}$, $\varphi(X) \in \overrightarrow{AB}$. Since $\varphi(\overline{AX}) = \overline{A\varphi(X)}$, by Definition NEUT.3(B) $\overline{AX} \cong \overline{A\varphi(X)}$. By Property R.4 of Definition NEUT.2, $\varphi(X) = X$.

(II) By Property B.3 of Definition IB.1 there exists a point B' such that B–A–B'. Let X be any member of $\overrightarrow{AB'}$. Reasoning as in (I) we get $\varphi(X) = X$.

By (I) and (II) every point on $\overleftrightarrow{AB}$ is a fixed point of φ. □

Theorem NEUT.24. *Let φ be an isometry of the neutral plane $\mathcal{P}$. If φ has three noncollinear fixed points, then $\varphi = \iota$.*

Proof. Let A, B, and C be noncollinear fixed points of φ. By Theorem NEUT.23 every member of $\overleftrightarrow{AB} \cup \overleftrightarrow{AC} \cup \overleftrightarrow{BC}$ is a fixed point of φ.

Let X be any member of $\mathcal{P} \setminus (\overleftrightarrow{AB} \cup \overleftrightarrow{AC} \cup \overleftrightarrow{BC})$. By Theorem PSH.22 (Denseness property for betweenness) there exists a point D between A and B. By Theorem PSH.6 (Pasch) there exists a point E such that $\overleftrightarrow{XD} \cap \overline{AC} = \{E\}$ or $\overleftrightarrow{XD} \cap \overline{BC} = \{E\}$. Since both D and E are fixed points of φ, by Theorem NEUT.23 every point of $\overleftrightarrow{DE} = \overleftrightarrow{XD}$ is a fixed point of φ, and X is a fixed point of $\mathcal{P}$. □

Theorem NEUT.25. *Let each of α or β be an isometry of the neutral plane $\mathcal{P}$. If α and β are equal at three noncollinear points of $\mathcal{P}$, then $\alpha = \beta$.*

Proof. Since the set of isometries is a group (*cf* Corollary NEUT.12) under composition of mappings and since α and β are isometries, $\alpha^{-1} \circ \beta$ is an isometry. Since there exist three noncollinear points A, B, and C, such that $\alpha(A) = \beta(A)$, $\alpha(B) = \beta(B)$, and $\alpha(C) = \beta(C)$, each of A, B, and C is a fixed point of

$\alpha^{-1} \circ \beta$. By Theorem NEUT.24 $\alpha^{-1} \circ \beta = \iota$ (the identity mapping). But then $\alpha \circ (\alpha^{-1} \circ \beta) = (\alpha \circ \alpha^{-1}) \circ \beta = \iota \circ \beta = \beta$ and $\alpha \circ \iota = \alpha$. Hence $\alpha = \beta$. □

Corollary NEUT.25.1. *Two distinct isometries which agree at two points A and B cannot agree at any point off* $\overleftrightarrow{AB}$.

In Theorem NEUT.15 we showed that every isometry is a collineation and a belineation. In Chapter 19 (Theorems AA.10 and AA.11), Theorems NEUT.24 and NEUT.25 will be generalized to all belineations of a Euclidean/LUB plane.

Theorem NEUT.26 (Uniqueness of angle reflection and line of symmetry). *For any angle* $\angle AOB$ *in a neutral plane, there exists at most one angle reflection mapping the angle onto itself, only one line of symmetry, and only one bisecting ray. Thus it is proper to speak of THE reflection mapping* $\angle AOB$ *to* $\angle AOB$*, THE line of symmetry of* $\angle AOB$*, and THE bisecting ray of* $\angle AOB$.

Proof. Suppose there exist two reflections $\mathcal{R}_{\mathcal{L}}$ and $\mathcal{R}_{\mathcal{M}}$ (the lines $\mathcal{L}$ and $\mathcal{M}$ may or may not be the same) such that $\mathcal{R}_{\mathcal{L}}(\angle AOB) = \angle AOB$ and $\mathcal{R}_{\mathcal{M}}(\angle AOB) = \angle AOB$. By Theorem NEUT.20, $\mathcal{R}_{\mathcal{L}}(O) = \mathcal{R}_{\mathcal{M}}(O) = O$, $\mathcal{R}_{\mathcal{L}}(A) \in \overrightarrow{OB}$ and $\mathcal{R}_{\mathcal{M}}(A) \in \overrightarrow{OB}$. Since a reflection is its own inverse (cf Definition NEUT.1(C)),

$$(\mathcal{R}_{\mathcal{M}} \circ \mathcal{R}_{\mathcal{L}})(\mathcal{R}_{\mathcal{L}}(A)) = \mathcal{R}_{\mathcal{M}}(\mathcal{R}_{\mathcal{L}}(\mathcal{R}_{\mathcal{L}}(A))) = \mathcal{R}_{\mathcal{M}}(A);$$

thus, by Theorem NEUT.15(5) and Definition NEUT.3(B), $\overline{O\mathcal{R}_{\mathcal{L}}(A)} \cong \overline{O\mathcal{R}_{\mathcal{M}}(A)}$; by Property R.4 of Definition NEUT.2 (linear scaling), $\mathcal{R}_{\mathcal{L}}(A) = \mathcal{R}_{\mathcal{M}}(A)$. It follows that

$$\mathcal{R}_{\mathcal{M}}(\mathcal{R}_{\mathcal{L}}(A)) = \mathcal{R}_{\mathcal{M}}(\mathcal{R}_{\mathcal{M}}(A)) = A = \mathcal{R}_{\mathcal{L}}(\mathcal{R}_{\mathcal{L}}(A)),$$

so that $\mathcal{R}_{\mathcal{L}}$ and $\mathcal{R}_{\mathcal{M}}$ are isometries which agree at the three noncollinear points A, $\mathcal{R}_{\mathcal{L}}(A)$, and O; hence by Theorem NEUT.25 $\mathcal{R}_{\mathcal{L}} = \mathcal{R}_{\mathcal{M}}$. From Remark NEUT.1.1, $\mathcal{L} = \mathcal{M}$. There can be only one bisecting ray, since it is the intersection $\mathcal{L} \cap$ ins $\angle AOB$. □

8.6 Constructed mirror mappings

The following two theorems do not need Axiom REF and are the exception to the blanket invocation of the neutral plane in Remark NEUT.9.

Theorem NEUT.27. *Let $\mathcal{M}$ and $\mathcal{L}$ be lines on a Pasch plane, and let $\mathcal{R}_\mathcal{M}$ and $\mathcal{R}_\mathcal{L}$ be mirror mappings over these lines. Then the mapping $\varphi = \mathcal{R}_\mathcal{L} \circ \mathcal{R}_\mathcal{M} \circ \mathcal{R}_\mathcal{L}$ is a mirror mapping over the line $\mathcal{R}_\mathcal{L}(\mathcal{M})$.*

Proof. We show that φ satisfies Properties (A) through (D) of Definition NEUT.1.

(A) Suppose $Y \in \mathcal{R}_\mathcal{L}(\mathcal{M})$. Then there exists a point $X \in \mathcal{M}$ such that $Y = \mathcal{R}_\mathcal{L}(X)$. Then

$$\begin{aligned}\varphi(Y) &= \mathcal{R}_\mathcal{L}(\mathcal{R}_\mathcal{M}(\mathcal{R}_\mathcal{L}(Y))) = \mathcal{R}_\mathcal{L}(\mathcal{R}_\mathcal{M}(\mathcal{R}_\mathcal{L}(\mathcal{R}_\mathcal{L}(X))))\\ &= \mathcal{R}_\mathcal{L}(\mathcal{R}_\mathcal{M}(X)) = \mathcal{R}_\mathcal{L}(X) = Y.\end{aligned}$$

(B) Suppose $Y \notin \mathcal{R}_\mathcal{L}(\mathcal{M})$; we show that $\varphi(Y)$ and Y are on opposite sides of $\mathcal{R}_\mathcal{L}(\mathcal{M})$. There exists a point $X \notin \mathcal{M}$ such that $Y = \mathcal{R}_\mathcal{L}(X)$. By Property (B), $\mathcal{R}_\mathcal{M}(X)$ and X are on opposite sides of $\mathcal{M}$, and by Theorem PSH.12 there exists a point $D \in \mathcal{M}$ such that $D \in \overline{X(\mathcal{R}_\mathcal{M}(X))}$. Then $\mathcal{R}_\mathcal{L}(D) \in \mathcal{R}_\mathcal{L}(\mathcal{M})$ and

$$\begin{aligned}\mathcal{R}_\mathcal{L}(D) \in \mathcal{R}_\mathcal{L}(\overline{X(\mathcal{R}_\mathcal{M}(X))}) &= \overline{(\mathcal{R}_\mathcal{L}(X))(\mathcal{R}_\mathcal{L}(\mathcal{R}_\mathcal{M}(X)))}\\ &= \overline{(\mathcal{R}_\mathcal{L}(X))(\mathcal{R}_\mathcal{L}(\mathcal{R}_\mathcal{M}(\mathcal{R}_\mathcal{L}(\mathcal{R}_\mathcal{L}(X)))))}\\ &= \overline{Y(\mathcal{R}_\mathcal{L}(\mathcal{R}_\mathcal{M}(\mathcal{R}_\mathcal{L}(Y))))}\end{aligned}$$

so that Y and $\mathcal{R}_\mathcal{L}(\mathcal{R}_\mathcal{M}(\mathcal{R}_\mathcal{L}(Y)))$ are on opposite sides of $\mathcal{R}_\mathcal{L}(\mathcal{M})$.

(C) $\mathcal{R}_\mathcal{L} \circ \mathcal{R}_\mathcal{M} \circ \mathcal{R}_\mathcal{L} \circ \mathcal{R}_\mathcal{L} \circ \mathcal{R}_\mathcal{M} \circ \mathcal{R}_\mathcal{L} = \mathcal{R}_\mathcal{L} \circ \mathcal{R}_\mathcal{M} \circ \mathcal{R}_\mathcal{M} \circ \mathcal{R}_\mathcal{L} = \mathcal{R}_\mathcal{L} \circ \mathcal{R}_\mathcal{L} = \iota$.

(D) Since each of the mappings $\mathcal{R}_\mathcal{M}$ and $\mathcal{R}_\mathcal{L}$ preserves betweenness, so does $\varphi = \mathcal{R}_\mathcal{L} \circ \mathcal{R}_\mathcal{M} \circ \mathcal{R}_\mathcal{L}$. □

Corollary NEUT.27.1. *Let $\mathcal{M}$ be a line on a Pasch plane, $\mathcal{R}_\mathcal{M}$ be a mirror mapping over $\mathcal{M}$, and suppose α is a composition of mirror mappings of the plane. Then the mapping $\varphi = \alpha \circ \mathcal{R}_\mathcal{M} \circ \alpha^{-1}$ is a mirror mapping over the line $\alpha(\mathcal{M})$.*

Proof. By Definition NEUT.3(A), either α is (1) the identity ι, (2) a mirror mapping, or (3) the composition $\mathcal{R}_{\mathcal{M}_1} \circ \cdots \circ \mathcal{R}_{\mathcal{M}_n}$ of a finite number of mirror mappings $\mathcal{R}_{\mathcal{M}_k}$ over lines $\mathcal{M}_k$ in the plane.

In case (1), $\varphi = \mathcal{R}_\mathcal{M}$ which is already a mirror mapping; case (2) follows from Theorem NEUT.27. For case (3), note first that $\alpha^{-1} = \mathcal{R}_{\mathcal{M}_n} \circ \cdots \circ \mathcal{R}_{\mathcal{M}_1}$; then

$$\begin{aligned}\varphi &= \alpha \circ \mathcal{R}_\mathcal{M} \circ \alpha^{-1}\\ &= \mathcal{R}_{\mathcal{M}_1} \circ \cdots \circ \left(\mathcal{R}_{\mathcal{M}_{n-1}} \circ \left(\mathcal{R}_{\mathcal{M}_n} \circ \mathcal{R}_\mathcal{M} \circ \mathcal{R}_{\mathcal{M}_n}\right) \circ \mathcal{R}_{\mathcal{M}_{n-1}}\right) \circ \cdots \circ \mathcal{R}_{\mathcal{M}_1}.\end{aligned}$$

By Theorem NEUT.27, $\mathcal{R}_{\mathcal{M}_n} \circ \mathcal{R}_\mathcal{M} \circ \mathcal{R}_{\mathcal{M}_n}$ is a mirror mapping over $\mathcal{R}_{\mathcal{M}_n}(\mathcal{M})$; applying the same theorem again

$$\mathcal{R}_{\mathcal{M}_{n-1}} \circ \left(\mathcal{R}_{\mathcal{M}_n} \circ \mathcal{R}_{\mathcal{M}} \circ \mathcal{R}_{\mathcal{M}_n}\right) \circ \mathcal{R}_{\mathcal{M}_{n-1}}$$

is a mirror mapping over $\mathcal{R}_{\mathcal{M}_{n-1}}(\mathcal{R}_{\mathcal{M}_n}(\mathcal{M}))$; successive repetitions of this process produce the final result. □

8.7 Complementary mappings and perpendicularity

Remark NEUT.28. Up to this point we have not invoked Property R.2 of Definition NEUT.2, which states that there can be only one reflection over a line. We now invoke this property in the following definition, when we speak of *the* reflection $\mathcal{R}_{\mathcal{M}}$ over $\mathcal{M}$. We also invoke Property R.3 of Definition NEUT.2 (closure) for the first time.

Definition NEUT.29. Let $\mathcal{M}$ be a line on a neutral plane, let $\mathcal{L} \neq \mathcal{M}$ be a fixed line for the reflection $\mathcal{R}_{\mathcal{M}}$ over $\mathcal{M}$, and let $\{O\} = \mathcal{M} \cap \mathcal{L}$. By Theorem NEUT.22(D) $\mathcal{L}$ is the only fixed line (other than $\mathcal{M}$) for $\mathcal{R}_{\mathcal{M}}$ that passes through O. Let A and B be two points distinct from O, such that $\overrightarrow{OA} \subseteq \mathcal{M}$ and $\overrightarrow{OB} \subseteq \mathcal{L}$. By Theorem NEUT.26 there is exactly one line of symmetry $\mathcal{S}$ for $\angle AOB$, and exactly one reflection $\mathcal{R}_{\mathcal{S}}$ over $\mathcal{S}$ such that $\mathcal{R}_{\mathcal{S}}(A) \in \overrightarrow{OB}$.

The mapping $CO(\mathcal{R}_{\mathcal{M}}) = \mathcal{R}_{\mathcal{S}} \circ \mathcal{R}_{\mathcal{M}} \circ \mathcal{R}_{\mathcal{S}}$ is called a **complement** or a **complementary mapping of** $\mathcal{R}_{\mathcal{M}}$ **at the point** O. By Property R.2 of Definition NEUT.2, $\mathcal{R}_{\mathcal{M}}$ is uniquely determined by the choice of $\mathcal{M}$; by Theorem NEUT.26, $\mathcal{S}$ and $\mathcal{R}_{\mathcal{S}}$ are uniquely determined by the choice of $\mathcal{M}$, $\mathcal{L}$, $\overrightarrow{OA}$, and $\overrightarrow{OB}$. Thus by Theorem NEUT.27, $CO(\mathcal{R}_{\mathcal{M}})$ is completely determined by the choices of $\mathcal{M}$, $\mathcal{L}$, $\overrightarrow{OA}$, and $\overrightarrow{OB}$.

Theorem NEUT.30. *Let $\mathcal{M}$ be a line on a neutral plane, let $\mathcal{L} \neq \mathcal{M}$ be a fixed line for the reflection $\mathcal{R}_{\mathcal{M}}$ over $\mathcal{M}$; and let $\{O\} = \mathcal{M} \cap \mathcal{L}$. Let A and B be points, not O, such that, $\overrightarrow{OA} \subseteq \mathcal{M}$ and $\overrightarrow{OB} \subseteq \mathcal{L}$, and let $\mathcal{S}$ be the line of symmetry of $\angle AOB$.*

(I) *The complementary mapping $CO(\mathcal{R}_{\mathcal{M}}) = \mathcal{R}_{\mathcal{S}} \circ \mathcal{R}_{\mathcal{M}} \circ \mathcal{R}_{\mathcal{S}}$ is the reflection over $\mathcal{L}$ having $\mathcal{M}$ as a fixed line. Hence $CO(\mathcal{R}_{\mathcal{M}}) = \mathcal{R}_{\mathcal{L}}$.*

(II) *$CO(CO(\mathcal{R}_{\mathcal{M}})) = \mathcal{R}_{\mathcal{M}}$. That is, $CO(\mathcal{R}_{\mathcal{L}}) = \mathcal{R}_{\mathcal{M}}$, the reflection over $\mathcal{M}$ having $\mathcal{L}$ as a fixed line.*

Proof. (I) Let A' and B' be points such that A–O–A' and B–O–B'. By Exercise NEUT.10 $\mathcal{S}$ is a line of symmetry for $\angle A'OB'$, so that $\mathcal{R}_{\mathcal{S}}(\overrightarrow{OA}) = \overrightarrow{OB}$, $\mathcal{R}_{\mathcal{S}}(\overrightarrow{OB}) = \overrightarrow{OA}$, $\mathcal{R}_{\mathcal{S}}(\overrightarrow{OA'}) = \overrightarrow{OB'}$, and $\mathcal{R}_{\mathcal{S}}(\overrightarrow{OB'}) = \overrightarrow{OA'}$.

Since $\mathcal{R}_{\mathcal{M}}$ is a reflection with fixed line $\mathcal{L}$, Property (B) of Definition NEUT.1 says that $\mathcal{R}_{\mathcal{M}}(\overrightarrow{OB}) = \overrightarrow{OB'}$ and $\mathcal{R}_{\mathcal{M}}(\overrightarrow{OB'}) = \overrightarrow{OB}$. If $X \in \overrightarrow{OA}$, $\mathcal{R}_{\mathcal{S}}(X) \in \overrightarrow{OB}$ so $\mathcal{R}_{\mathcal{M}}(\mathcal{R}_{\mathcal{S}}(X)) \in \overrightarrow{OB'}$ and

$$CO(\mathcal{R}_{\mathcal{M}})(X) = \mathcal{R}_{\mathcal{S}}(\mathcal{R}_{\mathcal{M}}(\mathcal{R}_{\mathcal{S}}(X))) \in \overrightarrow{OA'}.$$

Similarly, if $X \in \overrightarrow{OA'}$,

$$CO(\mathcal{R}_{\mathcal{M}})(X) = \mathcal{R}_{\mathcal{S}}(\mathcal{R}_{\mathcal{M}}(\mathcal{R}_{\mathcal{S}}(X)) \in \overrightarrow{OA}.$$

This shows that $CO(\mathcal{R}_{\mathcal{M}})$ maps $\mathcal{M}$ onto $\mathcal{M}$, which is then a fixed line for $CO(\mathcal{R}_{\mathcal{M}})$.

By Theorem NEUT.27, $CO(\mathcal{R}_{\mathcal{M}}) = \mathcal{R}_{\mathcal{S}} \circ \mathcal{R}_{\mathcal{M}} \circ \mathcal{R}_{\mathcal{S}}$ is a mirror mapping over $\mathcal{R}_{\mathcal{S}}(\mathcal{M}) = \mathcal{L}$. By Property R.3 of Definition NEUT.2 (closure) $CO(\mathcal{R}_{\mathcal{M}})$ is a reflection over $\mathcal{L}$. By Property R.2 of Definition NEUT.2 there is only one reflection $\mathcal{R}_{\mathcal{L}}$ over $\mathcal{L}$, so $CO(\mathcal{R}_{\mathcal{M}}) = \mathcal{R}_{\mathcal{L}}$, proving (I).

(II) $CO(CO(\mathcal{R}_{\mathcal{M}})) = CO(\mathcal{R}_{\mathcal{S}} \circ \mathcal{R}_{\mathcal{M}} \circ \mathcal{R}_{\mathcal{S}}) = \mathcal{R}_{\mathcal{S}} \circ \mathcal{R}_{\mathcal{S}} \circ \mathcal{R}_{\mathcal{M}} \circ \mathcal{R}_{\mathcal{S}} \circ \mathcal{R}_{\mathcal{S}} = \mathcal{R}_{\mathcal{M}}$ since $\mathcal{R}_{\mathcal{S}} \circ \mathcal{R}_{\mathcal{S}} = \iota$ by Remark NEUT.1.3. □

Definition NEUT.31. Two lines $\mathcal{L}$ and $\mathcal{M}$ on the neutral plane $\mathcal{P}$ are **perpendicular** to each other (notation $\mathcal{L} \perp \mathcal{M}$) iff $\mathcal{L}$ and $\mathcal{M}$ are distinct and each of them is a line of symmetry of the other.

Theorem NEUT.32. *Let $\mathcal{M}$ and $\mathcal{L}$ be distinct lines on the neutral plane. Then the following are equivalent:*

(a) *$\mathcal{M}$ is a line of symmetry for $\mathcal{L}$;*
(b) *$\mathcal{L}$ is a line of symmetry for $\mathcal{M}$;*
(c) *$\mathcal{R}_{\mathcal{M}}(\mathcal{L}) = \mathcal{L}$ is a fixed line for $\mathcal{R}_{\mathcal{M}}$ which is not $\mathcal{M}$;*
(d) *$\mathcal{R}_{\mathcal{L}}(\mathcal{M}) = \mathcal{M}$ is a fixed line for $\mathcal{R}_{\mathcal{L}}$ which is not $\mathcal{L}$;*
(e) *$\mathcal{M}$ and $\mathcal{L}$ are perpendicular.*

Proof. By Definition NEUT.4 and Definition CAP.0(C), (a) $\Leftrightarrow$ (c) and (b) $\Leftrightarrow$ (d). By Theorem NEUT.30(I), (c) $\Rightarrow$ (d); by part (II) of the same theorem, (d) $\Rightarrow$ (c). Now (a) $\Leftrightarrow$ (b) so (a) $\Leftrightarrow$ ((a) and (b)) $\Leftrightarrow$ $\mathcal{L} \perp \mathcal{M}$, by Definition NEUT.31. □

Remark NEUT.32.1 (On temptations). (A) It is tempting to think that if O is a point on $\mathcal{M}$ there should be a complementary mapping $CO(\mathcal{R}_{\mathcal{M}})$ for $\mathcal{R}_{\mathcal{M}}$ which is a reflection over a line through O. However, at this point, we do not know that there is a fixed line for $\mathcal{R}_{\mathcal{M}}$ through an arbitrary point O on $\mathcal{M}$—this will be proved in Theorem NEUT.47.

(B) Let $\mathcal{M}$ and $\mathcal{L}$ be lines on the neutral plane, $\mathcal{R}_{\mathcal{M}}$ a reflection over $\mathcal{M}$ with fixed line $\mathcal{L}$, and $CO(\mathcal{R}_{\mathcal{M}})$ a reflection over $\mathcal{L}$, so that by Theorem NEUT.30, $\mathcal{M}$ is a fixed line for $CO(\mathcal{R}_{\mathcal{M}})$. It is tempting to speculate that $CO(\mathcal{R}_{\mathcal{M}})$ maps fixed lines of $\mathcal{R}_{\mathcal{M}}$ into fixed lines of $\mathcal{R}_{\mathcal{M}}$. That is, for every fixed line $\overleftrightarrow{A\mathcal{R}_{\mathcal{M}}(A)}$ of $\mathcal{R}_{\mathcal{M}}$, $CO(\mathcal{R}_{\mathcal{M}})(\overleftrightarrow{A(\mathcal{R}_{\mathcal{M}}(A))})$ is a fixed line for $\mathcal{R}_{\mathcal{M}}$. The following argument shows that this is indeed the case if Axiom PS holds.

For notational convenience, we denote $CO(\mathcal{R}_{\mathcal{M}})$ as $\mathcal{R}_{\mathcal{L}}$. By Theorem NEUT.22, $\mathcal{L} \parallel \overleftrightarrow{A\mathcal{R}_{\mathcal{M}}(A)}$, because both $\mathcal{L}$ and $\overleftrightarrow{A\mathcal{R}_{\mathcal{M}}(A)}$ are fixed lines for $\mathcal{R}_{\mathcal{M}}$. Then by Theorem CAP.3 and Theorem NEUT.15(1),

$$\mathcal{L} = \mathcal{R}_{\mathcal{L}}(\mathcal{L}) \parallel \mathcal{R}_{\mathcal{L}}(\overleftrightarrow{A\mathcal{R}_{\mathcal{M}}(A)}) = \overleftrightarrow{\mathcal{R}_{\mathcal{L}}(A)(\mathcal{R}_{\mathcal{L}}(\mathcal{R}_{\mathcal{M}}(A)))}.$$

The right side of this is a line containing the point $\mathcal{R}_{\mathcal{L}}(A)$, and is the image of $\overleftrightarrow{A\mathcal{R}_{\mathcal{M}}(A)}$ under mapping by $\mathcal{R}_{\mathcal{L}}$. Applying $\mathcal{R}_{\mathcal{M}}$ to $\overleftrightarrow{\mathcal{R}_{\mathcal{L}}(A)(\mathcal{R}_{\mathcal{M}}(\mathcal{R}_{\mathcal{L}}(A)))}$ and using Theorem NEUT.15(1), we again get $\overleftrightarrow{\mathcal{R}_{\mathcal{L}}(A)(\mathcal{R}_{\mathcal{M}}(\mathcal{R}_{\mathcal{L}}(A)))}$ so this is a fixed line for $\mathcal{R}_{\mathcal{M}}$, hence is parallel to $\mathcal{L}$ by Theorem NEUT.22(E).

Thus both $\overleftrightarrow{\mathcal{R}_{\mathcal{L}}(A)(\mathcal{R}_{\mathcal{L}}(\mathcal{R}_{\mathcal{M}}(A)))}$ and $\overleftrightarrow{\mathcal{R}_{\mathcal{L}}(A)(\mathcal{R}_{\mathcal{M}}(\mathcal{R}_{\mathcal{L}}(A)))}$ are parallel to $\mathcal{L}$ and contain the point $\mathcal{R}_{\mathcal{L}}(A)$, so by Axiom PS, they are the same line. Thus the image under $\mathcal{R}_{\mathcal{L}}$ of a fixed line for $\mathcal{R}_{\mathcal{M}}$ is a fixed line for $\mathcal{R}_{\mathcal{M}}$.

Theorem NEUT.33. (A) *Let $\mathcal{M}$ be a line on the neutral plane $\mathcal{P}$, and let O be a point of $\mathcal{P}$. Then there is at most one line $\mathcal{L}$ through O which is perpendicular to $\mathcal{M}$.*

(B) *If $\mathcal{L}$ is a line in the neutral plane, and $O \in \mathcal{L}$, there is no more than one line $\mathcal{M} \neq \mathcal{L}$ containing O such that $\mathcal{L}$ is a fixed line for $\mathcal{R}_{\mathcal{M}}$.*

Proof. (A) By Theorem NEUT.32, $\mathcal{L} \perp \mathcal{M}$ iff $\mathcal{L}$ is a fixed line for $\mathcal{R}_{\mathcal{M}}$ and $\mathcal{L} \neq \mathcal{M}$. By Theorem NEUT.22(D) there is at most one fixed line $\mathcal{L}$ for $\mathcal{R}_{\mathcal{M}}$ through a point O. Therefore there is at most one perpendicular to $\mathcal{M}$ through O.

(B) If $\mathcal{L}$ is a fixed line for $\mathcal{R}_{\mathcal{M}}$, then by Theorem NEUT.32 $\mathcal{L} \perp \mathcal{M}$; thus if both $\mathcal{M}$ and $\mathcal{M}'$ contain O and $\mathcal{L}$ is a fixed line for both $\mathcal{R}_{\mathcal{M}}$ and $\mathcal{R}_{\mathcal{M}'}$, both $\mathcal{M}$ and $\mathcal{M}'$ are perpendicular to $\mathcal{L}$ at O, contradicting part (A). □

8.8 Properties of certain isometries; *Pons Asinorum*

The following lemma is used to prove Theorem NEUT.35; it will also be used for Theorem NEUT.50, which proves uniqueness of midpoints of arbitrary segments.

Lemma NEUT.34 (Midpoints of fixed segments of reflections are unique). *Let $\mathcal{L}$ be a line on a neutral plane and let A and B be distinct points such that $\mathcal{R}_{\mathcal{L}}(A) = B$.*

(A) *Let $\{D\} = \mathcal{L} \cap \overleftrightarrow{AB}$, and let D' be a point such that A–D'–B and $\overline{AD'} \cong \overline{BD'}$. Then $D = D'$.*

(B) *If $\mathcal{L}'$ is any line such that $\mathcal{R}_{\mathcal{L}'}(A) = B$, then $\mathcal{L}' = \mathcal{L}$ so that by Property R.2 of Definition NEUT.2, $\mathcal{R}_{\mathcal{L}'} = \mathcal{R}_{\mathcal{L}}$.*

Proof. (A) By Definition NEUT.1(D), since A–D'–B, $\mathcal{R}_{\mathcal{L}}(A)$–$\mathcal{R}_{\mathcal{L}}(D')$–$\mathcal{R}_{\mathcal{L}}(B)$, that is, B–$\mathcal{R}_{\mathcal{L}}(D')$–$A$. By Theorem NEUT.15(5) $\mathcal{R}_{\mathcal{L}}(\overline{AD'}) = \overline{\mathcal{R}_{\mathcal{L}}(A)\mathcal{R}_{\mathcal{L}}(D')} = \overline{B\mathcal{R}_{\mathcal{L}}(D')}$. By Definition NEUT.3(B) $\overline{AD'} \cong \overline{B\mathcal{R}_{\mathcal{L}}(D')}$. By hypothesis, $\overline{AD'} \cong \overline{BD'}$ so that $\overline{BD'} \cong \overline{B\mathcal{R}_{\mathcal{L}}(D')}$, by Theorem NEUT.14 (congruence is an equivalence relation).

Also $\mathcal{R}_{\mathcal{L}}(D') \in \overrightarrow{BD'}$. Since $\overline{B\mathcal{R}_{\mathcal{L}}(D')} \cong \overline{BD'}$, $\mathcal{R}_{\mathcal{L}}(D') = D'$ by Property R.4 of Definition NEUT.2. Therefore D' is a fixed point for $\mathcal{R}_{\mathcal{L}}$, so that $D' \in \mathcal{L}$. Since $D' \in \overleftrightarrow{AB}$ and $\{D\} = \mathcal{L} \cap \overleftrightarrow{AB}$, it follows that $D' = D$.

(B) If $\mathcal{L}'$ is a line such that $\mathcal{R}_{\mathcal{L}'}(A) = B$, $\overleftrightarrow{AB}$ is a fixed line for $\mathcal{R}_{\mathcal{L}'}$, as well as for $\mathcal{R}_{\mathcal{L}}$. By Theorem NEUT.33(B), $\mathcal{L}' = \mathcal{L}$, hence by Property R.2 of Definition NEUT.2, $\mathcal{R}_{\mathcal{L}'} = \mathcal{R}_{\mathcal{L}}$. □

The proof of the following theorem contains our first use of Property R.5 of Definition NEUT.2, which says that every angle has an angle reflection and a line of symmetry.

Theorem NEUT.35 (Side-preserving isometry (A)). *Suppose α is an isometry of a neutral plane $\mathcal{P}$, A and B are fixed points of α, and $C \notin \overleftrightarrow{AB}$. If $\alpha(C) \in \overrightarrow{ABC}$ (that is, $\alpha(C)$ is on the C-side of $\overleftrightarrow{AB}$), then $\alpha = \iota$.*

Proof. By Theorem NEUT.23 every point on $\overleftrightarrow{AB}$ is a fixed point for α.

(Case 1: $\alpha(C) = C$.) By Theorem NEUT.24, $\alpha = \iota$.

(Case 2: $\alpha(C) \neq C$ and the line $\overleftrightarrow{C\alpha(C)}$ intersects $\overleftrightarrow{AB}$ at some point X.) Then by Theorem NEUT.15(5) $\alpha(\overline{XC}) = \overline{\alpha(X)\alpha(C)} = \overline{X\alpha(C)}$. By Definition NEUT.3(B)

$\overline{XC} \cong \overline{X\alpha(C)}$, and by definition $\alpha(C) \in \overrightarrow{XC}$. Then by Property R.4 of Definition NEUT.2 $\alpha(C) = C$ and by Theorem NEUT.24 $\alpha = \iota$.

(Case 3: $\alpha(C) \neq C$ and the line $\overleftrightarrow{C\alpha(C)}$ does not intersect $\overleftrightarrow{AB}$.) Then $\overleftrightarrow{C\alpha(C)} \parallel \overleftrightarrow{AB}$. In particular, the line $\overleftrightarrow{AC}$ does not contain $\alpha(C)$. By Property R.5 of Definition NEUT.2, $\angle CA\alpha(C)$ has a line $\mathcal{L}$ of symmetry, and there exists a reflection $\mathcal{R}_{\mathcal{L}}$ such that $\mathcal{R}_{\mathcal{L}}(\alpha(C)) \in \overrightarrow{AC}$.

If $B \in \mathcal{L}$, then $\overleftrightarrow{AB} = \mathcal{L}$ and then C and $\alpha(C)$ would be on opposite sides of $\overleftrightarrow{AB}$ which is false by hypothesis; therefore $B \notin \mathcal{L}$.

By Definition NEUT.3(A) $\gamma = \mathcal{R}_{\mathcal{L}} \circ \alpha$ is an isometry of $\mathcal{P}$. By Definition NEUT.1(A) $\gamma(A) = \mathcal{R}_{\mathcal{L}}(\alpha(A)) = \mathcal{R}_{\mathcal{L}}(A) = A$. By Theorem NEUT.15(5) $\gamma(\overline{AC}) = \overline{\gamma(A)\gamma(C)} = \overline{A\gamma(C)}$. By Definition NEUT.3(B) (congruence) $\overline{AC} \cong \overline{A\gamma(C)}$. By Property R.4 of Definition NEUT.2, $C = \gamma(C)$, *i.e.*, $(\mathcal{R}_{\mathcal{L}} \circ \alpha)(C) = C$. Thus $\mathcal{R}_{\mathcal{L}}$ maps $\alpha(C)$ to C, so that $\overleftrightarrow{C\alpha(C)}$ is a fixed line for $\mathcal{R}_{\mathcal{L}}$.

Now choose $A' \neq A$ to be any other point on $\overleftrightarrow{AB}$. The same argument shows that there is another line $\mathcal{L}'$ such that $\mathcal{R}_{\mathcal{L}'}$ maps $\alpha(C)$ to C, so that $\overleftrightarrow{C\alpha(C)}$ is a fixed line for $\mathcal{R}_{\mathcal{L}'}$.

Thus we have two distinct lines $\mathcal{L}$ and $\mathcal{L}'$, such that both $\mathcal{R}_{\mathcal{L}}(\alpha(C)) = C$ and $\mathcal{R}_{\mathcal{L}'}(\alpha(C)) = C$. By Lemma NEUT.34(B) $\mathcal{L} = \mathcal{L}'$, so that the lines are not distinct, but the same line, a contradiction. Thus case 3 is ruled out, and $\alpha = \iota$. □

Theorem NEUT.36 (Side-preserving isometry (B)). *Let A, B, C, and D be points on the neutral plane $\mathcal{P}$ such that C and D are on the same side of $\overleftrightarrow{AB}$ and $\angle BAC \cong \angle BAD$. Then $\overrightarrow{AC} = \overrightarrow{AD}$.*

Proof. By Definition NEUT.3 there exists an isometry α of $\mathcal{P}$ such that $\alpha(\angle BAC) = \angle BAD$. Statements (1) through (13) of Theorem NEUT.15 are true for α. We will use these without further reference in this proof.

Then $\angle BAD = \alpha(\angle BAC) = \angle\alpha(B)\alpha(A)\alpha(C)$, and by Theorem PSH.33 $\alpha(A) = A$. By Definition PSH.29 there are exactly two possibilities: $\alpha(B) \in \overrightarrow{AB}$ or $\alpha(B) \in \overrightarrow{AD}$.

(Case 1: $\alpha(B) \in \overrightarrow{AB}$.) In this case, $\alpha(C) \in \overrightarrow{AD}$. Since α is an isometry, by Definition NEUT.3, $\overline{AB} \cong \overline{A\alpha(B)}$ and by Property R.4 of Definition NEUT.2, $\alpha(B) = B$. Thus A and B are fixed points for α. By hypothesis $\alpha(C)$ is on the C-side of $\overleftrightarrow{AB}$, so by Theorem NEUT.35 $\alpha = \iota$ and $C = \alpha(C) \in \overrightarrow{AD}$. Therefore $\overrightarrow{AC} = \overrightarrow{AD}$, again using Theorem PSH.16.

(Case 2: $\alpha(B) \in \overrightarrow{AD}$.) In this case, $\alpha(C) \in \overrightarrow{AB}$, so that by Theorem PSH.16,

$$\alpha(\overrightarrow{AB}) = \overrightarrow{\alpha(A)\alpha(B)} = \overrightarrow{A\alpha(B)} = \overrightarrow{AD} \text{ and}$$
$$\alpha(\overrightarrow{AC}) = \overrightarrow{\alpha(A)\alpha(C)} = \overrightarrow{A\alpha(C)} = \overrightarrow{AB}.$$

By Property R.5 of Definition NEUT.2 there exists an angle reflection $\mathcal{R}_{\mathcal{M}}$ and line $\mathcal{M}$ of symmetry for $\angle BAC$, so that $\mathcal{R}_{\mathcal{M}}(A) = A$, $\mathcal{R}_{\mathcal{M}}(\overrightarrow{AB}) = \overrightarrow{AC}$ and $\mathcal{R}_{\mathcal{M}}(\overrightarrow{AC}) = \overrightarrow{AB}$. The mapping $\alpha \circ \mathcal{R}_{\mathcal{M}}$ is an isometry (by Definition NEUT.3) which maps $\angle BAC$ onto $\angle BAD$, such that

$$(\alpha \circ \mathcal{R}_{\mathcal{M}})(\overrightarrow{AB}) = \alpha(\overrightarrow{AC}) = \overrightarrow{AB} \text{ and}$$
$$(\alpha \circ \mathcal{R}_{\mathcal{M}})(\overrightarrow{AC}) = \alpha(\overrightarrow{AB}) = \overrightarrow{AD}.$$

Thus $(\alpha \circ \mathcal{R}_{\mathcal{M}})(B) \in \overrightarrow{AB}$, and we may apply Case 1 to conclude that $\alpha \circ \mathcal{R}_{\mathcal{M}} = \iota$ and $C = \alpha \circ \mathcal{R}_{\mathcal{M}}(C) \in \overrightarrow{AD}$. Therefore $\overrightarrow{AC} = \overrightarrow{AD}$. □

Theorem NEUT.37 (An isometry with two fixed points is the identity or a reflection). *Let α be an isometry of the neutral plane such that A and B are distinct fixed points of α; then either $\alpha = \iota$ or $\alpha = \mathcal{R}_{\overleftrightarrow{AB}}$.*

Proof. (I) If α has a fixed point C not belonging to $\overleftrightarrow{AB}$, then by Theorem NEUT.24 $\alpha = \iota$.

(II) If α has no fixed point off of $\overleftrightarrow{AB}$, let X be any member of $\mathcal{P} \setminus \overleftrightarrow{AB}$, then by Theorem NEUT.15(8) $\alpha(\angle BAX) = \angle\alpha(B)\alpha(A)\alpha(X) = \angle BA\alpha(X)$. By Definition NEUT.3(B) $\angle BAX \cong \angle BA\alpha(X)$. By the contrapositive of Theorem NEUT.35, X and $\alpha(X)$ are on opposite sides of $\overleftrightarrow{AB}$. Let $\mathcal{R}_{\overleftrightarrow{AB}}$ be a reflection over $\overleftrightarrow{AB}$ and define $\gamma = \mathcal{R}_{\overleftrightarrow{AB}} \circ \alpha$. Then γ is an isometry of $\mathcal{P}$ with distinct fixed points A and B, and X and $\gamma(X)$ are on the same side of $\overleftrightarrow{AB}$; by Theorem NEUT.35 $\gamma = \iota$. By elementary mapping theory, $\alpha = \mathcal{R}_{\overleftrightarrow{AB}}$. □

Theorem NEUT.38 (Isometry construction for angles). *Let A, B, C, D, E, and F be points on the neutral plane $\mathcal{P}$ such that A, B, and C are noncollinear, D, E, and F are noncollinear, and $\angle BAC \cong \angle EDF$. Then there exists an isometry φ of $\mathcal{P}$ such that $\varphi(\overrightarrow{AB}) = \overrightarrow{DE}$ and $\varphi(\overrightarrow{AC}) = \overrightarrow{DF}$.*

Proof. By Definition NEUT.3(B) there exists an isometry α of $\mathcal{P}$ such that $\alpha(\angle BAC) = \angle EDF$. By Theorem NEUT.15(8), $\alpha(\angle BAC) = \angle\alpha(B)\alpha(A)\alpha(C)$ so that $\angle\alpha(B)\alpha(A)\alpha(C) = \angle EDF$. By Theorem PSH.33 $\alpha(A) = D$ so $\angle\alpha(B)D\alpha(C) = \angle EDF$. By Definition PSH.29 there are two cases.

(Case 1: $\alpha(B) \in \overrightarrow{DE}$ and $\alpha(C) \in \overrightarrow{DF}$.) Let $\varphi = \alpha$. Then by Theorem PSH.16,

$$\varphi(\overrightarrow{AB}) = \alpha(\overrightarrow{AB}) = \overrightarrow{\alpha(A)\alpha(B)} = \overrightarrow{D\alpha(B)} = \overrightarrow{DE}$$

and

$$\varphi(\overrightarrow{AC}) = \alpha(\overrightarrow{AC}) = \overrightarrow{\alpha(A)\alpha(C)} = \overrightarrow{D\alpha(C)} = \overrightarrow{DF}.$$

(Case 2: $\alpha(B) \in \overrightarrow{DF}$ and $\alpha(C) \in \overrightarrow{DE}$.) By Theorem PSH.16 we have $\overrightarrow{D\alpha(B)} = \overrightarrow{DF}$ and $\overrightarrow{D\alpha(C)} = \overrightarrow{DE}$. Then let $\mathcal{M}$ be the line of symmetry of $\angle EDF$, and $\mathcal{R}_{\mathcal{M}}$ its angle reflection, so that $D \in \mathcal{M}$, $\mathcal{R}_{\mathcal{M}}(\overrightarrow{DE}) = \overrightarrow{DF}$ and $\mathcal{R}_{\mathcal{M}}(\overrightarrow{DF}) = \overrightarrow{DE}$. Let $\varphi = \mathcal{R}_{\mathcal{M}} \circ \alpha$. Then

$$\begin{aligned}\varphi(\overrightarrow{AB}) &= \mathcal{R}_{\mathcal{M}}(\alpha(\overrightarrow{AB})) = \mathcal{R}_{\mathcal{M}}(\overrightarrow{\alpha(A)\alpha(B)}) \\ &= \mathcal{R}_{\mathcal{M}}(\overrightarrow{D\alpha(B)}) = \mathcal{R}_{\mathcal{M}}(\overrightarrow{DF}) = \overrightarrow{DE},\end{aligned}$$

and

$$\begin{aligned}\varphi(\overrightarrow{AC}) &= \mathcal{R}_{\mathcal{M}}(\alpha(\overrightarrow{AC})) = \mathcal{R}_{\mathcal{M}}(\overrightarrow{\alpha(A)\alpha(C)}) \\ &= \mathcal{R}_{\mathcal{M}}(\overrightarrow{D\alpha(C)}) = \mathcal{R}_{\mathcal{M}}(\overrightarrow{DE}) = \overrightarrow{DF}.\end{aligned}$$

□

Theorem NEUT.39 (Line of symmetry angle criterion). *Let A, B, and C be noncollinear points on the neutral plane $\mathcal{P}$, and let D be a member of* ins $\angle BAC$. *Then $\overleftrightarrow{AD}$ is the line of symmetry of $\angle BAC$ (so that $\overrightarrow{AD}$ is the bisecting ray) iff $\angle BAD \cong \angle CAD$.*

Proof. (I) If $\overleftrightarrow{AD} = \mathcal{M}$ is the line of symmetry of $\angle BAC$, then by Theorem NEUT.20(E) $\angle BAD \cong \angle CAD$.

(II) Conversely, let $\angle BAD \cong \angle CAD$. By Theorem NEUT.38 there exists an isometry φ mapping $\angle BAD$ to $\angle CAD$ such that $\varphi(\overrightarrow{AD}) = \overrightarrow{AD}$ and $\varphi(\overrightarrow{AB}) = \overrightarrow{AC}$.

A is a fixed point for φ and by Theorem NEUT.15(3) $\overrightarrow{AD} = \varphi(\overrightarrow{AD}) = \overrightarrow{\varphi(A)\varphi(D)} = \overrightarrow{A\varphi(D)}$ so by Theorem PSH.24 $\varphi(D) \in \overrightarrow{AD}$. Now $\varphi(\overline{AD}) = \overline{\varphi(A)\varphi(D)} = \overline{A\varphi(D)}$ so by Definition NEUT.3 $\overline{AD} \cong \overline{A\varphi(D)}$. By Property R.4 of Definition NEUT.2 $\varphi(D) = D$ so D is a fixed point for φ.

Let $\mathcal{M}$ be the line of symmetry of $\angle BAC$. By Definition NEUT.3, $\mathcal{R}_{\mathcal{M}}(\overrightarrow{AB}) = \overrightarrow{AC}$ and $\mathcal{R}_{\mathcal{M}}(\overrightarrow{AC}) = \overrightarrow{AB}$.

Let $\gamma = \mathcal{R}_{\mathcal{M}} \circ \varphi$, which is an isometry by Definition NEUT.3. Note that $\gamma(A) = A$ because A is a fixed point for both $\mathcal{R}_{\mathcal{M}}$ and φ. Now $\varphi(B) \in \overrightarrow{AC}$ and $\mathcal{R}_{\mathcal{M}}(\overrightarrow{AC}) \subseteq \overrightarrow{AB}$, so $\gamma(B) = \mathcal{R}_{\mathcal{M}}(\varphi(B)) \in \overrightarrow{AB}$. By Theorem NEUT.15(5) $\gamma(\overline{AB}) = \overline{\gamma(A)\gamma(B)} = \overline{A\gamma(B)}$ so that by Definition NEUT.3(B) $\overline{A\gamma(B)} \cong \overline{AB}$ and by Property R.4 of Definition NEUT.2, $\gamma(B) = B$.

Recapitulating, we see that A and B are both fixed points for γ, and $\gamma(C) = \mathcal{R}_{\mathcal{M}}(\varphi(C)) \in \mathcal{R}_{\mathcal{M}}(\overrightarrow{AB}) \subseteq \overrightarrow{AC}$ so that $\gamma(C)$ and C are on the same side of the line $\overleftrightarrow{AB}$. Then by Theorem NEUT.35, $\gamma = \iota$, and by elementary mapping theory, $\varphi = \mathcal{R}_{\mathcal{M}}$.

Since D is a fixed point for $\mathcal{R}_{\mathcal{M}}$, by Remark NEUT.1.1 $D \in \mathcal{M}$ so by Exercise I.2 $\overleftrightarrow{AD} = \mathcal{M}$. □

Theorem NEUT.40. (A) (*Pons Asinorum* **or Isosceles Triangle theorem**)[5] *If A, B, and C are noncollinear points on the neutral plane $\mathcal{P}$ such that $\overline{AB} \cong \overline{AC}$, then $\angle ABC \cong \angle ACB$.*

(B) (**Converse of** *Pons Asinorum*) *If A, B, and C are noncollinear points on the neutral plane $\mathcal{P}$ such that $\angle ABC \cong \angle ACB$, then $\overline{AB} \cong \overline{AC}$.*

Proof. In this proof we will apply Theorem NEUT.15 without reference.

(A) Let $\mathcal{M}$ be the line of symmetry of $\angle BAC$, so that $A \in \mathcal{M}$, $\mathcal{R}_{\mathcal{M}}(\overrightarrow{AB}) = \overrightarrow{AC}$, and $\mathcal{R}_{\mathcal{M}}(B) \in \overrightarrow{AC}$. Since $\mathcal{R}_{\mathcal{M}}(\overline{AB}) = \overline{A(\mathcal{R}_{\mathcal{M}}(B))}$, $\overline{AB} \cong \overline{A(\mathcal{R}_{\mathcal{M}}(B))}$. Since $\overline{AB} \cong \overline{A(\mathcal{R}_{\mathcal{M}}(B))}$ and $\overline{AB} \cong \overline{AC}$, $\overline{A(\mathcal{R}_{\mathcal{M}}(B))} \cong \overline{AC}$. By Property R.4 of Definition NEUT.2 $\mathcal{R}_{\mathcal{M}}(B) = C$ so by Definition NEUT.1(C), $\mathcal{R}_{\mathcal{M}}(C) = B$. Therefore $\mathcal{R}_{\mathcal{M}}(\angle ABC) = \angle(\mathcal{R}_{\mathcal{M}}(A))(\mathcal{R}_{\mathcal{M}}(B))(\mathcal{R}_{\mathcal{M}}(C)) = \angle ACB$, and by Definition NEUT.3(B), $\angle ABC \cong \angle ACB$.

(B) By Theorem NEUT.38 there exists an isometry φ of $\mathcal{P}$ such that $\varphi(\angle ABC) = \angle ACB$,

$$\varphi(\overrightarrow{BA}) = \overrightarrow{CA} \quad (1) \quad \text{and} \quad \varphi(\overrightarrow{BC}) = \overrightarrow{CB}. \quad (2)$$

Then $\varphi(\angle ABC) = \angle\varphi(A)\varphi(B)\varphi(C) = \angle ACB$, so by Theorem PSH.32

$$\varphi(B) = C. \quad (3)$$

Combining (1) and (3), $\overrightarrow{CA} = \varphi(\overrightarrow{BA}) = \overrightarrow{\varphi(B)\varphi(A)} = \overrightarrow{C\varphi(A)}$. By Theorem PSH.24, $\varphi(A) \in \overrightarrow{CA}$.

Likewise, combining (2) and (3), $\overrightarrow{CB} = \varphi(\overrightarrow{BC}) = \overrightarrow{\varphi(B)\varphi(C)} = \overrightarrow{C\varphi(C)}$ so that by Theorem PSH.24 $\varphi(C) \in \overrightarrow{CB}$. Then $\varphi(\overline{CB}) = \overline{\varphi(C)\varphi(B)} = \overline{\varphi(C)C}$ so that by Definition NEUT.3 $\overline{CB} \cong \overline{C\varphi(C)}$. By Property R.4 of Definition NEUT.2

$$\varphi(C) = B. \quad (4)$$

[5] Literally, bridge for donkeys in Latin. This term has come to mean any problem that severely tests the ability of a person, such as Euclid's 5th postulate. The name may have come from the bridgelike appearance of Euclid's figure, including the construction lines, used in his complicated proof of this theorem.

Now $\angle ABC \cong \angle ACB \cong \varphi(\angle ACB) = \angle\varphi(A)\varphi(C)\varphi(B) = \angle\varphi(A)BC$, so that by Theorem NEUT.14 (congruence is an equivalence relation) $\angle ABC \cong \angle\varphi(A)BC$. A and $\varphi(A)$ are on the same side of $\overleftrightarrow{BC}$ because $\varphi(A) \in \overrightarrow{CA}$. Then by Theorem NEUT.36, $\overrightarrow{BA} = \overrightarrow{B\varphi(A)}$ and by Theorem PSH.24 $\varphi(A) \in \overrightarrow{BA}$. Hence $\varphi(A) \in \overrightarrow{BA} \cap \overrightarrow{CA} = \{A\}$ and $\varphi(A) = A$. Combining this with (3) above, $\varphi(\overline{AB}) = \overline{\varphi(A)\varphi(B)} = \overline{\varphi(A)C} = \overline{AC}$. By Definition NEUT.3(B), $\overline{AB} \cong \overline{AC}$. □

8.9 Vertical and supplementary angles; more perpendicularity

Definition NEUT.41. Let $\mathcal{D}$ and $\mathcal{E}$ be angles on the neutral plane $\mathcal{P}$.

(A) $\mathcal{D}$ and $\mathcal{E}$ are **vertical to each other** iff there exist points A, B, C, B', and C' such that A, B, and C are noncollinear, B–A–B', C–A–C', $\mathcal{D} = \angle BAC$ and $\mathcal{E} = \angle B'AC'$.

(B) $\mathcal{D}$ and $\mathcal{E}$ are **supplementary angles** iff there exist points A, B, C, and D on $\mathcal{P}$ such that B–A–C, $D \in (\mathcal{P} \setminus \overleftrightarrow{AB})$, $\mathcal{D} = \angle BAD$ and $\mathcal{E} = \angle CAD$. We may also say that the angles are **supplemental** or that each is a **supplement** of the other.

(C) An angle on the neutral plane $\mathcal{P}$ is **right** iff it is congruent to a supplement of itself. That is to say, if A, O, and A' are collinear points on the plane, and $C \notin \overleftrightarrow{AA'}$, then $\angle AOC$ ($\angle A'OC$) is a right angle iff $\angle AOC \cong \angle A'OC$.

Theorem NEUT.42 (Vertical angles). *Let $\mathcal{D}$ and $\mathcal{E}$ be angles on the neutral plane $\mathcal{P}$. If $\mathcal{D}$ and $\mathcal{E}$ are vertical to each other, then $\mathcal{D} \cong \mathcal{E}$.*

Proof. By Definition NEUT.41 there exist points A, B, C, B', and C' such that A, B, and C are noncollinear, B–A–B', C–A–C', $\mathcal{D} = \angle BAC$ and $\mathcal{E} = \angle B'AC'$. By Theorem NEUT.26 $\angle B'AC$ has a unique line $\mathcal{M}$ of symmetry and $\mathcal{R}_{\mathcal{M}}(\overrightarrow{AC}) = \overrightarrow{AB'}$. By Exercise NEUT.10 $\mathcal{M}$ is the line of symmetry of $\angle BAC'$. Then $\mathcal{R}_{\mathcal{M}}(\overrightarrow{AB}) = \overrightarrow{AC'}$.

By Definition PSH.29 $\angle BAC = \overrightarrow{AB} \cup \overrightarrow{AC}$. Hence $\mathcal{R}_{\mathcal{M}}(\angle BAC) = \angle B'AC'$ and so $\angle BAC \cong \angle B'AC'$. □

Remark NEUT.42.1. If E, F, and O are noncollinear points on the neutral plane $\mathcal{P}$ and if E' and F' are points on $\mathcal{P}$ such that E'–O–E and F–O–F', then each of the angles $\angle EOF'$ or $\angle FOE'$ is a supplement of $\angle EOF$. By Theorem NEUT.42, $\angle EOF' \cong \angle FOE'$, since they are vertical angles.

Theorem NEUT.43 (Supplements of congruent angles are congruent). *Let $\mathcal{C}$, $\mathcal{G}$, $\mathcal{C}'$, and $\mathcal{G}'$ be angles on the neutral plane $\mathcal{P}$ such that $\mathcal{C}$ and $\mathcal{C}'$ are supplements of each other and $\mathcal{G}$ and $\mathcal{G}'$ are supplements of each other. If $\mathcal{C} \cong \mathcal{G}$, then $\mathcal{C}' \cong \mathcal{G}'$.*

Proof. By Definition NEUT.41(B) there exist points A, B, B', C, E, F, F' and G on $\mathcal{P}$ such that $B\text{–}C\text{–}B'$, $A \notin \overleftrightarrow{BC}$, $F\text{–}G\text{–}F'$, $E \notin \overleftrightarrow{FG}$, $\mathcal{C} = \angle BCA$, $\mathcal{G} = \angle FGE$, $\mathcal{C}' = \angle B'CA$, and $\mathcal{G}' = \angle F'GE$. Since $\angle BCA \cong \angle FGE$, by Theorem NEUT.38 there exists an isometry α of $\mathcal{P}$ such that $\alpha(\angle BCA) = \angle FGE$, $\alpha(\overrightarrow{CB}) = \overrightarrow{GF}$, $\alpha(\overrightarrow{CA}) = \overrightarrow{GE}$, and $\alpha(\overleftrightarrow{CB}) = \overleftrightarrow{GF}$. By Theorem NEUT.15(1) and (3) $\alpha(\overrightarrow{CB}) = \overrightarrow{\alpha(C)\alpha(B)}$, $\alpha(\overrightarrow{CA}) = \overrightarrow{\alpha(C)\alpha(A)}$, and $\alpha(\overleftrightarrow{CB}) = \overleftrightarrow{\alpha(C)\alpha(B)}$.

By Theorem PSH.24 $\alpha(C) = G$, $\alpha(B) \in \overrightarrow{GF}$, $\alpha(B') \in \overrightarrow{GF'}$, and $\alpha(A) \in \overrightarrow{GE}$. By Theorem PSH.15 $\overleftrightarrow{GF}$ is the union of the disjoint sets $\overrightarrow{GF}$, $\{G\}$, and $\overrightarrow{GF'}$. Since $B'\text{–}C\text{–}B$ and the fact that $\alpha(C) = G$, $\alpha(B)\text{–}G\text{–}\alpha(B')$. By Theorem PSH.16 $\overrightarrow{G\alpha(B)} = \overrightarrow{GF}$, and $\overrightarrow{G\alpha(B')} = \overrightarrow{GF'}$. Thus

$$\begin{aligned}\alpha(\angle B'CA) &= \angle\alpha(B')\alpha(C)\alpha(A) = \angle\alpha(B')G\alpha(A) = \overrightarrow{G\alpha(A)} \cup \overrightarrow{G\alpha(B')}\\ &= \overrightarrow{GE} \cup \overrightarrow{GF'} = \angle F'GE \text{ (cf Definition PSH.29.)}\end{aligned}$$

By Definition NEUT.3 $\mathcal{C}' = \angle B'CA \cong \angle F'GE = \mathcal{G}'$. □

We now deal with perpendicularity from the point of view of angles. The next theorem is an extension of Theorem NEUT.32 to include the concept of right angle.

Theorem NEUT.44. *Suppose $\mathcal{M}$ and $\mathcal{L}$ are distinct lines on a neutral plane. Then the following statements are equivalent.*

(A) *There exists a point O such that $\mathcal{L} \cap \mathcal{M} = \{O\}$; if $Q \in \mathcal{L}\setminus\{O\}$ and $P \in \mathcal{M}\setminus\{O\}$, then $\angle POQ$ is a right angle.*
(B) *There exists a point O such that $\mathcal{L} \cap \mathcal{M} = \{O\}$; if $Q \in \mathcal{L}\setminus\{O\}$ and $P \in \mathcal{M}\setminus\{O\}$, and $P\text{–}O\text{–}P'$ and $Q\text{–}O\text{–}Q'$, then all the angles $\angle POQ$, $\angle P'OQ$, $\angle P'OQ'$, and $\angle POQ'$ are congruent, and are all right angles.*
(C) *$\mathcal{M}$ is a line of symmetry for $\mathcal{L}$.*
(D) *$\mathcal{L}$ is a line of symmetry for $\mathcal{M}$.*
(E) *$\mathcal{L}$ is a fixed line for $\mathcal{R}_{\mathcal{M}}$ which is not $\mathcal{M}$.*
(F) *$\mathcal{M}$ is a fixed line for $\mathcal{R}_{\mathcal{L}}$ which is not $\mathcal{L}$.*
(G) *$\mathcal{M} \perp \mathcal{L}$.*

Proof. First, note that if (B) is true then (A) is true.

By Theorem NEUT.32, statements (C) through (G) are equivalent. We will show that (E) implies both (B) and (A), and conversely, that statement (A) implies statement (D). First note that if any of (C) through (G) is true, then by Theorem NEUT.22(F) $\mathcal{L}$ intersects $\mathcal{M}$ at some point O.

Let $Q \in \mathcal{L} \setminus \{O\}$ and $P \in \mathcal{M} \setminus \{O\}$, and $P\text{–}O\text{–}P'$ and $Q\text{–}O\text{–}Q'$.

(I) Assuming (E) is true, $\mathcal{R}_{\mathcal{M}}$ is a reflection which has $\mathcal{L}$ as a fixed line. Then $\mathcal{R}_{\mathcal{M}}(\overrightarrow{OQ}) = \overrightarrow{OQ'}$ and $\mathcal{R}_{\mathcal{M}}(\angle POQ) = \angle POQ'$ so that $\angle POQ \cong \angle POQ'$, its supplement. Therefore by Definition NEUT.41(C) both $\angle POQ$ and $\angle POQ'$ are right angles. By Theorem NEUT.42 $\angle P'OQ' \cong \angle POQ \cong \angle POQ' \cong \angle P'OQ$ so all the angles are right; hence both (A) and (B) are true.

(II) To prove the converse it suffices to prove that (A) implies (D). If (A) is true, then $\angle POQ \cong \angle P'OQ$ or $\angle POQ \cong \angle POQ'$. If the latter is true, by Definition NEUT.41(A), $\angle P'OQ$ is vertical to $\angle POQ'$ and hence by Theorem NEUT.42 and the transitivity of congruence (*cf* Theorem NEUT.14),

$$\angle P'OQ \cong \angle POQ' \cong \angle POQ.$$

In either case, there exists an isometry α mapping $\mathcal{P}$ onto $\mathcal{P}$ such that

$$\alpha(\angle POQ) = \angle\alpha(P)\alpha(O)\alpha(Q) = \angle P'OQ.$$

Here we have used Theorem PSH.33 to show that $\alpha(O) = O$.

Either $\alpha(\overrightarrow{OQ}) = \overrightarrow{OQ}$ or $\alpha(\overrightarrow{OQ}) = \overrightarrow{OP'}$. If the latter holds, we may let $\mathcal{N}$ be the line of symmetry (*cf* Property R.5 of Definition NEUT.2) of $\angle P'OQ$; then the mapping $\beta = \mathcal{R}_{\mathcal{N}} \circ \alpha$ is an isometry that satisfies $\beta(\overrightarrow{OQ}) = \overrightarrow{OQ}$. Thus there is no loss of generality to assume that $\alpha(\overrightarrow{OQ}) = \overrightarrow{OQ}$ and $\alpha(\overrightarrow{OP}) = \overrightarrow{OP'}$.

Then $\overrightarrow{OQ} = \alpha(\overrightarrow{OQ}) = \overrightarrow{\alpha(O)\alpha(Q)}) = \overrightarrow{O\alpha(Q)}$ so by Theorem PSH.24 $\alpha(Q) \in \overrightarrow{OQ}$. Also, $\alpha(\overline{OQ}) = \overline{\alpha(O)\alpha(Q)} = \overline{O\alpha(Q)}$ so that $\overline{OQ} \cong \overline{O\alpha(Q)}$ and by Property R.4 of Definition NEUT.2, $\alpha(Q) = Q$.

Therefore both Q and O are fixed points of α. By Theorem NEUT.37, since $\alpha \neq \iota$, $\alpha = \mathcal{R}_{\overleftrightarrow{OQ}} = \mathcal{R}_{\mathcal{L}}$. Since P, O, and P' are members of $\mathcal{M}$, $\mathcal{L} \neq \mathcal{M}$, and $\mathcal{R}_{\mathcal{L}}(\overrightarrow{OP}) = \alpha(\overrightarrow{OP}) = \overrightarrow{OP'}$, $\mathcal{R}_{\mathcal{L}}(\mathcal{M}) = \mathcal{M}$ and $\mathcal{L}$ is a line of symmetry for $\mathcal{M}$, so that (D) holds. □

Corollary NEUT.44.1. *Let $\mathcal{L}$ and $\mathcal{M}$ be lines on the neutral plane $\mathcal{P}$ and let α be an isometry of $\mathcal{P}$. Then $\mathcal{L} \perp \mathcal{M}$ iff $\alpha(\mathcal{L}) \perp \alpha(\mathcal{M})$.*

Proof. If $\mathcal{L} \perp \mathcal{M}$ by Theorem NEUT.44, there exists a point O such that $\mathcal{L} \cap \mathcal{M} = \{O\}$; let $Q \in \mathcal{L} \setminus \{O\}$ and $P \in \mathcal{M} \setminus \{O\}$, and let Q–O–Q'; then $\angle POQ \cong \angle POQ'$. By Theorem NEUT.13 $\alpha(\angle POQ) \simeq \alpha(\angle POQ')$ and by Theorem NEUT.15(8)

$$\alpha(\angle POQ) \cong \angle\alpha(P)\alpha(O)\alpha(Q) \text{ and } \alpha(\angle POQ') \cong \angle\alpha(P)\alpha(O)\alpha(Q'),$$

Therefore

$$\angle\alpha(P)\alpha(O)\alpha(Q) \cong \alpha(\angle POQ) \cong \alpha(\angle POQ') \cong \angle\alpha(P)\alpha(O)\alpha(Q').$$

Now $\alpha(P) \in \alpha(\mathcal{M})$, $\alpha(Q)$ and $\alpha(Q') \in \alpha(\mathcal{L})$, and $\{\alpha(O)\} = \alpha(\mathcal{M}) \cap \alpha(\mathcal{L})$; therefore by Theorem NEUT.44 $\alpha(\mathcal{L}) \perp \alpha(\mathcal{M})$. The converse is proved by a similar proof, applying α^{-1} (an isometry by Theorem NEUT.11) to $\alpha(\mathcal{L}) \perp \alpha(\mathcal{M})$. □

Corollary NEUT.44.2. *Let $\angle BAC \cong \angle EDF$; then $\angle BAC$ is right iff $\angle EDF$ is right.*

Proof. By Definition NEUT.3(B) $\angle BAC \cong \angle EDF$ means that there exists an isometry α such that $\alpha(\angle BAC) = \angle EDF$, and hence α maps the set $\{\overleftrightarrow{AB}, \overleftrightarrow{AC}\}$ onto the set $\{\overleftrightarrow{DE}, \overleftrightarrow{DF}\}$. By Theorem NEUT.44 $\angle BAC$ is a right angle iff $\overleftrightarrow{AB} \perp \overleftrightarrow{AC}$, which by Corollary NEUT.44.1 is true iff $\overleftrightarrow{DE} \perp \overleftrightarrow{DF}$ which by Theorem NEUT.44 is true iff $\angle EDF$ is right. □

Theorem NEUT.45 (Lines of symmetry of supplementary angles are perpendicular.). *Let $\mathcal{P}$ be a neutral plane, O, P, and Q noncollinear points on $\mathcal{P}$, P' a point such that P'–O–P; let $\mathcal{M}$ be the line of symmetry of $\angle POQ$ and $\mathcal{L}$ be the line of symmetry of $\angle P'OQ$. Then $\mathcal{L} \perp \mathcal{M}$.*

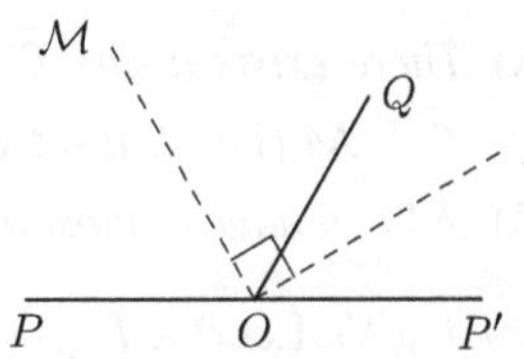

Fig. 8.2 For Theorem NEUT.45; dashed lines are the lines of symmetry.

Proof. See Figure 8.2 for a visualization. Let Q' be a point such that Q'–O–Q. By Exercise NEUT.10 $\mathcal{M}$ is the line of symmetry of $\angle P'OQ'$ and $\mathcal{L}$ is the line of symmetry of $\angle POQ'$. By Theorem NEUT.20 and Corollary PSH.39.2, P and Q are on opposite sides of $\mathcal{M}$ and P' and Q are on opposite sides of $\mathcal{L}$. By Theorem PSH.12 (plane separation) there exist points R and S such that $\overline{PQ} \cap \mathcal{M} = \{R\}$ and $\overline{P'Q} \cap \mathcal{L} = \{S\}$.

By Theorem NEUT.20 and Definition NEUT.1(A), $\mathcal{R}_{\mathcal{M}}(\overrightarrow{OQ}) = \overrightarrow{OP}$ and $\mathcal{R}_{\mathcal{M}}(\overrightarrow{OP'}) = \overrightarrow{OQ'}$. By the elementary theory of mappings, Theorem NEUT.15(8), and the definition of an angle (*cf* Definition PSH.29),

$$\begin{aligned}\mathcal{R}_{\mathcal{M}}(\angle P'OQ) &= \mathcal{R}_{\mathcal{M}}(\overrightarrow{OP'} \cup \overrightarrow{OQ}) = \mathcal{R}_{\mathcal{M}}(\overrightarrow{OP'}) \cup \mathcal{R}_{\mathcal{M}}(\overrightarrow{OQ})\\ &= \overrightarrow{OQ'} \cup \overrightarrow{OP} = \angle Q'OP.\end{aligned}$$

Since $\mathcal{L}$ is the line of symmetry of $\angle P'OQ$, by Theorem NEUT.39 $\angle P'OS \cong \angle QOS$. Then

$$\mathcal{R}_{\mathcal{M}}(\overrightarrow{OS}) = \overrightarrow{(\mathcal{R}_{\mathcal{M}}(O))(\mathcal{R}_{\mathcal{M}}(S))} = \overrightarrow{O(\mathcal{R}_{\mathcal{M}}(S))}.$$

By Definition PSH.29, Definition NEUT.1(A), and the elementary theory of mappings,

$$\begin{aligned}\mathcal{R}_{\mathcal{M}}(\angle QOS) &= \mathcal{R}_{\mathcal{M}}(\overrightarrow{OQ} \cup \overrightarrow{OS}) = \mathcal{R}_{\mathcal{M}}(\overrightarrow{OQ}) \cup \mathcal{R}_{\mathcal{M}}(\overrightarrow{OS})\\ &= \overrightarrow{OP} \cup \overrightarrow{O(\mathcal{R}_{\mathcal{M}}(S))} = \angle PO(\mathcal{R}_{\mathcal{M}}(S)), \text{ and}\\ \mathcal{R}_{\mathcal{M}}(\angle P'OS) &= \mathcal{R}_{\mathcal{M}}(\overrightarrow{OP'} \cup \overrightarrow{OS}) = \mathcal{R}_{\mathcal{M}}(\overrightarrow{OP'}) \cup \mathcal{R}_{\mathcal{M}}(\overrightarrow{OS})\\ &= \overrightarrow{OQ'} \cup \overrightarrow{O(\mathcal{R}_{\mathcal{M}}(S))} = \angle Q'O(\mathcal{R}_{\mathcal{M}}(S)).\end{aligned}$$

Since $\angle QOS \cong \angle P'OS$, by Theorem NEUT.13 $\mathcal{R}_{\mathcal{M}}(\angle QOS) \cong \mathcal{R}_{\mathcal{M}}(\angle P'OS)$. Combining this with the last two equalities, $\angle PO(\mathcal{R}_{\mathcal{M}}(S)) \cong \angle Q'O(\mathcal{R}_{\mathcal{M}}(S))$. By Theorem NEUT.39, $\overrightarrow{O(\mathcal{R}_{\mathcal{M}}(S))}$ is the line of symmetry of $\angle POQ'$, which we know already to be $\mathcal{L}$. Therefore $\mathcal{R}_{\mathcal{M}}(\mathcal{L}) = \mathcal{L}$, so by Theorem NEUT.44 (or NEUT.32) $\mathcal{L}$ is a fixed line for $\mathcal{R}_{\mathcal{M}}$ and $\mathcal{L} \perp \mathcal{M}$. □

Theorem NEUT.46 (A line has a unique perpendicular through each of its points). *Let $\mathcal{M}$ be a line on the neutral plane $\mathcal{P}$ and let O be any point on $\mathcal{M}$.*

(A) *There exists a line $\mathcal{L}$ containing O which is a line of symmetry for $\mathcal{M}$, and $\mathcal{L} \perp \mathcal{M}$ (i.e., $\mathcal{L}$ is a fixed line for $\mathcal{R}_{\mathcal{M}}$).*

(B) *$\mathcal{M}$ can have at most one line of symmetry $\mathcal{L}$ containing the point O.*

Proof. (A) Let $P \in \mathcal{P}\backslash\mathcal{M}$. If $\mathcal{R}_{\mathcal{M}}(P)$ is collinear with O and P, then $\mathcal{R}_{\mathcal{M}}(P)$–$O$–$P$, and $O \in \overleftrightarrow{P\mathcal{R}_{\mathcal{M}}(P)}$. By Theorem NEUT.22 this is a fixed line for $\mathcal{R}_{\mathcal{M}}$. Hence by Theorem NEUT.44 $\mathcal{L} \perp \mathcal{M}$.

Otherwise, let $Q = \mathcal{R}_{\mathcal{M}}(P)$, and let P' and Q' be points such that P–O–P' and Q–O–Q'. $\mathcal{M}$ is the line of symmetry of $\angle POQ$ since $\mathcal{R}_{\mathcal{M}}(P) = Q$ and $\mathcal{R}_{\mathcal{M}}(\overrightarrow{OP}) = \overrightarrow{OQ}$ by Theorem NEUT.15(3). By Property R.5 of Definition NEUT.2 there exists a line of symmetry $\mathcal{L}$ for $\angle POQ'$.

By Theorem NEUT.45 since $\angle POQ'$ and $\angle POQ$ are supplementary, $\mathcal{L} \perp \mathcal{M}$. By Theorem NEUT.44 each of $\mathcal{L}$ and $\mathcal{M}$ is a line of symmetry of the other.[6]

(B) By Theorem NEUT.44 (or for that matter Theorem NEUT.32) a line $\mathcal{L}$ (not equal to $\mathcal{M}$) is a line of symmetry for $\mathcal{M}$ iff it is a fixed line for $\mathcal{R}_{\mathcal{M}}$, which is true iff $\mathcal{L} \perp \mathcal{M}$. By Theorem NEUT.22(D) there can be only one such line through any point $O \in \mathcal{M}$. □

Corollary NEUT.46.1. *If O, P, and Q are noncollinear points on the neutral plane $\mathcal{P}$, there exists a point R on* the Q-side of $\overleftrightarrow{OP}$ *such that $\angle ROP$ is right.*

Proof. By Theorem NEUT.46 there exists a unique line $\mathcal{L}$ such that $\mathcal{L} \cap \overleftrightarrow{OP} = \{O\}$ and $\mathcal{L} \perp \overleftrightarrow{OP}$. By Exercise IB.17 there exists a point R on $\mathcal{L}$ which is on the Q-side of $\overleftrightarrow{OP}$. By Theorem NEUT.44, $\angle POR$ is a right angle. □

Theorem NEUT.47. *Let $\mathcal{P}$ be a neutral plane, and let $\mathcal{L}$ be a line on $\mathcal{P}$.*

(A) *If two distinct lines $\mathcal{M}$ and $\mathcal{N}$ are perpendicular to $\mathcal{L}$, then they are parallel.*
(B) *Given a point $A \in \mathcal{P}$ there can be only one line through A which is perpendicular to $\mathcal{L}$.*

Proof. (A) By Theorem NEUT.44 (or Theorem NEUT.32) both $\mathcal{M}$ and $\mathcal{N}$ are fixed lines for the reflection $\mathcal{R}_{\mathcal{L}}$. By Theorem NEUT.22, if these lines are distinct, they are parallel.

(B) If there are two lines $\mathcal{M}$ and $\mathcal{N}$ containing A, both perpendicular to $\mathcal{L}$, by part (A) they are parallel and therefore cannot intersect, a contradiction. □

Theorem NEUT.48. *Let $\mathcal{M}$ be a line on a neutral plane $\mathcal{P}$, and let P be any member of $\mathcal{P}$.*

(A) *There exists a unique line $\mathcal{L}$ such that $P \in \mathcal{L}$ and $\mathcal{M} \perp \mathcal{L}$. If $P \notin \mathcal{M}$, $\mathcal{L} = \overleftrightarrow{P(\mathcal{R}_{\mathcal{M}}(P))}$.*
(B) *(Property PE) For every point Q belonging to $\mathcal{P} \setminus \mathcal{M}$, there exists a line $\mathcal{L}$ through Q which is parallel to $\mathcal{M}$.*

Proof. (A) If $P \in \mathcal{M}$, this is Theorem NEUT.46. If $P \notin \mathcal{M}$, by Theorem NEUT.22, $\mathcal{L} = \overleftrightarrow{P\mathcal{R}_{\mathcal{M}}(P)}$ is a fixed line for $\mathcal{R}_{\mathcal{M}}$. Hence by Theorem NEUT.44,

[6] If Axiom PS were in force, we could replace this paragraph by the observation that $\overleftrightarrow{P\mathcal{R}_{\mathcal{M}}(P)}$ is a fixed line for $\mathcal{R}_{\mathcal{M}}$ that does not contain O; then by Theorem CAP.5, there exists a unique fixed line of $\mathcal{R}_{\mathcal{M}}$ through O.

$\mathcal{L}$ is perpendicular to $\mathcal{M}$. The uniqueness follows immediately from Theorem NEUT.47(B).

(B) By part (A) there exists a line $\mathcal{N}$ such that $\mathcal{N} \perp \mathcal{M}$; then again by part (A) there exists a line $\mathcal{L}$ containing Q which is perpendicular to $\mathcal{N}$. Since $Q \notin \mathcal{M}$, $\mathcal{L} \neq \mathcal{M}$. By Theorem NEUT.47 $\mathcal{L} \parallel \mathcal{M}$. □

Remark NEUT.49. (A) From Theorem NEUT.48(B), it might be tempting to think that $\mathcal{L}$ is the only line through Q which is parallel to $\mathcal{M}$. But this is not the case; using the notation of the proof of part (B), the possibility exists that there could be a different line $\mathcal{L}'$ through Q which is also parallel to $\mathcal{M}$. What is ruled out is that $\mathcal{L}'$ could be *perpendicular* to $\mathcal{N}$. We still have ambiguity as to parallelism, because we have not invoked either Axiom PS or PW.

(B) It is a big step forward to eliminate ambiguity about perpendicularity (which is defined entirely by our notion of reflection), by proving that there can be only one line perpendicular to another at a point. We will soon confront many of the standard congruence theorems of geometry, and a well-defined notion of perpendicularity will be essential.

From the basic properties of mirror mappings and the first five properties of Definition NEUT.2, we have proved that reflections (and isometries) behave in respectably decent ways.

In Theorem NEUT.24 we proved that an isometry having three noncollinear fixed points is the identity; in Theorem NEUT.26 that angle reflections are unique; Theorem NEUT.36 showed that if an isometry maps an angle $\angle AOB$ to an angle $\angle AOC$, where B and C are on the same side of $\overleftrightarrow{AO}$, then $\overrightarrow{OB} = \overrightarrow{OC}$.

We have now added to this list a usable notion of perpendicularity, and proceed forthwith to deal with midpoints.

8.10 Midpoints of segments

Before this point we have not invoked Property R.6 of Definition NEUT.2, stating the existence of midpoints of segments. We now do so, and this property will remain in force throughout the rest of the chapter.

Theorem NEUT.50. *If A and B are distinct points on the neutral plane $\mathcal{P}$, then $\overline{AB}$ has a unique midpoint. Thus it is legitimate to speak of THE midpoint of a segment.*

Proof. Existence: by Property R.6 of Definition NEUT.2, there exists a midpoint of $\overline{AB}$.

Uniqueness: suppose there are two midpoints M and N for $\overline{AB}$. By Definition NEUT.3(C) $\overline{MA} \cong \overline{MB}$. Using Theorem NEUT.48 let $\mathcal{L}$ be the line on $\mathcal{P}$ such that $M \in \mathcal{L}$ and $\mathcal{L} \perp \overleftrightarrow{AB}$; then $\{M\} = \mathcal{L} \cap \overleftrightarrow{AB}$. By Theorem NEUT.15(5) and Definition NEUT.1(A), $\mathcal{R}_{\mathcal{L}}(\overline{AM}) = \overline{(\mathcal{R}_{\mathcal{L}}(A))(\mathcal{R}_{\mathcal{L}}(M))} = \overline{(\mathcal{R}_{\mathcal{L}}(A))M}$ so that by Definition NEUT.3(B), $\overline{MA} \cong \overline{M(\mathcal{R}_{\mathcal{L}}(A))}$. By Definition NEUT.1(B), A and $\mathcal{R}_{\mathcal{L}}(A)$ are on opposite sides of $\mathcal{L}$ so that A–M–$\mathcal{R}_{\mathcal{L}}(A)$. By Theorem PSH.13, since A–M–B, $\mathcal{R}_{\mathcal{L}}(A) \in \overrightarrow{MB}$. Since $\overline{M(\mathcal{R}_{\mathcal{L}}(A))} \cong \overline{MA}$ and $\overline{MA} \cong \overline{MB}$, by Theorem NEUT.14 $\overline{M(\mathcal{R}_{\mathcal{L}}(A))} \cong \overline{MB}$. By Property R.4 of Definition NEUT.2, $\mathcal{R}_{\mathcal{L}}(A) = B$.

We have already seen that $\{M\} = \mathcal{L} \cap \overleftrightarrow{AB}$. By assumption, N is also a midpoint of $\overline{AB}$, so by Definition NEUT.3(C), $\overline{AN} \cong \overline{BN}$ and A–N–B. By Lemma NEUT.34(A), $N = M$. □

Definition NEUT.51. Let A and B be distinct points on the neutral plane $\mathcal{P}$ and let M be the midpoint of $\overline{AB}$. The **perpendicular bisector** of $\overline{AB}$ is the line $\mathcal{M}$ such that $M \in \mathcal{M}$ and $\mathcal{M} \perp \overleftrightarrow{AB}$.

Theorem NEUT.52. *Let A and B be distinct points on the neutral plane $\mathcal{P}$, and let M be the midpoint of $\overline{AB}$. Then*

(A) *there exists a unique line $\mathcal{M}$ containing M which is perpendicular to $\overleftrightarrow{AB}$ (the perpendicular bisector of $\overline{AB}$);*
(B) *$\mathcal{R}_{\mathcal{M}}(A) = B$ and $\mathcal{R}_{\mathcal{M}}(B) = A$;*
(C) *$\mathcal{R}_{\mathcal{M}}(\overline{AB}) = \overline{BA}$, so that $\mathcal{M}$ is a line of symmetry of $\overline{AB}$; and*
(D) *$\mathcal{M}$ is the unique line of symmetry of $\overline{AB}$.*

Proof. (A) follows immediately from Theorem NEUT.48. Then by Theorem NEUT.44, $\overleftrightarrow{AB}$ is a fixed line for $\mathcal{R}_{\mathcal{M}}$, and $\mathcal{M}$ is a line of symmetry for $\overleftrightarrow{AB}$.

Then $\mathcal{R}_{\mathcal{M}}(A) \in \overleftrightarrow{AB}$, and $\mathcal{R}_{\mathcal{M}}(A)$ is on the opposite side of $\mathcal{M}$ from A, that is, A–M–$\mathcal{R}_{\mathcal{M}}(A)$, and since A–M–B, $\mathcal{R}_{\mathcal{M}}(A) \in \overrightarrow{MB}$. Also, by Theorem NEUT.15,

$$\mathcal{R}_{\mathcal{M}}(\overline{AM}) = \overline{(\mathcal{R}_{\mathcal{M}}(A))(\mathcal{R}_{\mathcal{M}}(M))} = \overline{(\mathcal{R}_{\mathcal{M}}(A))M}$$

so that by Definition NEUT.3(C) $\overline{BM} \cong \overline{AM} \cong \overline{(\mathcal{R}_{\mathcal{M}}(A))M}$. By Property R.4 of Definition NEUT.2 (linear scaling), $\mathcal{R}_{\mathcal{M}}(A) = B$, and by Definition NEUT.1(C) $\mathcal{R}_{\mathcal{M}}(B) = A$. This proves part (B).

By Theorem NEUT.15(5),

$$\mathcal{R}_{\mathcal{M}}(\overline{AB}) = \overline{(\mathcal{R}_{\mathcal{M}}(A))(\mathcal{R}_{\mathcal{M}}(B))} = \overline{AB}$$

so that $\overline{AB} \cong \overline{BA}$, proving (C).

To prove (D), suppose that $\mathcal{L}$ is a line of symmetry for $\overline{AB}$; then $\overline{AB} \not\subseteq \mathcal{L}$, and $\mathcal{R}_{\mathcal{L}}(\overline{AB}) = \overline{AB}$. By Remark NEUT.16 $\mathcal{R}_{\mathcal{L}}(A) = B$, and by Definition NEUT.1(B) A and B are on opposite sides of $\mathcal{L}$; by Axiom PSA there exists a point $D \in \overline{AB}$ such that $D \in \mathcal{L}$; then

$$\mathcal{R}_{\mathcal{L}}(\overline{AD}) = \overline{(\mathcal{R}_{\mathcal{L}}(A))(\mathcal{R}_{\mathcal{L}}(D))} = \overline{BD}$$

so that $\overline{BD} \cong \overline{AD}$, and D is a midpoint of $\overline{AB}$. By Theorem NEUT.50, there is only one midpoint of $\overline{AB}$, so that $D = M$. Since $\mathcal{L} \perp \overleftrightarrow{AB}$ and $\mathcal{M} \perp \overleftrightarrow{AB}$ and both $\mathcal{L}$ and $\mathcal{M}$ contain the point M, by Theorem NEUT.48, $\mathcal{L} = \mathcal{M}$, which is therefore the unique line of symmetry for $\overline{AB}$, proving part (D). □

Theorem NEUT.53. *Let A and B be distinct points on the neutral plane $\mathcal{P}$, and let $\mathcal{M}$ be the perpendicular bisector of $\overline{AB}$. Then for every $C \in \mathcal{M}$, $\overline{AC} \cong \overline{BC}$.*

Proof. If $C \in \mathcal{M}$, C is a fixed point for $\mathcal{R}_{\mathcal{M}}$; by Theorem NEUT.52(B), $\mathcal{R}_{\mathcal{M}}(A) = B$; then by Theorem NEUT.15(5),

$$\mathcal{R}_{\mathcal{M}}(\overline{AC}) = \overline{(\mathcal{R}_{\mathcal{M}}(A))(\mathcal{R}_{\mathcal{M}}(C))} = \overline{BC};$$

thus $\overline{AC} \cong \overline{BC}$, proving the theorem. □

Theorem NEUT.54. *Let $\mathcal{M}$ be a line on the neutral plane $\mathcal{P}$ and let Q be a member of $\mathcal{P} \setminus \mathcal{M}$; then $\mathcal{M}$ is the perpendicular bisector of $\overline{Q(\mathcal{R}_{\mathcal{M}}(Q))}$.*

Proof. By Theorem NEUT.22(A), $\overleftrightarrow{Q(\mathcal{R}_{\mathcal{M}}(Q))}$ is a fixed line for $\mathcal{R}_{\mathcal{M}}$. By part (E) of the same theorem, $\overleftrightarrow{Q(\mathcal{R}_{\mathcal{M}}(Q))}$ intersects $\mathcal{M}$ at exactly one point D, and D is the midpoint of $\overline{Q(\mathcal{R}_{\mathcal{M}}(Q))}$. By Theorem NEUT.32 (or NEUT.44) $\overleftrightarrow{Q(\mathcal{R}_{\mathcal{M}}(Q))} \perp \mathcal{M}$, so $\mathcal{M}$ is its perpendicular bisector. □

Theorem NEUT.55. *Let A, B, and C be noncollinear points on the neutral plane $\mathcal{P}$ such that $\overline{AB} \cong \overline{AC}$, and let $\mathcal{M}$ be the line of symmetry of $\angle BAC$. Then $\mathcal{M}$ is the line of symmetry of $\overline{BC}$. Furthermore, there exists a point D such that $\{D\} = \mathcal{M} \cap \overline{BC}$, and D is the midpoint of $\overline{BC}$.*

Proof. By Theorem NEUT.20 $A \in \mathcal{M}$ and $\mathcal{R}_{\mathcal{M}}(\overrightarrow{AB}) = \overrightarrow{AC}$, so $\mathcal{R}_{\mathcal{M}}(B) \in \overrightarrow{AC}$. Since $\overline{AB} \cong \overline{AC}$, by Property R.4 of Definition NEUT.2, $\mathcal{R}_{\mathcal{M}}(B) = C$. By Theorem NEUT.22(A) $\overleftrightarrow{A(\mathcal{R}_{\mathcal{M}}(B))} = \overleftrightarrow{BC}$ is a fixed line for $\mathcal{R}_{\mathcal{M}}$, and by Theorem NEUT.32 (or NEUT.44) $\overleftrightarrow{BC} \perp \mathcal{M}$.

By Theorem NEUT.20 there exists a point D such that $\overline{B\mathcal{R}_{\mathcal{M}}(B)} \cap \mathcal{M} = \{D\}$ and (by part (E)(2)) $\overline{BD} \cong \overline{CD}$. Therefore, by Definition NEUT.51, $\mathcal{M}$ is the perpendicular bisector and the line of symmetry of $\overline{BC}$. □

Theorem NEUT.56. *Let A, B, C, and D be points on the neutral plane $\mathcal{P}$ such that $A \neq B$, $C \neq D$, and $\overline{AB} \cong \overline{CD}$. Then there exists an isometry α of $\mathcal{P}$ such that $\alpha(\overline{AB}) = \overline{CD}$, $\alpha(A) = C$, and $\alpha(B) = D$.*

Proof. By Definition NEUT.3(B) there exists an isometry λ of $\mathcal{P}$ such that $\lambda(\overline{AB}) = \overline{CD}$. By Remark NEUT.16, $\{\lambda(A), \lambda(B)\} = \{C, D\}$. Hence either $\lambda(A) = C$ and $\lambda(B) = D$ or $\lambda(A) = D$ and $\lambda(B) = C$. In the first case we take $\alpha = \lambda$. In the second case using Theorem NEUT.52 let $\mathcal{L}$ be the line of symmetry of $\overline{CD}$ and take $\alpha = \mathcal{R}_{\mathcal{M}} \circ \lambda$. Since α is an isometry of $\mathcal{P}$ (*cf* Theorem NEUT.11) and since $\alpha(A) = C$ and $\alpha(B) = D$, the theorem is proved. □

Theorem NEUT.57. *Let A, B, C, and D be points on the neutral plane $\mathcal{P}$ such that $A \neq B$, $C \neq D$, and $\overline{AB} \cong \overline{CD}$.*

(A) *If E and F are points such that A–E–B, $F \in \overrightarrow{CD}$ and $\overline{AE} \cong \overline{CF}$, then C–F–D.*
(B) *If G and H are points such that A–B–G, $H \in \overrightarrow{CD}$ and $\overline{AG} \cong \overline{CH}$, then C–D–H.*

Proof. Since $\overline{AB} \cong \overline{CD}$, by Theorem NEUT.56 there exists an isometry α of $\mathcal{P}$ such that $\alpha(\overline{AB}) = \overline{CD}$, $\alpha(A) = C$, and $\alpha(B) = D$. By Definition NEUT.1(D), $\alpha(A)$–$\alpha(E)$–$\alpha(B)$ and $\alpha(A)$–$\alpha(B)$–$\alpha(G)$, so that C–$\alpha(E)$–D and C–D–$\alpha(G)$. By Theorem NEUT.15(5) $\alpha(\overline{AE}) = \overline{\alpha(A)\alpha(E)} = \overline{C\alpha(E)}$ and $\alpha(\overline{AG}) = \overline{\alpha(A)\alpha(G)} = \overline{C\alpha(G)}$. By Definition IB.4 $\alpha(E) \in \overrightarrow{CD}$ and $\alpha(G) \in \overrightarrow{CD}$. By Property R.4 of Definition NEUT.2, $\alpha(E) = F$ and $\alpha(G) = H$. Thus C–F–D and C–D–H. □

Theorem NEUT.58. *Let A, B, C, D, E, and F be points on the neutral plane $\mathcal{P}$ such that A, B, and C are noncollinear, D, E, and F are noncollinear, and $\overline{AB} \cong \overline{DE}$. Then there exists an isometry α of $\mathcal{P}$ such that $\alpha(\overline{AB}) = \overline{DE}$, $\alpha(A) = D$, $\alpha(B) = E$, and $\alpha(C) \in \overrightarrow{DEF}$.*

Proof. By Theorem NEUT.56 there exists an isometry φ of $\mathcal{P}$ such that $\varphi(\overline{AB}) = \overline{DE}$, $\varphi(A) = D$, and $\varphi(B) = E$. If $\varphi(C) \in \overrightarrow{DEF}$, then we take $\alpha = \varphi$. If $\varphi(C)$ does not belong to the F-side of $\overleftrightarrow{DE}$, then since $\varphi(C)$ does not belong to $\overleftrightarrow{DE}$, it belongs to (the side of $\overleftrightarrow{DE}$ opposite the F-side). In this case we take $\alpha = \mathcal{R}_{\overleftrightarrow{DE}} \circ \varphi$. By Theorem NEUT.11 α is an isometry of $\mathcal{P}$ and furthermore $\alpha(C) \in \overrightarrow{DEF}$. □

Definition NEUT.59. A **kite** is a quadrilateral with two distinct pairs of consecutive edges that are congruent.

Theorem NEUT.60 (Kite). *Let A, B, C, and D be points on the neutral plane $\mathcal{P}$ such that $\overline{AB} \cup \overline{BC} \cup \overline{CD} \cup \overline{DA}$ is a quadrilateral (cf Definition PSH.31), $\overline{AB} \cong \overline{AD}$ and $\overline{BC} \cong \overline{DC}$. Then $\overleftrightarrow{AC}$ is the line of symmetry of $\angle BAD$, $\angle BCD$, $\overline{BD}$, and $\square ABCD$. Furthermore, $\angle BAC \cong \angle DAC$, $\angle BCA \cong \angle DCA$, $\angle ABC \cong \angle ADC$, and $\triangle ABC \cong \triangle ADC$.*

Fig. 8.3 Showing a kite.

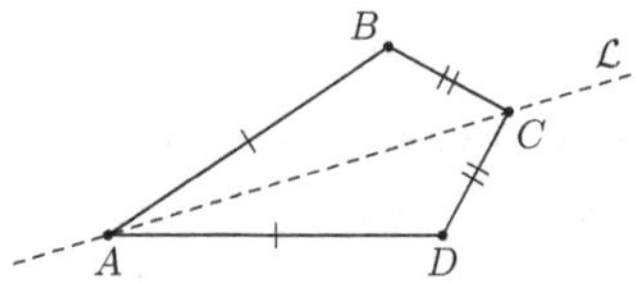

Proof. See Figure 8.3. Using Theorem NEUT.26 let $\mathcal{L}$ be the line of symmetry of $\angle BAD$ and let $\mathcal{L}'$ be the line of symmetry of $\angle BCD$. By Theorem NEUT.55 each of these lines is the line of symmetry of $\overline{BD}$, so $\mathcal{L} = \mathcal{L}' = \overleftrightarrow{AC}$ and $\mathcal{R}_{\mathcal{L}}(B) = D$. Furthermore, $\mathcal{R}_{\mathcal{L}}(\angle BAC) = \angle DAC$, $\mathcal{R}_{\mathcal{L}}(\overline{AB}) = \overline{AD}$, $\mathcal{R}_{\mathcal{L}}(\overline{CB}) = \overline{CD}$, and $\mathcal{R}_{\mathcal{L}}(\angle ABC) = \angle ADC$. By Definition NEUT.3(B) $\angle BAC \cong \angle DAC$, $\angle BCA \cong \angle DCA$, and $\angle ABC \cong \angle ADC$. Moreover, $\mathcal{R}_{\mathcal{L}}(\square ABCD) = \square ADCB = \square ABCD$. □

8.11 Congruence of triangles and angles

Remark NEUT.61. Let $\mathcal{P}$ be a neutral plane. According to Definition NEUT.3(B), two triangles $\triangle ABC$ and $\triangle DEF$ on $\mathcal{P}$ are congruent iff there exists an isometry φ of $\mathcal{P}$ such that $\varphi(\triangle ABC) = \triangle DEF$. This means, referring back to Definition IB.7, that

$$\varphi(\overline{AB} \cup \overline{BC} \cup \overline{CA}) = \overline{DE} \cup \overline{EF} \cup \overline{FD}.$$

However, this equality simply means the set on the left side is equal to the set on the right; it gives no information about which edge of $\triangle ABC$ maps to which edge of $\triangle DEF$. We used this equality to prove Theorem NEUT.15(10).

Stating that there is an isometry φ mapping a triangle to another triangle implies a pairing of the corners and of the edges of the respective triangles. That is to say, the mapping φ maps each of the corners A, B, and C to one of the corners E, F, or G. We express this by saying that if two triangles are congruent, then corresponding edges are congruent and the corresponding angles are congruent. Sometimes it is necessary to specify which corners map to which corners.

If we choose the notation so that $\varphi(A) = D$, $\varphi(B) = E$, and $\varphi(C) = F$, then

$$\varphi(\angle BAC) = \angle EDF, \varphi(\angle ABC) = \angle DEF, \text{ and } \varphi(\angle ACB) = \angle DFE,$$

so that

$$\angle BAC \cong \angle EDF, \angle ABC \cong \angle DEF, \text{ and } \angle ACB \cong \angle DFE$$

and

$$\varphi(\overline{AB}) = \overline{DE}, \varphi(\overline{BC}) = \overline{EF}, \text{ and } \varphi(\overline{CA}) = \overline{FD}.$$

Then

$$\overline{AB} \cong \overline{DE}, \overline{BC} \cong \overline{EF}, \text{ and } \overline{CA} \cong \overline{FD}.$$

In order to make the "congruence theorems" (NEUT.62, NEUT.64, and NEUT.65) completely clear, we will express their conclusions not only in terms of congruence but also in terms of specific corner pairings.

Theorem NEUT.62 (EEE congruence theorem for triangles). *Let A, B, C, D, E, and F be points on the neutral plane $\mathcal{P}$ such that A, B, and C are noncollinear and D, E, and F are noncollinear, $\overline{AB} \cong \overline{DE}$, $\overline{AC} \cong \overline{DF}$, and $\overline{BC} \cong \overline{EF}$. Then there exists an isometry α such that $\alpha(\triangle ABC) = \triangle DEF$, $\alpha(A) = D$, $\alpha(B) = E$, and $\alpha(C) = F$. Thus $\triangle ABC \cong \triangle DEF$, and corresponding angles are congruent. (Corresponding edges are congruent by hypothesis).*

Proof. By Theorem NEUT.58 there exists an isometry β of $\mathcal{P}$ such that $\beta(\overline{AB}) = \overline{DE}$, $\beta(A) = D$, $\beta(B) = E$, and $\beta(C) = F'$, where F' is a member of the side of $\overleftrightarrow{DE}$ opposite $\overrightarrow{DEF}$ (cf Definition IB.11).

Then by Theorem NEUT.15(5) $\beta(\overline{AC}) = \overline{\beta(A)\beta(C)} = \overline{DF'}$ so by hypothesis and Definition NEUT.3(B) $\overline{DF} \cong \overline{AC} \cong \overline{DF'}$. Likewise $\beta(\overline{BC}) = \overline{\beta(B)\beta(C)} = \overline{EF'}$ so $\overline{EF} \cong \overline{BC} \cong \overline{EF'}$. Then by Theorem NEUT.60 (Kite) $\overleftrightarrow{DE}$ is the line of symmetry of $\overline{FF'}$, so that $F = \mathcal{R}_{\overleftrightarrow{DE}}(F')$.

Let $\alpha = \mathcal{R}_{\overleftrightarrow{DE}} \circ \beta$. Then $\alpha(A) = D$, $\alpha(B) = E$, and $\alpha(C) = F$. By hypothesis we know that $\overline{AB} \cong \overline{DE}$, $\overline{AC} \cong \overline{DF}$, and $\overline{BC} \cong \overline{EF}$. By Theorem NEUT.15(8)

$$\alpha(\angle BAC) = \angle\alpha(B)\alpha(A)\alpha(C) = \angle EDF,$$
$$\alpha(\angle ABC) = \angle\alpha(A)\alpha(B)\alpha(C) = \angle DEF, \text{ and}$$
$$\alpha(\angle ACB) = \angle\alpha(A)\alpha(C)\alpha(B) = \angle DFE, \text{ so that}$$

$$\angle BAC \cong \angle EDF, \angle ABC \cong \angle DEF, \text{ and } \angle ACB \cong \angle DFE.$$

Finally, by Theorem NEUT.15(10),

$$\alpha(\triangle ABC) = \triangle\alpha(A)\alpha(B)\alpha(C) = \triangle DEF, \text{ and } \triangle ABC \cong \triangle DEF. \qquad \square$$

Theorem NEUT.63. *Let $\mathcal{P}$ be a neutral plane, A and B be distinct points on $\mathcal{P}$, M be the midpoint of $\overline{AB}$ and Q be a member of $\mathcal{P} \setminus \overleftrightarrow{AB}$ such that $\overline{AQ} \cong \overline{BQ}$. Then $\overleftrightarrow{MQ}$ is the line of symmetry of $\angle AQB$ and of $\overline{AB}$.*

Proof. Since M is the midpoint of $\overline{AB}$, by Definition NEUT.3(C) $\overline{AM} \cong \overline{BM}$. Moreover $\overline{MQ} \cong \overline{MQ}$. Since $\overline{AQ} \cong \overline{BQ}$, by Theorem NEUT.62 (EEE) $\angle AMQ \cong \angle BMQ$ and $\angle AQM \cong \angle BQM$ (cf Remark NEUT.61). By Theorem NEUT.39 $\overleftrightarrow{MQ}$ is the line of symmetry of $\angle AQB$.

By Property B.3 of Definition IB.1 let Q' be a point such that Q–M–Q'; by Theorem NEUT.42 (vertical angles), since $\angle AMQ \cong \angle BMQ$, $\angle BMQ' \cong \angle AMQ \cong \angle BMQ \cong \angle AMQ'$. By Theorem NEUT.44 (parts (B) and (G)) $\overleftrightarrow{MQ} \perp \overleftrightarrow{AB}$. By Theorem NEUT.52 $\overleftrightarrow{MQ}$ is the line of symmetry of $\overline{AB}$. $\square$

Theorem NEUT.64 (Isometry construction for angles). *Let A, B, C, D, E, and F be points on the neutral plane $\mathcal{P}$ such that A, B, and C are noncollinear, D, E, and F are noncollinear, $\overline{AB} \cong \overline{DE}$, $\overline{AC} \cong \overline{DF}$, and $\angle BAC \cong \angle EDF$. Then there exists an isometry α such that $\alpha(\triangle ABC) = \triangle DEF$, $\alpha(A) = D$, $\alpha(B) = E$, and $\alpha(C) = F$. Thus $\triangle ABC \cong \triangle DEF$, and corresponding edges and angles are congruent.*

Proof. By Theorem NEUT.58 there exists an isometry α of $\mathcal{P}$ such that $\alpha(\overline{AB}) = \overline{DE}$, $\alpha(A) = D$, $\alpha(B) = E$, and $\alpha(C)$ is a member of $\overrightarrow{DEF}$. By Theorem NEUT.15(8) $\alpha(\angle BAC) = \angle\alpha(B)\alpha(A)\alpha(C) = \angle ED\alpha(C)$. By Definition NEUT.3(B) $\angle BAC \cong \angle ED\alpha(C)$. By Theorem NEUT.14 (congruence is an equivalence relation), $\angle EDF \cong \angle ED\alpha(C)$. By Theorem NEUT.36 $\overrightarrow{DF} = \overrightarrow{D\alpha(C)}$. By Theorem PSH.24 $\alpha(C) \in \overrightarrow{DF}$. By Theorem NEUT.15(8) $\alpha(\overline{AC}) = \overline{\alpha(A)\alpha(C)} = \overline{D\alpha(C)}$. By Definition NEUT.3(B) $\overline{AC} \cong \overline{D\alpha(C)}$. Since congruence is an equivalence relation (Theorem NEUT.14) $\overline{D\alpha(C)} \cong \overline{DF}$. By Property R.4 of Definition NEUT.2 $\alpha(C) = F$. Since $\alpha(A) = D$, $\alpha(B) = E$, and $\alpha(C) = F$, by Theorem NEUT.15(5) $\alpha(\overline{BC}) = \overline{EF}$ so that $\overline{BC} \cong \overline{EF}$; the other corresponding edges are congruent by hypothesis.

We may now apply Theorem NEUT.62 (EEE Congruence theorem), completing the proof. $\square$

Theorem NEUT.65 (AEA congruence theorem for triangles). *Let A, B, C, D, E, and F be points on the neutral plane $\mathcal{P}$ such that A, B, and C are noncollinear, D, E, and F are noncollinear, $\overline{AB} \cong \overline{DE}$, $\angle BAC \cong \angle EDF$, and $\angle ABC \cong \angle DEF$. Then there exists an isometry α such that $\alpha(\triangle ABC) = \triangle DEF$, $\alpha(A) = D$, $\alpha(B) = E$, and $\alpha(C) = F$. Thus $\triangle ABC \cong \triangle DEF$, and corresponding edges and angles are congruent.*

Proof. By Theorem NEUT.58 there exists an isometry α of $\mathcal{P}$ such that $\alpha(\overline{AB}) = \overline{DE}$, $\alpha(A) = D$, $\alpha(B) = E$, and $\alpha(C) \in \overrightarrow{DEF}$. By Theorem NEUT.15(8)

$$\alpha(\angle BAC) = \angle\alpha(B)\alpha(A)\alpha(C) = \angle ED\alpha(C)$$

and

$$\alpha(\angle ABC) = \angle\alpha(A)\alpha(B)\alpha(C) = \angle DE\alpha(C).$$

By Definition NEUT.3(B) $\angle BAC \cong \angle ED\alpha(C)$ and $\angle ABC \cong \angle DE\alpha(C)$.

Since congruence is an equivalence relation (Theorem NEUT.14) and since $\angle BAC \cong \angle EDF$ and $\angle ABC \cong \angle DEF$, $\angle ED\alpha(C) \cong \angle EDF$ and $\angle DEF \cong \angle DE\alpha(C)$. By Theorem NEUT.36 $\overrightarrow{D\alpha(C)} = \overrightarrow{DF}$ and $\overrightarrow{E\alpha(C)} = \overrightarrow{EF}$. By Theorem PSH.24(B) $\alpha(C) \in \overrightarrow{DF}$ and $\alpha(C) \in \overrightarrow{EF}$. By Theorem I.5 $\overrightarrow{DF}$ is a subset of $\overleftrightarrow{DF}$ and $\overrightarrow{EF}$ is a subset of $\overleftrightarrow{EF}$. By Exercise I.1 $\overleftrightarrow{DF} \cap \overleftrightarrow{EF} = \{F\}$. Therefore $\alpha(C) = F$. Since $\alpha(A) = D$, $\alpha(B) = E$, and $\alpha(C) = F$, by Theorem NEUT.15(5) $\alpha(\overline{AC}) = \overline{DF}$ and $\alpha(\overline{BC}) = \overline{EF}$ so that $\overline{AC} \cong \overline{DF}$ and $\overline{BC} \cong \overline{EF}$; $\overline{AB} \cong \overline{DE}$ by hypothesis. We may now apply Theorem NEUT.62 (EEE Congruence theorem), completing the proof. □

Theorem NEUT.66. *On a neutral plane $\mathcal{P}$ every angle congruent to a right angle is a right angle.*

Proof. Let A, B, C, D, E, and F be points on $\mathcal{P}$ such that A, B, and C are noncollinear, D, E, and F are noncollinear, $\angle EDF$ is a right angle and $\angle BAC \cong \angle EDF$. Using Property B.3 of Definition IB.1 let B' and E' be points such that B–A–B' and E–D–E'. By Definition NEUT.41(C) $\angle EDF \cong \angle E'DF$.

Since $\angle BAC \cong \angle EDF$, $\angle E'DF \cong \angle B'AC$ by Theorem NEUT.43 (supplements of congruent angles are congruent). Since congruence is an equivalence relation, $\angle B'AC \cong \angle E'DF \cong \angle EDF \cong \angle BAC$, and by Definition NEUT.41(C) $\angle BAC$ is a right angle. □

Theorem NEUT.67 (segment construction). *Let A, B, C, and D be points on the neutral plane $\mathcal{P}$ such that $A \neq B$ and $C \neq D$. Then there exists a unique point P belonging to $\overrightarrow{CD}$ such that $\overline{CP} \cong \overline{AB}$.*

Proof. (I: Existence) There are four cases:

(Case 1: $A = C$ and $B \in \overrightarrow{CD}$.) Then $P = B$.

(Case 2: $A = C$ and B–C–D.) Let Q be a member of $\mathcal{P} \setminus \overleftrightarrow{AB}$. Using Theorem NEUT.26 let $\mathcal{M}$ be the line of symmetry of $\angle BCQ$ and let $\mathcal{N}$ be the line of symmetry of $\angle DCQ$. Let $\alpha = \mathcal{R}_{\mathcal{N}} \circ \mathcal{R}_{\mathcal{M}}$. By Theorem NEUT.20, Theorem NEUT.15(3), and Definition NEUT.1(A), $\alpha(\overrightarrow{CB}) = \mathcal{R}_{\mathcal{N}}(\mathcal{R}_{\mathcal{M}}(\overrightarrow{CB})) = \mathcal{R}_{\mathcal{N}}(\overrightarrow{CQ}) = \overrightarrow{CD}$ and $\alpha(\overline{CB}) = \overline{\alpha(C)\alpha(B)} = \overline{C\alpha(B)}$. By Definition NEUT.3(B) $\overline{C\alpha(B)} \cong \overline{CB}$. Letting $P = \alpha(B)$ we have $P \in \overrightarrow{CD}$ and $\overline{CP} \cong \overline{CB} \cong \overline{AB}$.

(Case 3: $A = C$ and $D \in (\mathcal{P} \setminus \overleftrightarrow{AB})$.) Using Theorem NEUT.26 let $\mathcal{L}$ be the line of symmetry of $\angle BAD$. By Theorem NEUT.20 $C \in \mathcal{L}$ and $\mathcal{R}_{\mathcal{L}}(\overrightarrow{CB}) = \overrightarrow{CD}$. By Theorem NEUT.15(5) and Definition NEUT.1(A),

$$\mathcal{R}_{\mathcal{L}}(\overline{CB}) = \overline{\mathcal{R}_{\mathcal{L}}(C)\mathcal{R}_{\mathcal{L}}(B)} = \overline{C\mathcal{R}_{\mathcal{L}}(B)}.$$

By Theorem NEUT.20 $\overrightarrow{C\mathcal{R}_{\mathcal{L}}(B)} = \overrightarrow{CD}$ so by Theorem PSH.24 $\mathcal{R}_{\mathcal{L}}(B) \in \overrightarrow{CD}$. By Theorem NEUT.15(5) and Definition NEUT.1(A),

$$\mathcal{R}_{\mathcal{L}}(\overline{CB}) = \overline{\mathcal{R}_{\mathcal{L}}(C)\mathcal{R}_{\mathcal{L}}(B)} = \overline{C\mathcal{R}_{\mathcal{L}}(B)}.$$

By Definition NEUT.3(B) $\overline{C\mathcal{R}_{\mathcal{L}}(B)} \cong \overline{CB}$. Let $P = \mathcal{R}_{\mathcal{L}}(B)$, then $P \in \overrightarrow{CD}$ and $\overline{CP} \cong \overline{CB} \cong \overline{AB}$.

(Case 4: $A \neq C$.) Using Theorem NEUT.52 let $\mathcal{J}$ be the line of symmetry of $\overline{AC}$, then $\mathcal{R}_{\mathcal{J}}(A) = C$ and $\mathcal{R}_{\mathcal{J}}(C) = A$. Let $\mathcal{R}_{\mathcal{J}}(B) = B'$. By Theorem NEUT.15(5) $\mathcal{R}_{\mathcal{J}}(\overline{AB}) = \overline{\mathcal{R}_{\mathcal{J}}(A)\mathcal{R}_{\mathcal{J}}(B)} = \overline{CB'}$. By Definition NEUT.3(B) $\overline{AB} \cong \overline{CB'}$. By whichever applies, case 1, case 2, or case 3, there exists a point P such that $P \in \overrightarrow{CD}$ and $\overline{CB'} \cong \overline{CP}$. By Theorem NEUT.14 (congruence is an equivalence relation) $\overline{AB} \cong \overline{CP}$.

(II: Uniqueness) Let P and P' be points such that $P \in \overrightarrow{CD}$, $P' \in \overrightarrow{CD}$, $\overline{CP} \cong \overline{AB}$, and $\overline{CP'} \cong \overline{AB}$. By Theorem NEUT.14 $\overline{CP'} \cong \overline{CP}$. By Property R.4 of Definition NEUT.2, $P' = P$. □

Theorem NEUT.68 (Angle construction). *Let A, B, and C be noncollinear points on the neutral plane $\mathcal{P}$, P and Q be distinct points on $\mathcal{P}$, $\mathcal{H}$ be a side of $\overleftrightarrow{PQ}$. Then there exists a point R belonging to $\mathcal{H}$ such that $\angle QPR \cong \angle BAC$. Furthermore, if S is a member of $\mathcal{H}$ such that $\angle QPS \cong \angle BAC$, then $\overrightarrow{PR} \cong \overrightarrow{PS}$ and $\angle QPR = \angle QPS$.*

Proof. (I: Existence) Let $\mathcal{H}'$ be the side of $\overleftrightarrow{PQ}$ opposite $\mathcal{H}$. There are six cases.

(Case 1: $\overrightarrow{PQ} = \overrightarrow{AB}$ and $C \in \mathcal{H}$.) Take $R = C$.

(Case 2: $\overrightarrow{PQ} = \overrightarrow{AB}$ and $C \in \mathcal{H}'$.) Take $R = \mathcal{R}_{\overleftrightarrow{AB}}(C)$; then by Theorem NEUT.14, Definition NEUT.1(A) and Definition PSH.29,

$$\begin{aligned}\mathcal{R}_{\overleftrightarrow{AB}}(\angle BAC) &= \angle \mathcal{R}_{\overleftrightarrow{AB}}(B)\mathcal{R}_{\overleftrightarrow{AB}}(A)\mathcal{R}_{\overleftrightarrow{AB}}(C)\\ &= \angle BAR = \overrightarrow{AB} \cup \overrightarrow{AR} = \overrightarrow{PQ} \cup \overrightarrow{PR} = \angle QPR.\end{aligned}$$

By Definition NEUT.3(B) $\angle BAC \cong \angle QPR$. Furthermore, by Definition NEUT.1(B), R and C are on opposite sides of $\overleftrightarrow{AB} = \overleftrightarrow{PQ}$. By Theorem PSH.12 (plane separation) $R \in \mathcal{H}$.

(Case 3: $P = A$, Q–P–B, and $C \in \mathcal{H}'$.) Then $\overrightarrow{AB}$ and $\overrightarrow{PQ}$ are opposite rays. Using Theorem NEUT.26 let $\mathcal{M}$ be the line of symmetry of $\angle QPC$. Let C' be a point such that C–P–C'. Then $\angle QPC$ and $\angle BPC'$ are vertical angles and by Exercise NEUT.10 $\mathcal{M}$ is the line of symmetry for $\angle BPC'$. Then by Theorem NEUT.20 $\mathcal{R}_{\mathcal{M}}(\overrightarrow{PB}) = \overrightarrow{PC'}$ and $\mathcal{R}_{\mathcal{M}}(\overrightarrow{PC}) = \overrightarrow{PQ}$.

Let $R = \mathcal{R}_{\mathcal{M}}(B)$; then $\overrightarrow{PR} = \overrightarrow{PC'}$, and by Theorem PSH.24 $R \in \overrightarrow{PC'}$. By Definition IB.11 C and C' are on opposite sides of $\overleftrightarrow{PQ}$. Since $C \in \mathcal{H}'$, $C' \in \mathcal{H}$. By Theorem PSH.38 $\overrightarrow{PC} = \overleftrightarrow{CC'} \cap \mathcal{H}'$ and $\overrightarrow{PC'} = \overleftrightarrow{CC'} \cap \mathcal{H}$. Therefore $R = \mathcal{R}_{\mathcal{M}}(B) \in \mathcal{H}$.

By Definition PSH.29, Definition NEUT.1(A), and Theorem NEUT.15(3),

$$\begin{aligned}\mathcal{R}_{\mathcal{M}}(\angle BPC) &= \mathcal{R}_{\mathcal{M}}(\overrightarrow{PC} \cup \overrightarrow{PB}) = \mathcal{R}_{\mathcal{M}}(\overrightarrow{PC}) \cup \mathcal{R}_{\mathcal{M}}(\overrightarrow{PB})\\ &= \mathcal{R}_{\mathcal{M}}(\overrightarrow{PC}) \cup \overrightarrow{\mathcal{R}_{\mathcal{M}}(P)\mathcal{R}_{\mathcal{M}}(B)} = \mathcal{R}_{\mathcal{M}}(\overrightarrow{PC}) \cup \overrightarrow{PR}\\ &= \overrightarrow{PQ} \cup \overrightarrow{PR} = \angle QPR,\end{aligned}$$

so that $\angle BAC = \angle BPC \cong \angle QPR$.

(Case 4: $P = A$, Q–P–B, and $C \in \mathcal{H}$.) By Case 3 there exists a point R' belonging to $\mathcal{H}'$ such that $\angle BAC \cong \angle QPR'$. Let $R = \mathcal{R}_{\overleftrightarrow{PQ}}(R')$. Then by Theorem NEUT.15(8) and Definition NEUT.1(A),

$$\mathcal{R}_{\overleftrightarrow{PQ}}(\angle QPR') = \angle \mathcal{R}_{\overleftrightarrow{PQ}}(Q)\mathcal{R}_{\overleftrightarrow{PQ}}(P)\mathcal{R}_{\overleftrightarrow{PQ}}(R') = \angle QPR.$$

By Definition NEUT.3(B) $\angle QPR' \cong \angle QPR$. Since $\angle BAC \cong \angle QPR'$, by Theorem NEUT.14 (congruence is an equivalence relation), $\angle BAC \cong \angle QPR$. By Definition NEUT.1(B), R and R' are on opposite sides of $\overleftrightarrow{PQ}$. Since $R' \in \mathcal{H}'$ by Theorem PSH.12 (plane separation), $R \in \mathcal{H}$.

(Case 5: $P = A$ and $Q \in (\mathcal{P} \setminus \overleftrightarrow{AB})$.) Using Theorem NEUT.26 let $\mathcal{L}$ be the line of symmetry of $\angle BAQ$. By Theorem NEUT.20 $A = P \in \mathcal{L}$, and $\mathcal{R}_\mathcal{L}(\overrightarrow{PB}) = \overrightarrow{PQ}$. By Theorem NEUT.15(5) and Definition NEUT.1(A),

$$\mathcal{R}_\mathcal{L}(\overrightarrow{PB}) = \overrightarrow{\mathcal{R}_\mathcal{L}(P)\mathcal{R}_\mathcal{L}(B)} = \overrightarrow{P\mathcal{R}_\mathcal{L}(B)}.$$

By Theorem NEUT.20 $\overrightarrow{P\mathcal{R}_\mathcal{L}(B)} = \overrightarrow{PQ}$, by Theorem PSH.24(B), $\mathcal{R}_\mathcal{L}(B) \in \overrightarrow{AQ}$. By Definition PSH.29, Theorem NEUT.15(3), and Definition NEUT.1(A),

$$\begin{aligned}\mathcal{R}_\mathcal{L}(\angle BPC) &= \mathcal{R}_\mathcal{L}(\overrightarrow{PB} \cup \overrightarrow{PC}) = \mathcal{R}_\mathcal{L}(\overrightarrow{PB}) \cup \mathcal{R}_\mathcal{L}(\overrightarrow{PC}) \\ &= \overrightarrow{PQ} \cup \overrightarrow{\mathcal{R}_\mathcal{L}(P)\mathcal{R}_\mathcal{L}(C)} = \overrightarrow{PQ} \cup \overrightarrow{P\mathcal{R}_\mathcal{L}(C)} = \angle QP\mathcal{R}_\mathcal{L}(C).\end{aligned}$$

By Definition NEUT.3(B) $\angle BPC \cong \angle QP\mathcal{R}_\mathcal{L}(C)$. If $\mathcal{R}_\mathcal{L}(C) \in \mathcal{H}$, then by case 1 we take $R = \mathcal{R}_\mathcal{L}(C)$. If $\mathcal{R}_\mathcal{L}(C) \in \mathcal{H}'$, then by case 2 there exists a point R belonging to $\mathcal{H}$ such that $\angle QPR \cong \angle QP\mathcal{R}_\mathcal{L}(C)$. Since $\angle QP\mathcal{R}_\mathcal{L}(C) \cong \angle BPC$, by Theorem NEUT.14 $\angle QPR = \angle BPC = \angle BAC$.

(Case 6: $P \neq A$.) Using Theorem NEUT.51 let $\mathcal{N}$ be the line of symmetry of $\overline{PA}$. By Theorem NEUT.15(8) $\mathcal{R}_\mathcal{N}(\angle BAC) = \angle\mathcal{R}_\mathcal{N}(B)\mathcal{R}_\mathcal{N}(A)\mathcal{R}_\mathcal{N}(C) = \angle\mathcal{R}_\mathcal{N}(B)P\mathcal{R}_\mathcal{N}(C)$. By Definition NEUT.3(B) $\angle BAC \cong \angle\mathcal{R}_\mathcal{N}(B)P\mathcal{R}_\mathcal{N}(C)$. By whichever of the cases 1, 2, 3, 4, or 5 applies, there exists a point R belonging to $\mathcal{H}$ such that $\angle QPR \cong \mathcal{R}_\mathcal{N}(B)P\mathcal{R}_\mathcal{N}(C)$. Since $\angle\mathcal{R}_\mathcal{N}(B)P\mathcal{R}_\mathcal{N}(C) \cong \angle BAC$, by Theorem NEUT.14 $\angle QPR \cong \angle BAC$.

(II: Uniqueness) If R and S are points belonging to $\mathcal{H}$ such that $\angle BAC \cong \angle QPR$ and $\angle BAC \cong \angle QPS$, then $\angle QPR \cong \angle QPS$. By Theorem NEUT.36 $\overrightarrow{PR} = \overrightarrow{PS}$, so $\angle QPR = \angle QPS$. □

We are finally ready to prove Euclid's Fourth Postulate.

Theorem NEUT.69. *On a neutral plane $\mathcal{P}$ any two right angles are congruent.*

Proof. Let A, B, C, D, E, and F be points on the neutral plane such that A, B, and C are noncollinear, D, E, and F are noncollinear, $\angle BAC$ is right and $\angle EDF$ is right. By Theorem NEUT.68 there exists a point P belonging to the C-side of $\overleftrightarrow{AB}$ such that $\angle EDF \cong \angle BAP$. By Theorem NEUT.66 $\angle BAP$ is right. By Theorem NEUT.44 $\overleftrightarrow{AB} \perp \overleftrightarrow{AP}$, and $\overleftrightarrow{AB} \perp \overleftrightarrow{AC}$. By Theorem NEUT.47(B), $\overleftrightarrow{AC} = \overleftrightarrow{AP}$. Since C and P belong to the same side of $\overleftrightarrow{AB}$ by Theorem PSH.38(A), $\overrightarrow{AC} = \overrightarrow{AP}$ so $\angle BAC = \angle BAP$. Since $\angle BAP \cong \angle EDF$, $\angle BAC \cong \angle EDF$. □

8.12 Ordering segments and angles

Definition NEUT.70. (A) Let A, B, C, and D be points on the neutral plane $\mathcal{P}$ such that $A \neq B$ and $C \neq D$. Then $\overline{AB} < \overline{CD}$ iff there exists a point P such that $C\text{–}P\text{–}D$ and $\overline{AB} \cong \overline{CP}$. $\overline{AB} > \overline{CD}$ iff $\overline{CD} < \overline{AB}$.

(B) Let A, B, C, D, E, and F be points on the neutral plane $\mathcal{P}$ such that A, B, and C are noncollinear, and D, E, and F are noncollinear. Then $\angle BAC < \angle EDF$ iff there exists a point P belonging to ins $\angle EDF$ such that $\angle BAC \cong \angle EDP$. $\angle BAC > \angle EDF$ iff $\angle EDF < \angle BAC$.

(C) The symbol $<$ is read "is **smaller than**."

(D) The symbol $>$ is read "is **larger than**."

(E) Let A, B, C, and D be points on the neutral plane $\mathcal{P}$ such that $A \neq B$ and $C \neq D$. Then

(1) $\overline{AB} \leq \overline{CD}$ iff $\overline{AB} < \overline{CD}$ or $\overline{AB} \cong \overline{CD}$, and

(2) $\overline{AB} \geq \overline{CD}$ iff $\overline{CD} \leq \overline{AB}$.

(F) Let A, B, C, D, E, and F be points on the neutral plane $\mathcal{P}$ such that A, B, and C are noncollinear and D, E, and F are noncollinear.

(1) $\angle BAC \leq \angle EDF$ iff $\angle BAC < \angle EDF$ or $\angle BAC \cong \angle EDF$.

(2) $\angle BAC \geq \angle EDF$ iff $\angle EDF \leq \angle BAC$.

(G) The symbol $\leq$ is read "is **smaller than or congruent to**."

(H) The symbol $\geq$ is read "is **larger than or congruent to**."

Theorem NEUT.71. *Let A, B, C, and D be points on the neutral plane $\mathcal{P}$ such that $A \neq B$ and $C \neq D$. Then $\overline{AB} > \overline{CD}$ iff there exists a point V such that $C\text{–}D\text{–}V$ and $\overline{AB} \cong \overline{CV}$.*

Proof. (I) If $\overline{AB} > \overline{CD}$, then by Definition NEUT.70 $\overline{CD} < \overline{AB}$ and there exists a point U such that $A\text{–}U\text{–}B$ and $\overline{CD} \cong \overline{AU}$. By Theorem NEUT.67 (segment construction) there exists a point V belonging to $\overrightarrow{CD}$ such that $\overline{AB} \cong \overline{CV}$. By Theorem NEUT.57 $C\text{–}D\text{–}V$.

(II) Suppose that there exists a point V such that $C\text{–}D\text{–}V$. and $\overline{AB} \cong \overline{CV}$. By Theorem NEUT.67 (segment construction) there exists a point U such that $U \in \overrightarrow{AB}$ and $\overline{AU} \cong \overline{CD}$. By Theorem NEUT.57 $A\text{–}U\text{–}B$. By Definition NEUT.70 $\overline{CD} < \overline{AB}$, i.e., $\overline{AB} > \overline{CD}$. □

Theorem NEUT.72 (Trichotomy for segments). *Let A, B, C, and D be points on the neutral plane $\mathcal{P}$ such that $A \neq B$ and $C \neq D$. Then one and only one of the following statements holds:*

$$(1)\ \overline{AB} \cong \overline{CD}, \qquad (2)\ \overline{AB} < \overline{CD}, \qquad (3)\ \overline{AB} > \overline{CD}.$$

Proof. Using Theorem NEUT.67 (segment construction) let P be the point on $\overrightarrow{CD}$ such that $\overline{AB} \cong \overline{CP}$. By Definition IB.4 one and only one of the following statements (A), (B), or (C) holds:

(A: $P = D$) Then $\overline{AB} \cong \overline{CP} = \overline{CD}$, i.e., $\overline{AB} \cong \overline{CD}$.
(B: C–P–D) Then by Definition NEUT.70 $\overline{AB} < \overline{CD}$.
(C: C–D–P) Then by Definition NEUT.70 $\overline{AB} > \overline{CD}$. □

Theorem NEUT.73 (Transitivity for segments). *Let A, B, C, D, E, and F be points on the neutral plane $\mathcal{P}$ such that $A \neq B$, $C \neq D$, and $E \neq F$.*

(A) *If $\overline{AB} < \overline{CD}$ and $\overline{AB} \cong \overline{EF}$, then $\overline{EF} < \overline{CD}$.*
(B) *If $\overline{AB} < \overline{CD}$ and $\overline{CD} \cong \overline{EF}$, then $\overline{AB} < \overline{EF}$.*
(C) *If $\overline{AB} < \overline{CD}$ and $\overline{CD} < \overline{EF}$, then $\overline{AB} < \overline{EF}$.*
(D) *If $\overline{AB} > \overline{CD}$ and $\overline{AB} \cong \overline{EF}$, then $\overline{EF} > \overline{CD}$.*
(E) *If $\overline{AB} > \overline{CD}$ and $\overline{CD} \cong \overline{EF}$, then $\overline{AB} > \overline{EF}$.*
(F) *If $\overline{AB} > \overline{CD}$ and $\overline{CD} > \overline{EF}$, then $\overline{AB} > \overline{EF}$.*

Proof. (A) If $\overline{AB} < \overline{CD}$, then by Definition NEUT.70 there exists a point J such that C–J–D and $\overline{AB} \cong \overline{CJ}$. Since $\overline{AB} \cong \overline{CJ}$ and $\overline{AB} \cong \overline{EF}$, $\overline{EF} \cong \overline{CJ}$. By Definition NEUT.70 $\overline{EF} < \overline{CD}$.

(B) By Theorem NEUT.72 (trichotomy for segments) one and only one of the following statements is true:

$$(1)\ \overline{AB} \cong \overline{EF}, \qquad (2)\ \overline{AB} < \overline{EF}, \qquad (3)\ \overline{AB} > \overline{EF}.$$

If $\overline{AB} \cong \overline{EF}$ were true, then since $\overline{CD} \cong \overline{EF}$, we would have $\overline{AB} \cong \overline{CD}$, contrary to the fact that $\overline{AB} < \overline{CD}$. If $\overline{AB} > \overline{EF}$ were true, then by part (A) $\overline{CD} < \overline{AB}$ would be true, contrary to the fact that $\overline{AB} < \overline{CD}$. Since $\overline{AB} \cong \overline{EF}$ is false and $\overline{AB} > \overline{EF}$ is false, by Theorem NEUT.72 (trichotomy for segments), $\overline{AB} < \overline{EF}$.

(C) Since $\overline{AB} < \overline{CD}$ and $\overline{CD} < \overline{EF}$, by Definition NEUT.70 there exists a point P such that C–P–D and $\overline{AB} \cong \overline{CP}$ and there exists a point Q such that E–Q–F and $\overline{CD} \cong \overline{EQ}$. By Theorem NEUT.56 there exists an isometry α of $\mathcal{P}$ such that $\alpha(\overline{CD}) = \overline{EQ}$, $\alpha(C) = E$, and $\alpha(D) = Q$. Since

C–P–D by Definition NEUT.1(D) E–$\alpha(P)$–Q. Since E–$\alpha(P)$–Q and E–Q–F, by Theorem PSH.12 E–$\alpha(P)$–Q–F so that E–$\alpha(P)$–F. By Theorem NEUT.15(5) $\alpha(\overline{CP}) = \overline{\alpha(C)\alpha(P)} = \overline{E\alpha(P)}$. By Definition NEUT.3(B) $\overline{CP} \cong \overline{E\alpha(P)}$. Since $\overline{AB} \cong \overline{CP}$ and $\overline{CP} \cong \overline{E\alpha(P)}$, $\overline{AB} \cong \overline{E\alpha(P)}$. By Definition NEUT.70 $\overline{AB} < \overline{EF}$. This completes the proof of (C).

In the next three parts, we use the fact, from Definition NEUT.70(A), that for any $A \neq B$ and $C \neq D$, $\overline{AB} > \overline{CD}$ iff $\overline{CD} < \overline{AB}$.

(D) Interchange $\overline{AB}$ and $\overline{CD}$ in part (B) and we have (D).
(E) Interchange $\overline{AB}$ and $\overline{CD}$ in part (A) and we have (E).
(F) Interchange $\overline{AB}$ and $\overline{EF}$ in part (C) and we have (F). □

Theorem NEUT.74. *Let A and B be distinct points on a neutral plane $\mathcal{P}$, and let O and Q be distinct points on $\mathcal{P}$ such that $\overline{OQ} \cong \overline{AB}$; then for any $X \in \overrightarrow{OQ}$,*

(1) O–X–Q iff $\overline{OX} < \overline{OQ} \cong \overline{AB}$ and (2) O–Q–X iff $\overline{OX} > \overline{OQ} \cong \overline{AB}$.

Proof. (1) If O–X–Q, then by Definition NEUT.70 $\overline{OX} < \overline{OQ}$. (Theorem NEUT.73 (transitivity for segments) says that $\overline{OX} < \overline{AB}$.)

Conversely, if $\overline{OX} < \overline{OQ} \cong \overline{AB}$ (cf Definition NEUT.70), there exists a point U such that O–U–Q and $\overline{OX} \cong \overline{OU}$; by Property 4 of Definition NEUT.2, $X = U$, so O–X–Q.

(2) If O–Q–X, then by Theorem NEUT.71 $\overline{OX} > \overline{OQ} \cong \overline{AB}$. (By transitivity for segments $\overline{OX} > \overline{AB}$.)

Conversely, if $\overline{OX} > \overline{OQ} \cong \overline{AB}$ by Theorem NEUT.71 there exists a point V such that O–Q–V and $\overline{OX} \cong \overline{OV}$; by Property 4 of Definition NEUT.2, $X = V$, so O–Q–X. □

Theorem NEUT.75 (Trichotomy for angles). *Let A, B, C, D, E, and F be points on the neutral plane $\mathcal{P}$ such that A, B, and C are noncollinear and D, E, and F are noncollinear. Then one and only one of the following statements is true:*

(1) $\angle BAC \cong \angle EDF$, (2) $\angle BAC < \angle EDF$, (3) $\angle BAC > \angle EDF$.

Proof. Using Theorem NEUT.68 (angle construction) let P be a point such that $\overrightarrow{EP}$ is a subset of the F-side of $\overleftrightarrow{DE}$ and $\angle BAC \cong \angle EDP$. By Exercise PSH.32 one and only one of the following statements is true:

(A) $P \in \overrightarrow{DF}$, (B) $P \in \text{ins}\,\angle EDF$, (C) $F \in \text{ins}\,\angle EDP$.

If $P \in \overrightarrow{DF}$, then by Theorem PSH.30 $\angle EDP = \angle EDF$ and by Remark NEUT.7(A) $\angle EDP \cong \angle EDF$. By Theorem NEUT.14, since $\angle BAC \cong \angle EDP$, $\angle BAC \cong \angle EDF$.

If $P \in \text{ins}\,\angle EDF$, then since $\angle BAC \cong \angle EDP$, by Definition NEUT.70, $\angle BAC < \angle EDF$.

If $F \in \text{ins}\,\angle EDP$, then since $\angle BAC \cong \angle EDP$, by Definition NEUT.70, $\angle EDF < \angle BAC$, i.e. $\angle BAC > \angle EDF$. □

Theorem NEUT.76 (Transitivity for angles). *Let A, B, C, D, E, P, Q, and R be points on the neutral plane $\mathcal{P}$ such that A, B, and C are noncollinear, D, E, and F are noncollinear, and P, Q, and R are noncollinear.*

(A) *If $\angle BAC < \angle EDF$ and $\angle BAC \cong \angle QPR$, then $\angle QPR < \angle EDF$.*
(B) *If $\angle BAC < \angle EDF$ and $\angle EDF \cong \angle QPR$, then $\angle BAC < \angle QPR$.*
(C) *If $\angle BAC < \angle EDF$ and $\angle EDF < \angle QPR$, then $\angle BAC < \angle QPR$.*
(D) *If $\angle BAC > \angle EDF$ and $\angle BAC \cong \angle QPR$, then $\angle QPR > \angle EDF$.*
(E) *If $\angle BAC > \angle EDF$ and $\angle EDF \cong \angle QPR$, then $\angle BAC > \angle QPR$.*
(F) *If $\angle BAC > \angle EDF$ and $\angle EDF > \angle QPR$, then $\angle BAC > \angle QPR$.*

Proof. (A) If $\angle BAC < \angle EDF$, then by Definition NEUT.70 there exists a point S such that $S \in \text{ins}\,\angle EDF$ and $\angle BAC \cong \angle EDS$. Since $\angle BAC \cong \angle QPR$, $\angle QPR \cong \angle EDS$. Hence by Definition NEUT.70 $\angle QPR < \angle EDF$.

(B) If $\angle BAC < \angle EDF$, then there exists a point T such that $T \in \text{ins}\,\angle EDF$ and $\angle BAC \cong \angle EDT$. Since $\angle EDF \cong \angle QPR$, by Theorem NEUT.38 there exists an isometry φ such that $\varphi(\overrightarrow{DE}) = \overrightarrow{PQ}$ and $\varphi(\overrightarrow{DF}) = \overrightarrow{PR}$. Then by Theorem NEUT.15(11) $\varphi(\text{ins}\,\angle EDF) = \text{ins}\,\angle\varphi(E)\varphi(D)\varphi(F) = \text{ins}\,\angle QPR$. Let $U = \varphi(T)$. Then since $T \in \text{ins}\,\angle EDF$, $U \in \text{ins}\,\angle QPR$, and $\varphi(\angle EDT) = \angle QPU$ so that $\angle EDT \cong \angle QPU$ by Definition NEUT.3(B). Since congruence is an equivalence relation (Theorem NEUT.14), $\angle BAC \cong \angle QPU$, thus $\angle BAC < \angle QPR$.

(C) Since $\angle BAC < \angle EDF$, by Definition NEUT.70, there exists a point V such that $V \in \text{ins}\,\angle EDF$ and $\angle BAC \cong \angle EDV$. Since $\angle EDF < \angle QPR$, by Definition NEUT.70 there exists a point W such that $W \in \text{ins}\,\angle QPR$ and $\angle EDF \cong \angle QPW$.

By Theorem NEUT.38 there exists an isometry φ such that $\varphi(\overrightarrow{DE}) = \overrightarrow{PQ}$, $\varphi(\overrightarrow{DF}) = \overrightarrow{PW}$. By Theorem NEUT.15(11)

$$\varphi(\text{ins}\,\angle EDF) = \text{ins}\,\angle\varphi(E)\varphi(D)\varphi(F) = \text{ins}\,\angle QPW.$$

Let $X = \varphi(V)$. Then since $V \in \text{ins}\,\angle EDF$, $X \in \text{ins}\,\angle QPW$ which is a subset of ins $\angle QPR$ by Theorem PSH.41(D) (or Exercise PSH.18), so that $X \in \text{ins}\,\angle QPR$.

Now by Definition PSH.29 and Theorem NEUT.15(3)

$$\varphi(\angle EDV) = \varphi(\overrightarrow{DE} \cup \overrightarrow{DV}) = \varphi(\overrightarrow{DE}) \cup \varphi(\overrightarrow{DV}) = \overrightarrow{PQ} \cup \overrightarrow{PX} = \angle QPX.$$

Therefore by Definition NEUT.3(B) $\angle EDV \cong \angle QPX$, and since $\angle BAC \cong \angle EDV$, by Theorem NEUT.14 $\angle BAC \cong \angle QPX$.

We have seen that $X \in \text{ins}\,\angle QPR$ so that by Definition NEUT.70 $\angle BAC < \angle QPR$; this completes the proof of (C).

In the next three parts we will use the fact, from Definition NEUT.70(B), that for any noncollinear points A, B, and C, and any noncollinear points D, E, and F, $\angle BAC > \angle EDF$ iff $\angle EDF < \angle BAC$.

(D) Interchange $\angle BAC$ and $\angle EDF$ in part (B) and we have (D).

(E) Interchange $\angle BAC$ and $\angle EDF$ in part (A) and we have (E).

(F) Interchange $\angle BAC$ and $\angle QPR$ in part (C) and we have (F). □

Theorem NEUT.77. *Let A, B, C, A', B', and C' be points on the neutral plane $\mathcal{P}$ such that A, B, and C are noncollinear and A', B', and C' are noncollinear. Then $\angle B'A'C' > \angle BAC$ iff there exists a point D such that $B \in$ ins $\angle CAD$ and $\angle CAD \cong \angle C'A'B'$.*

Proof. (I) If $\angle C'A'B' > \angle CAB$, then by Definition NEUT.70 $\angle CAB < \angle C'A'B'$ and there exists a point E belonging to ins $\angle C'A'B'$ such that $\angle CAB \cong \angle C'A'E$. By Theorem NEUT.68 (angle construction) there exists a point D belonging to the B-side of $\overleftrightarrow{AC}$ such that $\angle CAD \cong \angle C'A'B'$. By Exercise NEUT.14 $B \in \text{ins}\,\angle CAD$.

(II) Conversely, if such a point D exists, then by Theorem NEUT.38 there exists an isometry φ such that $\varphi(\angle CAD) = \angle C'A'B'$, $\varphi(\overrightarrow{AC}) = \overrightarrow{A'C'}$ and $\varphi(\overrightarrow{AD}) = \overrightarrow{A'B'}$. $B \in \text{ins}\,\angle CAD$ so by Theorem NEUT.15(11) $\varphi(B) \in \text{ins}\,\angle C'A'B'$, and $\varphi(\angle CAB) = \angle C'A'\varphi(B)$ so that $\angle CAB \cong \angle C'A'\varphi(B)$. By Definition NEUT.70 $\angle CAB < \angle C'A'B'$, i.e., $\angle C'A'B' > \angle CAB$ or $\angle B'A'C' > \angle BAC$. □

Theorem NEUT.78. *Let A, B, C, and D be points on the neutral plane $\mathcal{P}$ such that A, B, and C are noncollinear, $D \in \overrightarrow{ABC}$ and $\angle BAD < \angle BAC$. Then $D \in$ ins $\angle BAC$.*

Proof. By Definition NEUT.70 there exists a point D' belonging to ins $\angle BAC$ such that $\angle BAD' \cong \angle BAD$. By Definition PSH.36 D and D' both belong to $\overrightarrow{ABC}$. By Theorem NEUT.36 $\overrightarrow{AD} = \overrightarrow{AD'}$. By Theorem PSH.24 $D \in \overrightarrow{AD'}$. By Exercise PSH.17(A) $\overrightarrow{AD} = \overrightarrow{AD'} \subseteq$ ins $\angle BAC$. Thus $D \in$ ins $\angle BAC$. □

Definition NEUT.79. Let $\mathcal{T} = \triangle ABC$ be a triangle on the neutral plane $\mathcal{P}$; let B' and C' be points such that B–A–B' and C–A–C'. Then each of the angles $\angle CAB'$ or $\angle BAC'$ is said to be an **outside angle with corner** A of $\mathcal{T}$.

Remark NEUT.79.1. By Theorem NEUT.41 (vertical angles), the pair of outside angles of $\mathcal{T}$ with corner A are congruent to each other. By Definition NEUT.79, if $\mathcal{D}$ is an angle of triangle $\mathcal{T}$ and $\mathcal{E}$ is an outside angle of $\mathcal{T}$ with the same corner as $\mathcal{D}$, then $\mathcal{D}$ and $\mathcal{E}$ are *supplementary angles* (cf Definition NEUT.41(B)).

Theorem NEUT.80 (Outside angles). *Let $\mathcal{P}$ be a neutral plane and let $\mathcal{T}$ be a triangle on $\mathcal{P}$. Any angle $\mathcal{E}$ of $\mathcal{T}$ is smaller than an outside angle whose corner is not the corner of $\mathcal{E}$.*

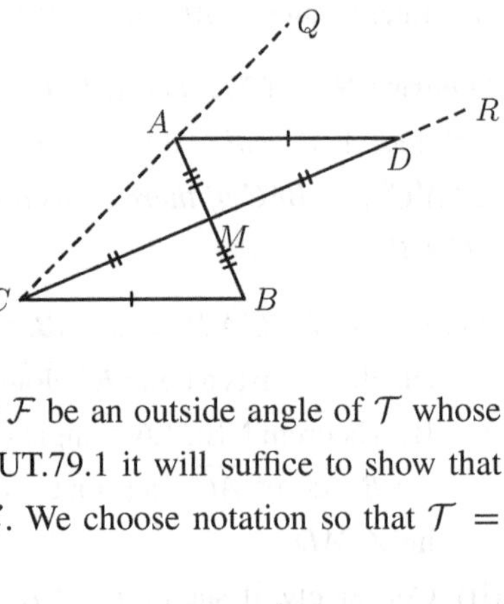

Fig. 8.4 For Theorem NEUT.80.

Proof. For a visualization, see Figure 8.4. Let $\mathcal{F}$ be an outside angle of $\mathcal{T}$ whose corner is not the corner of $\mathcal{E}$. By Remark NEUT.79.1 it will suffice to show that either $\mathcal{F}$ or its vertical angle is greater than $\mathcal{E}$. We choose notation so that $\mathcal{T} = \triangle ABC$, $\mathcal{E} = \angle ABC$, and $\mathcal{F}$ has A as its corner.

In the following we will use Definition IB.4 and associated facts without further reference. By Theorem NEUT.50 let M be the midpoint of $\overline{AB}$; by Property B.3 of Definition IB.1 let D' be a point on $\overrightarrow{CM}$ such that C–M–D', and let Q be a point such that C–A–Q, so that $Q \in \overrightarrow{CA}$.

By Theorem NEUT.67 (segment construction) there exists a point $D \in \overrightarrow{MD'}$ such that $\overline{MD} \cong \overline{MC}$. Again by Property B.3 let R be a point such that M–D–R. Then D, M, and R are members of $\overrightarrow{CM} = \overrightarrow{CR}$, and $D \in \overrightarrow{MD'} = \overrightarrow{MR}$.

By Theorem PSH.45 $R \in \text{ins}\,\angle QAB$; by Definition PSH.36 $R \in \overrightarrow{CAB} = \overrightarrow{QAB}$ and by Theorem PSH.38 $\overrightarrow{CR} \subseteq \overrightarrow{QAB}$. Also, $R \in \overrightarrow{ABQ}$ so by the same theorem $\overrightarrow{MR} \subseteq \overrightarrow{ABQ}$. Since $\overrightarrow{MR} \subseteq \overrightarrow{CR}$ then $\overrightarrow{MR} \subseteq \overrightarrow{AQB}$ and hence by Definition PSH.36 $\overrightarrow{MR} \subseteq \text{ins}\,\angle BAQ$. Therefore $D \in \text{ins}\,\angle BAQ$.

By Definition NEUT.3(C) $\overline{MB} \cong \overline{MA}$. By Theorem NEUT.42 (vertical angles), $\angle CMB \cong \angle DMA$. By Theorem NEUT.64 (EAE) $\triangle BMC \cong \triangle AMD$, so that corresponding angles are congruent; since C corresponds to D and A to B, $\angle ABC = \angle MBC \cong \angle MAD$. Since $D \in \text{ins}\,\angle BAQ$, $\angle ABC < \angle MAQ$ (cf Definition NEUT.70). □

8.13 Acute and obtuse angles

Definition NEUT.81. On a neutral plane an angle is **acute** iff it is smaller than a right angle, an angle is **obtuse** iff it is larger than a right angle.

By Theorem NEUT.75 (trichotomy for angles) an angle is one and only one of the following: (1) right, (2) acute, (3) obtuse.

Theorem NEUT.82. *Let $\mathcal{P}$ be a neutral plane.*

(A) *If an angle on $\mathcal{P}$ is acute, then its supplement is obtuse.*
(B) *If an angle on $\mathcal{P}$ is obtuse, then its supplement is acute.*

Proof. Let A, B, and C be noncollinear points on the neutral plane $\mathcal{P}$ and let B' be a point such that B–A–B'. By Corollary NEUT.46.1 there exists a point R on the C-side of $\overleftrightarrow{BB'}$ such that $\angle BAR$ is right. By Definition NEUT.41(C) $\angle BAR \cong \angle B'AR$ and $\angle B'AR$ is right.

(A) If $\angle BAC$ is acute, then by Definition NEUT.81 $\angle BAC < \angle BAR$, and by Theorem NEUT.78 $C \in \text{ins}\,\angle BAR$. By Exercise PSH.51 $R \in \text{ins}\,\angle B'AC$. By Definition NEUT.70 $\angle B'AR < \angle B'AC$. By Definition NEUT.81, since $\angle B'AR$ is right, $\angle B'AC$ is obtuse.
(B) If $\angle BAC$ is obtuse, then by Definition NEUT.81 $\angle BAC > \angle BAR$, and by Theorem NEUT.78 $R \in \text{ins}\,\angle BAC$. By Exercise PSH.51 $C \in \text{ins}\,\angle B'AR$. By Definition NEUT.70 $\angle B'AC < \angle B'AR$. By Definition NEUT.81, since $\angle B'AR$ is right, $\angle B'AC$ is acute. □

Theorem NEUT.83. *Let $\mathcal{P}$ be a neutral plane.*

(A) *Every angle on $\mathcal{P}$ congruent to an acute angle is acute.*
(B) *Every angle on $\mathcal{P}$ congruent to an obtuse angle is obtuse.*
(C) *Every angle on $\mathcal{P}$ smaller than an acute angle is acute.*
(D) *Every acute angle on $\mathcal{P}$ is smaller than every obtuse angle on $\mathcal{P}$.*

Proof. To prove (A), let $\mathcal{D}$ be an acute angle on $\mathcal{P}$, $\mathcal{E}$ an angle on $\mathcal{P}$ congruent to $\mathcal{D}$ and let $\mathcal{G}$ be a right angle on $\mathcal{P}$. By Definition NEUT.81 $\mathcal{D} < \mathcal{G}$. Since $\mathcal{D} < \mathcal{G}$ and $\mathcal{D} \cong \mathcal{E}$, by Theorem NEUT.76 (transitivity for angles) $\mathcal{E} < \mathcal{G}$. By Definition NEUT.81, $\mathcal{E}$ is acute.

The proof of assertions (B), (C), and (D) is Exercise NEUT.81. □

Theorem NEUT.84. *Let $\mathcal{T}$ be a triangle on the neutral plane $\mathcal{P}$. If an angle of $\mathcal{T}$ is right or is obtuse, then the other angles of $\mathcal{T}$ are acute.*

Proof. Let $\mathcal{E}$ be an angle of $\mathcal{T}$. If $\mathcal{E}$ is right, then by Definition NEUT.41(C) and Remark NEUT.79.1, an outside angle of $\mathcal{T}$ with the same corner as $\mathcal{E}$ is also right. By Theorem NEUT.80 (outside angles) the other angles are smaller than a right angle and are therefore acute by Definition NEUT.81. If $\mathcal{E}$ is obtuse, then the outside angle with the same corner is acute (*cf* Theorem NEUT.82). By Theorem NEUT.80 (outside angles) each of the other angles of $\mathcal{T}$ is acute. □

Definition NEUT.85. Let $\mathcal{T}$ be a triangle on the neutral plane $\mathcal{P}$.

(A) Triangle $\mathcal{T}$ is **right** iff an angle of $\mathcal{T}$ is right.
(B) Triangle $\mathcal{T}$ is **obtuse** iff an angle of $\mathcal{T}$ is obtuse.
(C) Triangle $\mathcal{T}$ is **acute** iff each angle of $\mathcal{T}$ is acute.
(D) An edge of $\mathcal{T}$ is **maximal** iff each of the other edges is smaller than or congruent to it.
(E) An angle of $\mathcal{T}$ is **maximal** iff each of the other two angles is congruent to or smaller than it.

Theorem NEUT.86. *Let A, B, and C be noncollinear points on the neutral plane $\mathcal{P}$. If $\angle ACB$ is a maximal angle of $\triangle ABC$, then each of $\angle ABC$ or $\angle CAB$ is acute.*

Proof. If $\angle ABC$ or $\angle CAB$ were right or obtuse, then $\angle ACB$ would be acute by Theorem NEUT.84 and thus by Definition NEUT.81 and Theorem NEUT.83 would not be a maximal angle of $\triangle ABC$. □

Theorem NEUT.87 (Alternate interior angles). *Let $\mathcal{P}$ be a neutral plane, P and Q be distinct points on $\mathcal{P}$, and T and R be distinct points on $\mathcal{P}$ which are on opposite sides of $\overleftrightarrow{PQ}$ such that $\angle RQP \cong \angle TPQ$. Then $\overleftrightarrow{PT} \parallel \overleftrightarrow{QR}$.*

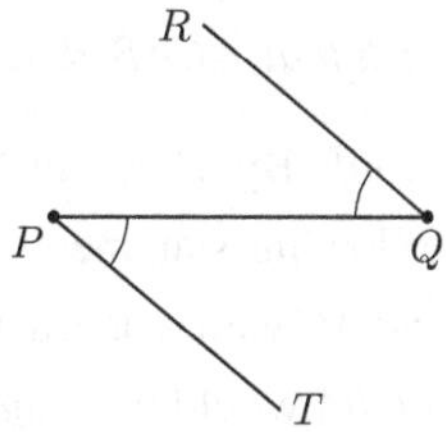

Fig. 8.5 Alternate interior angles.

Proof. For a visualization, see Figure 8.5. If $\overleftrightarrow{PT}$ and $\overleftrightarrow{QR}$ were to intersect at a point U which is on the R-side of $\overleftrightarrow{PQ}$, then by Theorem NEUT.80 (outside angles) $\angle RQP$ would be smaller than $\angle TPQ$ contrary to the fact that these two angles are congruent. If $\overleftrightarrow{PT}$ and $\overleftrightarrow{QR}$ were to intersect at a point U which is on the T-side of $\overleftrightarrow{PQ}$, then by Theorem NEUT.80 (outside angles) $\angle TPQ$ would be smaller than $\angle RQP$, contrary to the fact that $\angle RQP \cong \angle TPQ$. Hence $\overleftrightarrow{PT} \parallel \overleftrightarrow{QR}$. □

Remark NEUT.88. Theorem NEUT.87 is a generalization of Theorem NEUT.47(A). We next use Theorem NEUT.87 to give a second proof for Theorem NEUT.48(B) (Property PE), which is Theorem NEUT.89. See also Remark NEUT.49.

Theorem NEUT.89. *Let $\mathcal{P}$ be a neutral plane, $\mathcal{L}$ be a line on $\mathcal{P}$, and let Q be a member of $\mathcal{P} \setminus \mathcal{L}$. Then there exists a line $\mathcal{M}$ such that $Q \in \mathcal{M}$ and $\mathcal{M} \parallel \mathcal{L}$.*

Proof. Let P and T be distinct points on $\mathcal{L}$ and let $\mathcal{H}$ be the side of $\overleftrightarrow{PQ}$ opposite T. By Theorem NEUT.68 (angle construction) there exists a point R belonging to $\mathcal{H}$ such that $\angle PQR \cong \angle QPT$. Let $\mathcal{M} = \overleftrightarrow{QR}$. By Theorem NEUT.87 $\mathcal{M} \parallel \mathcal{L}$. □

Theorem NEUT.90. *Let A, B, and C be noncollinear points on the neutral plane $\mathcal{P}$ such that $\overline{BC} < \overline{AC}$; then $\angle CAB < \angle CBA$.*

Proof. Using Theorem NEUT.67 (segment construction), let B' be the point on $\overrightarrow{CB}$ such that $\overline{B'C} \cong \overline{AC}$, so that $\overline{BC} < \overline{AC} \cong \overline{B'C}$ and by Theorem NEUT.73 (transitivity for segments), $\overline{BC} < \overline{B'C}$; by Definition NEUT.70(A) C–B–B', so that by Theorem PSH.37 $B \in \text{ins}\,\angle CAB'$, and by Definition NEUT.70(B), $\angle CAB < \angle CAB'$.

By Theorem NEUT.40(A) (*Pons Asinorum*) $\angle CAB' \cong \angle CB'A$. By Definition NEUT.79 $\angle ABC$ is an outside angle at corner B for $\triangle ABB'$. By Theorem NEUT.80 (outside angles) $\angle CB'A = \angle BB'A < \angle ABC$. Then $\angle CAB < \angle CAB' \cong \angle CB'A < \angle ABC$ so that by Theorem NEUT.76 (transitivity for angles) $\angle CAB < \angle ABC$. □

Theorem NEUT.91. *If A, B, and C are noncollinear points on a neutral plane $\mathcal{P}$ such that $\angle CAB < \angle CBA$, then $\overline{BC} < \overline{AC}$.*

Proof. By Theorem NEUT.72 (trichotomy for segments) one and only one of the following statements holds: (1) $\overline{BC} \cong \overline{AC}$, (2) $\overline{BC} < \overline{AC}$, (3) $\overline{BC} > \overline{AC}$. If $\overline{BC}$ and $\overline{AC}$ were congruent, then by Theorem NEUT.40A (*Pons Asinorum*), $\angle CAB$ and $\angle CBA$ would be congruent contrary to the fact that $\angle CAB$ is smaller than $\angle CBA$ (cf Theorem NEUT.75 (trichotomy for angles)). If $\overline{BC}$ were larger than $\overline{AC}$, then $\overline{AC}$ would be smaller than $\overline{BC}$ and by Theorem NEUT.90 $\angle CBA$ would be smaller than $\angle CAB$ contrary to the fact that $\angle CAB$ is smaller than $\angle CBA$. Hence $\overline{BC} < \overline{AC}$. □

Theorem NEUT.92. *Let $\mathcal{P}$ be a neutral plane and let A, B, and C be noncollinear points on $\mathcal{P}$. Then $\overline{AB}$ is a maximal edge of $\triangle ABC$ iff $\angle ACB$ is a maximal angle of $\triangle ABC$.*

Proof. (I: If $\overline{AB}$ is a maximal edge of $\triangle ABC$, then $\angle ACB$ is a maximal angle of $\triangle ABC$.) If $\overline{AB} > \overline{BC}$ by Theorem NEUT.90 $\angle ACB > \angle BAC$; if $\overline{AB} \cong \overline{BC}$ by Theorem NEUT.40(A) (*Pons Asinorum*) $\angle ACB \cong \angle BAC$. Therefore if $\overline{AB} \geq \overline{BC}$, then $\angle ACB \geq \angle BAC$. A similar argument shows that if $\overline{AB} \geq \overline{AC}$ then $\angle ACB \geq \angle ABC$. Therefore by Definition NEUT.85 if $\overline{AB}$ is maximal, $\angle ACB$ is maximal.

(II: If $\angle ACB$ is a maximal angle of $\triangle ABC$, then $\overline{AB}$ is a maximal edge of $\triangle ABC$.) If $\overline{AB}$ is a not a maximal edge of $\triangle ABC$, then $\overline{AB} < \overline{AC}$ or $\overline{AB} < \overline{BC}$. If $\overline{AB} < \overline{AC}$, then by Theorem NEUT.90 $\angle ACB < \angle ABC$ so that $\angle ACB$ is not a maximal angle of $\triangle ABC$. Similarly, if $\overline{AB} < \overline{BC}$, $\angle ACB < \angle BAC$ so that $\angle ACB$ is not maximal. By the contrapositive, if $\angle ACB$ is a maximal angle, $\overline{AB}$ is a maximal edge. □

Theorem NEUT.93. *Let $\mathcal{P}$ be a neutral plane and let A, B, and C be noncollinear points on $\mathcal{P}$. If $\angle ACB$ is right or is obtuse, then $\overline{AB} > \overline{AC}$ and $\overline{AB} > \overline{BC}$.*

Proof. Exercise NEUT.17. □

Definition NEUT.94. Let $\mathcal{P}$ be a neutral plane and let $\mathcal{T}$ be a right triangle on $\mathcal{P}$. The **hypotenuse** of $\mathcal{T}$ is the maximal edge of $\mathcal{T}$. The other two edges of $\mathcal{T}$ are its **legs**.

Theorem NEUT.95. *Let A, B, and C be noncollinear points on a neutral plane $\mathcal{P}$ such that $\overline{AC} \leq \overline{AB}$. If P is a member of $\overline{BC}$, then $\overline{AP} < \overline{AB}$.*

Proof. By Theorem NEUT.90 $\angle ABP = \angle ABC \leq \angle ACB$. By Theorem NEUT.80 (outside angles) $\angle APB > \angle ACB = \angle ACP$. Then $\angle APB > \angle ACB \geq \angle ABP$ and by Theorem NEUT.76 (transitivity for angles) $\angle ABP < \angle APB$. By Theorem NEUT.91, $\overline{AP} < \overline{AB}$. □

Theorem NEUT.96 (Hypotenuse-leg). *Let A, B, C, D, E, and F be points on a neutral plane $\mathcal{P}$ such that A, B, and C are noncollinear, D, E, and F are noncollinear, $\angle ACB$ is right, $\angle DFE$ is right, $\overline{AB} \cong \overline{DE}$ and $\overline{AC} \cong \overline{DF}$. Then there exists an isometry α of $\mathcal{P}$ such that $\alpha(A) = D$, $\alpha(B) = E$, and $\alpha(C) = F$ so that $\alpha(\triangle ABC) = \triangle DEF$, i.e. $\triangle ABC \cong \triangle DEF$.*

Proof. By Theorem NEUT.58 there exists an isometry α of $\mathcal{P}$ such that $\alpha(\overline{AC}) = \overline{DF}$, $\alpha(A) = D$, $\alpha(C) = F$, and $\alpha(B) \in \overrightarrow{DFE}$. By Theorem NEUT.15(8) $\alpha(\angle ACB) = \angle\alpha(A)\alpha(C)\alpha(B) = \angle DF\alpha(B)$. By Definition NEUT.3(B) $\angle ACB \cong \angle DF\alpha(B)$ and $\overline{AB} \cong \overline{D\alpha(B)}$. By Theorem NEUT.69 $\angle ACB \cong \angle DFE$. By Theorem NEUT.14 (congruence is an equivalence relation) $\angle DF\alpha(B) \cong \angle DFE$. By Theorem NEUT.36 $\overrightarrow{FE} = \overrightarrow{F\alpha(B)}$, so that $\alpha(B) \in \overrightarrow{FE}$. By Theorem NEUT.15(3) $\alpha(\overline{AB}) = \overline{\alpha(A)\alpha(B)} = \overline{D\alpha(B)}$ so that by Definition NEUT.3, $\overline{AB} \cong \overline{D\alpha(B)}$.

By Definition IB.4 there exist three and only three possibilities: F–$\alpha(B)$–E, F–E–$\alpha(B)$, or $\alpha(B) = E$.

(A) Suppose F–$\alpha(B)$–E. By Exercise NEUT.17 $\overline{DF} < \overline{DE}$. By Theorem NEUT.95 $\overline{D\alpha(B)} < \overline{DE} \cong \overline{AB}$, so $\overline{D\alpha(B)} < \overline{AB}$ by Theorem NEUT.73. This contradicts $\overline{AB} \cong \overline{D\alpha(B)}$. Thus F–$\alpha(B)$–E is false.

(B) If F–E–$\alpha(B)$, then By Exercise NEUT.17 $\overline{DF} < \overline{D\alpha(B)}$. By Theorem NEUT.95 $\overline{DE} < \overline{D\alpha(B)} \cong \overline{AB}$. This contradicts our hypothesis that $\overline{DE} \cong \overline{AB}$. Thus F–E–$\alpha(B)$ is false.

Since F–$\alpha(B)$–E and F–E–$\alpha(B)$ are both false, $\alpha(B) = E$. It follows that $\alpha(\overline{BC}) = \overline{(\alpha(B))(\alpha(C))} = \overline{EF}$ and $\alpha(\overline{AB}) = \overline{DE}$ so by Definition IB.7 $\alpha(\triangle ABC) = \triangle DEF$, and $\triangle ABC \cong \triangle DEF$. □

Theorem NEUT.97 (EAA). *Let A, B, C, D, E, and F be points on the neutral plane $\mathcal{P}$ such that A, B, and C are noncollinear, D, E, and F are noncollinear,* $\overline{AB} \cong \overline{DE}$, $\angle ABC \cong \angle DEF$, *and* $\angle ACB \cong \angle DFE$. *Then there exists an isometry* α *of* $\mathcal{P}$ *such that* $\alpha(A) = D$, $\alpha(B) = E$, *and* $\alpha(C) = F$ *so that* $\triangle ABC \cong \triangle DEF$.

Proof. By Theorem NEUT.67 (segment construction) there exists a point F' on $\overrightarrow{EF}$ such that $\overline{EF'} \cong \overline{BC}$. Then $\overline{AB} \cong \overline{DE}$, $\overline{BC} \cong \overline{EF'}$, and $\angle ABC \cong \angle DEF'$, so by Theorem NEUT.64 (EAE), there exists an isometry α such that $\alpha(\triangle ABC) = \triangle DEF'$, $\alpha(A) = D$, $\alpha(B) = E$ and $\alpha(C) = F'$. Then $\triangle ABC \cong \triangle DEF'$, hence $\angle ACB \cong \angle DF'E$. By hypothesis $\angle ACB \cong \angle DFE$, so by Theorem NEUT.14 $\angle DFE \cong \angle DF'E$.

By Definition IB.4 and Property B.2 of Definition IB.1, one and only one of the following possibilities holds: (1) $E–F'–F$, (2) $E–F–F'$, or (3) $F' = F$.

If $E–F'–F$, then $\angle DF'E$ would be an outside angle of $\triangle DF'F$ with corner F'. By Theorem NEUT.80 (outside angles) $\angle DFE < \angle DF'E$, contradicting $\angle DFE \cong \angle DF'E$.

If $E–F–F'$, then $\angle DFE$ would be an outside angle of $\triangle DFF'$ with corner F. By Theorem NEUT.80 (outside angles) $\angle DFE > \angle DF'E$, contradicting $\angle DFE \cong \angle DF'E$.

Hence neither $E–F'–F$ nor $E–F–F'$, so that $F = F'$, i.e. $\alpha(C) = F$ and $\triangle ABC \cong \triangle DEF$. □

Theorem NEUT.98 (Hinge). *Let* $\mathcal{P}$ *be a neutral plane, A, B, C, D, E, and F be points on* $\mathcal{P}$ *such that A, B, and C are noncollinear, D, E, and F are noncollinear,* $\overline{AB} \cong \overline{DE}$, *and* $\overline{AC} \cong \overline{DF}$. *Then* $\overline{EF} < \overline{BC}$ *iff* $\angle EDF < \angle BAC$.

Proof. In this proof we will freely use Theorem NEUT.14 (congruence is an equivalence relation), Theorem NEUT.73 (transitivity for segments), and Theorem NEUT.76 (transitivity for angles) without further reference. For visualizations see Figures 8.6 and 8.7.

(I: If $\angle EDF < \angle BAC$, then $\overline{EF} < \overline{BC}$.) Using Theorem NEUT.68 (angle construction), let V be a member of $\overrightarrow{DFE}$ such that $\angle FDV \cong \angle CAB$. Using Theorem NEUT.67 (segment construction), let G be the point on $\overrightarrow{DV}$ such that $\overline{DG} \cong \overline{AB}$. Let U be a point on $\overrightarrow{DV}$ such that $D–G–U$; then $\angle FDU \cong \angle FDV \cong \angle CAB$. Since $\overline{AB} \cong \overline{DE}$ it follows that $\overline{DG} \cong \overline{DE}$ and by Theorem NEUT.40 (Pons Asinorum) $\angle DGE \cong \angle DEG$. We will use this fact in both Case A and Case C below.

Since $\angle BAC \cong \angle GDF$, $\overline{AC} \cong \overline{DF}$, and $\overline{AB} \cong \overline{DG}$, by Theorem NEUT.64 (EAE) there exists an isomorphism α of $\mathcal{P}$ such that $\alpha(\triangle ABC) = \triangle DGF$, $\alpha(A) = D$, $\alpha(B) = G$, and $\alpha(C) = F$, and corresponding angles and edges are congruent.

By hypothesis, $\angle EDF < \angle BAC \cong \angle GDF$, so $\angle EDF < \angle GDF$. By Theorem NEUT.78 $E \in \text{ins}\,\angle GDF$. By Theorem PSH.39 (Crossbar) $\overrightarrow{DE}$ and $\overline{FG}$ intersect at a point H. By Theorem PSH.37 $\overline{FG} \subseteq \text{ins}\,\angle GEF$, so $H \in \text{ins}\,\angle GEF$. By Definition IB.4 there are three possibilities: (A) D–H–E, (B) $H = E$, (C) D–E–H.

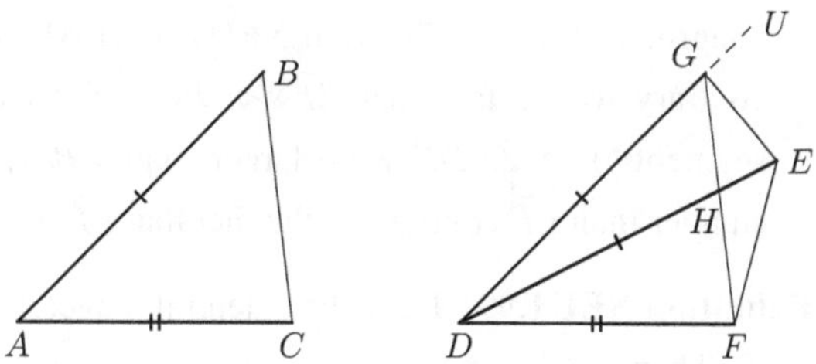

Fig. 8.6 For Theorem NEUT.98 (I) Case A.

(Case A: D–H–E.) By Theorem PSH.37 $H \in \overline{DE} \subseteq \text{ins}\,\angle DGE$, and by Definition NEUT.70 $\overline{DH} < \overline{DE}$ and $\angle HGE = \angle FGE < \angle DGE$.

Again by Theorem PSH.37 $H \in \overline{GF} \subseteq \text{ins}\,\angle FEG$, so by Exercise PSH.17, $\overrightarrow{EH} \subseteq \text{ins}\,\angle FEG$, so $D \in \text{ins}\,\angle FEG$ and by Definition NEUT.70 $\angle DEG < \angle FEG$.

Putting this together, we have $\angle FGE < \angle DGE \cong \angle DEG < \angle FEG$ so that $\angle FGE < \angle FEG$. Then by Theorem NEUT.91 $\overline{FE} < \overline{FG} \cong \overline{BC}$.

(Case B: $H = E$.) Then G–E–F and by Definition NEUT.70 $\overline{EF} < \overline{GF} \cong \overline{BC}$.

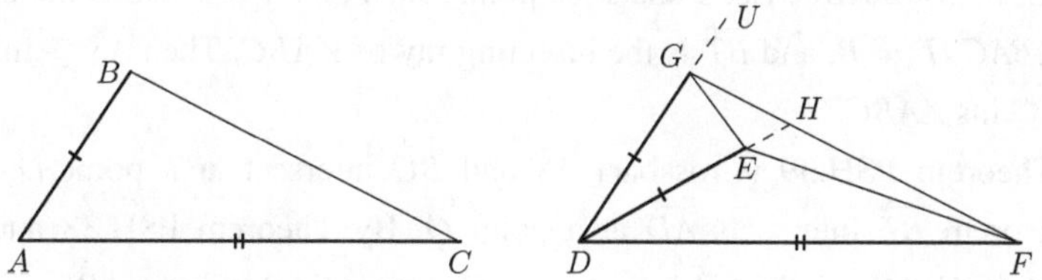

Fig. 8.7 For Theorem NEUT.98 (I) Case C.

(Case C: D–E–H.) By Theorem PSH.37 $H \in \overline{GF} \subseteq \text{ins}\,\angle GEF$, so by Definition NEUT.70 $\angle GEH < \angle GEF$.

Since $D–E–H$ and $D–G–U$, $H \in \overrightarrow{GEU}$. Also, by Theorem PSH.38(A) $H \in \overrightarrow{DE} \subseteq \overrightarrow{GUE}$ and hence $H \in \text{ins}\,\angle EGU = \overrightarrow{GEU} \cap \overrightarrow{GUE}$ and by Definition NEUT.70 $\angle EGF = \angle EGH < \angle EGU$.

Also, by Theorem NEUT.43 (congruence of supplements of congruent angles), since $\angle DGE \cong \angle DEG$ it follows that $\angle EGU \cong \angle GEH$.

Putting this together we have $\angle EGF < \angle EGU \cong \angle GEH < \angle GEF$ so that by Theorem NEUT.91 $\overline{EF} < \overline{GF} \cong \overline{BC}$.

(II: If $\overline{EF} < \overline{BC}$, then $\angle EDF < \angle BAC$.) By Theorem NEUT.75 (trichotomy for angles), one and only one of the following statements holds: (1) $\angle EDF \cong \angle BAC$, (2) $\angle EDF > \angle BAC$, or (3) $\angle EDF < \angle BAC$. If $\angle EDF$ and $\angle BAC$ were congruent, then by Theorem NEUT.64 (EAE) $\overline{EF}$ and $\overline{BC}$ would be congruent contrary to the fact that $\overline{EF} < \overline{BC}$ (cf Theorem NEUT.72 (trichotomy for segments)). If $\angle EDF$ were larger than $\angle BAC$, then by part (I) $\overline{BC}$ would be smaller than $\overline{EF}$ contrary to the fact that $\overline{EF} < \overline{BC}$. Hence $\angle EDF < \angle BAC$. □

Definition NEUT.99. Let $\mathcal{P}$ be a neutral plane, A be a point on $\mathcal{P}$, and $\mathcal{M}$ be a line on $\mathcal{P}$. Then

(A) $\text{pr}(A, \mathcal{M})$ denotes the line $\mathcal{L}$ such that $A \in \mathcal{L}$ and $\mathcal{L} \perp \mathcal{M}$, and
(B) $\text{ftpr}(A, \mathcal{M})$ denotes the point of intersection of $\text{pr}(A, \mathcal{M})$ and $\mathcal{M}$. This point is called the **foot** of the line $\mathcal{L}$.
(C) If A, B, and C are noncollinear points, the **altitude** of $\triangle ABC$ from corner A is $\overline{AD}$ where $D = \text{ftpr}(A, \overleftrightarrow{BC})$. Its **base** (belonging to that altitude) is the edge $\overline{BC}$.

Theorem NEUT.100. *If $\mathcal{P}$ is a neutral plane and $\mathcal{T}$ is any triangle on $\mathcal{P}$, then the bisecting rays of the angles of $\mathcal{T}$ intersect at a point inside the triangle.*

Proof. Let $\mathcal{T} = \triangle ABC$. Let S and T be points such that $S \neq A$, $\overrightarrow{AS}$ is the bisecting ray of $\angle BAC$, $T \neq B$, and $\overrightarrow{BT}$ is the bisecting ray of $\angle ABC$. Then $\overrightarrow{AS} \subseteq \text{ins}\,\angle CAB$ and $\overrightarrow{BT} \subseteq \text{ins}\,\angle ABC$.

By Theorem PSH.39 (Crossbar) $\overrightarrow{AS}$ and $\overline{BC}$ intersect at a point D. By the same theorem $\overrightarrow{BT}$ intersects $\overline{AD}$ at a point O. By Theorem PSH.38 and Theorem PSH.46(C) $\{O\} = \overrightarrow{BT} \cap \overrightarrow{AS} \subseteq \text{ins}\,\angle ABC \cap \text{ins}\,\angle CAB = \text{ins}\,\triangle ABC$.

Let $P = \text{ftpr}(O, \overleftrightarrow{BC})$, $Q = \text{ftpr}(O, \overleftrightarrow{AC})$, and $R = \text{ftpr}(O, \overleftrightarrow{AB})$. Then $\overline{OB} \cong \overline{OB}$ by Remark NEUT.7(A); by Theorem NEUT.69 $\angle ORB \cong \angle OPB$; since $\overrightarrow{BT}$ is the bisecting ray of $\angle ABC$, by Theorem NEUT.39 $\angle OBR \cong \angle OBP$. Thus $\triangle ORB$ and $\triangle OPB$ are right triangles which share a hypotenuse, and have a congruent angle. By Theorem NEUT.97(EAA), $\triangle ORB \cong \triangle OPB$ and hence $\overline{OR} \cong \overline{OP}$.

By similar reasoning, $\overline{OR} \cong \overline{OQ}$, so $\overline{OQ} \cong \overline{OP}$ by Theorem NEUT.14.

By Theorem NEUT.69 $\angle OQC \cong \angle OPC$ since they are both right angles. $\overline{OC} \cong \overline{OC}$ and $\overline{OQ} \cong \overline{OP}$ so by Theorem NEUT.96 (hypotenuse-leg) $\triangle OQC \cong \triangle OPC$. Thus $\angle OCQ \cong \angle OCD$ so by Theorem NEUT.39 $\overrightarrow{CO}$ is the bisecting ray for $\angle ACB$. Therefore all the bisecting rays intersect at the point O. □

In conclusion, we should mention that in the presence of Axiom PW it is possible to prove, independently of Property R.6 of Definition NEUT.2, that every segment $\overline{AB}$ has a midpoint. This proof is part of the online Supplement which may be accessed from the home page for this book at www.springer.com.

We would be most grateful if a reader with more perspicacity than we should come up with such a proof without invoking parallelism, thus making it possible to dispense with Property R.6 of Definition NEUT.2. Alternatively, creation of a model (in which parallelism does not hold) showing that Property R.6 is independent of the other properties would confirm that Property R.6 is essential. This is discussed more in Chapter 21, Section 21.7.3 (Independence of reflection properties) Remark RSI.4.

8.14 Exercises for neutral geometry

Answers to starred (*) exercises may be accessed from the home page for this book at www.springer.com.

Exercise NEUT.0*. There can be more than one mirror mapping over a line in the (real) coordinate plane $\mathbb{R}^2$. More specifically, if for each pair (u_1, u_2) of real numbers on the plane, we define $\Phi(u_1, u_2) = (u_1, -u_2)$ and $\Psi(u_1, u_2) = (u_1 - u_2, -u_2)$, both Φ and Ψ are mirror mappings over the x-axis.

Exercise NEUT.1*. Let $\mathcal{P}$ be a neutral plane and let $\mathcal{L}$ and $\mathcal{M}$ be parallel lines on $\mathcal{P}$, then $\mathcal{R}_{\mathcal{L}}(\mathcal{M})$ is a line which is contained in the side of $\mathcal{L}$ opposite the side containing $\mathcal{M}$ and $\mathcal{M} \parallel \mathcal{R}_{\mathcal{L}}(\mathcal{M})$.

Exercise NEUT.2*. Let $\mathcal{M}$ be any line on the neutral plane $\mathcal{P}$. If X is any point on $\mathcal{P}$ such that $\mathcal{R}_{\mathcal{M}}(X) = X$, then $X \in \mathcal{M}$.

Exercise NEUT.3*. Let $\mathcal{P}$ be a neutral plane and let $\mathcal{L}$ and $\mathcal{M}$ be lines on $\mathcal{P}$. If $\mathcal{R}_{\mathcal{L}} = \mathcal{R}_{\mathcal{M}}$, then $\mathcal{L} = \mathcal{M}$. This may be restated in its contrapositive form as follows: If $\mathcal{L} \neq \mathcal{M}$, then $\mathcal{R}_{\mathcal{L}} \neq \mathcal{R}_{\mathcal{M}}$.

Exercise NEUT.4*. Let A, B, and C be noncollinear points on the neutral plane $\mathcal{P}$, then neither $\overleftrightarrow{AB}$ nor $\overleftrightarrow{AC}$ is a line of symmetry of $\angle BAC$.

Exercise NEUT.5*. Let $\mathcal{S}$ be a nonempty subset of $\mathcal{P}$ which has a line $\mathcal{M}$ of symmetry, $\mathcal{H}_1$ and $\mathcal{H}_2$ be the sides of $\mathcal{M}$, $\mathcal{S}_1 = \mathcal{S} \cap \mathcal{H}_1$ and $\mathcal{S}_2 = \mathcal{S} \cap \mathcal{H}_2$, then $\mathcal{R}_{\mathcal{M}}(\mathcal{S}_2) = \mathcal{S}_1$.

Exercise NEUT.6*. (A) Let α be an isometry of the neutral plane $\mathcal{P}$ and let $\mathcal{L}$ be a line on $\mathcal{P}$ such that every point on $\mathcal{L}$ is a fixed point of α and no point off of $\mathcal{L}$ is a fixed point of α, then $\alpha = \mathcal{R}_{\mathcal{L}}$.

(B) Let α be an isometry of the neutral plane $\mathcal{P}$ which is also an axial affinity with axis $\mathcal{L}$. Then $\alpha = \mathcal{R}_{\mathcal{L}}$.

Exercise NEUT.7*. Let $\mathcal{L}$ and $\mathcal{M}$ be distinct lines on the neutral plane $\mathcal{P}$, then $\mathcal{R}_{\mathcal{M}} \circ \mathcal{R}_{\mathcal{L}} \neq \iota$ (the identity mapping of $\mathcal{P}$ onto itself).

Exercise NEUT.8*. If $\mathcal{L}$ and $\mathcal{M}$ are distinct lines on the neutral plane $\mathcal{P}$, then there exists a unique line $\mathcal{J}$ such that $\mathcal{R}_{\mathcal{L}} \circ \mathcal{R}_{\mathcal{M}} \circ \mathcal{R}_{\mathcal{L}} = \mathcal{R}_{\mathcal{J}}$. Moreover, $\mathcal{J} = \mathcal{R}_{\mathcal{L}}(\mathcal{M})$.

Exercise NEUT.9*. Let O, A, and B be noncollinear points on the neutral plane $\mathcal{P}$ and let $\mathcal{L}$ be a line such that $\mathcal{R}_{\mathcal{L}}(\overrightarrow{OA}) = \overrightarrow{OB}$. By Remark NEUT.6(B), $\mathcal{L}$ is a line of symmetry of $\angle AOB$, $\mathcal{R}_{\mathcal{L}}$ is an angle reflection for $\angle AOB$, and by Theorem NEUT.20, $\mathcal{R}_{\mathcal{L}}(O) = O$. Construct a proof that $\mathcal{R}_{\mathcal{L}}(O) = O$, using Theorem NEUT.15, but not Theorem NEUT.20 or Theorem PSH.33 (uniqueness of corners).

Exercise NEUT.10*. Let A, B, and C be noncollinear points on the neutral plane $\mathcal{P}$, B' and C' be points such that B–A–B', C–A–C', and $\mathcal{M}$ be a line of symmetry of $\angle BAC$, then $\mathcal{M}$ is a line of symmetry of $\angle B'AC'$.

Exercise NEUT.11*. Let O, P, and Q be noncollinear points on the neutral plane $\mathcal{P}$ such that $\overleftrightarrow{OP}$ is a line of symmetry of $\overleftrightarrow{OQ}$ and let Q' be a point such that Q'–O–Q. If we let $\mathcal{L} = \overleftrightarrow{OP}$, then $\mathcal{R}_{\mathcal{L}}(\overrightarrow{OQ}) = \overrightarrow{OQ'}$ and $\mathcal{R}_{\mathcal{L}}(Q) \in \overrightarrow{OQ'}$.

Exercise NEUT.12*. Let $\mathcal{P}$ be a neutral plane and let O, A, A', B, and B' be points such that: (1) A–O–A', (2) B and B' are on opposite sides of $\overleftrightarrow{OA}$ (so that $\{A, O, B\}$ and $\{A', O, B'\}$ are noncollinear), and (3) $\angle AOB \cong \angle A'OB'$. Then B–O–B'.

Exercise NEUT.13*. Let A, B, C, D, A', B', C', and D' be points on the neutral plane $\mathcal{P}$ such that A, B, and C are noncollinear, A', B', and C' are noncollinear, $D \in \text{ins}\,\angle BAC$, $D' \in \overrightarrow{B'A'C'}$, $\angle BAC \simeq \angle B'A'C'$, and $\angle BAD \cong \angle B'A'D'$; then $\overrightarrow{A'D'} \subseteq \text{ins}\,\angle B'A'C'$.

Exercise NEUT.14*. Let A, B, C, D, A', B', C', and D' be points on the neutral plane $\mathcal{P}$ such that A, B, and C are noncollinear, A', B', and C' are noncollinear, $B \in \text{ins}\,\angle CAD$ (so that by Corollary PSH.39.2 C and D are on opposite sides of $\overleftrightarrow{AB}$), $B' \in \overrightarrow{C'A'D'}$, $\angle CAB \cong \angle C'A'B'$, and $\angle CAD \cong \angle C'A'D'$, then $B' \in \text{ins}\,\angle C'A'D'$ (so that C' and D' are on opposite sides of $\overleftrightarrow{A'B'}$.

Exercise NEUT.15*. Let A and B be distinct points on the neutral plane $\mathcal{P}$, M be the midpoint of $\overline{AB}$, C and D be points on the same side of $\overleftrightarrow{AB}$ such that $\overleftrightarrow{AC} \perp \overleftrightarrow{AB}$ and $\overleftrightarrow{BD} \perp \overleftrightarrow{AB}$ and $\mathcal{M}$ be the perpendicular bisector of $\overline{AB}$, then $\mathcal{R}_{\mathcal{M}}(\overleftrightarrow{AC}) = \overleftrightarrow{BD}$ and $\mathcal{R}_{\mathcal{M}}(\overleftrightarrow{BD}) = \overleftrightarrow{AC}$.

Exercise NEUT.16*. Let O, P, Q, and R be points on the neutral plane $\mathcal{P}$ such that $\angle POQ$ is right, $\angle ROQ$ is right, and P and R are on opposite sides of $\overleftrightarrow{OQ}$, then P, O, and R are collinear.

Exercise NEUT.17*. Prove Theorem NEUT.93: let A, B, and C be noncollinear points on the neutral plane $\mathcal{P}$. If $\angle ACB$ is right or is obtuse, then $\overline{AC} < \overline{AB}$ and $\overline{BC} < \overline{AB}$.

Exercise NEUT.18*. Let O, P, and S be noncollinear points on the neutral plane $\mathcal{P}$ such that $\angle POS$ is acute, U be a member of $\overrightarrow{OP}$, and $V = \text{ftpr}(U, \overleftrightarrow{OS})$, then $V \in \overrightarrow{OS}$.

Exercise NEUT.19*. Let A, B, and C be noncollinear points on the neutral plane $\mathcal{P}$; by Definition NEUT.2 (Property R.5) there exists an angle reflection $\mathcal{R}_{\mathcal{M}}$ for $\angle BAC$, and by Theorem NEUT.20(E) a point $P \in \mathcal{M}$ such that $\overrightarrow{AP} \subseteq \text{ins}\,\angle BAC$. By Definition NEUT.3(D) $\overrightarrow{AP}$ is a bisecting ray for $\angle BAC$. Show that $\angle BAP$ is acute.

Exercise NEUT.20*. Let A, B, and C be noncollinear points on the neutral plane $\mathcal{P}$. If $\angle BAC$ and $\angle ABC$ are both acute, and if $D = \text{ftpr}(C, \overleftrightarrow{AB})$, then $D \in \overline{AB}$.

Exercise NEUT.21*. Let A, B, and C be noncollinear points on the neutral plane $\mathcal{P}$, if $\overline{AB}$ is the maximal edge of $\triangle ABC$ and if $D = \text{ftpr}(C, \overleftrightarrow{AB})$, then $D \in \overline{AB}$.

Exercise NEUT.22*. Let $\mathcal{L}$ be a line on the neutral plane $\mathcal{P}$ and let P be a point such that $P \notin \mathcal{L}$.

(I) Let $Q = \text{ftpr}(P, \mathcal{L})$; if X is any point on $\mathcal{L}$ distinct from Q, then $\overline{PQ} < \overline{PX}$.
(II) If Q is a point on $\mathcal{L}$ with the property that for every point X on $\mathcal{L}$ which is distinct from Q, $\overline{PQ} < \overline{PX}$, then $Q = \text{ftpr}(P, \mathcal{L})$.

Exercise NEUT.23*. Let $\mathcal{P}$ be a neutral plane, A, B, and C be noncollinear points on $\mathcal{P}$, P be a member of ins $\angle BAC$, $Q = \text{ftpr}(P, \overleftrightarrow{AB})$, and $R = \text{ftpr}(P, \overleftrightarrow{AC})$.

(1) If $\overrightarrow{AP}$ is the bisecting ray of $\angle BAC$, then $\overline{PQ} \cong \overline{PR}$.
(2) If $Q \in \overrightarrow{AB}$, $R \in \overrightarrow{AC}$, and $\overline{PQ} \cong \overline{PR}$, then $\overrightarrow{AP}$ is the bisecting ray of $\angle BAC$.

Exercise NEUT.24*. Let $\mathcal{P}$ be a neutral plane and let A, B, C, and D be points on $\mathcal{P}$ such that $\overline{AB} \cup \overline{BC} \cup \overline{CD} \cup \overline{DA}$ is a quadrilateral, and suppose that $\overleftrightarrow{AB} \perp \overleftrightarrow{AD}$ and $\overleftrightarrow{AB} \perp \overleftrightarrow{BC}$. Then

(1) $\square ABCD$ is rotund;
(2) $\overline{BC} \cong \overline{AD}$ iff $\angle ADC \cong \angle BCD$; and
(3) $\overline{BC} < \overline{AD}$ iff $\angle ADC < \angle BCD$.

Exercise NEUT.25*. Let A, B, C, A', B', and C' be points on the neutral plane $\mathcal{P}$ such that A, B, and C are noncollinear, A', B', and C' are noncollinear, both $\angle ACB$ and $\angle A'C'B'$ are right, $\overline{BC} \cong \overline{B'C'}$ and $\overline{AC} < \overline{A'C'}$, then $\angle ABC < \angle A'B'C'$, $\overline{AB} < \overline{A'B'}$, and $\angle B'A'C' < \angle BAC$.

Exercise NEUT.26*. Let A, B, C, A', B', and C' be points on the neutral plane $\mathcal{P}$ such that A, B, and C are noncollinear, A', B', and C' are noncollinear, $\angle ACB$ and $\angle A'C'B'$ are both right, $\overline{BC} < \overline{B'C'}$ and $\overline{AC} > \overline{A'C'}$, then $\angle ABC > \angle A'B'C'$ and $\angle BAC < \angle B'A'C'$.

Exercise NEUT.27*. Let A, B, C, A', B', and C' be points on the neutral plane $\mathcal{P}$ such that A, B, and C are noncollinear, A', B', and C' are noncollinear, both $\angle ACB$ and $\angle A'C'B'$ are right, $\overline{BC} < \overline{B'C'}$, and $\overline{AC} < \overline{A'C'}$, then $\overline{AB} < \overline{A'B'}$.

Exercise NEUT.28*. Let A, B, C, A', B', and C' be points on the neutral plane $\mathcal{P}$ such that A, B, and C are noncollinear, A', B', and C' are noncollinear, both $\angle ACB$ and $\angle A'C'B'$ are right, $\overline{BC} < \overline{B'C'}$ and $\overline{AB} \cong \overline{A'B'}$, then $\overline{A'C'} < \overline{AC}$, $\angle BAC < \angle B'A'C'$ and $\angle ABC > \angle A'B'C'$.

Exercise NEUT.29*. Let A, B, C, A', B', and C' be points on the neutral plane $\mathcal{P}$ such that both $\angle ACB$ and $\angle A'C'B'$ are right, $\overline{AB} \cong \overline{A'B'}$ and $\angle A'B'C' < \angle ABC$, then $\overline{A'C'} < \overline{AC}$, $\overline{BC} < \overline{B'C'}$ and $\angle BAC < \angle B'A'C'$.

Exercise NEUT.30*. Let A, B, C, A', B', and C' be points on the neutral plane $\mathcal{P}$ such that A, B, and C are noncollinear, A', B', and C' are noncollinear, both $\angle ACB$ and $\angle A'C'B'$ are right, $\overline{BC} \cong \overline{B'C'}$, and $\angle ABC < \angle A'B'C'$, then $\overline{AC} < \overline{A'C'}$, $\overline{AB} < \overline{A'B'}$, and $\angle B'A'C' < \angle BAC$.

Exercise NEUT.31*. Let P, O, and T be noncollinear points on the neutral plane $\mathcal{P}$, let S be a member of ins $\angle POT$ such that $\angle POS < \angle TOS$, and let M be a member of ins $\angle POT$ such that $\overrightarrow{OM}$ is the bisecting ray of $\angle POT$, then $M \in$ ins $\angle TOS$.

Exercise NEUT.32*. Let P, O, and T be noncollinear points on the neutral plane $\mathcal{P}$, S and V be members of ins $\angle POT$ such that $\angle POS < \angle TOS$ and $\angle POV \cong \angle TOS$, and M be a member of ins $\angle POT$ such that $\overrightarrow{OM}$ is the bisecting ray of $\angle POT$. Then

(1) $S \in$ ins $\angle POV$ and $V \in$ ins $\angle TOS$,
(2) $\overrightarrow{OM}$ is the bisecting ray of $\angle SOV$,
(3) $\angle TOV \cong \angle POS$, and
(4) $M \in$ ins $\angle TOS \cap$ ins $\angle POV$.

Exercise NEUT.33*. Let $\mathcal{P}$ be a neutral plane and let A_1, B_1, M_1, A_2, B_2, and M_2 be points on $\mathcal{P}$ such that $A_1 \neq B_1$ and $A_2 \neq B_2$, M_1 is the midpoint of $\overline{A_1B_1}$ and M_2 is the midpoint of $\overline{A_2B_2}$, then $\overline{A_1B_1} \cong \overline{A_2B_2}$ iff $\overline{A_1M_1} \cong \overline{A_2M_2}$.

Exercise NEUT.34*. Let $\mathcal{P}$ be a neutral plane, O and P be distinct points on $\mathcal{P}$, let the points on $\overleftrightarrow{OP}$ be ordered so that $O < P$, and let A and B be distinct points on $\overrightarrow{OP}$. Let M be the midpoint of $\overline{OA}$ and N be the midpoint of $\overline{OB}$, then $A < B$ iff $M < N$.

Exercise NEUT.35*. Let $\mathcal{P}$ be a neutral plane, O and P be distinct points on $\mathcal{P}$, A and B be distinct members of $\overrightarrow{OP}$, M be the midpoint of $\overline{OA}$, and N be the midpoint of $\overline{OB}$, then O–A–B iff O–M–N.

Exercise NEUT.36*. Let $\mathcal{P}$ be a neutral plane and let A_1, B_1, M_1, A_2, B_2, and M_2 be points on $\mathcal{P}$ such that $A_1 \neq B_1$, $A_2 \neq B_2$, M_1 be the midpoint of $\overline{A_1B_1}$ and M_2 be the midpoint of $\overline{A_2B_2}$, then $\overline{A_1B_1} < \overline{A_2B_2}$ iff $\overline{A_1M_1} < \overline{A_2M_2}$.

Exercise NEUT.37*. Let A_1, B_1, A_2, and B_2 be points on the neutral plane $\mathcal{P}$ such that $A_1 \neq B_1$, $A_2 \neq B_2$, and $\overline{A_1B_1} \cong \overline{A_2B_2}$ and let C_1 and C_2 be points such that A_1–C_1–B_1, $C_2 \in \overrightarrow{A_2B_2}$ and $\overline{A_1C_1} \cong \overline{A_2C_2}$, then A_2–C_2–B_2.

Exercise NEUT.38*. Let A_1, B_1, A_2, B_2, C_1, and C_2 be points on the neutral plane $\mathcal{P}$ such that $A_1 \neq B_1$, $A_2 \neq B_2$, $C_1 \in \overline{A_1B_1}$, and $C_2 \in \overline{A_2B_2}$.

(A) If $\overline{A_1C_1} \cong \overline{A_2C_2}$ and $\overline{C_1B_1} \cong \overline{C_2B_2}$, then $\overline{A_1B_1} \cong \overline{A_2B_2}$.
(B) If $\overline{A_1C_1} \cong \overline{A_2C_2}$ and $\overline{A_1B_1} \cong \overline{A_2B_2}$, then $\overline{C_1B_1} \cong \overline{C_2B_2}$.

Exercise NEUT.39*. Let $\mathcal{P}$ be a neutral plane and let A, B, C, D, A', B', C', and D' be points on $\mathcal{P}$ such that: (1) A, B, and C are noncollinear, (2) A', B', and C' are noncollinear, (3) $\overrightarrow{AD}$ is the bisecting ray of $\angle BAC$, (4) $\overrightarrow{A'D'}$ is the bisecting ray of $\angle B'A'C'$. Then $\angle BAC \cong \angle B'A'C'$ iff $\angle BAD \cong \angle B'A'D'$.

Exercise NEUT.40*. Let $\mathcal{P}$ be a neutral plane and let A, B, C, D, A', B', C', and D' be points on $\mathcal{P}$ such that: (1) A, B, and C are noncollinear, (2) A', B', and C' are noncollinear, (3) $D \in \text{ins}\,\angle BAC$ and $D' \in \text{ins}\,\angle B'A'C'$.

(A) If $\angle BAD \cong \angle B'A'D'$ and $\angle CAD \cong \angle C'A'D'$, then $\angle BAC \cong \angle B'A'C'$.
(B) If $\angle BAD \cong \angle B'A'D'$ and $\angle BAC \cong \angle B'A'C'$, then $\angle CAD \cong \angle C'A'D'$.

Exercise NEUT.41*. Let $\mathcal{P}$ be a neutral plane and let A_1, B_1, C_1, D_1, A_2, B_2, C_2, and D_2 be points on $\mathcal{P}$ such that: (1) A_1, B_1, and C_1 are noncollinear, (2) $D_1 \in \text{ins}\,\angle B_1A_1C_1$, (3) A_2, B_2, and C_2 are noncollinear, (4) $D_2 \in \text{ins}\,\angle B_2A_2C_2$, and (5) $\angle B_1A_1D_1 \cong \angle B_2A_2D_2$. Then $\angle B_1A_1C_1 < \angle B_2A_2C_2$ iff $\angle D_1A_1C_1 < \angle D_2A_2C_2$.

Exercise NEUT.42*. Let $\mathcal{P}$ be a neutral plane and let A_1, B_1, C_1, D_1, A_2, B_2, C_2, and D_2 be points on $\mathcal{P}$ such that: (1) A_1, B_1, and C_1 are noncollinear, (2) $D_1 \in \text{ins}\,\angle B_1A_1C_1$, (3) A_2, B_2, and C_2 are noncollinear, and (4) $D_2 \in \text{ins}\,\angle B_2A_2C_2$. Then $\angle B_1A_1C_1 < \angle B_2A_2C_2$ if $\angle B_1A_1D_1 < \angle B_2A_2D_2$ and $\angle D_1A_1C_1 < \angle D_2A_2C_2$.

Exercise NEUT.43*. Let $\mathcal{P}$ be a neutral plane and let A_1, B_1, C_1, D_1, A_2, B_2, C_2, and D_2 be points on $\mathcal{P}$ such that: (1) A_1, B_1, and C_1 are noncollinear, (2) $\overrightarrow{A_1D_1}$ is the bisecting ray of $\angle B_1A_1C_1$, (3) A_2, B_2, and C_2 are noncollinear, and (4) $\overrightarrow{A_2D_2}$ is the bisecting ray of $\angle B_2A_2C_2$. Then $\angle B_1A_1C_1 < \angle B_2A_2C_2$ iff $\angle B_1A_1D_1 < \angle B_2A_2D_2$.

Exercise NEUT.44*. Let $\mathcal{P}$ be a neutral plane and let A, B, C, P, and Q be points on $\mathcal{P}$ such that: (1) A, B, and C are noncollinear, (2) $P \in \text{ins}\,\angle BAC$, and (3) $Q \in \text{ins}\,\angle BAP$. Then $\angle QAP < \angle BAC$.

The reader will note that the next exercise is identical to Exercise NEUT.42, and at one point we thought to eliminate it. We decided to leave it in, since the method of proof is different from that for Exercise NEUT.42.

Exercise NEUT.45*. Use Exercise NEUT.44 to prove the following: Let $\mathcal{P}$ be a neutral plane and let A, B, C, D, A', B', C', and D' be points on $\mathcal{P}$ such that: (1) A, B, and C are noncollinear, (2) A', B', and C' are noncollinear, (3) $D \in \text{ins}\,\angle BAC$ and (4) $D' \in \text{ins}\,\angle B'A'C'$. If $\angle BAD < \angle B'A'D'$ and $\angle CAD < \angle C'A'D'$, then $\angle BAC < \angle B'A'C'$.

Exercise NEUT.46*. Let A and B be distinct points on the neutral plane $\mathcal{P}$, $\mathcal{L}$ be the perpendicular bisector of $\overline{AB}$, and α be an isometry of $\mathcal{P}$ such that $\alpha(\overline{AB}) = \overline{AB}$, then one and only one of the following statements is true: (A) α is the identity mapping ι of $\mathcal{P}$ onto itself, (B) $\alpha = \mathcal{R}_{\overleftrightarrow{AB}}$, (C) $\alpha = \mathcal{R}_{\mathcal{L}}$, or (D) $\alpha = \mathcal{R}_{\mathcal{L}} \circ \mathcal{R}_{\overleftrightarrow{AB}}$.

Exercise NEUT.47*. Let A, B, and C be distinct points on the neutral plane $\mathcal{P}$ and let α be an isometry of $\mathcal{P}$ such that A is a fixed point of α and B is not a fixed point of α. Then A is the midpoint of $\overline{BC}$ iff B–A–C and $\alpha(B) = C$.

Exercise NEUT.48*. Let $\mathcal{P}$ be a neutral plane, let $\mathcal{L}$ and $\mathcal{M}$ be distinct lines on $\mathcal{P}$ through the point O, and let $\mathcal{L}_1$ and $\mathcal{M}_1$ be lines on $\mathcal{P}$ such that $\mathcal{L}_1 \perp \mathcal{L}$ and $\mathcal{M}_1 \perp \mathcal{M}$, then $\mathcal{L}_1$ and $\mathcal{M}_1$ are distinct.

Exercise NEUT.49*. Let P, O, and T be noncollinear points on the neutral plane $\mathcal{P}$ and let S and V be members of $\text{ins}\,\angle POT$ such that $\angle POS < \angle TOS$ and $\angle POV \cong \angle TOS$. Furthermore, let X be any member of $\text{ins}\,\angle TOV$ and let W be a point such that $\angle POW < \angle POX$ and $\angle XOW \cong \angle POS$, then $W \in \text{ins}\,\angle POV$.

Exercise NEUT.50*. Let $\mathcal{P}$ be a neutral plane, $\mathcal{L}$ and $\mathcal{M}$ be lines on $\mathcal{P}$ such that $\mathcal{L} \perp \mathcal{M}$, and $\mathcal{E}$ be a side of $\mathcal{L}$. Then $\mathcal{M}$ is a line of symmetry of $\mathcal{E}$.

Exercise NEUT.51*. Let $\mathcal{P}$ be a neutral plane and let A, B, and C be noncollinear points on $\mathcal{P}$ such that $\angle ACB$ is a maximal angle of $\triangle ABC$.

(A) If D is any member of $\overrightarrow{BC}$, then $\overline{AD} < \overline{AB}$.
(B) If $\angle ACB$ is acute, then there exists a point $D \in \overrightarrow{BC}$ such that $\overline{AC} > \overline{AD}$.
(C) If $\angle ACB$ is right or obtuse, then for every $D \in \overrightarrow{BC}$, $\overline{AC} < \overline{AD}$.

Exercise NEUT.52*. Let $\mathcal{P}$ be a neutral plane, A, B, and C be points on $\mathcal{P}$ such that B–A–C, and D be a member of $\mathcal{P} \setminus \overleftrightarrow{AB}$ such that $\angle BAD < \angle CAD$, then $\angle BAD$ is acute and $\angle CAD$ is obtuse.

Exercise NEUT.53*. Let $\mathcal{P}$ be a neutral plane, A, B, and C be noncollinear points on $\mathcal{P}$ such that $\overline{AC} < \overline{AB}$ and D be the point of intersection of the bisecting ray of $\angle BAC$ and $\overline{BC}$ (so $\angle BAD \cong \angle CAD$), then $\angle ADC$ is acute, $\angle ADB$ is obtuse, and $\overline{DC} < \overline{DB}$.

Exercise NEUT.54*. Let $\mathcal{P}$ be a neutral plane and let A, B, and M be distinct collinear points on $\mathcal{P}$ such that $\overline{AM} \cong \overline{BM}$, then M is the midpoint of $\overline{AB}$.

Exercise NEUT.55*. Let $\mathcal{P}$ be a neutral plane, A and B be distinct points on $\mathcal{P}$, M be the midpoint of $\overline{AB}$, and C be a member of $\overset{\vdash\!\dashv}{AB}$. Then $C \in \overset{\vdash\!\dashv}{AM}$ iff $\overline{AC} < \overline{BC}$.

Exercise NEUT.56*. Let $\mathcal{P}$ be a neutral plane, A, B, and C be noncollinear points on $\mathcal{P}$, P be a member of ins $\angle BAC$ such that $\overrightarrow{AP}$ is the bisecting ray of $\angle BAC$, and let Q also be a member of ins $\angle BAC$. Then $Q \in$ ins $\angle BAP$ iff $\angle BAQ < \angle CAQ$.

Exercise NEUT.57*. Let $\mathcal{P}$ be a neutral plane, A, B, and C be noncollinear points on $\mathcal{P}$ such that $\overline{AC} < \overline{AB}$, and D be the midpoint of $\overline{BC}$.

(A) $\angle ADC$ is acute and $\angle ADB$ is obtuse.

(B) If E is the point of intersection of the bisecting ray of $\angle BAC$ and segment $\overset{\vdash\!\dashv}{BC}$, then C–E–D–B and $\angle BAD < \angle CAD$.

Exercise NEUT.58*. Let $\mathcal{P}$ be a neutral plane and let A, B, C, D, E, and F be points on $\mathcal{P}$ such that: (1) A, B, and C are noncollinear, (2) D, E, and F are noncollinear, (3) $\angle BAC \cong \angle EDF$ and $\angle CBA \cong \angle FED$, and (4) $\overline{AB} < \overline{DE}$. Then $\overline{AC} < \overline{DF}$ and $\overline{BC} < \overline{EF}$.

Exercise NEUT.59*. Let $\mathcal{P}$ be a neutral plane, A, B, and C be noncollinear points on $\mathcal{P}$, F be the midpoint of $\overline{AB}$, E be the midpoint of $\overline{AC}$, and O be the point of intersection of $\overset{\vdash\!\dashv}{BE}$ and $\overset{\vdash\!\dashv}{CF}$. If $\overline{AB} \cong \overline{AC}$, then $\overline{BE} \cong \overline{CF}$, $\angle CBE \cong \angle BCF$, $\angle ABE \cong \angle ACF$, $\overleftrightarrow{AO}$ is the perpendicular bisector of $\overline{BC}$ and $\overline{AO}$ is the bisecting ray of $\angle BAC$.

Exercise NEUT.60*. Let $\mathcal{P}$ be a neutral plane and let A, B, and C be noncollinear points on $\mathcal{P}$, E be the midpoint of $\overline{AC}$, and F be the midpoint of $\overline{AB}$. If $\overline{AC} < \overline{AB}$, then $\angle ABE < \angle ACF$.

Exercise NEUT.61*. Let $\mathcal{P}$ be a neutral plane and let A, B, C, E, and F be points on $\mathcal{P}$ such that: (1) A, B, and C are noncollinear, (2) E is the point where the bisecting ray of $\angle ABC$ and $\overset{\vdash\!\dashv}{AC}$ intersect, (3) F is the point where the bisecting ray of $\angle ACB$ and $\overset{\vdash\!\dashv}{AB}$ intersect. If $\overline{AB} < \overline{AC}$, then $\overline{BE} < \overline{CF}$.

Exercise NEUT.62 (Steiner-Lehmus)*. Let $\mathcal{P}$ be a neutral plane and let A, B, C, E, and F be points on $\mathcal{P}$ such that:

(1) A, B, and C are noncollinear,
(2) E is the point of intersection of the bisecting ray of $\angle ABC$, and $\overset{\circ}{\overline{AC}}$, and
(3) F is the point of intersection of the bisecting ray of $\angle ACB$ and $\overset{\circ}{\overline{AB}}$.

If $\overline{BE} \cong \overline{CF}$, then $\overline{AB} \cong \overline{AC}$.

Exercise NEUT.63*. (A) Let $\mathcal{P}$ be a neutral plane and let A, B, C, and D be points on $\mathcal{P}$ such that:

(1) A, B, and C are noncollinear,
(2) $\angle BAC$ is acute,
(3) B and D are on opposite sides of $\overleftrightarrow{AC}$,
(4) $\angle CAD \cong \angle CAB$.

Then D is on the C-side of $\overleftrightarrow{AB}$.

(B) Let $\mathcal{P}$ be a neutral plane and let A, B, C, and D be points on $\mathcal{P}$ such that:

(1) A, B, and C are noncollinear,
(2) $\angle BAC$ is acute,
(3) B and D are on opposite sides of $\overleftrightarrow{AC}$,
(4) $\angle CAD$ is acute or right.

Then D is on the C-side of $\overleftrightarrow{AB}$.

Exercise NEUT.64*. Let $\mathcal{P}$ be a neutral plane and let A_1, B_1, C_1, D_1, A_2, B_2, C_2, and D_2 be points on $\mathcal{P}$ such that:

(1) A_1, B_1, and C_1 are noncollinear,
(2) A_2, B_2, and C_2 are noncollinear,
(3) B_1 and D_1 are on opposite sides of $\overleftrightarrow{A_1C_1}$,
(4) B_2 and D_2 are on opposite sides of $\overleftrightarrow{A_2C_2}$,
(5) $\angle D_1A_1C_1 \cong \angle B_1A_1C_1$,
(6) $\angle D_2A_2C_2 \cong \angle B_2A_2C_2$,
(7) $\angle B_1A_1C_1 < \angle B_2A_2C_2$, and $\angle B_2A_2C_2$ is acute.

Then $\angle B_1A_1D_1 < \angle B_2A_2D_2$.

Exercise NEUT.65*. Let $\mathcal{P}$ be a neutral plane and let A, B, and C be noncollinear points on $\mathcal{P}$ such that each angle of $\triangle ABC$ is acute, $D = \text{ftpr}(B, \overleftrightarrow{AC})$ and $E = \text{ftpr}(C, \overleftrightarrow{AB})$, then $\overset{\circ}{\overline{BD}}$ and $\overset{\circ}{\overline{CE}}$ intersect at a point O which belongs to ins $\triangle ABC$.

Exercise NEUT.66*. Let $\mathcal{P}$ be a neutral plane and let A, B, C, D, E, and F be points on $\mathcal{P}$ such that: (1) A, B, and C are noncollinear, $\angle ABC$ and $\angle ACB$ are both acute, and $\overline{AC} < \overline{AB}$, (2) D is the midpoint of $\overline{BC}$, E is the point of intersection of the bisecting ray of $\angle BAC$ and $\overset{\supset\subset}{BC}$, and $F = \text{ftpr}(A, \overleftrightarrow{BC})$. If the points on $\overleftrightarrow{BC}$ are ordered so that $B < C$, then $B < D < E < F < C$. Moreover, $\overline{AF} < \overline{AE} < \overline{AD} < \overline{AB}$.

Exercise NEUT.67*. Let $\mathcal{P}$ be a neutral plane and let A, B, C, D, E, and F be points on $\mathcal{P}$ such that: (1) A, B, and C are noncollinear, (2) D is the midpoint of $\overline{BC}$, (3) E is the point of intersection of the bisecting ray of $\angle BAC$ and $\overset{\supset\subset}{BC}$, and (4) $F = \text{ftpr}(A, \overleftrightarrow{BC})$. If $\overline{AB} \cong \overline{AC}$, then $D = E = F$.

The following exercise will strike the reader as decidedly odd, because we can hardly imagine a triangle such that the perpendicular bisectors of the sides do not intersect. But this is all we can prove at this stage of our development. The issue will be resolved in Chapter 11, Theorem EUC.9.

Exercise NEUT.68*. Let $\mathcal{P}$ be a neutral plane, A, B, and C be noncollinear points on $\mathcal{P}$. Let $\mathcal{L}$, $\mathcal{M}$, and $\mathcal{N}$ be the perpendicular bisectors of $\overline{AB}$, $\overline{AC}$, and $\overline{BC}$ respectively. Then either (1) $\mathcal{L}$, $\mathcal{M}$, and $\mathcal{N}$ are concurrent at a point O or (2) $\mathcal{L} \parallel \mathcal{M}$, $\mathcal{L} \parallel \mathcal{N}$, and $\mathcal{M} \parallel \mathcal{N}$.

Exercise NEUT.69*. Let $\mathcal{L}$ be a line on a neutral plane $\mathcal{P}$; let A, B, and C be points on $\mathcal{L}$ such that B–A–C, and let $\mathcal{M}$ be the line such that $A \in \mathcal{M}$ and $\mathcal{M} \perp \mathcal{L}$. We order the points on $\mathcal{L}$ such that $A < B$. Let X and Y be points on $\mathcal{L}$. Then $X < Y$ iff $\mathcal{R}_{\mathcal{M}}(Y) < \mathcal{R}_{\mathcal{M}}(X)$.

Exercise NEUT.70*. Let $\mathcal{P}$ be a neutral plane, $\mathcal{L}$ and $\mathcal{M}$ be lines on $\mathcal{P}$ which intersect at the point O, A be a point on $\mathcal{L}$ distinct from O, and X and Y be points on $\mathcal{M}$ distinct from O such that X and Y are on the same side of $\mathcal{L}$. Let the points on $\mathcal{M}$ be ordered so that $O < X$. Then $O < X < Y$ iff $\angle OAX < \angle OAY$.

Exercise NEUT.71*. Let $\mathcal{P}$ be a neutral plane, A, B, and C be noncollinear points on $\mathcal{P}$, and D be a member of $\overleftrightarrow{BC} \setminus \{B, C\}$. Then B–D–C iff $\angle ACB < \angle ADB$ and $\angle ABC < \angle ADC$.

Exercise NEUT.72*. Let A, B, C, and M be points on the neutral plane $\mathcal{P}$ such that $A \neq B$, $A \neq C$, M is the midpoint of $\overline{AB}$ and M is the midpoint of $\overline{AC}$. Then $B = C$.

Exercise NEUT.73*. Let A and M be distinct points on the neutral plane $\mathcal{P}$. Then there exists a unique point B such that M is the midpoint of $\overline{AB}$.

Exercise NEUT.74*. Let $\mathcal{P}$ be a neutral plane, $\mathcal{L}$ be a line on $\mathcal{P}$, and θ be the mapping of $\mathcal{P}$ into $\mathcal{P}$ such that: (1) For every member X of $\mathcal{L}$, $\theta(X) = X$. (2) For every member X of $\mathcal{P} \setminus \mathcal{L}$, $\theta(X)$ is the point such that $\mathrm{ftpr}(X, \mathcal{L})$ is the midpoint of $\overline{X\theta(X)}$. Then $\theta = \mathcal{R}_{\mathcal{L}}$.

Exercise NEUT.75*. Let $\mathcal{P}$ be a neutral plane and let θ be an isometry of $\mathcal{P}$. Then:

(A) If A and B are distinct points of $\mathcal{P}$ and if M is the midpoint of $\overline{AB}$, then $\theta(M)$ is the midpoint of $\overline{\theta(A)\theta(B)}$.

(B) Let A, B, and C be noncollinear points on $\mathcal{P}$. If H is a member of ins $\angle BAC$ such that $\overrightarrow{AH}$ is the bisecting ray of $\angle BAC$, then $\overrightarrow{\theta(A)\theta(B)}$ is the bisecting ray of $\angle\theta(B)\theta(A)\theta(C)$ and if D is the point of intersection of $\overrightarrow{AH}$ and $\overline{BC}$, then $\theta(D)$ is the point of intersection of $\overrightarrow{\theta(A)\theta(H)}$ and $\overline{\theta(B)\theta(C)}$.

(C) If $\mathcal{L}$ is line on $\mathcal{P}$, Q is a member of $\mathcal{P} \setminus \mathcal{L}$, $\mathcal{M} = \mathrm{pr}(Q, \mathcal{L})$, and $F = \mathrm{ftpr}(Q, \mathcal{L})$, then $\theta(\mathcal{M}) = \mathrm{pr}(\theta(Q), \theta(\mathcal{L}))$ and $\theta(F) = \mathrm{ftpr}(\theta(Q), \theta(\mathcal{L}))$.

Exercise NEUT.76*. Let $\mathcal{P}$ be a neutral plane and let A_1, B_1, C_1, D_1, E_1, F_1, A_2, B_2, C_2, D_2, E_2, and F_2 be points on $\mathcal{P}$ such that:

(1) A_1, B_1, and C_1 are noncollinear; A_2, B_2, and C_2 are noncollinear; and $\triangle A_1B_1C_1 \cong \triangle A_2B_2C_2$.

(2) θ is an isometry of $\mathcal{P}$ such that $\theta(\triangle A_1B_1C_1) = \triangle A_2B_2C_2$, $\theta(A_1) = A_2$, $\theta(B_1) = B_2$, and $\theta(C_1) = C_2$.

(3) D_1 is the midpoint of $\overline{B_1C_1}$ and D_2 is the midpoint of $\overline{B_2C_2}$.

(4) E_1 is the point of intersection of the bisecting ray of $\angle B_1A_1C_1$ and $\overline{B_1C_1}$; and E_2 is the point of intersection of the bisecting ray of $\angle B_2A_2C_2$ and $\overline{B_2C_2}$.

(5) $F_1 = \mathrm{ftpr}(A_1, \overleftrightarrow{B_1C_1})$ and $F_2 = \mathrm{ftpr}(A_2, \overleftrightarrow{B_2C_2})$.

Then $\theta(D_1) = D_2$, $\theta(E_1) = E_2$, and $\theta(F_1) = F_2$.

Exercise NEUT.77*. Let A, B, C, D, and E be points on the neutral plane $\mathcal{P}$ such that A–B–C, A–B–D, A–D–E, and $\overline{BC} \cong \overline{DE}$, then A–C–E.

Exercise NEUT.78*. Let $\mathcal{P}$ be a neutral plane and let $\mathcal{F}$, $\mathcal{G}$, and $\mathcal{H}$ be distinct lines on $\mathcal{P}$ concurrent at the point O such that no two of them are perpendicular to each other, Q be a member of $\mathcal{F} \setminus \{O\}$, $R = \mathrm{ftpr}(Q, \mathcal{G})$, $S = \mathrm{ftpr}(R, \mathcal{H})$ and $T = \mathrm{ftpr}(Q, \mathcal{H})$. Then $S \neq T$.

Exercise NEUT.79*. Let A, B, and C be noncollinear points on the neutral plane $\mathcal{P}$ and Q be a member of ins $\angle BAC$. Then $\overset{[\rightarrow}{AQ}$ is the bisecting ray of $\angle BAC$ iff for every member T of $\overset{(\rightarrow}{AQ}$, $\overset{[\;]}{TD} \cong \overset{[\;]}{TE}$, where $D = \text{ftpr}(T, \overleftrightarrow{AB})$ and $E = \text{ftpr}(T, \overleftrightarrow{AC})$.

Exercise NEUT.80*. Prove parts (B), (C), and (D) in Theorem NEUT.83.

Exercise NEUT.81*. Without invoking Theorem NEUT.15 parts (4) through (7), prove that if $A \neq B$ are points in a neutral plane,

(A) $\overset{[\;]}{AB} \not\cong \overset{[\;)}{AB}$ and $\overset{[\;]}{AB} \not\cong \overset{(\;]}{AB}$;
(B) $\overset{[\;]}{AB} \not\cong \overset{(\;)}{AB}$; and
(C) $\overset{(\;)}{AB} \not\cong \overset{[\;)}{AB}$ and $\overset{(\;)}{AB} \not\cong \overset{(\;]}{AB}$.

Exercise NEUT.82*. Let A, B, and C be points on the neutral plane such that $A \neq B$, $C \in \overset{(\rightarrow}{AB}$, and $\overset{[\;]}{AB} \cong \overset{[\;]}{AC}$. Let φ be the isometry such that $\varphi(\overset{[\;]}{AB}) = \overset{[\;]}{AC}$. (A) Using only NEUT.1 through NEUT.20, show that if φ is its own inverse, then $B = C$. (B) Discuss why this type of proof will not work in the general case, where φ is not necessarily its own inverse. If it did, we could prove Property R.4 of Definition NEUT.2 as a theorem.

Exercise NEUT.83*. Let $\mathcal{L}$ be a line on a neutral plane $\mathcal{P}$. Let φ be a mapping obeying Properties (B) through (D) of Definition NEUT.1. Then if every point O of $\mathcal{L}$ is contained in some line $\overleftrightarrow{A\varphi(A)}$, where $A \notin \mathcal{L}$, Property (A) of Definition NEUT.1 holds for φ.

The following scrap came from some attempts to show that there is a Pasch plane in which there is a line over which there exists no reflection. It seemed, somehow, worth saving, as it gives some insight into the structure of fixed lines.

Exercise NEUT.84*. Let $\mathcal{L}$ be a line in a neutral plane $\mathcal{P}$, and let A and B be distinct points on the same side of $\mathcal{L}$. Then if φ is a reflection over $\mathcal{L}$, the lines $\overleftrightarrow{\varphi(A)B}$ and $\overleftrightarrow{A\varphi(B)}$ intersect at a point $P \in \mathcal{L}$.

Chapter 9
Free Segments of a Neutral Plane (FSEG)

Acronym: *FSEG*
Dependencies: *Chapters 1, 3 (definitions and Theorems CAP.1–CAP.4), 4, 5, 6, 7, and 8*
New Axioms: *none*
New Terms Defined: *free segment, sum, subtraction, ordering (of free segments)*

Abstract: Free segments are defined as congruence classes of segments; these are ordered in a natural way, and this ordering is shown to be transitive and to have the trichotomy property. Addition of free segments is defined, and its elementary properties and interactions with ordering are studied. These developments are sufficient to prove the triangle inequality, and provide a first step toward defining distance on a neutral plane.

Whether or not two segments on a neutral plane are congruent has nothing to do with their position on the plane or their orientation. In the previous chapter, Theorem NEUT.67 (segment construction) showed that given a segment on the neutral plane, another segment congruent to it can be constructed anywhere on the plane. Congruence, therefore, ignores position and orientation, and preserves what we would like to call "length," or "distance."

By Theorem NEUT.14, congruence is an *equivalence relation*. Thus, given a closed segment $\overline{AB}$, the collection of all segments which are congruent to $\overline{AB}$ is an *equivalence class*. (The behavior of equivalence classes is briefly outlined at the end of Section 1.4 of Chapter 1.) Thus each segment of the plane belongs to exactly one

E.J. Specht et al., *Euclidean Geometry and its Subgeometries*,
DOI 10.1007/978-3-319-23775-6_9

of these equivalence classes, which can be called *congruence classes*, and each of them can be named by any of its members.

We give such an equivalence class the slightly fanciful name of "free segment," suggesting that we think of all the segments in the class as the same segment, moved around to different locations and orientations. Literally, a free segment is not a segment at all, but a collection of them, and we could just as well call it a congruence class. But we got started with "free segment" and free segment it shall remain.

With appropriate definitions, free segments can be treated as algebraic objects. We can compare them using "<" and ">," we can add them, and we can subtract them, but only in the case where a smaller free segment is subtracted from a larger one. Thus, the algebra of free segments is very rudimentary; there is no additive identity (no zero segments), there is no additive inverse (no negative segments); and we do not yet have a definition of multiplication—that will come in Chapter 15 (SIM).

One might think that free segments aren't good for much, but they are actually quite useful as temporary surrogates for the concepts of "length" and "distance." This will be made clearer in Chapter 14, where we show that every line in a Euclidean plane is an ordered field, complete with positive and negative elements. Moreover we will show that the positive elements of such a field can be identified with free segments using a mapping Φ which we will define shortly in Definition FSEG.14. This mapping begins a process that will eventuate, in Chapter 14, Remark OF.14 and Definition OF.16, in a definition of "length" and "distance" for which our Cartesian selves yearn most earnestly.

9.1 Theorems for free segments

Theorem FSEG.1. *Let A, B, C, and D be points on the neutral plane $\mathcal{P}$ such that $A \neq B$ and $C \neq D$. Then there exist points E, F, and G on $\mathcal{P}$ such that E–F–G, $\overline{AB} \cong \overline{EF}$, and $\overline{CD} \cong \overline{FG}$.*

Proof. Let E and H be points on $\mathcal{P}$ such that $H \neq E$. Then there exists a unique point (*cf* Theorem NEUT.67 (Segment Construction)) F on $\overrightarrow{EH}$ such that $\overline{EF} \cong \overline{AB}$. Let I be a point such that E–F–I (*cf* Property B.3 of Definition IB.1). By Theorem PSH.13 $\{X| E\text{–}F\text{–}X\} = \overset{\rightrightarrows}{FI}$. By Theorem PSH.15 $\overrightarrow{EF}$ is the union of the disjoint sets $\overline{EF}$ and $\overset{\rightrightarrows}{FI}$. Let G be the unique point on $\overset{\rightrightarrows}{FI}$ such that $\overline{FG} \cong \overline{CD}$, then the proof is complete.

In the particular case where $B = C$ and A–B–D, we will satisfy the conclusion of this theorem by letting $E = A$, $F = B$, and $G = D$. □

Definition FSEG.2. Let $\mathcal{P}$ be a neutral plane, and let A and B be distinct points on $\mathcal{P}$. The set $\{\overset{\sqsubset\sqsupset}{XY} \mid \overset{\sqsubset\sqsupset}{XY} \cong \overset{\sqsubset\sqsupset}{AB}\}$ is denoted $[\overset{\sqsubset\sqsupset}{AB}]$ and called the **free segment** of $\overset{\sqsubset\sqsupset}{AB}$. $[\overset{\sqsubset\sqsupset}{AB}]$ is the *congruence class* of $\overset{\sqsubset\sqsupset}{AB}$, that is, its *equivalence class* under the equivalence relation $\cong$ (*cf* Theorem NEUT.14).

Definition FSEG.3. Let $\mathcal{P}$ be a neutral plane, A, B, C, and D be points on $\mathcal{P}$ such that $A \neq B$ and $C \neq D$.

(A) By Theorem FSEG.1, let E, F, and G be the points on $\mathcal{P}$ such that E–F–G, $\overset{\sqsubset\sqsupset}{EF} \cong \overset{\sqsubset\sqsupset}{AB}$, and $\overset{\sqsubset\sqsupset}{FG} \cong \overset{\sqsubset\sqsupset}{CD}$. Then $[\overset{\sqsubset\sqsupset}{EG}]$ is the **sum** of $[\overset{\sqsubset\sqsupset}{AB}]$ and $[\overset{\sqsubset\sqsupset}{CD}]$, and is denoted by $[\overset{\sqsubset\sqsupset}{AB}] \oplus [\overset{\sqsubset\sqsupset}{CD}]$.
(B) $[\overset{\sqsubset\sqsupset}{AB}] < [\overset{\sqsubset\sqsupset}{CD}]$ iff $\overset{\sqsubset\sqsupset}{AB} < \overset{\sqsubset\sqsupset}{CD}$ (cf. Definition NEUT.70) and $[\overset{\sqsubset\sqsupset}{AB}] > [\overset{\sqsubset\sqsupset}{CD}]$ iff $[\overset{\sqsubset\sqsupset}{CD}] < [\overset{\sqsubset\sqsupset}{AB}]$. $[\overset{\sqsubset\sqsupset}{AB}] \leq [\overset{\sqsubset\sqsupset}{CD}]$ iff $\overset{\sqsubset\sqsupset}{AB} < \overset{\sqsubset\sqsupset}{CD}$ or $\overset{\sqsubset\sqsupset}{AB} \cong \overset{\sqsubset\sqsupset}{CD}$. $[\overset{\sqsubset\sqsupset}{AB}] \geq [\overset{\sqsubset\sqsupset}{CD}]$ iff $[\overset{\sqsubset\sqsupset}{CD}] \leq [\overset{\sqsubset\sqsupset}{AB}]$.

Remark FSEG.3.1. (A) Given an equivalence relation on a set $\mathcal{E}$, it is traditional to write the equivalence class of an element $x \in \mathcal{E}$ as $[x]$; the notation $[\overset{\sqsubset\sqsupset}{AB}]$ is an amalgamation of this notation with the notation $\overset{\sqsubset\sqsupset}{AB}$ for closed segment.
(B) A free segment has many names—one for each segment belonging to it. Thus, $\overset{\sqsubset\sqsupset}{AB} \cong \overset{\sqsubset\sqsupset}{CD}$ iff $[\overset{\sqsubset\sqsupset}{AB}] = [\overset{\sqsubset\sqsupset}{CD}]$.
(C) We need to be able to give free segments names which do not refer to specific segments. We employ small capital script letters, such as $\mathcal{S}$, $\mathcal{T}$, $\mathcal{U}$, and $\mathcal{V}$ for this purpose.
(D) This chapter will deal exclusively with closed segments of the form $\overset{\sqsubset\sqsupset}{AB}$ and not with open or semi-open segments such as $\overset{][}{AB}$, $\overset{]\sqsupset}{AB}$, or $\overset{\sqsubset[}{AB}$.
(E) Notice carefully that Definition FSEG.3(A) does *not* define sums of segments. In this book we define sums of points, sums of free segments, but never sums of segments. Also, observe that Definition FSEG.3(B) defines the ordering of free segments, based on the ordering of segments already defined in Chapter 8.

Theorem FSEG.4. *Let $\mathcal{S}$, $\mathcal{T}$, $\mathcal{U}$, and $\mathcal{V}$ be free segments of the neutral plane $\mathcal{P}$ such that $\mathcal{S} = \mathcal{T}$ and $\mathcal{U} = \mathcal{V}$, then*

(A) $\mathcal{S} \oplus \mathcal{U} = \mathcal{T} \oplus \mathcal{V}$.
(B) $\mathcal{S} < \mathcal{U}$ *iff* $\mathcal{T} < \mathcal{V}$.

In other words, addition and ordering of free segments are well defined.

Proof. (A) By Definition FSEG.3 and Theorem FSEG.1 there exist points A, B, C, D, E, and F on $\mathcal{P}$ such that $A{-}B{-}C$, $D{-}E{-}F$, $\mathcal{S} = [\overline{AB}]$, $\mathcal{U} = [\overline{BC}]$, $\mathcal{T} = [\overline{DE}]$, $\mathcal{V} = [\overline{EF}]$, $\mathcal{S} \oplus \mathcal{U} = [\overline{AC}]$, and $\mathcal{T} \oplus \mathcal{V} = [\overline{DF}]$. By Exercise FSEG.1 $\overline{AB} \cong \overline{DE}$ and $\overline{BC} \cong \overline{EF}$. By Exercise NEUT.38(A) $\overline{AC} \cong \overline{DF}$ and using Exercise FSEG.1 again, $[\overline{AC}] = [\overline{DF}]$, so that $\mathcal{S} \oplus \mathcal{U} = \mathcal{T} \oplus \mathcal{V}$.

(B) By Definition FSEG.3 there exist points A, B, C, D, E, F, G, and H on $\mathcal{P}$ such that $\mathcal{S} = [\overline{AB}]$, $\mathcal{U} = [\overline{CD}]$, $\mathcal{T} = [\overline{EF}]$, $\mathcal{V} = [\overline{GH}]$. By Definition FSEG.2, since $\mathcal{S} = \mathcal{T}$, $\overline{AB} \cong \overline{EF}$; since $\mathcal{U} = \mathcal{V}$, $\overline{CD} \cong \overline{GH}$. If $\mathcal{S} < \mathcal{U}$ by Definition FSEG.3 $\overline{AB} < \overline{CD}$, and by Theorem NEUT.73 $\overline{EF} < \overline{GH}$ so that by Definition FSEG.3 again, $\mathcal{T} < \mathcal{V}$. Interchanging $\mathcal{S}$ with $\mathcal{T}$ and $\mathcal{U}$ with $\mathcal{V}$ proves the converse. □

Theorem FSEG.5 (Trichotomy property for free segments). *Let $\mathcal{S}$ and $\mathcal{T}$ be free segments of the neutral plane $\mathcal{P}$. Then one and only one of the following statements is true:*

$$(1)\ \mathcal{S} = \mathcal{T} \qquad (2)\ \mathcal{S} < \mathcal{T} \qquad (3)\ \mathcal{S} > \mathcal{T}.$$

Proof. Taking into account Definition FSEG.2 and Exercise FSEG.1 we need to show that if $\mathcal{S} \neq \mathcal{T}$, then either $\mathcal{S} < \mathcal{T}$, or $\mathcal{S} > \mathcal{T}$. There exist points A, B, C, and D on $\mathcal{P}$ such that $A \neq B$, $C \neq D$, $\mathcal{S} = [\overline{AB}]$ and $\mathcal{T} = [\overline{CD}]$. By Theorem NEUT.72 (trichotomy for segments), either $\overline{AB} < \overline{CD}$ or $\overline{AB} > \overline{CD}$. Since $\overline{AB} < \overline{CD}$ iff $[\overline{AB}] < [\overline{CD}]$ and $\overline{AB} > \overline{CD}$ iff $[\overline{AB}] > [\overline{CD}]$, the proof is complete. □

Remark FSEG.6. At this point it would be possible to develop the idea of rational multiples of free segments, and we do so partially in Exercise FSEG.3. In the interest of a more complete development which also would encompass rational multiples of points on a line, we have elected to defer the main part of this discussion to Chapter 17, *Rational Points on a Line*.

Theorem FSEG.7 (Transitivity property for free segments). *Let $\mathcal{S}$, $\mathcal{T}$, and $\mathcal{U}$ be free segments of the neutral plane $\mathcal{P}$. If $\mathcal{S} < \mathcal{T}$ and $\mathcal{T} < \mathcal{U}$, then $\mathcal{S} < \mathcal{U}$.*

Proof. There exist points A, B, C, D, E, and F on $\mathcal{P}$ (cf Definition FSEG.2) such that $\mathcal{S} = [\overline{AB}]$, $\mathcal{T} = [\overline{CD}]$, and $\mathcal{U} = [\overline{EF}]$. Furthermore, $\mathcal{S} < \mathcal{T}$ iff $\overline{AB} < \overline{CD}$ and $\mathcal{T} < \mathcal{U}$ iff $\overline{CD} < \overline{EF}$. Since Theorem NEUT.73 gives transitivity for segments, $\mathcal{S} < \mathcal{U}$. □

Theorem FSEG.8. *Let $\mathcal{S}$, $\mathcal{T}$, and $\mathcal{U}$ be free segments of the neutral plane $\mathcal{P}$, then*

(I) $\mathcal{S} \oplus \mathcal{T} = \mathcal{T} \oplus \mathcal{S}$ *and*

(II) $(\mathcal{S} \oplus \mathcal{T}) \oplus \mathcal{U} = \mathcal{S} \oplus (\mathcal{T} \oplus \mathcal{U})$.

That is to say, the operation $\oplus$ *for free segments of the neutral plane* $\mathcal{P}$ *is commutative and associative.*

Proof. By Theorem FSEG.1 there exist points A, B, and C and points E, F, and G such that A–B–C, E–F–G, $s = [\overline{AB}]$, $t = [\overline{BC}] = [\overline{EF}]$, and $u = [\overline{FG}]$. Thus $\overline{BC} \cong \overline{EF}$. By Theorem NEUT.56 there exists an isometry α such that $\alpha(E) = B$, $\alpha(F) = C$. Let $\alpha(G) = D$; since α preserves betweenness and is a collineation, $\alpha(E)$–$\alpha(F)$–$\alpha(G)$, that is B–C–D. Then $\overline{FG} \cong \overline{CD}$ so $u = [\overline{FG}] = [\overline{CD}]$.

Then A–B–C–D, $s = [\overline{AB}]$, $t = [\overline{BC}]$, and $u = [\overline{CD}]$. By Definition FSEG.3 $s \oplus t = [\overline{AB}] \oplus [\overline{BC}] = [\overline{AC}]$ and $t \oplus s = [\overline{CB}] \oplus [\overline{BA}] = [\overline{CA}] = [\overline{AC}]$, so that $s \oplus t = t \oplus s$. Moreover

$$(s \oplus t) \oplus u = ([\overline{AB}] \oplus [\overline{BC}]) \oplus [\overline{CD}] = [\overline{AC}] \oplus [\overline{CD}] = [\overline{AD}]$$

and

$$s \oplus (t \oplus u) = [\overline{AB}] \oplus ([\overline{BC}] \oplus [\overline{CD}]) = [\overline{AB}] \oplus [\overline{BD}] = [\overline{AD}],$$

so $(s \oplus t) \oplus u = s \oplus (t \oplus u)$. □

Theorem FSEG.9. *Let* s, t, u, *and* v *be free segments of the neutral plane* $\mathcal{P}$.

(I) *If* $s < t$, *then* $s \oplus u < t \oplus u$.

(II) *If* $s < t$ *and* $u < v$, *then* $s \oplus u < t \oplus v$.

Proof. (I) By Definition FSEG.3 there exist points A, B, C, D, and E on $\mathcal{P}$ such that A–B–C, A–B–D, A–D–E, $s = [\overline{AB}]$, $u = [\overline{BC}] = [\overline{DE}]$, and $t = [\overline{AD}]$. Furthermore, $s \oplus u = [\overline{AC}]$ and $t \oplus u = [\overline{AE}]$. Using Exercises FSEG.1 and NEUT.78 we have A–C–E. Thus $\overline{AC} < \overline{AE}$ and so $[\overline{AC}] < [\overline{AE}]$ (cf Definition FSEG.3) (*i.e.*, $s \oplus u < t \oplus u$).

(II) By Theorem FSEG.7 and part (I) $s \oplus u < t \oplus u$ and $t \oplus u = u \oplus t < v \oplus t = t \oplus v$. By Theorem FSEG.7 (transitivity) $s \oplus u < t \oplus v$. □

Theorem FSEG.10. *Let* s *and* t *be free segments of the neutral plane* $\mathcal{P}$. *If* $s < t$, *then there exists a unique free segment* u *of* $\mathcal{P}$ *such that* $t = s \oplus u$.

Proof. (I: Uniqueness.) If u_1 and u_2 are free segments such that $t = s \oplus u_1 = s \oplus u_2$, then if $u_1 < u_2$, by Theorem FSEG.9(I) $s \oplus u_1 < s \oplus u_2$, a contradiction. Similarly, if $u_1 > u_2$, $s \oplus u_1 > s \oplus u_2$, also a contradiction. By Theorem FSEG.5 (trichotomy), $u_1 = u_2$.

(II: Existence.) By Definition FSEG.2 there exist points A, B, C, and D on $\mathcal{P}$ such that $A \neq B$, $C \neq D$, $s = [\overline{AB}]$, $t = [\overline{CD}]$, and $\overline{AB} < \overline{CD}$. By

Definition NEUT.70 there exists a point E such that C–E–D and $\overline{AB} \cong \overline{CE}$. Let $\mathscr{u} = [\overline{ED}]$, then by Definition FSEG.3 $\mathscr{t} = \mathscr{s} \oplus \mathscr{u}$. □

Definition FSEG.11. The free segment $\mathscr{u}$ of Theorem FSEG.10 is denoted by $\mathscr{t} \ominus \mathscr{s}$ and is "the **subtraction** of $\mathscr{s}$ from $\mathscr{t}$." Less formally, $\mathscr{t} \ominus \mathscr{s}$ is $\mathscr{t}$ **minus** $\mathscr{s}$.

Theorem FSEG.12. *Let $\mathscr{u}$ be any free segment of the neutral plane $\mathcal{P}$. Then there exist free segments $\mathscr{s}$ and $\mathscr{t}$ of $\mathcal{P}$ such that $\mathscr{t} < \mathscr{s}$ and $\mathscr{u} = \mathscr{s} \ominus \mathscr{t}$.*

Proof. Suppose that $\mathscr{u} = [\overline{GH}]$; by Property B.3 of Definition IB.1 there exists a point I such that G–H–I. By Definition FSEG.3 $[\overline{GH}] \oplus [\overline{HI}] = [\overline{GI}]$. By Definition FSEG.11 $[\overline{GH}] = [\overline{GI}] \ominus [\overline{HI}]$. Let $\mathscr{s} = [\overline{GI}]$ and $\mathscr{t} = [\overline{HI}]$, then by Definition FSEG.3 $\mathscr{t} < \mathscr{s}$ and $\mathscr{u} = \mathscr{s} \ominus \mathscr{t}$. □

Theorem FSEG.13. *Let O and Q be distinct points on the neutral plane $\mathcal{P}$, $\mathcal{L} = \overleftrightarrow{OQ}$, and let $\mathbb{F}$ be the set of all free segments of $\mathcal{P}$. Then for each free segment $\mathscr{s} \in \mathbb{F}$, there exists a unique point $X \in \overrightarrow{OQ}$ such that $\overline{OX} \in \mathscr{s}$.*

Proof. Let $\overline{EF}$ be a segment belonging to $\mathscr{s}$. By Theorem NEUT.67 (segment construction), for each free segment $\mathscr{s} = [\overline{EF}] \in \mathbb{F}$ there exists a unique point $X \in \overrightarrow{OQ}$ such that $\overline{EF} \cong \overline{OX}$. □

Definition FSEG.14. Let O and Q be distinct points on the neutral plane $\mathcal{P}$, $\mathcal{L} = \overleftrightarrow{OQ}$, and let $\mathbb{F}$ be the set of all free segments of $\mathcal{P}$. For any points E and F in $\mathcal{P}$ define $\Phi[\overline{EF}]$ to be the point $X \in \overrightarrow{OQ}$ (whose existence and uniqueness is guaranteed by Theorem FSEG.13) such that $\overline{EF} \cong \overline{OX}$, that is, $\overline{EF} \cong \overline{O(\Phi[\overline{EF}])}$.

As a consequence of Definition FSEG.14, we have the rather odd-looking equality $[\overline{EF}] = \left[\overline{O(\Phi[\overline{EF}])}\right]$. If we let $[\overline{EF}]$ be denoted by $\mathscr{s}$, this becomes $\mathscr{s} = [\overline{O(\Phi(\mathscr{s}))}]$.

Theorem FSEG.15. *Let O and Q be distinct points on the neutral plane $\mathcal{P}$, $\mathcal{L} = \overleftrightarrow{OQ}$, and let $\mathbb{F}$ be the set of all free segments of $\mathcal{P}$.*

(A) *The mapping Φ defined in Definition FSEG.14 is a bijection of $\mathbb{F}$ onto $\overrightarrow{OQ}$.*
(B) *If the points on $\overleftrightarrow{OQ}$ are ordered so that $O < Q$, then $\mathscr{s} < \mathscr{t}$ iff O–$\Phi(\mathscr{s})$–$\Phi(\mathscr{t})$ iff $\Phi(\mathscr{s}) < \Phi(\mathscr{t})$.*

Proof. (A) By Theorem FSEG.13 Φ is well defined. If $\mathscr{s}$ and $\mathscr{t}$ are free segments in $\mathbb{F}$, and $\Phi(\mathscr{s}) = \Phi(\mathscr{t})$, then there exist segments in each of $\mathscr{s}$ and $\mathscr{t}$

both of which are congruent to the same $\overline{OX}$ and hence are congruent by Theorem NEUT.14, so that $\mathcal{S} = \mathcal{T}$; thus, Φ is one-to-one. Since every segment $\overline{OX} \in [\overline{OX}] \in \mathbb{F}$, Φ is onto $\overrightarrow{OQ}$.

(B) By Definition FSEG.14, $\mathcal{S} < \mathcal{T}$ iff $[\overline{O\Phi(\mathcal{S})}] < [\overline{O\Phi(\mathcal{T})}]$. By Definition FSEG.3 this is the same as saying that $\overline{O\Phi(\mathcal{S})} < \overline{O\Phi(\mathcal{T})}$. By Theorem NEUT.74 this is O–$\Phi(\mathcal{S})$–$\Phi(\mathcal{T})$. By Theorem ORD.6 this is true iff either $O < \Phi(\mathcal{S}) < \Phi(\mathcal{T})$ or $\Phi(\mathcal{T}) < \Phi(\mathcal{S}) < O$. Since $O < \Phi(\mathcal{S})$, $O < \Phi(\mathcal{S}) < \Phi(\mathcal{T})$. □

Remark FSEG.16. The mapping Φ in Definition FSEG.14 is very significant to the overall development. It allows us to associate any segment on the plane with a point on a given ray; eventually, in Chapter 14, this will enable us to associate each segment with a number, which will be its length (*cf* Definition OF.16).

Theorem FSEG.17 (Triangle inequality). *Let A, B, and C be noncollinear points on the neutral plane $\mathcal{P}$. Then $[\overline{AC}] < [\overline{AB}] \oplus [\overline{BC}]$. That is to say, any edge of a triangle is smaller than the sum of the other two edges.*

Proof. By Theorem NEUT.67 (segment construction) there exists a point D such that A–B–D and $\overline{BD} \cong \overline{BC}$. By Theorem PSH.37 $B \in \text{ins}\,\angle ACD$. By Definition NEUT.70 $\angle BCD < \angle ACD$. By Theorem NEUT.40(A) (*Pons Asinorum*) $\angle BCD \cong \angle BDC = \angle ADC$. By Theorem NEUT.76 (Transitivity for Angles) $\angle ADC < \angle ACD$. By Theorem NEUT.91 $\overline{AC} < \overline{AD}$. By Definition FSEG.3 $[\overline{AC}] < [\overline{AD}] = [\overline{AB}] \oplus [\overline{BD}]$. By Exercise FSEG.1 $[\overline{BD}] = [\overline{BC}]$. Thus $[\overline{AC}] < [\overline{AB}] \oplus [\overline{BC}]$. □

Theorem FSEG.18. *Let A, B, and C be distinct points on the neutral plane $\mathcal{P}$. Then $[\overline{AC}] \leq [\overline{AB}] \oplus [\overline{BC}]$.*

Proof. If A, B, and C are noncollinear, then by Theorem FSEG.17, $[\overline{AC}] < [\overline{AB}] \oplus [\overline{BC}]$. If A, B, and C are collinear, then by Property B.2 of Definition IB.1 one and only one of the following statements holds: A–B–C, A–C–B, or B–A–C. If A–B–C, then by Definition FSEG.3 $[\overline{AC}] = [\overline{AB}] \oplus [\overline{BC}]$. If A–C–B, then by Exercise FSEG.2 $[\overline{AC}] < [\overline{AB}] < [\overline{AB}] \oplus [\overline{BC}]$. If B–A–C, then $[\overline{AC}] < [\overline{BC}] < [\overline{AB}] \oplus [\overline{BC}]$. □

Theorem FSEG.19. *Let A, B, and C be noncollinear points on the neutral plane $\mathcal{P}$ such that $\overline{BC} < \overline{AC}$. Then $[\overline{AC}] \ominus [\overline{BC}] < [\overline{AB}]$.*

Proof. By Theorem FSEG.17 $[\overline{AC}] < [\overline{AB}] \oplus [\overline{BC}]$. By Exercise FSEG.6

$$[\overline{AC}] \ominus [\overline{BC}] < ([\overline{AB}] \oplus [\overline{BC}]) \ominus [\overline{BC}].$$

By Exercise FSEG.4 $([\overline{AB}] \oplus [\overline{BC}]) \ominus [\overline{BC}] = [\overline{AB}]$, so $[\overline{AC}] \ominus [\overline{BC}] < [\overline{AB}]$. □

Theorem FSEG.20. *Let A, B, and C be distinct points on the neutral plane $\mathcal{P}$. Then $[\overline{AC}] \oplus [\overline{CB}] = [\overline{AB}]$ iff A–C–B.*

Proof. (I: If A–C–B, then $[\overline{AC}] \oplus [\overline{CB}] = [\overline{AB}]$.) If A–C–B, then by Definition FSEG.3 $[\overline{AC}] \oplus [\overline{CB}] = [\overline{AB}]$.

(II: If $[\overline{AC}] \oplus [\overline{CB}] = [\overline{AB}]$, then A–C–B.) To prove this half we prove the equivalent statement (contrapositive): If $\neg(A$–C–$B)$, then $[\overline{AC}] \oplus [\overline{CB}] \neq [\overline{AB}]$. By Property B.2 of Definition IB.1, either B–A–C, or A–B–C. Using Theorem FSEG.5 (trichotomy for free segments), we need only show that if either B–A–C or A–B–C, then $[\overline{AC}] \oplus [\overline{BC}] > [\overline{AB}]$. However, this is done in the proof of Theorem FSEG.18. □

9.2 Exercises for free segments

Answers to starred (*) exercises may be accessed from the home page for this book at www.springer.com.

Exercise FSEG.1*. Let A, B, C, and D be points on the neutral plane $\mathcal{P}$ such that $A \neq B$ and $C \neq D$. Then $[\overline{AB}] = [\overline{CD}]$ iff $\overline{AB} \cong \overline{CD}$.

Exercise FSEG.2*. Let A, B, C, and D be points on the neutral plane $\mathcal{P}$ such that $A \neq B$ and $C \neq D$. Then $[\overline{AB}] < [\overline{AB}] \oplus [\overline{CD}]$.

Exercise FSEG.3*. Let A and B be distinct points on the neutral plane P and let m and n be natural numbers. For the purposes of this exercise, we use mathematical induction to make the following definitions:

(1): Define $1[\overline{AB}] = [\overline{AB}]$, and for any n, if a point C has been determined so that $n[\overline{AB}] = [\overline{AC}]$, define $(n+1)[\overline{AB}] = [\overline{AC}] \oplus [\overline{AB}]$.

(2): Using Theorem NEUT.50, let M be the midpoint of $\overline{AB}$. Then define $\frac{1}{2}[\overline{AB}] = [\overline{AM}]$, and if for any m, C has been determined so that $\frac{1}{2^m}[\overline{AB}] = [\overline{AC}]$, let D be the midpoint of $\overline{AC}$ and define $\frac{1}{2^{m+1}}[\overline{AB}] = [\overline{AD}]$.

(3): For any n and m, define $\frac{n}{2^m}[\overline{AB}] = \frac{1}{2^m}\left(n[\overline{AB}]\right)$.

Let A, B, C, and D be points on the neutral plane such that $A \neq B$ and $C \neq D$; using the definitions above show the following:

(I) If $[\overline{AB}] < [\overline{CD}]$, then for any natural numbers n and m,

(A) $n[\overline{AB}] < n[\overline{CD}]$,
(B) $\frac{1}{2^m}[\overline{AB}] < \frac{1}{2^m}[\overline{CD}]$, and
(C) $\frac{n}{2^m}[\overline{AB}] < \frac{n}{2^m}[\overline{CD}]$.

(II) $\frac{n}{2^m}([\overline{AB}] \oplus [\overline{CD}]) = \frac{n}{2^m}[\overline{AB}] \oplus \frac{n}{2^m}[\overline{CD}]$.

Exercise FSEG.4*. If $\mathcal{S}$ and $\mathcal{T}$ are any free segments of the neutral plane $\mathcal{P}$ such that $\mathcal{S} < \mathcal{T}$, then $(\mathcal{T} \oplus \mathcal{S}) \ominus \mathcal{S} = \mathcal{T}$ and $(\mathcal{T} \ominus \mathcal{S}) \oplus \mathcal{S} = \mathcal{T}$.

Exercise FSEG.5*. Let $\mathcal{S}$, $\mathcal{T}$, and $\mathcal{U}$ be free segments of the neutral plane $\mathcal{P}$.

(A) If $\mathcal{U} < \mathcal{S}$ and $\mathcal{U} < \mathcal{T}$, then $(\mathcal{S} \oplus \mathcal{T}) \ominus \mathcal{U} = (\mathcal{S} \ominus \mathcal{U}) \oplus \mathcal{T} = (\mathcal{T} \ominus \mathcal{U}) \oplus \mathcal{S}$.
(B) If $\mathcal{T} \oplus \mathcal{U} < \mathcal{S}$, then $\mathcal{S} \ominus (\mathcal{T} \oplus \mathcal{U}) = (\mathcal{S} \ominus \mathcal{T}) \ominus \mathcal{U} = (\mathcal{S} \ominus \mathcal{U}) \ominus \mathcal{T}$.

Exercise FSEG.6*. Let $\mathcal{S}$, $\mathcal{T}$, and $\mathcal{U}$ be free segments of the neutral plane $\mathcal{P}$ such that $\mathcal{U} < \mathcal{S}$ and $\mathcal{U} < \mathcal{T}$. If $\mathcal{S} < \mathcal{T}$, then $\mathcal{S} \ominus \mathcal{U} < \mathcal{T} \ominus \mathcal{U}$.

Exercise FSEG.7*. Let $\mathcal{S}$, $\mathcal{T}$, and $\mathcal{U}$ be free segments of the neutral plane $\mathcal{P}$ such that $\mathcal{S} \oplus \mathcal{U} < \mathcal{T} \oplus \mathcal{U}$, then $\mathcal{S} < \mathcal{T}$.

Exercise FSEG.8*. Let $\mathcal{S}$, $\mathcal{T}$, $\mathcal{U}$, and $\mathcal{V}$ be free segments of the neutral plane $\mathcal{P}$ such that $\mathcal{T} < \mathcal{S}$ and $\mathcal{V} < \mathcal{U}$, then $(\mathcal{S} \ominus \mathcal{T}) \oplus (\mathcal{U} \ominus \mathcal{V}) = (\mathcal{S} \oplus \mathcal{U}) \ominus (\mathcal{T} \oplus \mathcal{V})$.

Exercise FSEG.9*. Let $\mathcal{S}$, $\mathcal{T}$, $\mathcal{U}$, and $\mathcal{V}$ be free segments of the neutral plane $\mathcal{P}$ such that $\mathcal{T} < \mathcal{S}$ and $\mathcal{V} < \mathcal{U}$, then $\mathcal{S} \ominus \mathcal{T} = \mathcal{U} \ominus \mathcal{V}$ iff $\mathcal{S} \oplus \mathcal{V} = \mathcal{T} \oplus \mathcal{U}$.

Exercise FSEG.10*. If $\mathcal{S}$ and $\mathcal{T}$ are free segments of the neutral plane $\mathcal{P}$ such that $\mathcal{T} < \mathcal{S}$, then $\mathcal{S} \ominus \mathcal{T} < \mathcal{S}$ and $\mathcal{S} \ominus (\mathcal{S} \ominus \mathcal{T}) = \mathcal{T}$.

Exercise FSEG.11. If $\mathcal{S}$, $\mathcal{T}$, and $\mathcal{U}$ are any free segments of the neutral plane $\mathcal{P}$, then $(\mathcal{S} \oplus \mathcal{T}) \oplus \mathcal{U} = \mathcal{S} \oplus (\mathcal{T} \oplus \mathcal{S})$ (the operation $\oplus$ is *associative* on the set $\mathbb{F}$ of free segments.

Exercise FSEG.12. Construct a theory FANG of free angles analogous to that developed in this chapter for free segments, based on the following definition: the **free angle** $FA(\angle BAC) = \{\angle XYZ \mid \angle XYZ \cong \angle BAC\}$.

Chapter 10
Rotations About a Point of a Neutral Plane (ROT)

Acronym: *ROT*

Dependencies: *Chapters 1, 3 (definitions and Theorems CAP.1–CAP.4), 4, 5, 6, 7, 8, and 9*

New Axioms: *none*

New Terms Defined: *(point) rotation, point reflection, inverse of a rotation*

Abstract: This chapter defines point rotations and point reflections (about a point O) on a neutral plane, and derives their elementary properties to the extent possible without a parallel axiom. It ends with a classification of isometries of a neutral plane, and proof of the existence of a "square root" of a rotation.

A *rotation* of a neutral plane about a point O is the composition of two reflections over lines intersecting at O; thus every rotation is a collineation. Rotations were not discussed in Chapter 3 because reflections had not yet been defined there.

In this chapter we develop those properties of rotations which are not dependent on a parallel axiom. In Chapter 13, after the parallel axiom is invoked, rotations will be used to define *half-rotations* which are collineations but not isometries. These in turn will be used to prove the existence of *dilations* (*cf* Definition CAP.17) in the Euclidean plane, and in Chapter 14 these will be used to define multiplication of points on a line.

E.J. Specht et al., *Euclidean Geometry and its Subgeometries*,
DOI 10.1007/978-3-319-23775-6_10

10.1 Definitions and theorems for rotations

Definition ROT.1. (A) A mapping α from the neutral plane $\mathcal{P}$ onto itself is a **rotation** about the point O iff there exist distinct lines $\mathcal{L}$ and $\mathcal{M}$ which intersect at O, such that $\alpha = \mathcal{R}_{\mathcal{M}} \circ \mathcal{R}_{\mathcal{L}}$. A rotation may sometimes be referred to as a **point rotation**. See Figure 10.1.

(B) If, in part (A), $\mathcal{L} \perp \mathcal{M}$, then the rotation α is a **point reflection** about O. A point reflection about O is denoted by $\mathcal{R}_O$. See Figure 10.2.

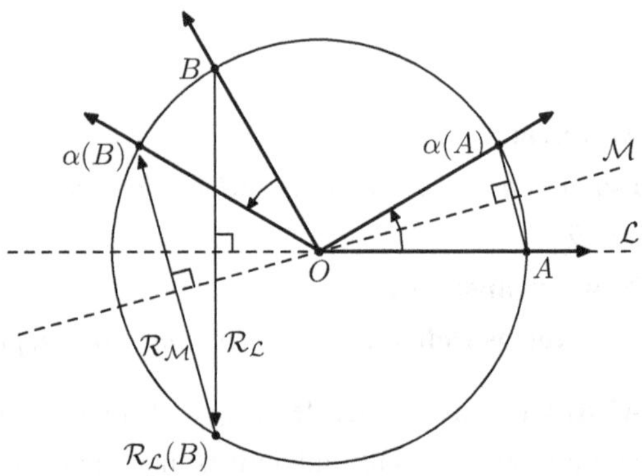

Fig. 10.1 Showing action of a rotation $\alpha = \mathcal{R}_{\mathcal{M}} \circ \mathcal{R}_{\mathcal{L}}$; $\mathcal{M}$ is the line of symmetry of $\angle AO\alpha(A)$.

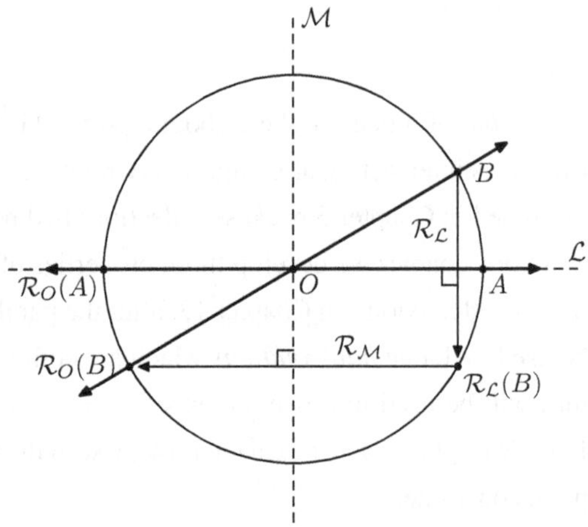

Fig. 10.2 Showing action of the point reflection $\mathcal{R}_O = \mathcal{R}_{\mathcal{M}} \circ \mathcal{R}_{\mathcal{L}}$, where $\mathcal{M}$ and $\mathcal{L}$ are perpendicular.

The rotation (or point reflection) α defined above is a composition of reflections and thus is a bijection and an isometry (*cf* Definition NEUT.3(A)). Since there is exactly one reflection over any line (*cf* Property R.2 of Definition NEUT.2) a rotation $\rho = \mathcal{R}_{\mathcal{L}} \circ \mathcal{R}_{\mathcal{M}}$ is completely determined by the two lines determining the reflections. It will follow from Theorem ROT.20(A) that the action of $\rho(X)$ at a single point $X \neq O$ determines its action at every point of the plane.

A rotation does not have a "sense"; it does not rotate a point either in the positive or negative (counterclockwise or clockwise) "direction." It is determined entirely by the final position of the points it rotates. Reverting back to the traditional measure of angle by "degrees": a 270 degree rotation counterclockwise is the same as a 90 degree rotation clockwise.

Theorem ROT.2. *If α is a rotation of the neutral plane $\mathcal{P}$ about the point O, then O is a fixed point of α and α has no other fixed points.*

Proof. By Definition ROT.1 there exist distinct lines $\mathcal{L}$ and $\mathcal{M}$ on $\mathcal{P}$ such that $\mathcal{L} \cap \mathcal{M} = \{O\}$ and $\mathcal{R}_O = \mathcal{R}_\mathcal{M} \circ \mathcal{R}_\mathcal{L}$. O is a fixed point of α since by Definition NEUT.1(A) $\alpha(O) = \mathcal{R}_\mathcal{M}(\mathcal{R}_\mathcal{L}(O)) = \mathcal{R}_\mathcal{M}(O) = O$.

Suppose $X \neq O$ is a fixed point for α. Then $\alpha(X) = \mathcal{R}_\mathcal{M}(\mathcal{R}_\mathcal{L}(X)) = X$. If $X \in \mathcal{L}$, then $\mathcal{R}_\mathcal{L}(X) = X$ and thus $\mathcal{R}_\mathcal{M}(X) = X$, so that $X \in \mathcal{M}$ and $X = O$, a contradiction; therefore, $X \notin \mathcal{L}$.

Then the fixed line $\overleftrightarrow{X\mathcal{R}_\mathcal{L}(X)}$ for $\mathcal{L}$ and the fixed line

$$\overleftrightarrow{\mathcal{R}_\mathcal{L}(X)\mathcal{R}_\mathcal{M}(\mathcal{R}_\mathcal{L}(X))} = \overleftrightarrow{\mathcal{R}_\mathcal{L}(X)\alpha(X)} = \overleftrightarrow{\mathcal{R}_\mathcal{L}(X)X}$$

of $\mathcal{M}$ are the same. By Theorem NEUT.44 both $\mathcal{M}$ and $\mathcal{L}$ are perpendicular to this line, and since they both contain O, by Theorem NEUT.48(A) $\mathcal{M} = \mathcal{L}$, contradicting our assumption that $\mathcal{M}$ and $\mathcal{L}$ are distinct. □

Theorem ROT.3. *Let $\mathcal{R}_O$ be a point reflection about the point O on a neutral plane $\mathcal{P}$.*

(A) *If $X \in \mathcal{P}\backslash\{O\}$, then X–O–$\mathcal{R}_O(X)$; also $\overline{OX} \cong \overline{O\mathcal{R}_O(X)}$ so that O is the midpoint of $\overline{X\mathcal{R}_O(X)}$.*

(B) *Every line $\mathcal{L}$ containing O is a fixed line for $\mathcal{R}_O$.*

Proof. (A) By Definition ROT.1 there exist lines $\mathcal{L}$ and $\mathcal{M}$ such that $\mathcal{L} \cap \mathcal{M} = \{O\}$, $\mathcal{L} \perp \mathcal{M}$, and $\mathcal{R}_O = \mathcal{R}_\mathcal{M} \circ \mathcal{R}_\mathcal{L}$. By Definition NEUT.3(A) $\mathcal{R}_O$ is an isometry, and by Theorem ROT.2 O is a fixed point for α; hence by Theorem NEUT.15(5), $\mathcal{R}_O(\overline{OX}) = \overline{\mathcal{R}_O(O)\mathcal{R}_O(X)} = \overline{O\mathcal{R}_O(X)}$, so that $\overline{OX} \cong \overline{O\mathcal{R}_O(X)}$. Thus, to show that O is the midpoint of $\overline{X\mathcal{R}_O(X)}$ all we need to show is that X–O–$\mathcal{R}_O(X)$ (*cf* Definition NEUT.3(C)).

(Case 1: $X \in (\mathcal{P} \setminus \mathcal{L} \cup \mathcal{M})$.) By Theorem NEUT.54 $\mathcal{L}$ is the perpendicular bisecting line of $\overline{X\mathcal{R}_\mathcal{L}(X)}$ so that $\mathcal{L} \perp \overleftrightarrow{X\mathcal{R}_\mathcal{L}(X)}$. By Theorem NEUT.47(A) $\mathcal{M} \parallel \overleftrightarrow{X\mathcal{R}_\mathcal{L}(X)}$. By Exercise PSH.14 $\overleftrightarrow{X\mathcal{R}_\mathcal{L}(X)} \subseteq$ (X-side of $\mathcal{M}$) so that X and

$\mathcal{R}_{\mathcal{L}}(X)$ belong to the same side of $\mathcal{M}$. By Definition NEUT.1(B) X and $\mathcal{R}_{\mathcal{L}}(X)$ are on opposite sides of $\mathcal{L}$; also $\mathcal{R}_{\mathcal{L}}(X)$ and $\mathcal{R}_O(X) = \mathcal{R}_{\mathcal{M}}(\mathcal{R}_{\mathcal{L}}(X))$ are on opposite sides of $\mathcal{M}$. They are also on the same side of $\mathcal{L}$, which is opposite X. By Theorem PSH.12 (plane separation) X and $\mathcal{R}_{\mathcal{M}}(\mathcal{R}_{\mathcal{L}}(X))$ are on opposite sides of $\mathcal{M}$ and on opposite sides of $\mathcal{L}$.

By Definition IB.11, there exists a point Q such that $\overline{X\mathcal{R}_{\mathcal{L}}(X)} \cap \mathcal{L} = \{Q\}$. By Definition NEUT.1(B) Q and $\mathcal{R}_{\mathcal{M}}(Q) = \mathcal{R}_O(Q)$ are on opposite sides of $\mathcal{M}$; thus, by Axiom PSA Q–O–$\alpha(Q)$. By Theorem NEUT.15(8) and the fact that $\mathcal{R}_O(O) = O$, $\mathcal{R}_O(\angle XOQ) = \angle\mathcal{R}_O(X)O\mathcal{R}_O(Q)$, so by Definition NEUT.3(B) (congruence) $\angle XOQ \cong \angle\mathcal{R}_O(X)O\mathcal{R}_O(Q)$. Applying Exercise NEUT.12 we get X–O–$\mathcal{R}_O(X)$.

(Case 2: $X \in \mathcal{L}$.) Note that $\mathcal{R}_O(X) = \mathcal{R}_{\mathcal{M}}(\mathcal{R}_{\mathcal{L}}(X)) = \mathcal{R}_{\mathcal{M}}(X)$; since $\mathcal{L}$ is a fixed line for $\mathcal{M}$, $\mathcal{R}_O(X) \in \mathcal{L}$, and by Definition NEUT.1(B) X and $\mathcal{R}_{\mathcal{M}}(X)$ are on opposite sides of $\mathcal{M}$. Then by Axiom PSA X–O–$\mathcal{R}_O(X)$.

(Case 3: $X \in \mathcal{M}$.) The proof is the same as Case 2, with the roles of $\mathcal{M}$ and $\mathcal{L}$ interchanged.

(B) Let $\mathcal{L}$ be any line containing O, and let $X \in \mathcal{L}$. Then by part (A), X–O–$\mathcal{R}_O(X)$, and by Definition IB.1, $\mathcal{L} = \overleftrightarrow{XO} = \overleftrightarrow{X\mathcal{R}_O(X)}$, so $\mathcal{R}_O(X) \in \mathcal{L}$. Therefore $\mathcal{L}$ is a fixed line for $\mathcal{R}_O$. □

The following theorem says that a point reflection about O is the composition of two line reflections over *any* two perpendicular lines containing O. That is, the point reflection does not depend on the choice of the perpendicular lines, and therefore there is only one point reflection about a given point.

Theorem ROT.4. *Let $\mathcal{P}$ be a neutral plane and let $\mathcal{J}$, $\mathcal{K}$, $\mathcal{L}$, and $\mathcal{M}$ be lines on $\mathcal{P}$ concurrent at O such that $\mathcal{J} \perp \mathcal{K}$ and $\mathcal{L} \perp \mathcal{M}$. Then $\mathcal{R}_{\mathcal{K}} \circ \mathcal{R}_{\mathcal{J}} = \mathcal{R}_{\mathcal{M}} \circ \mathcal{R}_{\mathcal{N}}$.*

Proof. Let $\alpha = \mathcal{R}_{\mathcal{K}} \circ \mathcal{R}_{\mathcal{J}}$ and Let $\beta = \mathcal{R}_{\mathcal{M}} \circ \mathcal{R}_{\mathcal{L}}$, then $\alpha(O) = O = \beta(O)$. Furthermore, if Q is any member of $\mathcal{P} \setminus \{O\}$, then by Theorem ROT.3 Q–O–$\alpha(Q)$, Q–O–$\beta(Q)$, $\overline{OQ} \cong \overline{O\alpha(Q)} \cong \overline{O\beta(Q)}$. By Theorem PSH.13 $\beta(Q) \in \overrightarrow{O\alpha(Q)}$, so that by Property R.3 of Definition NEUT.2 $\beta(Q) = \alpha(Q)$. Since Q is any member of $\mathcal{P} \setminus \{O\}$, $\beta = \alpha$. □

Corollary ROT.5. *If lines $\mathcal{L}$ and $\mathcal{M}$ on neutral plane $\mathcal{P}$ are concurrent at O and if $\mathcal{L} \perp \mathcal{M}$, then $\mathcal{R}_{\mathcal{M}} \circ \mathcal{R}_{\mathcal{L}} = \mathcal{R}_{\mathcal{L}} \circ \mathcal{R}_{\mathcal{M}}$.*

Proof. In Theorem ROT.4 take $\mathcal{K} = \mathcal{L}$ and $\mathcal{J} = \mathcal{M}$. □

Corollary ROT.6. *If $\mathcal{P}$ is a neutral plane and if O is any point on $\mathcal{P}$, then $\mathcal{R}_O \circ \mathcal{R}_O = \iota$ (the identity mapping of $\mathcal{P}$ onto itself).*

Proof. Let $\mathcal{L}$ and $\mathcal{M}$ be any lines on $\mathcal{P}$ concurrent at O, then $\mathcal{R}_O \circ \mathcal{R}_O = (\mathcal{R}_\mathcal{L} \circ \mathcal{R}_\mathcal{M}) \circ (\mathcal{R}_\mathcal{L} \circ \mathcal{R}_\mathcal{M}) = (\mathcal{R}_\mathcal{L} \circ \mathcal{R}_\mathcal{M}) \circ (\mathcal{R}_\mathcal{M} \circ \mathcal{R}_\mathcal{L}) = (\mathcal{R}_\mathcal{L} \circ (\mathcal{R}_\mathcal{M} \circ \mathcal{R}_\mathcal{M})) \circ \mathcal{R}_\mathcal{L} = \mathcal{R}_\mathcal{L} \circ \mathcal{R}_\mathcal{L} = \iota$. □

Corollary ROT.7. *Let O be a point on the neutral plane $\mathcal{P}$, and let $\mathcal{L}$ and $\mathcal{M}$ be perpendicular lines intersecting at O. Then for any $X \in \mathcal{L}$, $\mathcal{R}_O(X) = \mathcal{R}_\mathcal{M}(X)$. That is, the restriction of $\mathcal{R}_O$ to $\mathcal{L}$ is equal to the restriction of $\mathcal{R}_\mathcal{M}$ to $\mathcal{L}$.*

Proof. By Theorem ROT.4, $\mathcal{R}_O = \mathcal{R}_\mathcal{M} \circ \mathcal{R}_\mathcal{L}$. If X is any point of $\mathcal{L}$, $\mathcal{R}_O(X) = \mathcal{R}_\mathcal{M}(\mathcal{R}_\mathcal{L}(X)) = \mathcal{R}_\mathcal{M}(X)$. □

Theorem ROT.8. *Let $\mathcal{P}$ be a neutral plane, O be a point on $\mathcal{P}$, and α be a mapping of $\mathcal{P}$ such that $\alpha(O) = O$. If for every member of X of $\mathcal{P} \setminus \{O\}$, O is the midpoint of $\overline{X\alpha(X)}$, then $\alpha = \mathcal{R}_O$.*

Proof. Let Y be any member of $\mathcal{P} \setminus \overleftrightarrow{X\alpha(X)}$, $\mathcal{L} = \overleftrightarrow{OY}$, and $\mathcal{M} = \text{pr}(O, \mathcal{L})$. Let $\beta = \mathcal{R}_\mathcal{M} \circ \mathcal{R}_\mathcal{L}$. By Theorem ROT.4 O is the midpoint of $\overline{X\beta(X)}$ so that X–O–$\beta(X)$ and $\overline{OX} \cong \overline{O\beta(X)}$. Since O is the midpoint of $\overline{X\alpha(X)}$, X–O–$\alpha(X)$ and $\overline{OX} \cong \overline{O\alpha(X)}$. Since congruence is an equivalence relation for segments (Theorem NEUT.14), $\overline{O\alpha(X)} \cong \overline{O\beta(X)}$. By Theorem PSH.13 $\beta(X) \in \overrightarrow{O\alpha(X)}$ and so by Property R.3 of Definition NEUT.2 $\beta(X) = \alpha(X)$. Thus $\alpha = \mathcal{R}_O$. □

Theorem ROT.9. *Let $\mathcal{P}$ be a neutral plane.*

(A) *If $\mathcal{L}$ and $\mathcal{M}$ are lines on $\mathcal{P}$ such that $O \in \mathcal{L} \cap \mathcal{M}$ and $\mathcal{R}_\mathcal{M} \circ \mathcal{R}_\mathcal{L} = \iota$, then $\mathcal{L} = \mathcal{M}$.*

(B) *Let O and X be any points of $\mathcal{P}$ such that $X \neq O$; there is no rotation α about O such that $\alpha(\overrightarrow{OX}) = \overrightarrow{OX}$.*

Proof. (A) If $\mathcal{L}$ and $\mathcal{M}$ were distinct, then by Theorem ROT.2 the only fixed point of $\mathcal{R}_\mathcal{M} \circ \mathcal{R}_\mathcal{M}$ would be O. Hence $\mathcal{L} = \mathcal{M}$.

(B) If there were such a rotation α, then by Theorem NEUT.15(5) $\alpha(\overline{OX}) = \overline{O\alpha(X)}$ so that $\overline{OX} \cong \overline{O\alpha(X)}$. Since $\alpha(X) \in \overrightarrow{OX}$, by Property R.4 of Definition NEUT.2 $\alpha(X) = X$, so that X is a fixed point for α, contradicting Theorem ROT.2. □

Theorem ROT.10 (A rotation cannot be a line reflection). *Let $\mathcal{P}$ be a neutral plane, α be a rotation of $\mathcal{P}$ about O, and $\mathcal{L}$ be any line on $\mathcal{P}$ through O. Then $\alpha \neq \mathcal{R}_\mathcal{L}$.*

Proof. Since by Theorem ROT.2 O is the only fixed point of α, and whereas every point on $\mathcal{L}$ is a fixed point of $\mathcal{R}_{\mathcal{L}}$, $\alpha \neq \mathcal{R}_{\mathcal{L}}$. □

Theorem ROT.11. *Let $\mathcal{P}$ be a neutral plane, $\mathcal{L}$, $\mathcal{M}$, and $\mathcal{N}$ be distinct lines on $\mathcal{P}$ concurrent at O. Then there exists a unique line $\mathcal{J}$ such that $\mathcal{R}_{\mathcal{N}} \circ \mathcal{R}_{\mathcal{M}} \circ \mathcal{R}_{\mathcal{L}} = \mathcal{R}_{\mathcal{J}}$.*

Proof. We shall prove that the mapping $\alpha = \mathcal{R}_{\mathcal{N}} \circ \mathcal{R}_{\mathcal{M}} \circ \mathcal{R}_{\mathcal{L}}$ is a mirror mapping $\mathcal{R}_{\mathcal{J}}$. Once this is done, by Property R.3 (closure) of Definition NEUT.2 $\mathcal{R}_{\mathcal{J}}$ is a reflection. By Remark NEUT.1.1, $\mathcal{J}$ is the only possible axis for this mapping, proving uniqueness.

Let Q be any point on $\mathcal{L}$ distinct from O. By Definition NEUT.1(A) $\alpha(O) = O$ and $\mathcal{R}_{\mathcal{L}}(Q) = Q$. Since $Q \in \mathcal{L}$, $\alpha(Q) = \mathcal{R}_{\mathcal{N}}(\mathcal{R}_{\mathcal{M}}(Q))$; since $\mathcal{R}_{\mathcal{N}} \circ \mathcal{R}_{\mathcal{M}}$ is a rotation, $\alpha(Q) \neq Q$ by Theorem ROT.2. Let $\mathcal{J}$ be the line of symmetry of $\overline{Q\alpha(Q)}$ (*i.e.*, the perpendicular bisecting line of $\overline{Q\alpha(Q)}$, (*cf* Theorem NEUT.52(A)). Then Q is a fixed point of $\mathcal{R}_{\mathcal{J}} \circ \alpha$.

By Theorem NEUT.15(5) and the fact that $\alpha(O) = O$, $\alpha(\overline{OQ}) = \overline{O\alpha(Q)}$. By Definition NEUT.3(B) $\overline{OQ} \cong \overline{O\alpha(Q)}$. By Theorem NEUT.63 $\mathcal{J}$ is the line of symmetry of $\angle QO\alpha(Q)$ and in particular, $O \in \mathcal{J}$. Since O and Q are fixed points of $\mathcal{R}_{\mathcal{J}} \circ \alpha$, by Theorem NEUT.37 either $\mathcal{R}_{\mathcal{J}} \circ \alpha = \iota$ or $\mathcal{R}_{\mathcal{J}} \circ \alpha = \mathcal{R}_{\overleftrightarrow{OQ}} = \mathcal{R}_{\mathcal{L}}$.

If $\mathcal{R}_{\mathcal{J}} \circ \alpha$ were equal to $\mathcal{R}_{\mathcal{L}}$, then $\mathcal{R}_{\mathcal{L}} = \mathcal{R}_{\mathcal{J}} \circ \alpha = \mathcal{R}_{\mathcal{J}} \circ \mathcal{R}_{\mathcal{N}} \circ \mathcal{R}_{\mathcal{M}} \circ \mathcal{R}_{\mathcal{L}}$, and $\mathcal{R}_{\mathcal{J}} \circ \mathcal{R}_{\mathcal{N}} \circ \mathcal{R}_{\mathcal{M}} = \iota$, that is $\mathcal{R}_{\mathcal{N}} \circ \mathcal{R}_{\mathcal{M}} = \mathcal{R}_{\mathcal{J}}$. Thus the rotation $\mathcal{R}_{\mathcal{N}} \circ \mathcal{R}_{\mathcal{M}}$ is equal to the reflection $\mathcal{R}_{\mathcal{J}}$, which is impossible by Theorem ROT.10. Therefore $\mathcal{R}_{\mathcal{J}} \circ \alpha = \iota$, that is, $\alpha = \mathcal{R}_{\mathcal{J}}$ □

Theorem ROT.12. *If $\mathcal{L}$, $\mathcal{M}$, and $\mathcal{N}$ are distinct lines on the neutral plane $\mathcal{P}$ concurrent at O, then $\mathcal{R}_{\mathcal{M}} \circ \mathcal{R}_{\mathcal{N}} \circ \mathcal{R}_{\mathcal{L}} = \mathcal{R}_{\mathcal{L}} \circ \mathcal{R}_{\mathcal{N}} \circ \mathcal{R}_{\mathcal{M}}$.*

Proof. By Theorem ROT.11 there exists a unique line $\mathcal{J}$ such that $\mathcal{R}_{\mathcal{M}} \circ \mathcal{R}_{\mathcal{N}} \circ \mathcal{R}_{\mathcal{L}} = \mathcal{R}_{\mathcal{J}}$. By Remark NEUT.1.3 $\mathcal{R}_{\mathcal{J}}^{-1} = \mathcal{R}_{\mathcal{J}}$. By elementary algebra

$$\mathcal{R}_{\mathcal{J}}^{-1} = \mathcal{R}_{\mathcal{L}}^{-1} \circ \mathcal{R}_{\mathcal{N}}^{-1} \circ \mathcal{R}_{\mathcal{M}}^{-1} = \mathcal{R}_{\mathcal{L}} \circ \mathcal{R}_{\mathcal{N}} \circ \mathcal{R}_{\mathcal{M}}.$$

Thus $\mathcal{R}_{\mathcal{M}} \circ \mathcal{R}_{\mathcal{N}} \circ \mathcal{R}_{\mathcal{L}} = \mathcal{R}_{\mathcal{L}} \circ \mathcal{R}_{\mathcal{N}} \circ \mathcal{R}_{\mathcal{M}}$. □

The following theorem allows any rotation about a point to be written as the composition of two reflections over lines containing this point, where one of these lines has been chosen arbitrarily.

Theorem ROT.13. *Let $\mathcal{P}$ be a neutral plane, $\mathcal{L}$, $\mathcal{M}$, $\mathcal{N}$ be lines on $\mathcal{P}$ concurrent at O such that $\mathcal{M} \neq \mathcal{N}$. Then*

(A) *There exists a unique line $\mathcal{J}$ through O such that $\mathcal{R}_{\mathcal{L}} \circ \mathcal{R}_{\mathcal{J}} = \mathcal{R}_{\mathcal{N}} \circ \mathcal{R}_{\mathcal{M}}$.*
(B) *There exists a unique line $\mathcal{K}$ through O such that $\mathcal{R}_{\mathcal{K}} \circ \mathcal{R}_{\mathcal{L}} = \mathcal{R}_{\mathcal{N}} \circ \mathcal{R}_{\mathcal{M}}$.*

Proof. (A) (Case 1: $\mathcal{L} = \mathcal{N}$.) $\mathcal{R}_{\mathcal{N}} \circ \mathcal{R}_{\mathcal{J}} = \mathcal{R}_{\mathcal{N}} \circ \mathcal{R}_{\mathcal{M}}$ iff $\mathcal{R}_{\mathcal{J}} = \mathcal{R}_{\mathcal{M}}$. By Remark NEUT.1.1 $\mathcal{R}_{\mathcal{J}} = \mathcal{R}_{\mathcal{M}}$ iff $\mathcal{J} = \mathcal{M}$.

(Case 2: $\mathcal{L} = \mathcal{M}$.) $\mathcal{R}_{\mathcal{M}} \circ \mathcal{R}_{\mathcal{J}} = \mathcal{R}_{\mathcal{N}} \circ \mathcal{R}_{\mathcal{M}}$ iff $\mathcal{R}_{\mathcal{J}} = \mathcal{R}_{\mathcal{M}} \circ \mathcal{R}_{\mathcal{N}} \circ \mathcal{R}_{\mathcal{M}}$. Hence $\mathcal{J}$ is the line of Exercise NEUT.8, i.e. $\mathcal{J} = \mathcal{R}_{\mathcal{M}}(\mathcal{N})$.

(Case 3: $\mathcal{L}$, $\mathcal{M}$, and $\mathcal{N}$ are distinct.) $\mathcal{R}_{\mathcal{L}} \circ \mathcal{R}_{\mathcal{J}} = \mathcal{R}_{\mathcal{N}} \circ \mathcal{R}_{\mathcal{M}}$ iff $\mathcal{R}_{\mathcal{J}} = \mathcal{R}_{\mathcal{L}} \circ \mathcal{R}_{\mathcal{N}} \circ \mathcal{R}_{\mathcal{M}}$ Hence $\mathcal{J}$ is the line given by Theorem ROT.11.

(B) By part (A), there exists a unique line $\mathcal{K}$ through O such that $\mathcal{R}_{\mathcal{L}} \circ \mathcal{R}_{\mathcal{K}} = \mathcal{R}_{\mathcal{M}} \circ \mathcal{R}_{\mathcal{N}}$. Then taking inverses, we have $\mathcal{R}_{\mathcal{K}} \circ \mathcal{R}_{\mathcal{L}} = \mathcal{R}_{\mathcal{N}} \circ \mathcal{R}_{\mathcal{M}}$. □

Remark ROT.14. Parts (A) and (B) of the next theorem provide a standard way to construct a rotation $\mathcal{R}_{\mathcal{M}} \circ \mathcal{R}_{\mathcal{L}}$ that carries one ray of an angle into the other. Thus, specifying these two rays is all we need to determine the action of the rotation everywhere on the plane.

Theorem ROT.15. *Let O, A, and B be noncollinear points on the neutral plane $\mathcal{P}$, and let $\mathcal{L}$ be the line of symmetry of $\angle AOB$.*

(A) *There exists a unique rotation ρ about O such that $\rho(\overrightarrow{OA}) = \overrightarrow{OB}$; and $\rho = \mathcal{R}_{\mathcal{L}} \circ \mathcal{R}_{\overleftrightarrow{OA}} = \mathcal{R}_{\overleftrightarrow{OB}} \circ \mathcal{R}_{\mathcal{L}}$.*
(B) *If B′ is the point on $\overrightarrow{OB}$ such that $\overline{OB'} \cong \overline{OA}$, $\mathcal{L}$ is the line of symmetry (i.e., the perpendicular bisecting line) of $\overline{AB'}$.*
(C) *Let P be a point such that $P \notin \angle AOB$, where both $\angle AOP$ and $\angle POB$ are defined, with lines of symmetry $\mathcal{M}$ and $\mathcal{N}$, respectively. Then $\rho = \mathcal{R}_{\mathcal{N}} \circ \mathcal{R}_{\mathcal{M}}$ is the rotation that carries $\overrightarrow{OA}$ to $\overrightarrow{OB}$.*

Proof. (A) (I: Existence.) By Theorem NEUT.20 $\mathcal{R}_{\mathcal{L}}(\overrightarrow{OA}) = \overrightarrow{OB}$. Hence if we let $\rho = \mathcal{R}_{\mathcal{L}} \circ \mathcal{R}_{\overleftrightarrow{OA}}$ by Definition NEUT.1(A), we get

$$\rho(\overrightarrow{OA}) = \mathcal{R}_{\mathcal{L}}(\mathcal{R}_{\overleftrightarrow{OA}}(\overrightarrow{OA})) = \mathcal{R}_{\mathcal{L}}(\overrightarrow{OA}) = \overrightarrow{OB}.$$

(II: Uniqueness.) Suppose ρ is a rotation of $\mathcal{P}$ about O such that $\rho(\overrightarrow{OA}) = \overrightarrow{OB}$. By Theorem ROT.13 there exists a line $\mathcal{S}$ through O such that $\mathcal{R}_{\mathcal{S}} \circ \mathcal{R}_{\overleftrightarrow{OA}} = \rho$. By Definition NEUT.1(A)

$$\rho(\overrightarrow{OA}) = (\mathcal{R}_{\mathcal{S}} \circ \mathcal{R}_{\overleftrightarrow{OA}})(\overrightarrow{OA}) = \mathcal{R}_{\mathcal{S}}(\mathcal{R}_{\overleftrightarrow{OA}}(\overrightarrow{OA})) = \mathcal{R}_{\mathcal{S}}(\overrightarrow{OA}) = \overrightarrow{OB}.$$

By Exercise NEUT.9 $\mathcal{S}$ is a line of symmetry for $\angle AOB$ and by Theorem NEUT.26, there is only one such line; therefore, $\mathcal{S} = \mathcal{L}$ so $\rho = \mathcal{R}_{\mathcal{L}} \circ \mathcal{R}_{\overleftrightarrow{OA}}$.

Now $\mathcal{R}_{\overleftrightarrow{OB}} \circ \mathcal{R}_{\mathcal{L}}(\overrightarrow{OA}) = \overrightarrow{OB}$, so that by the uniqueness argument just above, $\mathcal{R}_{\overleftrightarrow{OB}} \circ \mathcal{R}_{\mathcal{L}} = \rho$.

(B) Follows immediately from Theorem NEUT.63.

(C) Note that the hypotheses of this part are satisfied if $P \in \text{ins}\,\angle AOB$. By Theorem NEUT.20 $\mathcal{R}_{\mathcal{M}}(\overrightarrow{OA}) = \overrightarrow{OP}$, and $\mathcal{R}_{\mathcal{N}}(\overrightarrow{OP}) = \overrightarrow{OB}$. Let $\rho = \mathcal{R}_{\mathcal{N}} \circ \mathcal{R}_{\mathcal{M}}$; then $\rho(\overrightarrow{OA}) = \overrightarrow{OB}$. $\mathcal{M} \neq \mathcal{N}$, for otherwise, A, B, and O would be collinear; thus ρ is a rotation. □

Theorem ROT.16. *Let α be an isometry of the neutral plane $\mathcal{P}$ which has one and only one fixed point O. Then there exist distinct lines $\mathcal{L}$ and $\mathcal{M}$ on $\mathcal{P}$ such that $\mathcal{L} \cap \mathcal{M} = \{O\}$ and $\alpha = \mathcal{R}_{\mathcal{L}} \circ \mathcal{R}_{\mathcal{M}}$ so that α is a rotation of $\mathcal{P}$ about O.*

Proof. Let X be any member of $\mathcal{P}\setminus\{O\}$. Since α has only one fixed point, $\alpha(X) \neq X$. By Property R.5 of Definition NEUT.2 (existence of angle reflection), we may let $\mathcal{L}$ be the line of symmetry for the angle $\angle XO(\alpha(X))$; then $O \in \mathcal{L}$. Then $\mathcal{R}_{\mathcal{L}}$ is the reflection such that $\mathcal{R}_{\mathcal{L}}(\overrightarrow{OX}) = \overrightarrow{O\alpha(X)}$, and let $Y = \mathcal{R}_{\mathcal{L}}(\alpha(X))$, which is a member of $\overrightarrow{OX}$. Let $\beta = \mathcal{R}_{\mathcal{L}} \circ \alpha$. Then $\beta(X) = Y$ and since X and Y are on the same ray, and β is an isometry, by Property R.4 of Definition NEUT.2 (linear scaling), $Y = X$; therefore X is a fixed point for β. Since O is a fixed point for both $\mathcal{R}_{\mathcal{L}}$ and α, it also is a fixed point for β.

By Theorem NEUT.37 either β is the identity mapping ι of $\mathcal{P}$ onto itself, or $\beta = \mathcal{R}_{\overleftrightarrow{OX}}$. If β were equal to ι, then α would be equal to $\mathcal{R}_{\mathcal{L}}$, contradicting Theorem ROT.10 (a rotation cannot be a reflection). Hence $\beta = \mathcal{R}_{\overleftrightarrow{OX}}$ and since we already know $\beta = \mathcal{R}_{\mathcal{L}} \circ \alpha$, $\mathcal{R}_{\overleftrightarrow{OX}} = \mathcal{R}_{\mathcal{L}} \circ \alpha$; applying $\mathcal{R}_{\mathcal{L}}$ to both sides, we have $\alpha = \mathcal{R}_{\mathcal{L}} \circ \mathcal{R}_{\overleftrightarrow{OX}}$, or $\alpha = \mathcal{R}_{\mathcal{L}} \circ \mathcal{R}_{\mathcal{M}}$, where $\mathcal{M} = \overleftrightarrow{OX}$. □

Theorem ROT.17. *Let $\mathcal{P}$ be a neutral plane, O be a point on $\mathcal{P}$, and α_1 and α_2 be rotations of $\mathcal{P}$ about O. Then either $\alpha_2 \circ \alpha_1$ is the identity mapping ι of $\mathcal{P}$ onto itself (in which case $\alpha_1 \circ \alpha_2 = \alpha_2 \circ \alpha_1 = \iota$) or $\alpha_2 \circ \alpha_1$ is a rotation of $\mathcal{P}$ about O.*

Proof. By Definition ROT.1 there exist distinct lines $\mathcal{L}_1$ and $\mathcal{M}_1$ as well as distinct lines $\mathcal{L}_2$ and $\mathcal{M}_2$ such that $\mathcal{L}_1 \cap \mathcal{M}_1 \cap \mathcal{L}_2 \cap \mathcal{M}_2 = \{O\}$, $\alpha_1 = \mathcal{R}_{\mathcal{M}_1} \circ \mathcal{R}_{\mathcal{L}_1}$ and $\alpha_2 = \mathcal{R}_{\mathcal{M}_2} \circ \mathcal{R}_{\mathcal{L}_2}$, so that $\alpha_2 \circ \alpha_1 = (\mathcal{R}_{\mathcal{M}_2} \circ \mathcal{R}_{\mathcal{L}_2}) \circ (\mathcal{R}_{\mathcal{M}_1} \circ \mathcal{R}_{\mathcal{L}_1}) = \mathcal{R}_{\mathcal{M}_2} \circ (\mathcal{R}_{\mathcal{L}_2} \circ (\mathcal{R}_{\mathcal{M}_1} \circ \mathcal{R}_{\mathcal{L}_1}))$. By Theorem ROT.11 there exists a unique line $\mathcal{J}$ on $\mathcal{P}$ such that $\mathcal{R}_{\mathcal{L}_2} \circ \mathcal{R}_{\mathcal{M}_1} \circ \mathcal{R}_{\mathcal{L}_1} = \mathcal{R}_{\mathcal{J}}$, thus $\alpha_2 \circ \alpha_1 = \mathcal{R}_{\mathcal{M}_2} \circ \mathcal{R}_{\mathcal{J}}$. If $\mathcal{J} = \mathcal{M}_2$, then by

Remark NEUT.1.3 $\alpha_2 \circ \alpha_1 = \iota$. If $\mathcal{J} \neq \mathcal{M}_2$, then by Definition ROT.1 $\alpha_2 \circ \alpha_1$ is a rotation of $\mathcal{P}$ about O. □

Theorem ROT.18. *Let O be a point on a neutral plane $\mathcal{P}$, and α a rotation of $\mathcal{P}$ about O. Then there exists a unique rotation β of $\mathcal{P}$ about O such that $\beta \circ \alpha = \iota$. That is, $\beta = \alpha^{-1}$, the inverse of α. Moreover, if $\alpha = \mathcal{R}_{\mathcal{M}} \circ \mathcal{R}_{\mathcal{L}}$, then $\beta = \mathcal{R}_{\mathcal{L}} \circ \mathcal{R}_{\mathcal{M}}$.*

Proof. (I: Existence.) By Definition ROT.1 there exist distinct lines $\mathcal{L}$ and $\mathcal{M}$ on $\mathcal{P}$ such that $\mathcal{L} \cap \mathcal{M} = \{O\}$ and $\alpha = \mathcal{R}_{\mathcal{M}} \circ \mathcal{R}_{\mathcal{L}}$. Let $\beta = \mathcal{R}_{\mathcal{L}} \circ \mathcal{R}_{\mathcal{M}}$. Then $\beta \circ \alpha = (\mathcal{R}_{\mathcal{L}} \circ \mathcal{R}_{\mathcal{M}}) \circ (\mathcal{R}_{\mathcal{M}} \circ \mathcal{R}_{\mathcal{L}}) = \mathcal{R}_{\mathcal{L}} \circ ((\mathcal{R}_{\mathcal{M}}) \circ \mathcal{R}_{\mathcal{M}}) \circ \mathcal{R}_{\mathcal{L}} = \mathcal{R}_{\mathcal{L}} \circ (\iota \circ \mathcal{R}_{\mathcal{L}}) = \mathcal{R}_{\mathcal{L}} \circ \mathcal{R}_{\mathcal{L}} = \iota$.

(II: Uniqueness.) This follows immediately from Theorem ROT.15(A), but here is a purely algebraic proof. If β and β' are rotations such that $\alpha \circ \beta = \iota$ and $\alpha \circ \beta' = \iota$, then $\alpha \circ \beta = \alpha \circ \beta'$. By part I there exists a rotation α' such that $\alpha' \circ \alpha = \iota$. Hence $\alpha' \circ (\alpha \circ \beta) = \alpha' \circ (\alpha \circ \beta')$, that is $(\alpha' \circ \alpha) \circ \beta = (\alpha' \circ \alpha) \circ \beta'$, so that $\beta = \beta'$. □

Theorem ROT.19. *Let α be a rotation about a point O of a neutral plane $\mathcal{P}$.*

(A) *If α has a fixed line, then α is a point reflection about O.*

(B) *If α is not a point reflection, then for every $X \neq O$, $\alpha(\overrightarrow{OX}) \neq \overrightarrow{OX}$.*

Proof. (A) We prove the contrapositive; that is, if α is not a point reflection, then α has no fixed line. Let $\mathcal{L}$ be any line on $\mathcal{P}$.

(Case 1: $O \in \mathcal{L}$.) Let X be any point on $\mathcal{L}$ distinct from O. Since α is not a point reflection about O, by Exercise ROT.2 $\alpha(X) \notin \overleftrightarrow{OX}$. Since $\alpha(O) = O$ by Theorem NEUT.15(1) $\alpha(\mathcal{L}) = \alpha(\overleftrightarrow{OX}) = \overleftrightarrow{O\alpha(X)}$. Thus $\alpha(\mathcal{L}) \neq \mathcal{L}$.

(Case 2: $O \notin \mathcal{L}$.) Let $\mathcal{M} = \text{pr}(O, \mathcal{L})$ and let $U = \text{ftpr}(O, \mathcal{L})$. (cf Theorem NEUT.48(A) and Definition NEUT.99), by case 1, $\alpha(\mathcal{M}) \neq \mathcal{M}$. By Corollary NEUT.44.1 $\alpha(\mathcal{M}) \perp \alpha(\mathcal{L})$. By Exercise NEUT.48 $\alpha(\mathcal{L}) \neq \mathcal{L}$.

(B) By the contrapositive of part (A), since α is not a point reflection, $\overleftrightarrow{OX} \neq \overleftrightarrow{O\alpha(X)}$ and $\overrightarrow{OX} \neq \overrightarrow{O\alpha(X)}$. □

Theorem ROT.20. *Let $\mathcal{P}$ be a neutral plane.*

(A) *If O, X, and Y are noncollinear points on $\mathcal{P}$ such that $\overline{OX} \cong \overline{OY}$, then there exists a unique rotation α of $\mathcal{P}$ about O such that $\alpha(X) = Y$.*

(B) *If X and Y are distinct points on $\mathcal{P}$ and if O is the midpoint of $\overline{XY}$, then there exists a unique point reflection β of $\mathcal{P}$ about O such that $\beta(X) = Y$.*

Proof. (A) By Theorem ROT.15 (A) there exists a unique rotation α such that $\alpha(\overrightarrow{OX}) = \overrightarrow{OY}$. Since α is an isometry, $\overline{OY} \cong \overline{OX} \cong \overline{O\alpha(X)}$ and since $\alpha(X) \in \overrightarrow{OY}$, by Property R.4 of Definition NEUT.2, $\alpha(X) = Y$.

If α and β are rotations of $\mathcal{P}$ about O such that $\alpha(X) = Y$ and $\beta(X) = Y$, then X is a fixed point of $\beta^{-1} \circ \alpha$. By Theorem ROT.2 $\beta^{-1} \circ \alpha$ is not a rotation; by Theorem ROT.17 $\beta^{-1} \circ \alpha = \iota$ so $\alpha = \beta$, thus showing uniqueness.

(B) If $\mathcal{J}$ and $\mathcal{K}$ are any lines through O which are perpendicular to each other and if $\beta = \mathcal{R}_{\mathcal{K}} \circ \mathcal{R}_{\mathcal{J}}$, then by Theorem ROT.3 X–O–$\beta(X)$ and $\overline{OX} \cong \overline{O\beta(X)}$. By the definition of midpoint, $\overline{OX} \cong \overline{OY}$ and X–O–Y so by Theorem PSH.15, $\beta(X) \in \overrightarrow{OY}$. By Theorem NEUT.14, $\overline{O\beta(X)} \cong \overline{OY}$; by Property R.4 of Definition NEUT.2 $\beta(X) = Y$. Then β is unique because by Theorem ROT.4 there is only one point reflection about any point O. □

Theorem ROT.21. *Let $\mathcal{P}$ be a neutral plane, O be a point on $\mathcal{P}$, and α and β be rotations of $\mathcal{P}$ about O. Then $\beta \circ \alpha = \alpha \circ \beta$.*

Proof. Let $\mathcal{M}$ be any line on $\mathcal{P}$ through O, then by Theorem ROT.13 there exist unique lines $\mathcal{L}$ and $\mathcal{N}$ on $\mathcal{P}$ through O such that $\alpha = \mathcal{R}_{\mathcal{M}} \circ \mathcal{R}_{\mathcal{L}}$ and $\beta = \mathcal{R}_{\mathcal{N}} \circ \mathcal{R}_{\mathcal{M}}$, so that $\beta \circ \alpha = (\mathcal{R}_{\mathcal{N}} \circ \mathcal{R}_{\mathcal{M}}) \circ (\mathcal{R}_{\mathcal{M}} \circ \mathcal{R}_{\mathcal{L}}) = \mathcal{R}_{\mathcal{N}} \circ \mathcal{R}_{\mathcal{L}}$. Using Theorem ROT.12 and Remark NEUT.1.3 we get

$$\begin{aligned}\alpha \circ \beta &= (\mathcal{R}_{\mathcal{M}} \circ \mathcal{R}_{\mathcal{L}}) \circ (\mathcal{R}_{\mathcal{N}} \circ \mathcal{R}_{\mathcal{M}}) = \mathcal{R}_{\mathcal{M}} \circ (\mathcal{R}_{\mathcal{L}} \circ \mathcal{R}_{\mathcal{N}} \circ \mathcal{R}_{\mathcal{M}})\\ &= \mathcal{R}_{\mathcal{M}} \circ (\mathcal{R}_{\mathcal{M}} \circ \mathcal{R}_{\mathcal{N}} \circ \mathcal{R}_{\mathcal{L}}) = (\mathcal{R}_{\mathcal{M}} \circ \mathcal{R}_{\mathcal{M}}) \circ (\mathcal{R}_{\mathcal{N}} \circ \mathcal{R}_{\mathcal{L}})\\ &= \iota \circ (\mathcal{R}_{\mathcal{N}} \circ \mathcal{R}_{\mathcal{L}}) = \mathcal{R}_{\mathcal{N}} \circ \mathcal{R}_{\mathcal{L}} = \beta \circ \alpha.\end{aligned}$$
□

The following theorem shows that a rotation is what we think of as a “rigid motion.”

Theorem ROT.22. *Let O, A, and B be points in the neutral plane $\mathcal{P}$, where $A \neq O$ and $B \neq O$, and let α be a rotation about O. Then $\angle AO\alpha(A) \cong \angle BO\alpha(B)$.*

Proof. By Theorem ROT.15 there exists a rotation β such that $\overrightarrow{OB} = \beta(\overrightarrow{OA}) = \overrightarrow{O\beta(A)}$. Then by Theorem NEUT.15(8) and Theorem ROT.21,

$$\begin{aligned}\beta(\angle AO(\alpha(A))) &= \angle \beta(A)O\beta(\alpha(A)) = \angle \beta(A)O\alpha(\beta(A))\\ &= \overrightarrow{O\beta(A)} \cup \overrightarrow{O\alpha(\beta(A))} = \overrightarrow{O\beta(A)} \cup \alpha(\overrightarrow{O\beta(A)})\\ &= \overrightarrow{OB} \cup \alpha(\overrightarrow{OB}) = \overrightarrow{OB} \cup \overrightarrow{O\alpha(B)} = \angle BO\alpha(B).\end{aligned}$$
□

Theorem ROT.23. *Under composition of mappings the set of rotations of a neutral plane $\mathcal{P}$ about the point O on $\mathcal{P}$, together with the identity mapping ι is an abelian group.*

Proof. By Theorem ROT.17 the set of rotations is closed under composition; by Theorem ROT.18 inverses exist and are rotations; therefore by *Bijections forming a group* in Section 1.5, the rotations form a group. By Theorem ROT.21, the group is abelian. □

Theorem ROT.24. *Let $\mathcal{P}$ be a neutral plane, $\mathcal{L}$ be a line on $\mathcal{P}$, A and B be distinct points on $\mathcal{L}$, $\mathcal{M} = \mathrm{pr}(A, \mathcal{L})$, and $\mathcal{N} = \mathrm{pr}(B, \mathcal{L})$, then $\mathcal{R}_{\mathcal{N}} \circ \mathcal{R}_{\mathcal{M}} = \mathcal{R}_B \circ \mathcal{R}_A$.*

Proof. By Definition ROT.1, Corollary ROT.5, and Remark NEUT.1.3,

$$\begin{aligned}\mathcal{R}_B \circ \mathcal{R}_A &= (\mathcal{R}_{\mathcal{N}} \circ \mathcal{R}_{\mathcal{L}}) \circ (\mathcal{R}_{\mathcal{L}} \circ \mathcal{R}_{\mathcal{M}})\\ &= (\mathcal{R}_{\mathcal{N}} \circ (\mathcal{R}_{\mathcal{L}} \circ \mathcal{R}_{\mathcal{L}}) \circ \mathcal{R}_{\mathcal{M}}) = \mathcal{R}_{\mathcal{N}} \circ \mathcal{R}_{\mathcal{M}}.\end{aligned}$$ □

Theorem ROT.25. *On the neutral plane $\mathcal{P}$, let α be an isometry which has no fixed points. Then either there exist three distinct lines $\mathcal{L}$, $\mathcal{M}$, $\mathcal{N}$ on $\mathcal{P}$ such that $\alpha = \mathcal{R}_{\mathcal{L}} \circ \mathcal{R}_{\mathcal{M}} \circ \mathcal{R}_{\mathcal{N}}$ or there exist distinct lines $\mathcal{L}$ and $\mathcal{M}'$ on $\mathcal{P}$ such that $\alpha = \mathcal{R}_{\mathcal{L}} \circ \mathcal{R}_{\mathcal{M}'}$.*

Proof. Let Q be any point on $\mathcal{P}$. Then $\alpha(Q) \neq Q$. Let $\mathcal{L}$ be the line of symmetry (*i.e.*, the perpendicular bisecting line of $\overline{Q\alpha(Q)}$ (*cf* Theorem NEUT.52(A)). Since $\mathcal{R}_{\mathcal{L}}(\alpha(Q)) = Q$, Q is a fixed point of $\mathcal{R}_{\mathcal{L}} \circ \alpha$. If $\mathcal{R} \circ \alpha$ has no fixed point different from Q, then by Theorem ROT.16 $\mathcal{R}_{\mathcal{L}} \circ \alpha$ is a rotation of $\mathcal{P}$ about Q and there exist distinct lines $\mathcal{M}$ and $\mathcal{N}$ which are concurrent at Q such that $\mathcal{R}_{\mathcal{L}} \circ \alpha = \mathcal{R}_{\mathcal{M}} \circ \mathcal{R}_{\mathcal{N}}$. Thus in this case $\alpha = \mathcal{R}_{\mathcal{L}} \circ \mathcal{R}_{\mathcal{M}} \circ \mathcal{R}_{\mathcal{N}}$. If $\mathcal{R}_{\mathcal{L}} \circ \alpha$ has a fixed point G distinct from Q but has no fixed point off of $\overleftrightarrow{QG}$, then by Theorem NEUT.37 $\mathcal{R}_{\mathcal{L}} \circ \alpha = \mathcal{R}_{\overleftrightarrow{QG}}$ so if $\mathcal{M}' = \overleftrightarrow{QG}$, then $\alpha = \mathcal{R}_{\mathcal{L}} \circ \mathcal{R}_{\mathcal{M}'}$. If $\mathcal{R}_{\mathcal{L}} \circ \alpha$ had a fixed point off of $\overleftrightarrow{QG}$, then by Theorem NEUT.24, $\mathcal{R}_{\mathcal{L}} \circ \alpha$ would be the identity mapping ι and α would be equal to $\mathcal{R}_{\mathcal{L}}$ and this would contradict the fact that α has no fixed points. □

Theorem ROT.26 (Classification of isometries). *Let θ be an isometry of the neutral plane $\mathcal{P}$, then one and only one of the following statements is true:*

(1) *θ is the identity mapping ι of $\mathcal{P}$ onto itself.*
(2) *There exists a line $\mathcal{H}$ of $\mathcal{P}$ such that $\theta = \mathcal{R}_{\mathcal{H}}$.*
(3) *There exist distinct lines $\mathcal{J}$ and $\mathcal{K}$ on $\mathcal{P}$ such that $\theta = \mathcal{R}_{\mathcal{K}} \circ \mathcal{R}_{\mathcal{J}}$.*
(4) *There exist distinct lines $\mathcal{L}$, $\mathcal{M}$, and $\mathcal{N}$ on $\mathcal{P}$ such that $\theta = \mathcal{R}_{\mathcal{N}} \circ \mathcal{R}_{\mathcal{M}} \circ \mathcal{R}_{\mathcal{L}}$.*

Proof. If θ has three noncollinear fixed points, then by Theorem NEUT.24 $\theta = \iota$. If θ has two distinct fixed points A and B but no fixed points off of $\overleftrightarrow{AB}$, then by

Theorem NEUT.37 $\alpha = \mathcal{R}_{\overleftrightarrow{AB}}$. If θ has one and only one fixed point O on $\mathcal{P}$, then by Theorem ROT.16 there exist distinct lines $\mathcal{J}$ and $\mathcal{K}$ such that $\mathcal{J} \circ \mathcal{K} = \{O\}$ and $\theta = \mathcal{R}_{\mathcal{K}} \circ \mathcal{R}_{\mathcal{J}}$. If θ has no fixed point then by Theorem ROT.25, either there exist distinct lines $\mathcal{L}$, $\mathcal{M}$, such that $\theta = \mathcal{R}_{\mathcal{M}'} \circ \mathcal{R}_{\mathcal{L}}$ or there exist three distinct lines $\mathcal{L}$, $\mathcal{M}$, and $\mathcal{N}$ on $\mathcal{P}$ such that $\theta = \mathcal{R}_{\mathcal{N}} \circ \mathcal{R}_{\mathcal{M}} \circ \mathcal{R}_{\mathcal{L}}$. The proof of Theorem ROT.25 makes clear the fact that these cases are mutually exclusive. □

Theorem ROT.27. *Let $\mathcal{P}$ be a neutral plane, O be a point on $\mathcal{P}$, and A, B, and C be points on $\mathcal{P}$ such that either (1) A–O–B and $\overleftrightarrow{OC} \perp \overleftrightarrow{OA}$, or (2) A, O, and B are noncollinear and $C \in$ ins $\angle AOB$ and $\angle AOC \cong \angle COB$; let α and β be rotations of $\mathcal{P}$ about O such that $\alpha(\overrightarrow{OA}) = \overrightarrow{OC}$ and $\beta(\overrightarrow{OC}) = \overrightarrow{OB}$; then $\alpha = \beta$.*

Proof. Since α is an isometry, $\alpha(\angle AOC) = \angle\alpha(A)\alpha(O)\alpha(C) = \angle\alpha(A)O\alpha(C) = \angle CO\alpha(C)$. By definition of congruence $\angle AOC \cong \angle CO\alpha(C)$; congruence is an equivalence relation (Theorem NEUT.14) so $\angle CO\alpha(C) \cong \angle COB$.

If A–O–B, by Definition IB.11 A and B are on opposite sides of $\overleftrightarrow{OC}$; if $C \in$ ins $\angle AOB$, this is true by Corollary PSH.39.2. By Theorem ROT.15, $\alpha = \mathcal{R}_{\overleftrightarrow{OC}} \circ \mathcal{R}_{\mathcal{L}}$ where $\mathcal{L}$ is the line of symmetry of $\angle AOC$. Then $\mathcal{R}_{\mathcal{L}}(C)$ is a point on $\overrightarrow{OA}$, which by Theorem IB.14 lies entirely on the A-side of $\overleftrightarrow{OC}$; by Property (B) of Definition NEUT.1, $\alpha(C) = \mathcal{R}_{\overleftrightarrow{OC}}(\mathcal{R}_{\mathcal{L}}(C))$ is on the side of $\overleftrightarrow{OC}$ opposite to A, that is, the B-side.

Then by Theorem NEUT.36 $\overrightarrow{OB} = \overrightarrow{O\alpha(C)} = \alpha(\overrightarrow{OC})$. By hypothesis $\overrightarrow{OB} = \beta(\overrightarrow{OC})$. By the uniqueness part of Theorem ROT.15(A), $\alpha = \beta$. □

Theorem ROT.28 (Existence of square root). *Let $\mathcal{P}$ be a neutral plane, O be a point on $\mathcal{P}$, and let α be a rotation of $\mathcal{P}$ about O. There exists a rotation β of $\mathcal{P}$ about O such that*

(A) *$\beta \circ \beta = \alpha$, and*
(B) *when α is not a point reflection, for every $X \in \mathcal{P} \setminus \{O\}$, $\beta(\overrightarrow{OX})$ is the bisecting ray of $\angle XO\alpha(X)$.*

Proof. (A) In this proof, all rotations will be rotations about the point O. Let A be a point on $\mathcal{P}$ distinct from O and let $B = \alpha(A)$. There are two possibilities (see Exercise ROT.2). Either α is the point reflection $\mathcal{R}_O$, or A, $\alpha(A)$, and O are noncollinear. If $\alpha = \mathcal{R}_O$, let C be a point such that $\angle AOC$ is right and $\angle COB$ is right. If $\alpha \neq \mathcal{R}_O$, let C be a point such that $\overrightarrow{OC}$ is the bisecting ray of $\angle AOB$. By Theorem ROT.15 let β be the rotation such that $\beta(\overrightarrow{OA}) = \overrightarrow{OC}$; let δ be the rotation such that $\delta(\overrightarrow{OC}) = \overrightarrow{OB}$. Then $(\delta \circ \beta)(\overrightarrow{OA}) = \overrightarrow{OB}$; by Theorem ROT.27 $\delta = \beta$. Hence $\beta \circ \beta = \alpha$.

(B) Since $\overrightarrow{OC}$ is the bisecting ray of $\angle AOB$, by Theorem NEUT.39 $\angle AOC \cong \angle COB$. Again using Theorem ROT.15, let γ be the rotation such that $\gamma(\overrightarrow{OA}) = \overrightarrow{OX}$. Without loss of generality we may assume that $\gamma(A) = X$ and $C = \beta(A)$. Then

$$\gamma(B) = \gamma(\alpha(A)) = \alpha(\gamma(A)) = \alpha(X) \text{ and}$$
$$\gamma(C) = \gamma(\beta(A)) = \beta(\gamma(A)) = \beta(X).$$

Therefore by Theorem NEUT.13 $\gamma(\angle AOC) \cong \gamma(\angle COB)$. By Theorem NEUT.15(8)

$$\angle XO\beta(X) = \angle\gamma(A)O\gamma(C) = \gamma(\angle AOC)$$
$$\cong \gamma(\angle COB) = \angle\gamma(C)O\gamma(B) = \angle\beta(X)O\alpha(X).$$

Since $C \in \text{ins}\,\angle AOB$, by Theorem NEUT.15(11)

$$\beta(X) = \gamma(C) \in \text{ins}\,\angle\gamma(A)O\gamma(B) = \text{ins}\,\angle XO\alpha(X)$$

and by Theorem NEUT.39, $\overrightarrow{O\beta(X)}$ is the angle bisector of $\angle XO\alpha(X)$. □

10.2 Exercises for rotations

Answers to starred (*) exercises may be accessed from the home page for this book at www.springer.com.

Exercise ROT.1*. Let $\mathcal{P}$ be a neutral plane.

(A) If O is a point on $\mathcal{P}$, and $\mathcal{R}_O$ is the point reflection about O, then $\mathcal{R}_O(O) = O$ and $\mathcal{R}_O \circ \mathcal{R}_O = \iota$.
(B) If $\mathcal{L}$ and $\mathcal{M}$ are distinct lines on $\mathcal{P}$ and if $\alpha = \mathcal{R}_\mathcal{M} \circ \mathcal{R}_\mathcal{L}$, then $\alpha^{-1} = \mathcal{R}_\mathcal{L} \circ \mathcal{R}_\mathcal{M}$.
(C) If G and H are points on $\mathcal{P}$ and if $\theta = \mathcal{R}_H \circ \mathcal{R}_G$, then $\theta^{-1} = \mathcal{R}_G \circ \mathcal{R}_H$.

Exercise ROT.2*. Let $\mathcal{P}$ be a neutral plane, O be a point on $\mathcal{P}$, and α be a rotation of $\mathcal{P}$ about O which is not a point reflection. If X is any member of $\mathcal{P} \setminus \{O\}$, then X, $\alpha(X)$, and O are noncollinear.

Exercise ROT.3*. Let $\mathcal{P}$ be a neutral plane, O be a point on $\mathcal{P}$, and α and β be rotations of $\mathcal{P}$ about O. If X is any member of $\mathcal{P} \setminus \{O\}$, then $O\alpha(X) \cong O\beta(X)$.

The following exercise shows that rotations (and half rotations, which we will meet in Chapter 13) behave as we expect them to—all points "rotate in the same direction."

Exercise ROT.4*. Let O, X, and Y be noncollinear points on the neutral plane $\mathcal{P}$ and let α be a rotation of $\mathcal{P}$ about O which is not a point reflection; we note that α cannot be the identity, as was proved in Theorem ROT.2.

(A1) α rotates X and Y through congruent angles; that is, $\angle XO\alpha(X) \cong \angle YO\alpha(Y)$.

(A2) Let α and β be rotations of $\mathcal{P}$ about O which are not point reflections. Let X be a point of $\mathcal{P} \setminus \{O\}$ such that

$$\alpha(X) \in \text{ins } \angle XO(\beta \circ \alpha(X)).$$

Then for any point $U \in \mathcal{P} \setminus \{O\}$,

$$\angle UO\alpha(U) \cong \angle XO\alpha(X);$$
$$\angle\alpha(U)O(\beta \circ \alpha)(U) \cong \angle\alpha(X)O(\beta \circ \alpha)(X);$$
$$\angle UO(\beta \circ \alpha)(U) \cong \angle XO(\beta \circ \alpha)(U); \text{ and}$$
$$\alpha(U) \in \text{ins } \angle UO(\beta \circ \alpha)(U).$$

(B) It cannot be true that both $\alpha(X) \in Y\text{-side } \overleftrightarrow{OX}$ and $\alpha(Y) \in X\text{-side } \overleftrightarrow{OY}$.

(C) It cannot be true that both $\alpha(X)$ is on the side of $\overleftrightarrow{OX}$ opposite Y and $\alpha(Y)$ is on the side of $\overleftrightarrow{OY}$ opposite X.

(D) $\alpha(X) \in Y$-side of $\overleftrightarrow{OX}$ iff $\alpha(Y)$ is on the side of $\overleftrightarrow{OY}$ opposite X; equivalently, $\alpha(Y) \in X$-side of $\overleftrightarrow{OY}$ iff $\alpha(X)$ is on the side of $\overleftrightarrow{OX}$ opposite Y.

(E) Let $W = \alpha(X)$, and $Z = \alpha(Y)$; let E be a point on the bisecting ray of $\angle XOW$ and F a point on the bisecting ray of $\angle YOZ$. Then $\angle EOX \cong \angle EOW \cong \angle FOY \cong \angle FOZ$.

(F) $E \in Y$-side of $\overleftrightarrow{OX}$ iff F is on the side of $\overleftrightarrow{OY}$ opposite X; equivalently, $F \in X$-side of $\overleftrightarrow{OY}$ iff E is on the side of $\overleftrightarrow{OX}$ opposite Y.

Exercise ROT.5*. Let $\mathcal{P}$ be a neutral plane, O be a point on $\mathcal{P}$, $\mathcal{L}$ and $\mathcal{M}$ be lines on $\mathcal{P}$ through O which are not perpendicular to each other, Q and R be points on $\mathcal{L}$ such that Q–O–R, S and T be points on $\mathcal{M}$ such that S–O–T and ρ be the rotation $\mathcal{R}_{\mathcal{M}} \circ \mathcal{R}_{\mathcal{L}}$ about O. If we choose the notation (using Theorem NEUT.82) so that $\angle QOS$ is acute, then $\rho(Q)$ is the member of ins $\angle ROS$ such that Q and $\rho(Q)$ are on opposite sides of $\mathcal{M}$, $\angle SO\rho(Q) \cong \angle QOS$ and $\overline{O\rho(Q)} \cong \overline{OQ}$.

Exercise ROT.6*. Let A, B, and C be noncollinear points on the neutral plane $\mathcal{P}$ and let $\mathcal{L} = \overleftrightarrow{AB}$, $\mathcal{M} = \overleftrightarrow{AC}$, and $\mathcal{N} = \overleftrightarrow{BC}$. Then there exists a unique point G and a unique line $\mathcal{J}$ such that $C \in \mathcal{J}$ and $\mathcal{R}_{\mathcal{N}} \circ \mathcal{R}_{\mathcal{M}} \circ \mathcal{R}_{\mathcal{L}} = \mathcal{R}_{\mathcal{J}} \circ \mathcal{R}_G$.

Exercise ROT.7*. Let A and B be distinct points on the neutral plane $\mathcal{P}$. If M is the midpoint of $\overline{AB}$, then $\mathcal{R}_M(A) = B$.

Exercise ROT.8*. Let $\mathcal{P}$ be a neutral plane, α be an isometry of $\mathcal{P}$ such that α has one and only one fixed point O, and for every member X of $\mathcal{P} \setminus \{O\}$, X–O–$\alpha(X)$, and $\overline{OX} \cong \overline{O\alpha(X)}$. Then α is the point reflection $\mathcal{R}_O$.

Exercise ROT.9*. Let $\mathcal{P}$ be a neutral plane, O be a point on $\mathcal{P}$, ρ be the reflection of $\mathcal{P}$ about O, $\mathcal{L}$ be a line on $\mathcal{P}$ through O which is ordered according to Definition ORD.1, and let X and Y be points on $\mathcal{L}$. Then $X < Y$ iff $\rho(Y) < \rho(X)$.

Exercise ROT.10*. Let $\mathcal{P}$ be a neutral plane, A, B, and O noncollinear points on $\mathcal{P}$. Then there exists a unique rotation α of $\mathcal{P}$ about O such that $\angle AO\alpha(A) = \angle AOB$.

Chapter 11
Euclidean Geometry Basics (EUC)

Acronym: *EUC*
Dependencies: *all prior Chapters 1 through 10*
New Axioms: *Axiom PS recalled from Chapter 2*
New Terms Defined: *Euclidean plane, Euclidean space, parallel (between a segment and a line), parallelogram, circumcenter (of a triangle), center (of a parallelogram), adjacent angles (of a quadrilateral), rectangle; orthocenter, median, centroid (of a triangle); complementary angles, complete triple of angles*

Abstract: This chapter combines the axioms of neutral geometry (incidence, betweenness, plane separation, and reflection) with the strong form of the Parallel Axiom to arrive at Euclidean geometry. It explores many well-known elementary results from plane geometry involving parallel lines, perpendicularity, adjacent and complementary angles, parallelograms and rectangles.

We have completed the development possible without a parallel axiom, either Axiom PS or PW, and now invoke Axiom PS to arrive at Euclidean geometry. Since Euclidean geometry includes Pasch and neutral geometry, we may now use any result from any previous chapter of the book.

The key theorems in this chapter are Theorems EUC.11, EUC.17, and EUC.22. Several theorems in this chapter are marked with double asterisks (for instance, Theorem EUC.3**). We encourage the reader to try to construct proofs for these before reading the proofs we give.

E.J. Specht et al., *Euclidean Geometry and its Subgeometries*,
DOI 10.1007/978-3-319-23775-6_11

It should be noted that in the presence of Property PE (which was proved as Theorem NEUT.48(B)), Axiom PW is equivalent to Axiom PS. For the reader's convenience we repeat Axiom PS, which was originally stated in Chapter 2.

Axiom PS (Strong Form of the Parallel Axiom). Given a line $\mathcal{L}$ and a point P not belonging to $\mathcal{L}$, there exists *exactly* one line $\mathcal{M}$ such that $P \in \mathcal{M}$ and $\mathcal{L} \parallel \mathcal{M}$. This line is denoted $\text{par}(P, \mathcal{L})$.

Theorem EUC.2 and Theorem EUC.3 below use Exercises IP.2 and IP.4 which we re-state here.

Exercise IP.2. Let $\mathcal{L}$, $\mathcal{M}$, and $\mathcal{N}$ be distinct lines contained in a single plane. Then (A) if $\mathcal{L} \parallel \mathcal{M}$ and $\mathcal{M} \parallel \mathcal{N}$, then $\mathcal{L} \parallel \mathcal{N}$, and (B) if $\mathcal{L}$ intersects $\mathcal{M}$, then $\mathcal{N}$ must intersect either $\mathcal{L}$ or $\mathcal{M}$.

Exercise IP.4. Let $\mathcal{L}$, $\mathcal{M}$, and $\mathcal{N}$ be distinct lines in a plane $\mathcal{E}$ such that $\mathcal{L} \parallel \mathcal{M}$. Then if $\mathcal{L} \cap \mathcal{N} \neq \emptyset$, $\mathcal{M} \cap \mathcal{N} \neq \emptyset$.

11.1 Definitions and theorems for Euclidean geometry

Definition EUC.1. Euclidean space is IB space in which Axiom PS holds and in which every plane is a neutral plane. Every such plane is a **Euclidean plane**, and the resulting geometry is **Euclidean geometry**.

Theorem EUC.2. *Let A, B, and C be noncollinear points in a Euclidean plane $\mathcal{E}$, and let $P \in$ ins $\angle BAC$. Then every line $\mathcal{L}$ containing P must intersect $\angle BAC$ (cf Theorem PSH.44).*

Proof. By Exercise IP.2(B) $\mathcal{L}$ intersects one or the other of the lines $\overleftrightarrow{AB}$ or $\overleftrightarrow{AC}$. If for some point Q, $\{Q\} = \mathcal{L} \cap \overleftrightarrow{AB}$, either $Q \in \overrightarrow{AB}$ or Q–A–B (from Theorem PSH.15). If $Q \in \overrightarrow{AB}$, then $Q \in \angle BAC$. If Q–A–B, then by Theorem IB.14 $\overrightarrow{QP}$ is a subset of the C-side of $\overleftrightarrow{AB}$, because P, being in ins $\angle BAC$, is on that side of $\overleftrightarrow{AB}$. Q and P are on opposite sides of $\overleftrightarrow{AC}$ because $P \in$ ins $\angle BAC$ and therefore is on the B-side of $\overleftrightarrow{AC}$. By Theorem PSH.12 there exists a point Q' such that $\{Q'\} = \overleftrightarrow{AC} \cap \overrightarrow{QP}$ and Q' is on the C-side of $\overleftrightarrow{AB}$. Hence $Q' \in \overrightarrow{AC}$, so $\mathcal{L}$ intersects $\angle BAC$. Similar reasoning holds if $\mathcal{L}$ intersects $\overleftrightarrow{AC}$. □

Theorem EUC.3.** *If $\mathcal{L}$, $\mathcal{M}$, and $\mathcal{N}$ are lines on the Euclidean plane $\mathcal{P}$ such that $\mathcal{L} \parallel \mathcal{M}$ and $\mathcal{L} \perp \mathcal{N}$, then $\mathcal{M} \perp \mathcal{N}$.*

Proof. By Definition NEUT.29 and Theorem NEUT.44, $\mathcal{L}$ and $\mathcal{N}$ intersect at a point B. By Exercise IP.4 there exists a point A such that $\mathcal{M} \cap \mathcal{N} = \{A\}$. By Theorem NEUT.46(A) there exists a unique line $\mathcal{M}'$ such that $A \in \mathcal{M}'$ and $\mathcal{M}' \perp \mathcal{N}$. By Theorem NEUT.47(A), $\mathcal{M}' \parallel \mathcal{L}$. By Axiom PS, $\mathcal{M}' = \mathcal{M}$, so that $\mathcal{M} \perp \mathcal{N}$. □

Corollary EUC.3.1. *Let $\mathcal{R}_{\mathcal{M}}$ be the line reflection over $\mathcal{M}$, and let $\mathcal{L}$ be a fixed line for $\mathcal{R}_{\mathcal{M}}$. Then $\mathcal{N} \parallel \mathcal{L}$ iff $\mathcal{N}$ is a fixed line for $\mathcal{R}_{\mathcal{M}}$.*

Proof. The proof is Exercise EUC.7. □

Corollary EUC.4. *Let $\mathcal{L}$, $\mathcal{M}$, and $\mathcal{N}$ be distinct lines on the Euclidean plane $\mathcal{P}$ such that $\mathcal{M}$ and $\mathcal{N}$ intersect at a point O, $\mathcal{M}$ and $\mathcal{N}$ are not perpendicular to each other, and $\mathcal{L} \perp \mathcal{N}$, then $\mathcal{L}$ and $\mathcal{M}$ intersect at a point Q. That is to say, if two lines intersect but are not perpendicular, then a line perpendicular to one of them must intersect the other.*

Proof. The proof is Exercise EUC.1. □

Definition EUC.5. (A) Two segments $\mathcal{D}$ and $\mathcal{E}$ on the Euclidean plane $\mathcal{P}$ are **parallel** iff there exist lines $\mathcal{L}$ and $\mathcal{M}$ on $\mathcal{P}$ which are parallel, and $\mathcal{D} \subseteq \mathcal{L}$ and $\mathcal{E} \subseteq \mathcal{M}$.

(B) A quadrilateral is a **parallelogram** iff its opposite edges are parallel.

Theorem EUC.6. *A parallelogram is a rotund quadrilateral.*

Proof. This follows immediately from Theorem PSH.53.1; see also Exercise EUC.2. □

Theorem EUC.7.** *Let $\mathcal{J}$, $\mathcal{K}$, $\mathcal{L}$, and $\mathcal{M}$ be distinct lines on the Euclidean plane $\mathcal{P}$ such that $\mathcal{J} \parallel \mathcal{K}$, $\mathcal{L} \perp \mathcal{J}$, and $\mathcal{M} \perp \mathcal{K}$, then $\mathcal{L} \parallel \mathcal{M}$.*

Proof. By Corollary EUC.4, $\mathcal{L} \perp \mathcal{K}$; by Theorem NEUT.47(A), $\mathcal{L} \parallel \mathcal{M}$. □

Corollary EUC.8. *Let $\mathcal{J}$, $\mathcal{K}$, $\mathcal{L}$, and $\mathcal{M}$ be distinct lines on the Euclidean plane $\mathcal{P}$ such that $\mathcal{L}$ and $\mathcal{M}$ intersect at the point O, $\mathcal{L} \perp \mathcal{J}$, and $\mathcal{M} \perp \mathcal{K}$, then $\mathcal{J}$ and $\mathcal{K}$ intersect at a point Q.*

Proof. The proof is Exercise EUC.3. □

Theorem EUC.9.** *Let A, B, and C be noncollinear points on the Euclidean plane $\mathcal{P}$ and let $\mathcal{L}$, $\mathcal{M}$, and $\mathcal{N}$ be the lines of symmetry (i.e., the perpendicular bisecting lines) of $\overline{AB}$, $\overline{AC}$, and $\overline{BC}$, respectively. Then $\mathcal{L}$, $\mathcal{M}$, and $\mathcal{N}$ are concurrent at a point O.*

Proof. $\mathcal{L} \perp \overleftrightarrow{AB}$ and $\mathcal{M} \perp \overleftrightarrow{AC}$ and $\overleftrightarrow{AB}$ intersects $\overleftrightarrow{AC}$; by Corollary EUC.8, $\mathcal{L}$ and $\mathcal{M}$ intersect at a point O. The conclusion follows from Exercise NEUT.68. □

Definition EUC.10. The point O of Theorem EUC.9 is the **circumcenter** of $\triangle ABC$.

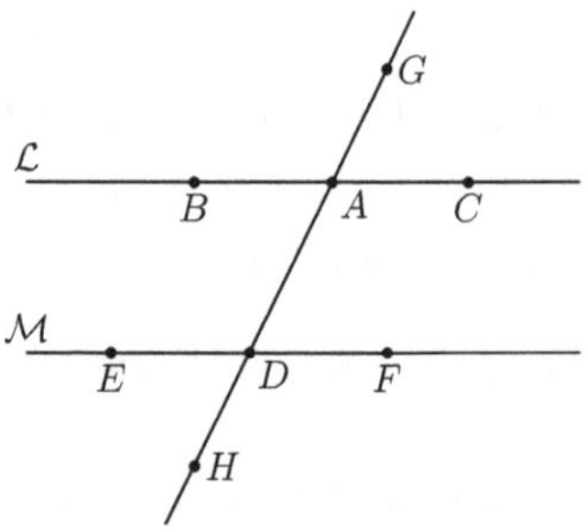

Fig. 11.1 For Theorem EUC.11.

Theorem EUC.11 (Congruence of alternate angles of parallel lines). *Let $\mathcal{L}$ and $\mathcal{M}$ be distinct lines on the Euclidean plane $\mathcal{P}$, and let A, B, C, D, E, F, G, and H be points on the plane such that*

(A) *D, E, and F are on $\mathcal{M}$ and E–D–F;*
(B) *A, B, and C are on $\mathcal{L}$ and B–A–C;*
(C) *G and H are on $\overleftrightarrow{AD}$ and G–A–D–H; and*
(D) *E and B are on the same side of $\overleftrightarrow{AD}$.*

Then the following statements are equivalent:

(1) $\mathcal{L} \parallel \mathcal{M}$,
(2) $\angle DAB \cong \angle ADF$,
(3) $\angle DAC \cong \angle ADE$,
(4) $\angle GAC \cong \angle ADF$,
(5) $\angle GAB \cong \angle HDF$, *and*
(6) $\angle GAB \cong \angle ADE$.

Proof. See Figure 11.1. We will proceed by proving statements (1) and (2) equivalent and then showing in turn that statements (3), (4), (5), and (6) are equivalent to (2).

(A) ((2) $\Rightarrow$ (1)) *cf* Theorem NEUT.87.

(B) ((1) $\Rightarrow$ (2)) By Theorem NEUT.68 (Angle Construction) there exists a point B' on the E-side of $\overleftrightarrow{AB}$ such that $\angle DAB' \cong \angle ADF$. By part (A) $\overleftrightarrow{AB'} \parallel \mathcal{M}$. By Axiom PS, $\overleftrightarrow{AB'} = \mathcal{L}$.

(C) ((2) $\Longleftrightarrow$ (3)) Since $\angle DAB$ and $\angle DAC$ are supplementary and $\angle ADE$ and $\angle ADF$ are supplementary, (2) $\Longleftrightarrow$ (3) is a consequence of Theorem NEUT.43.

(D) ((2) $\Longleftrightarrow$ (4)) Since $\angle DAB$ and $\angle GAC$ are vertical angles, this is a consequence of Theorem NEUT.42.

(E) ((2) $\Longleftrightarrow$ (5)) Since $\angle DAB$ and $\angle GAB$ are supplements, and $\angle ADF$ and $\angle HDF$ are supplements, this is a consequence of Theorem NEUT.43.

(F) ((2) $\Longleftrightarrow$ (6)) Since $\angle GAB$ and $\angle DAB$ are supplements, $\angle ADE$ and $\angle ADF$ are supplements, this is a consequence of Theorem NEUT.43. □

Theorem EUC.12.** (A) *If a quadrilateral $\square ABCD$ is a parallelogram, then its opposite edges are congruent and its opposite angles are congruent.*

(B) *If a quadrilateral $\square ABCD$ has a pair of opposite edges which are parallel (that is, the lines containing those edges are parallel) and congruent, then it is a parallelogram.*

Proof. (A) $\square ABCD$ is a trapezoid, and by Theorem PSH.53.1, is rotund; by Theorem PSH.54(A), its diagonals $\overline{AC}$ and $\overline{BD}$ intersect at some point O; by Theorem PSH.12 B and D are on opposite sides of $\overleftrightarrow{AC}$. Since $\overleftrightarrow{AB} \parallel \overleftrightarrow{CD}$, by Theorem EUC.11 $\angle BAC \cong \angle ACD$. Since $\overleftrightarrow{AD} \parallel \overleftrightarrow{BC}$, by Theorem EUC.11 $\angle ACB \cong \angle CAD$. By Theorem NEUT.65 (AEA) $\overline{AB} \cong \overline{CD}$, $\overline{AD} \cong \overline{BC}$, and $\angle ADC \cong \angle ABC$. A similar proof (using the diagonal $\overline{BD}$ instead of $\overline{AC}$) shows that $\angle BAD \cong \angle BCD$.

(B) $\square ABCD$ is a trapezoid; if we assume that $\overleftrightarrow{AB} \parallel \overleftrightarrow{CD}$, we can apply the first part of the proof for part (A) to get $\angle BAC \cong \angle ACD$. Assuming also that $\overline{AB} \cong \overline{CD}$, we may apply Theorem NEUT.64 (EAE) to get $\angle CAD \cong \angle ACB$. By Theorem EUC.11 $\overleftrightarrow{AD} \parallel \overleftrightarrow{BC}$. By Definition EUC.5(B) $\square ABCD$ is a parallelogram. If $\overleftrightarrow{AD} \parallel \overleftrightarrow{BC}$, a similar proof will apply. □

Theorem EUC.13. *A quadrilateral is a parallelogram iff there exists a point O which is the midpoint of both its diagonals.*

Proof. (A) Let $\square ABCD$ be a parallelogram. It follows from Theorem EUC.6 that $\square ABCD$ is rotund. By Theorem PSH.54(A) there exists a point O such that $\overline{AC} \cap \overline{BD} = \{O\}$. By Theorem EUC.12 $\overline{AB} \cong \overline{CD}$. By Theorem NEUT.65 (AEA) $\overline{AO} \cong \overline{CO}$ and $\overline{BO} \cong \overline{DO}$. By Definition NEUT.3(C) O is the midpoint of $\overline{AC}$ and of $\overline{BD}$.

(B) Let $\square ABCD$ be a quadrilateral for which there exists a point O such that O is the midpoint of $\overline{AC}$ and of $\overline{BD}$. By Theorem PSH.54(A) $\square ABCD$ is rotund. By Definition NEUT.3(C) $\overline{OA} \cong \overline{OC}$ and $\overline{OB} \cong \overline{OD}$. By Theorem NEUT.42 (Vertical Angles) $\angle AOB \cong \angle DOC$ and $\angle AOD \cong \angle BOC$. By Theorem NEUT.64 (EAE) $\overline{AB} \cong \overline{CD}$ and $\overline{AD} \cong \overline{BC}$. Also by Theorem NEUT.64 (EAE) $\angle OAB \cong \angle OCD$ and $\angle OBA \cong \angle ODC$. By Theorem EUC.11 $\overleftrightarrow{AB} \parallel \overleftrightarrow{DC}$ and $\overleftrightarrow{AD} \parallel BC$. By Definition EUC.5(B) $\square ABCD$ is a parallelogram. □

Definition EUC.14. (A) The **center** of a parallelogram is the point of intersection of its diagonals.

(B) The angles of a quadrilateral $\square ABCD$ are $\angle DAB$, $\angle ABC$, $\angle BCD$, and $\angle CDA$. Two angles of a quadrilateral are **adjacent** iff their corners are endpoints of an edge of the quadrilateral.

(C) A quadrilateral $ABCD$ is a **rectangle** iff of each the angles $\angle BAD$, $\angle ADC$, $\angle DCB$, and $\angle CBA$ is right.

Remark EUC.15. (A) If $\square ABCD$ is a rectangle, it follows from Theorem NEUT.47(A) that $\overleftrightarrow{AD} \parallel \overleftrightarrow{BC}$ and $\overleftrightarrow{AB} \parallel \overleftrightarrow{CD}$ so by Definition EUC.5(B) $\square ABCD$ is a parallelogram.

(B) For any rectangle $\square ABCD$, by Theorem EUC.12 $\overline{AB} \cong \overline{CD}$ and $\overline{AD} \cong \overline{BC}$, that is to say, opposite edges are congruent.

(C) Any rectangle is completely determined by three of its corners. That is, if $\square ABCD$ and $\square ABCD'$ are rectangles, then $D = D'$. For both $\overleftrightarrow{AB} \perp \overleftrightarrow{AD}$ and $\overleftrightarrow{AB} \perp \overleftrightarrow{AD'}$ so that by Theorem NEUT.47(B) $\overleftrightarrow{AD} = \overleftrightarrow{AD'}$; similarly, $\overleftrightarrow{CD} = \overleftrightarrow{CD'}$; by Exercise I.1, $D = D'$.

Theorem EUC.16.** *Let α be a belineation of the Euclidean plane $\mathcal{P}$ and let $\square ABCD$ be a parallelogram on $\mathcal{P}$. Then*

(1) *$\alpha(\square ABCD)$ is a parallelogram on $\mathcal{P}$, and*

(2) *$\alpha(\square ABCD) = \square\alpha(A)\alpha(B)\alpha(C)\alpha(D)$, where $\overleftrightarrow{\alpha(A)\alpha(B)} \parallel \overleftrightarrow{\alpha(C)\alpha(D)}$ and $\overleftrightarrow{\alpha(A)\alpha(D)} \parallel \overleftrightarrow{\alpha(B)\alpha(C)}$.*

Proof. In Chapter 7 Theorem COBE.5(13) we showed that if α is a belineation, $\alpha(\square ABCD)$ is a quadrilateral. Since $\square ABCD$ is a parallelogram $\overleftrightarrow{AB} \parallel \overleftrightarrow{CD}$ and $\overleftrightarrow{AD} \parallel \overleftrightarrow{BC}$. By Theorem CAP.3 $\alpha(\overleftrightarrow{AB}) \parallel \alpha(\overleftrightarrow{CD})$ and $\alpha(\overleftrightarrow{AD}) \parallel \alpha(\overleftrightarrow{BC})$. By Theorem COBE.5(13) $\alpha(\square ABCD) = \square\alpha(A)\alpha(B)\alpha(C)\alpha(D)$. So $\overleftrightarrow{\alpha(A)\alpha(B)} \parallel \overleftrightarrow{\alpha(C)\alpha(D)}$ and $\overleftrightarrow{\alpha(A)\alpha(D)} \parallel \overleftrightarrow{\alpha(B)\alpha(C)}$. $\square$

We are now ready to prove an important theorem in Euclidean geometry involving every collineation of the plane $\mathcal{P}$. Recall that $\text{par}(C, \mathcal{M})$ denotes the line through a point C which is parallel to the line $\mathcal{M}$.

Theorem EUC.17 (Collineations preserve midpoints). *Let A and C be distinct points on the Euclidean plane $\mathcal{P}$, φ be a collineation of $\mathcal{P}$, and M be the midpoint of $\overline{AC}$, then $\varphi(M)$ is the midpoint of $\overline{\varphi(A)\varphi(C)}$.*

Proof. Let B be any point off of $\overleftrightarrow{AC}$, $\mathcal{L} = \text{par}(C, \overleftrightarrow{AB})$, and $\mathcal{M} = \text{par}(A, \overleftrightarrow{BC})$. By Exercise IP.4, $\mathcal{L}$ and $\mathcal{M}$ intersect at a point D. Since $\overleftrightarrow{AB} \parallel \overleftrightarrow{CD}$ and $\overleftrightarrow{BC} \parallel \overleftrightarrow{AD}$, $\overline{AB} \cap \overline{CD} = \emptyset$ and $\overline{BC} \cap \overline{AD} = \emptyset$, so that by Definition PSH.31, $\square ABCD = \overline{AB} \cup \overline{BC} \cup \overline{CD} \cup \overline{DA}$ is a quadrilateral. By Definition EUC.5(B) it is a parallelogram, hence by Theorem EUC.6, it is rotund. By Theorem EUC.13, the diagonals $\overline{AC}$ and $\overline{BD}$ intersect at the midpoint of both diagonals, that is, at the point M.

Using Theorem CAP.1, we get

$$\varphi(\overleftrightarrow{AB}) = \overleftrightarrow{\varphi(A)\varphi(B)},\ \varphi(\overleftrightarrow{DC}) = \overleftrightarrow{\varphi(D)\varphi(C)},$$
$$\varphi(\overleftrightarrow{AC}) = \overleftrightarrow{\varphi(A)\varphi(C)}, \text{ and } \varphi(\overleftrightarrow{BD}) = \overleftrightarrow{\varphi(B)\varphi(D)}.$$

By Theorem CAP.3,

$$\overleftrightarrow{\varphi(A)\varphi(B)} \parallel \overleftrightarrow{\varphi(D)\varphi(C)} \text{ and}$$
$$\overleftrightarrow{\varphi(A)\varphi(D)} \parallel \overleftrightarrow{\varphi(B)\varphi(C)},$$

hence these pairs of lines do not intersect. Then

$$\overline{\varphi(A)\varphi(B)} \cup \overline{\varphi(B)\varphi(C)} \cup \overline{\varphi(C)\varphi(D)} \cup \overline{\varphi(D)\varphi(A)}$$

is a parallelogram by Definition EUC.5(B), and by Theorem EUC.6, it is rotund.

Notice that here all we are claiming is that the closed segments whose endpoints are $\varphi(A)$, $\varphi(B)$, $\varphi(C)$, and $\varphi(D)$ form a parallelogram; but this parallelogram is not necessarily the same as $\varphi(\square ABCD)$, and we are not invoking Theorem EUC.16.

Therefore by Theorem PSH.54(A), $\overline{\varphi(A)\varphi(C)} \cap \overline{\varphi(B)\varphi(D)} = \{N\}$ which by Theorem EUC.13 is the midpoint of $\overline{\varphi(A)\varphi(C)}$. By Theorem CAP.1

$$\overleftrightarrow{\varphi(A)\varphi(B)} \cap \overleftrightarrow{\varphi(B)\varphi(D)} = \{\varphi(M)\}$$

so that $N = \varphi(M)$, by Exercise I.1. Thus $\varphi(M)$ is the midpoint of $\overline{\varphi(A)\varphi(C)}$. □

Corollary EUC.17.1. *Let A, B, and C be points on the Euclidean plane $\mathcal{P}$ such that A–B–C and $\overline{AB} \cong \overline{BC}$ and let φ be a collineation of $\mathcal{P}$; then $\overline{\varphi(A)\varphi(B)} \cong \overline{\varphi(B)\varphi(C)}$.*

Proof. Since B is the midpoint of $\overline{AC}$ (cf Definition NEUT.3(C)), by Theorem EUC.17, $\varphi(B)$ is the midpoint of $\overline{\varphi(A)\varphi(C)}$. Hence $\overline{\varphi(A)\varphi(B)} \cong \overline{\varphi(B)\varphi(C)}$. □

Corollary EUC.17.2. *Let A, B, and M be points on the Euclidean plane $\mathcal{P}$, M the midpoint of $\overline{AB}$, and let φ be a collineation of $\mathcal{P}$. Then if A and B are fixed points of φ, M is a fixed point of φ.*

Proof. By Definition NEUT.3(C), A–M–B and $\overline{AM} \cong \overline{MB}$. Then by Corollary EUC.17.1, $\overline{A\varphi(M)} = \overline{\varphi(A)\varphi(M)} \cong \overline{\varphi(M)\varphi(B)} = \overline{\varphi(M)B}$ and A–$\varphi(M)$–B. Again by Definition NEUT.3(C), $\varphi(M)$ is the midpoint of $\overline{AB}$, and by uniqueness of midpoints (cf Theorem NEUT.50) $\varphi(M) = M$. □

Corollary EUC.17.3. *Let A, B, and M be points on the Euclidean plane $\mathcal{P}$, M the midpoint of $\overline{AB}$, and let φ be a belineation of $\mathcal{P}$. Then if A and M are fixed points of φ, B is a fixed point of φ.*

Proof. By Definition NEUT.3(C), $\overline{AM} \cong \overline{MB}$; by Corollary EUC.17.1, $\overline{AM} = \overline{\varphi(A)\varphi(M)} \cong \overline{\varphi(M)\varphi(B)} = \overline{M\varphi(B)}$. Then by Definition NEUT.3(C), M is the midpoint of $\overline{A\varphi(B)}$, and $\overline{M\varphi(B)} \cong \overline{AM} \cong \overline{MB}$. Since A–M–B, $\varphi(A)$–$\varphi(M)$–$\varphi(B)$ and thus A–M–$\varphi(B)$ (because φ is a belineation) and $\varphi(B) \in \overrightarrow{MB}$. By Property R.4 of Definition NEUT.2, $\varphi(B) = B$. □

Theorem EUC.18 (Criteria for a rectangle (A)).** *Let □ABCD be a quadrilateral such that $\overleftrightarrow{AB} \perp \overleftrightarrow{AD}$ and $\overleftrightarrow{AB} \perp \overleftrightarrow{BC}$. If any of the conditions (A) through (D) hold, then □ABCD is a rectangle.*

(A) $\overleftrightarrow{AB} \parallel \overleftrightarrow{CD}$;
(B) $\overleftrightarrow{AD} \perp \overleftrightarrow{CD}$ *(that is, the quadrilateral has three right angles);*
(C) $\overline{AD} \cong \overline{BC}$; *or*
(D) $\overline{AB} \cong \overline{CD}$.

Proof. By Theorem NEUT.47(A) $\overleftrightarrow{AD} \parallel \overleftrightarrow{BC}$.

(A) By Definition EUC.5(B) $\square ABCD$ is a parallelogram. By Theorem EUC.12(A) its opposite angles are congruent. By hypothesis, $\angle ABC$ and $\angle BAD$ are right angles, hence by Corollary NEUT.44.2 all the angles of $\square ABCD$ are right angles. By Definition EUC.14(C) $\square ABCD$ is a rectangle.

(B) By Theorem NEUT.47(A) $\overleftrightarrow{AB} \parallel \overleftrightarrow{DC}$. By part (A), $\square ABCD$ is a rectangle.

(C) By Theorem EUC.12(B) $\square ABCD$ is a parallelogram and hence by Theorem EUC.12(A) its opposite angles are congruent. The rest of the proof is the same as in part (A).

(D) Let C' be a point on $\overleftrightarrow{BC}$ such that $\overleftrightarrow{DC'} \parallel \overleftrightarrow{AB}$ and $C' \neq C$. By part (A) $\square ABC'D$ is a rectangle and $\angle DC'B$ is a right angle; that is, $\overleftrightarrow{DC'} \perp \overleftrightarrow{BC'} = \overleftrightarrow{BC}$. By Theorem EUC.12(A), $\overline{DC} \cong \overline{AB} \cong \overline{DC'}$. By Theorem NEUT.40(A) (*Pons Asinorum*) $\angle DCC' \cong \angle DC'C$; that is, both $\overleftrightarrow{DC} \perp \overleftrightarrow{BC}$ and $\overleftrightarrow{DC'} \perp \overleftrightarrow{BC}$. But this contradicts Theorem NEUT.48(A). Therefore $C' = C$, $\overleftrightarrow{DC} \parallel \overleftrightarrow{AB}$, and by part (A) $\square ABCD$ is a rectangle. □

Theorem EUC.19 (Criteria for a rectangle (B)).** *If one of the angles of a parallelogram $\square ABCD$ is right, then $\square ABCD$ is a rectangle.*

Proof. Without loss of generality, we can select any angle to be a right angle. Let $\angle DAB$ be a right angle; by Theorem NEUT.66 and Theorem EUC.12(A) $\angle BCD$ is right. Since $\overleftrightarrow{AD} \perp \overleftrightarrow{AB}$ and $\overleftrightarrow{AD} \parallel \overleftrightarrow{BC}$ by Theorem EUC.3 $\overleftrightarrow{AB} \perp \overleftrightarrow{BC}$ so that $\angle ABC$ is right. By Theorem EUC.18(B), $\square ABCD$ is a rectangle. □

Recall from Definition NEUT.99(C) that the *altitude* of $\triangle ABC$ through the point A is $\text{pr}\,(A, \overleftrightarrow{BC})$

Theorem EUC.20 (Concurrence of altitudes of a triangle).** *Let $\triangle ABC$ be a triangle on the Euclidean plane $\mathcal{P}$. Then* $\text{pr}\,(A, \overleftrightarrow{BC})$, $\text{pr}\,(B, \overleftrightarrow{AC})$, *and* $\text{pr}\,(C, \overleftrightarrow{AB})$ *are concurrent at a point O, i.e., the altitudes of a triangle are concurrent at a point O.*

Proof. Let $\mathcal{L} = \text{par}\,(A, \overleftrightarrow{BC})$, $\mathcal{M} = \text{par}\,(B, \overleftrightarrow{AC})$, and $\mathcal{N} = \text{par}\,(C, \overleftrightarrow{AB})$. By Theorem IP.5 points A', B', and C' exist such that $\mathcal{M} \cap \mathcal{N} = \{A'\}$, $\mathcal{L} \cap \mathcal{N} = \{B'\}$, and $\mathcal{L} \cap \mathcal{M} = \{C'\}$. By Definition EUC.5(B) both quadrilaterals $\square AB'CB$ and $\square AC'BC$ are parallelograms, so that by Theorem EUC.12(A), $\overline{AB'} \cong \overline{BC} \cong \overline{AC'}$. Then by Definition NEUT.3(C) A is the midpoint of $\overline{B'C'}$. Similar arguments show that B is the midpoint of $\overline{A'C'}$ and C is the midpoint of $\overline{A'B'}$. Thus $\text{pr}\,(A, \overleftrightarrow{BC})$ is the perpendicular bisecting line of $\overleftrightarrow{B'C'}$, $\text{pr}\,(B, \overleftrightarrow{AC})$ is the perpendicular bisecting line of

$\overleftrightarrow{A'C'}$, and pr $(C, \overleftrightarrow{AB})$ is the perpendicular bisecting line of $\overleftrightarrow{A'B'}$. By Theorem EUC.9 $\mathcal{L}$, $\mathcal{M}$, and $\mathcal{N}$ are concurrent at a point O. □

Definition EUC.21. The point O of Theorem EUC.20 is the **orthocenter** of $\triangle ABC$.

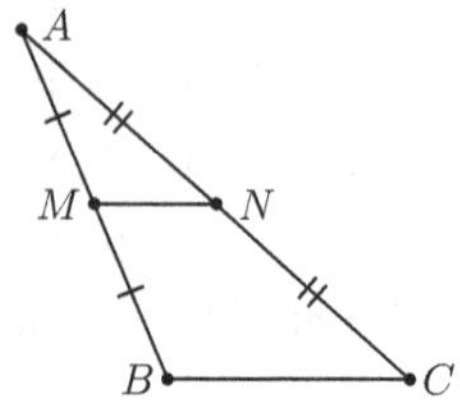

Fig. 11.2 For Theorem EUC.22.

Theorem EUC.22 (Parallel projection preserves midpoints). *Let A, B, and C be noncollinear points on the Euclidean plane $\mathcal{P}$. Let M be the midpoint of $\overline{AB}$ and $\mathcal{L}$ = par $(M, \overleftrightarrow{BC})$. Then $\mathcal{L}$ and $\overleftrightarrow{AC}$ intersect at a point N which is the midpoint of $\overline{AC}$.*

Proof. See Figure 11.2. By Theorem PSH.6 we may invoke the Postulate of Pasch. Therefore, since $\mathcal{L}$ intersects $\overline{AB}$, it must intersect exactly one of $\overline{AC}$, $\overline{BC}$, or $\{C\}$. The last two possibilities are ruled out because $\mathcal{L} \parallel \overleftrightarrow{BC}$. Hence $\mathcal{L}$ intersects $\overline{AC}$ at some point N.

Using Axiom PS let $\mathcal{M}$ = par $(N, \overleftrightarrow{AB})$; then by the same kind of reasoning as for the point N, $\mathcal{M}$ intersects $\overleftrightarrow{BC}$ at a point Q between B and C. By Definition EUC.5(B) $\square BMNQ$ is a parallelogram. We leave it to the reader (as Exercise EUC.4) to prove that $\triangle NQC \cong \triangle AMN$, so that $\overline{AN} \cong \overline{CN}$, proving (*cf* Definition NEUT.3(C)) that N is the midpoint of $\overline{AC}$. □

Corollary EUC.23. *Let $\mathcal{P}$ be a Euclidean plane and A, B, and C be noncollinear points on $\mathcal{P}$. If M is the midpoint of $\overline{AB}$ and N is the midpoint of $\overline{AC}$, then $\overleftrightarrow{MN} \parallel \overleftrightarrow{BC}$. Moreover, if L is the midpoint of $\overline{BC}$, then $\overline{BL} \cong \overline{MN}$.*

Proof. The proof is Exercise EUC.5. □

Corollary EUC.24.** *Let $\square ABCD$ be a parallelogram on the Euclidean plane $\mathcal{P}$. If M is the midpoint of $\overline{AB}$ and O is the center of the parallelogram, then $\overleftrightarrow{MO}$ and $\overleftrightarrow{DC}$ intersect at the midpoint N of $\overline{DC}$.*

Proof. Since M is the midpoint of $\overline{AB}$ and O is the midpoint of $\overline{AC}$ (*cf* Theorem EUC.13), by Corollary EUC.23 $\overleftrightarrow{OM} \parallel \overleftrightarrow{BC}$. By Theorem EUC.22 N is the midpoint of $\overline{DC}$. □

Corollary EUC.25. *Let A, B, and C be noncollinear points on the Euclidean plane* $\mathcal{P}$, *F be the midpoint of* $\overline{AB}$, *G the midpoint of* $\overline{AC}$, *H the midpoint of* $\overline{BC}$, *and M the midpoint of* $\overline{FG}$. *Then M is the midpoint of* $\overline{AH}$.

Proof. By Corollary EUC.23 $\overleftrightarrow{FA} \parallel \overleftrightarrow{HG}$ and $\overleftrightarrow{FH} \parallel \overleftrightarrow{AG}$. By Definition EUC.5(B) $\square AGHF$ is a parallelogram. By Theorem EUC.13 $\overline{FG}$ and $\overline{AH}$ intersect at their common midpoint. □

Definition EUC.26. The **median** of triangle ABC through A is the line $\overleftrightarrow{AM}$, where M is the midpoint of $\overline{BC}$.

Theorem EUC.27 (Concurrence of medians of a triangle). *Let A, B, and C be noncollinear points on the Euclidean plane* $\mathcal{P}$, *F the midpoint of* $\overline{AB}$, *G the midpoint of* $\overline{AC}$, *H the midpoint of* $\overline{BC}$, *O the point of intersection of* $\overline{BG}$ *and* $\overline{CF}$, *I the midpoint of* $\overline{BO}$, *and J the midpoint of* $\overline{CO}$. *Then* $\square IFGJ$ *is a parallelogram and A, O, and H are collinear. That is to say, the medians of a triangle on a Euclidean plane intersect at a point inside the triangle.*

Proof. Applying Corollary EUC.23 to $\triangle ABC$ and also to $\triangle OBC$ we find that $\overleftrightarrow{FG}$ and $\overleftrightarrow{IJ}$ are both parallel to $\overleftrightarrow{BC}$. Applying Corollary EUC.23 to $\triangle AOC$ and to $\triangle AOB$ we have $\overleftrightarrow{GJ}$ and $\overleftrightarrow{FI}$ are both parallel to $\overleftrightarrow{AO}$. Therefore by Theorem IP.6 the opposite edges of $\square GFIJ$ are parallel and by Definition EUC.5(B), $\square GFIJ$ is a parallelogram. Let M be the midpoint of $\overline{FG}$ and N be the midpoint of $\overline{IJ}$. Note that O is the center of $\square GFIJ$, so that by Corollary EUC.24, M, O, and N are collinear.

Applying Corollary EUC.25 to both $\triangle ABC$ and $\triangle OBC$, we find that A, M, N, and H are collinear. $\overleftrightarrow{AH}$ is the median of $\triangle ABC$ through A. Since $\overline{BG}$ and $\overline{CF}$ are subsets of the inside of $\triangle ABC$, O belongs to ins $\triangle ABC$. This completes the proof. □

Definition EUC.28. The point of intersection of the medians of a triangle is the **centroid** of the triangle.

Theorem EUC.29 (Fixed points of a belineation).** *Let φ be a belineation of the Euclidean plane $\mathcal{P}$, A and B be distinct fixed points of φ, M be the midpoint of $\overline{AB}$ and C and D be the points such that C–A–B–D and $\overline{CA} \cong \overline{BD} \cong \overline{AB}$, then M, C, and D are fixed points of φ.*

Proof. By Corollary EUC.17.2, M is a fixed point for φ. Since B is the midpoint of $\overline{AD}$, and A is the midpoint of $\overline{BC}$, by Corollary EUC.17.3 both C and D are fixed points for φ. □

See also Chapter 20 "Belineations of a Euclidean/LUB plan" for additional results related to Theorem EUC.29.

Definition EUC.30. Let $\angle BAC$ and $\angle DEF$ be acute angles on a Euclidean plane $\mathcal{P}$. Each of these angles is a **complement** of the other (and they are said to be **complementary** angles) iff there exist noncollinear points G, H, I, and J such that $I \in \text{ins}\,\angle GHJ$, $\angle BAC \cong \angle GHI$, $\angle DEF \cong \angle JHI$, and $\angle GHJ$ is right.

Theorem EUC.31. *Let A, B, and C be noncollinear points on the Euclidean plane $\mathcal{P}$ such that $\angle ACB$ is right, then $\angle ABC$ and $\angle BAC$ are complements of each other.*

Proof. Let $\mathcal{L} = \text{pr}(B, \overleftrightarrow{BC})$ and $\mathcal{M} = \text{pr}(A, \overleftrightarrow{AC})$. By Theorem NEUT.44, $\overleftrightarrow{BC} \perp \overleftrightarrow{AC}$, and by Theorem NEUT.47(A), $\mathcal{M} \parallel \overleftrightarrow{BC}$ and $\mathcal{L} \parallel \overleftrightarrow{AC}$. By Theorem NEUT.44 there exists a point D such that $\mathcal{L} \cap \mathcal{M} = \{D\}$ and by Theorem EUC.3 $\mathcal{L} \perp \mathcal{M}$, so that $\angle DBC$ is a right angle.

By Theorem EUC.11, $\angle DBA \cong \angle CAB$. Since opposite sides of the quadrilateral $\square ACBD$ are parallel, it is rotund, and by Theorem PSH.54(B), $A \in \text{ins}\,\angle DBC$. Then by Definition EUC.30, $\angle ABC$ and $\angle BAC$ are complementary. □

Theorem EUC.32. *Complements of acute congruent angles are congruent.*

Proof. The proof is left to the reader as Exercise EUC.6. □

Definition EUC.33. A triple of angles $\{\angle BAC, \angle DEF, \angle GHI\}$ on the Euclidean plane $\mathcal{P}$ is **complete** iff there exists a line $\mathcal{L}$ on $\mathcal{P}$, points Q, O, and R on $\mathcal{L}$ with Q–O–R, and distinct points S and T on a side of $\mathcal{L}$ such that $\angle QOS \cong \angle BAC$, $\angle SOT \cong \angle DEF$, and $\angle GHI \cong \angle ROT$.

Theorem EUC.34. *Let A, B, and C be noncollinear points on the Euclidean plane $\mathcal{P}$, then the triple of angles of $\triangle ABC$ is complete.*

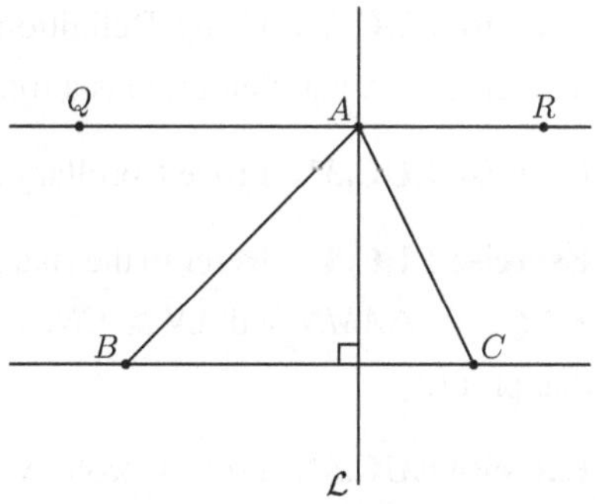

Fig. 11.3 For Theorem EUC.34.

Proof. See Figure 11.3. Let $\mathcal{L} = \text{pr}(A, \overleftrightarrow{BC})$ and let Q and R be points on $\mathcal{L}$ such that Q is on the side of $\overleftrightarrow{AB}$ opposite the C-side, and R is on the C-side of $\overleftrightarrow{AB}$. By Axiom PSA Q–A–R. By Theorem EUC.11 $\angle ABC \cong \angle BAQ$, and $\angle ACB \cong \angle CAR$. By Definition EUC.33 $\{\angle BAC, \angle ABC, \angle ACB\}$ is complete. □

Suppose a triangle $\mathcal{T}$ has two angles that are congruent to two angles of another triangle $\mathcal{S}$. Theorem EUC.34 suggests that the third angle of each of these triangles must also be congruent to each other. The next theorem proves this.

Theorem EUC.35. *On a Euclidean plane, let A, B, and C be noncollinear points and A', B', and C' be noncollinear points. If $\angle ABC \cong \angle A'B'C'$ and $\angle BAC \cong \angle B'A'C'$, then $\angle ACB \cong \angle A'C'B'$.*

Proof. By Theorem NEUT.67 (segment construction) there exist points $A'' \in \overrightarrow{BA}$ and $C'' \in \overrightarrow{BC}$ such that $\overline{BA''} \cong \overline{B'A'}$ and $\overline{BC''} \cong \overline{B'C'}$. Since $\angle ABC \cong \angle A'B'C'$, by Theorem NEUT.64 (EAE) $\Delta A'B'C' \cong \Delta A''BC''$ and corresponding angles are congruent. That is, $\angle BA''C'' \cong \angle B'A'C'$ and $\angle A''C''B \cong \angle A'C'B'$.

Then $\angle BAC \cong \angle B'A'C' \cong \angle BA''C''$. Using the equivalence of (1) and (4) in Theorem EUC.11, we have $\overleftrightarrow{A''C''} \parallel \overleftrightarrow{AC}$. Again by Theorem EUC.11, $\angle ACB \cong \angle A''C''B \cong \angle A'C'B'$. □

11.2 Exercises for Euclidean geometry

Answers to starred (*) exercises may be accessed from the home page for this book at www.springer.com.

Exercise EUC.1*. Prove Corollary EUC.4, using Theorem EUC.3.

Exercise EUC.2*. Using Definition PSH.31 and Theorem PSH.12, prove Theorem EUC.6: A parallelogram is a rotund quadrilateral.

Exercise EUC.3*. Prove Corollary EUC.8.

Exercise EUC.4*. Refer to the statement and proof of Theorem EUC.22; show that $\triangle NQC \cong \triangle AMN$ and $\overline{AN} \cong \overline{CN}$, so that N is the midpoint of $\overline{AC}$, thus completing that proof.

Exercise EUC.5*. Prove Corollary EUC.23.

Exercise EUC.6*. Prove Theorem EUC.32: Complements of acute congruent angles are congruent.

Exercise EUC.7*. Prove Corollary EUC.3.1: let $\mathcal{R}_{\mathcal{M}}$ be the line reflection over $\mathcal{M}$, and let $\mathcal{L}$ be a fixed line for $\mathcal{R}_{\mathcal{M}}$. Then $\mathcal{N} \parallel \mathcal{L}$ iff $\mathcal{N}$ is a fixed line for $\mathcal{R}_{\mathcal{M}}$.

Chapter 12
Isometries of a Euclidean Plane (ISM)

Acronym: *ISM*
Dependencies: *all prior Chapters 1 through 11*
New Axioms: *none*
New Terms Defined: *glide reflection*

Abstract: This chapter gives a complete classification of isometries on a Euclidean plane, proves a technical theorem to be used later to develop the properties of dilations, and describes a method for constructing a translation with a given action.

Until now, we have studied isometries mainly in the context of neutral geometry, where they were defined. In Euclidean geometry, we can give a complete classification of isometries; we do this in Theorem ISM.17. Theorem ISM.19 is the technical theorem referred to in the abstract, which will be used in the proof of the properties of half-rotations, which, in turn, are used in Chapter 13 to prove properties of dilations (*cf* Theorems DLN.4 and DLN.7).

In this chapter we will loosely follow J. Diller and J. Boczeck, in *Euclidean Planes*, Chapter 4 in *Fundamentals of Mathematics*, Volume 2, H. Behnke, F. Bachmann, K. Fladt, and H. Kunle, eds, translated by S. Gould, MIT Press, 1974 [2]. See also F. Bachmann, *Aufbau der Geometrie aus dem Spiegelungsbegriff*, 2nd ed., *Grundlehren der mathematischen Wissenschaften*, Springer (1973) [1].

E.J. Specht et al., *Euclidean Geometry and its Subgeometries*,
DOI 10.1007/978-3-319-23775-6_12

12.1 Properties and classification of isometries

Remark ISM.1. In this chapter a plane is a Euclidean plane, that is, a neutral plane for which Axiom PS holds. We will occasionally use the notation (from Definition CAP.10) $\mathcal{L} \;\mathbb{PE}\; \mathcal{M}$ to mean that either $\mathcal{L} \parallel \mathcal{M}$ or $\mathcal{L} = \mathcal{M}$.

Remark ISM.2. If α is an isometry of the Euclidean plane, and has three non-collinear fixed points, then by Theorem NEUT.24 $\alpha = \iota$, the identity mapping of $\mathcal{P}$ onto $\mathcal{P}$.

If α has distinct fixed points A and B, but no fixed point off of $\overleftrightarrow{AB}$, then by Exercise NEUT.6, $\alpha = \mathcal{R}_{\overleftrightarrow{AB}}$. Moreover, if α has one and only one fixed point O, then by Theorem ROT.16, α is a rotation of $\mathcal{P}$ about O.

The parallel Axiom PS is required for an adequate treatment of the case where α has no fixed point, and it is this case we particularly address in this chapter. First we do some preliminary explorations.

Theorem ISM.3. *Let O be a point on the Euclidean plane $\mathcal{P}$, and let $\mathcal{R}_O$ be a point reflection about O.*

(A) *O is a fixed point of $\mathcal{R}_O$, which has no other fixed point.*
(B) *Every line through O is a fixed line of $\mathcal{R}_O$.*
(C) *If $\mathcal{L}$ is a line and $O \notin \mathcal{L}$, then $\mathcal{R}_O(\mathcal{L}) \parallel \mathcal{L}$, so that by Definition CAP.17, $\mathcal{R}_O$ is a dilation of $\mathcal{P}$.*

Proof. (A) The proof is Theorem ROT.2.
(B) The proof is Theorem ROT.3.
(C) Let $\mathcal{L}$ be any line on $\mathcal{P}$ such that $O \notin \mathcal{L}$ and let X and Y be distinct points on $\mathcal{L}$. By Theorem ROT.3 X–O–$\mathcal{R}_O(X)$, Y–O–$\mathcal{R}_O(Y)$, $\overline{OX} \cong \overline{O\mathcal{R}_O(X)}$, and $\overline{OY} \cong \overline{O\mathcal{R}_O(Y)}$. By Theorem NEUT.42 (vertical angles) and Theorem NEUT.64 (EAE) $\angle XYO \cong \angle \mathcal{R}_O(X)\mathcal{R}_O(Y)O$. By Theorem EUC.11 $\mathcal{L} \parallel \mathcal{R}_O(\mathcal{L})$. By Definition CAP.17 $\mathcal{R}_O$ is a dilation of $\mathcal{P}$. □

Theorem ISM.4. (A) *Let $\mathcal{P}$ be a Euclidean plane and let A and B be distinct points on $\mathcal{P}$, then $\mathcal{R}_B \circ \mathcal{R}_A$ is a translation of $\mathcal{P}$.*
(B) *Let $\mathcal{P}$ be a Euclidean plane and let $\mathcal{L}$ and $\mathcal{M}$ be parallel lines on $\mathcal{P}$. Then $\alpha = \mathcal{R}_\mathcal{M} \circ \mathcal{R}_\mathcal{L}$ is a translation of $\mathcal{P}$. Moreover, the set of fixed lines of α is the set of lines each of which is perpendicular to $\mathcal{L}$ (and $\mathcal{M}$).*

(C) *If $\mathcal{P}$ is a Euclidean plane and if A and B are points on $\mathcal{P}$ such that $\mathcal{R}_A = \mathcal{R}_B$, then $A = B$.*

Proof. (A) Let $\mathcal{L}$ be any line on $\mathcal{P}$. By Theorem ISM.3 $\mathcal{R}_A$ and $\mathcal{R}_B$ are dilations of $\mathcal{P}$. By Definition CAP.17 $\mathcal{R}_A(\mathcal{L})$ $\mathbb{PE}$ $\mathcal{L}$ and $(\mathcal{R}_B \circ \mathcal{R}_A)(\mathcal{L}) = \mathcal{R}_B(\mathcal{R}_A(\mathcal{L}))$ $\mathbb{PE}$ $\mathcal{R}_A(\mathcal{L})$. Therefore by Theorem IP.6 $(\mathcal{R}_B \circ \mathcal{R}_A)(\mathcal{L})$ $\mathbb{PE}$ $\mathcal{L}$.

Let $\mathcal{M} = \text{pr}(A, \overleftrightarrow{AB})$ (*cf* Definition NEUT.99) and $\mathcal{N} = \text{pr}(B, \overleftrightarrow{AB})$. Using Definition ROT.1, by Theorem ROT.4 $\mathcal{R}_A = \mathcal{R}_{\overleftrightarrow{AB}} \circ \mathcal{R}_\mathcal{M}$ and $\mathcal{R}_A = \mathcal{R}_\mathcal{N} \circ \mathcal{R}_{\overleftrightarrow{AB}}$, so that

$$\mathcal{R}_B \circ \mathcal{R}_A = \mathcal{R}_\mathcal{N} \circ \mathcal{R}_{\overleftrightarrow{AB}} \circ \mathcal{R}_{\overleftrightarrow{AB}} \circ \mathcal{R}_\mathcal{M} = \mathcal{R}_\mathcal{N} \circ \mathcal{R}_\mathcal{M}.$$

By Theorem NEUT.47(A) $\mathcal{M} \parallel \mathcal{N}$.

If $\mathcal{R}_\mathcal{N} \circ \mathcal{R}_\mathcal{M}(X) = X$, $\mathcal{R}_\mathcal{N}(X) = \mathcal{R}_\mathcal{M}(X)$. By Theorem NEUT.48(A) and Theorem EUC.3, $\overleftrightarrow{X\mathcal{R}_\mathcal{N}(X)}$ and $\overleftrightarrow{X\mathcal{R}_\mathcal{M}(X)}$ are both perpendicular to both $\mathcal{N}$ and $\mathcal{M}$. Hence by Theorem NEUT.48(A) these are the same line, which we will call $\mathcal{J}$.

Let $\{C\} = \mathcal{J} \cap \mathcal{N}$ and let $\{D\} = \mathcal{J} \cap \mathcal{M}$. If X is either C or D, X would be a fixed point for $\mathcal{R}_\mathcal{N}$ or $\mathcal{R}_\mathcal{M}$ but not both, so that $\mathcal{R}_\mathcal{N}(X) \neq \mathcal{R}_\mathcal{M}(X)$. If $X \neq C$ and $X \neq D$, then since $\mathcal{R}_\mathcal{N}(X) = \mathcal{R}_\mathcal{M}(X)$, $\overline{X\mathcal{R}_\mathcal{N}(X)} = \overline{X\mathcal{R}_\mathcal{M}(X)}$. Since $\mathcal{R}_\mathcal{N}(C) = C$ by Theorem NEUT.15(5) we have $\mathcal{R}_\mathcal{N}(\overline{CX}) = \overline{\mathcal{R}_\mathcal{N}(C)\mathcal{R}_\mathcal{N}(X)} = \overline{C\mathcal{R}_\mathcal{N}(X)}$ so that $\overline{CX} \cong \overline{C\mathcal{R}_\mathcal{N}(X)}$ and C is therefore the midpoint of $\overline{X\mathcal{R}_\mathcal{N}(X)}$. Similarly, D is the midpoint of $\overline{X\mathcal{R}_\mathcal{M}(X)} = \overline{X\mathcal{R}_\mathcal{N}(X)}$. By Theorem NEUT.50, $C = D$; but $C \in \mathcal{N}$ and $D \in \mathcal{M}$ so that $\mathcal{N}$ and $\mathcal{M}$ are not parallel, a contradiction. Therefore $\mathcal{R}_\mathcal{N} \circ \mathcal{R}_\mathcal{M}$ has no fixed point, and by Definition CAP.6 is a translation.

(B) Let A be any point on $\mathcal{L}$, let $B \in \mathcal{M}$ be a point such that $\mathcal{N} = \overleftrightarrow{AB}$ is perpendicular to both $\mathcal{L}$ and $\mathcal{M}$. (Here we have used Theorem NEUT.48(A) and Theorem EUC.3.) By Theorem ROT.24, $\mathcal{R}_\mathcal{M} \circ \mathcal{R}_\mathcal{L} = \mathcal{R}_B \circ \mathcal{R}_A$, which by part (A) above, is a translation.

Let $\mathcal{J}$ be any line perpendicular to $\mathcal{L}$ (and hence by Theorem EUC.3 to $\mathcal{M}$). By Theorem NEUT.44 $\mathcal{J}$ is a fixed line of $\mathcal{R}_\mathcal{L}$ and $\mathcal{R}_\mathcal{M}$. Hence $\mathcal{J}$ is a fixed line of $\mathcal{R}_\mathcal{M} \circ \mathcal{R}_\mathcal{L}$.

Note that $(\mathcal{R}_\mathcal{M} \circ \mathcal{R}_\mathcal{L})(A) = \mathcal{R}_\mathcal{M}(\mathcal{R}_\mathcal{L}(A)) = \mathcal{R}_\mathcal{M}(A) = C$, where C is the point on $\mathcal{N}$ such that A–B–C and $\overline{BC} \cong \overline{AB}$. By Theorem CAP.8(B) $\overleftrightarrow{AC}$ is a fixed line of $\mathcal{R}_\mathcal{M} \circ \mathcal{R}_\mathcal{L}$. Let $\mathcal{J}$ be any fixed line of $\mathcal{R}_\mathcal{M} \circ \mathcal{R}_\mathcal{L}$. By Theorem CAP.8(A) there exists a point Q on $\mathcal{P}$ such that $\mathcal{J} = \overleftrightarrow{Q(\mathcal{R}_\mathcal{M} \circ \mathcal{R}_\mathcal{L}(Q))}$. By

Theorem CAP.8(C) $\mathcal{J} \parallel \overleftrightarrow{AC}$. Therefore the set of fixed lines of $\alpha = \mathcal{R}_{\mathcal{M}} \circ \mathcal{R}_{\mathcal{L}}$ is $\{\mathcal{J} \mid \mathcal{J}$ is a line on $\mathcal{P}$ and $\mathcal{J}\ \mathbb{PE}\ \overleftrightarrow{AC}\} = \{\mathcal{J} \mid \mathcal{J}$ is a line on $\mathcal{P}$ and $\mathcal{J} \perp \mathcal{L}\}$.

(C) If $\mathcal{R}_A = \mathcal{R}_B$, then B would be a fixed point of $\mathcal{R}_A$; by Theorem ISM.3, $\mathcal{R}_A$ has no fixed point other than A; therefore $A = B$. □

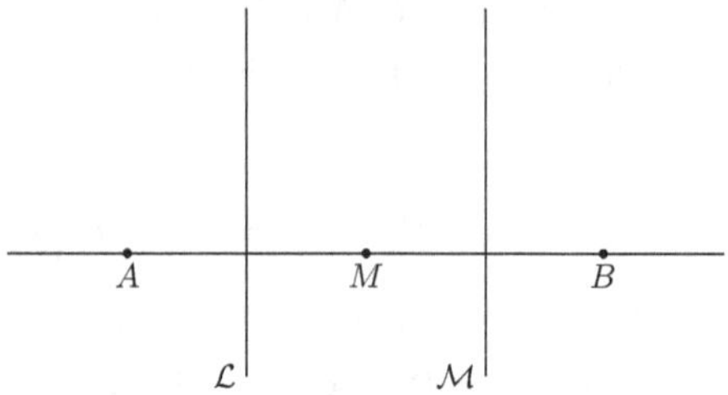

Fig. 12.1 For Theorem ISM.5.

Theorem ISM.5. *Let $\mathcal{P}$ be a Euclidean plane and let A and B be distinct points on $\mathcal{P}$. Then there exists a unique translation τ of $\mathcal{P}$ such that $\tau(A) = B$. Moreover, there exist parallel lines $\mathcal{L}$ and $\mathcal{M}$ on $\mathcal{P}$ such that $\tau = \mathcal{R}_{\mathcal{M}} \circ \mathcal{R}_{\mathcal{L}}$.*

Proof. (A: Uniqueness.) This is Theorem CAP.9. See Figure 12.1.

(B: Existence.) Let M be the midpoint of $\overline{AB}$ (by Theorem NEUT.50), C be the midpoint of $\overline{AM}$, and D be the midpoint of $\overline{MB}$. Let $\mathcal{L} = \text{pr}(C, \overleftrightarrow{AB})$ and $\mathcal{M} = \text{pr}(D, \overleftrightarrow{AB})$. By Theorem NEUT.47(A) $\mathcal{L} \parallel \mathcal{M}$. Let $\tau = \mathcal{R}_{\mathcal{M}} \circ \mathcal{R}_{\mathcal{L}}$. By Theorem NEUT.52 $\mathcal{R}_{\mathcal{L}}(A) = M$ and $\mathcal{R}_{\mathcal{M}}(M) = B$ so that $\tau(A) = B$. By Theorem ISM.4 τ is a translation of $\mathcal{P}$. □

Theorem ISM.6. *Let $\mathcal{P}$ be a Euclidean plane and let τ be a translation of $\mathcal{P}$. Then τ is an isometry of $\mathcal{P}$.*

Proof. For any translation τ, and any point A in the plane, $\tau(A)$ is a point on the plane; once this is specified, Theorem ISM.5 constructs τ as the composition of two reflections. Hence τ is an isometry by Definition NEUT.3(A). □

Theorem ISM.7. *Let $\mathcal{P}$ be a Euclidean plane.*

(A) *For every line $\mathcal{L}$ on $\mathcal{P}$ there exists a translation τ of $\mathcal{P}$ whose set of fixed lines is*

$$\{\mathcal{J} \mid \mathcal{J} \subseteq \mathcal{P} \text{ and } \mathcal{J}\ \mathbb{PE}\ \mathcal{L}\}.$$

(B) *If τ and σ are translations of $\mathcal{P}$, then $\sigma \circ \tau = \tau \circ \sigma$.*

Proof. (A) Let A and B be distinct points on $\mathcal{L}$, $\mathcal{M} = \text{pr}(A, \mathcal{L})$, and $\mathcal{N} = \text{pr}(B, \mathcal{L})$. By Theorem ISM.4 $\tau = \mathcal{R}_{\mathcal{N}} \circ \mathcal{R}_{\mathcal{M}}$ is a translation of $\mathcal{P}$ whose set of fixed lines is $\{\mathcal{J} \mid \mathcal{J}$ is a line on $\mathcal{P}$ and $\mathcal{J}$ PE $\mathcal{L}\}$.

(B) By part (A) and Theorem CAP.15(B), $\sigma \circ \tau = \tau \circ \sigma$. □

Theorem ISM.8. *Let $\mathcal{P}$ be a Euclidean plane.*

(A) *The set of translations of $\mathcal{P}$, together with ι is an abelian group with respect to composition of mappings.*

(B) *Let $\mathcal{M}$ be a line on $\mathcal{P}$ and let $\Pi_{\mathcal{M}} = \{\mathcal{J} \mid \mathcal{J}$ is a line on $\mathcal{P}$* and *$\mathcal{J}$ PE $\mathcal{M}\}$. Then $\Gamma_{\mathcal{M}} = \{\tau \mid \tau$ is a translation of $\mathcal{P}$ whose set of fixed lines is $\Pi_{\mathcal{M}}$ or $\tau = \iota\}$ is an abelian group with respect to composition of mappings.*

Proof. Follows from Theorem CAP.12 and Theorem ISM.7. □

Theorem ISM.9. *Let $\mathcal{P}$ be a Euclidean plane, τ be a translation of $\mathcal{P}$, and A and B be distinct points on $\mathcal{P}$. Then $\overline{A\tau(A)} \cong \overline{B\tau(B)}$.*

Proof. By Theorem ISM.5 there exists a translation σ of $\mathcal{P}$ such that $\sigma(A) = B$. Thus

$$\overline{B\tau(B)} = \overline{\sigma(A)\tau(\sigma(A))} = \overline{\sigma(A)\sigma(\tau(A))} = \sigma(\overline{A\tau(A)}).$$

Here we have used Theorems COBE.5(5) and ISM.8. By Definition NEUT.3(B) $\overline{B\tau(B)} \cong \overline{A\tau(A)}$. □

Theorem ISM.10. *Let $\mathcal{P}$ be a Euclidean plane and let A, B, and C be distinct points on $\mathcal{P}$. Then there exists a unique point D on $\mathcal{P}$ such that $\mathcal{R}_C \circ \mathcal{R}_B \circ \mathcal{R}_A = \mathcal{R}_D$. Moreover, $D \in \overleftrightarrow{A(\mathcal{R}_C \circ \mathcal{R}_B(A))}$.*

Proof. (A: Uniqueness.) If D and D' are points on $\mathcal{P}$ such that $\mathcal{R}_C \circ \mathcal{R}_B \circ \mathcal{R}_A = \mathcal{R}_D$ and $\mathcal{R}_C \circ \mathcal{R}_B \circ \mathcal{R}_A = \mathcal{R}_{D'}$, then $\mathcal{R}_D = \mathcal{R}_{D'}$ and by Theorem ISM.4(C) $D = D'$.

(B: Existence.) Since $\mathcal{R}_C \circ \mathcal{R}_B$ is a translation of $\mathcal{P}$ (cf Theorem ISM.4(A)) and since a translation has no fixed points (Definition CAP.6), by Exercise ROT.1(A) $(\mathcal{R}_C \circ \mathcal{R}_B \circ \mathcal{R}_A)(A) = (\mathcal{R}_C \circ \mathcal{R}_B)(A) \neq A$. By Theorem NEUT.50 there exists a unique midpoint D of $\overline{A(\mathcal{R}_C(\mathcal{R}_B(A)))}$. By Exercise ROT.7 $\mathcal{R}_D(A) = \mathcal{R}_C(\mathcal{R}_B(A))$, so that $\mathcal{R}_D(\mathcal{R}_A(A)) = \mathcal{R}_C(\mathcal{R}_B(A))$. By Theorem ISM.4(A) $\mathcal{R}_D \circ \mathcal{R}_A$ and $\mathcal{R}_C \circ \mathcal{R}_B$ are translations of $\mathcal{P}$. By Theorem ISM.5 $\mathcal{R}_D \circ \mathcal{R}_A = \mathcal{R}_C \circ \mathcal{R}_B$. Multiplying on the right by $\mathcal{R}_A$ and using Exercise ROT.1(A) again, $\mathcal{R}_D = \mathcal{R}_C \circ \mathcal{R}_B \circ \mathcal{R}_A$. □

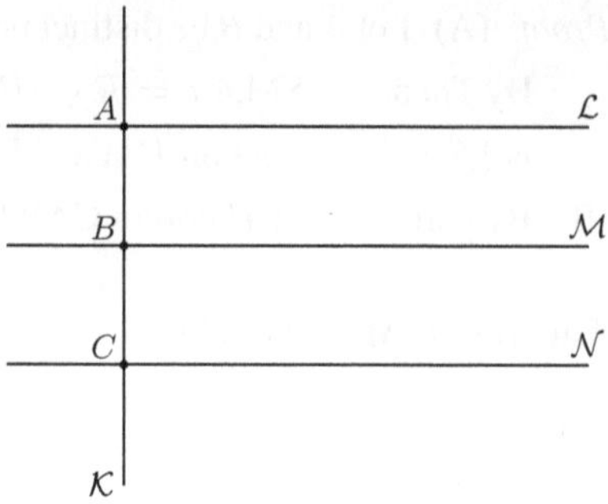

Fig. 12.2 For Theorem ISM.11.

Theorem ISM.11. *Let $\mathcal{P}$ be a Euclidean plane and let $\mathcal{L}$, $\mathcal{M}$, and $\mathcal{N}$ be lines on $\mathcal{P}$ such that $\mathcal{L} \parallel \mathcal{M} \parallel \mathcal{N}$. Then there exists a unique line $\mathcal{J}$ such that $\mathcal{J} \parallel \mathcal{L}$ and $\mathcal{R}_{\mathcal{N}} \circ \mathcal{R}_{\mathcal{M}} \circ \mathcal{R}_{\mathcal{L}} = \mathcal{R}_{\mathcal{J}}$.*

Proof. See Figure 12.2.

(I: Existence.) Let A be a point on $\mathcal{L}$ and let $\mathcal{K} = \text{pr}(A, \mathcal{L})$. By Corollary EUC.4 $\mathcal{K} \perp \mathcal{M}$ and $\mathcal{K} \perp \mathcal{N}$. Let $B = \text{ftpr}(A, \mathcal{M})$ and $C = \text{ftpr}(A, \mathcal{N})$. By Definition ROT.1, Remark NEUT.1.3, and Theorem ROT.4,

$$\begin{aligned}\mathcal{R}_{\mathcal{N}} \circ \mathcal{R}_{\mathcal{M}} \circ \mathcal{R}_{\mathcal{L}} &= \mathcal{R}_{\mathcal{N}} \circ \mathcal{R}_{\mathcal{K}} \circ \mathcal{R}_{\mathcal{K}} \circ \mathcal{R}_{\mathcal{M}} \circ \mathcal{R}_{\mathcal{L}} \circ \mathcal{R}_{\mathcal{K}} \circ \mathcal{R}_{\mathcal{K}} \\ &= \mathcal{R}_C \circ \mathcal{R}_B \circ \mathcal{R}_A \circ \mathcal{R}_{\mathcal{K}}.\end{aligned}$$

By Theorem ISM.10 there exists a point Q on $\mathcal{K}$ such that $\mathcal{R}_C \circ \mathcal{R}_B \circ \mathcal{R}_A = \mathcal{R}_Q$. Thus $\mathcal{R}_{\mathcal{N}} \circ \mathcal{R}_{\mathcal{M}} \circ \mathcal{R}_{\mathcal{L}} = \mathcal{R}_Q \circ \mathcal{R}_{\mathcal{K}}$. Let $\mathcal{J} = \text{pr}(Q, \mathcal{K})$. By Theorem NEUT.47(A) $\mathcal{J} \parallel \mathcal{L}$. Thus $\mathcal{R}_Q \circ \mathcal{R}_{\mathcal{K}} = \mathcal{R}_{\mathcal{J}} \circ \mathcal{R}_{\mathcal{K}} \circ \mathcal{R}_{\mathcal{K}} = \mathcal{R}_{\mathcal{J}}$ so $\mathcal{R}_{\mathcal{N}} \circ \mathcal{R}_{\mathcal{M}} \circ \mathcal{R}_{\mathcal{L}} = \mathcal{R}_{\mathcal{J}}$.

(II: Uniqueness.) If $\mathcal{J}$ and $\mathcal{J}'$ are lines on $\mathcal{P}$ such that $\mathcal{R}_{\mathcal{N}} \circ \mathcal{R}_{\mathcal{M}} \circ \mathcal{R}_{\mathcal{L}} = \mathcal{R}_{\mathcal{J}}$ and $\mathcal{R}_{\mathcal{N}} \circ \mathcal{R}_{\mathcal{M}} \circ \mathcal{R}_{\mathcal{L}} = \mathcal{R}_{\mathcal{J}'}$, then $\mathcal{R}_{\mathcal{J}} = \mathcal{R}_{\mathcal{J}'}$; by Remark NEUT.1.1 $\mathcal{J} = \mathcal{J}'$. □

Definition ISM.12. Let $\mathcal{P}$ be a Euclidean plane, α a mapping of $\mathcal{P}$ into $\mathcal{P}$, and $\mathcal{L}$ a line on $\mathcal{P}$. α is a **glide reflection** of $\mathcal{P}$ over $\mathcal{L}$ iff there exists a translation τ of $\mathcal{P}$ such that $\mathcal{L}$ is a fixed line of τ and $\alpha = \mathcal{R}_{\mathcal{L}} \circ \tau$.

A glide reflection is an isometry because it is the composition of two isometries.

Theorem ISM.13. *Let $\mathcal{P}$ be a Euclidean plane, and let $\alpha = \mathcal{R}_{\mathcal{L}} \circ \tau$ be a glide reflection of $\mathcal{P}$, where τ is a translation and $\mathcal{L}$ is a fixed line for τ. Then $\alpha = \mathcal{R}_{\mathcal{L}} \circ \tau = \tau \circ \mathcal{R}_{\mathcal{L}}$, α has no fixed point, and $\mathcal{L}$ is the only fixed line of α.*

Proof. (A) Let X be any point on $\mathcal{P}$. (Case 1: $X \in \mathcal{L}$.) Since every point on $\mathcal{L}$ is a fixed point of $\mathcal{R}_{\mathcal{L}}$, $\tau(\mathcal{R}_{\mathcal{L}}(X)) = \tau(X)$. Since $\mathcal{L}$ is a fixed line of τ, $\tau(X) \in \mathcal{L}$, so that $\mathcal{R}_{\mathcal{L}}(\tau(X)) = \tau(X) = \tau(\mathcal{R}_{\mathcal{L}}(X))$. This also shows that no point on $\mathcal{L}$ can be a fixed point of $\mathcal{R}_{\mathcal{L}} \circ \tau$, because τ has no fixed point.

(Case 2: $X \in (\mathcal{P} \setminus \mathcal{L})$.) By Theorem CAP.8 $\overleftrightarrow{X\tau(X)}$ and $\overleftrightarrow{\mathcal{R}_{\mathcal{L}}(X)\tau(\mathcal{R}_{\mathcal{L}}(X))}$ are fixed lines of τ and since $\mathcal{L}$ is a fixed line of τ, each is parallel to $\mathcal{L}$. By Theorem IP.6 $\overleftrightarrow{X\tau(X)} \parallel \overleftrightarrow{\mathcal{R}_{\mathcal{L}}(X)\tau(\mathcal{R}_{\mathcal{L}}(X))}$. By Theorem NEUT.15(1) $\tau(\overleftrightarrow{X\mathcal{R}_{\mathcal{L}}(X)}) = \overleftrightarrow{\tau(X)\tau(\mathcal{R}_{\mathcal{L}}(X))}$. By Theorem CAP.8(C) and Definition CAP.6 $\overleftrightarrow{X\mathcal{R}_{\mathcal{L}}(X)} \parallel \overleftrightarrow{\tau(X)\tau(\mathcal{R}_{\mathcal{L}}(X))}$. Thus $\square X\tau(X)\tau(\mathcal{R}_{\mathcal{L}}(X))\mathcal{R}_{\mathcal{L}}(X)$ is a parallelogram, by Definition EUC.5(B).

By Theorem NEUT.48(A) $\mathcal{L} \perp \overleftrightarrow{X\mathcal{R}_{\mathcal{L}}(X)}$ and $\mathcal{L} \perp \overleftrightarrow{\tau(X)\mathcal{R}_{\mathcal{L}}(\tau(X))}$. By Theorem NEUT.47(A) $\overleftrightarrow{X\mathcal{R}_{\mathcal{L}}(X)} \parallel \overleftrightarrow{\tau(X)\mathcal{R}_{\mathcal{L}}(\tau(X))}$. Since $\overleftrightarrow{X\tau(X)} \parallel \mathcal{L}$, by Exercise NEUT.1 $\overleftrightarrow{X\tau(X)} \parallel \overleftrightarrow{\mathcal{R}_{\mathcal{L}}(X)\mathcal{R}_{\mathcal{L}}(\tau(X))}$. By Definition EUC.5(B), $\square X\tau(X)\mathcal{R}_{\mathcal{L}}(\tau(X))\mathcal{R}_{\mathcal{L}}(X)$ is a parallelogram.

Since $\mathcal{L} \perp \overleftrightarrow{X\mathcal{R}_{\mathcal{L}}(X)}$ and $\mathcal{L} \parallel \overleftrightarrow{X\tau(X)}$, by Theorem EUC.3 $\overleftrightarrow{X\mathcal{R}_{\mathcal{L}}(X)} \perp \overleftrightarrow{X\tau(X)}$, and by Theorem NEUT.44 $\angle\mathcal{R}_{\mathcal{L}}(X)X\tau(X)$ is right. By Theorem EUC.19 both $\square X\tau(X)\tau(\mathcal{R}_{\mathcal{L}}(X))\mathcal{R}_{\mathcal{L}}(X)$ and $\square X\tau(X)\mathcal{R}_{\mathcal{L}}(\tau(X))\mathcal{R}_{\mathcal{L}}(X)$ are rectangles. By Remark EUC.15(C), $\mathcal{R}_{\mathcal{L}}(\tau(X)) = \tau(\mathcal{R}_{\mathcal{L}}(X))$. This construction also shows that no point of $\mathcal{P} \setminus \mathcal{L}$ can be a fixed point of $\mathcal{R}_{\mathcal{L}} \circ \tau$, because $\mathcal{R}_{\mathcal{L}}(\tau(X))$ is on the side of $\mathcal{L}$ opposite X.

(B) Since $\mathcal{L}$ is a fixed line of both τ and $\mathcal{R}_{\mathcal{L}}$, it is a fixed line of $\mathcal{R}_{\mathcal{L}} \circ \tau$. If $\mathcal{M}$ is a line parallel to $\mathcal{L}$, then by Theorem CAP.8 $\mathcal{M}$ is a fixed line of τ. Since $\mathcal{R}_{\mathcal{L}}(\mathcal{M})$ is a line which is a subset of the side of $\mathcal{L}$ which is opposite the side containing $\mathcal{M}$, $\mathcal{M}$ is not a fixed line of $\mathcal{R}_{\mathcal{L}} \circ \tau$. Let $\mathcal{J}$ be any line on $\mathcal{P}$ such that $\mathcal{J}$ and $\mathcal{L}$ intersect at the point Q. Since $\tau(Q) \neq Q$ and $\mathcal{R}_{\mathcal{L}}(\tau(Q)) = \tau(Q) \in \mathcal{L}$, $\mathcal{R}_{\mathcal{L}}(\tau(\mathcal{J})) \neq \mathcal{J}$. Thus $\mathcal{J}$ is not a fixed line of $\mathcal{R}_{\mathcal{L}} \circ \tau$. □

Corollary ISM.13.1. *A mapping α on a Euclidean plane $\mathcal{P}$ is a glide reflection iff there exists a translation τ of $\mathcal{P}$ such that $\mathcal{L}$ is a fixed line of τ and $\alpha = \tau \circ \mathcal{R}_{\mathcal{L}}$.*

Theorem ISM.14. *Let $\mathcal{P}$ be a Euclidean plane and let $\mathcal{L}$, $\mathcal{M}$, and $\mathcal{N}$ be distinct lines on $\mathcal{P}$ which are nonconcurrent and nonparallel (i.e., any two of the three lines intersect at a point and that point is not on the third line), then there exists a line $\mathcal{J}$ and a point Q on $\mathcal{P}$ such that $\mathcal{R}_{\mathcal{N}} \circ \mathcal{R}_{\mathcal{M}} \circ \mathcal{R}_{\mathcal{L}} = \mathcal{R}_Q \circ \mathcal{R}_{\mathcal{J}}$. Furthermore, $\mathcal{R}_{\mathcal{N}} \circ \mathcal{R}_{\mathcal{M}} \circ \mathcal{R}_{\mathcal{L}}$ is a glide reflection of $\mathcal{P}$.*

Proof. Any two of the three lines intersect at a point and that point is not on the third line. Thus we have three cases:

(Case 1: $\mathcal{M}$ and $\mathcal{L}$ intersect at the point G and $G \notin \mathcal{N}$.) We will refer to $\mathcal{N}$ as the "odd" line that does not contain G. Let $\mathcal{K} = \text{pr}(G, \mathcal{N})$ and $Q = \text{ftpr}(G, \mathcal{N})$. By Theorem ROT.13 there exists a unique line $\mathcal{J}$ such that $G \in \mathcal{J}$ and $\mathcal{R}_{\mathcal{K}} \circ \mathcal{R}_{\mathcal{J}} = \mathcal{R}_{\mathcal{M}} \circ \mathcal{R}_{\mathcal{L}}$. By Definition ROT.1 $\mathcal{R}_{\mathcal{N}} \circ \mathcal{R}_{\mathcal{K}} = \mathcal{R}_{Q}$. Hence $\mathcal{R}_{\mathcal{N}} \circ \mathcal{R}_{\mathcal{M}} \circ \mathcal{R}_{\mathcal{L}} = \mathcal{R}_{\mathcal{N}} \circ \mathcal{R}_{\mathcal{K}} \circ \mathcal{R}_{\mathcal{J}} = \mathcal{R}_{Q} \circ \mathcal{R}_{\mathcal{J}}$.

To show that this mapping is a glide reflection, let $\mathcal{S} = \text{par}(Q, \mathcal{J})$ and $\mathcal{T} = \text{pr}(Q, \mathcal{J})$. By Theorem EUC.3 $\mathcal{S} \perp \mathcal{T}$. By Definition ROT.1 and Theorem ROT.4 $\mathcal{R}_{Q} = \mathcal{R}_{\mathcal{T}} \circ \mathcal{R}_{\mathcal{S}} = \mathcal{R}_{\mathcal{N}} \circ \mathcal{R}_{\mathcal{K}}$. Thus

$$\begin{aligned} \mathcal{R}_{\mathcal{N}} \circ \mathcal{R}_{\mathcal{M}} \circ \mathcal{R}_{\mathcal{L}} &= \mathcal{R}_{Q} \circ \mathcal{R}_{\mathcal{J}} \\ &= \mathcal{R}_{\mathcal{T}} \circ \mathcal{R}_{\mathcal{S}} \circ \mathcal{R}_{\mathcal{J}} = \mathcal{R}_{\mathcal{T}} \circ (\mathcal{R}_{\mathcal{S}} \circ \mathcal{R}_{\mathcal{J}}) = \mathcal{R}_{\mathcal{T}} \circ \tau \end{aligned}$$

where $\tau = \mathcal{R}_{\mathcal{S}} \circ \mathcal{R}_{\mathcal{J}}$ is a translation, since $\mathcal{S} \parallel \mathcal{J}$. Moreover, $\mathcal{T}$ is perpendicular to both $\mathcal{S}$ and $\mathcal{J}$, so is a fixed line for τ. Therefore $\mathcal{R}_{\mathcal{N}} \circ \mathcal{R}_{\mathcal{M}} \circ \mathcal{R}_{\mathcal{L}}$ is a glide reflection by Definition ISM.12.

If we interchange $\mathcal{M}$ and $\mathcal{L}$ throughout, the proof also shows that $\mathcal{R}_{\mathcal{N}} \circ \mathcal{R}_{\mathcal{L}} \circ \mathcal{R}_{\mathcal{M}}$ is a glide reflection; we will refer to this result as the "alternate" result of Case 1.

(Case 2: $\mathcal{N}$ and $\mathcal{L}$ intersect at the point G and $G \notin \mathcal{M}$.) By Exercise IP.2, $\mathcal{M}$ intersects either $\mathcal{N}$ or $\mathcal{L}$, possibly both. If $\mathcal{M}$ intersects $\mathcal{L}$ at some point H, we have Case 1 again, with H substituted for G. If $\mathcal{M}$ intersects $\mathcal{N}$ at some point H, then $\mathcal{L}$ is the "odd" line and substituting H for G, and interchanging $\mathcal{N}$ and $\mathcal{L}$ in Case 1, we see that $\mathcal{R}_{\mathcal{L}} \circ \mathcal{R}_{\mathcal{M}} \circ \mathcal{R}_{\mathcal{N}}$ is a glide reflection. That is, there exists a line $\mathcal{T}$ and a translation τ, where $\mathcal{T}$ is a fixed line for τ, such that $\mathcal{R}_{\mathcal{L}} \circ \mathcal{R}_{\mathcal{M}} \circ \mathcal{R}_{\mathcal{N}} = \mathcal{R}_{\mathcal{T}} \circ \tau$. The inverse of this mapping is

$$\mathcal{R}_{\mathcal{N}} \circ \mathcal{R}_{\mathcal{M}} \circ \mathcal{R}_{\mathcal{L}} = \mathcal{R}_{\mathcal{N}}^{-1} \circ \mathcal{R}_{\mathcal{M}}^{-1} \circ \mathcal{R}_{\mathcal{L}}^{-1} = \tau^{-1} \circ \mathcal{R}_{\mathcal{T}}^{-1} = \tau^{-1} \circ \mathcal{R}_{\mathcal{T}}.$$

By Theorem CAP.12(A), τ^{-1} is a translation σ, and the fixed lines of τ and σ are the same, so by Corollary ISM.13.1, this mapping is $\mathcal{R}_{\mathcal{T}} \circ \sigma$, a glide reflection.

(Case 3: $\mathcal{M}$ and $\mathcal{N}$ intersect at the point G and $G \notin \mathcal{L}$.) By Exercise IP.2, $\mathcal{L}$ intersects either $\mathcal{M}$ or $\mathcal{N}$, possibly both. If $\mathcal{L}$ intersects $\mathcal{M}$ at some point H, we have Case 1 again, with H substituted for G. If $\mathcal{L}$ intersects $\mathcal{N}$ at some point H, then $\mathcal{M}$ is the "odd" line and substituting H for G, $\mathcal{L}$ for $\mathcal{N}$, $\mathcal{N}$ for $\mathcal{M}$, and $\mathcal{M}$ for $\mathcal{L}$, and using the alternate conclusion for Case 1, we have $\mathcal{R}_{\mathcal{L}} \circ \mathcal{R}_{\mathcal{M}} \circ \mathcal{R}_{\mathcal{N}}$ is a glide reflection; reasoning as at the end of Case 2, $\mathcal{R}_{\mathcal{N}} \circ \mathcal{R}_{\mathcal{M}} \circ \mathcal{R}_{\mathcal{L}}$ is a glide reflection.

□

Theorem ISM.15. *Let $\mathcal{P}$ be a Euclidean plane and let α be an isometry of $\mathcal{P}$ which has no fixed point. Then α is either a translation of $\mathcal{P}$, or α is a glide reflection of $\mathcal{P}$.*

Proof. By Theorem ROT.25 either there exist two distinct lines $\mathcal{L}$ and $\mathcal{M}$ on $\mathcal{P}$ such that $\alpha = \mathcal{R}_{\mathcal{M}} \circ \mathcal{R}_{\mathcal{L}}$, or there exist three distinct lines $\mathcal{H}$, $\mathcal{J}$, and $\mathcal{K}$ on $\mathcal{P}$ such that $\alpha = \mathcal{R}_{\mathcal{K}} \circ \mathcal{R}_{\mathcal{J}} \circ \mathcal{R}_{\mathcal{H}}$.

(Case 1: $\alpha = \mathcal{R}_{\mathcal{M}} \circ \mathcal{R}_{\mathcal{L}}$.) If $\mathcal{M}$ and $\mathcal{N}$ were concurrent at O on $\mathcal{P}$, then by Definition NEUT.1(A) and Definition CAP.0, O would be a fixed point of α, contrary to the given fact that α has no fixed point. Hence $\mathcal{L} \parallel \mathcal{M}$ and by Theorem ISM.4(B) α is a translation of $\mathcal{P}$.

(Case 2: $\alpha = \mathcal{R}_{\mathcal{K}} \circ \mathcal{R}_{\mathcal{J}} \circ \mathcal{R}_{\mathcal{H}}$.) If $\mathcal{H}$, $\mathcal{J}$, and $\mathcal{K}$ were concurrent at the point O, then O would be a fixed point for α contrary to our assumption that it has no fixed point. If $\mathcal{K} \parallel \mathcal{J} \parallel \mathcal{H}$, then by Theorem ISM.11 there would exist a line $\mathcal{T}$ such that $\mathcal{R}_{\mathcal{T}} = \mathcal{R}_{\mathcal{K}} \circ \mathcal{R}_{\mathcal{J}} \circ \mathcal{R}_{\mathcal{H}}$ and every point on $\mathcal{T}$ would be a fixed point of α, contrary to the given fact that α has no fixed point. Thus two of the three lines intersect at a point and that point is not on the third line and by Theorem ISM.14, α is a glide reflection of $\mathcal{P}$. □

Theorem ISM.16. *Let $\mathcal{P}$ be a Euclidean plane and let $\mathcal{L}$, $\mathcal{M}$, and $\mathcal{N}$ be distinct lines on $\mathcal{P}$. $\{\mathcal{L}, \mathcal{M}, \mathcal{N}\}$ is a pencil by Definition IP.1(D) iff there exists a line $\mathcal{J}$ on $\mathcal{P}$ such that $\mathcal{R}_{\mathcal{N}} \circ \mathcal{R}_{\mathcal{M}} \circ \mathcal{R}_{\mathcal{L}} = \mathcal{R}_{\mathcal{J}}$.*

Proof. (I: If $\{\mathcal{L}, \mathcal{M}, \mathcal{N}\}$ is a pencil, then there exists a line $\mathcal{J}$ on $\mathcal{P}$ such that $\mathcal{R}_{\mathcal{N}} \circ \mathcal{R}_{\mathcal{M}} \circ \mathcal{R}_{\mathcal{L}} = \mathcal{R}_{\mathcal{J}}$.) If there exists a point O on $\mathcal{P}$ such that $\mathcal{L} \cap \mathcal{M} \cap \mathcal{N} = \{O\}$, then by Theorem ROT.4 $\mathcal{J}$ exists. If $\mathcal{L} \parallel \mathcal{M} \parallel \mathcal{N}$, then by Theorem ISM.11, $\mathcal{J}$ exists.

(II: If there exists a line $\mathcal{J}$ on $\mathcal{P}$ such that $\mathcal{R}_{\mathcal{N}} \circ \mathcal{R}_{\mathcal{M}} \circ \mathcal{R}_{\mathcal{L}} = \mathcal{R}_{\mathcal{J}}$, then $\{\mathcal{L}, \mathcal{M}, \mathcal{N}\}$ is a pencil.) We will proceed by proving the contrapositive: If $\{\mathcal{L}, \mathcal{M}, \mathcal{N}\}$ is not a pencil, then two of the three lines intersect at a point and that point is not on the third line. By Theorem ISM.14 $\alpha = \mathcal{R}_{\mathcal{N}} \circ \mathcal{R}_{\mathcal{M}} \circ \mathcal{R}_{\mathcal{L}}$ is a glide reflection. By Theorem ISM.13 α has no fixed point and thus is not a reflection. □

Theorem ISM.17. *Let $\mathcal{P}$ be a Euclidean plane and let α be an isometry of $\mathcal{P}$. Then α is one and only one of the following:*

(1) *the identity,*
(2) *a line reflection,*
(3) *a rotation of $\mathcal{P}$ about a point,*
(4) *a translation of $\mathcal{P}$,*
(5) *a glide reflection of $\mathcal{P}$.*

Proof. If α has three noncollinear fixed points, then by Theorem NEUT.24 $\alpha = \iota$. If α has distinct fixed points A and B, but no fixed points off of $\overleftrightarrow{AB}$, then by Exercise NEUT.6(A) α is the line reflection $\mathcal{R}_{\overleftrightarrow{AB}}$. If α has one and only one fixed point O, then by Theorem ROT.16 α is a rotation of $\mathcal{P}$ about O. If α has no fixed point, then by Theorem ISM.15 α is either a translation of $\mathcal{P}$, or α is a glide reflection of $\mathcal{P}$. □

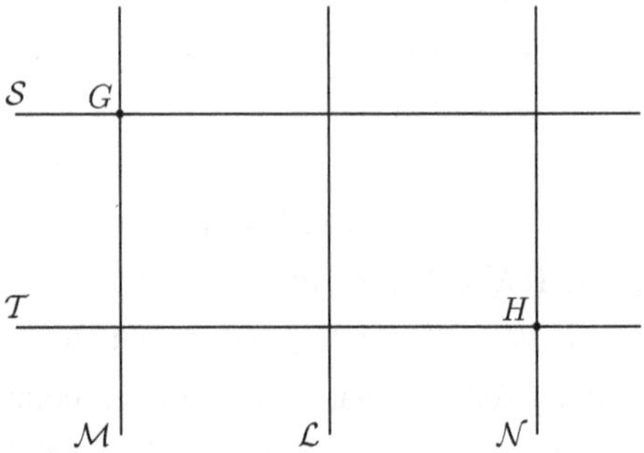

Fig. 12.3 For Theorem ISM.18.

Theorem ISM.18. *Let $\mathcal{P}$ be a Euclidean plane, $\mathcal{L}$ be a line on $\mathcal{P}$, and G and H be distinct points on $\mathcal{P}$ neither of which is on $\mathcal{L}$. Furthermore, let $\mathcal{S} = \text{pr}(G, \mathcal{L})$ and $\mathcal{T} = \text{pr}(H, \mathcal{L})$. Then there exists a line $\mathcal{F}$ such that $\mathcal{R}_H \circ \mathcal{R}_\mathcal{L} \circ \mathcal{R}_G = \mathcal{R}_\mathcal{F}$ iff $\mathcal{S} = \mathcal{T}$.*

Proof. See Figure 12.3. By Theorem NEUT.47(A) and Definition CAP.10 $\mathcal{S} \mathbb{PE} \mathcal{T}$. Let $\mathcal{M} = \text{par}(G, \mathcal{L})$ and $\mathcal{N} = \text{par}(H, \mathcal{L})$. By Theorem EUC.3 $\mathcal{N} \perp \mathcal{S}$, $\mathcal{N} \perp \mathcal{T}$, $\mathcal{M} \perp \mathcal{S}$, and $\mathcal{M} \perp \mathcal{T}$. By Theorem IP.6 $\mathcal{M} \parallel \mathcal{N}$. By Theorem ISM.11 there exists a line $\mathcal{J}$ such that $\mathcal{J} \parallel \mathcal{L}$ and $\mathcal{R}_\mathcal{J} = \mathcal{R}_\mathcal{N} \circ \mathcal{R}_\mathcal{L} \circ \mathcal{R}_\mathcal{M}$. By Definition ROT.1 and Theorem ROT.4, $\mathcal{R}_H = \mathcal{R}_\mathcal{T} \circ \mathcal{R}_\mathcal{N}$ and $\mathcal{R}_G = \mathcal{R}_\mathcal{M} \circ \mathcal{R}_\mathcal{S}$. Therefore

$$\mathcal{R}_H \circ \mathcal{R}_\mathcal{L} \circ \mathcal{R}_G = \mathcal{R}_\mathcal{T} \circ \mathcal{R}_\mathcal{N} \circ \mathcal{R}_\mathcal{L} \circ \mathcal{R}_\mathcal{M} \circ \mathcal{R}_\mathcal{S} = \mathcal{R}_\mathcal{T} \circ \mathcal{R}_\mathcal{J} \circ \mathcal{R}_\mathcal{S}.$$

But since $\mathcal{T} \perp \mathcal{J}$, by Corollary ROT.5 $\mathcal{R}_\mathcal{T} \circ \mathcal{R}_\mathcal{J} = \mathcal{R}_\mathcal{J} \circ \mathcal{R}_\mathcal{T}$, and

$$\mathcal{R}_H \circ \mathcal{R}_\mathcal{L} \circ \mathcal{R}_G = \mathcal{R}_\mathcal{J} \circ \mathcal{R}_\mathcal{T} \circ \mathcal{R}_\mathcal{S}.$$

If $\mathcal{T} \neq \mathcal{S}$, then $\mathcal{T}$ and $\mathcal{S}$ are parallel and so by Theorem ISM.4 $\mathcal{R}_\mathcal{T} \circ \mathcal{R}_\mathcal{S}$ is a translation of $\mathcal{P}$. Since $\mathcal{T} \perp \mathcal{J}$, $\mathcal{J}$ is a fixed line for this translation, and by Definition ISM.12 $\mathcal{R}_\mathcal{J} \circ \mathcal{R}_\mathcal{T} \circ \mathcal{R}_\mathcal{S}$ is a glide reflection, which has no fixed points (cf Theorem ISM.13) and hence is not a reflection.

If $\mathcal{T} = \mathcal{S}$, then by Property R.2 of Definition NEUT.2 and Definition NEUT.1(C), $\mathcal{R}_\mathcal{T} \circ \mathcal{R}_\mathcal{S} = \mathcal{R}_\mathcal{T} \circ \mathcal{R}_\mathcal{T} = \iota$ so $\mathcal{R}_H \circ \mathcal{R}_\mathcal{L} \circ \mathcal{R}_G = \mathcal{R}_\mathcal{J}$. □

The following theorem is essential to the proof of Theorem DLN.4, in Chapter 13, which is important in the development of half-rotations and dilations.

The reader may wish to get a feeling for what it says by choosing an arbitrary point on the figure, then reflecting it successively in $\mathcal{M}$, $\mathcal{L}$, and $\mathcal{N}$, and observing that the result is the same as reflecting it in $\mathcal{J}$. We have printed the figure a bit larger than usual to facilitate such an exercise.

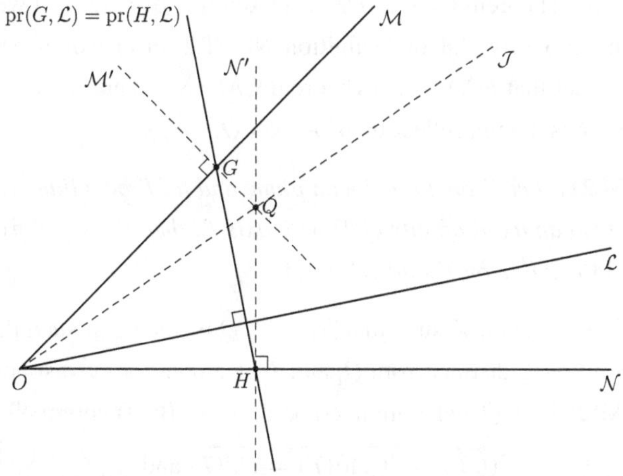

Fig. 12.4 For Theorem ISM.19.

Theorem ISM.19. *Let $\mathcal{P}$ be a Euclidean plane and let $\mathcal{L}$, $\mathcal{M}$, and $\mathcal{N}$ be distinct lines on $\mathcal{P}$ which are concurrent at O. Moreover, let G be a point on $\mathcal{M}$ distinct from O and H be a point on $\mathcal{N}$ distinct from O such that* $\text{pr}(G,\mathcal{L}) = \text{pr}(H,\mathcal{L})$. *Finally, let* $\mathcal{M}' = \text{pr}(G,\mathcal{M})$, $\mathcal{N}' = \text{pr}(H,\mathcal{N})$, *and let J be the line (cf Theorem ROT.11) such that $\mathcal{R}_{\mathcal{J}} = \mathcal{R}_{\mathcal{N}} \circ \mathcal{R}_{\mathcal{L}} \circ \mathcal{R}_{\mathcal{M}}$. Then $\mathcal{M}'$, $\mathcal{J}$, and $\mathcal{N}'$ are concurrent at a point Q.*

Proof. See Figure 12.4. By Theorem ISM.18 there exists a line $\mathcal{F}$ such that $\mathcal{R}_H \circ \mathcal{R}_{\mathcal{L}} \circ \mathcal{R}_G = \mathcal{R}_{\mathcal{F}}$. By Definition ROT.1 and Theorem ROT.4 $\mathcal{R}_H \circ \mathcal{R}_{\mathcal{L}} \circ \mathcal{R}_G = \mathcal{R}_{\mathcal{N}'} \circ \mathcal{R}_{\mathcal{N}} \circ \mathcal{R}_{\mathcal{L}} \circ \mathcal{R}_{\mathcal{M}} \circ \mathcal{R}_{\mathcal{M}'}$. Thus $\mathcal{R}_{\mathcal{F}} = \mathcal{R}_H \circ \mathcal{R}_{\mathcal{L}} \circ \mathcal{R}_G = \mathcal{R}_{\mathcal{N}'} \circ \mathcal{R}_{\mathcal{J}} \circ \mathcal{R}_{\mathcal{M}'}$. If $\mathcal{M}'$, $\mathcal{J}$, and $\mathcal{N}'$ were not concurrent at a point Q, then by Theorem ISM.14 $\mathcal{R}_{\mathcal{N}'} \circ \mathcal{R}_{\mathcal{J}} \circ \mathcal{R}_{\mathcal{M}'}$ would be a glide reflection and would not be equal to $\mathcal{R}_{\mathcal{F}}$, a contradiction. Therefore $\mathcal{M}'$, $\mathcal{J}$, and $\mathcal{N}'$ are concurrent at Q. □

Theorem ISM.20. *Let $\mathcal{P}$ be a Euclidean plane and let δ be a dilation of $\mathcal{P}$ with fixed point O, and also an isometry of $\mathcal{P}$. Then δ is a point reflection of $\mathcal{P}$ over O.*

Proof. Let X be any member of $\mathcal{P} \setminus \{O\}$, then by Property B.3 of Definition IB.1 (extension property) there exists a point X' such that $X\text{–}O\text{–}X'$. By Theorem PSH.15 $\overleftrightarrow{OX}$ is the union of the disjoint sets $\overrightarrow{OX}, \{O\}$, and $\overrightarrow{OX'}$. By Theorem NEUT.15(2) $\delta(\overrightarrow{OX}) = \overrightarrow{\delta(O)\delta(X)} = \overrightarrow{O\delta(X)}$. By Definition NEUT.3(B) $\overline{OX} \cong \overline{O\delta(X)}$. By Theorem CAP.18 $\overleftrightarrow{OX}$ is a fixed line of δ. If $\delta(X)$ were to belong to $\overrightarrow{OX}$, then by Property R.4 of Definition NEUT.2, $\delta(X)$ would be equal to X, contrary to the fact that $\delta(X) \neq X$ (Theorem CAP.18). Therefore $\delta(X) \in \overrightarrow{OX'}$. By Exercise ROT.8 δ is a point reflection of $\mathcal{P}$ over O. □

Theorem ISM.21. *Let $\mathcal{P}$ be a Euclidean plane and let $\mathcal{L}$ be a line on $\mathcal{P}$. If θ is an isometry and also an axial affinity of $\mathcal{P}$ with axis $\mathcal{L}$, then the set of fixed lines of θ is $\mathcal{L} \cup \{\mathcal{M} \mid \mathcal{M}$ is a line on $\mathcal{P}$ and $\mathcal{M} \perp \mathcal{L}\}$.*

Proof. Let $\mathcal{N}$ be a line on $\mathcal{P}$ such that $\mathcal{N} \perp \mathcal{L}$, Q be the point such that $\mathcal{L} \cap \mathcal{N} = \{Q\}$, S be a point on $\mathcal{L}$ distinct from Q, and T be a point on $\mathcal{N}$ distinct from Q. By Definition CAP.25 both Q and S are fixed points of θ. By Theorem NEUT.15(1)

$$\theta(\overleftrightarrow{QT}) = \overleftrightarrow{\theta(Q)\theta(T)} = \overleftrightarrow{Q\theta(T)} \text{ and}$$
$$\theta(\angle SQT) = \angle\theta(S)\theta(Q)\theta(T) = \angle SQ\theta(T).$$

By Definition NEUT.3(B) $\overline{QT} \cong \overline{Q\theta(T)}$ and $\angle SQT \cong \angle SQ\theta(T)$. By Theorem NEUT.44, $\angle SQT$ is right. By Theorem NEUT.66 $\angle SQ\theta(T)$ is right. By Theorem NEUT.44 $\overleftrightarrow{Q\theta(T)} \perp \mathcal{L}$. By Theorem NEUT.48(A) $\overleftrightarrow{Q\theta(T)} = \overleftrightarrow{QT}$. By Definition CAP.0 $\mathcal{N}$ is a fixed line of θ. By Theorem CAP.27(B), the set of fixed lines of θ is $\{\mathcal{L}\} \cup \{\mathcal{M} \mid \mathcal{M}$ is a line on $\mathcal{P}$ and $\mathcal{M} \perp \mathcal{L}\}$. □

Theorem ISM.22. *Let $\mathcal{P}$ be a Euclidean plane and let $\mathcal{L}$ be a line on $\mathcal{P}$.*

(A) *$\mathcal{R}_{\mathcal{L}}$ is an axial affinity of $\mathcal{P}$ with axis $\mathcal{L}$.*
(B) *If θ is an axial affinity of $\mathcal{P}$ with axis $\mathcal{L}$ such that θ is also an isometry of $\mathcal{P}$, then $\theta = \mathcal{R}_{\mathcal{L}}$.*

That is, the set of all reflections is identical to the set of all isometries which are axial affinities.

Proof. (A) By Remark NEUT.1.3 $\mathcal{R}_{\mathcal{L}}$ is a bijection of $\mathcal{P}$ onto itself. By Definition NEUT.1(D) and Remark NEUT.1.5 $\mathcal{R}_{\mathcal{L}}$ is a collineation of $\mathcal{P}$. By Definition NEUT.1(A), every point on $\mathcal{L}$ is a fixed point of $\mathcal{R}_{\mathcal{L}}$. By Definition CAP.25 $\mathcal{R}_{\mathcal{L}}$ is an axial affinity of $\mathcal{P}$ with axis $\mathcal{L}$.

(B) If X is any member of $\mathcal{L}$, then by Definition CAP.25 and Definition NEUT.1(A) $\theta(X) = \mathcal{R}_{\mathcal{L}}(X) = X$.

If X is any member of $\mathcal{P} \setminus \mathcal{L}$, let $\mathcal{M} = \operatorname{pr}(X, \mathcal{L})$ and let $U = \operatorname{ftpr}(X, \mathcal{L})$. By Theorem ISM.21 $\mathcal{M}$ is a fixed line of θ, so $\theta(X) \in \mathcal{M}$. If $\theta(X)$ were on the X-side of $\mathcal{L}$, then by Theorem PSH.38(A) $\theta(X)$ would belong to $\overrightarrow{UX}$. By Theorem NEUT.15(5) and Definition NEUT.1(A) we would have $\theta(\overline{UX}) = \overline{\theta(U)\theta(X)} = \overline{U\theta(X)}$ and by Definition NEUT.3(B) we would have $\overline{UX} \cong \overline{U\theta(X)}$ and by Property R.4 of Definition NEUT.2, $\theta(X)$ would equal X. This would contradict the fact from Theorem CAP.26 that X is not a fixed point of θ. By Theorem PSH.12 (Plane Separation), $\theta(X)$ and X are on opposite sides of $\mathcal{L}$. Using Property B.3 of Definition IB.1 let X' be a point such that $X{-}U{-}X'$. By Theorems PSH.15 and PSH.38(A) $\theta(X) \in \overrightarrow{UX'}$. By Definition NEUT.3(C) U is the midpoint of $\overline{X\theta(X)}$. By Exercise NEUT.74 $\theta = \mathcal{R}_{\mathcal{L}}$. □

Theorem ISM.23 (Construction of a translation). *Let $\mathcal{P}$ be a Euclidean plane, O a member of $\mathcal{P}$, A a member of $\mathcal{P} \setminus \{O\}$, τ_A the translation of $\mathcal{P}$ such that $\tau_A(O) = A$, and let X be any member of $\mathcal{P} \setminus \{O\}$. Then $\tau_A(X)$ is constructed as follows:*

(Case 1: $X \in \mathcal{P} \setminus \overleftrightarrow{OA}$.) *$\tau_A(X)$ is the point of intersection of* $\operatorname{par}(X, \overleftrightarrow{OA})$ *and* $\operatorname{par}(A, \overleftrightarrow{OX})$. *Furthermore A and $\tau_A(X)$ are on the same side of $\overleftrightarrow{OX}$.*

(Case 2: $X \in \overleftrightarrow{OA} \setminus \{O\}$.) *Let Y be any member of $(\mathcal{P} \setminus \overleftrightarrow{OA})$ and using case 1 find $\tau_A(Y)$. Then $\tau_A(X)$ is the point of intersection of $\overleftrightarrow{OA}$ and* $\operatorname{par}(\tau_A(Y), \overleftrightarrow{XY})$. *Moreover, if the points on $\overleftrightarrow{OA}$ are ordered so that $O < A$ (see Remark ORD.2), then for every X belonging to $\overleftrightarrow{OA}$, $X < \tau_A(X)$.*

Proof. (Case 1: $X \in (\mathcal{P} \setminus \overleftrightarrow{OA})$.) We will freely use Theorem CAP.8(B) and (C) without further reference. Since $\overleftrightarrow{OA} = \overleftrightarrow{O\tau_A(O)}$, it is a fixed line for τ_A, as is $\overleftrightarrow{X\tau_A(X)}$, and these lines are parallel. By definition, $\overleftrightarrow{OA} \parallel \operatorname{par}(X, \overleftrightarrow{OA})$; both $\operatorname{par}(X, \overleftrightarrow{OA})$ and $\overleftrightarrow{X\tau_A(X)}$ contain X, so by Axiom PS, $\operatorname{par}(X, \overleftrightarrow{OA}) = \overleftrightarrow{X\tau_A(X)}$ and this is a fixed line. The lines $\overleftrightarrow{OX}$ and $\overleftrightarrow{OA}$ are not parallel because they intersect at O, so $\overleftrightarrow{OX}$ is not a fixed line for τ_A and by Definition CAP.6 and Theorem CAP.1 (or Theorem NEUT.15(1)), $\tau_A(\overleftrightarrow{OX}) = \overleftrightarrow{A\tau_A(X)} \parallel \overleftrightarrow{OX}$. Then $\overleftrightarrow{A\tau_A(X)} = \operatorname{par}(A, \overleftrightarrow{OX})$; this line intersects $\operatorname{par}(X, \overleftrightarrow{OA}) = \overleftrightarrow{X\tau_A(X)}$ at $\tau_A(X)$. By Exercise PSH.14 A and $\tau_A(X)$ are on the same side of $\overleftrightarrow{OX}$.

(Case 2: $X \in (\overleftrightarrow{OA} \setminus \{O\}$.) By Theorem CAP.8(B) $\overleftrightarrow{Y\tau_A(Y)}$ and $\overleftrightarrow{OA} = \overleftrightarrow{O\tau_A(O)}$ are both fixed lines of τ_A, and by Theorem CAP.8(C) they are parallel. Since $X \in \overleftrightarrow{OA}$, $\tau_A(X) \in \overleftrightarrow{OA}$. $\overleftrightarrow{XY}$ is not a fixed line of τ_A so by Definition CAP.6, $\tau_A(\overleftrightarrow{XY}) =$

$\overleftrightarrow{\tau_A(X)\tau_A(Y)}$ is parallel to $\overleftrightarrow{XY}$, so is the same line as par $(\tau_A(Y), \overleftrightarrow{XY})$. Therefore $\tau_A(X)$ is the point of intersection of par $(\tau_A(Y), \overleftrightarrow{XY})$ and $\overleftrightarrow{OA}$, as required.

We now show that $X < \tau_A(X)$ in case 2 above. The following may seem like a lot of fuss to prove something so intuitively obvious, but it seems to be what is required. There are four subcases: since $X \neq O$ either O–A–X, $X = A$, O–X–A, or X–O–A.

(Case 2A: O–A–X.) By Exercise PSH.14 (which in the following we will use many times without further reference), $Y \in O$-side of $\overleftrightarrow{A\tau_A(Y)}$, which is opposite the X-side by Definition IB.11. Therefore $\overline{XY} \cap \overleftrightarrow{A\tau_A(Y)} \neq \emptyset$. Hence by Definition IB.11 $\tau_A(Y)$ is on the side of $\overleftrightarrow{XY}$ opposite to O and A. Let $\{W\} = \text{par}\,(O, \overleftrightarrow{XY}) \cap \overleftrightarrow{Y\tau_A(Y)}$, and let $\{Q\} = \text{par}\,(W, \overleftrightarrow{OY}) \cap \overleftrightarrow{OA}$. Because W is on the O-side of $\overleftrightarrow{XY}$, W–Y–$\tau_A(Y)$; and since Q is on the W-side of $\overleftrightarrow{OY}$, Q–O–A–X. $\tau_A(X)$ is on the $\tau_A(Y)$-side of $\overleftrightarrow{XY}$ so A–X–$\tau_A(X)$, and hence Q–O–A–X–$\tau_A(X)$. Since $O < A$ it follows from Theorem ORD.6 that $A < X$ and hence that $X < \tau_A(X)$.

(Case 2B: $X = A$.) $\square OY\tau_A(Y)X$ is a parallelogram by Definition EUC.5(B) and is rotund by Theorem EUC.6. By Theorem PSH.54(A) its diagonals intersect, so that O and $\tau_A(Y)$ are on opposite sides of $\overleftrightarrow{XY}$, by Definition IB.11. Let $\{W\} = \text{par}\,(O, \overleftrightarrow{XY}) \cap \overleftrightarrow{Y\tau_A(Y)}$; then $W \in O$-side of $\overleftrightarrow{XY}$ by Exercise PSH.14, so that W–Y–$\tau_A(Y)$ and hence by Exercise PSH.57 O–X–$\tau_A(X)$. Then since $O < A = X$ by Theorem ORD.6 $X < \tau_A(X)$.

(Case 2C: O–X–A.) Since $O < A$, $O < X$ by Theorem ORD.6. Let $\{W\} = \text{par}\,(O, \overleftrightarrow{XY}) \cap \overleftrightarrow{Y\tau_A(Y)}$, and let $\{Z\} = \text{par}\,(X, \overleftrightarrow{OY}) \cap \overleftrightarrow{Y\tau_A(Y)}$. $\square OYZX$ is a parallelogram and reasoning as in case 2B, O and Z are on opposite sides of $\overleftrightarrow{XY}$. Since W and O are on the same side of $\overleftrightarrow{XY}$, W is on the side opposite Z of $\overleftrightarrow{XY}$, and by Definition IB.11 W–Y–Z. Since O–X–A, by Exercise PSH.57 Y–Z–$\tau_A(Y)$. Therefore W–Y–Z–$\tau_A(Y)$. Again by Exercise PSH.57, O–X–$\tau_A(X)$, and $X < \tau_A(X)$ by Theorem ORD.6.

(Case 2D: X–O–A.) Since $O < A$ by Theorem ORD.6, $X < O$. Let $\{W\} = \text{par}\,(X, \overleftrightarrow{OY}) \cap \overleftrightarrow{Y\tau_A(Y)}$, and let $\{Q\} = \text{par}\,(W, \overleftrightarrow{XY}) \cap \overleftrightarrow{OA}$. Then $\square OYWX$ is a parallelogram and reasoning as in case 2B, O and W are on opposite sides of $\overleftrightarrow{XY}$. Since W and Q are on the same side of $\overleftrightarrow{XY}$, Q and O are on opposite sides of $\overleftrightarrow{XY}$, and by Definition IB.11 Q–X–O. Since $X < O$, $Q < X$ by Theorem ORD.6.

Since X–O–A, by Exercise PSH.57W–Y–$\tau_A(Y)$; again by Exercise PSH.57, Q–X–$\tau_A(X)$. Since $Q < X$, $X < \tau_A(X)$ by Theorem ORD.6. □

12.2 Exercises for isometries

Answers to starred (*) exercises may be accessed from the home page for this book at www.springer.com.

Exercise ISM.1*. Let $\mathcal{P}$ be a Euclidean plane.

(A) There is no translation τ of $\mathcal{P}$ such that $\tau \circ \tau = \iota$.
(B) For any translation τ of $\mathcal{P}$, $\tau \circ \tau$ is a translation, having no fixed point.

Exercise ISM.2*. Let $\mathcal{P}$ be a Euclidean plane, σ and τ be translations of $\mathcal{P}$ such that $\mathcal{L}$ is a fixed line of σ, $\mathcal{M}$ is a fixed line of τ, $\mathcal{L}$ and $\mathcal{M}$ are not parallel, and let Q be any point on $\mathcal{P}$. Then $\square Q(\sigma(Q))(\tau(\sigma(Q)))(\tau(Q))$ is a parallelogram.

Exercise ISM.3*. Let $\mathcal{P}$ be a Euclidean plane, A and B be distinct points on $\mathcal{P}$, and τ be a translation of $\mathcal{P}$ such that $\overleftrightarrow{AB}$ is not a fixed line of τ. Then $\overline{A\tau(A)}$ and $\overline{B\tau(B)}$ are opposite edges of a parallelogram.

Exercise ISM.4*. Let $\mathcal{P}$ be a Euclidean plane, $\mathcal{L}_1$ and $\mathcal{L}_2$ be parallel lines on $\mathcal{P}$, A_1 be a point on $\mathcal{L}_1$, A_2 be the point of intersection of $\mathrm{pr}\,(A_1, \mathcal{L}_1)$ and $\mathcal{L}_2$, and τ be the translation of $\mathcal{P}$ such that $\tau(A_1) = A_2$ (cf Theorem ISM.5). Then $\tau(\mathcal{L}_1) = \mathcal{L}_2$.

Exercise ISM.5*. Let $\mathcal{M}$ be a line on a Euclidean plane $\mathcal{P}$ and let σ be a translation along $\mathcal{M}$; that is, $\mathcal{M}$ is a fixed line for σ. Let $\mathcal{R}_{\mathcal{M}}$ be the reflection with axis $\mathcal{M}$. Then $\mathcal{R}_{\mathcal{M}} \circ \sigma = \sigma \circ \mathcal{R}_{\mathcal{M}}$.

Exercise ISM.6. Prove, disprove, or improve: let $\mathcal{P}$ be a Euclidean plane, τ a translation, and $\mathcal{L}$ a line on $\mathcal{P}$. Then $\mathcal{R}_{\tau(\mathcal{L})} \circ \tau = \tau \circ \mathcal{R}_{\mathcal{L}}$.

Exercise ISM.7. In Theorem ISM.23 Case 2, create a simpler proof of the fact that $X < \tau_A(X)$.

Exercise ISM.8*. Let $\mathcal{P}$ be a Euclidean plane, and let $\alpha = \mathcal{R}_{\mathcal{L}} \circ \tau$ be a glide reflection, where τ is a translation and $\mathcal{L}$ is the single fixed line for α according to Theorem ISM.13.

(A) If $\mathcal{N} \parallel \mathcal{L}$, then $\alpha(\mathcal{N}) \parallel \mathcal{L}$.
(B) If $\mathcal{N} \perp \mathcal{L}$, then $\alpha(\mathcal{N}) \perp \mathcal{L}$ and $\alpha(\mathcal{N}) \parallel \mathcal{N}$.

Chapter 13
Dilations of a Euclidean Plane (DLN)

Acronym: *DLN*

Dependencies: *all prior Chapters 1 through 12*

New Axioms: *none*

New Terms Defined: *half-rotation, associated rotation, group generated by a union of three sets*

Abstract: This chapter establishes a rich array of properties for dilations, which were defined in Chapter 3. These play a key role in the development of Euclidean geometry, both in the definition of multiplication and in the development of similarity. Half-rotations are defined and their properties developed in an intricate process; these, in turn, are used to define dilations, which are shown to be belineations. A method is provided for point-wise construction of a dilation having a given action. A classical proposition attributed to Pappus of Alexandria is proved.

Intuitively, a dilation is a uniform expansion (or contraction) of the plane in all directions from (or toward) a fixed point O. Dilations are not isometries, as are most of the mappings we've worked with so far.[1]

It is fairly simple to construct a dilation that *contracts* the plane; this is done (and illustrated) in Theorem DLN.5, using *half-rotations* which, like dilations, are not isometries, although they are derived from rotations. For the general case

[1]Dilations would be easy to construct if we had a notion of distance, but we don't; indeed, we need dilations in order to construct a definition of distance, which we do in Chapter 14, Definition OF.16.

E.J. Specht et al., *Euclidean Geometry and its Subgeometries*,
DOI 10.1007/978-3-319-23775-6_13

(both expansion and contraction) we must deal both with half-rotations and with their inverses. Theorem DLN.7 gives the details for this process, and includes a figure which, hopefully, will be helpful in clarifying the proof. The most difficult part of the whole process of defining and developing the properties of dilations is proving the properties of half-rotations; this is done mainly in Theorems DLN.2 and Theorem DLN.4.

In this chapter we will generally follow J. Diller and J. Boczeck, in *Euclidean Planes*, Chapter 4 in *Fundamentals of Mathematics*, Volume 2, H. Behnke, F. Bachmann, K. Fladt, and H. Kunle, eds, translated by S. Gould, MIT Press, 1974 [2].

13.1 Half-rotations and dilations

Recall from Definition CAP.17 that a collineation $\alpha \neq \iota$ is a *dilation* iff it has a fixed point, and for every line $\mathcal{L}$ on $\mathcal{P}$, either $\alpha(\mathcal{L}) \parallel \mathcal{L}$, or $\alpha(\mathcal{L}) = \mathcal{L}$. Alternatively, by Theorem CAP.22, a collineation is a *dilation* if O is its only fixed point, and every line containing O is a fixed line. Recall also from Theorem CAP.21 that the set of dilations with fixed point O (together with the identity) form a group under composition. In particular, the inverse of a dilation with fixed point O is a dilation with fixed point O.

Definition DLN.1. Let $\mathcal{P}$ be a Euclidean plane, O a point on $\mathcal{P}$, and ρ a rotation about O which is not a point reflection. Define $\alpha(O) = O$ and for every $X \in \mathcal{P} \setminus \{O\}$ define $\alpha(X)$ to be the midpoint of $\overline{X\rho(X)}$. Then α is called the **half-rotation** of $\mathcal{P}$ about O **associated** with the rotation ρ, and ρ is the rotation **associated** with α. We will sometimes denote the associated rotation by ρ_α.

Theorem ROT.9(B) shows that there can be no half-rotation about O that maps a ray $\overrightarrow{OA}$ to itself, because no rotation maps $\overrightarrow{OA}$ to itself.

Theorem DLN.2. *Let O be a point on the Euclidean plane $\mathcal{P}$, ρ a rotation of $\mathcal{P}$ about O, and α the half-rotation about O associated with $\mathcal{P}$. For any member A of $\mathcal{P} \setminus \{O\}$, by Property R.5 of Definition NEUT.2, and Theorem NEUT.26 there exists a unique line of symmetry $\mathcal{G}_A$ for $\angle AO\rho(A)$. Then*

(A) $\rho = \mathcal{R}_{\mathcal{G}_A} \circ \mathcal{R}_{\overleftrightarrow{OA}}$, *and $\mathcal{G}_A$ is the only line with this property.*

(B) $\mathcal{G}_A \not\perp \overleftrightarrow{OA}$;

(C) *for every* $X \in \overrightarrow{OA}$, $\rho(X) = \mathcal{R}_{\mathcal{G}_A}(X)$; *therefore* $\rho(\overrightarrow{OA}) = \mathcal{R}_{\mathcal{G}_A}(\overrightarrow{OA})$ *and* $\rho(\overleftrightarrow{OA}) = \mathcal{R}_{\mathcal{G}_A}(\overleftrightarrow{OA})$;

(D) $\alpha(\overleftrightarrow{OA}) = \mathcal{G}_A$; *for every* $X \in \overleftrightarrow{OA}$, $\overleftrightarrow{X\alpha(X)} = \overleftrightarrow{X\rho(X)}$ *is perpendicular to* $\mathcal{G}_A$, *so that* $\alpha(A) = \text{ftpr}\,(O, \overleftrightarrow{A\rho(A)}) = \text{ftpr}\,(A, \mathcal{G}_A)$; *and*

(E) $\alpha(A) \in \text{ins}\,\angle AO\rho(A)$, $\angle AO\alpha(A)$ *is an acute angle, and* $\alpha(\overrightarrow{OA}) = \overrightarrow{O\alpha(A)}$ *is a subset of the bisecting ray of* $\angle AO\rho(A)$.

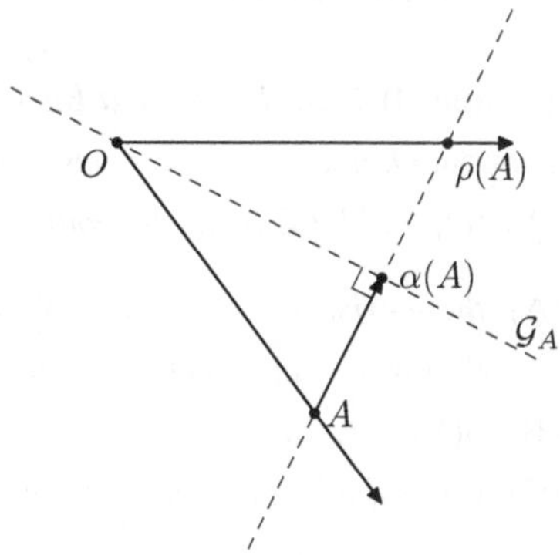

Fig. 13.1 For Theorem DLN.2 showing action of a half-rotation α.

Proof. See Figure 13.1.

(A) The decomposition $\rho = \mathcal{R}_{\mathcal{G}_A} \circ \mathcal{R}_{\overleftrightarrow{OA}}$ follows immediately from Theorem ROT.15. If $\mathcal{D}$ were a line through O such that $\mathcal{R}_{\mathcal{D}} \circ \mathcal{R}_{\overleftrightarrow{OA}} = \rho = \mathcal{R}_{\mathcal{G}_A} \circ \mathcal{R}_{\overleftrightarrow{OA}}$, then since $\mathcal{R}_{\overleftrightarrow{OA}} \circ \mathcal{R}_{\overleftrightarrow{OA}} = \iota$, $\mathcal{R}_{\mathcal{D}} = \mathcal{R}_{\mathcal{G}_A}$, and so by Remark NEUT.1.1 $\mathcal{D} = \mathcal{G}_A$.

(B) If $\mathcal{G}_A \perp \overleftrightarrow{OA}$, then by Definition ROT.1 and part (A), ρ is a point reflection, which contradicts Definition DLN.1.

(C) By part (A), if $X \in \overrightarrow{OA}$, $\rho(X) = \mathcal{R}_{\mathcal{G}_A}(\mathcal{R}_{\overleftrightarrow{OA}}(X)) = \mathcal{R}_{\mathcal{G}_A}(X)$. Thus by Theorem NEUT.15(3), $\rho(\overrightarrow{OA}) = \overrightarrow{\rho(O)\rho(A)} = \overrightarrow{O\mathcal{R}_{\mathcal{G}_A}(A)} = \mathcal{R}_{\mathcal{G}_A}(\overrightarrow{OA})$ and likewise $\rho(\overleftrightarrow{OA}) = \mathcal{R}_{\mathcal{G}_A}(\overleftrightarrow{OA})$.

(D) For every $X \in \overrightarrow{OA}$, by Definition DLN.1 and Theorem NEUT.20, the midpoint $\alpha(X)$ of $\overline{X\rho(X)}$ belongs to $\mathcal{G}_A$, the line of symmetry of $\angle AO\rho(A)$, so that $\alpha(\overrightarrow{OA}) \subseteq \mathcal{G}_A$. Since $\mathcal{G}_A$ is also the line of symmetry of the angle vertical to $\angle AO\rho(A)$, $\alpha(\overleftrightarrow{OA}) \subseteq \mathcal{G}_A$.

Conversely, suppose X is any point of $\mathcal{G}_A$ which is on the $\alpha(A)$-side of O, and let $\mathcal{L}_X = \text{pr}(X, \mathcal{G}_A)$ (that is, the line through X perpendicular to $\mathcal{G}_A$). By part (B) above and Corollary EUC.4, $\mathcal{L}_X$ must intersect $\overrightarrow{OA}$ at some point Y, since $\overrightarrow{OA}$ intersects $\mathcal{G}_A$ (at O) and $\mathcal{L}_X$ is perpendicular to $\mathcal{G}_A$. By Definition DLN.1,

$\alpha(Y) = X$. A similar argument will show the same thing for X on the opposite side of O, so that $\mathcal{G}_A \subseteq \alpha(\overleftrightarrow{OA})$ and hence $\mathcal{G}_A = \alpha(\overleftrightarrow{OA})$.

Finally, by Theorem NEUT.52, $\overleftrightarrow{X\alpha(X)} = \overleftrightarrow{X\rho(X)}$ is perpendicular to $\overleftrightarrow{O\alpha(A)} = \mathcal{G}_A$. This is true for $X = A$, so by Definition NEUT.99, $\alpha(A) = \text{ftpr}(O, \overleftrightarrow{A\rho(A)}) = \text{ftpr}(A, \mathcal{G}_A)$.

(E) By Theorem PSH.37 $\alpha(A) \in \text{ins} \angle AO\rho(A)$ and by Exercise NEUT.19, $\angle AO\alpha(A)$ is an acute angle. The last assertion follows from Definition NEUT.3(D). □

Lemma DLN.3. *Let $\mathcal{P}$ be a Euclidean plane, O a point on $\mathcal{P}$, α a half-rotation of $\mathcal{P}$ about O, and let ρ be the associated rotation of $\mathcal{P}$ about O. Let X be any point in $\mathcal{P} \setminus \{O\}$ and let $\mathcal{L}$ be a line containing X such that $\overleftrightarrow{OX} \perp \mathcal{L}$. Then:*

(A) *there exists a unique point Y of $\mathcal{P}$ such that $\alpha(Y) = X$, and this is a point of $\mathcal{L}$; thus α has an inverse mapping α^{-1};*
(B) *$\rho(Y) \in \mathcal{L}$; and*
(C) *$\overleftrightarrow{OX}$ is the line of symmetry of $\angle YO\rho(Y)$.*

Proof. Let $\mathcal{H}$ be the line of symmetry of $\angle XO\rho^{-1}(X)$. By Corollary EUC.4 $\mathcal{L}$ must intersect $\mathcal{H}$; let Y be this point of intersection.

By Theorem ROT.18, ρ^{-1} is a rotation so by Theorem DLN.2 $\rho^{-1} = \mathcal{R}_{\mathcal{H}} \circ \mathcal{R}_{\overleftrightarrow{OX}}$ and since reflections are their own inverses, $\rho = \mathcal{R}_{\overleftrightarrow{OX}} \circ \mathcal{R}_{\mathcal{H}}$. Then since $Y \in \mathcal{H}$, $\rho(Y) = \mathcal{R}_{\overleftrightarrow{OX}}(\mathcal{R}_{\mathcal{H}}(Y)) = \mathcal{R}_{\overleftrightarrow{OX}}(Y)$ and $\overleftrightarrow{OX}$ is thus the line of symmetry for $\angle YO\rho(Y)$, proving (C).

By Theorem DLN.2, $\alpha(Y) = \text{ftpr}(Y, \overleftrightarrow{OX})$. Both $\overleftrightarrow{Y\alpha(Y)}$ and $\mathcal{L}$ are lines containing Y that are perpendicular to $\overleftrightarrow{OX}$, so by Theorem NEUT.48(A), $\overleftrightarrow{Y\alpha(Y)} = \mathcal{L}$, and $\alpha(Y) = X$, proving the existence part of (A).

Thus $\mathcal{L}$ is a fixed line for $\mathcal{R}_{\overleftrightarrow{OX}}$, and by Theorem NEUT.22, $\rho(Y) = \mathcal{R}_{\overleftrightarrow{OX}}(Y) \in \mathcal{L}$, proving (B).

Finally, we show that Y is the only point of $\mathcal{P}$ such that $\alpha(Y) = X$, thus completing the proof of (A). If $\alpha(Z) = X$ and $Z \neq Y$, then $\overleftrightarrow{Z\alpha Z}$ contains X and is perpendicular to $\overleftrightarrow{OX}$, so equals $\mathcal{L}$ by Theorem NEUT.48(A). Thus $Z \in \mathcal{L}$. By Theorem DLN.2(D), $\rho = \mathcal{R}_{\overleftrightarrow{OX}} \circ \mathcal{R}_{\overleftrightarrow{OZ}}$, and since $\rho = \mathcal{R}_{\overleftrightarrow{OX}} \circ \mathcal{R}_{\mathcal{H}}$, $\mathcal{R}_{\overleftrightarrow{OZ}} = \mathcal{R}_{\mathcal{H}}$. By Remark NEUT.1.1, $\overleftrightarrow{OZ} = \mathcal{H}$, so that $Z = Y$. □

Theorem DLN.4. *Let $\mathcal{P}$ be a Euclidean plane, O a point on $\mathcal{P}$, α a half-rotation of $\mathcal{P}$ about O, and let ρ be the associated rotation of $\mathcal{P}$ about O. Then:*

(A) *α is a bijection of $\mathcal{P}$,*

(B) *α is a collineation of $\mathcal{P}$, and*

(C) *if $\mathcal{T}$ is any line in $\mathcal{P}$, then $\alpha(\text{pr}(O,\mathcal{T})) = \text{pr}(O,\alpha(\mathcal{T}))$; that is, if $\mathcal{L}$ is a line through O perpendicular to $\mathcal{T}$, $\alpha(\mathcal{L})$ is a line through O perpendicular to $\alpha(\mathcal{T})$.*

Proof. **(A) α is a bijection.** By Lemma DLN.3, if $X \neq O$ is any point, there is a unique point Y of $\mathcal{P}$ such that $\alpha(Y) = X$; therefore α is onto. Since Y is unique, α is one-to-one, and therefore a bijection.

(B) α is a collineation. This part introduces a quite technical proof; the reader may find Figure 13.2 useful for keeping things straight.

Let $\mathcal{T}$ be any line on $\mathcal{P}$. We show that $\alpha(\mathcal{T})$ is a line. If $O \in \mathcal{T}$, this is Theorem DLN.2(D). So we assume that $O \notin \mathcal{T}$. Let X be any point of $\mathcal{T}$; since α is a one-to-one mapping and $\alpha(O) = O$, $\alpha(X) \neq O$. Therefore $O \notin \alpha(\mathcal{T})$.

Let $P = \text{ftpr}(O,\mathcal{T})$, that is, the point of intersection of $\mathcal{T}$ with the line through O perpendicular to $\mathcal{T}$. By Lemma DLN.3, and since $\mathcal{T} \perp \overleftrightarrow{OP}$, there exists a point $Y \in \mathcal{T}$ such that $\alpha(Y) = P$ and $\overleftrightarrow{OP}$ is the line of symmetry for $\angle YO\rho(Y)$. Therefore $P \in \alpha(\mathcal{T})$ and P is the point of intersection of $\mathcal{T}$ and $\alpha(\mathcal{T})$.

By Theorem ROT.15 $\rho = \mathcal{R}_{\overleftrightarrow{O\rho(Y)}} \circ \mathcal{R}_{\overleftrightarrow{OP}}$. Then

$$\rho(P) = \mathcal{R}_{\overleftrightarrow{O\rho(Y)}}(\mathcal{R}_{\overleftrightarrow{OP}}(P)) = \mathcal{R}_{\overleftrightarrow{O\rho(Y)}}(P),$$

so that $\overleftrightarrow{O\rho(Y)}$ is the line of symmetry for $\angle PO\rho(P)$. Hence by Lemma DLN.3, $\overleftrightarrow{O\rho(Y)} = \overleftrightarrow{O\alpha(P)}$, which contains $\alpha(P)$. By Theorem DLN.2(D) $\overleftrightarrow{P\alpha(P)} \perp \overleftrightarrow{O\alpha(P)}$. We now begin the process of showing that α maps the line $\mathcal{T}$ onto the line $\overleftrightarrow{P\alpha(P)}$.

(B_1) In this part we prove that $\overleftrightarrow{O\alpha(X)}$ intersects $\overleftrightarrow{P\alpha(P)}$. We can show this much, even though we can't yet prove directly that $\alpha(X) \in \overleftrightarrow{P\alpha(P)}$.

Continuing our assumption that X is an arbitrary point of $\mathcal{T}$, we apply α to X. If $X = Y$, $\alpha(X) = P$; if $X = P$, $\alpha(P)$ belongs to both $\overleftrightarrow{P\alpha(P)}$ and $\overleftrightarrow{O\alpha(X)}$. In either case, $\overleftrightarrow{O\alpha(X)}$ intersects $\overleftrightarrow{P\alpha(P)}$.

For the general case, let $\mathcal{G}_X$ be the line of symmetry for $\angle XO\rho(X)$. By Theorem DLN.2(D), $\alpha(\overleftrightarrow{OX}) = \mathcal{G}_X = \overleftrightarrow{O\alpha(X)}$. By Theorem ROT.28 there exists a rotation σ such that for every $X \neq O$, $\sigma(\overrightarrow{OX}) = \overrightarrow{O\alpha(X)} \subseteq \mathcal{G}_X$. This holds true for $X = P$.

Thus, by Theorem NEUT.15(8), for any $X \neq P$ in $\mathcal{T}$, $\sigma(\angle XOP) = \angle\sigma(X)O\sigma(P) = \angle\alpha(X)O\alpha(P)$ showing that $\angle\alpha(X)O\alpha(P) \cong \angle XOP$. Since $\angle OPX$ is a right angle, by Theorem NEUT.84 $\angle XOP$ is acute, and by Theorem NEUT.83 $\angle\alpha(X)O\alpha(P)$ is acute. Since $\overleftrightarrow{O\alpha(X)}$ intersects $\overleftrightarrow{O\alpha(P)}$ at the point O and $\overleftrightarrow{O\alpha(P)} \perp \overleftrightarrow{P\alpha(P)}$, by Corollary EUC.4, $\overleftrightarrow{O\alpha(X)}$ intersects $\overleftrightarrow{P\alpha(P)}$ at some point we shall call Z, for now.

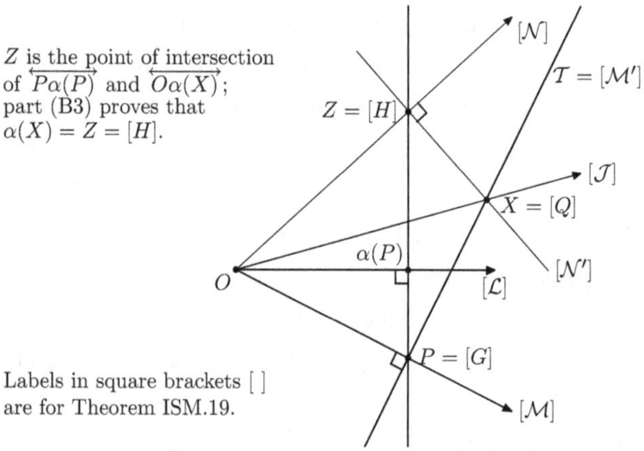

Fig. 13.2 For Theorem DLN.4.

(B₂) In this part we show that $\alpha(\mathcal{T}) \subseteq \overleftrightarrow{P\alpha(P)}$. The first set of computations sets the stage for a rather subtle application of Theorem ISM.19.

By Theorem DLN.2(D) the line of symmetry of $\angle PO\rho(P)$ is $\overleftrightarrow{O\alpha(P)}$, and $\overleftrightarrow{P\alpha(P)} \perp \overleftrightarrow{O\alpha(P)}$. Moreover, if X is any point of $\mathcal{T}$ other than P, the line of symmetry of $\angle XO\rho(X)$ is $\overleftrightarrow{O\alpha(X)}$, and $\overleftrightarrow{X\alpha(X)} \perp \overleftrightarrow{O\alpha(X)}$. By Theorem DLN.2(A) we can write ρ in two different ways:

$$\rho = \mathcal{R}_{\overleftrightarrow{O\alpha(X)}} \circ \mathcal{R}_{\overleftrightarrow{OX}} \text{ and } \rho = \mathcal{R}_{\overleftrightarrow{O\alpha(P)}} \circ \mathcal{R}_{\overleftrightarrow{OP}}$$

so that

$$\mathcal{R}_{\overleftrightarrow{O\alpha(X)}} \circ \mathcal{R}_{\overleftrightarrow{OX}} = \mathcal{R}_{\overleftrightarrow{O\alpha(P)}} \circ \mathcal{R}_{\overleftrightarrow{OP}} \quad (*)$$

and "multiplying" on the left by $\mathcal{R}_{\overleftrightarrow{O\alpha(X)}}$ we have

$$\mathcal{R}_{\overleftrightarrow{OX}} = \mathcal{R}_{\overleftrightarrow{O\alpha(X)}} \circ \mathcal{R}_{\overleftrightarrow{O\alpha(P)}} \circ \mathcal{R}_{\overleftrightarrow{OP}}. \quad (**)$$

Moreover, if there were a second line $\mathcal{K}$ such that

$$\rho = \mathcal{R}_{\overleftrightarrow{O\alpha(X)}} \circ \mathcal{R}_{\mathcal{K}} = \mathcal{R}_{\overleftrightarrow{O\alpha(X)}} \circ \mathcal{R}_{\overleftrightarrow{OX}}$$

then multiplying on the left by $\mathcal{R}_{\overleftrightarrow{O\alpha(X)}}$ yields $\mathcal{R}_{\mathcal{K}} = \mathcal{R}_{\overleftrightarrow{OX}}$, so that $\overleftrightarrow{OX}$ is the only line such that (*) above is true.

We now re-label the points and lines in this proof to correspond with points and lines in Theorem ISM.19, as in the table below:

DLN.4	ISM.19		DLN.4	ISM.19
Z	H		$\mathcal{T} = \overleftrightarrow{PX}$	$\overleftrightarrow{GQ} = \mathcal{M}'$
P	G		$\overleftrightarrow{O\alpha(P)}$	$\mathcal{L}$
X	Q		$\overleftrightarrow{OZ} = \overleftrightarrow{O\alpha(X)}$	$\overleftrightarrow{OH} = \mathcal{N}$
$\overleftrightarrow{OX}$	$\overleftrightarrow{OQ} = \mathcal{J}$		$\overleftrightarrow{XZ}$	$\overleftrightarrow{QH} = \mathcal{N}'$
$\overleftrightarrow{OP}$	$\overleftrightarrow{OG} = \mathcal{M}$		$\overleftrightarrow{PZ} = \overleftrightarrow{P\alpha(P)}$	$\overleftrightarrow{GH}$

Using these new labels, equality (**) above becomes $\mathcal{R}_{\mathcal{J}} = \mathcal{R}_{\mathcal{N}} \circ \mathcal{R}_{\mathcal{L}} \circ \mathcal{R}_{\mathcal{M}}$, as in Theorem ISM.19.

(B$_3$) At this stage in the proof, we know from (B$_1$) that the lines $\overleftrightarrow{O\alpha(X)}$ and $\overleftrightarrow{P\alpha(P)}$ intersect at some point Z; however, we don't know whether this point is $\alpha(X)$. The following argument will prove that it is.

By Theorem NEUT.48(A) let $\mathcal{N}'$ be the perpendicular to $\overleftrightarrow{O\alpha(X)} = \mathcal{N}$ at the point Z. We summarize the perpendicularity relations between the various lines in the following table:

DLN.4	ISM.19
$\overleftrightarrow{OP} \perp \mathcal{T}$	$\mathcal{M} \perp \mathcal{M}'$
$\overleftrightarrow{O\alpha(P)} \perp \overleftrightarrow{PZ}$	$\mathcal{L} \perp \overleftrightarrow{GH}$
$\overleftrightarrow{OZ} = \overleftrightarrow{O\alpha(X)} \perp \overleftrightarrow{XZ}$	$\mathcal{N} \perp \mathcal{N}'$

From Theorem ISM.19, the three lines $\mathcal{N}'$, $\mathcal{M}'$, and $\mathcal{J}$ intersect at a single point. Since $\mathcal{M}' = \mathcal{T}$ and $\mathcal{J} = \overleftrightarrow{OX}$ intersect at $Q = X$, this point is X.

So now we have two lines, $\overleftrightarrow{X\alpha(X)}$ and $\overleftrightarrow{XZ}$, both of which contain X and both of which are perpendicular to $\mathcal{N} = \overleftrightarrow{O\alpha(X)}$. One of them intersects $\mathcal{N}$ at $\alpha(X)$ and the other at Z. By Theorem NEUT.48(A) there can only be one such line; hence $Z = \alpha(X)$ and thus $\alpha(X) \in \overleftrightarrow{P\alpha(P)}$.

This completes the proof that $\alpha(\mathcal{T}) \subseteq \overleftrightarrow{P\alpha(P)}$.

(B$_4$) $\overleftrightarrow{P\alpha(P)} \subseteq \alpha(\mathcal{T})$. Here, P, $\alpha(P)$, $\mathcal{L}$, $\mathcal{M}$, and $\mathcal{T} = \mathcal{M}'$ are defined as before. Again we use Theorem ISM.19.

Suppose Z is any point of $\overleftrightarrow{P\alpha(P)}$ other than P or $\alpha(P)$. Then define $\mathcal{N} = \overleftrightarrow{OZ}$, and let $\mathcal{N}' = \text{pr}(Z, \mathcal{N})$, so that $Z \in \mathcal{N}'$ and $\mathcal{N}' \perp \mathcal{N}$. By Lemma DLN.3, there exists a point $Y \in \mathcal{N}'$ such that $\mathcal{N}$ is the line of symmetry of $\angle YO\rho(Y)$. Define $\mathcal{J} = \overleftrightarrow{OY}$. Then $\mathcal{J} \cap \mathcal{N}' = \{Y\}$.

By Theorem DLN.2(A), $\mathcal{R}_{\mathcal{N}} \circ \mathcal{R}_{\mathcal{J}} = \rho = \mathcal{R}_{\mathcal{L}} \circ \mathcal{R}_{\mathcal{M}}$. Multiplying on the left by $\mathcal{R}_{\mathcal{N}}$ we have $\mathcal{R}_{\mathcal{J}} = \mathcal{R}_{\mathcal{N}} \circ \mathcal{R}_{\mathcal{L}} \circ \mathcal{R}_{\mathcal{M}}$. Since $\mathcal{T} = \mathcal{M}'$, $\mathcal{M}' \perp \mathcal{M}$; by Theorem ISM.19, $\mathcal{M}'$, $\mathcal{N}'$, and $\mathcal{J}$ intersect at a point Q. But $\mathcal{J} \cap \mathcal{N}' = \{Y\}$, so $Q = Y$. $Y \in \mathcal{T}$ and $\alpha(Y) = Z$, showing that Z is a member of $\alpha(\mathcal{T})$. Thus $\overleftrightarrow{P\alpha(P)} = \alpha(\mathcal{T})$.

(C) follows immediately from the construction used in part (B). □

Theorem DLN.5 (Dilation contracting the plane). *Let $\mathcal{P}$ be a Euclidean plane, O a point on $\mathcal{P}$, and let α be a half-rotation of $\mathcal{P}$ about O with associated rotation ρ. Using Theorem ROT.28 let θ be the rotation of $\mathcal{P}$ about O such that $\theta \circ \theta = \rho$. Then $\theta^{-1} \circ \delta$ is a dilation of $\mathcal{P}$ with fixed point O. Moreover, for every $X \neq O$, $\theta(\overrightarrow{OX}) = \alpha(\overrightarrow{OX})$.*

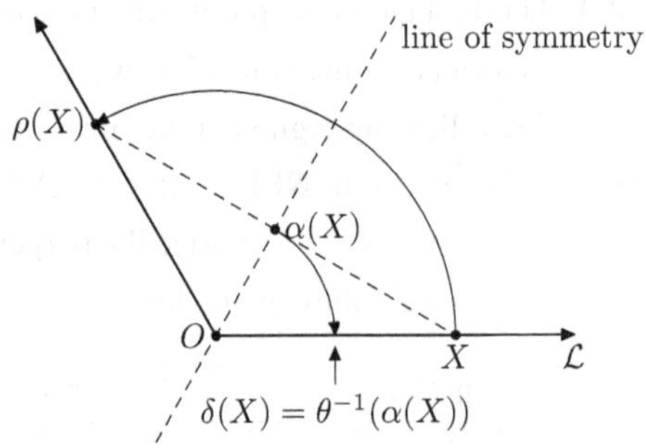

Fig. 13.3 For Theorem DLN.5 showing a dilation contracting the plane.

Proof. See Figure 13.3. Let $\mathcal{L}$ be any line such that $O \in \mathcal{L}$, and let $X \in \mathcal{L}$ and $X \neq O$. By Theorem DLN.2(D) $\alpha(\mathcal{L})$ is the line of symmetry of $\angle XO\rho(X)$. By Exercise ROT.4(A), $\angle XO\theta(X) \cong \angle\theta(X)O\theta(\theta(X)) = \angle\theta(X)O\rho(X)$ so that by Theorem NEUT.39, $\theta(\mathcal{L})$ is the line of symmetry of $\angle XO\rho(X)$. Therefore $\theta(\mathcal{L}) = \alpha(\mathcal{L})$ and $(\theta^{-1} \circ \alpha)(\mathcal{L}) = \mathcal{L}$. Therefore every line through O is a fixed line for $\theta^{-1} \circ \alpha$.

Let $Y = \theta^{-1} \circ \alpha(X)$. Then $Y \neq X$, because $\overline{OY} \cong \overline{O\theta(Y)} = \overline{O\alpha(X)}$ and $\overline{O\alpha(X)} < \overline{OX}$ by Theorem NEUT.93 (a leg of a right triangle is smaller than the hypotenuse). Hence X is not a fixed point of $\theta^{-1} \circ \alpha$, which mapping is not the identity.

Since $\theta^{-1} \circ \alpha$ is the composition of two collineations, by Theorem CAP.1(C) it is a collineation with no fixed point other than O; since every line through O is a fixed line, we may apply Theorem CAP.22, showing that $\theta^{-1} \circ \alpha$ is a dilation of $\mathcal{P}$ with fixed point O.

This last might also be proved as follows: if $\mathcal{T}$ is a line not containing O and $\mathcal{L}$ is a line through O such that $\mathcal{L} \perp \mathcal{T}$, then by Theorem DLN.4(C), $\alpha(\mathcal{T}) \perp \alpha(\mathcal{L})$. Now θ^{-1} is an isometry, so $\theta^{-1}(\alpha(\mathcal{T})) \perp \theta^{-1}(\alpha(\mathcal{L})) = \mathcal{L}$ so both $\mathcal{T}$ and $\theta^{-1}(\alpha(\mathcal{T}))$ are perpendicular to $\mathcal{L}$; hence by Theorem NEUT.47(A), if these lines are distinct, they are parallel. By Definition CAP.17 $\theta^{-1} \circ \alpha$ is a dilation. □

Theorem DLN.6. *Let $\mathcal{P}$ be a Euclidean plane and O be a point on $\mathcal{P}$.*

(A) *If α and β are half-rotations of $\mathcal{P}$ about O, then $\beta \circ \alpha = \alpha \circ \beta$.*

(B) *If α is a half-rotation and π a rotation of $\mathcal{P}$ about O, then $\pi \circ \alpha = \alpha \circ \pi$.*

(C) *If α and β are bijections of $\mathcal{P}$ (in particular, rotations or half-rotations) and $\beta \circ \alpha = \alpha \circ \beta$, then $\beta \circ \alpha^{-1} = \alpha^{-1} \circ \beta$.*

(D) *Here we anticipate the proof of Theorem DLN.7 below. If α, β, and γ are half-rotations of $\mathcal{P}$ about O, and π is a rotation of $\mathcal{P}$ about O, then $\pi \circ \gamma^{-1} \circ \beta \circ \alpha = \gamma^{-1} \circ \beta \circ \alpha \circ \pi$.*

Proof. (A) (Two half-rotations commute.) Let X be any member of $\mathcal{P} \setminus \{O\}$ and let $\mathcal{K} = \overleftrightarrow{OX}$. If ρ_α is the rotation of $\mathcal{P}$ about O associated with α and if ρ_β is the rotation of $\mathcal{P}$ about O associated with β, then by Theorem DLN.2(A) and (D), $\rho_\alpha = \mathcal{R}_{\alpha(\mathcal{K})} \circ \mathcal{R}_\mathcal{K}$ and $\rho_\beta = \mathcal{R}_{\beta(\mathcal{K})} \circ \mathcal{R}_\mathcal{K}$. These are true for any line $\mathcal{K}$ containing the point O, so in particular they are true for the lines $\beta(\mathcal{K})$, $\alpha(\mathcal{K})$, etc. Therefore, $\mathcal{R}_{\alpha(\mathcal{K})} \circ \mathcal{R}_\mathcal{K} = \rho_\alpha = \mathcal{R}_{\alpha(\beta(\mathcal{K}))} \circ \mathcal{R}_{\beta(\mathcal{K})}$ and $\mathcal{R}_{\beta(\mathcal{K})} \circ \mathcal{R}_\mathcal{K} = \rho_\beta = \mathcal{R}_{\beta(\alpha(\mathcal{K}))} \circ \mathcal{R}_{\alpha(\mathcal{K})}$. Thus $\mathcal{R}_{\alpha(\beta(\mathcal{K}))} = \mathcal{R}_{\alpha(\mathcal{K})} \circ \mathcal{R}_\mathcal{K} \circ \mathcal{R}_{\beta(\mathcal{K})}$ and

$$\mathcal{R}_{\beta(\alpha(\mathcal{K}))} = \mathcal{R}_{\beta(\mathcal{K})} \circ \mathcal{R}_\mathcal{K} \circ \mathcal{R}_{\alpha(\mathcal{K})}. \quad (*)$$

By Theorem ROT.12 $\mathcal{R}_{\alpha(\beta(\mathcal{K}))} = \mathcal{R}_{\beta(\alpha(\mathcal{K}))}$. By Remark NEUT.1.1 $\alpha(\beta(\mathcal{K})) = \beta(\alpha(\mathcal{K}))$.

It remains to prove that $(\beta \circ \alpha)(X) = (\alpha \circ \beta)(X)$. To accomplish this, first rewrite equation (*) above by multiplying on the left by $\mathcal{R}_{\beta(\mathcal{K})}$ and on the right by $\mathcal{R}_{\alpha(\mathcal{K})}$ to get

$$\mathcal{R}_{\beta(\mathcal{K})} \circ \mathcal{R}_{\beta(\alpha(\mathcal{K}))} \circ \mathcal{R}_{\alpha(\mathcal{K})} = \mathcal{R}_\mathcal{K}.$$

In Theorem ISM.19, let $\mathcal{M} = \alpha(\mathcal{K})$ so that $\alpha(X) \in \mathcal{M}$, and $\mathcal{N} = \beta(\mathcal{K})$ so that $\beta(X) \in \mathcal{N}$; and let $\mathcal{L} = \beta(\alpha(\mathcal{K})) = \alpha(\beta(\mathcal{K}))$. By Definition DLN.1, the line $\overleftrightarrow{\alpha(X)\beta(\alpha(X))} \perp \mathcal{L}$ at the point $\beta(\alpha(X))$. By Corollary EUC.4, $\overleftrightarrow{\alpha(X)\beta(\alpha(X))}$ intersects $\mathcal{N}$ at some point P. Let $\mathcal{N}'$ be the line perpendicular to $\mathcal{N}$ at the point P. Let $\mathcal{M}' = \overleftrightarrow{X\alpha(X)}$; this is perpendicular to $\mathcal{M}$ at the point $\alpha(X)$.

By Theorem ISM.19, $\mathcal{N}'$, $\mathcal{M}'$, and $\mathcal{K}$ intersect at a point Q; since the point of intersection of $\mathcal{M}'$ and $\mathcal{K}$ is the point X, $Q = X$. Since $\mathcal{N}'$ is perpendicular to $\mathcal{N}$, as is $\overleftrightarrow{X\beta(X)}$, and both $\mathcal{N}'$ and $\overleftrightarrow{X\beta(X)}$ contain X, by Theorem NEUT.48(A), $\mathcal{N}' = \overleftrightarrow{X\beta(X)}$ and $P = \beta(X)$.

We know that $\alpha(\beta(X))$ lies on $\mathcal{L}$, and since $\overleftrightarrow{\alpha(X)\beta(\alpha(X))}$ is perpendicular to $\mathcal{L}$, $\alpha(\beta(X))$ is the point of intersection of these two lines; but we already know this is $\beta(\alpha(X))$. Therefore $\alpha(\beta(X)) = \beta(\alpha(X))$, and since X is any member of $\mathcal{P} \setminus \{O\}$, $\beta \circ \alpha = \alpha \circ \beta$, completing the proof of part (A).

(B) (Half-rotations commute with rotations.) Let α be any half-rotation about O with associated rotation ρ and let $X \in \mathcal{P} \setminus \{O\}$. By Theorem ROT.15(A) (or Theorem DLN.5) let θ be the rotation about O such that $\theta(X) \in \overrightarrow{O\alpha(X)}$. Since for any $X \in \mathcal{P} \setminus \{O\}$, both $\alpha(X)$ and $\theta(X)$ lie in the angle bisector of $\angle XO\rho(X)$, $\overrightarrow{O\alpha(X)} = \overrightarrow{O\theta(X)}$.

Let π be a rotation of the plane about O. Then by Theorem ROT.21, for any $X \in \mathcal{P} \setminus \{O\}$, $\pi(\theta(X)) = \theta(\pi(X))$ and by Theorem NEUT.15(2)

$$\pi(\alpha(X)) \in \pi(\overrightarrow{O(\theta(X))}) = \overrightarrow{O(\pi(\theta(X)))} = \overrightarrow{O(\theta(\pi(X)))} = \overrightarrow{O(\alpha(\pi(X)))}.$$

Therefore $\overrightarrow{O(\pi(\alpha(X)))} = \overrightarrow{O(\alpha(\pi(X)))}$ and $\alpha(\pi(X)) \in \overrightarrow{O(\pi(\alpha(X)))}$.

Consider now triangle $\pi(\triangle XO\alpha(X)) = \triangle\pi(X)O\pi(\alpha(X))$. By Definition NEUT.3(B) $\triangle XO\alpha(X) \cong \triangle\pi(X)O\pi\alpha(X)$.

Since α is a half-rotation, we know that $\angle O\alpha(X)X$ is a right angle; hence by Corollary NEUT.44.1, $\angle O\pi(\alpha(X))\pi(X)$ is also right. Since $\alpha(\pi(X))$ is a member of $\overrightarrow{O\pi(\alpha(X))}$, $\alpha(\pi(X)) = \text{ftpr}\,(\pi(X), \overleftrightarrow{O\pi(\alpha(X))})$, and (by Theorem NEUT.47(B)) there can be only one perpendicular from a point to a line, it follows that $\alpha(\pi(X)) = \pi(\alpha(X))$. Again, since X is any member of $\mathcal{P} \setminus \{O\}$, $\alpha \circ \pi = \pi \circ \alpha$, completing the proof of part (B).

(C) If $\beta \circ \alpha = \alpha \circ \beta$, then since $\alpha^{-1} \circ \alpha = \iota$,

$$\beta \circ \alpha^{-1} = \alpha^{-1} \circ \alpha \circ \beta \circ \alpha^{-1} = \alpha^{-1} \circ \beta \circ \alpha \circ \alpha^{-1} = \alpha^{-1} \circ \beta.$$

(D) Let α, β, and γ be half-rotations about O and let π be a rotation about O. By part (B), $\pi \circ \gamma = \gamma \circ \pi$, and by part (C) $\pi \circ \gamma^{-1} = \gamma^{-1} \circ \pi$. Then, applying this, and part (B) twice, we have

$$\pi \circ \gamma^{-1} \circ \beta \circ \alpha = \gamma^{-1} \circ \pi \circ \beta \circ \alpha = \gamma^{-1} \circ \beta \circ \pi \circ \alpha = \gamma^{-1} \circ \beta \circ \alpha \circ \pi. \quad \square$$

Theorem DLN.7 (Structure of dilations). *Let O, A, and B be distinct collinear points on a Euclidean plane $\mathcal{P}$.*

(A) *There exist half-rotations α, β, and γ about O such that the mapping $\delta = \gamma^{-1} \circ \beta \circ \alpha$ is a collineation and maps A to B.*
(B) *For any point $A' \neq O$ in $\mathcal{P}$, the mapping δ defined in part (A) maps $\overleftrightarrow{OA'}$ to itself, so that every line through O is a fixed line for δ.*
(C) *There exists exactly one dilation δ with fixed point O such that $\delta(A) = B$, and it is the mapping defined in part (A).*
(D) *Every dilation δ of $\mathcal{P}$ with fixed point O can be written as in part (A).*
(E) *If δ is a dilation of $\mathcal{P}$ with fixed point O, and π is a rotation about O, then $\pi \circ \delta = \delta \circ \pi$.*

Proof. (A) Note that there may be many possible choices for α, β, and γ such that the mapping $\gamma^{-1} \circ \beta \circ \alpha$ carries A to B. This will become obvious in the following construction. We now undertake the basic construction for the proof. All rotations and half-rotations have fixed point O. See Figure 13.4.

Let Q be a member of $\mathcal{P} \setminus \overleftrightarrow{OA} = \mathcal{P} \setminus \mathcal{J}$, $\mathcal{K} = \overleftrightarrow{OQ}$, $C = \text{ftpr}(A, \mathcal{K})$, α be the half-rotation associated with the rotation $\rho_\alpha = \mathcal{R}_\mathcal{K} \circ \mathcal{R}_\mathcal{J}$. Then by Theorem DLN.2(D) $\alpha(A) = C$.

Let $\mathcal{M} = \overleftrightarrow{BC}$, $\mathcal{N} = \text{pr}(O, \mathcal{M})$, $D = \text{ftpr}(O, \mathcal{M})$, and β be the half-rotation associated with the rotation $\rho_\beta = \mathcal{R}_\mathcal{N} \circ \mathcal{R}_\mathcal{K}$. Then by Theorem DLN.2(D) $\beta(C) = D$.

Let γ be the half-rotation associated with the rotation $\rho_\gamma = \mathcal{R}_\mathcal{N} \circ \mathcal{R}_\mathcal{J}$. Then by Theorem DLN.2(D) $\gamma(B) = D$. By Theorem DLN.2(D) $\gamma(\mathcal{J}) = \mathcal{N}$. Thus $\gamma^{-1}(D) = B$, and $\gamma^{-1}(\mathcal{N}) = \mathcal{J}$. Therefore, if we let $\delta = \gamma^{-1} \circ \beta \circ \alpha$, then $\delta(A) = B$.

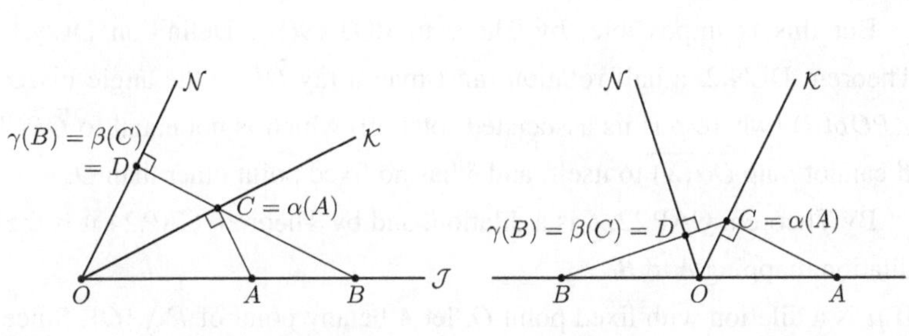

Fig. 13.4 For the construction of Theorem DLN.7. The left-hand figure is for the case O–A–B; the right-hand figure is for the case B–O–A. It might be instructive to construct a figure for the case where O–B–A, as in Theorem DLN.5.

By Theorem DLN.4, each of α, β, and γ is a bijection and a collineation, and by Theorem CAP.1(D') γ^{-1} is a collineation. By elementary mapping theory $\delta = \gamma^{-1} \circ \beta \circ \alpha$ is a bijection and collineation.

(B) We now show that for every point $A' \neq O$, $\delta(A') \in \overleftrightarrow{OA'}$. Let $C' = \alpha(A')$, $D' = \beta(C')$, and let $B' = \gamma^{-1}(D)$.

(Case 1: $A' \in \overleftrightarrow{OA}$.) By Theorem DLN.2, both $C = \alpha(A)$ and $C' = \alpha(A')$ are members of the line of symmetry for the rotation ρ associated with α. Thus $C' = \alpha(A') \in \overleftrightarrow{OC}$. Similarly, $D' = \beta(C') \in \overleftrightarrow{OD}$ and $D' = \gamma(B') \in \overleftrightarrow{OD}$. Since γ maps $\overleftrightarrow{OA}$ onto $\overleftrightarrow{OD}$, γ^{-1} maps $\overleftrightarrow{OD}$ onto $\overleftrightarrow{OA}$. Therefore $\delta = \gamma^{-1} \circ \beta \circ \alpha$ maps A' to a point of $\overleftrightarrow{OA}$, which is therefore a fixed line.

(Case 2: $A' \notin \overleftrightarrow{OA}$.) By Theorem ROT.15 there exists a rotation π such that $\pi(\overrightarrow{OA}) = \overrightarrow{OA'}$, so that $\pi(A) \in \overrightarrow{OA'}$. Without loss of generality we may choose A so that $\pi(A) = A'$. Then by Theorem DLN.6(B), $C' = \alpha(A') = \alpha(\pi(A)) = \pi(\alpha(A)) = \pi(C)$. Similarly, $D' = \beta(C') = \beta(\pi(C)) = \pi(\beta(C)) = \pi(D)$. Since $\gamma \circ \pi = \pi \circ \gamma$, by Theorem DLN.6(B) and (C), $\pi \circ \gamma^{-1} = \gamma^{-1} \circ \pi$, so that we have also $B' = \gamma^{-1}(D') = \gamma^{-1}(\pi(D)) = \pi(\gamma^{-1}(D)) = \pi(B)$. Since $B \in \overleftrightarrow{OA}$, $\pi(B) \in \pi(\overleftrightarrow{OA}) = \overleftrightarrow{O\pi(A)} = \overleftrightarrow{OA'}$, so that $B' = \delta(A') \in \overleftrightarrow{OA'}$. Therefore δ maps every point A' on the plane into $\overleftrightarrow{OA'}$, which is a fixed line for δ.

(C) To show that $\delta = \gamma^{-1} \circ \beta \circ \alpha$ is a dilation, we need only show that it has no fixed point other than O. Suppose that for some $X \in \mathcal{P} \backslash \{O\}$, $\delta(X) = X$. Then X is both the argument for δ (shown in the figure for part (A) as A) and the image $\delta(X)$ (shown as B), and $\overleftrightarrow{X\alpha(X)}$ and $\overleftrightarrow{X\gamma(X)}$ are the same line, because they both contain $\alpha(X)$. By Theorem NEUT.47(B) there is only one line containing O which is perpendicular to $\overleftrightarrow{X\alpha(X)} = \overleftrightarrow{X\gamma(X)}$; moreover, both $\alpha(X)$ and $\gamma(X)$ belong to this line, so that $\alpha(X) = \gamma(X)$. Since $(\beta \circ \alpha)(X) = \gamma(X)$ and $\alpha(X) = \gamma(X)$, β must map $\overrightarrow{O\alpha(X)}$ to itself.

But this is impossible; by Theorem ROT.19(B), Definition DLN.1, and Theorem DLN.2, a half-rotation must map a ray $\overrightarrow{OP}$ to the angle bisector of $\angle PO\rho(P)$ (where ρ is its associated rotation) which is not equal to $\overrightarrow{OP}$. Thus, β cannot map $\overrightarrow{O\alpha(X)}$ to itself, and δ has no fixed point other than O.

By Theorem CAP.22, δ is a dilation, and by Theorem CAP.24 it is the only dilation mapping A to B.

(D) If μ is a dilation with fixed point O, let A be any point of $\mathcal{P} \setminus \{O\}$. Since O is the only fixed point, $B = \mu(A) \neq A$. By Theorem CAP.18(D), B, A, and O are collinear. By parts (A) and (C) there exists a dilation $\delta = \gamma^{-1} \circ \beta \circ \alpha$ where α, β, and γ are half-rotations about O, and $\delta(A) = B$, and this is the only dilation mapping A to B, so $\delta = \mu$.

(E) Follows directly from part (D) above and Theorem DLN.6(D). □

Remark DLN.7.1. The preceding two results (Theorems DLN.6 and DLN.7) show that half-rotations, their inverses, as well as dilations commute with rotations. Thus, speaking intuitively, if we define one of these mappings by its action on a specific line, rotating that line to a new position rotates the entire "picture" and replicates it in the new position. This assures us that such a construction is a global one, even if we specify it on a specific line.

This last result, Theorem DLN.7(E) shows that a dilation δ expands or shrinks the plane equally in all directions. (It may also "mirror" it about the point O, in the case where X–O–$\delta(X)$.) Applying a dilation to a point X moves it along the line $\overleftrightarrow{OX}$ to some point; rotating X to another line through O, then applying the dilation, then rotating back to the original line, accomplishes exactly the same thing.

Theorem DLN.8. *Every dilation δ of a Euclidean plane $\mathcal{P}$*

(A) *is a collineation, and*

(B) *is a belineation; that is, if A, B, and C are any points of $\mathcal{P}$, and A–B–C, then $\delta(A)$–$\delta(B)$–$\delta(C)$.*

Proof. From Theorem DLN.7, $\delta = \gamma^{-1} \circ \beta \circ \alpha$ where each of γ, β, and α is half-rotations about O.

(A) By Theorem DLN.4, half-rotations are collineations. By Theorem CAP.1(D'), γ^{-1} is a collineation. Since a composition of collineations is a collineation, δ is a collineation.

(B) Assume that O is a fixed point for δ, and let A, B, and C be any points on $\mathcal{P}$ such that A–B–C.

(Case 1: A, B, C, and O are collinear.) Since α is a collineation, and has fixed point O, $\alpha(A)$, $\alpha(B)$, $\alpha(C)$, and O are collinear, and all these points are members of $\overleftrightarrow{O\alpha(A)}$. By Theorem NEUT.47(A), the lines $\overleftrightarrow{A\alpha(A)}$, $\overleftrightarrow{B\alpha(B)}$, and $\overleftrightarrow{C\alpha(C)}$ are all parallel, since they are perpendicular to $\overleftrightarrow{O\alpha(A)}$ by Theorem DLN.2(D). (See Figure 13.4, for Theorem DLN.7.) By Exercises PSH.57 and PSH.58, $\alpha(A)$–$\alpha(B)$–$\alpha(C)$. Thus α is a belineation. Similar proofs show that β and γ

preserve betweenness and are belineations. By Theorem COBE.3, γ^{-1} is a belineation. Thus

$$\gamma^{-1}(\beta(\alpha(A)))\text{–}\gamma^{-1}(\beta(\alpha(B)))\text{–}\gamma^{-1}(\beta(\alpha(C))).$$

That is, $\delta(A)$–$\delta(B)$–$\delta(C)$, showing that δ is a belineation.

(Case 2: A, B, and C are not collinear with O.) By Theorem CAP.18(D), $\delta(A) \in \overleftrightarrow{OA}$, $\delta(B) \in \overleftrightarrow{OB}$, and $\delta(C) \in \overleftrightarrow{OC}$.

Since A–B–C and $O \notin \overleftrightarrow{AC}$, by Theorem IB.14 $B \in \overrightarrow{AC}$ and $B \in$ the C-side of $\overleftrightarrow{OA}$). Similarly, $B \in$ the A-side of $\overleftrightarrow{OC}$, so that by Definition PSH.36 $B \in$ ins $\angle AOC$, and by Theorem PSH.38(B) $\overrightarrow{OB} \subseteq$ ins $\angle AOC$.

There are two subcases. In subcase (a), $\delta(A) \in \overrightarrow{OA}$; by Exercise DLN.5(II) $\delta(B) \in \overrightarrow{OB}$ and $\delta(C) \in \overrightarrow{OC}$. In subcase (b), by the same exercise, each of $\delta(A)$, $\delta(B)$, and $\delta(C)$ belongs to the opposing ray.

If subcase (a) holds, $\delta(A) \in \overrightarrow{OA}$ and $\delta(C) \in \overrightarrow{OC}$, so that $\angle\delta(A)O\delta(C) = \angle AOC$, and $\delta(B) \in \overrightarrow{OB} \in$ ins $\angle AOC$. By Corollary PSH.39.2 $\delta(A)$ and $\delta(C)$ are on opposite sides of the line $\overleftrightarrow{OB}$ and by Definition IB.11 $\overleftrightarrow{OB}$ intersects $\overline{\delta(A)\delta(C)} \subseteq \overleftrightarrow{\delta(A)\delta(C)}$ at some point P. By part (A) dilations are collineations, so that $\delta(B) \in \overleftrightarrow{\delta(A)\delta(C)}$, and $\overleftrightarrow{O\delta(B)} = \overleftrightarrow{OB}$ intersects $\overleftrightarrow{\delta(A)\delta(C)}$ at the point $\delta(B)$. By Exercise I.1, $\delta(B) = P \in \overline{\delta(A)\delta(C)}$, or $\delta(A)$–$\delta(B)$–$\delta(C)$.

If subcase (2b) holds, then $\angle\delta(A)O\delta(C)$ is vertical to $\angle AOC$, and $\delta(B) \in \overrightarrow{OB} \in$ ins $\angle\delta(A)O\delta(C)$. An argument similar to that for subcase (2a) above shows the same result. □

Theorem DLN.9 (Point-wise construction of the dilation of Theorem DLN.7.). *Let $\mathcal{P}$ be a Euclidean plane, O, A, and B be distinct collinear points on $\mathcal{P}$ and δ be the dilation of $\mathcal{P}$ such that $\delta(A) = B$.*

(A) *If X is any member of $\mathcal{P} \setminus \overleftrightarrow{OA}$, then $\delta(X)$ is the point such that* $\operatorname{par}(B, \overleftrightarrow{AX}) \cap \overleftrightarrow{OX} = \{\delta(X)\}$.

(B) *Let X be any member of $\overleftrightarrow{OA} \setminus \{O, A\}$. Let Q be a member of $\mathcal{P} \setminus \overleftrightarrow{OA}$, so that $\delta(Q) \in \overleftrightarrow{OQ}$. Then $\delta(X)$ is the point such that* $\operatorname{par}(\delta(Q), \overleftrightarrow{QX}) \cap \overleftrightarrow{OA} = \{\delta(X)\}$.

Proof. (A) By Theorem CAP.18 $\overleftrightarrow{OX}$ is a fixed line of δ but $\overleftrightarrow{AX}$ is not. By Theorem CAP.1(A) $\delta(\overleftrightarrow{AX}) = \overleftrightarrow{\delta(A)\delta(X)} = \overleftrightarrow{B\delta(X)}$. By Remark CAP.11 $\overleftrightarrow{B\delta(X)} \parallel \overleftrightarrow{AX}$ so that $\overleftrightarrow{B\delta(X)} = \operatorname{par}(B, \overleftrightarrow{AX})$. (cf Axiom PS.) By Exercises IP.4 and I.1, $\delta(X)$ is the point such that $\overleftrightarrow{OX} \cap \operatorname{par}(B, \overleftrightarrow{AX}) = \{\delta(X)\}$.

(B) By part (A) we may locate $\delta(Q)$ on $\overleftrightarrow{OQ}$. Then since neither A or X is in $\overleftrightarrow{OQ}$ we may apply part (A) again to locate $\delta(X)$ as the point of intersection of par $(\delta(Q), \overleftrightarrow{QX})$ and $\overleftrightarrow{OA}$. Since we know already that δ is a well-defined mapping, this completes the proof. □

Remark DLN.10. If we were using the construction in the proof of part (B) of Theorem DLN.9 as a definition of the mapping δ, it would be necessary, in order for the mapping to be well-defined, to prove that two different choices for Q would yield the same value for $\delta(X)$. However, Theorem DLN.9 assumes that δ is a well-defined mapping and is a dilation. Parts (A) and (B) of the proof merely use the properties of the dilation to show that for a given X, these constructions give the correct value of $\delta(X)$. This is true for the proof of part (B), even though Q is chosen arbitrarily. It follows that the end result of the construction of part (B) is independent of the choice of Q (and of the line $\overleftrightarrow{OQ}$).

The following theorem is credited to Pappus of Alexandria (c. 290–350).

Theorem DLN.11 (The Proposition of Pappus). *Let $\mathcal{P}$ be a Euclidean plane, O a point on $\mathcal{P}$, and let $\mathcal{L}$ and $\mathcal{L}'$ be lines on $\mathcal{P}$ such that $\mathcal{L} \cap \mathcal{L}' = \{O\}$. Let Q, R, and S be points on $\mathcal{L}$ and let Q', R', and S' be points on $\mathcal{L}'$ such that:*

(1) *the points O, Q, R, S, Q', R', and S' are distinct;*
(2) *there exist points T and V such that $\overleftrightarrow{QR'} \cap \overleftrightarrow{RS'} = \{T\}$ and $\overleftrightarrow{RQ'} \cap \overleftrightarrow{SR'} = \{V\}$; and*
(3) *$\overleftrightarrow{QR'} \parallel \overleftrightarrow{RQ'}$ and $\overleftrightarrow{RS'} \parallel \overleftrightarrow{SR'}$.*

Then $\overleftrightarrow{QS'} \parallel \overleftrightarrow{SQ'}$.

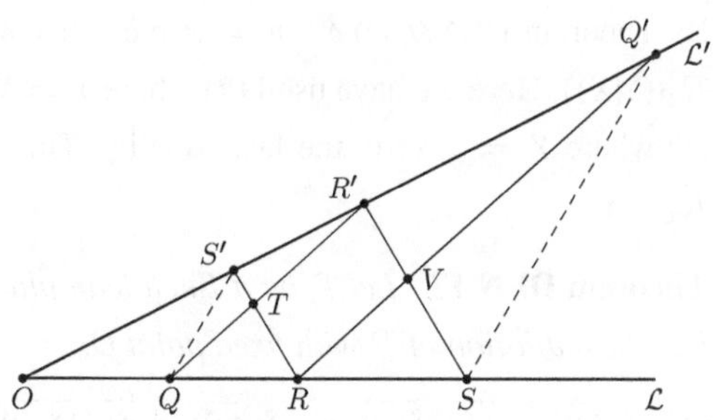

Fig. 13.5 For Theorem DLN.11.

Proof. See Figure 13.5. Using Theorem DLN.7 let δ be the dilation of $\mathcal{P}$ with fixed point O such that $\delta(Q) = R$ and let ϵ be the dilation of $\mathcal{P}$ with fixed point O such that $\epsilon(R) = S$. Then $S = (\epsilon \circ \delta)(Q)$. Since $\delta(Q) = R$ and $\overleftrightarrow{QR'} \parallel \overleftrightarrow{RQ'}$, by Definition CAP.17, $\delta(\overleftrightarrow{QR'}) = \overleftrightarrow{RQ'}$. Thus $\delta(R') = Q'$. Likewise since $\epsilon(R) = S$, $\epsilon(S') = R'$; combining these results, $(\delta \circ \epsilon)(S') = Q'$. By Exercise DLN.3 $\epsilon \circ \delta = \delta \circ \epsilon$ and thus $(\epsilon \circ \delta)(S') = (\delta \circ \epsilon)(S') = Q'$. By this equality and Theorem CAP.1

$$(\epsilon \circ \delta)(\overleftrightarrow{QS'}) = \overleftrightarrow{(\epsilon \circ \delta)(Q)(\epsilon \circ \delta)(S')} = \overleftrightarrow{SQ'}.$$

Definition CAP.17 says that a dilation maps a line to a line parallel to it, so the composition of two dilations does the same. Therefore $\overleftrightarrow{QS'} \parallel \overleftrightarrow{SQ'}$. □

13.2 Properties of dilations

Theorem DLN.12. *Let $\mathcal{P}$ be a Euclidean plane and let δ be a dilation of $\mathcal{P}$ with fixed point O. If $\mathcal{L}$ is a line on $\mathcal{P}$ through O, then $\delta \circ \mathcal{R}_{\mathcal{L}} = \mathcal{R}_{\mathcal{L}} \circ \delta$.*

Proof. $\mathcal{L}$ is a fixed line of δ and is pointwise fixed for $\mathcal{R}_{\mathcal{L}}$. Thus if $X \in \mathcal{L}$ $\mathcal{R}_{\mathcal{L}}(\delta(X)) = \delta(X) = \delta(\mathcal{R}_{\mathcal{L}}(X))$. Let X be any member of $\mathcal{P} \setminus \mathcal{L}$. By Theorem NEUT.15(3) and Definition NEUT.1(A) $\mathcal{R}_{\mathcal{L}}(\overrightarrow{OX}) = \overrightarrow{O\mathcal{R}_{\mathcal{L}}(X)}$. By Theorem CAP.18 both $\overleftrightarrow{OX}$ and $\overleftrightarrow{O\mathcal{R}_{\mathcal{L}}(X)}$ are fixed lines of δ.

By Theorem ROT.15 there exists a unique rotation π about O such that $\pi(\overrightarrow{OX}) = \overrightarrow{O\mathcal{R}_{\mathcal{L}}(X)}$. For any point $Y \in \overrightarrow{OX}$, $\pi(Y) \in \overrightarrow{O\mathcal{R}_{\mathcal{L}}(X)}$, and since π is an isometry, $\pi(\overline{OY}) = \overline{O\pi(Y)}$, so that $\overline{OY} \cong \overline{O\pi(Y)}$. Also, $\overline{OY} \cong \overline{O\mathcal{R}_{\mathcal{L}}(Y)}$, so that $\overline{O\pi(Y)} \cong \overline{O\mathcal{R}_{\mathcal{L}}(Y)}$, and by Property R.4 of Axiom NEUT.2,

$$\pi(Y) = \mathcal{R}_{\mathcal{L}}(Y) \quad (*).$$

By Theorem DLN.7(D) $\delta \circ \pi = \pi \circ \delta$. Then $\delta(\mathcal{R}_{\mathcal{L}}(X)) = \delta(\pi(X)) = \pi(\delta(X)) = \mathcal{R}_{\mathcal{L}}(\delta(X))$. Here we have used (*) where $Y = X$ for the first equality, and have used (*) where $Y = \delta(X)$ in the last equality. This completes the proof that $\delta \circ \mathcal{R}_{\mathcal{L}} = \mathcal{R}_{\mathcal{L}} \circ \delta$. □

Theorem DLN.13. *Let $\mathcal{P}$ be a Euclidean plane, O, A, B, and C points on $\mathcal{P}$, and let δ be a dilation of $\mathcal{P}$ with fixed point O.*

(A) $\delta(\overrightarrow{AB}) = \overrightarrow{\delta(A)\delta(B)}$ *and* $\delta(\overline{AB}) = \overline{\delta(A)\delta(B)}$ *(corresponding statements are true for open rays and open and half-open intervals)*

(B) *If A, B, and C are noncollinear,* $\delta(\angle BAC) = \angle\delta(B)\delta(A)\delta(C)$.

(C) *If A, B, and C are noncollinear,* $\delta(\text{ins}\,\angle BAC) = \text{ins}\,\angle\delta(B)\delta(A)\delta(C)$.

Proof. By Theorem DLN.8, δ is a belineation; the results follow immediately from Theorem COBE.5, parts (2) through (10), and part (11). □

Theorem DLN.14. *Let $\mathcal{P}$ be a Euclidean plane, O, A, B, and C be points on $\mathcal{P}$ such that A, B, and C are noncollinear, δ be a dilation of $\mathcal{P}$ with fixed point O. Then* $\delta(\angle BAC) \cong \angle BAC$.

Proof. (Case 1: $A = O$, $\delta(B) \in \overrightarrow{OB}$.) By Exercise DLN.5(III)(A) $\delta(C) \in \overrightarrow{OC}$. By Theorem PSH.16 $\overrightarrow{O\delta(B)} = \overrightarrow{OB}$ and $\overrightarrow{O\delta(C)} = \overrightarrow{OC}$. Hence by Definition PSH.29 $\angle BOC = \angle\delta(B)O\delta(C)$. By Theorem DLN.13(B), $\delta(\angle BOC) = \angle\delta(B)\delta(O)\delta(C) = \angle\delta(B)O\delta(C)$. Since angles that are equal are congruent, $\angle BOC \cong \delta(\angle BOC)$.

(Case 2: $A = O$, B' is a point such that B'–O–B and $\delta(B) \in \overrightarrow{OB'}$.) Let C' be a point such that C'–O–C. By Exercise DLN.5(III)(B) $\delta(C) \in \overrightarrow{OC'}$. By Theorem DLN.13(B), $\delta(\angle BOC) = \angle\delta(B)O\delta(C)$. By the reasoning in Case 1, $\angle\delta(B)O\delta(C) = \angle B'OC'$. By Theorem NEUT.42 (vertical angles) $\angle B'OC' \cong \angle BOC$. By Theorem NEUT.14 (congruence is an equivalence relation) $\angle BOC \cong \delta(\angle BOC)$.

(Case 3: $A \neq O$; O, A, and B are collinear; and O, A, and C are noncollinear.) By Theorem CAP.18 $\overleftrightarrow{OA}$ is a fixed line of δ. By Theorem DLN.13(A) $\delta(\overrightarrow{AB}) = \overrightarrow{\delta(A)\delta(B)} \subseteq \overleftrightarrow{OA}$. By Theorem CAP.1(A) $\delta(\overleftrightarrow{AC}) = \overleftrightarrow{\delta(A)\delta(C)}$. By Definition CAP.17 and Theorem CAP.18 $\delta(\overleftrightarrow{AC}) \parallel \overleftrightarrow{AC}$, so that $\overleftrightarrow{\delta(A)\delta(C)} \parallel \overleftrightarrow{AC}$. By Property B.2 of Definition IB.1 one and only one of the following statements is true: O–A–B, O–B–A, or A–O–B.

(I) If O–A–B, then by Theorem DLN.8 and the fact that O is a fixed point of δ, O–$\delta(A)$–$\delta(B)$.

(A) If $\delta(A) \in \overrightarrow{OA}$, then by Exercise DLN.5(III)(A), $\delta(C)$ belongs to the C-side of $\overleftrightarrow{AB}$. By Theorem CAP.18 $\overleftrightarrow{AC}$ is not a fixed line of δ. By Theorem EUC.11 $\angle BAC \cong \angle\delta(B)\delta(A)\delta(C)$. By Theorem DLN.13(B) $\angle\delta(B)\delta(A)\delta(C) = \delta(\angle BAC)$, so that by Theorem NEUT.14 $\angle BAC \cong \delta(\angle BAC)$.

(B) If A' is a point such that A'–O–A and if $\delta(A) \in \overrightarrow{OA'}$, then by Exercise DLN.5(III)(B) $\delta(C) \in$ the side of $\overleftrightarrow{AB}$ opposite the C-side. As in part (A), $\overleftrightarrow{\delta(A)\delta(C)} \parallel \overleftrightarrow{AC}$, $\angle\delta(B)\delta(A)\delta(C) \simeq \angle BAC$ and $\angle\delta(B)\delta(A)\delta(C) = \delta(\angle BAC)$.

(II) If O–B–A, then as in part (I) O–$\delta(B)$–$\delta(A)$, $\delta(C) \in C$-side of $\overleftrightarrow{AB}$ and $\angle BAC \cong \angle\delta(B)\delta(A)\delta(C) = \delta(\angle BAC)$.

(III) If A–O–B, then $\delta(A)$–O–$\delta(B)$.

(A) If $\delta(A) \in \overrightarrow{OA}$, then by Exercise DLN.5(III)(A) $\delta(C)$ belongs to the C-side of $\overleftrightarrow{AB}$ and as in parts (I) and (II), $\angle BAC \cong \delta(\angle BAC)$.

(B) If $\delta(A)$–O–A, then $\delta(C)$ belongs to the side of $\overleftrightarrow{OA} = \overleftrightarrow{AB}$ opposite the C-side. Again as in parts (I) and (II), $\angle BAC \cong \delta(\angle BAC)$.

(Case 4: O, A, and B are noncollinear and O, A, and C are collinear.) Interchange "B" and "C" in Case 3.

(Case 5: O, A, and B are noncollinear, O, A, and C are noncollinear, and B and C are on the same side of $\overleftrightarrow{OA}$.) Let A' be a point such that A'–O–A. We choose the notation so that $B \in \text{ins}\,\angle A'AC$. By cases 3 and 4 and Theorem DLN.13(B)

$$\angle A'AB \cong \delta(\angle A'AB) = \angle\delta(A')\delta(A)\delta(B), \text{ and}$$
$$\angle A'AC \cong \delta(\angle A'AC) = \angle\delta(A')\delta(A)\delta(C).$$

By Theorem DLN.13(C) $\delta(B)$ is a member of $\text{ins}\,\angle\delta(A')\delta(A)\delta(C)$. By Exercise NEUT.40(B)

$$\angle BAC \cong \angle\delta(B)\delta(A)\delta(C) = \delta(\angle BAC).$$

(Case 6: O, A, and B are noncollinear, O, A, and C are noncollinear, and B and C are on opposite sides of $\overleftrightarrow{OA}$.) By Theorem PSH.12 (Plane Separation) there exists a point Q such that $\overline{BC} \cap \overleftrightarrow{OA} = \{Q\}$. By Theorem PSH.37 $Q \in \text{ins}\,\angle BAC$. By cases 3 and 4 $\angle QAB \cong \delta(Q)\delta(A)\delta(B)$ and $\angle QAC \cong \delta(Q)\delta(A)\delta(C)$. By Exercise NEUT.40(A) and Theorem DLN.13(B),

$$\angle BAC \cong \angle\delta(B)\delta(A)\delta(C) = \delta(\angle BAC). \qquad \square$$

The next theorem is a generalization of Theorem DLN.12.

Theorem DLN.15. *Let $\mathcal{P}$ be a Euclidean plane, O a point on $\mathcal{P}$, $\mathcal{L}$ any line on $\mathcal{P}$, and let δ be a dilation of $\mathcal{P}$ with fixed point O. Then $\delta \circ \mathcal{R}_{\mathcal{L}} = \mathcal{R}_{\delta(\mathcal{L})} \circ \delta$.*

Proof. If $O \in \mathcal{L}$, then by Theorem CAP.18, $\mathcal{L}$ is a fixed line for δ so this is Theorem DLN.12.

If $O \notin \mathcal{L}$, let $\mathcal{M} = \text{pr}\,(O, \mathcal{L})$, $Q = \text{ftpr}\,(O, \mathcal{L})$, and let X be any point on $\mathcal{P}$.

(Case 1: $X = Q$.) $(\delta \circ \mathcal{R}_{\mathcal{L}})(Q) = \delta(\mathcal{R}_{\mathcal{L}}(Q)) = \delta(Q)$ and $\mathcal{R}_{\delta(\mathcal{L})}(\delta(Q)) = \delta(Q)$. Here we have used Definition NEUT.1(A).

(Case 2: $X \in (\mathcal{M}\setminus\{Q\})$.) By Theorem NEUT.54 $\mathcal{L}$ is the perpendicular bisecting line of $\overleftrightarrow{X\mathcal{R}_{\mathcal{L}}(X)}$ so that Q is the midpoint of $\overleftrightarrow{X\mathcal{R}_{\mathcal{L}}(X)}$. By Theorem DLN.4(B) and Theorem EUC.17 $\delta(Q)$ is the midpoint of $\delta(\overleftrightarrow{X\mathcal{R}_{\mathcal{L}}(X)}) = \overleftrightarrow{\delta(X)\delta(\mathcal{R}_{\mathcal{L}}(X))}$ (here we have used Theorem DLN.13(A)).

Now $\delta(\mathcal{L})$ is the perpendicular bisecting line of $\overleftrightarrow{\delta(X)\mathcal{R}_{\delta(\mathcal{L})}(\delta(X))}$ (by Theorem NEUT.54), and the intersection is the midpoint of the segment. By Theorem CAP.18 $\mathcal{L}$ is not a fixed line of δ, and by Definition CAP.17 $\delta(\mathcal{L}) \parallel \mathcal{L}$, so by Theorem EUC.3 $\delta(\mathcal{L}) \perp \mathcal{M}$. Then $\delta(X)$ is a member of both $\mathcal{M}$ and $\overleftrightarrow{\delta(X)\mathcal{R}_{\delta(\mathcal{L})}(\delta(X))}$ both of which are perpendicular to $\delta(\mathcal{L})$. By Theorem NEUT.47(B), $\overleftrightarrow{\delta(X)\mathcal{R}_{\delta(\mathcal{L})}(\delta(X))} \subseteq \mathcal{M}$. Hence $\delta(Q)$ is the point of intersection of $\delta(\mathcal{L})$ with both $\overleftrightarrow{\delta(X)\delta(\mathcal{R}_{\mathcal{L}}(X))}$ and $\overleftrightarrow{\delta(X)\mathcal{R}_{\delta(\mathcal{L})}(\delta(X))}$ and hence is the midpoint of both. By Exercise NEUT.72 $\delta(\mathcal{R}_{\mathcal{L}}(X)) = \mathcal{R}_{\delta(\mathcal{L})}(\delta(X))$.

(Case 3: $X \in (\mathcal{L}\setminus\{Q\})$.) By Definition NEUT.1(A) $(\delta \circ \mathcal{R}_{\mathcal{L}})(X) = \delta(\mathcal{R}_{\mathcal{L}}(X)) = \delta(X)$ and $(\mathcal{R}_{\delta(\mathcal{L})} \circ \delta)(X) = \mathcal{R}_{\delta(\mathcal{L})}(\delta(X)) = \delta(X)$. Therefore $(\delta \circ \mathcal{R}_{\mathcal{L}})(X) = (\mathcal{R}_{\delta(\mathcal{L})} \circ \delta)(X)$.

(Case 4: $X \in (\mathcal{P} \setminus (\mathcal{L} \cup \mathcal{M}))$.) Let Y be the midpoint of $\overleftrightarrow{X\mathcal{R}_{\mathcal{L}}(X)}$. By Theorem NEUT.54 $\mathcal{L}$ is the perpendicular bisecting line of $\overleftrightarrow{X\mathcal{R}_{\mathcal{L}}(X)}$ and Y is the point of intersection. By Theorem EUC.17 $\delta(Y)$ is the midpoint of $\delta(\overleftrightarrow{X\mathcal{R}_{\mathcal{L}}(X)}) = \overleftrightarrow{\delta(X)\delta(\mathcal{R}_{\mathcal{L}}(X))}$.

As in Case 2, $\delta(\mathcal{L})$ is the perpendicular bisecting line of $\overleftrightarrow{\delta(X)\mathcal{R}_{\delta(\mathcal{L})}(\delta(X))}$ (by Theorem NEUT.54), and the intersection is the midpoint of the segment. By an argument similar to that in Case 2, both $\overleftrightarrow{X\mathcal{R}_{\mathcal{L}}(X)}$ and $\overleftrightarrow{\delta(X)\mathcal{R}_{\delta(\mathcal{L})}(\delta(X))}$ are perpendicular to $\delta(\mathcal{L})$ and since both contain $\delta(X)$, their lines are the same, so that $\delta(Y)$ is the point of intersection of $\delta(\mathcal{L})$ with both $\delta(\overleftrightarrow{X\mathcal{R}_{\mathcal{L}}(X)})$ and $\overleftrightarrow{\delta(X)\mathcal{R}_{\delta(\mathcal{L})}(\delta(X))}$, and hence is the midpoint of both. By Exercise NEUT.72 $\delta(\mathcal{R}_{\mathcal{L}}(X)) = \mathcal{R}_{\delta(\mathcal{L})}(\delta(X))$. □

Theorem DLN.16. *Let $\mathcal{P}$ be a Euclidean plane, O be a point on $\mathcal{P}$, δ be a dilation of $\mathcal{P}$ with fixed point O, and θ be an isometry of $\mathcal{P}$. Then there exist isometries ω and ψ of $\mathcal{P}$ such that $\delta \circ \theta = \omega \circ \delta$ and $\theta \circ \delta = \delta \circ \psi$.*

Proof. First we show that there exists an isometry ω such that $\delta \circ \theta = \omega \circ \delta$. By Theorem ROT.26 there are four cases.

(Case 1: $\theta = \iota$, the identity mapping.) If $\theta = \iota$, take $\omega = \iota$. Then $\delta \circ \theta = \delta = \omega \circ \delta$.

(Case 2: There exists a line $\mathcal{H}$ on $\mathcal{P}$ such that $\theta = \mathcal{R}_{\mathcal{H}}$.) Using Theorem DLN.15 let $\omega = \mathcal{R}_{\delta(\mathcal{H})}$. Then $\delta \circ \mathcal{R}_{\mathcal{H}} = \mathcal{R}_{\delta(\mathcal{H})} \circ \delta$.

(Case 3: There exist distinct lines $\mathcal{J}$ and $\mathcal{K}$ on $\mathcal{P}$ such that $\theta = \mathcal{R}_{\mathcal{K}} \circ \mathcal{R}_{\mathcal{J}}$.) Using Theorem DLN.15

$$\begin{aligned}\delta \circ (\mathcal{R}_{\mathcal{K}} \circ \mathcal{R}_{\mathcal{J}}) &= (\delta \circ \mathcal{R}_{\mathcal{K}}) \circ \mathcal{R}_{\mathcal{J}} = (\mathcal{R}_{\delta(\mathcal{K})} \circ \delta) \circ \mathcal{R}_{\mathcal{J}} = \mathcal{R}_{\delta(\mathcal{K})} \circ (\delta \circ \mathcal{R}_{\mathcal{J}}) \\ &= \mathcal{R}_{\delta(\mathcal{K})} \circ (\mathcal{R}_{\delta(\mathcal{J})} \circ \delta) = (\mathcal{R}_{\delta(\mathcal{K})} \circ \mathcal{R}_{\delta(\mathcal{J})}) \circ \delta.\end{aligned}$$

Hence we may take $\omega = \mathcal{R}_{\delta(\mathcal{K})} \circ \mathcal{R}_{\delta(\mathcal{J})}$.

(Case 4: There exist three distinct lines $\mathcal{L}$, $\mathcal{M}$, and $\mathcal{N}$ such that $\theta = \mathcal{R}_{\mathcal{N}} \circ \mathcal{R}_{\mathcal{M}} \circ \mathcal{R}_{\mathcal{L}}$.) The proof of this case is Exercise DLN.8.

The proof that there exists an isometry ψ such that $\theta \circ \delta = \delta \circ \psi$ is Exercise DLN.9. □

Theorem DLN.17. *Let $\mathcal{P}$ be a Euclidean plane, $\mathcal{D}$ and $\mathcal{E}$ be nonempty subsets of $\mathcal{P}$ such that $\mathcal{D} \cong \mathcal{E}$, and δ be a dilation of $\mathcal{P}$ with fixed point O. Then $\delta(\mathcal{D}) \cong \delta(\mathcal{E})$.*

Proof. By Definition NEUT.3(B) there exists an isometry θ of $\mathcal{P}$ such that $\mathcal{E} = \theta(\mathcal{D})$. Hence $\delta(\mathcal{E}) = (\delta \circ \theta)(\mathcal{D})$. By Theorem DLN.16 there exists an isometry ω of $\mathcal{P}$ such that $\delta \circ \theta = \omega \circ \delta$ so that $\delta(\mathcal{E}) = \omega(\delta(\mathcal{D}))$. By Definition NEUT.3(B) $\delta(\mathcal{E}) \cong \delta(\mathcal{D})$. □

Corollary DLN.18. *Let $\mathcal{P}$ be a Euclidean plane, δ be a dilation of $\mathcal{P}$ with fixed point O, and let O, A, and B be distinct points on $\mathcal{P}$ such that $\overline{OA} \cong \overline{OB}$. Then*

(A) *$\delta(\overline{OA}) \cong \delta(\overline{OB})$ and $\overline{O\delta(A)} \cong \overline{O\delta(B)}$.*

(B) *$\overline{A\delta(A)} \cong \overline{B\delta(B)}$.*

Proof. (A) By Theorem DLN.17 since $\overline{OA} \cong \overline{OB}$, $\delta(\overline{OA}) \cong \delta(\overline{OB})$. By Theorem COBE.5(5) $\delta(\overline{OA}) = \overline{\delta(O)\delta(A)} = \overline{O\delta(A)}$ and $\delta(\overline{OB}) = \overline{\delta(O)\delta(B)} = \overline{O\delta(B)}$. Hence by Theorem NEUT.14 $\overline{O\delta(A)} \cong \overline{O\delta(B)}$.

(B) If O–A–$\delta(A)$, then by Exercise DLN.5(II)(A), O–B–$\delta(B)$. By Exercise NEUT.38(B) $\overline{A\delta(A)} \cong \overline{A\delta(A)}$.

If O–$\delta(A)$–A, then by Exercise DLN.5(II)(B), O–$\delta(B)$–B. By Exercise NEUT.38(B) $\overline{A\delta(A)} \cong \overline{A\delta(A)}$.

If A–O–$\delta(A)$, then by Exercise DLN.5(II)(C), B–O–$\delta(B)$. By Exercise NEUT.38(A) $\overline{A\delta(A)} \cong \overline{A\delta(A)}$. □

Theorem DLN.19. *Let $\mathcal{P}$ be a Euclidean plane, O be a point on $\mathcal{P}$, ρ be a rotation of $\mathcal{P}$ about O that is not the identity ι, and let δ be a dilation of $\mathcal{P}$ with fixed point O. Then $\delta \circ \rho$ is a collineation of $\mathcal{P}$ whose sole fixed point is O.*

Proof. Let X be any member of $\mathcal{P} \setminus O$. By Exercise ROT.2 O, X, and $\rho(X)$ are noncollinear. By Theorem CAP.18 $\overleftrightarrow{O\rho(X)}$ is a fixed line of δ and thus $\delta(\rho(X)) \in \overleftrightarrow{O\rho(X)}$. Since $\delta \circ \rho$ is a one-to-one mapping, $\delta(\rho(X)) \neq O$ and so O, X, and $\delta(\rho(X))$ are noncollinear. Thus $\delta(\rho(X)) \neq X$. Since O is a fixed point of $\delta \circ \rho$, it is the sole fixed point of $\delta \circ \rho$. □

Theorem DLN.20. *Let $\mathcal{P}$ be a Euclidean plane, O a point on $\mathcal{P}$, $\mathcal{R}^*$ the set of rotations of $\mathcal{P}$ about O and let $\mathcal{D}^*$ be the set of dilations of $\mathcal{P}$ with fixed point O. Then*

(A) $\mathbb{G} = \{\rho \circ \delta \mid \rho \in (\mathcal{R}^* \cup \{\iota\}) \text{ and } \delta \in (\mathcal{D}^* \cup \{\iota\})\}$ *is an abelian group under composition of mappings;*
(B) $\mathcal{R}^* \cup \mathcal{D}^* \cup \{\iota\} \subseteq \mathbb{G}$*; and*
(C) *if $\mathbb{H}$ is any group such that $\mathcal{R}^* \cup \mathcal{D}^* \cup \{\iota\} \subseteq \mathbb{H}$, then $\mathbb{G} \subseteq \mathbb{H}$. That is, $\mathbb{G}$ is the minimal group that contains $\mathcal{R}^* \cup \mathcal{D}^* \cup \{\iota\}$.*

Proof. (A) By Theorem ROT.23 $\mathcal{R}^* \cup \{\iota\}$ is an abelian group under composition of mappings so if $\rho \in \mathcal{R}^*$ then $\rho^{-1} \in \mathcal{R}^*$. By Theorem ROT.2 O is the sole fixed point of every member of $\mathcal{R}^*$.

By Theorem CAP.21 $\mathcal{D}^* \cup \{\iota\}$ is a group under composition and by Exercise DLN.3 it is abelian (cf Exercise DLN.4). Therefore if $\delta \in \mathcal{D}^*$, $\delta^{-1} \in \mathcal{D}^*$. By Theorem CAP.18 O is the sole fixed point of every member of $\mathcal{D}^*$.

If $\rho \in \mathcal{R}^* \cup \{\iota\}$ and $\delta \in \mathcal{D}^* \cup \{\iota\}$, then by Theorem DLN.7(E), $\rho \circ \delta = \delta \circ \rho$ so that $\mathbb{G}$ is the set of all $\delta \circ \rho$ as well as the set of all $\rho \circ \delta$, where $\rho \in (\mathcal{R}^* \cup \{\iota\})$ and $\delta \in (\mathcal{D}^* \cup \{\iota\})$.

If ρ_1 and ρ_2 are any members of $\mathcal{R}^* \cup \{\iota\}$ and if δ_1 and δ_2 are any members of $\mathcal{D}^* \cup \{\iota\}$, then $(\rho_1 \circ \delta_1) \circ (\rho_2 \circ \delta_2) = (\rho_1 \circ \rho_2) \circ (\delta_1 \circ \delta_2) \in \mathbb{G}$ since $(\rho_1 \circ \rho_2) \in \mathcal{R}^* \cup \{\iota\})$ and $(\delta_1 \circ \delta_2) \in \mathcal{D}^* \cup \{\iota\}$. Also by Theorem DLN.7(E),

$$\begin{aligned}(\rho_1 \circ \delta_1) \circ (\rho_2 \circ \delta_2) &= \rho_1 \circ (\delta_1 \circ \rho_2) \circ \delta_2 = \rho_1 \circ (\rho_2 \circ \delta_1) \circ \delta_2 \\ &= (\rho_1 \circ \rho_2) \circ (\delta_1 \circ \delta_2) = (\rho_2 \circ \rho_1) \circ (\delta_2 \circ \delta_1) \\ &= \rho_2 \circ (\rho_1 \circ \delta_2) \circ \delta_1 = \rho_2 \circ (\delta_2 \circ \rho_1) \circ \delta_1 \\ &= (\rho_2 \circ \delta_2) \circ (\rho_1 \circ \delta_1)\end{aligned}$$

so that any two elements of $\mathbb{G}$ commute.

If $\rho \in (\mathcal{R}^* \cup \{\iota\})$ and $\delta \in (\mathcal{D}^* \cup \{\iota\})$, then $(\rho^{-1} \circ \delta^{-1}) \in \mathbb{G}$ because $\rho^{-1} \in (\mathcal{R}^* \cup \{\iota\})$ and $\delta^{-1} \in (\mathcal{D}^* \cup \{\iota\})$. Then by Theorem DLN.7(E)

$$(\rho^{-1} \circ \delta^{-1}) \circ (\rho \circ \delta) = (\delta^{-1} \circ \rho^{-1}) \circ (\rho \circ \delta) = \delta^{-1} \circ (\rho^{-1} \circ \rho) \circ \delta$$
$$= \delta^{-1} \circ \iota \circ \delta = \delta^{-1} \circ \delta = \iota$$

so that $\rho^{-1} \circ \delta^{-1} = (\rho \circ \delta)^{-1}$; hence any member of $\mathbb{G}$ has an inverse in $\mathbb{G}$. It follows that $\mathbb{G}$ is an abelian group under composition of mappings.

(B) If $\rho \in \mathcal{R}^* \cup \{\iota\}$ and $\delta \in \mathcal{D}^* \cup \{\iota\}$, then $\rho = \rho \circ \iota \in \mathbb{G}$, $\delta = \delta \circ \iota \in \mathbb{G}$, and $\iota = \iota \circ \iota \in \mathbb{G}$, proving (B).

(C) If $\mathbb{H}$ is any group containing $\mathcal{R}^* \cup \mathcal{D}^* \cup \{\iota\}$, then $\mathbb{H}$ must contain all the compositions of elements of $\mathcal{R}^* \cup \{\iota\}$ and $\mathcal{D}^* \cup \{\iota\}$, and all possible compositions and inverses of those elements, that is, all elements comprising $\mathbb{G}$, so that $\mathbb{G} \subseteq \mathbb{H}$. □

Definition DLN.21. The group $\mathbb{G}$ defined in the statement of Theorem DLN.20 is called the **group generated by the set** $(\mathcal{R}^* \cup \mathcal{D}^* \cup \{\iota\})$.

Theorem DLN.22. *Let $\mathcal{P}$ be a Euclidean plane, O be a point on $\mathcal{P}$, and let $\mathcal{R}^*$, $\mathcal{D}^*$, and $\mathbb{G}$ be the sets defined in Theorem DLN.20. For any distinct members A and B of $\mathcal{P} \setminus \{O\}$, there is a unique member α of $\mathbb{G}$ such that $\alpha(A) = B$.*

Proof. (I: Existence.)

(Case 1: O, A, and B are noncollinear and $\overline{OA} \cong \overline{OB}$.) Then by Theorem ROT.15(A) there exists a rotation α of $\mathcal{P}$ about O such that $\alpha(\overrightarrow{OA}) = \overrightarrow{OB}$. Since a rotation is an isometry $\overline{OA} \cong \overline{O\alpha(A)}$, and since congruence is an equivalence relation $\overline{OB} \cong \overline{O\alpha(A)}$. By Property R.4 of Definition NEUT.2 $\alpha(A) = B$, and since $\mathcal{R}^* \subseteq \mathbb{G}$, $\alpha \in \mathbb{G}$.

(Case 2: O, A, and B are collinear.) Then by Theorem DLN.7 there exists a dilation α of $\mathcal{P}$ with fixed point O such that $\alpha(A) = B$. Since $\mathcal{D}^* \subseteq \mathbb{G}$, $\alpha \in \mathbb{G}$.

Note here that in the case where A–O–B and $\overline{OA} \cong \overline{OB}$, the point reflection $\mathcal{R}_O$ maps A to B; by Theorem ISM.3, $\mathcal{R}_O$ is a dilation, hence is the unique dilation guaranteed by Theorem DLN.7 which maps A to B.

(Case 3: O, A, and B are noncollinear and $\overline{OA} \not\cong \overline{OB}$.) Then by Theorem ROT.15(A) there exists a rotation ρ of $\mathcal{P}$ about O such that $\rho(\overrightarrow{OA}) = \overrightarrow{OB}$. Since $\overline{OA} \not\cong \overline{OB}$, $\rho(A) \neq B$. By Theorem DLN.7 there exists a dilation δ of $\mathcal{P}$ such that $\delta(\rho(A)) = B$. If we let $\alpha = \delta \circ \rho$, then $\alpha(A) = B$, and since $\alpha = \delta \circ \rho = \rho \circ \delta$, $\alpha \in \mathbb{G}$.

(II: Uniqueness.) Let α and β be members of $\mathbb{G}$ (cf Theorem DLN.20), such that $\alpha(A) = B$ and $\beta(A) = B$. Then $\alpha(A) = \beta(A)$ and $(\alpha \circ \beta^{-1})(A) = A$. By

Theorem DLN.20 $\alpha \circ \beta^{-1}$ is a member of $\mathbb{G}$, so $\alpha \circ \beta^{-1} = \delta \circ \rho$ where δ is either a dilation or ι, and ρ is either a rotation or ι.

Now ρ is an isometry so by Definition NEUT.3(B) $\overline{O\rho(A)} \cong \overline{OA}$.

Also, $\delta(\rho(A)) = A$. Since every line through O is a fixed line for δ, O, $\rho(A)$, and A must be collinear. Then either $\overrightarrow{O\rho(A)} = \overrightarrow{OA}$ or $\overrightarrow{O\rho(A)} = \overrightarrow{OA'}$ where A–O–A'.

In the first case, by Property R.4 of Definition NEUT.2, $\rho(A) = A$ and hence $\delta(A) = A$, and A is a fixed point for δ, a contradiction to Theorem CAP.18. In the second case, $\rho(A)$ is the point such that (i) $\overline{O\rho(A)} \cong \overline{OA}$ and (ii) A–O–$\rho(A)$.

By Theorem ROT.3 the point reflection $\mathcal{R}_O$ has properties (i) and (ii) just above, and since by Theorem ROT.15(A) there is a unique rotation mapping A to $\overrightarrow{O\rho(A)}$ (that is, property (ii)), $\rho = \mathcal{R}_O$.

By Corollary ROT.6 $\rho \circ \rho = \mathcal{R}_O \circ \mathcal{R}_O = \iota$, so that $\rho(\rho(A)) = A$. By Theorem ISM.3, $\rho = \mathcal{R}_O$ is a dilation, and by Theorem DLN.7 there is a unique dilation δ such that $\delta(\mathcal{R}_O(A)) = \delta(\rho(A)) = A$. Therefore $\delta = \mathcal{R}_O$, and $\alpha \circ \beta^{-1} = \rho \circ \rho = \mathcal{R}_O \circ \mathcal{R}_O = \iota$, so that $\alpha = \beta$. □

13.3 Exercises for dilations

Answers to starred (*) exercises may be accessed from the home page for this book at www.springer.com.

Exercise DLN.1*. Let O be a point on a Euclidean plane $\mathcal{P}$, and let α be a half-rotation of $\mathcal{P}$ about O. If X and Y are members of $\mathcal{P} \setminus \{O\}$ such that O, X, and Y are noncollinear, then $\angle XO\alpha(X) \cong \angle YO\alpha(Y)$.

Exercise DLN.2*. Let O be a point on a Euclidean plane $\mathcal{P}$, and let α and β be half-rotations of $\mathcal{P}$ about O; let R, S, and T be members of $\mathcal{P} \setminus \{O\}$ such that $\alpha(R) = S$, $\beta(S) = T$, and $S \in \text{ins}\,\angle ROT$. Then for every member U of $\mathcal{P} \setminus \{O\}$ $\angle UO\alpha(U) \cong \angle ROS$, $\angle\alpha(U)O(\beta \circ \alpha)(U) \cong \angle SOT$ $\angle UO(\beta \circ \alpha)(U) \cong \angle ROT$, and $\alpha(U) \in \text{ins}\,\angle UO(\beta \circ \alpha)(U)$.

Exercise DLN.3*. Let O be a point on a Euclidean plane $\mathcal{P}$, and let δ_1 and δ_2 be dilations of $\mathcal{P}$ with fixed point O. Then $\delta_1 \circ \delta_2 = \delta_2 \circ \delta_1$, *i.e.* composition of dilations with a common fixed point is commutative.

Exercise DLN.4*. Let O be a point on a Euclidean plane $\mathcal{P}$, and let $\mathbb{D} = \{\alpha \mid \alpha$ be a dilation of $\mathcal{P}$ with fixed point O, or $\alpha = \iota\}$. Then under composition of mappings $\mathbb{D}$ is an abelian group.

Exercise DLN.5*. Let O be a point on a Euclidean plane $\mathcal{P}$, and let δ be a dilation of $\mathcal{P}$ with fixed point O.

(I) If X and Y are members of $\mathcal{P} \setminus \{O\}$ such that O, X, and Y are noncollinear, then $\delta(X)$ and $\delta(Y)$ are on the same side of $\overleftrightarrow{XY}$.

(II) Let A be any member of $\mathcal{P} \setminus \{O\}$ and let X be any member of $\mathcal{P} \setminus \{O, A\}$.

(A) If O–A–$\delta(A)$, then O–X–$\delta(X)$.
(B) If O–$\delta(A)$–A, then O–$\delta(X)$–X.
(C) If $\delta(A)$–O–A, then $\delta(X)$–O–X.

(III) Let A be any member of $\mathcal{P} \setminus \{O\}$ and let X be any member of $\mathcal{P} \setminus \{O, A\}$.

(A) If $\delta(A) \in \overrightarrow{OA}$, then $\delta(X) \in \overrightarrow{OX}$.
(B) If A' is a point such that A'–O–A, X' is a point such that X'–O–X, and if $\delta(A) \in \overrightarrow{OA'}$, then $\delta(X) \in \overrightarrow{OX'}$.

(IV) Let A be any member of $\mathcal{P} \setminus \{O\}$ and let C be any member of $\mathcal{P} \setminus \overleftrightarrow{OA}$.

(A) If $\delta(A) \in \overrightarrow{OA}$, then $\delta(C)$ is on the C-side of $\overleftrightarrow{OA}$.
(B) If $\delta(A)$–O–A, then $\delta(C)$ is on the side of $\overleftrightarrow{OA}$ opposite the C-side.

Exercise DLN.6*. Let O be a point on a Euclidean plane $\mathcal{P}$; let δ be a dilation of $\mathcal{P}$ with fixed point O and let ρ be a rotation of $\mathcal{P}$ about O. Then $\rho^{-1} \circ \delta \circ \rho = \delta$ and $\delta^{-1} \circ \rho \circ \delta = \rho$.

Exercise DLN.7*. Let O be a point on a Euclidean plane $\mathcal{P}$; let δ be a dilation on $\mathcal{P}$ with fixed point O, and let $\mathcal{L}$ be any line on $\mathcal{P}$. Then

$$\mathcal{R}_{\mathcal{L}} \circ \delta = \delta \circ \mathcal{R}_{\delta^{-1}(\mathcal{L})}.$$

Exercise DLN.8*. Let O be a point on a Euclidean plane $\mathcal{P}$, and let δ be a dilation of $\mathcal{P}$ with fixed point O. Let $\mathcal{L}$, $\mathcal{M}$, and $\mathcal{N}$ be distinct lines on $\mathcal{P}$. Then $\delta \circ (\mathcal{R}_{\mathcal{N}} \circ \mathcal{R}_{\mathcal{M}} \circ \mathcal{R}_{\mathcal{L}}) = (\mathcal{R}_{\delta(\mathcal{N})} \circ \mathcal{R}_{\delta(\mathcal{M})} \circ \mathcal{R}_{\delta(\mathcal{L})}) \circ \delta$.

Exercise DLN.9*. Let O be a point on a Euclidean plane $\mathcal{P}$; let δ be a dilation of $\mathcal{P}$ with fixed point O, and let θ be an isometry of $\mathcal{P}$. Then there exists an isometry ψ of $\mathcal{P}$ such that $\theta \circ \delta = \delta \circ \psi$.

Exercise DLN.10. Using the construction of Theorem DLN.4, prove that for any half-rotation α, if A–B–C, then $\alpha(A)$–$\alpha(B)$–$\alpha(C)$.

Chapter 14
Every Line in a Euclidean Plane Is an Ordered Field (OF)

Acronym: *OF*

Dependencies: *all prior Chapters 1 through 13*

New Axioms: *none*

New Terms Defined: *origin, zero, unit; sum, product (of points on a line); inverse (additive, multiplicative), subtraction, division; positive, negative points (on the line); the positive half (of the line), absolute value; distance, length*

Abstract: This chapter is concerned with an arbitrary line in a Euclidean plane. It uses translations to define an operation of addition, and dilations to define multiplication on such a line; when equipped with these operations, the line becomes a field (defined in Chapter 1 Section 1.5). An ordering of the line is defined, so that the line becomes an ordered field. These concepts are used to define distance between points, and the length of a segment.

When we try to think about how to make an arbitrary line into an ordered field, we naturally think of the real numbers—the archetypical ordered field. When adding the numbers 2 and -3, we might first "do" the translation that takes 0 to 2, then follow that with another translation that takes 0 to -3, so that the composite translation takes 0 to -1. Thinking about multiplication in this way is harder, since it involves stretching the real numbers outward from the origin, rather than translating them. Such intuitions suggest a way for defining addition and multiplication on a line in the Euclidean Plane. To accomplish addition, we invoke translation; to accomplish multiplication, we invoke dilation.

E.J. Specht et al., *Euclidean Geometry and its Subgeometries*,
DOI 10.1007/978-3-319-23775-6_14

Some formal difficulties arise from the fact that according to the definitions for translations and dilations (CAP.6 and CAP.17, respectively), the identity ι is neither a translation nor a dilation, and the "zero" mapping, which takes the whole line to the origin, is not a dilation because dilations are collineations and this mapping takes everything to a single point. But we need both these mappings in order to make the arithmetic work correctly.

One way around the problem would have been to anoint both the identity and the "zero" mapping as "honorary" translations or dilations, as the case might be, but that would have brought another set of problems. In the following definition we will solve our current problem by simply making special definitions for the identity and the "zero" mapping.

In this chapter we follow *Geometry: An Introduction*, Chapter 3, by Günter Ewald, Ishi Press, Wadsworth, 2013 [7].

14.1 Building a line into an ordered field

Definition OF.1. Let $\mathcal{P}$ be a Euclidean Plane, $\mathbb{L}$ a line on $\mathcal{P}$, and let O be a point on $\mathbb{L}$.

(A) For each $A \in \mathbb{L} \setminus \{O\}$ define τ_A to be the translation of $\mathcal{P}$ such that $\tau_A(O) = A$. Theorem ISM.5 (Chapter 12) says that such a translation exists and is unique. Define τ_O to be the identity mapping ι.

(B) Let U be a member of $\mathbb{L} \setminus \{O\}$; for each $A \in \mathbb{L} \setminus \{O, U\}$, define δ_A to be the dilation with fixed point O such that $\delta_A(U) = A$. Theorem DLN.7 says that such a dilation exists and is unique. Define $\delta_U = \iota$, and define δ_O to be the mapping such that for every $X \in \mathbb{L}$, $\delta_O(X) = O$.

(C) If A and B are members of $\mathbb{L}$, define

$$A \oplus B = (\tau_B \circ \tau_A)(O) = \tau_B(\tau_A(O)) = \tau_B(A);$$

The operation $\oplus$ is called **addition** and $A \oplus B$ is the **sum** of A and B.

(D) If A and B are any members of $\mathbb{L}$, define

$A \odot B = (\delta_B \circ \delta_A)(U) = \delta_B(\delta_A(U)) = \delta_B(A)$. The operation $\odot$ is called **multiplication** and $A \odot B$ is the **product** of A and B. The point U is called the **unit**, and O is called the **zero** or **origin** of $\mathbb{L}$.

In Theorem OF.2(A) and (B) it will be seen that we could just as well have defined $A \oplus B = (\tau_A \circ \tau_B)(O) = \tau_A(B)$, since the composition of translations is commutative, and $A \odot B = (\delta_A \circ \delta_B)(U) = \delta_A(B)$, since the composition of dilations is commutative.

Throughout this chapter, $\mathbb{L}$ is a line on the Euclidean plane $\mathcal{P}$; $O, U, \tau_A, \delta_A, \oplus$ and $\odot$ are as defined in Definition OF.1. In this and in the rest of the main development of this book (Chapters 15, 17, 18, the first section of Chapter 19, and Chapter 20) we acknowledge the newly exalted status of a line by changing its notation to $\mathbb{L}$, rather than $\mathcal{L}$, as previously.

Theorem OF.2. (A) $\mathbb{L}$ *is an abelian group under the operation* $\oplus$. *(See the definition of group in Chapter 1, Section 1.5.)*

(B) $\mathbb{L} \setminus \{O\}$ *is an abelian group under the operation* $\odot$.

Proof. In Theorem CAP.12(A) we showed that the set of all translations, together with the identity ι, is an abelian group under composition. By Theorem CAP.21 and Exercise DLN.3, the set of all dilations with fixed point O, together with the identity ι, is also an abelian group under composition. Associativity of $\oplus$ and $\odot$ follows immediately from the associativity of composition of mappings (*cf* Chapter 1, Sections 1.4 and 1.5.).

(A) Let A and B be points in $\mathbb{L}$. Then by Definition OF.1, $A \oplus B = \tau_B(A) \in \mathbb{L}$. For any $A \in \mathbb{L}$; the translation τ_A has an inverse, which is also a translation τ_C, for some $C \in \mathbb{L}$; then $A \oplus C = (\tau_C \circ \tau_A)(O) = \iota(O) = O$. Finally, $A \oplus B = (\tau_B \circ \tau_A)(O) = (\tau_A \circ \tau_B)(O) = B \oplus A$. This shows that $\mathbb{L}$ is an abelian group under $\oplus$.

(B) Let A and B be points in $\mathbb{L} \setminus \{O\}$. Then by Definition OF.1,

$$A \odot B = (\delta_B \circ \delta_A)(U) = \delta_B(\delta_A(U)) = \delta_B(A) \in \mathbb{L}.$$

Now $\delta_B(\delta_A(O)) = O$; both δ_B and δ_A are one-to-one mappings, so their composition is one-to-one, so that $\delta_B(\delta_A(U)) \neq O$. Therefore $A \odot B \in \mathbb{L} \setminus \{O\}$.

For any $A \in \mathbb{L} \setminus \{O\}$ the dilation δ_A has an inverse δ_C, for some $C \in \mathbb{L} \setminus \{O\}$. Then $A \odot C = (\delta_C \circ \delta_A)(U) = \iota(U) = U$. Finally,

$$A \odot B = (\delta_B \circ \delta_A)(U) = (\delta_A \circ \delta_B)(U) = B \odot A.$$

This shows that $\mathbb{L} \setminus \{O\}$ is an abelian group under $\odot$. □

Theorem OF.3. *For every $A \in \mathbb{L}$,*

$$A \oplus O = O \oplus A = A,\ O \oplus O = O,$$
$$U \odot A = A \odot U = A, \text{ and}$$
$$A \odot O = O \odot A = O \odot O = O.$$

Proof. By Definitions OF.1(A) and (C), for every $A \in \mathbb{L}$,

$$A \oplus O = (\iota \circ \tau_A)(O) = \iota(\tau_A(O)) = \iota(A) = A;$$
$$O \oplus A = (\tau_A \circ \iota)(O) = \tau_A(\iota(O)) = \tau_A(O) = A; \text{ and}$$
$$O \oplus O = (\iota \circ \iota)(O) = O.$$

By Definitions OF.1(B) and (D), for every $A \in \mathbb{L}$,

$$U \odot A = (\delta_A \circ \iota)(U) = \delta_A(U) = A \text{ and } A \odot U = (\iota \circ \delta_A)(U) = \iota(A) = A;$$
$$A \odot O = (\delta_O \circ \delta_A)(U) = \delta_O(\delta_A(U)) = \delta_O(A) = O$$

since δ_O maps $\mathbb{L}$ to O; likewise

$$O \odot A = (\delta_A \circ \delta_O)(U) = \delta_A(\delta_O(U)) = \delta_A(O) = O \text{ and } O \odot O = O.$$ □

Definition OF.4. (A) For every member A of $\mathbb{L}$, define ${}^{\ominus}A$ to be the unique member of $\mathbb{L}$, which is guaranteed by Theorem OF.2(A) above, such that $A \oplus {}^{\ominus}A = {}^{\ominus}A \oplus A = O$. ${}^{\ominus}A$ is the **additive inverse** of A. Note that ${}^{\ominus}O = O$ and ${}^{\ominus}({}^{\ominus}O) = O$.

(B) For every member A of $\mathbb{L}\setminus\{O\}$, define A^{-1} to be the unique member of $\mathbb{L}\setminus\{O\}$, guaranteed by Theorem OF.2(B) above, such that $A \odot A^{-1} = A^{-1} \odot A = U$. A^{-1} is the **multiplicative inverse** of A. Again, note that $U^{-1} = U$ and $(U^{-1})^{-1} = U$.

Theorem OF.5. *Let φ be a collineation of $\mathcal{P}$ with fixed line $\mathbb{L}$ such that $\varphi(O) = O$. (For example, φ may be a dilation of $\mathcal{P}$ with fixed point O.) Then for all points S and T on $\mathbb{L}$ $\varphi(S+T) = \varphi(S) + \varphi(T)$.*

Proof. (Case 1: $S = 0$.) By Definition OF.1 each side of the given equality is $\varphi(T)$.
(Case 2: $S \neq 0$.) By Theorem CAP.13, $\varphi \circ \tau_S \circ \varphi^{-1}$ is a translation of $\mathcal{P}$. Since

$$(\varphi \circ \tau_S \circ \varphi^{-1})(O) = \varphi(\tau_S(\varphi^{-1}(O))) = \varphi(\tau_S(O)) = \varphi(S),$$

$\varphi \circ \tau_S \circ \varphi^{-1} = \tau_{\varphi(S)}$, and multiplying on the right by φ, we have $\varphi \circ \tau_S = \tau_{\varphi(S)} \circ \varphi$, so that $\varphi(\tau_S(T)) = \tau_{\varphi(S)}(\varphi(T))$. But $\tau_S(T) = S+T$ and $\tau_{\varphi(S)}(\varphi(T)) = \varphi(S)+\varphi(T)$. This, together with the last equality yields $\varphi(S+T) = \varphi(\tau_S(T)) = \varphi(S)+\varphi(T)$. □

Theorem OF.6 (Distributive property). *For all points A, B, and C of $\mathbb{L}$, $A \odot (B \oplus C) = (A \odot B) \oplus (A \odot C)$ and $(A \oplus B) \odot C = (A \odot C) \oplus (B \odot C)$.*

Proof. (Case 1: $A = O$.) Each side of the given equality is O.

(Case 2: $A \neq O$.) By Theorem OF.2(B), $A \odot (B \oplus C) = \delta_A(B \oplus C)$. By Theorem OF.5 $\delta_A(B \oplus C) = \delta_A(B) \oplus \delta_A(C)$. By Definition OF.1 $\delta_A(B) \oplus \delta_A(C) = (A \odot B) \oplus (A \odot C)$. Thus $A \odot (B \oplus C) = (A \odot B) + (A \odot C)$. That $(A \oplus B) \odot C) = (A \odot C) \oplus (B \odot C)$ follows directly from commutativity of the operations $\oplus$ and $\odot$, which was proved in Theorem OF.2. □

Theorem OF.7. *$\mathbb{L}$ is a field under the operations $\oplus$ and $\odot$.*

Proof. The proof is simply a synthesis of Theorems OF.2 and OF.6. For the definition of field, see Chapter 1 Section 1.5. □

Definition OF.8. (A) For all members A and B of $\mathbb{L}$, $A \ominus B = A \oplus {}^{\ominus}B$. The operation $\ominus$ is called **subtraction** and "$A \ominus B$" is read "A less B" or "A minus B."

(B) For every member A of $\mathbb{L}$ and for every member B of $\mathbb{L}\setminus\{O\}$, $A \odiv B = A \odot B^{-1}$. The operation $\odiv$ is called **division**, and "$A \odiv B$" is read "A divided by B."

(C) For every member A of $\mathbb{L}$, A is **positive** iff $A \in \overrightarrow{OU}$ and A is **negative** iff A–O–U.

(D) The points on $\mathbb{L}$ are ordered (*cf* Definition ORD.1) so that $O < U$.

Theorem OF.9. *Let A be any member of $\mathbb{L} \setminus \{O\}$. Then A is positive iff $A > O$ and A is negative iff $A < O$.*

Proof. By Definition OF.8 $U > O$ and A is positive iff $A \in \overrightarrow{OU}$, which, by Theorem ORD.7, equals $\{X \mid X > O\}$; thus A is positive iff $A > O$. Again by Definition OF.8, A is negative iff A–O–U; by Theorem ORD.6 A–O–U iff either $A < O < U$ or $A > O > U$; but $O < U$ so the latter case is ruled out. Therefore, A is negative iff $A < O$. □

Theorem OF.10. *Let A and B be any members of $\mathbb{L} \setminus \{O\}$.*

(A) *(1) ${}^{\ominus}A = \mathcal{R}_O(A)$; (2) A–O–$({}^{\ominus}A)$; (3) ${}^{\ominus}({}^{\ominus}A) = \mathcal{R}_O(\mathcal{R}_O(A)) = A$; and (4) $\overline{O({}^{\ominus}A)} \cong \overline{OA}$.*

(B) *If A is positive, then ${}^{\ominus}A$ is negative and if A is negative, then ${}^{\ominus}A$ is positive.*

(C) *If A and B are positive, then $A \oplus B$ is positive and $A \odot B$ is positive.*

(D) *(1) $({}^{\ominus}U) \odot A = {}^{\ominus}A$; (2) ${}^{\ominus}(A \odot B) = ({}^{\ominus}A) \odot B = A \odot ({}^{\ominus}B)$; and (3) $A \odot B = ({}^{\ominus}A) \odot ({}^{\ominus}B)$.*

(E) *(1) If one of A or B is positive and the other negative, then $A \odot B$ is negative; and (2) A and A^{-1} are both positive or both negative.*

(F) $(^{\ominus}A) \oplus (^{\ominus}B) = {}^{\ominus}(A \oplus B)$.

(G) $A^{-1} \odot B^{-1} = (A \odot B)^{-1}$.

(H) *Let A and B be any members of $\mathbb{L}$; then $A \odot B = O$ iff $A = O$ or $B = O$.*

(I) *If A and B are negative, then $A \oplus B$ is negative and $A \odot B$ is positive.*

Proof. (A) First note that $O = A \oplus (^{\ominus}A) = (\tau_A \circ \tau_{\ominus A})(O) = \tau_A(^{\ominus}A)$, and $\tau_A(O) = A$. By Theorem ISM.6, every translation is an isometry, so by Theorem COBE.5(5) (or Theorem NEUT.15(5)), $\overline{O(^{\ominus}A)} \cong \overline{OA}$.

If $^{\ominus}A \in \overrightarrow{OA}$, then by Property R.4 of Definition NEUT.2, $A = {}^{\ominus}A$, so that $O = A \oplus A = \tau_A(\tau_A(O))$ and O is a fixed point of $\tau_A \circ \tau_A$ which is impossible by Exercise ISM.1(B). Then $^{\ominus}A \notin \overrightarrow{OA}$ and hence A–O–$^{\ominus}A$ and O is the midpoint of $\overline{A(^{\ominus}A)}$. By Theorem ROT.3 A–O–$\mathcal{R}_O(A)$ and O is the midpoint of $\overline{A\mathcal{R}_O(A)}$. By Exercise NEUT.33 $\overline{A\mathcal{R}_O(A)} \cong \overline{A(^{\ominus}A)}$, and since both $^{\ominus}A$ and $\mathcal{R}_O(A)$ are members of $\overrightarrow{AO}$, by Property R.4 of Definition NEUT.2, $^{\ominus}A = \mathcal{R}_O(A)$. Then $^{\ominus}(^{\ominus}A) = \mathcal{R}_O(\mathcal{R}_O(A)) = A$ by Corollary ROT.6.

(B) By Theorem OF.9, if A is positive, then $A > O$ and by Theorem ORD.6, since $(^{\ominus}A)$–O–A, $^{\ominus}A < O < A$, and thus $^{\ominus}A$ is negative. Similarly, if A is negative, then $A < O$ and $^{\ominus}A > O$ so that $^{\ominus}A$ is positive.

(C) If A and B are positive, by Theorem ISM.23, $\tau_A(B) > B > O$; by Theorem OF.2, $A \oplus B = \tau_A(B)$, so that $A \oplus B > O$ is positive. To prove the second assertion, note that $A \odot B = \delta_A(B)$. By Theorem DLN.8 δ_A is a belineation so by Theorem COBE.5(2), $\delta_A(\overrightarrow{OU}) = \overrightarrow{\delta_A(O)\delta_A(U)} = \overrightarrow{OA}$. By Theorem PSH.16, $\overrightarrow{OA} = \overrightarrow{OU} = \overrightarrow{OB}$; since $B \in \overrightarrow{OU}$, $A \odot B = \delta_A(B) \in \overrightarrow{OU}$ and $A \odot B > O$.

(D) By Theorem OF.3 and Theorem OF.6,

$$\begin{aligned}((^{\ominus}U) \odot A) \oplus A &= ((^{\ominus}U) \odot A) \oplus (U \odot A)\\ &= (^{\ominus}U \oplus U) \odot A = O \odot A = O,\end{aligned}$$

so that $(^{\ominus}U) \odot A = {}^{\ominus}A$.

Since $((^{\ominus}A) \odot B) \oplus (A \odot B) = ((^{\ominus}A) \oplus A) \odot B = O \odot B = O$, it follows that $^{\ominus}(A \odot B) = (^{\ominus}A) \odot B$. Applying Theorem OF.2(B) (commutativity), we have

$$^{\ominus}(A \odot B) = {}^{\ominus}(B \odot A) = (^{\ominus}B) \odot A = A \odot (^{\ominus}B).$$

The next equality follows from Theorem OF.2(B), part (A) of this theorem, and two applications of what we just proved:

$$(^{\ominus}A) \odot (^{\ominus}B) = {}^{\ominus}(A \odot (^{\ominus}B)) = {}^{\ominus}(^{\ominus}(A \odot B)) = A \odot B.$$

(E) Choose the notation so that A is negative and B is positive. Then by part (D) $(^{\ominus}A) \odot B = {}^{\ominus}(A \odot B)$. By part (B) ${}^{\ominus}(A \odot B)$ is positive so again by part (B) its additive inverse $A \odot B$ is negative. If one of A or A^{-1} were positive and the other negative, $A \odot A^{-1} = U$ would be negative, which is impossible since $U \in \overrightarrow{OU}$ and hence is positive.

(F) $$((^{\ominus}A) \oplus (^{\ominus}B)) \oplus (A \oplus B) = (^{\ominus}B) \oplus ((^{\ominus}A) \oplus A) \oplus B$$
$$= (^{\ominus}B) \oplus O \oplus B = {}^{\ominus}B + B = O.$$

Hence $(^{\ominus}A) \oplus (^{\ominus}B)$ is the additive inverse of $A \oplus B$ so
$$(^{\ominus}A) \oplus (^{\ominus}B) = {}^{\ominus}(A \oplus B).$$

(G) $$(A^{-1} \odot B^{-1}) \odot (A \odot B) = (A^{-1} \odot B^{-1}) \odot (B \odot A)$$
$$= A^{-1} \odot (B^{-1} \odot B) \odot A$$
$$= A^{-1} \odot U \odot A = A^{-1} \odot A = U.$$

Hence $A^{-1} \odot B^{-1}$ is the multiplicative inverse of $A \odot B$ and
$$A^{-1} \odot B^{-1} = (A \odot B)^{-1}.$$

(H) If $B = O$, by Definition OF.1(D), $A \odot O = \delta_O(A) = O$. If $A = O$, then $O \odot B = B \odot O = O$. Given $A \odot B = O$, if $A \neq O$ then $O = A^{-1} \odot (A \odot B) = (A^{-1} \odot A) \odot B = U \odot B = B$. Similarly if $B \neq O$, then $A = O$.

(I) From part (F), $A \oplus B = {}^{\ominus}(^{\ominus}A) \oplus {}^{\ominus}(^{\ominus}B) = {}^{\ominus}((^{\ominus}A) \oplus (^{\ominus}B))$. Since ${}^{\ominus}A$ and ${}^{\ominus}B$ are positive, by part (C), $(^{\ominus}A) \oplus (^{\ominus}B)$ is positive. So by part (B) ${}^{\ominus}((^{\ominus}A) \oplus (^{\ominus}B)) = A \oplus B$ is negative.

Since A is negative, ${}^{\ominus}A$ is positive by part (B). So ${}^{\ominus}A \odot B$ is negative by part (E). Then by part (B), ${}^{\ominus}((^{\ominus}A) \odot B)$ is positive. By parts (D) and (A), ${}^{\ominus}((^{\ominus}A) \odot B) = (^{\ominus}U) \odot ((^{\ominus}A) \odot B) = ((^{\ominus}U) \odot (^{\ominus}A)) \odot B = (^{\ominus}(^{\ominus}A)) \odot B = A \odot B$. □

Theorem OF.11. *Let $\mathcal{P}$ be a Euclidean plane, $\mathbb{L}$ be a line on $\mathcal{P}$, O be a point on $\mathbb{L}$, and A, B, and C be members of $\mathbb{L}$. Then*

(A) *$A < B$ iff $B \ominus A > O$ iff $(^{\ominus}B) < (^{\ominus}A)$.*
(B) *$A < B$ iff $A \oplus C < B \oplus C$; also $O < B$ iff $C < B \oplus C$ and $A < O$ iff $A \oplus C < C$.*
(C) *If $A < B$ and $C > O$, then $A \odot C < B \odot C$.*
(D) *If $A < B$ and $C < O$, then $B \odot C < A \odot C$.*

Proof. In this proof we will write $A \oplus (^{\ominus}B)$ as $A \ominus B$, as permitted by Definition OF.8(A).

(A) If $A < B$ and $B \ominus A \not> O$, by trichotomy for $<$, either $B \ominus A = O$ or $B \ominus A < O$. If $B \ominus A = O$, then $B = A$ which contradicts $A < B$. If $B \ominus A < O$, then by Theorem OF.10(F) and (A),

$$^{\ominus}(B \ominus A) = (^{\ominus}B) \oplus (^{\ominus}(^{\ominus}A)) = (^{\ominus}B) \oplus A = A \ominus B$$

which is greater than O by Theorem OF.10(B). Then $\tau_{A \ominus B}(B) = (A \ominus B) \oplus B = A \oplus (^{\ominus}B \oplus B) = A \oplus O = A$; by the last statement of Case 2 of Theorem ISM.23, $B < A$, which contradicts $A < B$. Therefore if $A < B$, $B \ominus A > O$.

Conversely, if $B \ominus A > O$, then

$$\begin{aligned}\tau_{B \ominus A}(A) &= (B \ominus A) \oplus A = (B \oplus (^{\ominus}A)) \oplus A \\ &= B \oplus ((^{\ominus}A) \oplus A) = B \oplus O = B\end{aligned}$$

so that again by Theorem ISM.23 $A < B$.

The proof that $B \ominus A > O$ iff $(^{\ominus}B) < (^{\ominus}A)$ is Exercise OF.9.

(B) Using Theorem OF.10(F),

$$\begin{aligned}(B \oplus C) \ominus (A \oplus C) &= (B \oplus C) \oplus {}^{\ominus}(A \oplus C) \\ &= (B \oplus C) \oplus ((^{\ominus}A) \oplus (^{\ominus}C)) \\ &= B \oplus C \oplus (^{\ominus}A) \oplus (^{\ominus}C) \\ &= (B \oplus (^{\ominus}A)) \oplus (C \oplus (^{\ominus}C)) \\ &= (B \oplus (^{\ominus}A)) \oplus O = B \ominus A.\end{aligned}$$

By part (A) this is greater than O iff $B > A$, so that

$$(B \oplus C) \ominus (A \oplus C) > O \text{ iff } B > A.$$

That $O < B$ iff $C < B \oplus C$ follows by substituting O for A in the above; that $A < O$ iff $A \oplus C < C$ follows by substituting O for B.

(C) If $A < B$, then by part (A) $B \ominus A > O$. Since $C > O$, by Theorem OF.10(C) $C \odot (B \ominus A) > O$. Then by Theorem OF.10(D) and Theorem OF.6,

$$\begin{aligned}C \odot (B \ominus A) &= C \odot (B \oplus {}^{\ominus}A) = (C \odot B) \oplus (C \odot {}^{\ominus}A) \\ &= (C \odot B) \ominus (C \odot A) > O\end{aligned}$$

Therefore by part (A) $C \odot A < C \odot B$.

(D) The proof of part (D) is Exercise OF.7. □

Theorem OF.12. *Let $\mathcal{P}$ be a Euclidean plane, $\mathbb{L}$ be a line on $\mathcal{P}$, O be a point on $\mathbb{L}$, and U be a member of $\mathbb{L} \setminus \{O\}$. If the operations $\oplus$ and $\odot$ are established on $\mathbb{L}$ in accordance with Definition OF.1 and if the relation $<$ is established on $\mathbb{L}$ with the properties listed in Theorem OF.11, then $\mathbb{L}$ is an ordered field.*

Proof. This is simply a synthesis of Theorems OF.7 and OF.11. □

Definition OF.13. (A) The ray $\overrightarrow{OU}$ is the **positive half** of $\mathbb{L} \setminus \{O\}$.
(B) For any member X of $\mathbb{L}$, the **absolute value** of X is

$$|X| = \begin{cases} X \text{ if } X \geq O \\ \ominus X \text{ if } X < O \end{cases}.$$

Remark OF.14 (Identification of $\mathbb{F}$ and $\overrightarrow{OU}$). Early in Chapter 9 (Definition FSEG.3) we defined addition and ordering of free segments. Doubtless the reader has noted that the arithmetic of free segments is very limited, since there is no "zero" segment and no "negative" segments, or any definition of multiplication. (This kind of algebraic system is known as a *semigroup*.)

So far in this chapter we have shown how a line $\mathbb{L}$ in the Euclidean plane can be "built" into a field, complete with addition $\oplus$ and multiplication $\odot$, so that it has a complete system of arithmetic. Also in Definition OF.8(D) we have specified an ordering for this field.

In Theorem FSEG.15, we showed that the mapping Φ (from Definition FSEG.14) is a bijection of the set $\mathbb{F}$ of all free segments onto $\overrightarrow{OU}$, and this mapping also preserves order. (Here O and U are distinct points of the line, O being the *origin*, and the *unit* U replaces Q in Theorem FSEG.15.) Thus it makes sense to define length and distance as positive members of $\mathbb{L}$; we will do this in Definition OF.16.

In Theorem OF.17, we will show that Φ preserves addition, and later on in Theorem SIM.8 (after products of free segments have been defined), it will be shown that Φ also preserves products. Thus Φ allows us to *identify* $\mathbb{F}$ with $\overrightarrow{OU}$, meaning that these two sets are algebraically indistinguishable.

Note carefully that we use the symbol $\oplus$ both for addition of free segments and for addition of points on the line $\mathbb{L}$; its meaning in these two situations is quite different, the first being derived from pasting two free segments together end-to-end, the second from translations.

Theorem OF.15. *Let A, B, C, and D be distinct points on the line $\mathbb{L}$ on the Euclidean plane $\mathcal{P}$ (where $\mathbb{L}$ is equipped with origin O and unit U as in Definition OF.1).*

(A) $\overline{AB} \cong \overline{O(B \ominus A)} \cong \overline{O|B \ominus A|}$.
(B) $|B \ominus A| = |D \ominus C|$ *iff* $\overline{AB} \cong \overline{CD}$.
(C) *Let Φ be the mapping defined in Definition FSEG.14, where U takes the place of Q in that definition. Then* $\Phi[\overline{AB}] = |B \ominus A|$.

Proof. (A) (I) For the moment, choose the notation so that $A < B$; by Theorem OF.11 (A), $B \ominus A > O$.

Then $\tau_{\ominus A}(A) = A \ominus A = O$ and $\tau_{\ominus A}(B) = B \ominus A$; since $\tau_{\ominus A}$ is an isometry, we may use Theorem NEUT.15(5) to get $\tau_{\ominus A}(\overline{AB}) = \overline{O(B \ominus A)}$, and $\overline{AB} \cong \overline{O(B \ominus A)}$.

(II) On the other hand, if $A > B$, by Theorem OF.11(A) $A \ominus B > O$, and by (I) above, $\overline{BA} \cong \overline{O(A \ominus B)}$. By Theorem NEUT.15(5) and Theorem OF.10(A)

$$\mathcal{R}_O(\overline{O(A \ominus B)}) = \overline{O(\mathcal{R}_O(A \ominus B))} = \overline{O(^{\ominus}(A \ominus B))} = \overline{O(B \ominus A)},$$

so that $\overline{O(A \ominus B)} \cong \overline{O(B \ominus A)}$. Again using (I), $\overline{BA} \cong \overline{O(A \ominus B)} \cong \overline{O(B \ominus A)}$, and since by Definition IB.3 $\overline{AB} = \overline{BA}$, $\overline{AB} \cong \overline{O(B \ominus A)}$.

(III) By the result just above and Definition IB.3,

$$\overline{O(B \ominus A)} \cong \overline{AB} = \overline{BA} \cong \overline{O(A \ominus B)} \cong \overline{O\,^{\ominus}(B \ominus A)},$$

and $|B \ominus A|$ is either $B \ominus A$ or $A \ominus B$. Hence $\overline{AB} \cong |\,\overline{O(B \ominus A)}\,|$. This completes the proof of part (A).

(B) If $|B \ominus A| = |D \ominus C|$, then $\overline{O|B \ominus A|} = \overline{O|D \ominus C|}$; by part (A) above,

$$\overline{AB} \cong \overline{O|B \ominus A|} = \overline{O|D \ominus C|} \cong \overline{CD},$$

and $\overline{AB} = \overline{CD}$.

Conversely, if $\overline{AB} \cong \overline{CD}$, then applying part (A) twice, we have

$$\overline{O|B \ominus A|} \cong \overline{AB} \cong \overline{CD} \cong \overline{O|D \ominus C|}.$$

Since both $|B \ominus A|$ and $|D \ominus C|$ are members of $\overrightarrow{OU}$, by Property R.4 of Definition NEUT.2, $|B \ominus A| = |D \ominus C|$.

(C) Since $X = \Phi[\overline{AB}]$ is defined to be the point of $\overrightarrow{OU}$ such that $[\overline{OX}] = [\overline{AB}]$, we know that $\overline{OX} \cong \overline{AB}$. By part (B), and using the fact that $X > O$,

$$|B \ominus A| = |X \ominus O| = |X| = X = \Phi[\overline{AB}]. \qquad \square$$

Definition OF.16. Define the **distance** between two points A and B, or the **length of** $\overline{AB}$ to be $|B \ominus A| = \Phi[\overline{AB}]$.

Theorem OF.17. *Let $\mathcal{P}$ be a Euclidean plane, $\mathbb{L}$ be the ordered field of Theorem OF.12, O be the origin of $\mathbb{L}$, and let S, T, and V be positive members of $\mathbb{L}$ (i.e., members of $\overrightarrow{OU}$). Let Φ be as in Definition FSEG.14. Then*

(A) $[\overline{OT}] = [\overline{S(S \oplus T)}]$, *that is,* $\overline{OT} \cong \overline{S(S \oplus T)}$*; this is true for any S and T in $\mathbb{L}$;*

(B) $[\overline{OS}] \oplus [\overline{S(S \oplus T)}] = [\overline{OS}] \oplus [\overline{OT}] = [\overline{O(S \oplus T)}]$*;*

(C) $\Phi([\overline{OS}] \oplus [\overline{OT}]) = \Phi[\overline{OS}] \oplus \Phi[\overline{OT}]$; *and*

(D) $[\overline{OS}] \oplus [\overline{OT}] = [\overline{OV}]$ *iff* $S \oplus T = V$.

Proof. (A) By Definition OF.1 and using Theorem COBE.5(5), $\tau_S(\overline{OT}) = \overline{\tau_S(O)\tau_S(T)} = \overline{S(S \oplus T)}$. Since τ_S is an isometry, by Definition NEUT.3(B) $\overline{OT} \cong \overline{S(S \oplus T)}$ (*cf* Definition FSEG.2); this proves part (A) for any T and S in $\mathbb{L}$.

(B) By Theorem OF.11(B) $S \oplus T > S$ so that O–S–$(S \oplus T)$. By Definition FSEG.3 and part (A), $[\overline{OS}] \oplus [\overline{OT}] = [\overline{OS}] \oplus [\overline{S(S \oplus T)}] = [\overline{O(S \oplus T)}]$. This proves part (B).

(C) In this part, $[\overline{OS}] \oplus [\overline{OT}]$ means addition of free segments, while $\Phi[\overline{OS}] \oplus \Phi[\overline{OT}]$ means addition of points on the line $\mathbb{L}$. Since S, T, and V are positive, so is $S \oplus T$. Then by part (B) $\Phi([\overline{OS}] \oplus [\overline{OT}]) = \Phi([\overline{O(S \oplus T)}]) = S \oplus T$, $\Phi[\overline{OS}] = S$, and $\Phi[\overline{OT}] = T$. Here we have used Definition FSEG.14 three times.

(D) The proof that if $S \oplus T = V$ then $[\overline{OS}] \oplus [\overline{OT}] = [\overline{OV}]$ follows immediately from part (B). Conversely, if $[\overline{OS}] \oplus [\overline{OT}] = [\overline{OV}]$, by part (B) $[\overline{OV}] = [\overline{O(S \oplus T)}]$ so that by Definition FSEG.2 $\overline{OV} \cong \overline{O(S \oplus T)}$; since V and $S \oplus T$ are positive, $V \in \overrightarrow{O(S \oplus T)}$ and by Definition NEUT.2 Property R.4 (linear scaling), $V = S \oplus T$. □

Corollary OF.18. *Assume the hypotheses of Theorem OF.17. In addition, assume that there is a second line* $\mathbb{M}$ *which has been made into an ordered field with unit* U', *and which intersects* $\mathbb{L}$ *at their common point* O *of origin. Let* S' *and* T' *be positive members of* $\mathbb{M}$ *(i.e., members of* $\overrightarrow{OU'}$*). If* $\overline{OS} \cong \overline{OS'}$ *and* $\overline{OT} \cong \overline{OT'}$, *then* $\overline{O(S \oplus T)} \cong \overline{O(S' \oplus T')}$.

Proof. $\overline{OS} \cong \overline{OS'}$ and $\overline{OT} \cong \overline{OT'}$ iff $[\overline{OS}] = [\overline{OS'}]$ and $[\overline{OT}] = [\overline{OT'}]$; by Theorem OF.17(B), if this is true then $[\overline{O(S \oplus T)}] = [\overline{O(S' \oplus T')}]$, that is, $\overline{O(S \oplus T)} \cong \overline{O(S' \oplus T')}$. □

Remark OF.19. Note that by Theorem OF.17 we may add two positive points A and B on $\mathbb{L}$ by first constructing the two segments $\overline{OA}$ and $\overline{OB}$, then constructing a point C such that O–A–C and $\overline{AC} \cong \overline{OB}$. Then $A \oplus B = C$. This observation will be pivotal in our development in Chapter 17 (QX) of rational multiples of points on a line.

14.2 Exercises for ordered fields

Answers to starred (*) exercises may be accessed from the home page for this book at www.springer.com.

Exercise OF.1*. Let $\mathcal{P}$ be a Euclidean plane; let $\mathbb{L}$ be an ordered field on $\mathcal{P}$ with origin O, and τ_A be the translation of $\mathcal{P}$ such that $\tau_A(O) = A$, where A is any member of $\mathbb{L} \setminus \{O\}$. Then for every member X of $\mathbb{L}$, $\tau_A(X) = X \oplus A$.

Exercise OF.2*. Let $\mathcal{P}$ be a Euclidean plane; let $\mathbb{L}$ be an ordered field on $\mathcal{P}$ with origin O and unit U (where $U \in (\mathbb{L} \setminus \{O\})$) and let δ_A be the dilation of $\mathcal{P}$ with fixed point O such that $\delta_A(U) = A$. Then for every member X of $\mathbb{L} \setminus \{O\}$, $\delta_A(X) = X \odot A$.

Exercise OF.3*. (A) If A, B, and C are members of the ordered field $\mathbb{L}$ (cf. Definition OF.1) such that $A \oplus C = B \oplus C$, then $A = B$.

(B) If A and B are members of $\mathbb{L}$ and if C is a member of $\mathbb{L} \setminus \{O\}$ such that $A \odot C = B \odot C$, then $A = B$.

Exercise OF.4*. (A) If A, B, and C are members of the field $\mathbb{L}$ such that $A \oplus B = A \oplus C$, then $B = C$.

(B) If A is a member of $\mathbb{L} \setminus \{O\}$, and if B and C are members of $\mathbb{L}$ such that $A \odot B = A \odot C$, then $B = C$.

Exercise OF.5*. Let A, B, and C be members of the field $\mathbb{L}$; then $(B \ominus A) \odot C = (B \odot C) \ominus (A \odot C)$.

Exercise OF.6*. Let δ be a dilation of the Euclidean plane $\mathcal{P}$ with fixed point O, and let $\mathbb{L}$ be an ordered field with origin O and unit U. If K and T are any members of $\mathbb{L}$, then $\delta(K \odot T) = K \odot \delta(T)$.

Exercise OF.7*. Let A and B be members of $\mathbb{L}$. Complete the proof of Theorem OF.11(A), by showing that $B \ominus A$ iff $({}^{\ominus}B) < ({}^{\ominus}A)$.

Exercise OF.8*. Prove part D of Theorem OF.11: If $A < B$ and $C < O$, then $B \odot C < A \odot C$.

Exercise OF.9*. Let A and B be negative members of $\mathbb{L}$. Then $A < B$ iff $|B| < |A|$.

Exercise OF.10*. (A) Let $\mathbb{T} = \{\tau_A \mid A \in \mathbb{L}\}$; then the mapping $\alpha{:}A \to \tau_A$ is a bijection of $\mathbb{L}$ onto $\mathbb{T}$.

(B) Let $\mathbb{M} = \{\delta_A \mid A \in \mathbb{L}\}$; then the mapping $\mu{:}A \to \delta_A$ is a bijection of $\mathbb{L}$ onto $\mathbb{M}$; furthermore μ maps $\mathbb{L} \setminus \{O\}$ onto $\mathbb{M} \setminus \{O\}$.

Exercise OF.11*. (This result is analogous to Theorem CAP.23.) Let $\mathcal{P}$ be a Euclidean plane, and let $\mathbb{L}_1$ and $\mathbb{L}_2$ be parallel lines on $\mathcal{P}$, where $\mathbb{L}_1$ has been built into an ordered field with origin O_1 and unit U_1. Let O_2 be a point of $\mathbb{L}_2$, and let σ be the translation of $\mathcal{P}$ such that $\sigma(O_1) = O_2$. (The existence and uniqueness of this translation is guaranteed by Theorem ISM.5.) Let $A \in \mathbb{L}_1 \setminus \{O_1, U_1\}$. Then $\sigma \circ \delta_A \circ \sigma^{-1}$ is a dilation of $\mathcal{P}$ with fixed point O_2. In fact, $\sigma \circ \delta_A \circ \sigma^{-1} = \delta_{\sigma(A)}$ so that $\sigma \circ \delta_A = \delta_{\sigma(A)} \circ \sigma$.

Exercise OF.12*. Let $\mathcal{P}$ be a Euclidean plane; let $\mathbb{L}$ be an ordered field on $\mathcal{P}$ with origin O and unit U, A be a member of $\mathbb{L} \setminus \{O, U\}$, and let τ_A and δ_A be as in Definition OF.1. Then $\delta_A \circ \tau_A = \tau_{\delta_A(A)} \circ \delta_A$.

Chapter 15
Similarity on a Euclidean Plane (SIM)

Acronym: *SIM*
Dependencies: *all prior Chapters 1 through 14*
New Axioms: *none*
New Terms Defined: *similar, similarity, similarity mapping, unit free segment; product, ratio (of free segments)*

Abstract: This chapter defines a similarity mapping on a Euclidean plane as a dilation, an isometry, or a composition of a dilation and an isometry. Such mappings are used to define the similarity of two sets. Similarity is shown to be an equivalence relation, and criteria are developed for similarity of triangles. The chapter concludes with a proof of the Pythagorean Theorem, and a proof that the product of the base and altitude of a triangle is constant.

15.1 Theorems on similarity

Definition SIM.1. (A) Let $\mathcal{P}$ be a Euclidean plane; if δ is either a dilation, or the identity ι, and φ is an isometry of $\mathcal{P}$, then $\delta \circ \varphi$ is a **similarity mapping** (or simply a **similarity**).

(B) If $\mathcal{D}$ and $\mathcal{E}$ are any nonempty subsets of $\mathcal{P}$, they are said to be **similar** (notation: $\mathcal{D} \sim \mathcal{E}$) iff there exists a similarity mapping $\delta \circ \varphi$ such that $(\delta \circ \varphi)(\mathcal{D}) = \delta(\varphi(\mathcal{D})) = \mathcal{E}$.

E.J. Specht et al., *Euclidean Geometry and its Subgeometries*,
DOI 10.1007/978-3-319-23775-6_15

Theorem SIM.2. (A) *Every dilation and every isometry (including the identity ι) is a similarity mapping.*

(B) *For any similarity mapping $\delta \circ \varphi$, where δ is a dilation and φ is an isometry, there exists an isometry ψ such that $\psi \circ \delta = \delta \circ \varphi$.*

Therefore a mapping is a similarity mapping iff it is either of the form $\varphi \circ \delta$ or $\delta \circ \varphi$, where δ is either a dilation or the identity, and φ is an isometry.

(C) *The inverse of a similarity mapping is a similarity mapping.*

(D) *The composition of two similarity mappings is a similarity mapping.*

(E) *Every similarity mapping is a belineation.*

Proof. (A) Since the identity ι is an isometry, by Definition SIM.1, $\iota \circ \iota = \iota$ is a similarity mapping; if φ is any isometry, then by the same definition, $\iota \circ \varphi$ is a similarity mapping. If δ is any dilation, then $\delta = \delta \circ \iota$ is a similarity mapping.

(B) This follows immediately from Theorem DLN.16.

(C) $(\varphi \circ \delta)^{-1} = \delta^{-1} \circ \varphi^{-1}$; by Theorem CAP.21, δ^{-1} is a dilation, and by Theorem NEUT.11 φ^{-1} is an isometry; by part (B) above, $\delta^{-1} \circ \varphi^{-1}$ is a similarity mapping.

(D) Let $\delta \circ \varphi$ and $\delta' \circ \varphi'$ be similarity mappings, where φ and φ' are isometries, δ is either a dilation with fixed point O or the identity ι, and δ' is either a dilation with fixed point O' or ι. We show that $\omega = \delta \circ \varphi \circ \delta' \circ \varphi'$ is a similarity mapping.

(I) If $\delta = \iota$, then $\omega = \varphi \circ \delta' \circ \varphi'$. By part (B) above, there exists an isometry ψ such that $\psi \circ \delta' = \delta' \circ \varphi'$, so that $\omega = (\varphi \circ \psi) \circ \delta'$; since $\varphi \circ \psi$ is an isometry, by part (B) ω is a similarity mapping.

If $\delta' = \iota$, $\omega = \delta \circ (\varphi \circ \varphi')$ and since $\varphi \circ \varphi'$ is an isometry, by part (B) ω is a similarity mapping.

(II) Now suppose that both δ and δ' are dilations, with, as noted above, fixed points O and O', respectively. By Theorem ISM.5, there exists a translation τ such that $\tau(O') = O$. By Theorem CAP.23(C), $\epsilon = \tau^{-1} \circ \delta' \circ \tau$ is a dilation with fixed point O. Thus

$$\begin{aligned}\omega = \delta \circ \varphi \circ \delta' \circ \varphi' &= \delta \circ \varphi \circ \tau \circ (\tau^{-1} \circ \delta' \circ \tau) \circ \tau^{-1} \circ \varphi' \\ &= \delta \circ (\varphi \circ \tau) \circ \epsilon \circ (\tau^{-1} \circ \varphi')\end{aligned}$$

where $\varphi \circ \tau$ and $\tau^{-1} \circ \varphi'$ are both isometries. By Theorem DLN.16 there exists an isometry φ'' such that $\varphi \circ \tau \circ \epsilon = \epsilon \circ \varphi''$, so that

$$\omega = \delta \circ \epsilon \circ (\varphi'' \circ \tau^{-1} \circ \varphi')$$

and $\varphi'' \circ \tau^{-1} \circ \varphi'$ is an isometry. By Theorem CAP.21, $\delta \circ \epsilon$ is either the identity or a dilation with fixed point O, so that by Definition SIM.1, ω is a similarity mapping.

(E) This is an immediate consequence of Theorem DLN.8 and the fact that every isometry is a belineation. □

Theorem SIM.3. *Similarity is an equivalence relation on the set $\mathbb{O}$ of nonempty subsets of the Euclidean plane $\mathcal{P}$.*

Proof. Let $\mathcal{D}$, $\mathcal{E}$, and $\mathcal{F}$ be members of $\mathbb{O}$.

(I: $\mathcal{D} \sim \mathcal{D}$.) $\iota(\mathcal{D}) = \mathcal{D}$, and by Theorem SIM.2(A), the identity is a similarity mapping, so $\mathcal{D} \sim \mathcal{D}$.

(II: If $\mathcal{D} \sim \mathcal{E}$, then $\mathcal{E} \sim \mathcal{D}$.) If $\mathcal{D} \sim \mathcal{E}$, there exists a similarity mapping ω such that $\omega(\mathcal{D}) = \mathcal{E}$; then $\omega^{-1}(\mathcal{E}) = \mathcal{D}$ and by Theorem SIM.2(C), ω^{-1} is a similarity mapping.

(III: If $\mathcal{D} \sim \mathcal{E}$ and $\mathcal{E} \sim \mathcal{F}$, then $\mathcal{D} \sim \mathcal{F}$.) By Definition SIM.1, there exist similarity mappings σ and ω such that $\sigma(\mathcal{D}) = \mathcal{E}$ and $\omega(\mathcal{E}) = \mathcal{F}$; By Theorem SIM.2(D), $\omega \circ \sigma$ is a similarity mapping, and $\omega \circ \sigma(\mathcal{D}) = \omega(\sigma(\mathcal{D})) = \omega(\mathcal{E}) = \mathcal{F}$ so that $\mathcal{D} \sim \mathcal{F}$. □

Remark SIM.4. (A) Definition SIM.1(B) can be stated as follows: if $\mathcal{D}$ and $\mathcal{E}$ are any nonempty sets, then $\mathcal{D} \sim \mathcal{E}$ iff there exists a nonempty set $\mathcal{G}$ and a dilation δ such that $\mathcal{G} \cong \mathcal{D}$ and $\delta(\mathcal{G}) = \mathcal{E}$. By Theorem SIM.2(B), this is the same as saying that there exist a nonempty set $\mathcal{G}$ and a dilation δ such that $\delta(\mathcal{D}) = \mathcal{G}$ and $\mathcal{G} \cong \mathcal{E}$.

(B) Notice that our definition of similarity provides a generalization of congruence; if two sets are congruent, they are similar; that is, an isometry is a similarity mapping. But the converse is not true.

Theorem SIM.5. *Let $\mathcal{P}$ be a Euclidean plane.*

(A) *Let A, B, and C be noncollinear points on $\mathcal{P}$ and let ω be a similarity mapping of $\mathcal{P}$. Then $\angle BAC \cong \angle\omega(B)\omega(A)\omega(C)$, $\angle CBA \cong \angle\omega(C)\omega(B)\omega(A)$, and $\angle ACB \cong \angle\omega(A)\omega(C)\omega(B)$.*

(B) *If A_1, B_1, C_1, A_2, B_2, and C_2 are points on $\mathcal{P}$ such that A_1, B_1, and C_1 are noncollinear, A_2, B_2, and C_2 are noncollinear, $\angle B_1A_1C_1 \cong \angle B_2A_2C_2$, $\angle C_1B_1A_1 \cong \angle C_2B_2A_2$, and $\angle A_1C_1B_1 \cong \angle A_2C_2B_2$, then*

$$\Delta A_1B_1C_1 \sim \Delta A_2B_2C_2.$$

Proof. (A) By Theorem SIM.2(A) and (B) either ω is an isometry of $\mathcal{P}$, or there exist an isometry φ and a dilation δ such that $\omega = \varphi \circ \delta$. If ω is an isometry, then by Theorem NEUT.15(8), the congruences stated are true. If $\omega = \varphi \circ \delta$, where φ is an isometry and δ is a dilation, then by Theorem DLN.14 and Theorem NEUT.15(8) the congruences stated are true.

(B) By Theorem NEUT.67 (Segment construction) there exist points B_2' and C_2' such that $B_2' \in \overrightarrow{A_2B_2}$, $C_2' \in \overrightarrow{A_2C_2}$, $\overline{A_1B_1} \cong \overline{A_2B_2'}$, and $\overline{A_1C_1} \cong \overline{A_2C_2'}$. By Theorem NEUT.64 (EAE), $\triangle A_1B_1C_1 \cong \triangle A_2B_2'C_2'$, $\angle A_2B_2'C_2' \cong \angle A_1B_1C_1$, $\angle A_2C_2'B_2' \cong \angle A_1C_1B_1$, and $\overline{B_2'C_2'} \cong \overline{B_1C_1}$. Since $\angle A_1B_1C_1 \cong \angle A_2B_2C_2$ by Theorem NEUT.14 (congruence is an equivalence relation), $\angle A_2B_2'C_2' \cong \angle A_2B_2C_2$.

If $B_2' = B_2$, then by Theorem NEUT.65 (AEA) $\overline{A_2C_2'} \cong \overline{A_2C_2}$. By Property R.4 of Definition NEUT.2, $C_2' = C_2$ so that $\triangle A_2B_2'C_2' = \triangle A_2B_2C_2$ and thus $\triangle A_1B_1C_1 \sim \triangle A_2B_2C_2$.

If $B_2' \neq B_2$, then using Theorem DLN.7 let δ be the dilation of $\mathcal{P}$ with fixed point A_2 such that $\delta(B_2') = B_2$. By Theorem CAP.1(A)

$$\delta(\overleftrightarrow{B_2'C_2'}) = \overleftrightarrow{\delta(B_2')\delta(C_2')} = \overleftrightarrow{B_2\delta(C_2')}.$$

By Definition CAP.17 $\overleftrightarrow{B_2\delta(C_2')} \parallel \overleftrightarrow{B_2'C_2'}$. Since

$$\begin{aligned}\angle A_2B_2\delta(C_2') &= \angle\delta(A_2)\delta(B_2')\delta(C_2') = \delta(\angle A_2B_2C_2)\\ &\cong \angle A_2B_2C_2 \cong \angle A_2B_2'C_2',\end{aligned}$$

by Theorem EUC.11, $\overleftrightarrow{B_2C_2} \parallel \overleftrightarrow{B_2'C_2'}$. By Axiom PS $\overleftrightarrow{B_2C_2} = \overleftrightarrow{B_2\delta(C_2')}$ so that $\delta(C_2') = C_2$.

Summarizing, we have $\delta(B_2') = B_2$ and $\delta(C_2') = C_2$ so that $\delta(\triangle A_2B_2'C_2') = \triangle A_2B_2C_2$. Also $\triangle A_1B_1C_1 \cong \triangle A_2B_2'C_2'$, so that $\triangle A_1B_1C_1 \sim \triangle A_2B_2C_2$, by Definition SIM.1. □

Theorem SIM.6. *Let $\mathcal{P}$ be a Euclidean plane and let A_1, B_1, C_1, A_2, B_2, and C_2 be points on $\mathcal{P}$ such that A_1, B_1, and C_1 are noncollinear, A_2, B_2, and C_2 are noncollinear, $\angle B_1A_1C_1 \cong \angle B_2A_2C_2$ and $\angle A_1B_1C_1 \cong \angle A_2B_2C_2$, then $\triangle A_1B_1C_1 \sim \triangle A_2B_2C_2$.*

Proof. An immediate consequence of Theorem SIM.5 and Theorem EUC.35. □

In the following, we will use smaller typeface script letters to denote free segments of $\mathcal{P}$, as in Chapter 9 (FSEG). We will use the symbol $<$ and its natural variants for the order relation on free segments that we defined in that chapter.

Definition SIM.7. Let $\mathcal{P}$ be a Euclidean plane, O a point on $\mathcal{P}$, and U a member of $\mathcal{P} \setminus \{O\}$. Let $\mathcal{A}$ and $\mathcal{B}$ be any free segments of $\mathcal{P}$, and let A and B be points on $\overrightarrow{OU}$ such that $[\overline{OA}] = \mathcal{A}$ and $[\overline{OB}] = \mathcal{B}$.

(A) Define the **unit free segment** to be $\mathcal{U} = [\overline{OU}]$. It will remain fixed throughout our development.

(B) Define the **product** $\mathcal{A} \odot \mathcal{B} = [\overline{OA}] \odot [\overline{OB}]$ of $\mathcal{A}$ and $\mathcal{B}$ as the free segment $[\overline{O(A \odot B)}]$, where (as in Definition OF.1) $A \odot B = \delta_B(\delta_A(U)) = \delta_B(A)$.

Theorem SIM.8. *Let $\mathcal{P}$ be a Euclidean plane.*

(A) *Under the operation $\odot$, the set of free segments of $\mathcal{P}$ is an abelian group with identity element $\mathcal{U}$.*

(B) *If $\mathcal{A}$, $\mathcal{B}$, and $\mathcal{C}$ are free segments of $\mathcal{P}$, then $\mathcal{A} \odot (\mathcal{B} \oplus \mathcal{C}) = (\mathcal{A} \odot \mathcal{B}) \oplus (\mathcal{A} \odot \mathcal{C})$.*

(C) *The mapping Φ which is an order and sum-preserving bijection of $\mathbb{F}$ onto $\overrightarrow{OU}$ (cf Definition FSEG.14 and Theorem OF.17) also preserves products. That is, for any free segments $\mathcal{A}$ and $\mathcal{B}$, $\Phi(\mathcal{A} \odot \mathcal{B}) = \Phi(\mathcal{A}) \odot \Phi(\mathcal{B})$.*

Proof. (A) Let A, B, and C be members of $\overrightarrow{OU}$ such that $[\overline{OA}] = \mathcal{A}$, $[\overline{OB}] = \mathcal{B}$, and $[\overline{OC}] = \mathcal{C}$. In the following reasoning we use Definition SIM.7 and Theorem OF.7.

(I) (Existence of identity) $\mathcal{A} \odot \mathcal{U} = [\overline{O(A \odot U)}] = [\overline{OA}] = \mathcal{A}$.

(II) (Existence of inverses) Let $\mathcal{D} = [\overline{OA^{-1}}]$. Then

$$\mathcal{A} \odot \mathcal{D} = [\overline{OA}] \odot [\overline{OA^{-1}}] = [\overline{O(A \odot A^{-1})}] = [\overline{OU}] = \mathcal{U}.$$

Hence $\mathcal{D} = \mathcal{A}^{-1}$.

(III) (Commutativity)

$$\mathcal{A} \odot \mathcal{B} = [\overline{OA}] \odot [\overline{OB}] = [\overline{O(A \odot B)}] = [\overline{O(B \odot A)}] = \mathcal{B} \odot \mathcal{A}.$$

(IV) (Associativity)

$$(\mathcal{A} \odot \mathcal{B}) \odot \mathcal{C} = [\overline{O((A \odot B) \odot C)}] = [\overline{O(A \odot (B \odot C))}] = \mathcal{A} \odot (\mathcal{B} \odot \mathcal{C}).$$

(B) (Distributivity) In addition to Definition SIM.7 and Theorem OF.7, in this part we use Theorem OF.17.

$$\begin{aligned} \mathcal{A} \odot (\mathcal{B} \oplus \mathcal{C}) &= [\overline{O(A \odot (B \oplus C))}] = [\overline{O((A \odot B) \oplus (A \odot C))}] \\ &= [\overline{O(A \odot B)}] \oplus [\overline{O(A \odot C)}] = (\mathcal{A} \odot \mathcal{B}) \oplus (\mathcal{A} \odot \mathcal{C}). \end{aligned}$$

(C) Let A and B be members of $\overrightarrow{OU}$ such that $\mathcal{A} = [\overline{OA}]$ and $\mathcal{B} = [\overline{OB}]$. By Definition SIM.7, $\mathcal{A} \odot \mathcal{B} = [\overline{O(A \odot B)}]$ so $\Phi(\mathcal{A} \odot \mathcal{B}) = A \odot B = \Phi(\mathcal{A}) \odot \Phi(\mathcal{B})$.

□

Remark SIM.8.1. (A) Theorem SIM.8, in combination with Theorem FSEG.15, Theorem OF.17, and Theorem OF.7 fulfills the promise we made in Remark OF.14. It shows that the product defined on the set of free segments does turn that set into a group under the operation $\odot$, and that the mapping Φ is a group isomorphism with respect to that operation. Thus the set of free segments may be identified with $\overrightarrow{OU}$ as to sums, products, and order.

(B) We should make clear exactly what the objects being defined here are. The product of two free segments is again a free segment; later on, in Definition SIM.12, we will define the ratio of two free segments, which again will be a free segment. Intuitively we are used to thinking of ratios as numbers. It is possible to do that here, too, since we have mapped the set of free segments onto $\overrightarrow{OU}$ using the mapping Φ defined in Definition FSEG.14.

Theorem SIM.9. *Let $\mathcal{A}$, $\mathcal{B}$, and $\mathcal{C}$ be free segments of $\mathcal{P}$.*

(I) *If $\mathcal{B} < \mathcal{C}$, then $\mathcal{A} \odot \mathcal{B} < \mathcal{A} \odot \mathcal{C}$.*
(II) *If $\mathcal{B} > \mathcal{C}$, then $\mathcal{A} \odot \mathcal{B} > \mathcal{A} \odot \mathcal{C}$.*

Proof. (I) Let A, B, and C be points on $\overrightarrow{OU}$ such that $[\overline{OA}] = \mathcal{A}$, $[\overline{OB}] = \mathcal{B}$, and $[\overline{OC}] = \mathcal{C}$; and let δ, ϵ, and θ be dilations of $\mathcal{P}$ with fixed point O such that $\delta(U) = A$, $\epsilon(U) = B$, and $\theta(U) = C$.

By Definitions SIM.7 and OF.1, $\mathcal{A} \circ \mathcal{B} = [\overline{O((\epsilon \circ \delta)(U))}]$ and $\mathcal{A} \circ \mathcal{C} = [\overline{O((\theta \circ \delta)(U))}]$. By Exercise DLN.3 $\mathcal{A} \circ \mathcal{B} = \mathcal{B} \circ \mathcal{A} = [\overline{O((\delta \circ \epsilon)(U))}] = [\overline{O\delta(B)}]$ and $\mathcal{A} \circ \mathcal{C} = \mathcal{C} \circ \mathcal{A} = [\overline{O((\delta \circ \theta)(U))}] = [\overline{O\delta(C)}]$. If $\mathcal{B} < \mathcal{C}$, then by Definition FSEG.3, $\overline{OB} < \overline{OC}$. By Theorem NEUT.74 O–B–C. By Theorem SIM.2(E) (every similarity is a belineation), and the fact that O is a fixed point of δ, O–$\delta(B)$–$\delta(C)$. By Definition NEUT.70 $\overline{O\delta(B)} < \overline{O\delta(C)}$. By Definition FSEG.3 $[\overline{O\delta(B)}] < [\overline{O\delta(C)}]$. Thus $\mathcal{A} \odot \mathcal{B} < \mathcal{A} \odot \mathcal{C}$.

(II) If $\mathcal{B} > \mathcal{C}$, then by Definition FSEG.3, $\mathcal{C} < \mathcal{B}$. By part (I) $(\mathcal{A} \odot \mathcal{B}) > \mathcal{A} \odot \mathcal{C}$. □

Definition SIM.10. Let $\mathcal{P}$ be a Euclidean plane and let $\mathcal{T}$ be a free segment of $\mathcal{P}$, then $\mathcal{T}^2 = \mathcal{T} \odot \mathcal{T}$.

The reader may find the following theorem easier to visualize in its re-statement as Theorem SIM.13, which deals with ratios of "lengths" of segments.

Theorem SIM.11. *Let $\mathcal{P}$ be a Euclidean plane and let O, A, and B be noncollinear points on $\mathcal{P}$; let C and D be points such that $C \in (\overleftrightarrow{OA} \setminus \{A\})$ and $D \in (\overleftrightarrow{OB} \setminus \{B\})$, $\mathcal{A} = [\overline{OA}]$, $\mathcal{B} = [\overline{OB}]$, $\mathcal{C} = [\overline{OC}]$, and $\mathcal{D} = [\overline{OD}]$.*

(A) *If* $\overleftrightarrow{AB} \parallel \overleftrightarrow{CD}$, *then* $\mathcal{A} \odot \mathcal{D} = \mathcal{B} \odot \mathcal{C}$.
(B) *If* $\mathcal{A} \odot \mathcal{D} = \mathcal{B} \odot \mathcal{C}$, *then* $\overleftrightarrow{AB} \parallel \overleftrightarrow{CD}$.

Proof. Using Theorem NEUT.67 and Exercise FSEG.1, define U as the unit for $\overleftrightarrow{OA}$ and U' as the unit for $\overleftrightarrow{OB}$; that is, U and U' are those points on their respective lines such that $[\overline{OU'}] = [\overline{OU}] = u$, where u is the unit free segment as in Definition SIM.7(A).

Using Theorem DLN.7 let α, β, γ, and δ be the dilations of $\mathcal{P}$ with fixed point O such that $\alpha(U) = A$, $\beta(U') = B$, $\gamma(U) = C$, and $\delta(U') = D$. By Definitions SIM.7 and OF.1

$$\mathcal{A} \odot \mathcal{D} = [\overline{O(\delta \circ \alpha)(U)}] = [\overline{O\delta(\alpha(U))}] \text{ and}$$
$$\mathcal{B} \odot \mathcal{C} = [\overline{O(\gamma \circ \beta)(U')}] = [\overline{O\gamma(\beta(U'))}].$$

Since $A = \alpha(U)$, $U = \alpha^{-1}(A)$ and so $C = \gamma(U) = (\gamma \circ \alpha^{-1})(A)$. Since $B = \beta(U')$, $U' = \beta^{-1}(B)$ and so $D = \delta(U') = (\delta \circ \beta^{-1})(B)$. By Theorem CAP.21 $\gamma \circ \alpha^{-1}$ is either ι (the identity mapping), or it is a dilation of $\mathcal{P}$. If $\gamma \circ \alpha^{-1}$ were equal to ι, then $\gamma = \alpha$, contrary to the fact that $\alpha(U) = A \neq C = \gamma(U)$. Hence $\gamma \circ \alpha^{-1}$ is a dilation of $\mathcal{P}$. The same kind of reasoning shows that $\delta \circ \beta^{-1}$ is a dilation of $\mathcal{P}$.

(A) (If $\overleftrightarrow{AB} \parallel \overleftrightarrow{CD}$, then $\mathcal{A} \odot \mathcal{D} = \mathcal{C} \odot \mathcal{B}$.) First note that $(\gamma \circ \alpha^{-1})(A) = C$ and by Definition CAP.17, $(\gamma \circ \alpha^{-1})(\overleftrightarrow{AB}) \parallel \overleftrightarrow{AB}$. By Axiom PS there is only one line through C parallel to $\overleftrightarrow{AB}$, so that $(\gamma \circ \alpha^{-1})(\overleftrightarrow{AB}) = \overleftrightarrow{CD}$ and $(\gamma \circ \alpha^{-1})(B) = D$. Since $(\delta \circ \beta^{-1})(B) = D$, by Theorem CAP.24 $\gamma \circ \alpha^{-1} = \delta \circ \beta^{-1}$. Multiplying both sides on the right by $\alpha \circ \beta = \beta \circ \alpha$ we have $\beta \circ \gamma = \gamma \circ \beta = \delta \circ \alpha$. Here we have used Exercise DLN.3. By Definitions SIM.7 and OF.1, $\mathcal{A} \odot \mathcal{D} = \mathcal{C} \odot \mathcal{B}$.

(B) (If $\mathcal{A} \odot \mathcal{D} = \mathcal{C} \odot \mathcal{B}$, then $\overleftrightarrow{AB} \parallel \overleftrightarrow{CD}$.) By Definition SIM.7 $\mathcal{A} \odot \mathcal{D} = [\overline{O(\delta \circ \alpha)(U)}]$ and $\mathcal{B} \odot \mathcal{C} = [\overline{O(\gamma \circ \beta)(U')}]$. If $\mathcal{A} \odot \mathcal{D} = \mathcal{C} \odot \mathcal{B}$, we have $\overline{O(\delta \circ \alpha)(U)} \cong \overline{O(\gamma \circ \beta)(U')}$, and since $\overline{OU} \cong \overline{OU'}$, by Corollary DLN.18(A) $\overline{O(\gamma \circ \beta)(U')} \cong \overline{O(\gamma \circ \beta)(U)}$, so that $\overline{O(\delta \circ \alpha)(U)} \cong \overline{O(\gamma \circ \beta)(U)}$. Since $\delta(\alpha(U))$ and $\gamma(\beta(U))$ both belong to $\overrightarrow{OU}$, we may apply Property R.4 of Definition NEUT.2 to get $(\delta \circ \alpha)(U) = (\gamma \circ \beta)(U)$. By Theorem CAP.24, $\delta \circ \alpha = \gamma \circ \beta$. Multiplying on the right by $\alpha^{-1} \circ \beta^{-1} = \beta^{-1} \circ \alpha^{-1}$, we have $\gamma \circ \alpha^{-1} = \delta \circ \beta^{-1}$. Define θ to be this dilation, which has fixed point O.

Then $\theta(A) = (\gamma \circ \alpha^{-1})(A) = C$ and $\theta(B) = (\delta \circ \beta^{-1})(B) = D$; by Theorem CAP.1(A) and Definition CAP.17 $\theta(\overleftrightarrow{AB}) = \overleftrightarrow{\theta(A)\theta(B)} = \overleftrightarrow{CD}$, and $\theta(\overleftrightarrow{AB}) \parallel \overleftrightarrow{AB}$. Thus $\overleftrightarrow{AB} \parallel \overleftrightarrow{CD}$. □

Definition SIM.12. Let $\mathcal{P}$ be a Euclidean plane and let $\mathcal{A}$ and $\mathcal{B}$ be free segments of $\mathcal{P}$. Then $\mathcal{A} \oplus \mathcal{B} = \mathcal{A} \odot \mathcal{B}^{-1} = \mathcal{B}^{-1} \odot \mathcal{A}$. $\mathcal{A} \oplus \mathcal{B}$ is the **ratio** of $\mathcal{A}$ to $\mathcal{B}$.

Theorem SIM.13 (Restatement of Theorem SIM.11.). *Let $\mathcal{P}$ be a Euclidean plane, O, A, and B be noncollinear points on $\mathcal{P}$, C and D be points such that $C \in (\overleftrightarrow{OA} \setminus \{A\})$ and $D \in (\overleftrightarrow{OB} \setminus \{B\})$, $\mathcal{A} = [\overline{OA}]$, $\mathcal{B} = [\overline{OB}]$, $\mathcal{C} = [\overline{OC}]$, and $\mathcal{D} = [\overline{OD}]$. Then $\mathcal{A} \oplus \mathcal{B} = \mathcal{C} \oplus \mathcal{D}$ iff $\overleftrightarrow{AB} \parallel \overleftrightarrow{CD}$.*

(A) *If $\overleftrightarrow{AB} \parallel \overleftrightarrow{CD}$, then $\mathcal{A} \oplus \mathcal{B} = \mathcal{C} \oplus \mathcal{D}$.*
(B) *If $\mathcal{A} \oplus \mathcal{B} = \mathcal{C} \oplus \mathcal{D}$, then $\overleftrightarrow{AB} \parallel \overleftrightarrow{CD}$.*

Theorem SIM.14. *Let $\mathcal{P}$ be a Euclidean plane; O and A be distinct points on $\mathcal{P}$; $B \in (\overrightarrow{OA} \setminus \{A\})$; $\mathcal{U}_1$ and $\mathcal{U}_2$ be free segments of $\mathcal{P}$; U_1 and U_2 be the points on $\overrightarrow{OA}$ such that $[\overline{OU_1}] = \mathcal{U}_1$ and $[\overline{OU_2}] = \mathcal{U}_2$; and δ_1, δ_2, ϵ_1, and ϵ_2 be the dilations of $\mathcal{P}$ such that $\delta_1(U_1) = A$, $\delta_2(U_2) = A$, $\epsilon_1(U_1) = B$, and $\epsilon_2(U_2) = B$. Then $\epsilon_1 \circ \delta_1^{-1} = \epsilon_2 \circ \delta_2^{-1}$ (i.e. $\epsilon_1 \oplus \delta_1 = \epsilon_2 \oplus \delta_2$).*

Proof. Since $\delta_1(U_1) = A$, $U_1 = \delta_1^{-1}(A)$. Thus $B = \epsilon_1(U_1) = \epsilon_1(\delta_1^{-1}(A)) = (\epsilon_1 \circ \delta_1^{-1})(A)$. In the same manner $B = (\epsilon_2 \circ \delta_2^{-1})(A)$. By Theorem CAP.24 $\epsilon_1 \circ \delta_1^{-1} = \epsilon_2 \circ \delta_2^{-1}$ (i.e., $\epsilon_1 \oplus \delta_1 = \epsilon_2 \oplus \delta_2$). □

Remark SIM.15. Theorem SIM.14 means that the ratio of two free segments is independent of the unit free segment that has been chosen. To see this most clearly (using Theorem SIM.13 and its notation), let $\mathcal{A} = [\overline{OA}]$ and $\mathcal{B} = [\overline{OB}]$. Then by Theorem SIM.14 $\mathcal{A} \oplus \mathcal{B} = [\overline{O(\epsilon_1 \circ \delta_1^{-1})(U_1)}] = [\overline{O(\epsilon_2 \circ \delta_2^{-1})(U_2)}]$. Thus the ratio $\mathcal{A} \oplus \mathcal{B}$ is the same for either U_1 or U_2.

Theorem SIM.16. *Let $\mathcal{P}$ be a Euclidean plane and let A_1, B_1, C_1, A_2, B_2, and C_2 be points on $\mathcal{P}$ such that A_1, B_1, and C_1 are noncollinear and A_2, B_2, and C_2 are noncollinear. Furthermore, let $\mathcal{A}_1 = [\overline{B_1C_1}]$, $\mathcal{B}_1 = [\overline{A_1C_1}]$, $\mathcal{C}_1 = [\overline{A_1B_1}]$, $\mathcal{A}_2 = [\overline{B_2C_2}]$, $\mathcal{B}_2 = [\overline{A_2C_2}]$, and $\mathcal{C}_2 = [\overline{A_2B_2}]$. Then $\angle B_1A_1C_1 \cong \angle B_2A_2C_2$ and $\angle A_1B_1C_1 \cong \angle A_2B_2C_2$ iff $\mathcal{A}_1 \oplus \mathcal{A}_2 = \mathcal{B}_1 \oplus \mathcal{B}_2 = \mathcal{C}_1 \oplus \mathcal{C}_2$.*

Proof. (I: If $\angle B_1A_1C_1 \cong \angle B_2A_2C_2$ and $\angle A_1B_1C_1 \cong \angle A_2B_2C_2$, then $\mathcal{A}_1 \oplus \mathcal{A}_2 = \mathcal{B}_1 \oplus \mathcal{B}_2 = \mathcal{C}_1 \oplus \mathcal{C}_2$.) By Theorem EUC.35

$$\angle A_1C_1B_1 \cong \angle A_2C_2B_2.$$

(Case 1: $\overline{A_1B_1} \cong \overline{A_2B_2}$.) By Theorem NEUT.65 (AEA) $\overline{A_1C_1} \cong \overline{A_2C_2}$ and $\overline{B_1C_1} \cong \overline{B_2C_2}$. By Exercise FSEG.1 $\mathcal{A}_1 = \mathcal{A}_2$, $\mathcal{B}_1 = \mathcal{B}_2$, and $\mathcal{C}_1 = \mathcal{C}_2$. Hence $\mathcal{A}_1 \oplus \mathcal{A}_2 = \mathcal{B}_1 \oplus \mathcal{B}_2 = \mathcal{C}_1 \oplus \mathcal{C}_2 = \mathcal{U}$.

(Case 2: $\overline{A_1B_1} \not\cong \overline{A_2B_2}$.) Using Theorem NEUT.67 (segment construction) let B'_1 and C'_1 be the points such that $B'_1 \in \overrightarrow{A_1B_1}$, $C'_1 \in \overrightarrow{A_1C_1}$, $\overline{A_1B'_1} \cong \overline{A_2B_2}$ and $\overline{A_1C'_1} \cong \overline{A_2C_2}$. By Theorem NEUT.64 (EAE), $\angle A_1B'_1C'_1 \cong \angle A_2B_2C_2$. Since $\angle A_1B_1C_1 \cong \angle A_2B_2C_2$, by Theorem NEUT.14 (congruence is an equivalence relation), $\angle A_1B'_1C'_1 \cong \angle A_1B_1C_1$. By Theorem EUC.11 $\overleftrightarrow{B'_1C'_1} \parallel \overleftrightarrow{B_1C_1}$. By Theorem SIM.13 $\mathcal{B}_1 \oplus \mathcal{B}_2 = \mathcal{C}_1 \oplus \mathcal{C}_2$. Using Theorem NEUT.67 (segment construction), let A'_1 and C''_1 be the points such that $A'_1 \in \overrightarrow{B_1A_1}$, $C''_1 \in \overrightarrow{B_1C_1}$, $\overline{B_1A'_1} \cong \overline{B_2A_2}$ and $\overline{B_1C''_1} \cong \overline{B_2C_2}$. By Theorem NEUT.64 (EAE), $\angle B_1A'_1C''_1 \cong \angle B_2A_2C_2$. Since $\angle B_1A_1C_1 \cong \angle B_2A_2C_2$, by Theorem NEUT.14, $\angle B_1A'_1C''_1 \cong \angle B_1A_1C_1$. By Theorem EUC.11 $\overleftrightarrow{A'_1C''_2} \parallel \overleftrightarrow{A_1C_1}$. By Theorem SIM.11 $\mathcal{A}_1 \oplus \mathcal{A}_2 = \mathcal{C}_1 \oplus \mathcal{C}_2$. Hence $\mathcal{A}_1 \oplus \mathcal{A}_2 = \mathcal{B}_1 \oplus \mathcal{B}_2 = \mathcal{C}_1 \oplus \mathcal{C}_2$.

(II: If $\mathcal{A}_1 \oplus \mathcal{A}_2 = \mathcal{B}_1 \oplus \mathcal{B}_2 = \mathcal{C}_1 \oplus \mathcal{C}_2$, then $\angle B_1A_1C_1 \cong \angle B_2A_2C_2$ and $\angle A_1B_1C_1 \cong \angle A_2B_2C_2$.) Using Theorem NEUT.67 (segment construction) let B'_1 be the point on $\overrightarrow{C_1B_1}$ such that $\overline{C_1B'_1} \cong \overline{C_2B_2} \in \mathcal{A}_2$. By Exercise FSEG.1 $\mathcal{A}_2 = [\overline{C_1B'_1}]$. Using Axiom PS, let $\mathcal{J} = \text{par}\,(B'_1, \overleftrightarrow{A_1B_1})$. By Exercise IP.4 $\mathcal{J}$ and $\overleftrightarrow{A_1C_1}$ intersect at a point A'_1. By Theorem EUC.11 $\angle C_1B'_1A'_1 \cong \angle C_1B_1A_1$ and $\angle C_1A'_1B'_1 \cong \angle C_1A_1B_1$. By part (I)

$$\mathcal{A}_1 \oplus [\overline{C_1B'_1}] = \mathcal{B}_1 \oplus [\overline{C_1A'_1}] = \mathcal{C}_1 \oplus [\overline{A'_1B'_1}].$$

By assumption

$$\mathcal{A}_1 \oplus \mathcal{A}_2 = \mathcal{B}_1 \oplus \mathcal{B}_2 = \mathcal{C}_1 \oplus \mathcal{C}_2,$$

and we have already seen that $[\overline{C_1B'_1}] = \mathcal{A}_2$. Therefore

$$\mathcal{B}_1 \oplus [\overline{C_1A'_1}] = \mathcal{A}_1 \oplus [\overline{C_1B'_1}] = \mathcal{A}_1 \oplus \mathcal{A}_2 = \mathcal{B}_1 \oplus \mathcal{B}_2$$

so that $\mathcal{B}_1 \odot \mathcal{B}_2 = \mathcal{B}_1 \odot [\overline{C_1A'_1}]$ and hence $[\overline{C_1A'_1}] = \mathcal{B}_2$. A similar calculation shows that $\mathcal{C}_1 \oplus [\overline{A'_1B'_1}] = \mathcal{C}_1 \oplus \mathcal{C}_2$ and $[\overline{A'_1B'_1}] = \mathcal{C}_2$.

By Exercise FSEG.1 $\overline{C_1A'_1} \cong \overline{C_2A_2}$ and $\overline{A'_1B'_1} \cong \overline{A_2B_2}$. By Theorem NEUT.62 (EEE), $\triangle A'_1B'_1C_1 \cong \triangle A_2B_2C_2$, $\angle B'_1A'_1C_1 \cong \angle B_2A_2C_2$, and $\angle A'_1B'_1C_1 \cong \angle A_2B_2C_2$. By Theorem EUC.11 $\angle A'_1B'_1C_1 \cong \angle A_1B_1C_1$ and $\angle B'_1A'_1C_1 \cong \angle B_1A_1C_1$. By Theorem NEUT.14 (congruence is an equivalence relation), $\angle B_1A_1C_1 \cong \angle B_2A_2C_2$ and $\angle A_1B_1C_1 \cong \angle A_2B_2C_2$. □

Theorem SIM.17. *Let $\mathcal{P}$ be a Euclidean plane, let A_1, B_1, C_1, A_2, B_2, and C_2 be points on $\mathcal{P}$ such that A_1, B_1, and C_1 are noncollinear, and A_2, B_2, and C_2 are noncollinear. Furthermore, let $\mathcal{B}_1 = [\overline{A_1C_1}]$, $\mathcal{B}_2 = [\overline{A_2C_2}]$, $\mathcal{C}_1 = [\overline{A_1B_1}]$, and $\mathcal{C}_2 = [\overline{A_2B_2}]$. If $\angle B_1A_1C_1 \cong \angle B_2A_2C_2$, $\mathcal{B}_1 \neq \mathcal{B}_2$, and $\mathcal{B}_1 \oplus \mathcal{B}_2 = \mathcal{C}_1 \oplus \mathcal{C}_2$, then $\angle A_1B_1C_1 \cong \angle A_2B_2C_2$ and $\angle A_1C_1B_1 \cong \angle A_2C_2B_2$.*

Proof. Using Theorem NEUT.67 (segment construction) let B_1' and C_1' be the points such that $B_1' \in \overrightarrow{A_1B_1}$, $C_1' \in \overrightarrow{A_2C_2}$, $\overline{A_1B_1'} \cong \overline{A_2B_2}$ and $\overline{A_1C_1'} \cong \overline{A_2C_2}$. By Exercise FSEG.1 $[\overline{A_1B_1'}] = \mathcal{C}_2$ and $[\overline{A_1C_1'}] = \mathcal{B}_2$. Since $\mathcal{B}_1 \neq \mathcal{B}_2$ and $\mathcal{B}_1 \oplus \mathcal{B}_2 = \mathcal{C}_1 \oplus \mathcal{C}_2$, $\mathcal{C}_1 \neq \mathcal{C}_2$. Multiplying both sides of $\mathcal{B}_1 \oplus \mathcal{B}_2 = \mathcal{C}_1 \oplus \mathcal{C}_2$ by $\mathcal{B}_2 \odot \mathcal{C}_1{}^{-1}$, (or by Exercise SIM.4) we have $\mathcal{B}_1 \oplus \mathcal{C}_1 = \mathcal{B}_2 \oplus \mathcal{C}_2$.

We may now apply Theorem SIM.13, by substituting, in the statement of that theorem, A_1 for O, C_1 for A, B_1 for B, C_1' for C, and B_1' for D. Then in the statement of Theorem SIM.13, $\mathcal{A}$ becomes $[\overline{A_1C_1}] = \mathcal{B}_1$, $\mathcal{B}$ becomes $[\overline{A_1B_1}] = \mathcal{C}_1$, $\mathcal{C}$ becomes $[\overline{A_1C_1'}] = [\overline{A_2C_2}] = \mathcal{B}_2$, and $\mathcal{D}$ becomes $[\overline{A_1B_1'}] = [\overline{A_1B_2}] = \mathcal{C}_2$. Then $\mathcal{A} \oplus \mathcal{B} = \mathcal{C} \oplus \mathcal{D}$ in Theorem SIM.13 becomes $\mathcal{B}_1 \oplus \mathcal{C}_1 = \mathcal{B}_2 \oplus \mathcal{C}_2$, which we have seen earlier to be true.

Thus, by Theorem SIM.13, $\overleftrightarrow{AB} \parallel \overleftrightarrow{CD}$, that is, $\overleftrightarrow{B_1C_1} \parallel \overleftrightarrow{B_1'C_1'}$. By Theorem EUC.11 $\angle A_1B_1'C_1' \cong \angle A_1B_1C_1$ and $\angle A_1C_1'B_1' \cong \angle A_1C_1B_1$. By Theorem NEUT.64 (EAE) $\angle A_1B_1'C_1' \cong \angle A_2B_2C_2$ and $\angle A_1C_1'B_1' \cong \angle A_2C_2B_2$. By Theorem NEUT.14 (congruence is an equivalence relation) $\angle A_1B_1C_1 \cong \angle A_2B_2C_2$ and $\angle A_1C_1B_1 \cong \angle A_2C_2B_2$. □

Theorem SIM.18. *Let $\mathcal{P}$ be a Euclidean plane and let $\mathcal{T}_1$ and $\mathcal{T}_2$ be triangles on $\mathcal{P}$. Then all the following statements are equivalent:*

(I) $\mathcal{T}_1 \sim \mathcal{T}_2$

(II) *There exists a matching of the angles of $\mathcal{T}_1$ with the angles of $\mathcal{T}_2$ such that pairs of matched angles are congruent to each other.*

(III) *There exists a matching of the edges of $\mathcal{T}_1$ with the edges of $\mathcal{T}_2$ such that the ratios of matched edges are equal.*

(IV) *There exists a matching of the corners A_1, B_1, C_1 of $\mathcal{T}_1$ with the corners A_2, B_2, C_2 of $\mathcal{T}_2$ such that $\angle B_1A_1C_1 \cong \angle B_2A_2C_2$ and*

$$[\overline{A_1B_1}] \oplus [\overline{A_2B_2}] = [\overline{A_1C_1}] \oplus [\overline{A_2C_2}].$$

(V) *There exists a matching of the corners A_1, B_1, C_1 of $\mathcal{T}_1$ with the corners A_2, B_2, C_2 of $\mathcal{T}_2$ such that $\angle B_1A_1C_1 \cong \angle B_2A_2C_2$ and $\angle A_1B_1C_1 \cong \angle A_2B_2C_2$.*

Proof. By Theorem SIM.5 part (A) and Theorem SIM.6 statements (I) and (II) are equivalent.

By Theorem SIM.16 statements (II) and (III) are equivalent. By Theorems SIM.17 and SIM.16 statements (II) and (IV) are equivalent. Therefore statements (I), (II), (III), and (IV) are equivalent. By Theorem EUC.35, statement (V) implies statement (II), which implies statement (V). □

Theorem SIM.19. *Let $\mathcal{P}$ be a Euclidean plane and let O, A, B, C, and D be points on $\mathcal{P}$ such that O–A–C, $\overleftrightarrow{BD}$ and $\overleftrightarrow{AC}$ are concurrent at O, and $\overleftrightarrow{AB} \parallel \overleftrightarrow{CD}$. Then $[\overline{OA}] \oplus [\overline{OB}] = [\overline{OC}] \oplus [\overline{OD}]$ and $[\overline{OA}] \oplus [\overline{OB}] = [\overline{AC}] \oplus [\overline{BD}]$.*

Proof. (Case 1: O–A–C.) See Figure 15.1. Let $\mathcal{A} = [\overline{OA}]$, $\mathcal{B} = [\overline{OB}]$, $\mathcal{C} = [\overline{OC}]$, and $\mathcal{D} = [\overline{OD}]$. By Exercise PSH.56 O–B–D. By Definition FSEG.2 $\mathcal{A} < \mathcal{C}$ and $\mathcal{B} < \mathcal{D}$. By Definition FSEG.11 $\mathcal{C} \ominus \mathcal{A} = [\overline{AC}]$ and $\mathcal{D} \ominus \mathcal{B} = [\overline{BD}]$. By Theorem SIM.13 $\mathcal{A} \oplus \mathcal{B} = \mathcal{C} \oplus \mathcal{D}$ so that $[\overline{OA}] \oplus [\overline{OB}] = [\overline{OC}] \oplus [\overline{OD}]$. By Exercise SIM.5 $\mathcal{A} \oplus \mathcal{B} = (\mathcal{C} \ominus \mathcal{A}) \oplus (\mathcal{D} \ominus \mathcal{B})$, so that $[\overline{OA}] \oplus [\overline{OB}] = [\overline{AC}] \oplus [\overline{BD}]$.

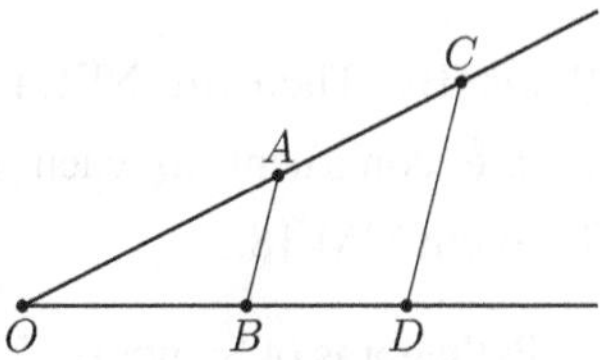

Fig. 15.1 For Theorem SIM.19 Case 1.

(Case 2: A–O–C.) By Exercise PSH.56 B–O–D. By Theorem SIM.13 $[\overline{OA}] \oplus [\overline{OB}] = [\overline{OC}] \oplus [\overline{OD}]$.

Let $\mathbb{L} = \text{par}\,(D, (\overleftrightarrow{OC}))$, and let B' be the point such that $\{B'\} = \mathbb{L} \cap \overleftrightarrow{AB}$. Then $\square ACDB'$ is a parallelogram, and therefore by Theorem EUC.12(A) $\overline{AC} \cong \overline{B'D}$. By Theorem SIM.18(II), $\triangle BDB' \sim \triangle DOC$, because $\angle BDB' \cong \angle DOC$, $\angle DB'B \cong \angle OCD$, and $\angle DBB' \cong \angle ODC$, all of these being true from Theorem EUC.11. Again, by Theorem SIM.18(III), $[\overline{B'D}] \oplus [\overline{OC}] = [\overline{BD}] \oplus [\overline{OD}]$, so that by Exercise SIM.4 $[\overline{B'D}] \oplus [\overline{BD}] = [\overline{OC}] \oplus [\overline{OD}]$; since we already know that $\overline{AC} \cong \overline{B'D}$,

$$[\overline{AC}] \oplus [\overline{BD}] = [\overline{B'D}] \oplus [\overline{BD}] = [\overline{OC}] \oplus [\overline{OD}].$$

By an argument similar to that above, $\triangle OAB \sim \triangle OCD$ and by Theorem SIM.18(III) and Exercise SIM.4, $[\overline{OC}] \oplus [\overline{OD}] = [\overline{OA}] \oplus [\overline{OB}]$ so that

$$[\overline{AC}] \oplus [\overline{BD}] = [\overline{OC}] \oplus [\overline{OD}] = [\overline{OA}] \oplus [\overline{OB}],$$

which was to be proved. □

Corollary SIM.20. *With the assumptions and notation of Theorem SIM.19*

$$\frac{[\overline{OA}]}{[\overline{OC}]} = \frac{[\overline{OB}]}{[\overline{OD}]} \text{ and } \frac{[\overline{OA}]}{[\overline{AC}]} = \frac{[\overline{OB}]}{[\overline{BD}]},$$

where $\dfrac{[\overline{OA}]}{[\overline{OC}]}$ is equal to $[\overline{OA}] \oplus [\overline{OC}]$ and $\dfrac{[\overline{OB}]}{[\overline{OD}]}$ is equal to $[\overline{OB}] \oplus [\overline{OD}]$.

Proof. Follows immediately from Exercise SIM.4. □

Corollary SIM.21. *With the assumptions and notation of Theorem SIM.19, if A is the midpoint of* $\overline{OC}$, *then B is the midpoint of* $\overline{OD}$.

Proof. If A is the midpoint of $\overline{OC}$, then by Definition NEUT.3(C), $\overline{OA} \cong \overline{AC}$ and O–A–C. Then if u is the unit free segment, $\dfrac{[\overline{OA}]}{[\overline{AC}]} = u = \dfrac{[\overline{OB}]}{[\overline{BD}]}$; so that $\overline{OB} \cong \overline{BD}$; by Exercise PSH.56, O–B–D, so B is the midpoint of $\overline{OD}$. □

Theorem SIM.22. *Let* $\mathcal{P}$ *be a Euclidean plane and let* $\mathcal{T}_1$ *and* $\mathcal{T}_2$ *be right triangles on* $\mathcal{P}$. *If an acute angle of* $\mathcal{T}_1$ *and an acute angle of* $\mathcal{T}_2$ *are congruent to each other, then* $\mathcal{T}_1 \sim \mathcal{T}_2$.

Proof. By Theorem NEUT.69 the right angle of $\mathcal{T}_1$ and the right angle of $\mathcal{T}_2$ are congruent to each other. Hence $\mathcal{T}_1 \sim \mathcal{T}_2$ by Theorem SIM.16 and Theorem SIM.18. □

Pythagoras of Samos (c. 570–495 BC), for whom the following theorem is named (although it is disputed whether he had anything at all to do with it), was an Ionian Greek philosopher and mathematician. He also was the founder of a religious sect called Pythagoreanism. In 1955 the town of Tigani, located on the south side of the island of Samos, was renamed Pithagoreio in his honor. On the jetty extending into its harbor there is a statue of Pythagoras "illustrating" this theorem.

Theorem SIM.23 (Pythagorean Theorem). *Let A, B, and C be noncollinear points on the Euclidean plane* $\mathcal{P}$, $\mathcal{A} = [\overline{BC}]$, $\mathcal{B} = [\overline{AC}]$, *and* $\mathcal{C} = [\overline{AB}]$. *Then* $\angle ACB$ *is right iff* $\mathcal{A}^2 \oplus \mathcal{B}^2 = \mathcal{C}^2$.

Proof. (I: If $\angle ACB$ is right, then $\mathcal{A}^2 \oplus \mathcal{B}^2 = \mathcal{C}^2$.) Let $\mathbb{L} = \mathrm{pr}\,(C, \overleftrightarrow{AB})$ and let $D = \mathrm{ftpr}\,(C, \overleftrightarrow{AB})$. By Exercise NEUT.20 $D \in \overline{AB}$. Let $\mathcal{T} = [\overline{AD}]$ and $\mathcal{S} = [\overline{BD}]$. Then by Definition FSEG.3 $\mathcal{S} \oplus \mathcal{T} = \mathcal{C}$. By Theorem SIM.22, since $\triangle ABC$ and $\triangle ACD$ have $\angle BAC$ in common, they are similar, and since $\triangle ABC$ and $\triangle CBD$ have $\angle ABC$ in common, they are similar. By Theorem SIM.16 $\mathcal{A} \oslash \mathcal{C} = \mathcal{S} \oslash \mathcal{A}$ and $\mathcal{B} \oslash \mathcal{C} = \mathcal{T} \oslash \mathcal{B}$. Hence $\mathcal{A}^2 = \mathcal{C} \odot \mathcal{S}$ and $\mathcal{B}^2 = \mathcal{C} \odot \mathcal{T}$. By Theorem SIM.8 $\mathcal{A}^2 \oplus \mathcal{B}^2 = \mathcal{C} \odot \mathcal{S} \oplus \mathcal{C} \odot \mathcal{T} = \mathcal{C} \odot (\mathcal{S} \oplus \mathcal{T}) = \mathcal{C}^2$.

(II: If $\mathcal{A}^2 \oplus \mathcal{B}^2 = \mathcal{C}^2$, then $\angle ACB$ is right.) Let D, E, and F be points on $\mathcal{P}$ such that $\overline{DF} \cong \overline{AC}$ and $\overline{EF} \cong \overline{BC}$ and $\angle DFE$ is a right angle. Let $\mathcal{D} = [\overline{EF}]$, $\mathcal{E} = [\overline{DF}]$, and $\mathcal{F} = [\overline{DE}]$. By part (I), $\mathcal{D}^2 \oplus \mathcal{E}^2 = \mathcal{F}^2$.

By assumption, $\mathcal{A}^2 \oplus \mathcal{B}^2 = \mathcal{C}^2$; then $\mathcal{F}^2 = \mathcal{D}^2 \oplus \mathcal{E}^2 = \mathcal{A}^2 \oplus \mathcal{B}^2 = \mathcal{C}^2$. By Exercise SIM.2 $\mathcal{C} = \mathcal{F}$, so that by Exercise FSEG.1 $\overline{AB} \cong \overline{DE}$. By

Theorem NEUT.62 (EEE) $\triangle ABC \cong \triangle DEF$, and therefore $\angle ACB \cong \angle DFE$. Since $\angle DFE$ is right, by Theorem NEUT.66 $\angle ACB$ is right. □

Theorem SIM.23.1 (Second form of the Pythagorean Theorem). *Let A, B, and C be noncollinear points on the Euclidean plane $\mathcal{P}$, Let $\hat{A} = \Phi[\overline{BC}]$, $\hat{B} = \Phi[\overline{AC}]$, and $\hat{C} = \Phi[\overline{AB}]$ be the points of $\overrightarrow{OU}$ such that $[\overline{BC}] = [\overline{O\hat{A}}]$, $[\overline{AC}] = [\overline{O\hat{B}}]$, and $[\overline{AB}] = [\overline{O\hat{C}}]$. Then $\angle ACB$ is right iff $\hat{A}^2 \oplus \hat{B}^2 = \hat{C}^2$.*

Proof. By Theorem SIM.23 $\angle ACB$ is right iff $[\overline{O\hat{A}}]^2 \oplus [\overline{O\hat{B}}]^2 = [\overline{O\hat{C}}]^2$. By Definition SIM.7 this is $[\overline{O(\hat{A}^2)}] \oplus [\overline{O(\hat{B}^2)}] = [\overline{O(\hat{C}^2)}]$. By Theorem OF.17 this is $[\overline{O(\hat{A}^2 \oplus \hat{B}^2)}] = [\overline{O(\hat{C}^2)}]$. By Property R.4 of Definition NEUT.2, since $\hat{A}$, $\hat{B}$, and $\hat{C}$ are all in $\overrightarrow{OU}$, this is true iff $\hat{A}^2 \oplus \hat{B}^2 = \hat{C}^2$. □

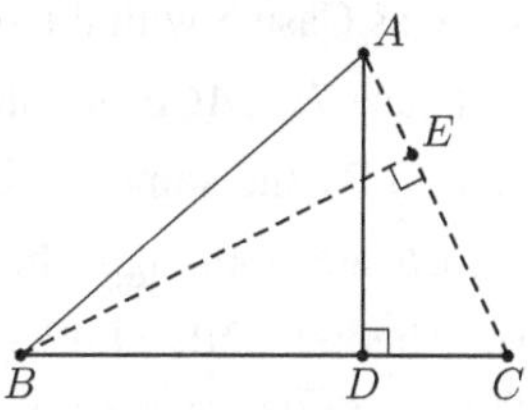

Fig. 15.2 For Theorem SIM.24: the product of any altitude and its base is constant.

Theorem SIM.24 (Product of base and altitude). *Let $\mathcal{P}$ be a Euclidean plane, A, B, and C be noncollinear points on $\mathcal{P}$, and $\overline{AD}$, $\overline{BE}$, and $\overline{CF}$ be the altitudes (see Definition NEUT.99) of $\triangle ABC$, respectively, from A, B, and C. Then $[\overline{AD}] \odot [\overline{BC}] = [\overline{BE}] \odot [\overline{AC}]$.*

Proof. See Figure 15.2.

(Case 1: $\angle ACB$ is right.) By Theorem NEUT.44 $\overleftrightarrow{AC} \perp \overleftrightarrow{BC}$. Thus $D = E = C$. Therefore $[\overline{AD}] \odot [\overline{BC}] = [\overline{AC}] \odot [\overline{BC}]$ and $[\overline{BE}] \odot [\overline{AC}] = [\overline{AC}] \odot [\overline{BC}]$ and so $[\overline{AD}] \odot [\overline{BC}] = [\overline{BE}] \odot [\overline{AC}]$.

(Case 2: $\angle ACB$ is acute and $\angle ABC$ is right.) By Exercise NEUT.20 and Definition IB.3 A–E–C and by Theorem NEUT.48(A) $B = D$. Then $\angle ABC \cong \angle BEC$ are both right angles, and $\angle ACB$ is an angle of both $\triangle BEC$ and $\triangle ADC = \triangle ABC$. By Theorem SIM.16

$$[\overline{BE}] \oslash [\overline{AB}] = [\overline{BE}] \oslash [\overline{AD}] = [\overline{BC}] \oslash [\overline{AC}],$$

so that $[\overline{BC}] \odot [\overline{AD}] = [\overline{BE}] \odot [\overline{AC}]$.

(Case 3: $\angle ACB$ is right and $\angle ABC$ is acute.) The proof is the same as Case 2, with the roles of A and B interchanged.

(Case 4: Each angle of $\triangle ABC$ is acute.) By Exercise NEUT.20 and Definition IB.3 B–D–C and A–E–C. Since $\triangle BEC$ and $\triangle ADC$ are right and since $\angle BCE = \angle ACD$, by Theorem SIM.16 $[\overline{BE}] \oplus [\overline{AD}] = [\overline{BC}] \oplus [\overline{AC}]$, so that $[\overline{BC}] \odot [\overline{AD}] = [\overline{BE}] \odot [\overline{AC}]$.

(Case 5: $\angle ACB$ is acute, $\angle BAC$ is obtuse, and $\angle ABC$ is acute.) By Exercise NEUT.20 B–D–C. Let C' be a point of $\overrightarrow{AC}$ such that C'–A–C. Since $\angle BAC$ is obtuse, $\angle BAC'$ is acute, by Theorem NEUT.82. Now $E = \text{ftpr}\,(B, \overleftrightarrow{AC})$ and by Exercise NEUT.15, since $B \in \overrightarrow{AB}$, $E \in \overrightarrow{AC'}$ and hence C–A–E.

Since $\triangle BEC$ and $\triangle ADC$ are right and since $\angle ACB$ is common to them, they are similar to each other. By Theorem SIM.16 $[\overline{BC}] \oplus [\overline{AC}] = [\overline{BE}] \oplus [\overline{AD}]$. Thus $[\overline{AD}] \odot [\overline{BC}] = [\overline{BE}] \odot [\overline{AC}]$.

(Case 6: $\angle ACB$ is acute, $\angle ABC$ is obtuse, and $\angle BAC$ is acute.) The proof is the same as Case 5 with the roles of A and B interchanged.

(Case 7: $\angle ACB$ is obtuse.) By Theorem NEUT.84, both $\angle BAC$ and $\angle ABC$ are acute.) By the same reasoning as in Case 5, both B–C–D and A–C–E. $\angle ADC$ and $\angle CEB$ are both right, hence congruent by Theorem NEUT.69. $\angle ACD \cong \angle BCE$ by Theorem NEUT.42 (vertical angles). By Theorem SIM.16 $[\overline{BC}] \oplus [\overline{AC}] = [\overline{BE}] \oplus [\overline{AD}]$. Thus $[\overline{AD}] \odot [\overline{BC}] = [\overline{BE}] \odot [\overline{AC}]$. □

Remark SIM.25. The standard definition of area of a triangle is "$1/2$ the product of the altitude and the base," where the base is the length of the edge which is a subset of the line perpendicular to the altitude. Theorem SIM.24 shows that such a definition is a "good" definition. Our only problem is the "$1/2$"—to which we so far have not given meaning. We will develop the topic of rational multiples of segments and of points on the line in Chapter 17, and then will state the definition for area of a triangle.

15.2 Exercises for similarity

Answers to starred (*) exercises may be accessed from the home page for this book at www.springer.com.

Exercise SIM.1*. Let $\mathcal{P}$ be a Euclidean plane and let $\mathcal{A}$ and $\mathcal{B}$ be free segments of $\mathcal{P}$.

(I) If $\mathcal{A} < \mathcal{B}$, then $\mathcal{A}^2 < \mathcal{B}^2$.
(II) If $\mathcal{A} > \mathcal{B}$, then $\mathcal{A}^2 > \mathcal{B}^2$.

Exercise SIM.2*. Let $\mathcal{P}$ be a Euclidean plane and let $\mathcal{A}$ and $\mathcal{B}$ be free segments of $\mathcal{P}$. If $\mathcal{A}^2 = \mathcal{B}^2$, then $\mathcal{A} = \mathcal{B}$.

Exercise SIM.3*. Let $\mathcal{P}$ be a Euclidean plane and let $\mathcal{A}$, $\mathcal{B}$, and $\mathcal{C}$ be free segments of $\mathcal{P}$ such that $\mathcal{C} < \mathcal{B}$. Then $\mathcal{A} \odot (\mathcal{B} \ominus \mathcal{C}) = (\mathcal{A} \odot \mathcal{B}) \ominus (\mathcal{A} \odot \mathcal{C})$.

Exercise SIM.4*. Let $\mathcal{P}$ be a Euclidean plane and let $\mathcal{A}$, $\mathcal{B}$, $\mathcal{C}$, and $\mathcal{D}$ be free segments on $\mathcal{P}$. Then the following statements are equivalent to each other.

(1) $\mathcal{A} \odot \mathcal{D} = \mathcal{B} \odot \mathcal{C}$.
(2) $\mathcal{A} \odiv \mathcal{B} = \mathcal{C} \odiv \mathcal{D}$.
(3) $\mathcal{A} \odiv \mathcal{C} = \mathcal{B} \odiv \mathcal{D}$.
(4) $\mathcal{B} \odiv \mathcal{A} = \mathcal{D} \odiv \mathcal{C}$.
(5) $(\mathcal{A} \oplus \mathcal{B}) \odiv \mathcal{B} = (\mathcal{C} \oplus \mathcal{D}) \odiv \mathcal{D}$.

Exercise SIM.5*. Let $\mathcal{P}$ be a Euclidean plane and let $\mathcal{A}$, $\mathcal{B}$, $\mathcal{C}$, and $\mathcal{D}$ be free segments on $\mathcal{P}$ such that $\mathcal{A} < \mathcal{C}$, $\mathcal{B} < \mathcal{D}$, and $\mathcal{A} \odiv \mathcal{B} = \mathcal{C} \odiv \mathcal{D}$. Then $\mathcal{A} \odiv \mathcal{B} = (\mathcal{C} \ominus \mathcal{A}) \odiv (\mathcal{D} \ominus \mathcal{B})$.

Exercise SIM.6*. Let $\mathcal{P}$ be a Euclidean plane and let A_1, B_1, C_1, A_2, B_2, and C_2 be points on $\mathcal{P}$ such that A_1, B_1, and C_1 are noncollinear and A_2, B_2, and C_2 are noncollinear. Furthermore, let $\mathcal{A}_1 = [\overline{B_1C_1}]$, $\mathcal{B}_1 = [\overline{A_1C_1}]$, $\mathcal{C}_1 = [\overline{A_1B_1}]$, $\mathcal{A}_2 = [\overline{B_2C_2}]$, $\mathcal{B}_2 = [\overline{A_2C_2}]$, and $\mathcal{C}_2 = [\overline{A_2B_2}]$. Then: $\angle B_1A_1C_1 \cong \angle B_2A_2C_2$ and $\angle C_1B_1A_1 \cong \angle C_2B_2A_2$ iff $\mathcal{A}_1 \odiv \mathcal{B}_1 = \mathcal{A}_2 \odiv \mathcal{B}_2$, $\mathcal{A}_1 \odiv \mathcal{C}_1 = \mathcal{A}_2 \odiv \mathcal{C}_2$, and $\mathcal{B}_1 \odiv \mathcal{C}_1 = \mathcal{B}_2 \odiv \mathcal{C}_2$.

Exercise SIM.7*. Let $\mathcal{P}$ be a Euclidean plane, O be a point on $\mathcal{P}$, δ be a dilation of $\mathcal{P}$ with fixed point O, and X and Y be distinct members of $\mathcal{P} \setminus \{O\}$ such that O, X, and Y are collinear. Then $[\overline{OX}] \odiv [\overline{O\delta(X)}] = [\overline{OY}] \odiv [\overline{O\delta(Y)}]$.

Chapter 16
Axial Affinities of a Euclidean Plane (AX)

Acronym: *AX*

Dependencies: *all prior Chapters 1 through 15*

New Axioms: *none*

New Terms Defined: *projection map, stretch, shear*

Abstract: The main results of this short chapter are Theorems AX.3 and AX.4. The first of these shows that every axial affinity (defined in Chapter 3, Definition CAP.25) on a Euclidean plane is either a stretch or a shear; the second proves that every axial affinity is a belineation.

16.1 Theorems for axial affinities

We begin by defining stretches and shears, temporarily reverting back to the use of script letters $\mathcal{L}$ and $\mathcal{M}$ for lines.

Definition AX.0. (A) A **stretch** φ of a plane $\mathcal{P}$ is an axial affinity of $\mathcal{P}$ with axis $\mathcal{M}$ such that there exists a line $\mathbb{L}$ on $\mathcal{P}$ which is a fixed line for φ, is not parallel to $\mathcal{M}$, and the set of fixed lines of φ is $\{\mathcal{M}\} \cup \{\mathcal{J} \mid \mathcal{J} \text{ PE } \mathcal{L}\}$.

It is easy to see that a *reflection* is a *stretch*. In Theorem AX.1 we will prove the existence of stretches other than reflections.

(B) A **shear** ψ of the affine plane $\mathcal{P}$ is an axial affinity of $\mathcal{P}$ with axis $\mathcal{M}$ such that the set of fixed lines of ψ is $\{\mathcal{J} \mid \mathcal{J} \text{ PE } \mathcal{M}\}$.

E.J. Specht et al., *Euclidean Geometry and its Subgeometries*,
DOI 10.1007/978-3-319-23775-6_16

Theorem AX.1. *Let $\mathcal{P}$ be a Euclidean plane, $\mathcal{M}$ be a line on $\mathcal{P}$, and A and B be distinct members of $\mathcal{P}\setminus\mathcal{M}$ such that $\overleftrightarrow{AB}\cap\mathcal{M}\neq\emptyset$. Then there exists a unique axial affinity φ of $\mathcal{P}$ with axis $\mathcal{M}$, such that $\varphi(A)=B$; this axial affinity is also a stretch.*

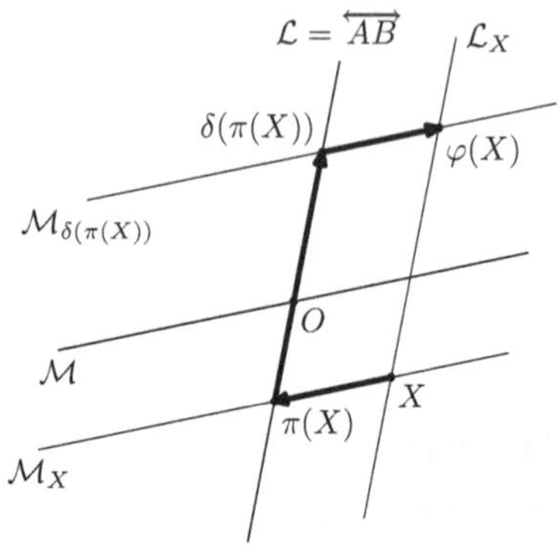

Fig. 16.1 For Theorem AX.1: showing action of a stretch φ, where A–O–$\varphi(A)$.

Proof. For a visualization see Figure 16.1. The reader should not be put off by the length of this proof; the bulk of the proof is in parts (VII) through (VII(c)), showing that the mapping φ is a collineation.

We first construct an axial affinity φ of $\mathcal{P}$ with axis $\mathcal{M}$ such that $\varphi(A)=B$.

(I: There can be no more than one such axial affinity.) This is Theorem CAP.29.

(II: Notation.) We first adopt some notation which is specific to the following construction. Let $\mathcal{L}=\overleftrightarrow{AB}$ and let O be the point such that $\{O\}=\mathcal{L}\cap\mathcal{M}$. For every $X\in\mathcal{P}$, define $\mathcal{L}_X=\text{par}(X,\mathcal{L})$ (in case $X\in\mathcal{L}$, let $\mathcal{L}_X=\mathcal{L}$), and define $\mathcal{M}_X=\text{par}(X,\mathcal{M})$ (in case $X\in\mathcal{M}$, let $\mathcal{M}_X=\mathcal{M}$). Then for every $X\in\mathcal{P}$, $\{X\}=\mathcal{L}_X\cap\mathcal{M}_X$.

Define the **projection map** π so that for each $X\in\mathcal{P}$, $\{\pi(X)\}=\mathcal{M}_X\cap\mathcal{L}$. Then for any $X'\in\mathcal{M}_X$, $\pi(X')=\pi(\mathcal{M}_X)=\pi(X)$. That is, π has the same value everywhere on each line parallel to $\mathcal{M}$.

(III: The construction.) Let δ be the unique dilation on $\mathcal{P}$ with fixed point O, such that $\delta(A)=B$. The existence and uniqueness of δ are guaranteed by Theorem DLN.7. For every $X\in\mathcal{P}$, define $\varphi(X)$ to be the point of intersection of $\mathcal{M}_{\delta(\pi(X))}$ and $\mathcal{L}_X$; that is, $\{\varphi(X)\}=\mathcal{M}_{\delta(\pi(X))}\cap\mathcal{L}_X$. Then in particular, if $X\in\mathcal{M}$, $\{\varphi(X)\}=\mathcal{M}_{\delta(\pi(X))}\cap\mathcal{L}_X=\mathcal{M}\cap\mathcal{L}_X=\{X\}$; and if $X\in\overleftrightarrow{AB}=\mathcal{L}$, then $\{\varphi(X)\}=\mathcal{M}_{\delta(X)}\cap\mathcal{L}=\{\delta(X)\}$, that is, $\varphi(X)=\delta(X)$. In particular, $\varphi(A)=\delta(A)=B$.

(IV: $\varphi(\mathcal{M}_X) = \mathcal{M}_{\delta(\pi(X))}$.) From the construction, if $X' \in \mathcal{M}_X$, then $\pi(X') = \pi(X)$ and $\delta(\pi(X')) = \delta(\pi(X))$, so that $\mathcal{M}_{\delta(\pi(X'))} = \mathcal{M}_{\delta(\pi(X))}$ and $\varphi(X') \in \mathcal{M}_{\delta(\pi(X))}$. Therefore $\varphi(\mathcal{M}_X) \subseteq \mathcal{M}_{\delta(\pi(X))}$.

Now let Y be any point of $\mathcal{M}_{\delta(\pi(X))}$. Then $\pi(Y) = \delta(\pi(X))$ and

$$\delta^{-1}(\pi(Y)) = \delta^{-1}(\delta(\pi(X))) = \pi(X).$$

Let $\{X'\} = \mathcal{M}_{\pi(X)} \cap \mathcal{L}_Y$ so that $\pi(X') = \pi(X)$. Then

$$\{\varphi(X')\} = \mathcal{M}_{\delta(\pi(X))} \cap \mathcal{L}_Y = \{Y\},$$

and $\mathcal{M}_{\delta(\pi(X))} \subseteq \varphi(\mathcal{M}_X)$.

Since we can choose X arbitrarily, this also shows that for every line $\mathcal{N}$ which is parallel to $\mathcal{M}$, $\varphi(\mathcal{N})$ is a line parallel to $\mathcal{M}$. It also shows that φ is onto $\mathcal{P}$, since Y can be chosen arbitrarily.

(V: $\mathcal{L}_X = \varphi(\mathcal{L}_X)$.) By the construction, both X and $\varphi(X)$ are members of $\mathcal{L}_X$ so that $\varphi(\mathcal{L}_X) \subseteq \mathcal{L}_X$. The argument in part (IV) shows that if $Y \in \mathcal{P}$, there exists a point $X' \in \mathcal{L}_Y$ such that $\varphi(X') = Y$. Therefore, $\mathcal{L}_X \subseteq \varphi(\mathcal{L}_X)$ and hence $\mathcal{L}_X = \varphi(\mathcal{L}_X)$. Thus every line $\mathcal{L}_X$ is a fixed line for φ.

(VI: φ is one-to-one, hence a bijection.) If X and X' are points of $\mathcal{P}$, and $\varphi(X) = \varphi(X')$, then both $X \in \mathcal{L}_X$ and $X' \in \mathcal{L}_X$, since $\mathcal{L}_X$ is a fixed line. Also, $\delta(\pi(X)) = \delta(\pi(X'))$, since $\varphi(X) = \varphi(X')$ belongs both to $\mathcal{M}_{\delta\pi(X))}$ and $\mathcal{M}_{\delta\pi(X'))}$. Since δ is one-to-one, $\pi(X) = \pi(X')$, and hence both X and X' are in the intersection $\mathcal{M}_X \cap \mathcal{L}_X$, which is a single point. Hence $X = X'$, and φ is shown to be one-to-one.

(VII: φ is a collineation.) We already know that the lines $\mathcal{M}_X$ map into other lines parallel to themselves, and that the lines $\mathcal{L}_X$ map into themselves.

Now let $\mathcal{J}$ be any line on $\mathcal{P}$ which is neither parallel to $\overleftrightarrow{AB}$ nor to $\mathcal{M}$ and let Q and R be the points such that $\mathcal{J} \cap \mathcal{M} = \{Q\}$ and $\mathcal{J} \cap \overleftrightarrow{AB} = \{R\}$. Since $\varphi(Q) = Q$, $Q \in \varphi(\mathcal{J})$. In this part we prove that $\varphi(\mathcal{J}) \subseteq \overleftrightarrow{Q\varphi(R)}$.

Let X be any member of $\mathcal{J} \setminus \{Q, R\}$. In part (V) we showed $\mathcal{L}_X$ is a fixed line of φ, so that $\varphi(X) \in \mathcal{L}_X$. Let S be the point such that $\mathcal{L}_X \cap \mathcal{M} = \{S\}$ and let Y be the point such that $\overleftrightarrow{Q\varphi(R)} \cap \mathcal{L}_X = \{Y\}$.

(VII(a): $\varphi(X)$ and Y are on the same side of $\mathcal{M}$.) Since $\varphi(R) = \delta(R)$, we may apply Exercise DLN.5(II) as follows:

if R–O–$\varphi(R)$, then $\pi(X)$–O–$\delta(\pi(X))$, hence X–S–$\varphi(X)$

by Exercise PSH.56 and the fact that $\pi(S) = O$. By similar reasoning,

if O–R–$\varphi(R)$, then O–$\pi(X)$–$\delta(\pi(X))$ hence S–X–$\varphi(X)$; and

if R–$\varphi(R)$–O, then $\pi(X)$–$\delta(\pi(X))$–O hence X–$\varphi(X)$–S.

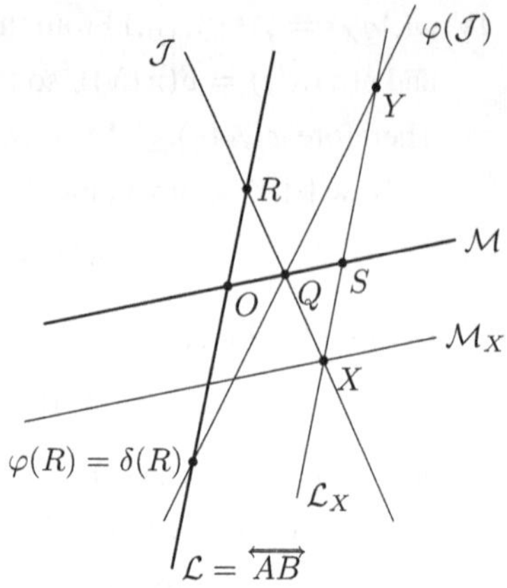

Fig. 16.2 For the proof of Theorem AX.1, part (VII(a)) Case 1, where R and $\varphi(R)$ are on opposite sides of $\mathcal{M}$.

Using Definition IB.11 we see that if R and $\varphi(R)$ are on opposite sides of $\mathcal{M}$, then X and $\varphi(X)$ are on opposite sides of $\mathcal{M}$. Using Definition IB.4 and Theorem IB.14 we see that if R and $\varphi(R)$ are on the same side of $\mathcal{M}$, then X and $\varphi(X)$ are on the same side of $\mathcal{M}$. Since O, Q, and S are distinct points on $\mathcal{M}$, by Property B.2 of Definition IB.1 there are three cases.

(Case 1: O–Q–S.) For a visualization of this case see Figure 16.2. By Exercise PSH.56 R–Q–X and $\varphi(R)$–Q–Y. By Definition IB.11 R and X are on opposite sides of $\mathcal{M}$ and $\varphi(R)$ and Y are on opposite sides of $\mathcal{M}$. Thus by Theorem PSH.12 (plane separation), if R and $\varphi(R)$ are on the same side of $\mathcal{M}$, then X and Y are on the same side of $\mathcal{M}$, and from the argument above, X and $\varphi(X)$ are on the same side, so that Y and $\varphi(X)$ are on the same side of $\mathcal{M}$.

On the other hand, if R and $\varphi(R)$ are on opposite sides of $\mathcal{M}$, then X and Y are on opposite sides of $\mathcal{M}$; from the argument above, X and $\varphi(X)$ are on opposite sides of $\mathcal{M}$, so that again $\varphi(X)$ and Y are on the same side of $\mathcal{M}$. By Theorem PSH.38(A) $\varphi(X) \in \overrightarrow{SY}$.

(Case 2: O–S–Q.) By Exercise PSH.56 R–X–Q and $\varphi(R)$–Y–Q. By Definition IB.4 and Theorem IB.14 R and X are on the same side of $\mathcal{M}$ and Y and $\varphi(R)$ are on the same side of $\mathcal{M}$. Using Theorem PSH.12 (plane separation), we see that if R and $\varphi(R)$ are on the same side of $\mathcal{M}$, then X and Y are on the same side of $\mathcal{M}$, whereas if R and $\varphi(R)$ are on opposite sides of $\mathcal{M}$, then X and Y are on opposite sides of $\mathcal{M}$. Reasoning as in Case 2, either way Y and $\varphi(X)$ are on the same side of $\mathcal{M}$ and by Theorem PSH.38(A) $\varphi(X) \in \overrightarrow{SY}$.

(Case 3: S–O–Q.) By Exercise PSH.56 X–R–Q and Y–$\varphi(R)$–Q. By Definition IB.4 and Theorem IB.14 R and X are on the same side of $\mathcal{M}$ and $\varphi(R)$ and Y are on the same side of $\mathcal{M}$. Using Theorem PSH.12 (plane separation) we see that if R and $\varphi(R)$ are on the same side of $\mathcal{M}$, then X and Y are on the same side of $\mathcal{M}$, whereas if R and $\varphi(R)$ are on opposite sides of $\mathcal{M}$, then X and Y are on opposite sides of $\mathcal{M}$. Either way Y and $\varphi(X)$ are on the same side of $\mathcal{M}$ and by Theorem PSH.38(A) $\varphi(X) \in \overrightarrow{SY}$.

(VII(b): $\varphi(X) = Y$.) In all of Cases 1 through 3 in part (VII(a)) above, by Theorem EUC.11 $\angle QXS \cong \angle QRO$ and $\angle QYS \cong \angle Q\varphi(R)O$. Since $\angle RQO = \angle XQS$ and $\angle\varphi(R)QO = \angle YQS$, by Theorem SIM.6 $\triangle XQS \sim \triangle RQO$ and $\triangle YQS \sim \triangle\varphi(R)QO$. By Theorem SIM.16

$$[\overline{SX}] \oplus [\overline{OR}] = [\overline{QS}] \oplus [\overline{QO}] \text{ and } [\overline{SY}] \oplus [\overline{O\varphi(R)}] = [\overline{QS}] \oplus [\overline{QO}].$$

We may restate these equalities as

$$\frac{[\overline{SX}]}{[\overline{OR}]} = \frac{[\overline{QS}]}{[\overline{QO}]} \text{ and } \frac{[\overline{SY}]}{[\overline{O\varphi(R)}]} = \frac{[\overline{QS}]}{[\overline{QO}]}.$$

From these, by Exercise SIM.4 we have

$$\frac{[\overline{QO}]}{[\overline{QS}]} = \frac{[\overline{OR}]}{[\overline{SX}]} \text{ and } \frac{[\overline{QO}]}{[\overline{QS}]} = \frac{[\overline{O\varphi(R)}]}{[\overline{SY}]}.$$

Therefore

$$\frac{[\overline{OR}]}{[\overline{SX}]} = \frac{[\overline{O\varphi(R)}]}{[\overline{SY}]},$$

that is,

$$\frac{[\overline{SY}]}{[\overline{SX}]} = \frac{[\overline{O\varphi(R)}]}{[\overline{OR}]}.$$

Since $\varphi(R) = \delta(R)$ and $\pi(X)$ is collinear with O and R, by Exercise SIM.7

$$\frac{[\overline{O\delta(\pi(X))}]}{[\overline{O\pi(X)}]} = \frac{[\overline{O\varphi(R)}]}{[\overline{OR}]}.$$

The quadrilaterals $\square OSX\pi(X)$ and $\square OS\varphi(X)\delta(\pi(X))$ are parallelograms, so by Theorem EUC.12(A)

$$\overline{O\pi(X)} \cong \overline{SX} \text{ and } \overline{O\delta(\pi(X))} \cong \overline{S\varphi(X)}.$$

Thus

$$\frac{[\overline{S\varphi(X)}]}{[\overline{SX}]} = \frac{[\overline{O\delta(\pi(X))}]}{[\overline{O\pi(X)}]} = \frac{[\overline{O\varphi(R)}]}{[\overline{OR}]} = \frac{[\overline{SY}]}{[\overline{SX}]}$$

and multiplying both sides by $[\overline{SX}]$ we have $[\overline{S\varphi(X)}] = [\overline{SY}]$, that is, $\overline{S\varphi(X)} \cong \overline{SY}$. In all three cases above, $\varphi(X) \in \overrightarrow{SY}$, so by Property R.4 of Definition NEUT.2, $\varphi(X) = Y$.

This completes the proof that $\varphi(\mathcal{J}) \subseteq \overleftrightarrow{Q\varphi(R)}$.

(VII(c): $\varphi(\mathcal{J}) = \overleftrightarrow{Q\varphi(R)}$.) Let Z be any member of $\overleftrightarrow{Q\varphi(R)} \setminus \{Q, \varphi(R)\}$, and let W be the point such that $\{W\} = \mathcal{J} \cap \mathcal{L}_Z$. Since by part (VII) $\varphi(W) \in \overleftrightarrow{Q\varphi(R)}$ and also $\varphi(W) \in \mathcal{L}_Z$, by Exercise I.1 $\varphi(W) = Z$. Therefore $\overleftrightarrow{Q\varphi(R)} \subseteq \varphi(\mathcal{J})$ and hence $\varphi(\mathcal{J}) = \overleftrightarrow{Q\varphi(R)}$.

Summarizing, parts (VII) through (VII(c)) show that φ is a collineation; by the construction in part (III), all points of $\mathcal{M}$ are fixed points for φ and $\varphi(A) = B$; thus $\varphi \neq \iota$, since $A \neq B$. This proves that φ is an axial affinity, as defined in Definition CAP.25.

It remains only to show that φ is a stretch. By part (V) all the lines $\mathcal{L}_X$, where $X \in \mathcal{P}$ are fixed lines for φ. If $X \notin \mathcal{M}$ then $\mathcal{M}_X$ is not a fixed line, because δ has only O as a fixed point. If $\mathcal{J} = \overleftrightarrow{QR}$ is any line not parallel to either $\mathcal{L}$ or $\mathcal{M}$ (here we are using the notation of part (II) and the construction of part (VII)), $\varphi(\mathcal{J})$ contains the point $\varphi(R)$ which is not in $\mathcal{J}$, so $\mathcal{J}$ is not a fixed line. Hence the set of fixed lines for φ is $\{\mathcal{L}_X \mid X \in \mathcal{P}\} \cup \{M\}$, as required by Definition AX.0, and φ is a stretch. □

Theorem AX.2. *Let $\mathcal{P}$ be a Euclidean plane, $\mathcal{M}$ be a line on $\mathcal{P}$, and A and B be distinct points such that $\overleftrightarrow{AB} \parallel \mathcal{M}$. Then there exists a shear ψ of $\mathcal{P}$ with axis $\mathcal{M}$ such that $\psi(A) = B$, and the set of fixed lines for ψ is $\{\mathcal{L}_X \mid X \in \mathcal{P}\} \cup \{M\}$, where $\mathcal{L}_X$ is the line through X parallel to $\mathcal{M}$.*

Proof. For a visualization see Figure 16.3.

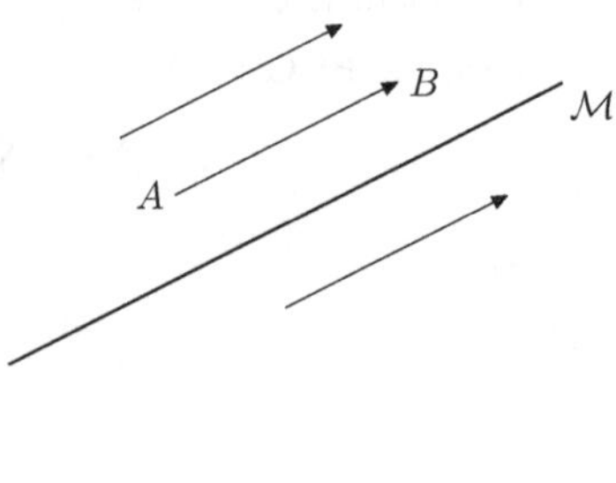

Fig. 16.3 For Theorem AX.2: showing action of a shear.

(I: Construction of ψ.) For every X belonging to $\mathcal{M}$ let $\psi(X) = X$. Let $C = \operatorname{ftpr}(A, \mathcal{M})$. For every member X of $\mathcal{P} \setminus (\mathcal{M} \cup \overleftrightarrow{AC})$, let $T = \operatorname{ftpr}(X, \mathcal{M})$, and define $\psi(X)$ as the point such that $\operatorname{par}(X, \mathcal{M}) \cap \operatorname{par}(T, \overleftrightarrow{BC}) = \{\psi(X)\}$. If $X \in \overleftrightarrow{AC} \setminus \{C\}$, then let $\psi(X)$ be the point such that $\operatorname{par}(X, \mathcal{M}) \cap \overleftrightarrow{BC} = \{\psi(X)\}$.

(II: Lines parallel to $\mathcal{M}$ are fixed lines of ψ.) Let $\mathcal{L}$ be any line which is parallel to $\mathcal{M}$ and let X be any member of $\mathcal{L}$. By the construction, $\psi(X) \in \operatorname{par}(X, \mathcal{M}) = \mathcal{L}$ so $\psi(\mathcal{L}) \subseteq \mathcal{L}$.

Let Y be any member of $\mathcal{L}$, S be the point such that $\operatorname{par}(Y, \overleftrightarrow{BC}) \cap \mathcal{M} = \{S\}$, and let X be the point such that $\operatorname{pr}(S, \mathcal{M}) \cap \mathcal{L} = \{X\}$. By the construction of part (I) $\psi(X) = Y$ so that $\mathcal{L} \subseteq \psi(\mathcal{L})$. Therefore $\psi(\mathcal{L}) = \mathcal{L}$ and $\mathcal{L}$ is a fixed line of ψ.

(III: ψ is a bijection.) Since Y can be chosen arbitrarily, the second argument of part (II) also shows that ψ is onto $\mathcal{P}$.

Now suppose X and X' are points of $\mathcal{P} \setminus \overleftrightarrow{AC}$ such that $\psi(X) = \psi(X')$; let $T = \operatorname{ftpr}(X, \mathcal{M})$ and $T' = \operatorname{ftpr}(X', \mathcal{M})$. Then $\operatorname{par}(X, \mathcal{M}) \cap \operatorname{par}(T, \overleftrightarrow{BC}) = \{\psi(X)\} = \operatorname{par}(X', \mathcal{M}) \cap \operatorname{par}(T', \overleftrightarrow{BC})$ so both $\operatorname{par}(T, \overleftrightarrow{BC})$ and $\operatorname{par}(T', \overleftrightarrow{BC})$ contain the point $\psi(X)$ and by Axiom PS must be the same line. Hence $T = T'$ and again by Axiom PS, $X = X'$. (If $X \in \overleftrightarrow{AC}$ modify the proof by substituting C for T and $\overleftrightarrow{BC}$ for $\operatorname{par}(T, \overleftrightarrow{BC})$ to show the same result.) Thus ψ is one-to-one, and a bijection.

(IV: ψ is a collineation.) Let $\mathcal{J}$ be any line on $\mathcal{P}$ such that $\mathcal{J}$ and $\mathcal{M}$ are not parallel, Q be the point such that $\mathcal{M} \cap \mathcal{J} = \{Q\}$, and R be any member of $\mathcal{J} \setminus \{Q\}$. Let $U = \operatorname{ftpr}(R, \mathcal{M})$, so that $S = \psi(R)$ is the point of intersection of $\operatorname{par}(R, \mathcal{M})$ and $\operatorname{par}(U, \overleftrightarrow{BC})$.

First we prove that $\psi(\mathcal{J}) \subseteq \overleftrightarrow{QS}$. Let $X \neq R$ be any member of $\mathcal{J} \setminus \{Q, R\}$. Let $T = \operatorname{ftpr}(X, \mathcal{M})$, and let Y be the point such that $\overleftrightarrow{QS} \cap \operatorname{par}(T, \overleftrightarrow{BC}) = \{Y\}$. (In case $X \in \overleftrightarrow{AC}$ so that $T = C$, let Y be the point such that $\overleftrightarrow{QS} \cap \overleftrightarrow{BC} = \{Y\}$.) By the construction we know that $\{\psi(X)\} = \operatorname{par}(X, \mathcal{M}) \cap \operatorname{par}(T, \overleftrightarrow{BC})$.

(Case 1: R and X are on the same side of $\mathcal{M}$.) We will use the following facts:

(i) $\overleftrightarrow{TY} \parallel \overleftrightarrow{US}$ since they are both parallel to $\overleftrightarrow{BC}$.
(ii) $\overleftrightarrow{RS} \parallel \overleftrightarrow{X\psi(X)}$ since they are both parallel to $\mathcal{M}$.
(iii) $\angle URS$, $\angle TX\psi(X)$, $\angle QUR$, and $\angle QTX$ are right angles, hence are congruent by Theorem NEUT.69.

By Theorem SIM.6, $\triangle QSU \sim \triangle QYT$, since $\angle UQS$ is shared by both triangles and by Theorem EUC.11 $\angle QUS \cong \angle QTY$. By the same theorem, $\triangle TQX \sim \triangle UQR$, since $\angle QUR \cong \angle QTX$ and $\angle UQR$ is shared by both triangles. And finally, $\triangle RSU \sim \triangle X\psi(X)T$ since $\angle URS \cong \angle TX\psi(X)$ and by Theorem EUC.11 $\angle SUR \cong \angle\psi(X)TX$.

Using Theorem SIM.16, we have from the first of these similarities that

$$\frac{[\overline{TY}]}{[\overline{SU}]} = \frac{[\overline{TQ}]}{[\overline{UQ}]}.$$

from the second similarity, we have

$$\frac{[\overline{TQ}]}{[\overline{UQ}]} = \frac{[\overline{XT}]}{[\overline{RU}]}$$

and from the third,

$$\frac{[\overline{XT}]}{[\overline{RU}]} = \frac{[\overline{T\psi(X)}]}{[\overline{SU}]}.$$

Combining these three equalities we have

$$\frac{[\overline{TY}]}{[\overline{SU}]} = \frac{[\overline{T\psi(X)}]}{[\overline{SU}]}$$

and multiplying both sides by $[\overline{SU}]$ we have $[\overline{TY}] = [\overline{T\psi(X)}]$, that is $\overline{TY} \cong \overline{T\psi(X)}$. $\psi(X)$ and Y are on the same side of T, for by Exercise PSH.14, R and S are on the same side of $\mathcal{M}$, and X and $\psi(X)$ are on the same side of $\mathcal{M}$. By Theorem PSH.38(A), $Y \in \overrightarrow{T\psi(X)}$ and by Property R.4 of Definition NEUT.2, $\psi(X) = Y \in \overleftrightarrow{SQ}$.

(Case 2: R and X are on opposite sides of $\mathcal{M}$.) Then (i), (ii), and (iii) hold as in Case 1. By Theorem NEUT.42, $\angle UQS \cong \angle TQY$ and $\angle UQR \cong \angle TQX$, because they are vertical angles. Then the triangles listed in Case 1 are congruent, and we conclude that $[\overline{TY}] = [\overline{T\psi(X)}]$, just as before.

Now by assumption X–Q–R, so by Exercise PSH.56 T–Q–U and Y–Q–S; by Exercise PSH.14, S and R are on the same side of $\mathcal{M}$ and X and $\psi(X)$ are on the same side of $\mathcal{M}$. Then S and R are on the opposite side from X and $\psi(X)$, and on the opposite side from Y, so that $\psi(X)$ and Y are on the same side of $\mathcal{M}$. By Theorem PSH.38(A), $Y \in \overrightarrow{T\psi(X)}$. By Property R.4 of Definition NEUT.2, $\psi(X) = Y \in \overleftrightarrow{SQ}$. This completes the proof that $\psi(\mathcal{J}) \subseteq \overleftrightarrow{QS}$.

Now let Z be any member of $\overleftrightarrow{QS} \setminus \{Q, S\}$, let $\mathcal{K} = \text{par}(Z, \overleftrightarrow{RS})$, and let W be the point such that $\mathcal{K} \cap \mathcal{J} = \{W\}$. Since $\psi(W) \in \overleftrightarrow{QS}$, by Exercise I.1 $\psi(W) = Z$. Therefore $\overleftrightarrow{QS} \subseteq \psi(\mathcal{J})$, and $\psi(\mathcal{J}) = \overleftrightarrow{QS}$.

(V: ψ is a shear of $\mathcal{P}$.) By part (II) of this proof, Theorem CAP.27, and Definition AX.0, ψ is a shear of $\mathcal{P}$. □

Theorem AX.3. *Let $\mathcal{M}$ be a line on a Euclidean plane $\mathcal{P}$.*

(A) *Every axial affinity φ on $\mathcal{P}$ with axis $\mathcal{M}$ is either a stretch or a shear, but not both.*

(B) *If A and B are any two points of $\mathcal{P} \setminus \mathcal{M}$, there exists a unique axial affinity φ on $\mathcal{P}$ with axis $\mathcal{M}$ such that $\varphi(A) = B$.*

Proof. (A) If an axial affinity has a fixed line $\mathcal{L} \neq \mathcal{M}$ that intersects $\mathcal{M}$, by Theorem CAP.27 all its fixed lines (other than $\mathcal{M}$) are parallel to $\mathcal{L}$, and all intersect $\mathcal{M}$. By the same theorem, if it has a fixed line parallel to $\mathcal{M}$ all its fixed lines are parallel (or equal) to $\mathcal{M}$, hence none except $\mathcal{M}$ can intersect $\mathcal{M}$. By Definition AX.0, in the first case φ is a stretch and not a shear; in the second it is a shear and not a stretch. This proves part (A).

(B) Note first that if $A \in \mathcal{P} \setminus \mathcal{M}$, and φ is an axial affinity on $\mathcal{P}$, $B = \varphi(A) \notin \mathcal{M}$, since φ is a one-to-one mapping.

If $\overleftrightarrow{AB}$ is not parallel to $\mathcal{M}$, by Theorem AX.1 there exists an axial affinity φ of $\mathcal{P}$ such that $\varphi(A) = B$ and $\overleftrightarrow{AB}$ is a fixed line of φ, so that φ is a stretch.

If $\overleftrightarrow{AB}$ is parallel to $\mathcal{M}$, then by Theorem AX.2 there exists an axial affinity φ of $\mathcal{P}$ such that $\varphi(A) = B$, $\overleftrightarrow{AB}$ is a fixed line of φ, which is a shear.

In either case, by Theorem CAP.29 there is only one axial affinity φ such that $\varphi(A) = B$. □

Theorem AX.4. *Let φ be an axial affinity with axis $\mathcal{M}$ on a Euclidean plane $\mathcal{P}$; then for any A, B, and C in $\mathcal{P}$, if A–B–C, then $\varphi(A)$–$\varphi(B)$–$\varphi(C)$. Thus φ preserves betweenness and is a belineation.*

Proof. By Theorem AX.3, φ is either a stretch or a shear. By Theorem CAP.27, if it is a stretch, there is a fixed line $\mathcal{L}$ that intersects $\mathcal{M}$, and the fixed lines of φ (other than $\mathcal{M}$) are the lines parallel to $\mathcal{L}$, all of which intersect $\mathcal{M}$. If it is a shear, its fixed lines (other than $\mathcal{M}$) are the lines parallel to $\mathcal{M}$. Now assume that A–B–C.

(A) If $\overleftrightarrow{AC} = \mathcal{M}$, the theorem is trivially true since all members of $\mathcal{M}$ are fixed points.

(B) Suppose now that $\overleftrightarrow{AC}$ is not a fixed line for φ.

(Case 1: φ is a stretch.) Then there exists a fixed line $\mathcal{L}$ which intersects $\mathcal{M}$, and all fixed lines other than $\mathcal{M}$ are parallel to $\mathcal{L}$. Thus $\overleftrightarrow{AC} \not\parallel \mathcal{L}$ (else it would be a fixed line) so intersects all fixed lines other than $\mathcal{M}$, and possibly $\mathcal{M}$ as well. By Axiom PS there are lines $\mathcal{L}_A$, $\mathcal{L}_B$, and $\mathcal{L}_C$ containing A, B, and C, respectively, which are parallel to $\mathcal{L}$ and therefore are fixed lines.

(Case 2: φ is a shear.) Then $\overleftrightarrow{AC}$ is not parallel to $\mathcal{M}$ or to any fixed line, and so intersects all of them. By Axiom PS there are lines $\mathcal{L}_A$, $\mathcal{L}_B$, and $\mathcal{L}_C$ containing A, B, and C, respectively, which are parallel to $\mathcal{M}$ and therefore are fixed lines.

In either case $\varphi(A) \in \mathcal{L}_A$, $\varphi(B) \in \mathcal{L}_B$, and $\varphi(C) \in \mathcal{L}_C$, and since φ is a collineation, all these points belong to the line $\varphi(\overleftrightarrow{AC})$ which by Theorem CAP.1 is $\overleftrightarrow{\varphi(A)\varphi(C)}$. If one of the points A, B, or C is a member of $\mathcal{M}$, it is the common point of $\overleftrightarrow{AC}$ and $\overleftrightarrow{\varphi(A)\varphi(C)}$, so by Exercise PSH.56, $\varphi(A)$–$\varphi(B)$–$\varphi(C)$; if none of A, B, or C is a member of $\mathcal{M}$, the same result follows from Exercise PSH.57.

(C) Suppose $\overleftrightarrow{AC}$ is a fixed line for φ but $\overleftrightarrow{AC} \neq \mathcal{M}$. Define $\mathcal{L} = \overleftrightarrow{AC}$. Then $\mathcal{L} \neq \mathcal{M}$, and all the points A, B, C, $\varphi(A)$, $\varphi(B)$, and $\varphi(C)$ are members of $\mathcal{L}$. Let A' be a point off of $\mathcal{L}$ and define $\mathcal{N}_A = \overleftrightarrow{AA'}$; by Axiom PS let $\mathcal{N}_B$ and $\mathcal{N}_C$ be lines parallel to $\mathcal{N}_A$ containing B and C, respectively. Let O be a point of $\mathcal{L}$ such that O–A–B–C and $O \notin \mathcal{M}$, and let $\mathcal{K} = \overleftrightarrow{OA'}$. Since $\mathcal{K}$ intersects $\mathcal{N}_A$, by Exercise IP.4 it must intersect $\mathcal{N}_B$ at a point B' and $\mathcal{N}_C$ at a point C', so that $\mathcal{N}_B = \overleftrightarrow{BB'}$ and $\mathcal{N}_C = \overleftrightarrow{CC'}$. By Exercise PSH.57 A'–B'–C'.

Since $\mathcal{K}$ intersects $\mathcal{L}$, $\mathcal{K}$ is not parallel to $\mathcal{L}$, nor to any fixed line parallel to $\mathcal{L}$, and $\mathcal{K} \neq \mathcal{M}$ because $\mathcal{K}$ contains the point O which is not in $\mathcal{M}$. Thus $\mathcal{K} = \overleftrightarrow{A'C'}$ is not a fixed line; moreover, not all of A', B', and C' are in $\mathcal{M}$, for if they were, $\mathcal{K}$ would be equal to $\mathcal{M}$. Thus we may apply part (B) to the points A', B', and C' giving us $\varphi(A')$–$\varphi(B')$–$\varphi(C')$.

Since $\overleftrightarrow{AA'} \parallel \overleftrightarrow{BB'} \parallel \overleftrightarrow{CC'}$ it follows from Theorem CAP.1(A) and Theorem CAP.3 that $\overleftrightarrow{\varphi(A)\varphi(A')} \parallel \overleftrightarrow{\varphi(B)\varphi(B')} \parallel \overleftrightarrow{\varphi(C)\varphi(C')}$ so that by Exercise PSH.57, $\varphi(A)$–$\varphi(B)$–$\varphi(C)$. □

Remark AX.5. In a Euclidean/LUB plane, Theorem AA.8 in Chapter 20 is a converse for Theorem AX.4. This is stated formally as Theorem AA.11.

16.2 Exercises for axial affinities

Answers to starred (*) exercises may be accessed from the home page for this book at www.springer.com.

Exercise AX.1*. Let $\mathcal{M}$ be a line on a Euclidean plane $\mathcal{P}$; let A and B be distinct points such that $\overleftrightarrow{AB} \parallel \mathcal{M}$. By Theorem AX.2 there exists a shear ψ with axis $\mathcal{M}$ such $\psi(A) = B$. Let $\mathcal{L}$ be a line parallel to $\mathcal{M}$; either $\mathcal{L} = \overleftrightarrow{AB}$ or $\mathcal{L} \parallel \overleftrightarrow{AB}$. Let

$C = \text{ftpr}(A, \mathcal{M})$, let D be the point of intersection of $\overleftrightarrow{AC}$ and $\mathcal{L}$, and let E be the point of intersection of $\overleftrightarrow{BC}$ and $\mathcal{L}$. Then by Theorem AX.2 $\psi(D) = E$. Using Theorem ISM.5 let τ be the translation of $\mathcal{P}$ such that $\tau(D) = E$. Show that for every $X \in \mathcal{L}$, $\psi(X) = \tau(X)$. This shows that the action of a shear on a line parallel to its axis is the same as that of a translation.

Exercise AX.2*. Let $\mathcal{P}$ be a Euclidean plane, and let φ be an axial affinity with axis $\mathcal{M}$ on $\mathcal{P}$, and let $\mathcal{L}$ be a line distinct from $\mathcal{M}$. Then $\mathcal{L}$ is a fixed line for φ iff for some $Q \notin \mathcal{M}$, $\mathcal{L} = \overleftrightarrow{Q\varphi(Q)}$.

Chapter 17
Rational Points on a Line (QX)

Acronym: *QX*
Dependencies: *all prior Chapters 1 through 16*
New Axioms: *none*
New Terms Defined: *dilation δ_n, rational point, polygonal domain*

Abstract: This chapter is concerned with an arbitrary line in a Euclidean plane, where this line has been built into an ordered field. It defines the meaning of a rational multiple of a point on this line, develops the arithmetical properties of such multiples, and uses these to show the existence of an order-preserving isomorphism between the set of all rational numbers and a subset of the line.

At this point we begin in earnest the process of identifying a line in a Euclidean plane with a number system. In Chapter 18, we will complete this process by defining real multiples of points on a line, and establishing an order-preserving isomorphism between the set of real numbers and any line in the plane. (Isomorphism is defined in Chapter 1, Section 1.5.)

In Chapter 14 we proved that a line in a Euclidean plane could be built into an ordered field, but this did not provide a way to relate points on the line to integers or to rational numbers. Thus, at that stage of the development, we could not create a correspondence between the set of rational numbers and some subset of the line.

To construct such a correspondence, we define integral multiples, then rational multiples of an arbitrary point $A \neq O$ of the ordered field. The greater part of the chapter is given to showing that the algebraic properties of the rational multiples

E.J. Specht et al., *Euclidean Geometry and its Subgeometries*,
DOI 10.1007/978-3-319-23775-6_17

of A mimic those of rational numbers. Then, in Theorem QX.16, we show that the natural correspondence between a rational number r and rU is an order preserving isomorphism of the rational numbers and the set of all rational multiples of U.

17.1 Integral multiples of a point

In this entire chapter, $\mathcal{P}$ will be a Euclidean plane, and $\mathbb{L}$ will be a line in $\mathcal{P}$ which has been made into an ordered field as described in Definition OF.1 and Theorem OF.2, with origin O and unit U.

Definition QX.1. Let $\mathcal{P}$ be a Euclidean plane, $\mathbb{L}$ a line in $\mathcal{P}$ which has been developed as an ordered field according to Chapter 14, with origin O and unit U, and let $A \in \mathbb{L}$. Define multiples of A as follows:

(A) $0A = O$.
(B) For every rational number r, $rO = O$.
(C) Let n be a natural number and let $A \in \mathbb{L} \setminus \{O\}$. Define $1A = A$. Assuming nA has been defined, define $(n+1)A = nA \oplus A$.
(D) Let n be a negative integer ($n < 0$) so that $-n$ is a natural number, and let $A \in \mathbb{L} \setminus \{O\}$. Define $nA = {}^{\ominus}((-n)A)$.

Definition QX.1(C) inductively defines the product of a natural number n and a member A, and part (D) extends that definition to any integer n.

Theorem QX.2. *Let A and B be distinct points of $\mathbb{L} \setminus \{O\}$.*

(A) *If n is a natural number, and δ is the unique dilation with fixed point O such that $\delta(A) = nA$; then for any other $B \in \mathbb{L} \setminus \{O\}$, $\delta(B) = nB$.*
(B) *If n is a nonzero integer, and δ is the unique dilation with fixed point O such that $\delta(A) = nA$, then for any other $B \in \mathbb{L} \setminus \{O\}$, $\delta(B) = nB$.*

Proof. (A) Note first that the existence of the dilation δ is guaranteed by Theorem DLN.7. By Theorem CAP.24 δ is unique. For each natural number n, define $A_n = nA$ and $B_n = nB$; then $A_1 = A$ and $B_1 = B$. By Definition QX.1(C), for every natural number n, $A_n = A_{n-1} \oplus A_1$, and $B_n = B_{n-1} \oplus B_1$. By Theorem OF.17(A), for every natural number n, $\overline{OA_1} \cong \overline{A_{n-1}(A_{n-1} \oplus A_1)} = \overline{A_{n-1}A_n}$ and $\overline{OB_1} \cong \overline{B_{n-1}(B_{n-1} \oplus B_1)} = \overline{B_{n-1}B_n}$. If $A_1 > 0$, by Theorem OF.11(B) $A_n > A_{n-1} > 0$; if $A_1 < 0$, $A_n < A_{n-1} < 0$ (and likewise for B_1 and B_n). Thus by Theorem ORD.6, O–A_{n-1}–A_n and O–B_{n-1}–B_n.

Let α be the dilation with fixed point O such that $\alpha(A_1) = B_1$. By Theorem DLN.8 every dilation is a belineation, and by Theorem COBE.5(5) and Theorem DLN.19,

$$\overline{OB_1} = \overline{O\alpha(A_1)} = \alpha(\overline{OA_1}) \cong \alpha(\overline{A_{n-1}A_n}) = \overline{\alpha(A_{n-1})\alpha(A_n)}. \quad (*)$$

By Exercise DLN.3 $\delta \circ \alpha = \alpha \circ \delta$, so that $\alpha(A_n) = \alpha(nA_1) = \alpha(\delta(A_1)) = \delta(\alpha(A_1)) = \delta(B_1)$. Therefore by (*), $\overline{OB_1} \cong \overline{\alpha(A_{n-1})\delta(B_1)}$.

By definition, $\alpha(A_1) = B_1$; assume that we have proved that $\alpha(A_{n-1}) = B_{n-1}$; then the above becomes $\overline{OB_1} \cong \overline{B_{n-1}\delta(B_1)}$. We have already seen, by construction, that $\overline{OB_1} \cong \overline{B_{n-1}B_n}$, so that $\overline{B_{n-1}\delta(B_1)} \cong \overline{B_{n-1}B_n}$.

We know that O–A_{n-1}–A_n. By Theorem DLN.19, O–$\alpha(A_{n-1})$–$\alpha(A_n)$. Since $\alpha(A_{n-1}) = B_{n-1}$ and $A_n = \delta(A_1)$, this is O–B_{n-1}–$\alpha(\delta(A_1))$. By Exercise DLN.3, O–B_{n-1}–$\delta(\alpha(A_1))$ that is, O–B_{n-1}–$\delta(B_1)$.

Since O–B_{n-1}–B_n, $B_n = nB_1$ and $\delta(B_1)$ are on the same side of B_{n-1}. Thus $\delta(B_1) \in \overrightarrow{B_{n-1}B_n}$, and because $\overline{B_{n-1}\delta(B_1)} \cong \overline{B_{n-1}B_n}$, by Property R.4 of Definition R.2 (linear scaling), $\delta(B_1) = nB_1$ as required.

(B) If $n > 0$, this is Part (A). If $n < 0$, let δ be the dilation such that $\delta(A) = nA$ which is equal to ${}^{\ominus}((-n)A)$ by Definition QX.1(D). Let δ' be the dilation such that $\delta'(A) = (-n)A$. Then $\delta(A) = {}^{\ominus}(\delta'(A)) = \mathcal{R}_O \circ \delta'(A)$, where $\mathcal{R}_O$ is the point reflection about O, and by Theorem ISM.3, $\mathcal{R}_O$ is a dilation, so that by Theorem CAP.21 $\mathcal{R}_O \circ \delta$ is a dilation, and by uniqueness (Theorem CAP.24) $\delta(X) = {}^{\ominus}(\delta'(X))$ for every $X \in \mathbb{L} \setminus \{O\}$. By part (A) above, $\delta'(B) = (-n)B$, and $\delta(B) = {}^{\ominus}\delta'(B) = {}^{\ominus}((-n)B) = nB$, by Definition QX.1(D). □

Definition QX.3. For any integer $n \neq 0$, δ_n is the dilation with fixed point O such that $\delta_n(U) = nU$ (and hence, for every $A \neq O$, $\delta_n(A) = nA$).

Theorem QX.4. *Let $n \neq 0$ and $m \neq 0$ be integers, and let A and B be points of $\mathbb{L} \setminus \{O\}$.*

(A) $n(A \odot B) = nA \odot B = A \odot nB$.

(B) $nA = U \odot nA = A \odot nU = nU \odot A$.

(C) *If n is any integer, then* $(-n)A = {}^{\ominus}(nA)$.

(D) $(-1)A = {}^{\ominus}A$.

Proof. (A) By Theorem QX.2(B), Definition QX.3, the commutativity of dilations (Exercise DLN.3), and Definition OF.1(D),

$$n(A \odot B) = \delta_n(A \odot B) = \delta_n(\delta_A(B)) = \delta_A(\delta_n(B)) = A \odot nB.$$

By this result, and the commutativity of $\odot$ on $\mathbb{L}$ (Theorem OF.2(B)),

$$n(A \odot B) = n(B \odot A) = B \odot nA = nA \odot B.$$

(B) The first and last equalities of (B) follow immediately from Theorem OF.3. For the second equality, let $B = U$ in part (A).

(C) If $n > 0$, by Definition QX.1(D), $(-n)A = {}^{\ominus}((-(-n))A) = {}^{\ominus}(nA)$. If $n < 0$, by the same definition, $nA = {}^{\ominus}((-n)A)$; taking the negative of both sides, we have ${}^{\ominus}(nA) = {}^{\ominus}({}^{\ominus}((-n)A))$; by Theorem OF.10(A), this is the same as $(-n)A$.

(D) In part (C) let $n = 1$. □

Theorem QX.5 (Associativity for integer multiplication). *Let $n \neq 0$ and $m \neq 0$ be integers, and let $A \in \mathbb{L} \setminus \{O\}$.*

(A) $(nm)A = (mn)A = n(mA) = m(nA)$.

(B) $\delta_{mn} = \delta_m \circ \delta_n$.

Proof. (A) (Case 1: $m > 0$ and $n > 0$.) By Definition QX.1(C),

$$(mn)A = A \oplus A \oplus \ldots \oplus A$$

where there are $mn = nm$ terms. The corresponding sum for nA has n terms, and for $m(nA)$ the sum has $mn = m(n)$ terms, so that $(mn)A = m(nA)$. Reversing m and n yields $(nm)A = n(mA)$.

(Case 2: $m < 0$ and $n > 0$.) Since $mn < 0$,

$(mn)A = {}^{\ominus}(-(mn)A)$	by Definition QX.1(D)
$= {}^{\ominus}(((-m)n)A)$	by arithmetic
$= {}^{\ominus}((-m)(nA))$	by Case 1, since both $-m > 0$ and $n > 0$
$= m(nA)$	by Definition QX.1(D) applied to m and nA.

On the other hand,

$(nm)A = {}^{\ominus}(-(nm)A)$	by Definition QX.1(D)
$= {}^{\ominus}((n(-m))A)$	by arithmetic
$= {}^{\ominus}(n((-m)A))$	by Case 1, since both $-m > 0$ and $n > 0$
$= (-n)((-m)A)$	by Theorem QX.4(C) applied to n and $(-m)A$
$= n(-1)((-m)A)$	by arithmetic
$= n({}^{\ominus}((-m)A)$	by Theorem QX.4(D)
$= n(mA)$	by Definition QX.1(D) applied to m and A.

(Case 3: $m > 0$ and $n < 0$.) This is just Case 2 with m and n reversed.

(Case 4: $m < 0$ and $n < 0$.)

$$\begin{aligned}(mn)A &= ((-m)(-n))A) && \text{by arithmetic}\\ &= (-m)((-n)A) && \text{by Case 1}\\ &= (m)(-1)((-n)A) && \text{by arithmetic}\\ &= m(^{\ominus}((-n)A)) && \text{by Theorem QX.4(D)}\\ &= m(nA) && \text{by Definition QX.1(D).}\end{aligned}$$

Likewise, $(mn)A = (nm)A = n(mA)$.

(B) By part (A) and Definition QX.3, $\delta_{mn}(A) = (mn)A = m(nA) = m(\delta_n(A)) = \delta_m(\delta_n(A))$; hence $\delta_{mn} = \delta_m \circ \delta_n$, by Theorem DLN.7(C). □

Theorem QX.6 (Distributive property for integer multiplication). *Let $n \neq 0$ and $m \neq 0$ be integers, and let A be a point of $\mathbb{L} \setminus \{O\}$. Then $(n+m)A = nA \oplus mA$.*

Proof. (Case 1: $m > 0$ and $n > 0$.) By Definition QX.1(C),

$$mA = A \oplus A \oplus \ldots \oplus A$$

where there are m terms. The corresponding sum for nA has n terms and for $(m+n)A$ has $m+n$ terms, so that $(m+n)A = mA \oplus nA$.

(Case 2: $m < 0$ and $n > 0$, where $n + m > 0$.) Since $-m > 0$, by Case 1,

$(m+n)A \oplus (-m)A = (m+n+(-m))A = nA$, so that

$(m+n)A = {}^{\ominus}((-m)A) \oplus nA = mA \oplus nA$. This uses Definition QX.1(D) applied to m and A.

(Case 3: $m < 0$ and $n > 0$, where $n + m < 0$.) First, note that $-(n+m) = -n-m > 0$, and $-n < 0$ and $-m > 0$. We then have by Case 2

$(-n-m)A = (-n)A \oplus (-m)A$, so that

$${}^{\ominus}U \odot (-n-m)A = ({}^{\ominus}U \odot (-n)A) \oplus ({}^{\ominus}U \odot (-m)A)$$

by Theorem OF.10(D) and distributivity for $\mathbb{L}$ (Theorem OF.6). Applying Theorem QX.4(C) three times we have

$${}^{\ominus}U \odot ({}^{\ominus}(n+m)A) = ({}^{\ominus}U \odot ({}^{\ominus}nA)) \oplus ({}^{\ominus}U \odot ({}^{\ominus}mA))$$

which, by Theorem OF.10(D), yields

$${}^{\ominus}({}^{\ominus}(n+m)A) = {}^{\ominus}({}^{\ominus}nA) \oplus {}^{\ominus}({}^{\ominus}mA).$$

By Theorem OF.10(A) $(n+m)A = nA \oplus mA$.

(Case 4: $m < 0$ and $n < 0$.) By Case 1, $(-m-n)A = (-m)A \oplus (-n)A$. Then

$${}^{\ominus}U \odot (-m-n)A = ({}^{\ominus}U \odot (-m)A) \oplus ({}^{\ominus}U \odot (-n)A)$$

by Theorem OF.10(D) and distributivity for $\mathbb{L}$ (Theorem OF.6). Applying Theorem QX.4(C) three times we have

$$^{\ominus}U \odot (^{\ominus}(m+n)A) = (^{\ominus}U \odot (^{\ominus}mA)) \oplus (^{\ominus}U \odot (^{\ominus}nA))$$

which, by Theorem OF.10(D), yields

$$^{\ominus}(^{\ominus}(m+n)A) = (^{\ominus}(^{\ominus}mA)) \oplus (^{\ominus}(^{\ominus}nA)).$$

By Theorem OF.10(A) $(m+n)A = mA \oplus nA$. □

17.2 Rational multiples of a point

Recall from Definition QX.1(B) that for every rational number $r \in \mathbb{Q}$, $r \neq 0$, $rO = O$.

Definition QX.7. Let n be a natural number, $m \neq 0$ an integer, and $A \in \mathbb{L} \setminus \{O\}$. In this definition we shall refer to the dilation δ_n, the dilation defined by Definition QX.3 such that for every $A \in \mathcal{P} \setminus \{O\}$, $\delta_n(A) = nA$. The existence of δ_n is guaranteed by Theorem QX.2 and its uniqueness by Theorem CAP.24.

(A) Define $\delta_{\frac{1}{n}} = \delta_n^{-1}$; this mapping exists and is a dilation by Theorem CAP.21.
(B) Define $\frac{1}{n}A = \delta_{\frac{1}{n}}(A)$.
(C) Define $\delta_{\frac{m}{n}} = \delta_m \circ \delta_{\frac{1}{n}} = \delta_{\frac{1}{n}} \circ \delta_m$. This is a dilation by Theorem CAP.21.
(D) Define $\frac{m}{n}A = \delta_{\frac{m}{n}}(A)$.

Theorem QX.8. *Let n be a natural number, $m \neq 0$ an integer, and $A \in \mathbb{L} \setminus \{O\}$.*

(A) $n(\frac{1}{n}A) = A = \frac{1}{n}(nA)$.
(B) $m(\frac{1}{n}A) = \frac{m}{n}A = \frac{1}{n}(mA)$.

Proof. (A) $n(\frac{1}{n}A) = \delta_n(\delta_{\frac{1}{n}}(A)) = \delta_n(\delta_n^{-1}(A)) = A$ and $\frac{1}{n}(nA) = \delta_{\frac{1}{n}}(\delta_n(A)) = \delta_n^{-1}(\delta_n(A)) = A$.
(B) By Exercise DLN.3,

$$m(\tfrac{1}{n}A) = \delta_m(\delta_{\frac{1}{n}}(A)) = \delta_{\frac{m}{n}}(A)) = \delta_{\frac{1}{n}}(\delta_m(A)) = \tfrac{1}{n}(mA).$$ □

Theorem QX.9. *Let n be a natural number, $m \neq 0$ be an integer and let A and B be members of $\mathbb{L} \setminus \{O\}$.*

(A) $\frac{1}{n}(A \odot B) = (\frac{1}{n}A) \odot B = A \odot (\frac{1}{n}B)$.
(B) $\frac{1}{n}A = U \odot (\frac{1}{n}A) = (\frac{1}{n}U) \odot A$.

(C) $\frac{m}{n}A = U \odot (\frac{m}{n}A) = (\frac{m}{n}U) \odot A$.

(D) *If n and q are any natural numbers,* $\delta_{\frac{1}{nq}} = \delta_{\frac{1}{n}} \circ \delta_{\frac{1}{q}}$.

Proof. (A) By Theorem QX.4(A) and Theorem QX.8(A), $n((\frac{1}{n}A) \odot B) = (n(\frac{1}{n}A)) \odot B = A \odot B$. Multiplying both sides by $\frac{1}{n}$, then $\frac{1}{n}(n((\frac{1}{n}A) \odot B)) = (\frac{1}{n}A) \odot B = \frac{1}{n}(A \odot B)$. By this result and commutativity of $\odot$,

$$\tfrac{1}{n}(A \odot B) = \tfrac{1}{n}(B \odot A) = (\tfrac{1}{n}B) \odot A = A \odot (\tfrac{1}{n}B).$$

(B) The first equality is immediate from Theorem OF.3 and the second from part (A) and commutativity by setting $B = U$.

(C) The first equality follows immediately from Theorem OF.3. By Definition QX.7(D), Theorem QX.4(B), and part (A) of this theorem,

$$\tfrac{m}{n}A = \tfrac{1}{n}(mA) = \tfrac{1}{n}((mU) \odot A) = (\tfrac{1}{n}(mU)) \odot A = (\tfrac{m}{n}U) \odot A,$$

which proves the second equality.

(D) For every $A \neq O$,

$nq(\delta_{\frac{1}{n}}(\delta_{\frac{1}{q}}(A))) = nq(\delta_{\frac{1}{q}}(\delta_{\frac{1}{n}}(A)))$	by Exercise DLN.3
$= \delta_{nq}(\delta_{\frac{1}{q}}(\delta_{\frac{1}{n}}(A)))$	by Definition QX.3
$= \delta_n(\delta_q(\delta_{\frac{1}{q}}(\delta_{\frac{1}{n}}(A))))$	by Theorem QX.5(B)
$= \delta_n((\delta_q \circ \delta_{\frac{1}{q}})(\delta_{\frac{1}{n}}(A)))$	by associativity of functions
$= \delta_n(\delta_{\frac{1}{n}}(A)) = A$	by inverses.

Multiplying both sides by $\frac{1}{nq}$, $\delta_{\frac{1}{q}}(\delta_{\frac{1}{n}}(A)) = \frac{1}{nq}(A) = \delta_{\frac{1}{nq}}(A)$, and $\delta_{\frac{1}{q}} \circ \delta_{\frac{1}{n}} = \delta_{\frac{1}{nq}}$, by Definition QX.7(A). □

Theorem QX.10 (Associative properties for rational multiples). *Let* $r = \frac{m}{n}$ *and* $s = \frac{p}{q}$ *be nonzero rational numbers, where n and q are natural numbers, m and p are integers. Let A and B be members of* $\mathbb{L} \setminus \{O\}$.

(A) $r(A \odot B) = (rA) \odot B = A \odot (rB)$.

(B) $(rs)A = r(sA)$.[1]

(C) $rA \odot sB = (rs)(A \odot B)$.[2]

---

[1]This may appear to be obvious, but there is no *a priori* assurance that multiplication of a point on the plane by two rational numbers successively is the same as a single multiplication by their product. The proof will use the fact that successive multiplication of a point by two natural numbers *is* the same as multiplication by their product.

[2]Strictly speaking, this is not an associative property, but it seems to fit here.

Proof. (A) By Theorem QX.9(C) and the associative property for points of $\mathbb{L}$ (see Theorem OF.2), $r(A \odot B) = (rU) \odot (A \odot B) = (rU \odot A) \odot B = (rA) \odot B$. Using this result and commutativity, $r(A \odot B) = (rB) \odot A = A \odot (rB)$.

(B)
$$\begin{aligned}
(rs)A &= (\tfrac{m}{n}\tfrac{p}{q})A = \tfrac{mp}{nq}A && \text{by arithmetic}\\
&= \tfrac{1}{nq}(mpA) && \text{by Theorem QX.8(B)}\\
&= \tfrac{1}{n}(\tfrac{1}{q}(m(pA))) && \text{by arithmetic and Theorem QX.5(A)}\\
&= \delta_{\frac{1}{n}}(\delta_{\frac{1}{q}}(\delta_m(\delta_p(A)))) && \text{by Theorem QX.5(B) and part (A)}\\
&= \delta_{\frac{1}{n}}((\delta_{\frac{1}{q}} \circ \delta_m)\delta_p(A)) && \text{by associativity of mappings}\\
&= \delta_{\frac{1}{n}}((\delta_m \circ \delta_{\frac{1}{q}})\delta_p(A)) && \text{by Exercise DLN.3}\\
&= \delta_{\frac{1}{n}}(\delta_m(\delta_{\frac{1}{q}}(\delta_p(A)))) && \text{by associativity of mappings}\\
&= (\delta_{\frac{1}{n}}(\delta_m(\delta_{\frac{p}{q}}(A)) && \text{by Theorem QX.8(B)}\\
&= \delta_{\frac{m}{n}}(\delta_{\frac{p}{q}}(A)) && \text{by Theorem QX.8(B)}\\
&= \tfrac{m}{n}(\tfrac{p}{q}(A)) = r(sA) && \text{renaming numbers.}
\end{aligned}$$

(C) By two applications of part (A), one application of part (B), and commutativity of $\odot$ in $\mathbb{L}$ (Theorem OF.2(B)), we have

$$rA \odot sB = r(A \odot sB) = r(sB \odot A) = r(s(B \odot A)) =$$
$$(rs)(B \odot A) = (rs)(A \odot B).$$ □

Theorem QX.11 (Distributive properties for rational multiples). *Let $r = \frac{m}{n}$ and $s = \frac{p}{q}$ be nonzero rational numbers, where n and q are natural numbers, m and p are integers. Let A and B be members of $\mathbb{L} \setminus \{O\}$.*

(A) $(r + s)A = rA \oplus sA$.
(B) $r(A \oplus B) = rA \oplus rB$.

Proof. (A)
$$\begin{aligned}
nq(\tfrac{m}{n}A \oplus \tfrac{p}{q}A) &= nq(\tfrac{m}{n}A) \oplus nq(\tfrac{p}{q}A) && \text{by Theorem QX.6}\\
&= (nq\tfrac{m}{n})A \oplus (nq\tfrac{p}{q})A && \text{by Theorem QX.10}\\
&= (qm)A \oplus (np)A && \text{by arithmetic}\\
&= (qm + np)A && \text{by Theorem QX.6.}
\end{aligned}$$

Then by Theorem QX.8(A), Theorem QX.10 and arithmetic,
$$\begin{aligned}
(\tfrac{m}{n}A \oplus \tfrac{p}{q}A) &= \tfrac{1}{nq}((qm + np)A) = (\tfrac{1}{nq}(qm + np))A\\
&= (\tfrac{qm+np}{nq})A = (\tfrac{m}{n} + \tfrac{p}{q})A
\end{aligned}$$
so $(r + s)A = rA \oplus sA$.

(B)
$$\begin{aligned}
r(A \oplus B) &= rU \odot (A \oplus B) && \text{by Theorem QX.9(C)}\\
&= (rU \odot A) \oplus (rU \odot B) && \text{by Theorem OF.6}\\
&= rA \oplus rB && \text{by Theorem QX.9(C).}
\end{aligned}$$ □

Corollary QX.12. *For every A and B belonging to* $\mathbb{L} \setminus \{O\}$ *and every rational number r,*

(A) $(-1)A = {}^{\ominus}A$.
(B) $(-r)A = {}^{\ominus}(rA)$.
(C) $(-r)({}^{\ominus}A) = rA$.
(D) $rA = {}^{\ominus}(r({}^{\ominus}A))$, *that is,* ${}^{\ominus}(rA) = r({}^{\ominus}A)$.
(E) $r(A \ominus B) = rA \ominus rB$.

Proof. (A) $O = (1-1)A = A \oplus (-1)A$, hence ${}^{\ominus}A = (-1)A$.
(B) By Theorem QX.10(B) and part (A),

$$(-r)A = ((-1)r)A = (-1)(rA) = {}^{\ominus}(rA).$$

(C) By part (A), Theorem QX.10(B), and Theorem OF.10(A),

$$(-r)({}^{\ominus}A) = ((r)(-1))({}^{\ominus}A) = (r)((-1)({}^{\ominus}A)) = r({}^{\ominus}({}^{\ominus}A)) = rA.$$

(D) By part (A), part (C) and Theorem QX.10(B),

$$rA = (-r)({}^{\ominus}A) = ((-1)r)({}^{\ominus}A) = (-1)(r({}^{\ominus}A)) = {}^{\ominus}(r({}^{\ominus}A)).$$

(E) Using, in succession, Definition OF.8(A), Theorem OF.6, part (D) above, and again, Definition OF.8(A), we have $r(B \ominus A) = r(B \oplus {}^{\ominus}A) = rB \oplus r({}^{\ominus}A) = rB \oplus {}^{\ominus}rA$. □

Theorem QX.13. *Let* $r = \frac{m}{n}$ *be a nonzero rational number, where m is an integer and n is a natural number. Let* $A \in \mathbb{L} \setminus \{O\}$.

(A) *If r is positive and A is positive, then rA is positive.*
(B) *If r is negative and A is positive, then rA is negative.*
(C) *If r is positive and A is negative, then rA is negative.*
(D) *If r is negative and A is negative, then rA is positive.*

Proof. Since n is a natural number, by Definition QX.1(C)

$$n(\tfrac{1}{n}A) = \tfrac{1}{n}A \oplus \tfrac{1}{n}A \oplus \ldots \oplus \tfrac{1}{n}A = A$$

(where there are n terms in the sum). By Theorem OF.10(C) if $\frac{1}{n}A > O$ then $A > O$, and if $\frac{1}{n}A < O$, by Theorem OF.10(I) $A < O$. Hence $\frac{1}{n}A > O$ $(< O)$ iff $A > O$ $(< O)$.

By a similar argument, if $r = \frac{m}{n} > 0$, since $n > 0$, $m > 0$ (a natural number), and $mU > O$ because $U > O$. If $r = \frac{m}{n} < 0$, $m < 0$, $-m > 0$, $(-m)U > O$, so that by Definition QX.1(D), $mU = {}^{\ominus}(-m)U$ which is $< O$ by Theorem OF.10(B).

By Theorem QX.8(B) and Theorem QX.4(B)

$$rA = \tfrac{m}{n}A = m(\tfrac{1}{n}A) = mU \odot (\tfrac{1}{n}A).$$

If $r > 0$ and $A > O$, $mU > O$, by Theorem OF.10(C) $rA = mU \odot (\frac{1}{n}A) > O$.
If $r < 0$ and $A < O$, $mU < O$, by Theorem OF.10(I), $rA = mU \odot (\frac{1}{n}A) > O$.
If $r > 0$ and $A < O$, then $mU > O$; if $r < 0$ and $A > O$, then $mU < O$; in either case, by Theorem OF.10(E), $rA = mU \odot (\frac{1}{n}A) < O$. □

Corollary QX.14. *Let r be a nonzero rational number and $A < B$ be points other than O on $\mathbb{L}$. Then $r > 0$ iff $rA < rB$, and $r < 0$ iff $rA > rB$.*

Proof. By Theorem OF.11(A) since $A < B$, $B \ominus A > O$. By Theorem QX.13, $r > 0$ iff $O < r(B \ominus A) = rB \ominus rA$, which is true iff $rB > rA$. Again by Theorem QX.13, $r < 0$ iff $r(B \ominus A) = rB \ominus rA < O$ which is true iff $rB < rA$. □

Definition QX.15. Let U be the unit in the line $\mathbb{L}$. Define the set $\mathbb{L}_Q = \{rU \mid r \in \mathbb{Q}\}$ be the set of **rational points** of $\mathbb{L}$.

Theorem QX.16. (A) *$\mathbb{L}_Q$ is an ordered field under the operations $\oplus$ and $\odot$ and the order relation $<$, and is a subfield of $\mathbb{L}$.*

(B) *There exists an order-preserving field isomorphism Ψ mapping $\mathbb{Q}$, the field of rational numbers, onto $\mathbb{L}_Q$.*

Proof. (A) Let r and s be rational numbers. By Theorem QX.11, $rU \oplus sU = (r+s)U$ and by Theorem QX.10(C), $(rU) \odot (sU) = rs(U \odot U) = rsU$. This shows that $\mathbb{L}_Q$ is closed under both $\oplus$ and $\odot$.

$(-r)U$ is the additive inverse of rU, since $(-r)U \oplus rU = (-r + r)U = 0U = O$ (cf Definition QX.1(A)). If $r \neq 0$, then by Theorem QX.9(A), the commutativity of $\odot$, and Theorem QX.10(B) we have

$(\frac{1}{r}U) \odot (rU) = \frac{1}{r}(U \odot rU) = \frac{1}{r}(rU \odot U) = \frac{1}{r}r(U \odot U) = 1 \cdot U = U,$

so that $\frac{1}{r}U$ is the multiplicative inverse of rU.

Both $\oplus$ and $\odot$ are commutative and associative in $\mathbb{L}$ and therefore have these same properties in $\mathbb{L}_Q$. The distributive property holds in $\mathbb{L}_Q$ since it holds in $\mathbb{L}$; thus, if r, s,and t are rational numbers,

$rU \odot (sU \oplus tU) = (rU \odot sU) \oplus (rU \odot tU) = rsU \oplus rtU = (rs + rt)U.$

Therefore $\mathbb{L}_Q$ is a group under $\oplus$ and $\mathbb{L}_Q \setminus \{O\}$ is a group under $\odot$; since the distributive law holds, $\mathbb{L}_Q$ is a field, a subfield of $\mathbb{L}$. Since $\mathbb{L}$ is ordered by the ordering "$<$", so is its subset $\mathbb{L}_Q$.

(B) Define the mapping Ψ as follows: for each rational number r, let $\Psi(r) = rU$. It is obvious that Ψ maps onto $\mathbb{L}_Q$.

Let r be a rational number. Then by Definition QX.1, if $r = 0$ then $rU = O$. If $r \neq 0$ then either $r > 0$ or $r < 0$ so by Theorem QX.13(A) or (B), $rU > O$ or $rU < O$, and $rU \neq O$. By the contrapositive, if $rU = O$, then $r = 0$; therefore $rU = O$ iff $r = 0$. It follows that if $rU = sU$,

$$\begin{aligned}(r-s)U &= (r+(-s))U = rU \oplus (-s)U \\ &= rU \oplus {}^{\ominus}(sU) = rU \oplus {}^{\ominus}(rU) = O\end{aligned}$$

so that $r - s = 0$ and $r = s$. Hence Ψ is one-to-one, and is a bijection.

By Theorem QX.11(A), $\Psi(r+s) = (r+s)U = rU + sU = \Psi(r) \oplus \Psi(s)$. If r and s are both nonzero, $\Psi(rs) = (rs)U = rU \odot sU = \Psi(r) \odot \Psi(s)$ by Theorem QX.10(C). Furthermore $\Psi(0) = 0U = O$ and $\Psi(1) = 1U = U$. Thus Ψ fulfills all the requirements to be an *isomorphism* of $\mathbb{Q}$ onto $\mathbb{L}_Q$.

Finally, by Exercise QX.2, $r < s$ iff $rU < sU$, so Ψ is order-preserving. □

Corollary QX.16.1. *Let H be any member of $\mathbb{L}$ such that $H > O$, and let $\mathbb{L}_H = \{rH \mid r \in \mathbb{Q}\}$. Then*

(A) *$\mathbb{L}_H$ is an ordered field under the operations $\oplus$ and $\odot$ and the order relation $<$, and is a subfield of $\mathbb{L}$.*

(B) *There exists an order-preserving field isomorphism Ψ mapping $\mathbb{Q}$, the field of rational numbers, onto $\mathbb{L}_H$.*

Proof. The proof is exactly the proof of Theorem QX.16, where H has been substituted for U. That is, the unit U can be chosen arbitrarily to be any point greater than O. □

Remark QX.17. The existence of an isomorphism between $\mathbb{Q}$ and $\mathbb{L}_Q$ means that these cannot be distinguished as algebraic objects.

17.3 Applications of rational multiples

Theorem QX.18 (Midpoint of a segment). *Let A and B be distinct members of the ordered field $\mathbb{L}$, then $\frac{1}{2}(A \oplus B) = \frac{A \oplus B}{2}$ is the midpoint of $\overline{AB}$.*

Proof. Choose the notation so that $A < B$; by Corollary QX.14 $\frac{A}{2} < \frac{B}{2}$. Using Theorem OF.11(B) and Theorem QX.11(B),

$$A = \tfrac{A}{2} \oplus \tfrac{A}{2} < \tfrac{A}{2} \oplus \tfrac{B}{2} = \tfrac{A \oplus B}{2} < \tfrac{B}{2} \oplus \tfrac{B}{2} = B$$

so that by Theorem ORD.6, $\frac{A\oplus B}{2}$ is between A and B. Since $\frac{A\oplus B}{2} > A$ and $B > \frac{A\oplus B}{2}$,

$$|\tfrac{A\oplus B}{2} \ominus A| = \tfrac{B\ominus A}{2} = \tfrac{1}{2}(B \ominus A) = \tfrac{B\ominus A}{2}$$

and

$$|B \ominus \tfrac{A\oplus B}{2}| = \tfrac{1}{2}(B \ominus A) = \tfrac{B\ominus A}{2}.$$

Therefore by Theorem OF.15(B) $\overline{A\frac{A\oplus B}{2}} \cong \overline{B\frac{A\oplus B}{2}}$. Since $\frac{A\oplus B}{2} \in \overline{AB}$, by Definition NEUT.3(C) $\frac{A\oplus B}{2}$ is the midpoint of $\overline{AB}$. □

Remark QX.19 (On free segments). In Chapter 15 (Definition SIM.7) we gave meaning to the *product* of two free segments $\mathcal{A} = [\overline{OA}]$ and $\mathcal{B} = [\overline{OB}]$ by defining $\mathcal{A} \odot \mathcal{B} = [\overline{O(A \odot B)}]$. We can also define a (positive) rational multiple of a free segment, as follows.

Definition QX.20 (Rational multiple of a free segment). If $\mathcal{A} = [\overline{OA}]$ is any free segment and $r > 0$ is any rational number, define $r\mathcal{A} = [\overline{O(rA)}]$.

Remark QX.21 (On the product of altitude and base of a triangle). Recall from Definition NEUT.99(C) the definitions of *altitude* and *base* of a triangle. Theorem SIM.24 shows that if $\overline{AD}$ and $\overline{BE}$ are two altitudes and $\overline{BC}$ and $\overline{AC}$ are the respective bases for those altitudes, $[\overline{AD}] \odot [\overline{BC}] = [\overline{BE}] \odot [\overline{AC}]$.

Let X, X', Y, and Y' be points of $\overrightarrow{OU}$ such that $[\overline{AD}] = [\overline{OX}]$, $[\overline{BC}] = [\overline{OY}]$, $[\overline{BE}] = [\overline{OX'}]$, and $[\overline{AC}] = [\overline{OY'}]$. By Theorem OF.15 and Definition OF.16, the lengths of these free segments are, respectively, X, Y, X', and Y'.

We can re-state the result of Theorem SIM.24 as: If X is the length of an altitude of a triangle, and Y is the length of its base, and X' is the length of another altitude of the same triangle, and Y' is the length of its base, then $[\overline{OX}] \odot [\overline{OY}] = [\overline{OX'}] \odot [\overline{OY'}]$. By Definition SIM.7 this is $[\overline{O(X \odot Y)}] = [\overline{O(X' \odot Y')}]$, or $\overline{O(X \odot Y)} \cong \overline{O(X' \odot Y')}$, which by Property R.4 of Definition NEUT.2 is $X \odot Y = X' \odot Y'$.

Thus for any triangle, the product of the lengths of an altitude and its base is independent of the choice of altitude, so that the following definition is a "good" one.

Definition QX.22. The **area** of a triangle is half the product of the length of an altitude and its base. More formally, if $\overline{AD}$ is an altitude of a triangle, and if $\overline{BC}$ is the base of that altitude, and if $[\overline{AD}] = [\overline{OX}]$ and $[\overline{BC}] = [\overline{OY}]$, then the area of the triangle is $\frac{1}{2}(X \odot Y)$. For a visualization see Figure 17.1.

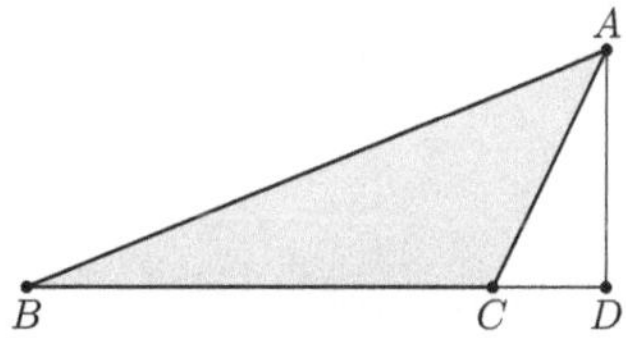

Fig. 17.1 For Definition QX.22.

Remark QX.23 (Area of a polygonal domain). In Definition QX.20 we defined the area of a triangle. This provides a basis for defining the area of a *polygonal domain*, a union $\bigcup_{\mathcal{H}\in\mathbb{T}} \operatorname{enc}\mathcal{H}$, where $\mathbb{T}$ is a finite set of triangles on the Euclidean plane, and the intersection of any two sets $\operatorname{enc}\mathcal{S}$ and $\operatorname{enc}\mathcal{T}$ is empty, or is a common corner or common edge of $\mathcal{S}$ and $\mathcal{T}$.

Chapter 10 of *Geometry, A Metric Approach with Models* (R. S. Millman and G. D. Parker, Springer-Verlag, 1981 [15]) develops this approach, using the positive real numbers; thus Axiom LUB is required. However, their development carries over verbatim to our situation if free segments and their lengths (defined in Chapter 14, Definition OF.16) are used in lieu of the positive real numbers. Thus the concept of area for polygonal domains does not depend on Axiom LUB.

17.4 Exercises for rational points on a line

Answers to starred (*) exercises may be accessed from the home page for this book at www.springer.com.

Exercise QX.1*. Let r be a nonzero rational number, and let $A \in \mathbb{L} \setminus \{O\}$.

(A) If A is positive, then rA is positive iff r is positive.

(B) If A is negative, then rA is negative iff r is positive.

Exercise QX.2*. Let A be a positive member of $\mathbb{L}$ and r and s be rational numbers. Then $rA < sA$ iff $r < s$.

Exercise QX.3*. Let A be a negative member of $\mathbb{L}$ and r and s be rational numbers. Then $rA > sA$ iff $r < s$.

Exercise QX.4*. Let $\mathcal{P}$ be a Euclidean plane, $\mathbb{L}$ be an ordered field on $\mathcal{P}$, T be a member of $\mathbb{L}$ and r be a rational number. Then $(-r)T = {}^{\ominus}(rT)$.

Exercise QX.5*. Let $\mathbb{L}$ be an ordered field with origin O on a Euclidean plane $\mathcal{P}$, and let X and Y be positive members of $\mathbb{L}$. Then there exist noncollinear points A, B, and C on $\mathcal{P}$ such that $\frac{1}{2}X \odot Y$ is the area of $\triangle ABC$.

Chapter 18
A Line as Real Numbers (REAL); Coordinatization of a Plane (RR)

Acronyms: *REAL, RR*

Dependencies: *all prior Chapters 1 through 17*

New Axioms: *Axiom LUB*

New Terms Defined: *least upper bound (lub), greatest lower bound (glb), complete (ordered field); rational, irrational members of a complete ordered field; Archimedean property; sequence, limit; sum of two subsets of an ordered field; addition of points on a plane; scalar multiple of a point on a plane; coordinatization, coordinatization map, axes, origin, first and second coordinates on a plane, right-handed system*

Abstract: This chapter derives basic properties of least upper bounds and explores their relationship with the Archimedean property. On an arbitrary line in a Euclidean/LUB plane (which has been built into an ordered field) real multiples of points are defined and their algebraic properties derived. These properties are used to show the existence of an order-preserving isomorphism between the set of all real numbers and the whole line. The chapter ends with coordinatization of a Euclidean/LUB plane.

We continue the project of embedding number systems in lines on a plane. In order to carry this out, it is necessary for the plane to have the *LUB property*, a property of the set of real numbers. We show that every line $\mathbb{L}$ in such a plane is order-isomorphic to the set $\mathbb{R}$ of real numbers, and finally, in a process called *coordinatization*, that the plane itself is a copy of the coordinate plane $\mathbb{R}^2$.

E.J. Specht et al., *Euclidean Geometry and its Subgeometries*,
DOI 10.1007/978-3-319-23775-6_18

Lines are built into ordered fields in Chapter 14, ordering is defined in Chapter 6, and in Definition ORD.8 of that chapter, upper and lower bounds are defined. Isomorphism is defined in Chapter 1 Section 1.5.

18.1 The basics of least upper bounds

Definition REAL.1. Let $\mathbb{L}$ be a line in a Euclidean plane $\mathcal{P}$. Suppose that $\mathbb{L}$ has been built into an ordered field, and that $\mathcal{E}$ is a nonempty subset of $\mathbb{L}$.

(A) If $\mathcal{E}$ is *bounded above*, and if the set of all *upper bounds* of $\mathcal{E}$ has a *minimum*, then this minimum is called the **least upper bound** of $\mathcal{E}$, and is denoted **lub** $\mathcal{E}$.

(B) If $\mathcal{E}$ is *bounded below*, and if the set of all *lower bounds* of $\mathcal{E}$ has a *maximum*, then this maximum is called the **greatest lower bound** of $\mathcal{E}$, and is denoted **glb** $\mathcal{E}$.

Axiom LUB. Let $\mathbb{L}$ be a line which is equipped with an order relation as defined in Definition ORD.1. Every nonempty subset $\mathcal{E}$ of $\mathbb{L}$ which is bounded above has a least upper bound lub $\mathcal{E}$.

Strictly speaking, Axiom LUB applies to lines, or ordered fields; however, we will freely speak of this axiom as being true "on space" or "on a plane," meaning that the Axiom is true for all lines in that space (or plane).

Definition REAL.2. (A) A Euclidean space or plane on which Axiom LUB is true is called a **Euclidean/LUB space (or plane)**; **Euclidean/LUB geometry** is the resulting geometry on such a plane or line.

(B) An ordered field for which Axiom LUB holds is a **complete** ordered field.

(C) For notational convenience, for any set $\mathcal{E} \subseteq \mathbb{L}$, where $\mathbb{L}$ is an ordered field, we define ${}^{\ominus}\mathcal{E} = \{{}^{\ominus}X \mid X \in \mathcal{E}\}$.

Remark REAL.3. (A) The least upper bound of a set $\mathcal{E}$ may or may not be a member of $\mathcal{E}$; if it is a member of $\mathcal{E}$, then it is the *maximum* element of $\mathcal{E}$ (*cf* Definition ORD.8).

(B) It is well known that the set $\mathbb{R}$ of real numbers is an ordered field for which Axiom LUB holds.

Theorem REAL.4 (GLB). *Let $\mathcal{E}$ be a subset of an ordered field $\mathbb{L}$ (which is a subset of a Euclidean/LUB plane) with origin O and unit U. If $\mathcal{E}$ is nonempty and is bounded below, then*

(A) *$\mathcal{E}$ has a greatest lower bound—that is,* $\text{glb}\,\mathcal{E}$ *exists, and*
(B) $\text{glb}\,\mathcal{E} =^{\ominus} \text{lub}(^{\ominus}\mathcal{E})$.
(C) $\text{glb}\,(^{\ominus}\mathcal{E}) =^{\ominus} \text{lub}\,\mathcal{E}$.

Proof. If B is a lower bound of $\mathcal{E}$, then $^{\ominus}B$ is an upper bound of $^{\ominus}\mathcal{E}$. By Axiom LUB the set of upper bounds of $^{\ominus}\mathcal{E}$ has a minimum $\text{lub}(^{\ominus}\mathcal{E})$. We show that $^{\ominus}\,\text{lub}(^{\ominus}\mathcal{E})$ is the maximum of the lower bounds of $\mathcal{E}$. If D is any lower bound of $\mathcal{E}$, then for every member X of $\mathcal{E}$, $D \leq X$ and so $^{\ominus}D \geq {}^{\ominus}X$. Since $^{\ominus}D$ is an upper bound of $^{\ominus}\mathcal{E}$, $\text{lub}(^{\ominus}\mathcal{E}) \leq {}^{\ominus}D$, but this means that $D \leq^{\ominus} \text{lub}(^{\ominus}\mathcal{E})$. So $^{\ominus}\,\text{lub}(^{\ominus}\mathcal{E})$ is the maximum of the lower bounds of $\mathcal{E}$, that is, $\text{glb}\,\mathcal{E}$. This proves both parts (A) and (B).

The proof of (C) follows from (B) by substituting $^{\ominus}\mathcal{E}$ for $\mathcal{E}$. □

Theorem REAL.5. *Let $\mathcal{P}$ be a Euclidean/LUB plane and let $\mathbb{L}$ be an ordered field on $\mathcal{P}$ with origin O and unit U, and let $\mathcal{E} \subseteq \mathbb{L}$.*

(A) *If* $\text{lub}\,\mathcal{E}$ *exists, it is unique.*
(B) *If* $\text{glb}\,\mathcal{E}$ *exists, it is unique.*

Proof. (A) Suppose A and B both are least upper bounds for $\mathcal{E}$. Since A is an upper bound, then $A \geq B$; since B is an upper bound, $B \geq A$. By Theorem ORD.5, if $A \neq B$ then either $A > B$ or $B > A$ both of which are impossible.

(B) The proof is similar to part (A) and is left to the reader. □

Definition REAL.6. Let $\mathcal{P}$ be a Euclidean/LUB plane, and let $\mathbb{L}$ be an ordered field on $\mathcal{P}$ with origin O and unit U.

(A) A function whose domain is the set $\mathbb{N}$ of all natural numbers and whose values are members of $\mathbb{L}$ is a **sequence**. The customary notation for a sequence whose value at each natural number n is S_n is $\{S_n\}$.
(B) A sequence $\{S_n\}$ is said to have a **limit** L iff for every $\epsilon > O$ which is a member of $\mathbb{L}$, there exists a natural number m such that for every $n > m$, $^{\ominus}\epsilon < S_n \ominus L < \epsilon$. If there exists a limit L for $\{S_n\}$, we write $L = \lim S_n$.

Theorem REAL.7. *Let $\mathcal{P}$ be a Euclidean/LUB plane, $\mathbb{L}$ an ordered field on $\mathcal{P}$, and $\{S_n\}$ a sequence with values in $\mathbb{L}$ such that for every natural number n, $S_{n+1} \geq S_n$. If $\{S_n \mid n$ is a natural number$\}$ is bounded above, and if $B = \text{lub}\{S_n \mid n$ is a natural number$\}$, then $\lim S$ exists and equals B.*

Proof. Let ϵ be any positive member of $\mathbb{L}$ such that $B \ominus \epsilon$ is positive. By Axiom LUB there exists a natural number m such that $S_m > B \ominus \epsilon$. Then ${}^{\ominus}\epsilon < S_m \ominus B < O$ (since B is an upper bound for $\{S_n\}$). If $n \geq m$, then $S_n \geq S_m$ so $B \ominus S_n \leq B \ominus S_m < \epsilon$ and ${}^{\ominus}\epsilon < S_n \ominus B < O$. But this means that $\lim S_n$ exists and equals B. □

18.2 Archimedes, Eudoxus, and least upper bounds

Remark REAL.8. Archimedes of Syracuse (c. 287–212 BC) is commonly listed with Newton and Gauss as one of the three greatest mathematicians of all time, having all but invented calculus to solve various problems. This next property still bears his name, although some say that Archimedes attributed it to Eudoxus of Cnidus, whom we will encounter shortly.

The Archimedean property has meaning (is either true or false) for any ordered field $\mathbb{L}$ that is equipped with an origin O and a unit U, where for any $A \in \mathbb{L}$ and any natural number n, nA is defined to be $A \oplus A \oplus \ldots \oplus A$ (with n terms in the summation), as in Definition QX.1(C).

Archimedean property For any $H > O$ and any $K > O$ there exists a natural number n such that $nH > K$. Alternatively, the set $\mathcal{D} = \{nH \mid n \text{ is a natural number}\}$ is unbounded above.

We will sometimes say "the plane $\mathcal{P}$ is Archimedean" to mean that the Archimedean property holds on every ordered field in the plane $\mathcal{P}$.

Theorem REAL.9 (Every Euclidean/LUB plane is Archimedean). *Let $\mathcal{P}$ be a Euclidean/ LUB plane, $\mathbb{L}$ an ordered field on $\mathcal{P}$, and suppose that for every $A \in \mathbb{L}$, nA has been defined as in Definition QX.1. Then if H is a positive member of $\mathbb{L}$, and $\mathcal{D} = \{nH \mid n$ is a natural number$\}$, $\mathcal{D}$ is unbounded above.*

Proof. Assume $\mathcal{D}$ is bounded above. By Axiom LUB the set of upper bounds of $\mathcal{D}$ has a minimum $B = \text{lub}\,\mathcal{D}$. By Theorem PSH.22 there exists a member V of $\mathbb{L}$ such that O–V–H and by Theorem ORD.6, $O < V < H$.

If all members of $\mathcal{D}$ were less than or equal to $B \ominus V$, this would be an upper bound for $\mathcal{D}$, and since $B \ominus V < B$, B would not be the least upper bound. Thus there exists a member of $\mathcal{D}$ which is greater than $B \ominus V$, that is, for some natural number m, $mH > B \ominus V$. Hence $(m+1)H = mH \oplus H > B \ominus V \oplus H = B \oplus (H \ominus V) > B$, because $H \ominus V > O$. But $(m+1)H \in \mathcal{D}$ so that $(m+1)H \leq B$, a contradiction. Therefore $\mathcal{D}$ is unbounded above. □

Corollary REAL.9.1. *Assuming that the hypotheses of Theorem REAL.9 are true, the set $\{rH \mid r \text{ is a rational number}\}$ is unbounded above.*

Proof. $\mathcal{D} \subseteq \{rH \mid r \text{ is a rational number}\}$. □

Remark REAL.10 (LUB and the Archimedean property). As we saw in Theorem REAL.9, the LUB property of an ordered field implies the Archimedean property. But in general, the Archimedean property does not imply the LUB property, as will be established in the next two theorems.

Theorem REAL.11 ($\mathbb{Q}$ is Archimedean). *The Archimedean property holds on the set $\mathbb{Q}$ of rational numbers.*

Proof. Let $r = \frac{p}{q}$ and $b = \frac{s}{t}$, where r and b are positive rational numbers and p, q, s, and t are natural numbers. We show that there exists a natural number n such that $nr > b$. Note first that $q \geq \frac{q}{p}$ and $s \geq \frac{s}{t}$. Let $n = sq + 1 > sq$; then $n > sq \geq \frac{q}{p}\frac{s}{t}$. Multiplying both sides by $r = \frac{p}{q}$, $nr = n\frac{p}{q} > \frac{s}{t} = b$. □

Theorem REAL.12 (Axiom LUB does not hold on $\mathbb{Q}$). *The set $\mathcal{E} = \{r \mid r \text{ is a rational number and } r^2 < 2\}$ has no least upper bound in $\mathbb{Q}$.*

Proof. Suppose the contrary, that $\frac{p}{q}$ is the least upper bound for $\mathcal{E}$, where p and q are natural numbers. By trichotomy for numbers, there are three cases; we show that all these cases lead to contradictions.

(Case 1: $\frac{p^2}{q^2} < 2$.) Then $2q^2 > p^2$ and $2q^2 - p^2 > 0$. By the Archimedean property, there exists a natural number n' such that $n' > \frac{4pq}{2q^2-p^2}$, so $\frac{1}{n'} < \frac{2q^2-p^2}{4pq}$. Also there exists a natural number n'' such that $n'' > \frac{2q^2}{2q^2-p^2}$, so $\frac{1}{(n'')^2} < \frac{1}{n''} < \frac{2q^2-p^2}{2q^2}$. Let $n = \max\{n', n''\}$. Then

$$\begin{aligned}\left(\tfrac{p}{q} + \tfrac{1}{n}\right)^2 &= \tfrac{p^2}{q^2} + \tfrac{2p}{qn} + \tfrac{1}{n^2} < \tfrac{p^2}{q^2} + \tfrac{2p}{q}\left(\tfrac{2q^2-p^2}{4pq}\right) + \tfrac{2q^2-p^2}{2q^2} \\ &= \tfrac{p^2}{q^2} + \tfrac{2q^2-p^2}{2q^2} + \tfrac{2q^2-p^2}{2q^2} = \tfrac{4q^2}{2q^2} = 2.\end{aligned}$$

Therefore $\frac{p}{q} + \frac{1}{n}$ is a member of $\mathcal{E}$, and is greater than $\frac{p}{q}$, and $\frac{p}{q}$ is not an upper bound for $\mathcal{E}$, contradicting our hypothesis.

(Case 2: $\frac{p^2}{q^2} > 2$.) Then $2q^2 < p^2$ and $p^2 - 2q^2 > 0$. Let n' be a natural number such that $n' > \frac{q}{p}$, so that $0 < \frac{1}{n'} < \frac{p}{q}$, $\frac{p}{q} - \frac{1}{n'} > 0$, and $\left(\frac{p}{q} - \frac{1}{n'}\right)^2 < \frac{p^2}{q^2}$. Also let n'' be a natural number such that $n'' > \frac{2qp}{p^2-2q^2}$, so $\frac{1}{n''} < \frac{p^2-2q^2}{2qp}$. Let $n = \max\{n', n''\}$. Then

$$\frac{p^2}{q^2} - \frac{2p}{qn} > \frac{p^2}{q^2} - \frac{2p}{q}\left(\frac{p^2-2q^2}{2qp}\right) = \frac{p^2}{q^2} - \frac{p^2-2q^2}{q^2} = \frac{2q^2}{q^2} = 2$$

so that $\left(\frac{p}{q} - \frac{1}{n}\right)^2 = \frac{p^2}{q^2} - \frac{2p}{qn} + \frac{1}{n^2} > 2 + \frac{1}{n^2} > 2$. Therefore, for every $r \in \mathcal{E}$, $\left(\frac{p}{q} - \frac{1}{n}\right)^2 > 2 > r^2$, and $\frac{p}{q} - \frac{1}{n} > r$; hence $\frac{p}{q} - \frac{1}{n}$ is an upper bound for $\mathcal{E}$. Since $\frac{p}{q} - \frac{1}{n} < \frac{p}{q}$, $\frac{p}{q}$ is not the least upper bound for $\mathcal{E}$, contradicting our hypothesis.

(Case 3: $\frac{p^2}{q^2} = 2$.) The following proof is well known: reduce $\frac{p}{q}$ to lowest terms, so that p and q have no common factor. Then $p^2 = 2q^2$ so that 2 is a factor of p^2, hence is a factor of p, and there exists a natural number a such that $p = 2a$. Then $4a^2 = 2q^2$ and $2a^2 = q^2$, so 2 is a factor of q^2, and therefore a factor of q. This shows that 2 is a common factor of both p and q, in contradiction to our original assumption that $\frac{p^2}{q^2}$ is in lowest terms. □

The geometries we develop from here on are Archimedean geometries. The study of non-Archimedean geometries has been an active field of research.

Theorem REAL.13. *Let $\mathcal{P}$ be a Euclidean/LUB plane, A, B, C, and D be points on $\mathcal{P}$ such that $A \neq B$ and $C \neq D$. Then there exists a natural number m such that $m[\overline{AB}] > [\overline{CD}]$, where $m[\overline{AB}]$ is defined as in Definition QX.20.*

Proof. In the following, let Φ be as in Definition FSEG.14. If $[\overline{AB}] > [\overline{CD}]$, then $m = 1$ satisfies the inequality. If $[\overline{AB}] \leq [\overline{CD}]$, then by Theorem FSEG.13 there exists a unique point $H = \Phi[\overline{AB}]$ on an ordered field $\mathbb{L}$ with origin O such that $H > O$ and $[\overline{OH}] = [\overline{AB}]$. Let $V > O$ be the point on $\mathbb{L}$ such that $V = \Phi[\overline{CD}]$ and $[\overline{OV}] = [\overline{CD}]$. By Theorem REAL.11 there exists a natural number m such that $mH > V$. Since by Theorem FSEG.15(B) Φ preserves order, $mH > V$ iff $m[\overline{AB}] > [\overline{CD}]$. □

Remark REAL.14. Eudoxus of Cnidus (c. 408–355 BC) lived between the times of Pythagoras and Euclid and was contemporary with Plato. Part of Eudoxus' work amounted to a rigorous definition of real numbers and Richard Dedekind (1831–1916) was inspired by his ideas. (This is the Dedekind who originated the now-standard method of "Dedekind cuts" for completing the set of real numbers to include irrational numbers.) The following theorem bears Eudoxus' name.

Theorem REAL.15 (Eudoxus). *Let H be a positive member of $\mathbb{L}$, and let A and B be members of $\mathbb{L}$ such that $B < A$. Then there exists a rational number r such that $B < rH < A$.*

Proof. (Case 1: $A > B > O$.) By the Archimedean property (Theorem REAL.5) choose a natural number q such that $q(A \ominus B) > H$. Then multiplying both sides by $\frac{1}{q}$, we have $\frac{1}{q}H < A \ominus B$. Adding B and subtracting $\frac{1}{q}H$ yields $B < A \ominus \frac{1}{q}H$.

By the Archimedean property choose a natural number p so that $\frac{p}{q}H > B$, and let p be the smallest such number.

Then $\frac{p}{q}H < A$, for otherwise, $\frac{p}{q}H \geq A$ and $\frac{p-1}{q}H = \frac{p}{q}H \ominus \frac{1}{q}H \geq A \ominus \frac{1}{q}H > B$, so that $\frac{p}{q}$ is not the smallest integer such that $\frac{p}{q}H > B$, contradicting the definition of p. Therefore $B < \frac{p}{q}H < A$.

(Case 2: $A > O > B$.) Let $r = 0$.

(Case 3: $O > A > B$.) By Theorem OF.10(B) and Exercise OF.9, ${}^{\ominus}B$ and ${}^{\ominus}A$ are both positive and $O < {}^{\ominus}A < {}^{\ominus}B$. By Case 1, let r be such that ${}^{\ominus}A < rH < {}^{\ominus}B$. Then $B < {}^{\ominus}(rH) = (-r)H < A$. Here we have used Corollary QX.12(B). □

18.3 Real multiples of members of $\mathbb{L}$

Remark REAL.16. (A) Unless explicitly stated otherwise, in the remainder of this chapter $\mathcal{P}$ will denote a Euclidean/LUB plane, $\mathbb{L}$ will be an ordered field on $\mathcal{P}$ with origin O and unit U, in which rational multiples of points have been defined as in Chapter 17.

(B) Unless explicitly stated otherwise, the letters r, s, and t will denote rational numbers. So we will routinely (but not always) omit the reminder that "$r \in \mathbb{Q}$".

Remark REAL.17. Let x be any real number, and $H > O$ a member of $\mathbb{L}$; let $\mathbb{L}_H = \{rH \mid r \in \mathbb{Q}\}$ (as in Corollary QX.16.1).

(A) Let $\mathcal{E}$ and $\mathcal{F}$ be subsets of $\mathbb{L}_H$. If $\mathcal{E}$ and $\mathcal{F}$ are bounded above (below) and have the same set of upper (lower) bounds in $\mathbb{L}_H$, then $\operatorname{lub}\mathcal{E} = \operatorname{lub}\mathcal{F}$ ($\operatorname{glb}\mathcal{E} = \operatorname{glb}\mathcal{F}$). This is obviously true by the plain meaning of the words upper and lower bound, least and greatest, *etc.*

(B) Let a, a', b, and b' be rational numbers such that $a < x < b$ and $a' < x < b'$.

(1) $\{rH \mid a < r < x\}$, $\{rH \mid a' < r < x\}$, and $\{rH \mid r < x\}$ are bounded above, have a common set of upper bounds, and a common least upper bound.

(2) $\{rH \mid x < r < b\}$, $\{rH \mid x < r < b'\}$, and $\{rH \mid x < r\}$ are bounded below, have a common set of lower bounds, and a common greatest lower bound.

(3) $\{r(^{\ominus}H) \mid a < r < x\}$, $\{r(^{\ominus}H) \mid a' < r < x\}$, and $\{r(^{\ominus}H) \mid r < x\}$ are bounded below, have a common set of lower bounds and a common greatest lower bound.

(4) $\{r(^{\ominus}H) \mid x < r < b\}$, $\{r(^{\ominus}H) \mid x < r < b'\}$, and $\{r(^{\ominus}H) \mid x < r\}$ are bounded above, have a common set of upper bounds and a common least upper bound.

To see part (B)(1), note that there exists a rational number s such that $a < r < x < s$. By Exercise QX.2, if $H > O$, $rH < xH < sH$ so sH is an upper bound for $\{rH \mid a < r < x\}$ and the other sets in the list, which obviously have the same set of upper bounds; part (A) says that their least upper bounds are the same. The proofs of the other parts of (B) are similar, and are left to the reader.

(C) Part (B) above shows that in most cases involving least upper bounds or greatest lower bounds, it is legitimate to write $\{rH \mid r < x\}$ instead of $\{rH \mid a < r < x\}$, or $\{rH \mid x < r\}$ instead of $\{rH \mid x < r < b\}$.

(D) Theorem REAL.18, which follows, provides the basis for general application of Definition REAL.19, which will define xH where x is an irrational number and H is any point of $\mathbb{L}$. It also facilitates the proofs of future theorems. The proof is long, but the reader should not be daunted by it.

Theorem REAL.18. *Let $H > O$ be a member of $\mathbb{L}$; let $\mathbb{L}_H = \{rH \mid r \in \mathbb{Q}\}$ (as in Corollary QX.16.1).*

(A) *If x is rational, then* $xH = \text{lub}\{rH \mid r < x\} = \text{glb}\{rH \mid x < r\}$.

(B) *If x is any real number,* $\text{lub}\{rH \mid r < x\} = \text{glb}\{rH \mid x < r\}$.

(C) *If x is any real number,* $\text{lub}\{r(^{\ominus}H) \mid x < r\} = \text{glb}\{r(^{\ominus}H) \mid r < x\}$.

(D) *If x is rational,* $x(^{\ominus}H) = \text{lub}\{r(^{\ominus}H) \mid x < r\} = \text{glb}\{r(^{\ominus}H) \mid r < x\}$.

Proof. Let x be any real number; if r and s are rational numbers such that $r < x < s$, by Exercise QX.2 $rH < sH$, so $\{rH \mid r < x\} < \{sH \mid x < s\}$ and $\{rH \mid r < x\} \cap \{sH \mid x < s\} = \emptyset$. Then every member of $\{rH \mid r < x\}$ is a lower bound for $\{sH \mid x < s\}$, and every member of $\{sH \mid x < s\}$ is an upper bound for $\{rH \mid r < x\}$.

Moreover, for any real x, $\{sH \mid x < s\}$ has no least (minimum) element and $\{rH \mid r < x\}$ has no greatest (maximum) element, because $\{s \mid x < s\}$ has no least element and $\{r \mid r < x\}$ has no greatest element. These facts will be used several times in the rest of this proof.

There can be at most one rational number t such that $r < t < s$ for all $r < x$ and all $s > x$; if there is such a t, then $t = x$, so that in this case x is rational. Therefore, if x is rational,

$$\{rH \mid r < x\} \cup \{sH \mid x < s\} \cup \{xH\} = \mathbb{L}_H \quad (*)$$

and the sets in this union are disjoint. If x is irrational, then every rational number is either greater or less than x, so that

$$\{rH \mid r < x\} \cup \{sH \mid x < s\} = \mathbb{L}_H.$$

Again, all r, s, and t are rational numbers, but for brevity we omit the reminders $r \in \mathbb{Q}$, *etc.*

Claim 1. For any real number x, neither $\text{lub}\{rH \mid r < x\}$ nor $\text{glb}\{sH \mid x < s\}$ can be a member of either $\{rH \mid r < x\}$ or $\{sH \mid x < s\}$. Also,

$$\{rH \mid r < x\} < \text{lub}\{rH \mid r < x\} < \{rH \mid x < r\}$$

and

$$\{rH \mid r < x\} < \text{glb}\{rH \mid x < r\} < \{rH \mid x < r\}.$$

Proof of Claim 1. $\text{lub}\{rH \mid r < x\}$ is an upper bound for $\{rH \mid r < x\}$, so $\{rH \mid r < x\} \le \text{lub}\{rH \mid r < x\}$. If $\text{lub}\{rH \mid r < x\} \in \{rH \mid r < x\}$, it would be an upper bound for that set and hence this set would contain a maximum element, which does not exist; therefore $\text{lub}\{rH \mid r < x\} \notin \{rH \mid r < x\}$, and $\{rH \mid r < x\} < \text{lub}\{rH \mid r < x\}$.

All the members of $\{rH \mid x < r\}$ are upper bounds for $\{rH \mid r < x\}$, so $\text{lub}\{rH \mid r < x\} \le \{rH \mid x < r\}$. If $\text{lub}\{rH \mid r < x\} \in \{rH \mid x < r\}$, it would be less than or equal to all members of $\{rH \mid x < r\}$ and thus its least element, which does not exist; therefore $\text{lub}\{rH \mid r < x\} \notin \{rH \mid x < r\}$, and $\text{lub}\{rH \mid r < x\} < \{rH \mid x < r\}$.

$\text{glb}\{rH \mid x < r\}$ is a lower bound for $\{rH \mid x < r\}$, so $\text{glb}\{rH \mid x < r\} \le \{rH \mid x < r\}$. If $\text{glb}\{rH \mid x < r\} \in \{rH \mid x < r\}$, it would be a lower bound for that set and hence its minimum element, which does not exist; so $\text{glb}\{rH \mid x < r\} \notin \{rH \mid x < r\}$, and $\text{glb}\{rH \mid x < r\} < \{rH \mid x < r\}$.

All the members of $\{rH \mid r < x\}$ are lower bounds for $\{rH \mid x < r\}$, so $\text{glb}\{rH \mid x < r\} \ge \{rH \mid r < x\}$. If $\text{glb}\{rH \mid x < r\} \in \{rH \mid r < x\}$, it would be greater than or equal to all members of $\{rH \mid r < x\}$ and thus would be its greatest element, which does not exist; so $\text{glb}\{rH \mid x < r\} \notin \{rH \mid r < x\}$, and $\{rH \mid r < x\} < \text{glb}\{rH \mid x < r\}$. This completes the proof of Claim 1.

Claim 2. For any real number x, $\text{lub}\{rH \mid r < x\} \le \text{glb}\{rH \mid x < r\}$.

Proof of Claim 2. As we have already noted, all points of $\{rH \mid x < r\}$ are upper bounds for $\{rH \mid r < x\}$; therefore they are all greater or equal to the least upper bound of $\{rH \mid x < r\}$; that is, $\{rH \mid x < r\} \geq \text{lub}\{rH \mid r < x\}$. Then $\text{lub}\{rH \mid r < x\}$ is a lower bound for $\{rH \mid x < r\}$ hence is less or equal to $\text{glb}\{rH \mid x < r\}$. Therefore $\text{lub}\{rH \mid r < x\} \leq \text{glb}\{rH \mid x < r\}$.

Note that up until this point we have assumed that x is an arbitrary real number. Now the proof splits.

(A) Assume that x is a rational number. If $\text{lub}\{rH \mid r < x\} \neq \text{glb}\{rH \mid x < r\}$, then by Claim 2 $\text{lub}\{rH \mid r < x\} < \text{glb}\{rH \mid x < r\}$. By Theorem REAL.15 (Eudoxus) there exists a rational number s such that $\text{lub}\{rH \mid r < x\} < sH < \text{glb}\{rH \mid x < r\}$; then neither sH nor xH is a member of either $\{rH \mid r < x\}$ or $\text{glb}\{rH \mid x < r\}$; therefore by (*), $sH = xH$, or $s = x$, showing that $\text{lub}\{rH \mid r < x\} < xH < \text{glb}\{rH \mid x < r\}$.

Again by Theorem REAL.15 there exists a rational number t such that $\text{lub}\{rH \mid r < x\} < tH < xH$. Then $t < x$ so $tH \in \{rH \mid r < x\}$ and $tH \leq \text{lub}\{rH \mid r < x\} < tH$, a contradiction. Therefore $\text{lub}\{rH \mid r < x\} = \text{glb}\{rH \mid x < r\}$ and by Claim 2, $\text{lub}\{rH \mid r < x\} = xH = \text{glb}\{rH \mid x < r\}$.

(B) If x is irrational, again from Claim 2 we know that $\text{lub}\{rH \mid r < x\} \leq \text{glb}\{rH \mid x < r\}$. If $\text{lub}\{rH \mid r < x\} < \text{glb}\{rH \mid x < r\}$, by Theorem REAL.15 there exists a rational number s such that $\text{lub}\{rH \mid r < x\} < sH < \text{glb}\{rH \mid x < r\}$. Since x is irrational either $s < x$ or $x < s$; hence sH is either a member of $\{rH \mid r < x\}$ or of $\{rH \mid x < r\}$, both of which are impossible by the definition of sH. Therefore $\text{lub}\{rH \mid r < x\} = \text{glb}\{rH \mid x < r\}$ where x is an irrational number. From part (A) this is true for all real numbers x, proving part (B).

(C) Applying, in order, Theorem QX.12(D), Theorem REAL.4(B), Part (B) above, Theorem REAL.4(C), and Theorem QX.12(D), we have

$$\begin{aligned}\text{lub}\{r({}^{\ominus}H) \mid x < r\} &= \text{lub}\{{}^{\ominus}(rH) \mid x < r\}\\ &= {}^{\ominus}\,\text{glb}\{rH \mid x < r\}\\ &= {}^{\ominus}\,\text{lub}\{rH \mid r < x\}\\ &= \text{glb}\ {}^{\ominus}\{rH \mid r < x\}\\ &= \text{glb}\{r({}^{\ominus}H) \mid r < x\}.\end{aligned}$$

(D) If x is rational, by part (A) $xH = \text{glb}\{rH \mid x < r\} = \text{lub}\{rH \mid r < x\}$. From Corollary QX.12(D)

$$x({}^{\ominus}H) = {}^{\ominus}(xH) = {}^{\ominus}\,\text{glb}\{rH \mid x < r\} = {}^{\ominus}\,\text{lub}\{rH \mid r < x\}.$$

Thus $x(^{\ominus}H)$ is equal to the second and third lines in the calculation above for part (C), so that

$$x(^{\ominus}H) = \text{glb}\{r(^{\ominus}) \mid r < x\} = \text{lub}\{r(^{\ominus}H) \mid x < r\}. \qquad \square$$

Definition REAL.19. (A) Let x be a real number and let H be a member of $\mathbb{L}$.

(1) If $x = 0$ or $H = O$, then $xH = O$.
(2) If r is a nonzero rational number and if $H \neq O$, then rH is given by Definition QX.1.
(3) If x is any irrational number and H is a positive member of $\mathbb{L}$, then $xH = \text{lub}\{rH \mid r \in \mathbb{Q} \textit{ and } b < r < x\}$, where b is any number such that $b < x$.
(4) If x is any irrational number and J is a negative member of $\mathbb{L}$, then $xJ = {}^{\ominus}(x(^{\ominus}J))$.

(B) H is a **rational** member of $\mathbb{L}$ iff there exists a rational number r such that $H = rU$.
(C) H is an **irrational** member of $\mathbb{L}$ iff there exists an irrational number a such that $H = aU$.

Theorem REAL.20 (Summary of Theorem REAL.18 and Definition REAL.19). *Let x be any real number, and let H be any positive member of $\mathbb{L}$.*

(A) $xH = \text{lub}\{rH \mid r < x\} = \text{glb}\{rH \mid x < r\}$.
(B) $x(^{\ominus}H) = {}^{\ominus}(xH) = \text{lub}\{r(^{\ominus}H) \mid x < r\} = \text{glb}\{r(^{\ominus}H) \mid r < x\}$.

Proof. (A) If x is rational, this is Theorem REAL.18(A). For irrational numbers the proof follows from Definition REAL.19(A)(3) and Theorem REAL.18(B).
(B) If x is rational, this follows from Theorem REAL.18(D) and Corollary QX.12(D). If x is irrational, by Definition REAL.19(A)(4) $x(^{\ominus}H) = {}^{\ominus}(xH))$; by Definition REAL.19(A)(3) this is

$$\begin{aligned} {}^{\ominus}\,\text{lub}\{rH \mid r < x\} &= \text{glb}^{\ominus}\{rH \mid r < x\} \\ &= \text{glb}\{^{\ominus}(rH) \mid r < x\} \\ &= \text{glb}\{r(^{\ominus}H) \mid r < x\} \\ &= \text{lub}\{r(^{\ominus}H) \mid x < r\}. \end{aligned}$$

The first equality is by Theorem REAL.4(C), the third by Theorem QX.12(D), the last by Theorem REAL.18(C). $\square$

Theorem REAL.21. *Let $\mathcal{P}$ be a Euclidean/LUB plane and let $\mathbb{L}$ be an ordered field on $\mathcal{P}$ with origin O and unit U, in which rational multiples of points have been defined as in Chapter 17. Let x be any real number and S any member of $\mathbb{L}$.*

(A) $(-x)S = {}^{\ominus}(xS)$ *(so* $(-1)S = {}^{\ominus}S$*, as in Theorem QX.4(D)).*

(B) $(-x)({}^{\ominus}S) = xS$.

(C) $xS = {}^{\ominus}(x({}^{\ominus}S))$*, that is,* ${}^{\ominus}(xS) = x({}^{\ominus}S)$.

Proof. If x is rational, this is Corollary QX.12, parts (B), (C), and (D). If $S = O$, they are all true by Definition REAL.19(A)(1).

(A) (Case 1: x is an irrational number and $S > O$.) Applying, in sequence, Definition REAL.19(A)(3), re-naming variable, Theorem QX.12(B), Definition REAL.2(C), Theorem REAL.4(C), and Theorem REAL.20(A), we have

$$\begin{aligned}(-x)S &= \text{lub}\{rS \mid r < -x\} = \text{lub}\{(-r)S \mid -r < -x\}\\ &= \text{lub}\{{}^{\ominus}(rS) \mid x < r\} = \text{lub}\ {}^{\ominus}\{rS \mid x < r\}\\ &= {}^{\ominus}\,\text{glb}\{rS \mid x < r\}.\\ &= {}^{\ominus}(xS).\end{aligned}$$

(A) (Case 2: x is an irrational number and $S < O$.) Applying, in sequence, Theorem REAL.20(B), re-naming variable, Theorem QX.12(B), Definition REAL.2(C), Theorem REAL.4(C), and Theorem REAL.20(B), we have

$$\begin{aligned}(-x)S &= \text{lub}\{rS \mid -x < r\} = \text{lub}\{(-r)S \mid -x < -r\}\\ &= \text{lub}\{{}^{\ominus}(rS) \mid r < x\} = \text{lub}\ {}^{\ominus}\{(rS) \mid r < x\}\\ &= {}^{\ominus}\,\text{glb}\{rS \mid r < x\} = {}^{\ominus}(xS).\end{aligned}$$

(B) (Case 1: x is an irrational number and $S > O$.) Applying, in sequence, Theorem REAL.20(B), re-naming variable, Theorem QX.12(C), and Definition REAL.19(A)(3), we have

$$\begin{aligned}(-x)({}^{\ominus}S) &= \text{lub}\{r({}^{\ominus}S) \mid -x < r\} = \text{lub}\{(-r)({}^{\ominus}S) \mid -x < -r\}\\ &= \text{lub}\{rS \mid r < x\} = xS.\end{aligned}$$

(B) (Case 2: x is any irrational number and $S < O$.) Applying, in sequence, Definition REAL.19(A)(3), re-naming variable, Theorem QX.12(C), and Theorem REAL.20(B), we have

$$\begin{aligned}(-x)({}^{\ominus}S) &= \text{lub}\{r({}^{\ominus}S) \mid r < -x\} = \text{lub}\{(-r)({}^{\ominus}S) \mid -r < -x\}\\ &= \text{lub}\{rS \mid x < r\} = xS.\end{aligned}$$

(C) We have now proved that (A) and (B) are true for all irrational x and all points $S \in \mathbb{L}$. Applying part (A) first, and then part (B) we have, for any $S \in \mathbb{L}$, ${}^{\ominus}(xS) = (-x)S = (-(-x))({}^{\ominus}S) = x({}^{\ominus}S)$. □

Corollary REAL.21.1. *Let x be any nonzero real number, and let $A \in \overrightarrow{OU}$. Then $\overline{O(xU)} \cong \overline{OA}$ iff $|x|U = A$. That is to say, $\Phi[\overline{OA}] = A = |x|U$, where Φ is as in Definition FSEG.14.*

Proof. Assume $\overline{O(xU)} \cong \overline{OA}$. If $x > 0$, then $xU \in \overrightarrow{OU}$ and by Property R.4 of Definition NEUT.2, $|x|U = xU = A$. If $x < 0$, then $(-x)U \in \overrightarrow{OU}$ and $|x|U = (-x)U = A$. Conversely, if $|x|U = A$ either $x > 0$ in which case $\overline{O(xU)} = \overline{OA}$, or $x < 0$ in which case $-x > 0$ and $A = (-x)U = {}^{\ominus}(xU)$ by Theorem REAL.21(A). Then ${}^{\ominus}A = xU$ and by Theorem OF.10(A), $\overline{O(xU)} = \overline{O({}^{\ominus}A)} \cong \overline{OA}$. □

Lemma REAL.22. *Suppose $x > 0$ and $y > 0$ are real numbers; then r is a rational number such that $0 < r < xy$ iff there exist rational numbers s and t such that $0 < s < x$, $0 < t < y$ and $st = r$.*

Proof. This proof uses the property that between every two distinct real numbers, there exists a rational number. Since $0 < r < xy$, $\frac{r}{y} < x$ so we may choose a rational number s such that $\frac{r}{y} < s < x$. Then $r = y\frac{r}{y} < ys < xy$ so that $\frac{r}{s} < y$. Let $t = \frac{r}{s}$. Then $0 < t < y$ and $st = s\frac{r}{s} = r$. Conversely, if $0 < s < x$ and $0 < t < y$, then $0 < st < xy$. □

We call Theorems REAL.23 and REAL.25 "Associative" properties, even though that stretches the language somewhat.

Theorem REAL.23 (Associative property I for scalar product). *If x is any real number, and S is any member of $\mathbb{L}$, then $x(yS) = (xy)S = y(xS)$.*

Proof. Again in this proof we will denote rational numbers by the letters r, s, or t; x and y may be any real numbers, rational or irrational.

(Case 1: Either $x = 0$, $y = 0$, or $S = O$.) The proof follows immediately from Definition REAL.19(A)(1).

(Case 2: $x > 0$, $y > 0$, and $S > O$.) By Lemma REAL.22, and Theorem QX.10(A), B is an upper bound for $\{rS \mid 0 < r < xy\}$ iff

$$\begin{aligned} B \geq \{rS \mid 0 < r < xy\} &= \{(st)S \mid 0 < s < x \textit{ and } 0 < t < y\} \\ &= \{s(tS) \mid 0 < s < x \textit{ and } 0 < t < y\}, \end{aligned}$$

Thus, for every s such that $0 < s < x$, $B \geq \{s(tS) \mid 0 < t < y\}$, and the following statements are equivalent:

(a) $B \geq \{s(tS) \mid 0 < t < y\} = s(\{(tS) \mid 0 < t < y\})$;
(b) $\frac{B}{s} \geq \{tS \mid 0 < t < y\}$;
(c) $\frac{B}{s} \geq \text{lub}\{tS \mid 0 < t < y\} = yS$ (by Definition REAL.19(A)(3));
(d) $B = s\frac{B}{s} \geq s(yS)$.

Therefore B is an upper bound for $\{rS \mid 0 < r < xy\}$ iff it is an upper bound for $\{s(yS) \mid 0 < s < x\}$. By Remark REAL.17(A), the least upper bounds for these sets are the same. Thus, using Definition REAL.19(A)(3),

$$(xy)S = \text{lub}\{rS \mid 0 < r < xy\} = \text{lub}\{s(yS) \mid 0 < s < x\} = x(yS).$$

(Case 3: $x > 0$, $y < 0$, and $S > O$.) Using, successively, arithmetic, Theorem REAL.21(A), Theorem REAL.21(C), Case 2 above, arithmetic, and Theorem REAL.21(A), we have

$$\begin{aligned} x(yS) &= x(-(-y))S = x(^{\ominus}((-y)S)) = {}^{\ominus}(x((-y)S)) \\ &= {}^{\ominus}((x(-y))S) = {}^{\ominus}((-(xy))S) = (xy)S. \end{aligned}$$

(Case 4: $x < 0$, $y > 0$, and $S > O$.) The proof is Exercise REAL.6.

(Case 5: $x < 0$ and $y < 0$ are any real numbers and $S > O$.) Using, in succession, arithmetic, Theorem REAL.21(A), Case 3 above, arithmetic, Theorem REAL.21(A), and Theorem OF.10(A), we have

$$\begin{aligned} x(yS) &= (-(-x))(yS) = {}^{\ominus}((-x)(yS)) = {}^{\ominus}(((-x)y)S) \\ &= {}^{\ominus}(-(xy)S) = {}^{\ominus}(^{\ominus}((xy)S)) = (xy)S. \end{aligned}$$

Cases 2 through 5 show that $(xy)S = x(yS)$ for all real numbers x and y where $S > O$.

(Case 6: x and y are any real numbers and $S < O$.) Using, in succession, two applications of Theorem REAL.21(C), Cases 2 through 5 above, and Theorem REAL.21(C), we have

$$x(yS) = x(^{\ominus}(y(^{\ominus}S))) = {}^{\ominus}(x(y(^{\ominus}S))) = {}^{\ominus}((xy)(^{\ominus}S)) = (xy)S.$$

This shows that $x(yS) = (xy)S$ for all real numbers x and y and all members S of $\mathbb{L}$. By commutativity, $x(yS) = (xy)S = (yx)S = y(xS)$. □

Lemma REAL.24. *If $\mathcal{E}$ is a subset of $\mathbb{L}$ which is bounded above, and $T > O$ is a member of $\mathbb{L}$, then* $(\text{lub}\,\mathcal{E}) \odot T = \text{lub}(\mathcal{E} \odot T)$.

Proof. The proof is Exercise REAL.4. □

Theorem REAL.25 (Associative property II for scalar product). *Let S and T be members of $\mathbb{L}$, and let x be any real number. Then $(xS) \odot T = x(S \odot T) = S \odot (xT)$. In particular, $(xU) \odot T = xT = U \odot (xT)$, where U is the unit for $\mathbb{L}$.*

Proof. In this proof we will use Theorem OF.10(A) ($^{\ominus}(^{\ominus}A) = A$) without reference. We will first show that $(xS) \odot T = x(S \odot T)$.

(Case 0: x is rational.) This is Theorem QX.10(A).

(Case 1: x is any irrational number, $S > O$, and $T > O$.) Applying, in succession, Definition REAL.19(A)(3), Lemma REAL.24, Theorem QX.10(A), and Definition REAL.19(A)(3), we have

$$\begin{aligned} xS \odot T &= \text{lub}\{rS \mid r < x\} \odot T = \text{lub}\{(rS) \odot T \mid r < x\} \\ &= \text{lub}\{r(S \odot T) \mid r < x\} = x(S \odot T). \end{aligned}$$

(Case 2: x is any irrational number, $S < O$ and $T < O$.) Applying, in succession, Theorem REAL.21(C), Theorem OF.10(D), Case 1 above, and Theorem OF.10(D), we have

$$\begin{aligned}(xS) \odot T &= {}^{\ominus}(x({}^{\ominus}S)) \odot T = (x({}^{\ominus}S)) \odot ({}^{\ominus}T) \\ &= (x(({}^{\ominus}S) \odot ({}^{\ominus}T)) = x(S \odot T).\end{aligned}$$

(Case 3: x is any irrational number, $S < O$ and $T > O$.) The proof is Exercise REAL.7.

(Case 4: x is any irrational number, $S > O$ and $T < O$.) Applying, in succession, Theorem OF.10(A), Theorem OF.10(D), Case 1 above, Theorem OF.10(D), and Theorem REAL.21(C).

$$\begin{aligned}(xS) \odot T &= (xS) \odot {}^{\ominus}({}^{\ominus}T) = {}^{\ominus}((xS) \odot ({}^{\ominus}T)) \\ &= {}^{\ominus}(x(S \odot ({}^{\ominus}T))) = {}^{\ominus}(x({}^{\ominus}(S \odot T))) \\ &= x(S \odot T).\end{aligned}$$

Cases 1 through 4 show that for all real numbers x and all members S and T of $\mathbb{L}$, $(xS) \odot T = x(S \odot T)$. Using this result, by commutativity, $x(S \odot T) = x(T \odot S) = (xT) \odot S = S \odot (xT)$, completing the proof. □

Theorem REAL.26. *If x and y are real numbers, and U is the unit for $\mathbb{L}$, then $(xU) \odot (yU) = (xy)U = x(yU) = y(xU)$.*

Proof. Applying Theorem REAL.25 twice, then Theorem REAL.23, we have

$$(xU) \odot (yU) = x(U \odot (yU)) = x(y(U \odot U)) = x(yU) = (xy)U = y(xU). \quad \square$$

Definition REAL.27 (Notation for addition of sets).

(A) If both $\mathcal{S}$ and $\mathcal{T}$ are subsets of $\mathbb{L}$, we will denote the set $\{X \oplus Y \mid X \in \mathcal{S}$ and $Y \in \mathcal{T}\}$ by $\mathcal{S} \oplus \mathcal{T}$.

(B) If $\mathcal{S}$ is a subset of, and H is a point of $\mathbb{L}$ we denote the set $\{X \oplus H \mid X \in \mathcal{S}\}$ by $\mathcal{S} \oplus H$.

Note that if either of the sets $\mathcal{S}$ or $\mathcal{T}$ is empty, then $\mathcal{S} \oplus \mathcal{T} = \emptyset$. Likewise, if $\mathcal{S} = \emptyset$, $\mathcal{S} \oplus H = \emptyset$.

Theorem REAL.28. (A) *If $\mathcal{S}$ and $\mathcal{T}$ are nonempty subsets of $\mathbb{L}$ which are bounded above, $\mathcal{S} \oplus \mathcal{T}$ is bounded above and* $\operatorname{lub}(\mathcal{S} \oplus \mathcal{T}) = \operatorname{lub}\mathcal{S} \oplus \operatorname{lub}\mathcal{T}$.

(B) *if $\mathcal{S}$ and $\mathcal{T}$ are nonempty subsets of $\mathbb{L}$ which are bounded below, $\mathcal{S} \oplus \mathcal{T}$ is bounded below and* $\operatorname{glb}(\mathcal{S} \oplus \mathcal{T}) = \operatorname{glb}\mathcal{S} \oplus \operatorname{glb}\mathcal{T}$.

Proof. (A) (I) Let B be an upper bound of $\mathcal{S}$ and let D be an upper bound of $\mathcal{T}$. For all X and Y, if $X \in \mathcal{S}$ and $Y \in \mathcal{T}$, then by two applications of Theorem OF.11(B), $X \oplus Y \leq B \oplus D$, so $\mathcal{S} \oplus \mathcal{T}$ is bounded above. By Axiom LUB, lub$(\mathcal{S} \oplus \mathcal{T})$ exists. Moreover, since for all members X of $\mathcal{S}$ and all members Y of $\mathcal{T}$, $X \leq \text{lub}\,\mathcal{S}$ and $Y \leq \text{lub}\,\mathcal{T}$. $X \oplus Y \leq \text{lub}\,\mathcal{S} \oplus \text{lub}\,\mathcal{T}$, and $\text{lub}\,\mathcal{S} \oplus \text{lub}\,\mathcal{T}$ is an upper bound for $\mathcal{S} \oplus \mathcal{T}$. Thus the least such upper bound $\text{lub}(\mathcal{S} \oplus \mathcal{T}) \leq \text{lub}\,\mathcal{S} \oplus \text{lub}\,\mathcal{T}$.

(II) Suppose now that $\text{lub}(\mathcal{S} \oplus \mathcal{T}) < \text{lub}\,\mathcal{S} \oplus \text{lub}\,\mathcal{T}$; there exists $\epsilon > O$ such that $\text{lub}(\mathcal{S} \oplus \mathcal{T}) < \text{lub}\,\mathcal{S} \oplus \text{lub}\,\mathcal{T} \ominus \epsilon$. By Exercise REAL.5, there is a member X of $\mathcal{S}$ such that $X > (\text{lub}\,\mathcal{S}) \ominus \frac{\epsilon}{2}$ and a member Y of $\mathcal{T}$ such that $Y > (\text{lub}\,\mathcal{T}) \ominus \frac{\epsilon}{2}$. Since $X \oplus Y \in \mathcal{S} \oplus \mathcal{T}$,
$\text{lub}(\mathcal{S} \oplus \mathcal{T}) \geq X \oplus Y > ((\text{lub}\,\mathcal{S}) \ominus \frac{\epsilon}{2}) \oplus ((\text{lub}\,\mathcal{T}) \ominus \frac{\epsilon}{2}) = \text{lub}\,\mathcal{S} \oplus \text{lub}\,\mathcal{T} \ominus \epsilon$,
a contradiction. Thus $\text{lub}(\mathcal{S} \oplus \mathcal{T}) \geq \text{lub}\,\mathcal{S} \oplus \text{lub}\,\mathcal{T}$. Together with part (I), this shows that $\text{lub}(\mathcal{S} \oplus \mathcal{T}) = \text{lub}\,\mathcal{S} \oplus \text{lub}\,\mathcal{T}$.

(B) Let B be a lower bound of $\mathcal{S}$ and let D be a lower bound of $\mathcal{T}$. For all X and Y, if $X \in \mathcal{S}$ and $Y \in \mathcal{T}$, then by Theorem OF.11(B) $X \oplus Y \geq B \oplus D$, $\mathcal{S} \oplus \mathcal{T}$ is bounded below, and by Theorem REAL.4(A) glb$(\mathcal{S} \oplus \mathcal{T})$ exists. Then

$$\begin{aligned}\text{glb}(\mathcal{S} \oplus \mathcal{T}) &= {}^{\ominus}\text{lub}({}^{\ominus}(\mathcal{S} \oplus \mathcal{T})) = {}^{\ominus}\text{lub}(({}^{\ominus}\mathcal{S}) \oplus ({}^{\ominus}\mathcal{T})) \\ &= {}^{\ominus}(\text{lub}({}^{\ominus}\mathcal{S}) \oplus \text{lub}({}^{\ominus}\mathcal{T})) = ({}^{\ominus}\text{lub}({}^{\ominus}\mathcal{S})) \oplus ({}^{\ominus}\text{lub}({}^{\ominus}\mathcal{T})) \\ &= \text{glb}\,\mathcal{S} \oplus \text{glb}\,\mathcal{T},\end{aligned}$$

where the first equality in this string is by Theorem REAL.4(B), the second by Theorem OF.10(F), the third by part (A), the fourth by Theorem OF.10(F), and the last is by Theorem REAL.4(B). □

Remark REAL.29. The proof of Theorem REAL.31 consists of several cases; we initially thought we might state it as a series of theorems, but finally decided on the structure given here. To complete this proof, we need the following numerical result, which is quite intuitive but oddly complex to prove.

Lemma REAL.30. *Let x and y be irrational numbers such that $x > 0$ and $y > 0$. Then r is a rational number such that $0 < r < x + y$ iff there exist rational numbers s and t such that $0 < s < x$, $0 < t < y$, and $r = s + t$.*

Proof. Suppose r is a rational number such that $0 < r < x+y$; then $-y < r-y < x$ so $r - y < x$; since $y > 0$, $r - y < r$. Therefore $\min\{r, x\} > r - y$.

Also, both $r > 0$ and $x > 0$, so $\min\{r, x\} > 0$ and $\min\{r, x\} > \max\{r - y, 0\}$. Choose s to be a rational number such that $\min\{r, x\} > s > \max\{r - y, 0\}$, and define $t = r - s$.

By definition, $s > 0$ and $s < x$. Since $s < r$, $-s > -r$ so that $t = r-s > r-r = 0$. Also $s > r - y$ implies that $-s < y - r$, so that $t = r - s < r + y - r = y$, that is, $t < y$. Therefore, $r = s + t$ where $0 < s < x$ and $0 < t < y$.

Conversely, suppose there exist rational numbers s and t such that $0 < s < x$, $0 < t < y$, and $r = s + t$; then $0 < r < x + y$. □

Theorem REAL.31 (Distributive property I). *Let x and y be any real numbers, and let H be any member of $\mathbb{L}$. Then $(x + y)H = xH \oplus yH$.*

Proof. (Case 0: $x = 0$ or $y = 0$ or $H = O$.) If $x = 0$, then $(x + y)H = yH = 0H \oplus yH = xH \oplus yH$. Similarly for $y = 0$. If $H = O$, then $(x + y)H = (x + y)O = O \oplus O = xH \oplus yH$. Here we have used Definition REAL.19(A)(1).

In the remainder of the proof we assume that x and y are nonzero, and $H \neq O$.

(Case 1: x and y are both rational numbers.) This is Theorem QX.11(A).

(Case 2: One of x or y is a rational number and the other is irrational, and $H > O$.) Without loss of generality we choose x to be the rational number s, and y to be irrational. As usual we denote rational numbers by r, s, and t. By Definition REAL.19(A)(3),

$$\begin{aligned}(s + y)H &= \text{lub}\{rH \mid r \in \mathbb{Q} \text{ and } 0 < r < s + y\}\\ &= \text{lub}\{(s + t)H \mid 0 < t < y\}.\end{aligned}$$

We now apply Theorem QX.11(A), Theorem REAL.28(A), and Definition REAL.19(A)(3) to get

$$\begin{aligned}&= \text{lub}\{sH \oplus tH \mid 0 < t < y\}\\ &= sH \oplus \text{lub}\{rH \mid 0 < r < y\} = sH \oplus yH.\end{aligned}$$

(Case 3: $x > 0$ and $y > 0$ are irrational numbers, and $H > O$.) By Definition REAL.19(A).3, $(x + y)H = \text{lub}\{rH \mid 0 < r < x + y\}$. By Lemma REAL.30,

$$\{rH \mid 0 < r < x + y\} = \{(s + t)H \mid 0 < s < x \text{ and } 0 < t < y\},$$

and by Theorem QX.11(A) this is

$$\{sH \oplus tH \mid 0 < s < x \text{ and } 0 < t < y\}.$$

By Definition REAL.27(A), this is

$$\{sH \mid 0 < s < x\} \oplus \{tH \mid 0 < t < y\}.$$

Thus by Theorem REAL.28(A),

$$\begin{aligned}(x+y)H &= \text{lub}\{rH \mid 0 < r < x+y\} \\ &= \text{lub}(\{sH \mid 0 < s < x\} \oplus \{tH \mid 0 < t < y\}) \\ &= \text{lub}(\{sH \mid 0 < s < x\} \oplus \{tH \mid 0 < t < y\}) \\ &= xH \oplus yH.\end{aligned}$$

(Case 4: $x > 0$ and $y > 0$ are irrational numbers, and $H < O$.) Applying Theorem REAL.21(C), Case 3 above, Theorem REAL.21(C), and Theorem OF.10(F), in that order, we have

$$\begin{aligned}{}^{\ominus}((x+y)H) &= (x+y)({}^{\ominus}H) = x({}^{\ominus}H) \oplus y({}^{\ominus}H) \\ &= {}^{\ominus}(xH) \oplus {}^{\ominus}(yH) = {}^{\ominus}(xH \oplus yH).\end{aligned}$$

By Theorem OF.10(A), $(x+y)H = xH \oplus yH$.

Cases 3 and 4 combined say that for irrational numbers $x > 0$ and $y > 0$, $(x+y)H = xH \oplus yH$ for any $H \in \mathbb{L}$.

(Case 5: $x < 0$ and $y < 0$ are irrational numbers, and H is any member of $\mathbb{L}$.) The proof is Exercise REAL.8.

(Case 6: x and y are irrational numbers, one of which is greater than 0 and the other less than 0, and H is any member of $\mathbb{L}$.) Without loss of generality, we assume that $x > 0$ and $y < 0$.

(Subcase 6A: $x + y > 0$.) Then $-y > 0$ so we may apply Cases 3 and 4 and Theorem REAL.21(A) to get

$$xH = (x+y-y)H = (x+y)H \oplus (-y)H = (x+y)H \oplus {}^{\ominus}(yH).$$

Adding yH to both sides we have, by Definition OF.4, $xH \oplus yH = (x+y)H$.

(Subcase 6B: $x + y < 0$.) Then $-x < 0$ so we may apply Case 5 and Theorem REAL.21(A) to get

$$yH = (-x+x+y)H = (-x)H \oplus (x+y)H = {}^{\ominus}(xH) \oplus (x+y)H.$$

Adding xH to both sides we have $xH \oplus yH = (x+y)H$. □

Theorem REAL.32 (Distributive property II). *Let x be any real number, and let S and T be members of $\mathbb{L}$. Then $x(S \oplus T) = xS \oplus xT$.*

Proof. By Theorem OF.3, $S \oplus T = U \odot (S \oplus T)$; then applying this, Theorem REAL.25, Theorem OF.6, Theorem REAL.25 again, and Theorem OF.3, we have

$$\begin{aligned}x(S \oplus T) &= x(U \odot (S \oplus T)) = (xU) \odot (S \oplus T)) \\ &= (xU \odot S) \oplus (xU \odot T) = x(U \odot S) \oplus x(U \odot T) \\ &= xS \oplus xT.\end{aligned}$$

□

The above theorem may also be proved directly from Definition REAL.19; the proof is Exercise REAL.9.

Theorem REAL.33. *Let $H > O$ be a member of $\mathbb{L}$, and let x and y be real numbers.*

(A) $x < y$ *iff* $xH < yH$ *iff* $y(^{\ominus}H) < x(^{\ominus}H)$.
(B) $x > 0$ *iff* $xH > O$ *iff* $x(^{\ominus}H) < O$.
(C) $x < 0$ *iff* $xH < O$ *iff* $x(^{\ominus}H) > O$.

Proof. (B) Let s be a rational number such that $0 < s < x$; by Theorem QX.13(A) $sH > O$. Then $sH \in \{rH \mid r < x\}$ so that $O < sH \leq \text{lub}\{rH \mid r < x\}$; by Definition REAL.19(A)(3) this is xH. If $xH = \text{lub}\{rH \mid r < x\} > O$, there must be some member $sH \in \{rH \mid r < x\}$ such that $sH > O$; by Theorem QX.13 $s > 0$, and therefore $x > s > 0$. By Theorem REAL.21(C) $x(^{\ominus}H) = {}^{\ominus}(xH)$ which is negative iff $xH > O$, by Theorem OF.10(B).

(A) $x < y$ iff $y - x > 0$ and by part (B) above, this is true iff $(y - x)H > O$ iff $(y - x)(^{\ominus}H) > O$. By Theorem REAL.31 and Theorem REAL.21(A), $(y - x)H > O$ is equivalent to
$O < (y - x)H = (yH) \oplus ((-x)H) = (yH) \oplus (^{\ominus}(xH)) = (yH) \ominus (xH)$.
By Theorem OF.11(A), this is true iff $(xH) < (yH)$. By the same theorem, it is also true iff $^{\ominus}(yH) < {}^{\ominus}(xH)$, and by Theorem REAL.21(C) this is $y(^{\ominus}H) < x(^{\ominus}H)$.

(C) $x < 0$ iff $-x > 0$ which by part (B) is true iff $(-x)H > O$ iff $(-x)(^{\ominus}H) < O$. By Theorem REAL.21(A) $^{\ominus}(xH) = (-x)H > O$ which by Theorem OF.10(B) is true iff $xH < O$. By Theorem REAL.21(C) and Theorem OF.10(B) $x(^{\ominus}H) = {}^{\ominus}(xH) > O$. □

Corollary REAL.34. *Let H be any member of $\mathbb{L}$ other than O, and let x and y be real numbers.*

(A) $x \neq 0$ *iff* $xH \neq O$.
(B) $x = 0$ *iff* $xH = O$.
(C) $x = y$ *iff* $xH = yH$.

Proof. (A) $x \neq 0$ iff either $x > 0$ (in which case, by Theorem REAL.33(B), $xH > O$ and $x(^{\ominus}H) < O$), or $x < 0$ (in which case $xH < O$ and $x(^{\ominus}H) > O$). In either case, $xH \neq O$ and $x(^{\ominus}H) \neq O$.

Conversely, if $xH \neq O$ or $x(^{\ominus}H) \neq O$, then $x \neq 0$ by Definition REAL.19(A)(1).

(B) If $x = 0$ by Definition REAL.19(A)(1) $xH = O$ and $x(^{\ominus}H) = O$. If $x \neq 0$ by part (A) $xH \neq O$.

(C) $x = y$ iff $x - y = 0$ iff $(x - y)H = xH \ominus yH = O$, by part (B). Since by Theorem OF.2(A) $\mathbb{L}$ is a group under "$\oplus$", $xH \ominus yH = O$ iff $xH = (xH \ominus yH) \oplus yH = yH$. □

Theorem REAL.35. *Let $\mathcal{P}$ be a Euclidean/LUB plane, and let $\mathbb{L} \subseteq \mathcal{P}$ be an ordered field with origin O and unit U. Define a mapping Θ from the set $\mathbb{R}$ of all real numbers to $\mathbb{L}$, as follows: for each real number x define $\Theta(x) = xU$. Then Θ is an order-preserving isomorphism of $\mathbb{R}$ onto $\mathbb{L}$. That is,*

(A) *For every $A \in \mathbb{L}$ there exists a unique real number x such that $xU = A$; and Θ is a bijection of $\mathbb{R}$ onto $\mathbb{L}$;*

(B) *for every x and y in $\mathbb{R}$, $x < y$ iff $\Theta(x) < \Theta(y)$;*

(C) *for every x and y in $\mathbb{R}$, $\Theta(x + y) = \Theta(x) \oplus \Theta(y)$; and*

(D) *for every x and y in $\mathbb{R}$, $\Theta(x \cdot y) = \Theta(x) \odot \Theta(y)$.*

Proof. (B) is Theorem REAL.33(A). (C) is Theorem REAL.31. (D) is Theorem REAL.26.

What remains to be proved is part (A). That Θ is one-to-one (and hence x is unique) is easy to see from part (B); for if $x \neq y$ then either $x < y$ or $y < x$ so that either $\Theta(x) < \Theta(y)$ or $\Theta(y) < \Theta(x)$, hence $\Theta(y) \neq \Theta(x)$.

To prove that Θ is onto $\mathbb{L}$ we must show that for every $A \in \mathbb{L}$ there exists a real number x such that $xU = A$.

(Case 1: $A > O$.) Define $\mathcal{D} = \{r \mid r \in \mathbb{Q}$ *and* $rU < A\}$; by Exercise REAL.3, $\mathcal{D}$ is bounded above. Let $x = \operatorname{lub} \mathcal{D}$. We will prove that $xU = A$.

(I) If $xU < A$, by Eudoxus' theorem there exists a rational number r such that $xU < rU < A$. By Theorem REAL.33(A) $x < r$, and by $rU < A$, x is not an upper bound for $\mathcal{D} = \{r \mid r \in \mathbb{Q}$ *and* $rU < A\}$, contradicting the definition of x.

(II) If $xU > A$, by Eudoxus' theorem there exists a rational number r such that $A < rU < xU$. If $s \in \mathcal{D}$ then $sU < A < rU$ and hence $s < r$ by Theorem REAL.33(A). Thus r is an upper bound for $\mathcal{D}$, which is smaller than x, and x is not the least upper bound for $\mathcal{D}$. Hence $A = xU$.

(Case 2: $A < O$.) By Case 1, there exists a real number x such that ${}^{\ominus}A = xU$. By Theorem REAL.21(A), $(-x)U = {}^{\ominus}(xU) = {}^{\ominus}({}^{\ominus}A) = A$. □

Corollary REAL.35.1. *For any two points A and B of $\mathbb{L}$ which are distinct from O, there exists a unique real number t such that $tA = B$.*

Proof. By Theorem REAL.35(A) for all A and B in $\mathbb{L}\backslash\{O\}$, there exist real numbers r and s such that $rU = A$ and $sU = B$. By Theorem REAL.23 $U = (\frac{1}{s}s)U = \frac{1}{s}(sU) = \frac{1}{s}B$, so

$$A = r(\tfrac{1}{s}B) = (r\tfrac{1}{s})B = \tfrac{r}{s}B.$$

Let $t = \frac{r}{s}$. Then $tA = B$.

To show uniqueness, suppose that u is any real number such that $uA = B$. Then by Theorem REAL.21 $(-u)A = {}^{\ominus}(uA) = {}^{\ominus}B$. By Theorem REAL.31

$$(t-u)A = (t + (-u))A = tA \oplus (-u)A = B \oplus {}^{\ominus}B = O;$$

by Corollary REAL.34, $t - u = 0$ so that $u = t$. □

Remark REAL.36. (A) Eudoxus' theorem (Theorem REAL.15) moves a well-known property of the real numbers (that between any two real numbers, there is a rational number) over to the line $\mathbb{L}$. If Theorem REAL.35 had been proved before Theorem REAL.15, the latter would have become a consequence of Theorem REAL.33(A) and the fact that Eudoxus' theorem holds in the real numbers.

As it is, Theorem REAL.15 is needed to prove Theorem REAL.35, by showing that the mapping Θ is onto $\mathbb{L}$.

(B) Theorem REAL.35 shows that $\mathbb{L}$ is isomorphic to the set $\mathbb{R}$ of real numbers, so that these two sets cannot be distinguished algebraically, and can be identified. Thus our axioms, even though they seem to have nothing to do with real numbers, provide a plane in which the set of all real numbers can be embedded.

Because of this isomorphism it would be possible, when discussing points on a line $\mathbb{L}$, to use lowercase italic letters $a, b, c, \ldots$ both for real numbers and for points on the line, treating them all as real numbers. In the interest of conceptual clarity in some circumstances, we will not do this, instead maintaining for the time being the notational distinction between real numbers and points, using lowercase letters for the former and capital letters for the latter. We do, however, at this point abandon the symbols $\oplus$, $\ominus$, and $\odot$ in favor of the ordinary $+$, $-$, and "$\cdot$" or juxtaposition, using the same symbols for operations on both points and real numbers. We will revert to the use of $\oplus$, $\ominus$, or $\odot$ only in cases where we need two symbols to distinguish between two different operations.

The product of two points on a line is defined using dilations; the product of a real number and a point on a line is defined differently. The following theorem shows that the product of a real number and a point can also be expressed using a dilation.

Theorem REAL.37. *If U is the unit in $\mathbb{L}$, T is any point of $\mathbb{L}$, x is any real number, and δ_x is a dilation of the plane $\mathcal{P}$ with fixed point O, then $\delta_x(U) = xU$ iff $\delta_x(T) = xT$. That is, there is a single dilation δ_x such that $\delta_x(T) = xT$ for every $T \in \mathbb{L}$.*

Proof. Let δ_x and δ_T be dilations with fixed point O such that $\delta_x(U) = xU$ and $\delta_T(U) = T$. Then by Exercise DLN.3 (commutativity of dilations)

$$\delta_x(T) = \delta_x(\delta_T(U)) = \delta_T(\delta_x(U)) = \delta_T(xU) = T \odot xU = xU \odot T,$$

and by Theorem REAL.25 this is xT.

Conversely, if $\delta_x(T) = xT$, and $\delta_{T^{-1}}$ is the dilation with fixed point O such that $\delta_{T^{-1}}(U) = T^{-1}$,

$$\begin{aligned}\delta_x(U) &= \delta_x(\delta_{T^{-1}}(T)) = \delta_{T^{-1}}(\delta_x(T))\\ &= \delta_{T^{-1}}(xT) = T^{-1} \odot xT = xT \odot T^{-1},\end{aligned}$$

which by Theorem REAL.25 is xU. □

Definition REAL.38. Let x be any real number. Define δ_x as the dilation on $\mathcal{P}$ with fixed point O such that for all $T \in \mathbb{L}$, $\delta_x(T) = xT$.

Remark REAL.39. In Theorem REAL.37 we established that *for points A and B on a given line through O*, if x is a real number, there is a single dilation δ_x (with fixed point O) such that both $xA = \delta_x(A)$ and $xB = \delta_x(B)$. This establishes that the dilation δ_x given by Definition REAL.38 is a "good" definition. In Theorems REAL.40, REAL.41, and REAL.42 we extend this result to the entire plane, showing that even if A and B *are on different lines* $\mathbb{L}_1$ and $\mathbb{L}_2$ through O, $xA = \delta_x(A)$ and $xB = \delta_x(B)$.

Theorem REAL.40. *Let $\mathcal{P}$ be a Euclidean/LUB plane, and let O be a point (the origin) on $\mathcal{P}$. Suppose that $\mathbb{L}_1$ and $\mathbb{L}_2$ are distinct lines which intersect at the point O, their common origin, and that each has been built into an ordered field, with units U_1 and U_2, respectively. Let ρ be the rotation with fixed point O such that*

$\rho(U_1) \in \overrightarrow{OU_2}$, and let $A \in \overrightarrow{OU_1}$ and $B \in \overrightarrow{OU_2}$ be two points such that $\rho(A) = B$. Then for any rational number $r = \frac{m}{n} > 0$, $\rho(rA) = rB$ and $\overline{OrA} \cong \overline{OrB}$.

Proof. (I) We prove that for any natural number n, $\rho(nA) = nB = n\rho(A)$, so that $\overline{OnA} \cong \overline{OnB}$. This is trivially true for $n = 1$ because $\rho(A) = B$. Suppose we have proved it for n. By Definition QX.1(C), $(n+1)A = nA+A$ and $(n+1)B = nB + B$. Then $\overline{O(n+1)A} = \overline{O(nA+A)}$ and $\overline{O(n+1)B} = \overline{O(nB+B)}$. By Corollary OF.18, $\overline{O(nA+A)} \cong \overline{O(nB+B)}$ since $\overline{OnA} \cong \overline{OnB}$ and $\overline{OA} \cong \overline{OB}$. Then $\overline{O(nB+B)} \cong \overline{O(nA+A)} \cong \overline{O\rho(nA+A)}$, and since both $\rho(nA+A)$ and $nB + B$ are members of $\overrightarrow{OU_2}$, $\rho(nA + A) = nB + B$ by Property R.4 of Definition NEUT.2. This proves (I) by induction.

(II) Next we show from $\rho(A) = B$, that $\rho(\frac{1}{n}A) = \frac{1}{n}B$. For if $\rho(\frac{1}{n}A) \neq \frac{1}{n}B$, by Theorem ORD.5 (Trichotomy) either $\rho(\frac{1}{n}A) < \frac{1}{n}B$ or $\rho(\frac{1}{n}A) > \frac{1}{n}B$.

Assume that $\rho(\frac{1}{n}A) < \frac{1}{n}B$. Claim: for every natural number m, $\rho(\frac{m}{n}A) < \frac{m}{n}B$. This is true by assumption for $m = 1$. Assume we have proved it for m. Applying (I) to $\frac{1}{n}A$ and using Theorem OF.11(C) and our assumption that $\rho(\frac{1}{n}A) < \frac{1}{n}B$,

$$\rho(\tfrac{m+1}{n}A) = \rho((m+1)\tfrac{1}{n}A) = (m+1)\rho(\tfrac{1}{n}A) < (m+1)\tfrac{1}{n}B = \tfrac{m+1}{n}B.$$

This proves the claim. The claim holds, in particular for $m = n$, so that $\rho(A) = \rho(\frac{n}{n}A) < \frac{n}{n}B = B$, contradicting our original assumption that $\rho(A) = B$.

By similar methods, a contradiction follows also from the other alternative. Therefore $\rho(\frac{1}{n}A) = \frac{1}{n}B$, and $\overline{O\rho(\frac{1}{n}A)} \cong \overline{O\frac{1}{n}B}$.

(III) It follows from (I) and (II) above, that if $\rho(A) = B$, for any rational number $r = \frac{m}{n} > 0$, $\rho(rA) = rB$ and $\overline{OrA} \cong \overline{OrB}$. □

Theorem REAL.41. *Assume the hypotheses of Theorem REAL.20. As in that theorem, let ρ be the rotation with fixed point O such that $\rho(U_1) \in \overrightarrow{OU_2}$, and let $A \in \overrightarrow{OU_1}$ and $B \in \overrightarrow{OU_2}$ be such that $\rho(A) = B$. Then for any real number x, $\rho(xA) = xB$, so that $\overline{OxA} \cong \overline{O\rho(xA)} = \overline{OxB}$.*

Proof. (Case 1: $x > 0$.) We consider only irrational x, since for rational x the result is already proved by Theorem REAL.40. Let

$\mathcal{E}_1 = \{rA \mid r \in \mathbb{Q} \text{ and } 0 < r < x\}$, and

$\mathcal{E}_2 = \{rB \mid r \in \mathbb{Q} \text{ and } 0 < r < x\}$.

By Definition REAL.19(A)(3), $xA = \operatorname{lub} \mathcal{E}_1$ and $xB = \operatorname{lub} \mathcal{E}_2$. By Theorem REAL.40, for every rational $r > 0$, $\rho(rA) = rB$, and therefore $\rho(\mathcal{E}_1) = \mathcal{E}_2$.

Suppose Y is any upper bound for $\rho(\mathcal{E}_1) = \mathcal{E}_2$, that is, for every rational r with $0 < r < x$, $Y > \rho(rA) > O$. Now ρ and ρ^{-1} are isometries, so by Theorem NEUT.15, each is a belineation; hence by Theorem ORD.6, $\rho^{-1}(Y) > \rho^{-1}(\rho(rA)) = rA > O$. It follows that $\rho^{-1}(Y)$ is an upper bound for $\mathcal{E}_1$ and therefore $\rho^{-1}(Y) \geq \text{lub}\,\mathcal{E}_1$. By definition of least upper bound, $\rho^{-1}(\text{lub}\,\mathcal{E}_2) \geq \text{lub}\,\mathcal{E}_1$.

By a similar argument, if X is an upper bound for $\mathcal{E}_1$, $\rho(X)$ is an upper bound for $\mathcal{E}_2$ and thus $\rho(X) \geq \text{lub}\,\mathcal{E}_2$, hence $\rho(\text{lub}\,\mathcal{E}_1) \geq \text{lub}\,\mathcal{E}_2$.

It follows that $\rho(\text{lub}\,\mathcal{E}_1) = \text{lub}\,\mathcal{E}_2$, that is, $\rho(xA) = xB$.

(Case 2: $x < 0$.) Then $-x > 0$ and by Case 1, $\rho((-x)A) = (-x)B$. By Theorem REAL.21(A) this is $\rho(-xA) = -xB$. By Theorem OF.10(A), $-(xA) = \mathcal{R}_O(xA)$ and $-(xB) = \mathcal{R}_O(xB)$. Because rotations commute, we have $\mathcal{R}_O(\rho(xA)) = \rho(\mathcal{R}_O(xA)) = \rho(-(xA)) = -xB = \mathcal{R}_O(xB)$, and applying $\mathcal{R}_O$ to both sides, we have $\rho(xA) = xB$. Since ρ is an isometry, $\overline{O(xA)} \cong \overline{O(xB)}$. □

Theorem REAL.42. *Let $\mathcal{P}$ be a Euclidean/LUB plane, and let O be a point (the origin) on $\mathcal{P}$. Let A and B be points of $\mathcal{P}$ such that A, B, and O are noncollinear, and suppose that $\overleftrightarrow{OA} = \mathbb{L}_1$ and $\overleftrightarrow{OB} = \mathbb{L}_2$ have been built into ordered fields with units U_1 and U_2, respectively. Suppose further that δ_x is the dilation of Definition REAL.38 such that for all $T \in \mathbb{L}_1$, $\delta_x(T) = xT$. Then $\delta_x(A) = xA$ iff $\delta_x(B) = xB$.*

Proof. By Theorem REAL.37, if $\delta_x(A) = xA$ then $\delta_x(U_1) = xU_1$. Let ρ be the rotation about O such that $\rho(U_1) \in \overrightarrow{OU_2}$, and let $C = \rho(U_1)$. Then by Theorem REAL.41 and the commutativity of dilations and rotations (Theorem DLN.7(E)),

$$xC = \rho(xU_1) = \rho(\delta_x(U_1)) = \delta_x(\rho(U_1)) = \delta_x(C)$$

and again by Theorem REAL.37, for every $B \in \mathbb{L}_2$, $\delta_x(B) = xB$. Thus if $\delta_x(A) = xA$ then $\delta_x(B) = xB$. The converse follows from reversing the roles of A and B. This establishes that multiplication by a real number x on the plane is implemented by a single dilation δ_x for all points on the plane. □

Remark REAL.43. Finally, we make an important connection between free segments and real numbers. Let $\mathcal{P}$ be a Euclidean/LUB plane, and let $\mathbb{L}$ be a line in $\mathcal{P}$ which has been built into an ordered field and identified with $\mathbb{R}$, the set of real numbers. If O is the origin and U the unit of $\mathbb{L}$, and if $A = aU$ and $B = bU$ are points of $\mathbb{L}$ such that a and b are both positive, then the following are true.

(A) $[\overline{O(aU)}] + [\overline{O(bU)}] = [\overline{O(aU + bU)}] = [\overline{O((a+b)U)}]$ by Theorem OF.17 and Theorem REAL.31.

(B) $[\overline{O(aU)}] \cdot [\overline{O(bU)}] = [\overline{O((aU)(bU))}] = [\overline{O((ab)U)}]$ by Definition SIM.7 and Theorems REAL.23 and REAL.25.

(C) By part (B) $[\overline{O(aU)}] \cdot [\overline{O(\frac{1}{a}U)}] = [\overline{O(a\frac{1}{a})U}] = [\overline{OU}]$ so that $[\overline{O(\frac{1}{a}U)}] = [\overline{O(aU)}]^{-1}$. Then by Definition SIM.12,

$$\frac{[\overline{O(aU)}]}{[\overline{O(bU)}]} = [\overline{O(aU)}] \cdot [\overline{O(\tfrac{1}{b}U)}] = [\overline{O((a\tfrac{1}{b})U)}] = [\overline{O(\tfrac{a}{b}U)}].$$

18.4 Coordinatizing the plane

So far in this chapter we have shown how to assign a real number to each point on a line in a Euclidean/LUB plane; this could be called *coordinatizing* the line. Now we go one step farther: we coordinatize the Euclidean plane, assigning to each point on it a pair (a, b) of numbers. It would be possible to coordinatize Euclidean space, assigning to each point a triple (a, b, c) of real numbers, but we do not pursue this.

Our treatment is necessarily somewhat sketchy, and we rely on the reader's prior familiarity with vector spaces, in particular with the vector space consisting of ordered pairs (a, b) of real numbers—that is, the *coordinate plane*. A summary of these matters is found in Chapter 1, Section 1.5; the reader who desires more detail may wish to consult the supplementary material online which may be accessed from the home page for this book at www.springer.com.

Here we shall use the acronym "*RR*" to suggest the coordinate plane, consisting of the Cartesian product of the real line with itself. We also remind the reader that we have abandoned the symbols $\oplus$, $\ominus$, and $\odot$ in favor of the ordinary $+$, $-$, and "$\cdot$" or juxtaposition, and will use the same symbols for operations on both points and real numbers.

Definition RR.1. (A) For each $A \in \mathcal{P} \setminus \{O\}$, define τ_A to be the translation of $\mathcal{P}$ such that $\tau_A(O) = A$. Theorem ISM.5 says that such a translation exists and is unique.

(B) Define $\tau_O = \iota$, the identity.

(C) For any A and B in $\mathcal{P}$, define

$$A + B = (\tau_B \circ \tau_A)(O) = \tau_B(\tau_A(O)) = \tau_B(A).$$

The operation $+$ is called **addition** and $A + B$ is the **sum** of A and B.

Remark RR.2. (A) The operation $+$ from Definition RR.1 applied to points on a line $\mathbb{L}$ through O is identical to the operation $\oplus$ from Definition OF.1(A) and (C). It is quite easy to see (from Theorem ISM.8(A)) that the Euclidean/LUB plane $\mathcal{P}$ is an abelian group under the operation $+$.

(B) It is also easy to see, from Exercise ISM.2, that if O, A, and B are noncollinear points, then $A+B$ is the fourth corner of the parallelogram whose other corners are O, A, and B.

(C) The translation τ_A not only maps O to A but also maps B to $A+B$, and $\tau_{A-B}(B) = (A-B)+B = A$ so τ_{A-B} maps B to A.

(D) If A and B are any two points, then $\tau_{-B}(B) = O$ and $\tau_{-B}(A) = A-B$. By Theorem NEUT.15(5) (since τ_{-B} is an isometry) $\tau_{-B}(\overline{AB}) = \overline{(A-B)O}$ and hence $\overline{AB} \cong \overline{O(A-B)}$.

(E) Since the line $\mathbb{L} = \overleftrightarrow{OA}$ is built into an ordered field using the machinery of Chapter 14, by Theorem OF.10(A), for each $A \in \mathcal{P}$, $-A = \mathcal{R}_O(A)$. Hence for any A, $\mathcal{R}_O(\overline{OA}) = \overline{O\mathcal{R}_O(A)} = \overline{O(-A)}$. It follows that $\overline{AB} \cong \overline{O(A-B)} \cong \overline{O(B-A)}$.

Definition RR.3. For every point $A \in \mathcal{P}$, and every real number x, define xA as in Definition REAL.19, where the line $\overleftrightarrow{OA}$ has been built into an ordered field. xA is called the **scalar product** of x and A, and the number x is called a **scalar**.

Theorem RR.4. (A) *For every $A \in \mathcal{P} \setminus \{O\}$, $\overleftrightarrow{OA} = \{xA \in \mathcal{P} \mid x \in \mathbb{R}\}$. That is, every line through the origin is the set of all scalar multiples of any point in that line which is distinct from O.*

Moreover, if A and B are any points in $\mathcal{P}$ and x and y are any real numbers,

(B) $x(yA) = (xy)A$ *(scalar multiplication is associative)*

(C) $x(A+B) = xA + xB$ *(scalar multiplication is distributive with respect to addition of points)*

(D) $(x+y)A = xA+yA$ *(scalar multiplication is distributive with respect to addition of scalars)*

(E) $1A = A$, *and*

(F) $xA = O$ *iff* $x = 0$ *or* $A = O$.

Proof. The proof is Exercise RR.2. □

Remark RR.5. The above result, together with our previous observation that $\mathcal{P}$ forms an abelian group under "+", shows that $\mathcal{P}$ forms a vector space over the field $\mathbb{R}$ of real numbers, when equipped with the addition and scalar multiplication operations specified in Definitions RR.1 and RR.3.

Theorem RR.6. *Let $\mathcal{P}$ be a Euclidean/LUB plane, and let O be its origin. Let $\mathbb{L}_1$ and $\mathbb{L}_2$ be lines in $\mathcal{P}$ such that $\mathbb{L}_1 \cap \mathbb{L}_2 = \{O\}$ and $\mathbb{L}_1 \perp \mathbb{L}_2$. Let $U_1 \in \mathbb{L}_1$ and $U_2 \in \mathbb{L}_2$ be points (distinct from O) chosen so that $\varphi(U_1) = U_2$, where φ is the angle reflection for $\angle U_1OU_2$.*[1]

Using the machinery of Chapter 14 (OF) and the earlier part of this chapter (REAL), build each of the lines $\mathbb{L}_1$ and $\mathbb{L}_2$ into an ordered field which is isomorphic to $\mathbb{R}$, the set of all real numbers, with U_1 and U_2, respectively, as their units, so that U_1 and U_2 correspond to the real number 1 under their respective isomorphisms.

(A) *For every $A \in \mathcal{P}$, there exist unique real numbers a and b such that $A = aU_1 + bU_2$.*
(B) *$aU_1 + bU_2 = O$ iff $a = b = 0$.*
(C) *If $A \notin \mathbb{L}_1 \cup \mathbb{L}_2$, so that both $a \neq 0$ and $b \neq 0$, $\overline{O(aU_1)} \cong \overline{(bU_2)A}$ and $\overline{O(bU_2)} \cong \overline{(aU_1)A}$.*

Proof. First, note that the requirement that $\varphi(U_1) = U_2$ is not needed for the algebraic proof (nor, for that matter, is the requirement that $\mathbb{L}_1 \perp \mathbb{L}_2$). But this is geometry, and it seems only reasonable that a *reflection* carrying $\mathbb{L}_1$ to $\mathbb{L}_2$ should carry a point one unit from the origin into another such point, thus establishing the same scale on both lines. Moreover, $\varphi(U_1) = U_2$ implies that $\overline{OU_1} \cong \overline{OU_2}$, as required for the development of complex numbers. For more detail, the reader may wish to consult the Supplemental materials which may be accessed from the home page for this book at www.springer.com.

If A is any point on $\mathcal{P}$, by Axiom PS there exists a unique line $\mathbb{M}_1$ containing the point A such that either $\mathbb{M}_1 = \mathbb{L}_1$ (in case $A \in \mathbb{L}_1$) or $\mathbb{M}_1 \parallel \mathbb{L}_1$; and there exists a unique line $\mathbb{M}_2$ such that either $\mathbb{M}_2 = \mathbb{L}_2$ (in case $A \in \mathbb{L}_2$) or $\mathbb{M}_2 \parallel \mathbb{L}_2$.

By Exercise I.1, $\mathbb{M}_1$ intersects $\mathbb{L}_2$ in exactly one point, which we shall call A_2, and $\mathbb{M}_2$ intersects $\mathbb{L}_1$ in exactly one point which we call A_1. By Theorem REAL.35, there exists a unique real number a such that $A_1 = aU_1$ and a unique real number b such that $A_2 = bU_2$. Since A uniquely determines $\mathbb{M}_1$ and $\mathbb{M}_2$, and these lines

[1] The existence of φ is guaranteed by Property R.5 of Definition NEUT.2 and Axiom REF.

uniquely determine the points A_1 and A_2, which in turn uniquely determine a and b, a and b are uniquely determined by A.

Moreover, $A \in \mathbb{L}_1$ iff $A_2 = O$ iff $b = 0$, in which case

$$A = A_1 + O = A_1 + A_2 = aU_1 + bU_2;$$

$A \in \mathbb{L}_2$ iff $A_1 = O$ iff $a = 0$, in which case

$$A = O + A_2 = A_1 + A_2 = aU_1 + bU_2;$$

and $A = O$ iff $A \in \mathbb{L}_1 \cap \mathbb{L}_2$ iff $a = b = 0$, and again in this case

$$A = O + O = aU_1 + bU_2.$$

If $A \in \mathcal{P} \setminus (\mathbb{L}_1 \cup \mathbb{L}_2)$, by Theorem RR.4, $aU_1 + bU_2$ is the fourth corner of the parallelogram of which O, aU_1, bU_2 are the other three corners. Since $\mathbb{M}_1$ contains the point A_2 and $\mathbb{M}_2$ contains the point A_1 and are parallel to $\mathbb{L}_1$ and $\mathbb{L}_2$, respectively, they are the same, respectively, as the sides $\overleftrightarrow{(aU_2)(aU_1 + bU_2)}$ and $\overleftrightarrow{(aU_1)(aU_1 + bU_2)}$ of this parallelogram. Since both $\mathbb{M}_1$ and $\mathbb{M}_2$ contain A, $A = aU_1 + bU_2$. This completes the proof of parts (A) and (B).

(C) The quadrilateral $\square O(aU_1)A(bU_2)$ is a parallelogram because $\mathbb{M}_1 \parallel \mathbb{L}_1$ and $\mathbb{M}_2 \parallel \mathbb{L}_2$. The result follows from Theorem EUC.12(A). □

Definition RR.7. (A) In Theorem RR.6, the two units U_1 and U_2, together with their lines $\mathbb{L}_1$ and $\mathbb{L}_2$ will be referred to as a **coordinatization** of $\mathcal{P}$. $\mathbb{L}_1$ and $\mathbb{L}_2$ are the **axes** of this coordinatization, and O is its **origin**.

(B) For every $A \in \mathcal{P}$, by Theorem RR.6(A) there exist unique real numbers a and b such that $A = aU_1 + bU_2$. For each such $A \in \mathcal{P}$, define $\lambda(A) = \lambda(aU_1 + bU_2) = (a, b)$. This mapping is called the **coordinatization map** belonging to the coordinatization (U_1, U_2) (*cf* part (A)).

Remark RR.8. It is a fairly routine matter to verify that the mapping λ defined just above is a (vector space) isomorphism of $\mathcal{P}$ onto

$$\mathbb{R} \times \mathbb{R} = \{(a, b) \mid a \text{ and } b \text{ are both members of } \mathbb{R}\},$$

that is, onto $\mathbb{R}^2$, the *Cartesian product* of $\mathbb{R}$ and $\mathbb{R}$ (*cf* Chapter 1 Section 1.3).

Definition RR.9. For any point $(a, b) \in \mathbb{R}^2$, we will refer to a as the **first coordinate** of (a, b), and to b as its **second coordinate**. The point $(0, 0)$ is the **origin**.

Remark RR.10. It is customary, when visualizing points of the plane, to show the first coordinate on the horizontal axis (commonly called the *x*-**axis**), with positive numbers to the right of the origin; and the second coordinate on the vertical axis (commonly called the *y*-**axis**), with positive numbers above the origin. This visualization yields what is termed a **right-handed** system. Then the rotation ρ of $\mathcal{P}$ such that $\rho(\overrightarrow{OU_1}) = \overrightarrow{OU_2}$ is *counterclockwise.*

At this point, we have established the identity of the Euclidean/LUB plane with the well-known coordinate plane, and we are free to invoke notions such as slope of a line and generally to indulge in what is called "analytic" geometry. Linear transformations (as well as affine mappings) on the plane may be characterized using *matrices*, and determinants may be used to study their properties. Several interesting results relating affine mappings, collineations, and isometries are set forth in the Supplementary materials, which may be accessed from the home page for this book at www.springer.com.

18.5 Exercises for real numbers and the coordinate plane

Answers to starred (*) exercises may be accessed from the home page for this book at www.springer.com.

Exercise REAL.1*. Let A, B, C, and D be points on the Euclidean plane $\mathcal{P}$ such that $A \neq B$ and $C \neq D$. Then there exists a natural number n such that $\frac{[\overline{AB}]}{2^n} < [\overline{CD}]$.

Note: in the following exercises REAL.2 through REAL.9, $\mathcal{P}$ will denote a Euclidean/LUB plane and $\mathbb{L}$ will be a line in $\mathcal{P}$ having origin O and unit U.

Exercise REAL.2*. If T and V are positive members of $\mathbb{L}$, there exists a natural number n such that $\frac{1}{n}T < V$.

Exercise REAL.3*. If T is a positive member of $\mathbb{L}$, $\{s \mid s \in \mathbb{Q} \textit{ and } sU < T\}$ is bounded above.

Exercise REAL.4*. Prove Lemma REAL.24: if $\mathcal{E}$ is a subset of $\mathbb{L}$ which is bounded above, and $T > O$ is a member of $\mathbb{L}$, then $(\operatorname{lub}\mathcal{E}) \odot T = \operatorname{lub}(\mathcal{E} \odot T)$.

Exercise REAL.5*. Prove Lemma REAL.24: let $\mathcal{S}$ be a subset of $\mathbb{L}$ which is bounded above, and suppose A is an upper bound for $\mathcal{S}$. Then $A = \operatorname{lub} S$ iff the following property holds: for every $\epsilon > O$ in $\mathbb{L}$, there exists $x \in \mathbb{L}$ such that $x > A \ominus \epsilon$.

Exercise REAL.6*. Complete the proof of Case 4 of Theorem REAL.23: let $S > O$ be a member of $\mathbb{L}$. Then if $x < 0$ and $y > 0$ are irrational numbers, $x(yS) = (xy)S$.

Exercise REAL.7*. Complete the proof of Theorem REAL.25, Case 3: let $S < O$ and $T > O$ be members of $\mathbb{L}$. If x is an irrational number, then $(xS) \odot T = x(S \odot T)$.

Exercise REAL.8*. Complete the proof of Case 5 of Theorem REAL.31: let $x < 0$ and $y < 0$ be irrational numbers, and let H be any member of $\mathbb{L}$; then $(x + y)H = xH \oplus yH$.

Exercise REAL.9 (Alternative proof of Theorem REAL.32)*. Let x be any real number, and let S and T be members of $\mathbb{L}$. Prove, using Definition REAL.19 and other theorems from this chapter and previous ones, including Theorem REAL.21, that $x(S \oplus T) = xS \oplus xT$.

Exercise RR.1*. Complete the computations necessary to prove Remark RR.2(A) from Theorem ISM.8(A), that is, show that $\mathcal{P}$ is an abelian group under the operation $+$.

Exercise RR.2*. Prove Theorem RR.4: (A) For every $A \in \mathcal{P} \setminus \{O\}$, $\overleftrightarrow{OA} = \{xA \in \mathcal{P} \mid x \in \mathbb{R}\}$. That is, every line through the origin is the set of all scalar multiples of any point in that line which is distinct from O.

Moreover, if A and B are any points in $\mathcal{P}$ and x and y are any real numbers, (B) $x(yA) = (xy)A$, (C) $x(A + B) = xA + xB$, (D) $(x + y)A = xA + yA$, (E) $1A = A$, (F) $xA = O$ iff $x = 0$ or $A = O$ (or both).

Chapter 19
Belineations on a Euclidean/LUB Plane (AA)

Acronym: *AA*
Dependencies: *all prior Chapters 1 through 18*
New Axioms: *none*
New Terms Defined: *set of midpoints generated by a segment*

Abstract: This brief chapter shows that on a Euclidean/LUB plane, any non-identity belineation which has more than one fixed point and is not the identity, is an axial affinity; it concludes with a classification of belineations. To prove the main result of this chapter we need Axiom LUB; this explains its placement after the chapter on real numbers.

19.1 Belineations with two fixed points are axial affinities

Theorem AA.1. *Let $\mathcal{P}$ be a Euclidean/LUB plane, $\mathbb{L}$ be an ordered field on $\mathcal{P}$ with origin O and unit U. If A and B are members of $\mathbb{L}$ such that $A < B$, for every integer n let $A_n = A + n(B - A)$, so that $A_0 = A$, $A_1 = B$, $A_{-1} = A - (B - A)$, $A_2 = A + 2(B - A)$, $A_{-2} = A - 2(B - A)$, etc.*
Let $\mathbb{J} = \{A_n \mid n \text{ is an integer}\} = \{\ldots, A_{-2}, A_{-1}, A_0, A_1, A_2, \ldots\}$.

(1) *If m and n are distinct integers, $m < n$ iff $A_m < A_n$.*
(2) *If n is any integer, then $X \in \overline{A_{n-1}A_n}$ iff $A_{n-1} < X < A_n$.*
(3) *For every integer n, $\overline{A_{n-1}A_n} \cap \overline{A_nA_{n+1}} = \{A_n\}$.*
(4) *For all integers m and n, $\overline{A_mA_{m+1}} \cong \overline{A_nA_{n+1}}$.*

E.J. Specht et al., *Euclidean Geometry and its Subgeometries*,
DOI 10.1007/978-3-319-23775-6_19

(5) *For all integers m and n such that* $m+1<n$, $\overline{A_mA_{m+1}} \cap \overline{A_nA_{n+1}} = \emptyset$.

(6) *Every member of* $\mathbb{J}$ *is the midpoint of a segment whose endpoints belong to* $\mathbb{J}$. *In particular,* A_n *is the midpoint of the segment* $\overline{A_{n-1}A_{n+1}}$.

(7) *For every natural number n,*

$$\bigcup_{k=1}^{n} \overline{A_{k-1}A_k} = \overline{A_0A_n} \text{ and } \bigcup_{k=1}^{n} \overline{A_{-k+1}A_{-k}} = \overline{A_0A_{-n}}.$$

Proof. (1) $A_m < A_n$ iff $A + m(B-A) < A + n(B-A)$ iff $m < n$.

(2) Follows immediately from Theorem ORD.7.

(3) Follows immediately from the observation that A_n is a member of both segments, and if $X \in \overrightarrow{A_{n-1}A_n}$ and $Y \in \overleftarrow{A_nA_{n+1}}$ then $X < A_n < Y$.

(4) For all integers n,

$$\begin{aligned} A_{n+1} - A_n &= A + (n+1)(B-A) - (A + n(B-A)) \\ &= A + (n+1)(B-A) - A - n(B-A)) = B - A, \end{aligned}$$

so that by Theorem OF.15(B), $\overline{A_mA_{m+1}} \cong \overline{AB} \cong \overline{A_nA_{n+1}}$.

(5) If $X \in \overline{A_mA_{m+1}}$ and $Y \in \overline{A_nA_{n+1}}$, then $X \leq A_{m+1} < A_n \leq Y$, showing that these two segments are disjoint.

(6) By Theorem QX.18,

$$\begin{aligned} \tfrac{1}{2}(A_{n-1} + A_{n+1}) &= \tfrac{1}{2}(A + (n-1)(B-A) + A + (n+1)(B-A)) \\ &= \tfrac{1}{2}(2A + ((n-1) + (n+1))(B-A)) \\ &= \tfrac{1}{2}(2A + 2n(B-A)) \\ &= \tfrac{1}{2}(2(A + n(B-A))) = A + n(B-A) = A_n \end{aligned}$$

is the midpoint of the segment $\overline{A_{n-1}A_{n+1}}$.

(7) We use induction on n. Both equalities are trivially true for $n = 1$. Assume that the equalities are true for any natural number n. Then

$$\begin{aligned} \bigcup_{k=1}^{n+1} \overline{A_{k-1}A_k} &= \bigcup_{k=1}^{n} \overline{A_{k-1}A_k} \cup \overline{A_nA_{n+1}} \\ &= \overline{A_0A_n} \cup \overline{A_nA_{n+1}} = \overline{A_0A_{n+1}}, \end{aligned}$$

and

$$\begin{aligned} \bigcup_{k=1}^{n+1} \overline{A_{-(k-1)}A_{-k}} &= \bigcup_{k=1}^{n} \overline{A_{-(k-1)}A_{-k}} \cup \overline{A_{-n}A_{-(n+1)}} \\ &= \overline{A_0A_{-n}} \cup \overline{A_{-n}A_{-(n+1)}} = \overline{A_0A_{-(n+1)}}. \end{aligned}$$

Hence statement (7) holds for every natural number n. □

Theorem AA.2. *Let* $\mathbb{L}$ *be an ordered field on a Euclidean/LUB plane* $\mathcal{P}$ *with origin O and unit U, and suppose A and B are members of* $\mathbb{L}$ *such that* $A < B$. *Let*

$$\mathbb{E} = \{A + \tfrac{m}{2^n}(B-A) \mid m \text{ is an integer and } n \text{ is a natural number}\}.$$

As in Theorem AA.1, for every integer n let $A_n = A + n(B - A)$, and let $\mathbb{J} = \{A_n \mid n$ is an integer$\}$. Then $\mathbb{E}$ has the following properties:

(1) $\mathbb{J} \subseteq \mathbb{E}$.

(2) *Every member of $\mathbb{E}$ is the midpoint of a segment whose endpoints belong to $\mathbb{E}$.*

(3) *If T_1 and T_2 are members of $\mathbb{E}$ such that $T_1 < T_2$, then the midpoint of $\overline{T_1T_2}$ belongs to $\mathbb{E}$.*

(4) *A is the midpoint of $\overline{(A + \frac{m}{2^n}(B - A))(A - \frac{m}{2^n}(B - A))}$. and thus $\mathbb{E}$ is symmetric with respect to A. Moreover, $\mathcal{R}_A(A + \frac{m}{2^n}(B - A)) = A - \frac{m}{2^n}(B - A)$, where $\mathcal{R}_A$ is the point reflection about A.*

Proof. (1) For every integer n, $A_n = A + \frac{n}{2^0}(B - A)$

(2) Let m be any integer and let n be any natural number. Since

$$A + \tfrac{m}{2^n}(B - A) = \tfrac{1}{2}(A + \tfrac{m-1}{2^n}(B - A) + A + \tfrac{m+1}{2^n}(B - A)),$$

by Theorem QX.18 $A + \frac{m}{2^n}(B - A)$ is the midpoint of

$$\overline{(A + \tfrac{m-1}{2^n}(B - A))(A + \tfrac{m+1}{2^n}(B - A))}.$$

(3) There exist integers m_1 and m_2 and there exist natural numbers n_1 and n_2 such that $T_1 = A + \frac{m_1}{2^{n_1}}(B - A)$ and $T_2 = A + \frac{m_2}{2^{n_2}}(B - A)$. By Theorem QX.18 the midpoint of $\overline{T_1T_2}$ is

$$H = \tfrac{1}{2}(A + \tfrac{m_1}{2^{n_1}}(B - A) + A + \tfrac{m_2}{2^{n_2}}(B - A)) = A + \tfrac{m_1 2^{n_2} + m_2 2^{n_1}}{2^{n_1+n_2}}(B - A).$$

Since $m_1 2^{n_2} + m_2 2^{n_1}$ is an integer and $n_1 + n_2$ is a natural number, H is a member of $\mathbb{E}$.

(4) By Theorem QX.18 the midpoint of

$$\overline{(A + \tfrac{m}{2^n}(B - A))(A + \tfrac{-m}{2^n}(B - A))}$$

is $\frac{1}{2}\big(A + \frac{m}{2^n}(B - A) + A - \frac{m}{2^n}(B - A)\big) = A$. Let $A^+ = A + \frac{m}{2^n}(B - A)$ and $A^- = A - \frac{m}{2^n}(B - A)$. Then by Definition NEUT.3(C)

$$\overline{AA^+} \cong \overline{AA^-} \text{ and } (A^-)\text{–}A\text{–}A^+.$$

By Theorem ROT.3, the point reflection $\mathcal{R}_A$ maps A^+ to a point Y such that $\overline{AA^+} \cong \overline{AY}$ and (A^+)–A–Y. Thus $Y \in \overrightarrow{AA^-}$ and by Property R.4 of Definition NEUT.2, $Y = A^-$. Therefore $\mathcal{R}_A(A^+) = A^-$ as required. □

Definition AA.3. The set $\mathbb{E}$ of Theorem AA.2 is the **set of midpoints generated by $\overline{AB}$.**

Theorem AA.4. *Let $\mathbb{L}$ be an ordered field on a Euclidean/LUB plane $\mathcal{P}$ with origin O and unit U, and suppose A and B are members of $\mathbb{L}$ such that $A < B$. As in*

Theorem AA.1, for every integer n let $A_n = A + n(B - A)$, *and let* $\mathbb{J} = \{A_n \mid n$ *is an integer*$\}$. *If* φ *is a belineation of* $\mathcal{P}$ *such that A and B are fixed points of* φ, *then every member of* $\mathbb{J}$ *is a fixed point of* φ.

Proof. We use mathematical induction. Let $\mathbb{T}_1 = \{A_{-1}, A_0, A_1\}$, and for every natural number $n > 1$ let $\mathbb{T}_n = \{A_{-n}, A_n\} \cup \mathbb{T}_{n-1}$. Since $A_0 = A$ and $A_1 = B$ are fixed points of φ, and since by Theorem AA.1(6), A_0 is the midpoint of $\overline{A_{-1}A_1}$, by Corollary EUC.17.3, A_{-1} is a fixed point, hence every member of $\mathbb{T}_1$ is a fixed point of φ.

Assume that we have shown that every member of $\mathbb{T}_{n-1}$ is a fixed point of φ. Then again by Theorem AA.1(6), $A_{-(n-1)}$ is the midpoint of $\overline{A_{-n}A_{-(n-2)}}$ and A_{n-1} is the midpoint of $\overline{A_nA_{n-2}}$, so that by Corollary EUC.17.3, A_{-n} and A_n are both fixed points, and every member of $\mathbb{T}_n$ is a fixed point of φ. Since $\mathbb{J}$ is the union of all the sets $\mathbb{T}_n$, every member of $\mathbb{J}$ is a fixed point of φ. □

Theorem AA.5. *Let* $\mathbb{L}$ *be an ordered field on a Euclidean/LUB plane* $\mathcal{P}$ *with origin O and unit U. Let* $A < B$ *be fixed points of a belineation* φ *of* $\mathcal{P}$. *As in Theorem AA.2 let* $\mathbb{E} = \{A + \frac{m}{2^n}(B - A) \mid m$ *is an integer and n is a natural number*$\}$. *Then every member of* $\mathbb{E}$ *is a fixed point of* φ.

Proof. For every natural number n let

$$\mathbb{F}_n = \{A + \tfrac{m}{2^{n-1}}(B - A) \mid m \text{ is an integer}\}.$$

By Theorem AA.4 every member of $\mathbb{F}_1$ is a fixed point of φ. Assume now that we have proved that every member of $\mathbb{F}_n$ is a fixed point of φ. By Corollary EUC.17.2, the midpoint of

$$\overline{(A + \tfrac{m}{2^{n-1}}(B - A))(A + \tfrac{m+1}{2^{n-1}}(B - A))}$$

is a fixed point of φ. By Theorem QX.18 that midpoint is $A + \frac{2m+1}{2^n}(B - A)$, so all points of this form are fixed points.

Now $\mathbb{F}_{n+1} = \{A + \frac{k}{2^n}(B - A) \mid k \in \mathbb{Z}\}$, ($\mathbb{Z}$ is the set of all integers) where k is either even or odd, that is, for some integer m, $k = 2m$ or $k = 2m + 1$. If $k = 2m$, then $A + \frac{2m}{2^n}(B - A) = A + \frac{m}{2^{n-1}}(B - A) \in \mathbb{F}_n$. Thus $\mathbb{F}_{n+1} = \mathbb{F}_n \cup \{A + \frac{2m+1}{2^n}(B - A) \mid m \in \mathbb{Z}\}$. We know already that every member of $\mathbb{F}_n$ is a fixed point, so every member of $\mathbb{F}_{n+1}$ is a fixed point of φ. By mathematical induction, this shows that for all natural numbers n, every member of $\mathbb{F}_n$ is a fixed point. Since $\mathbb{E} = \bigcup_{n \in \mathbb{N}} \mathbb{F}_n$, every member of $\mathbb{E}$ is a fixed point of φ. □

Theorem AA.6. *Let $\mathbb{L}$ be an ordered field on a Euclidean/LUB plane $\mathcal{P}$ with origin O and unit U, and suppose A and B are members of $\mathbb{L}$ such that $A < B$. As in Theorem AA.1, for every integer n let $A_n = A + n(B - A)$. Then* $\bigcup_{n\in\mathbb{Z}} \overline{A_{n-1}A_n} = \overleftrightarrow{AB}$.

Proof. Let $T \in \overleftrightarrow{AB}$. By Theorem IB.5 either $T \in \overrightarrow{AB}$ or T–A–B.

(I) If $T \in \overrightarrow{AB}$ either $T = A$ or by Theorem ORD.7, $T > A$. In this latter case, both $T - A > O$ and $B - A > O$ so by the Archimedean property Theorem REAL.9, there exists a natural number n such that $n(B - A) > T - A$. Thus $A_0 = A < T < A + n(B - A) = A_n$, and by Theorem AA.1(7) $T \in \overline{A_0A_n} = \bigcup_{k=1}^{n} \overline{A_{k-1}A_k}$. Therefore, for some integer k, $T \in \overline{A_{k-1}A_k}$.

(II) If T–A–B by Theorem ORD.6 $T < A$. Then both $A - T > O$ and $B - A > O$ so by the Archimedean property, there exists a natural number n such that $n(B - A) > A - T$, or $-n(B - A) < T - A$. Then
$$A_0 = A > T > A - n(B - A) = A_{-n},$$
and by Theorem AA.1(7)
$$T \in \overline{A_0A_{-n}} = \bigcup_{k=1}^{n} \overline{A_{-k+1}A_{-k}}.$$
Then for some natural number k, $T \in \overline{A_{-k+1}A_{-k}}$, and if we let $l = -k + 1$, we have $T \in \overline{A_lA_{l-1}} = \overline{A_{l-1}A_l}$, where l is an integer. Therefore
$$\bigcup_{n\in\mathbb{Z}} \overline{A_{n-1}A_n} = \overleftrightarrow{AB}.$$
□

Theorem AA.7. *Let $\mathcal{P}$ be a Euclidean/LUB plane, $\mathbb{L}$ be an ordered field on $\mathcal{P}$ with origin O and unit U. As in Theorem AA.1, for every integer k let $A_k = A + k(B - A)$, and as in Theorem AA.2 let $\mathbb{E} = \{A + \frac{m}{2^n}(B - A) \mid m$ is an integer and n is a natural number$\}$. Let $T_1 < T_2$ be members of $\mathbb{L}$. Then there exist members V_1, V_2, and V_3 of the set $\mathbb{E}$ such that $V_1 < T_1 < V_2 < T_2 < V_3$.*

Proof. (I) Existence of V_1 and V_3. By Theorem AA.6, for some integers k and l with $k \le l$, $T_1 \in \overline{A_{k-1}A_k}$ and $T_2 \in \overline{A_{l-1}A_l}$. Let $V_1 = A_{k-2}$ and $V_3 = A_{l+1}$. Then $V_1 = A_{k-2} < T_1 < T_2 < A_{l+1} = V_3$ and both V_1 and V_3 belong to $\mathbb{E}$.

(II) Existence of V_2.

(Case 1: there is no integer k such that both T_1 and T_2 belong to $\overline{A_{k-1}A_k}$.) Then for some integers k and l, $T_1 \in \overline{A_{k-1}A_k}$, $T_2 \in \overline{A_{l-1}A_l}$, and $k < l$. If $k + 1 = l$, neither T_1 nor T_2 can belong to $\overline{A_{k-1}A_k} \cap \overline{A_{l-1}A_l}$ because then both would belong to the same segment, either $\overline{A_{k-1}A_k}$ or $\overline{A_{l-1}A_l}$. Let $V_2 = V_k$. Then $T_1 < V_2 < T_2$, and $V_2 \in \mathbb{E}$.

(Case 2: there exists an integer k such that both T_1 and T_2 belong to $[A_{k-1}A_k]$.) Then $T_2 - T_1 \leq A_k - A_{k-1} = B - A$. By Theorem REAL.9 (Archimedean property) choose n so that $2^n(T_2 - T_1) > n(T_2 - T_1) > B - A$, so that

$$0 < \frac{1}{2^n}(B - A) < T_2 - T_1 < A_k - A_{k-1} = B - A.$$

Then $A_{k-1} + \frac{1}{2^n}(B - A) < A_k$ and $\frac{1}{2^n}(B - A) < T_2 - T_1$.

(Subcase A: $T_1 = A_{k-1}$.) Then

$$T_1 < T_1 + \frac{1}{2^n}(B - A) < T_1 + (T_2 - T_1) = T_2.$$

Choose $V_2 = A_{k-1} + \frac{1}{2^n}(B - A)$. Then $V_2 \in \mathbb{E}$ and $T_1 < V_2 < T_2$.

(Subcase B: $T_2 = A_k$.) Choose $V_2 = A_k - \frac{1}{2^n}(B - A)$, so that $V_2 < A_k$. Then

$$T_2 - V_2 = T_2 - (T_2 - \frac{1}{2^n}(B - A)) = \frac{1}{2^n}(B - A) < T_2 - T_1$$

so $-V_2 < -T_1$, that is $V_2 > T_1$. Then $V_2 \in \mathbb{E}$ and $T_1 < V_2 < T_2$.

(Subcase C: $T_1 > A_{k-1}$ and $T_2 < A_k$.) Then since

$$A_{k-1} + \frac{2^n}{2^n}(B - A) = A_{k-1} + (B - A) = A_k > T_1,$$

there exists a smallest natural number m such that $A_{k-1} + \frac{m}{2^n}(B - A) > T_1$. Choose $V_2 = A_{k-1} + \frac{m}{2^n}(B - A)$, so that $V_2 > T_1$.

Claim: $V_2 < T_2$. Otherwise, $V_2 \geq T_2$, and

$$\begin{aligned} A_{k-1} + \frac{m-1}{2^n}(B - A) &= (A_{k-1} + \frac{m}{2^n}(B - A)) - \frac{1}{2^n}(B - A) \\ &= V_2 - \frac{1}{2^n}(B - A) \\ &\geq T_2 - \frac{1}{2^n}(B - A) > T_2 - (T_2 - T_1) = T_1 \end{aligned}$$

(since $\frac{1}{2^n}(B - A) < T_2 - T_1$). Thus m is not the smallest integer such that $A_{k-1} + \frac{m}{2^n}(B - A) > T_1$, a contradiction. Thus $V_2 \in \mathbb{E}$ and $T_1 < V_2 < T_2$. □

Theorem AA.8. *Let $\mathcal{P}$ be a Euclidean/LUB plane and let φ be a belineation of $\mathcal{P}$ such that φ has distinct fixed points A and B and φ is not the identity of $\mathcal{P}$. Then φ is an axial affinity of $\mathcal{P}$ with axis $\overleftrightarrow{AB}$.*

Proof. Using Chapters 14 (ordered fields) and 18 (real numbers), build $\overleftrightarrow{AB} = \mathbb{L}$ into an ordered field where $A < B$. Assume there exists a point X on $\mathbb{L}$ which is not a fixed point of φ. Then $X \neq \varphi(X)$, and either $X < \varphi(X)$ or $\varphi(X) < X$. In the first case, use Theorem AA.7 to choose V_1 and V_2, both fixed points for φ, so that $V_1 < X < V_2 < \varphi(X)$; in the second case, choose V_1 and V_2 so that $\varphi(X) < V_2 < X < V_1$. By Theorem ORD.6, in either case we have V_1–X–V_2–$\varphi(X)$. Since φ is a belineation (it preserves betweenness), $\varphi(V_1)$–$\varphi(X)$–$\varphi(V_2)$; since V_1 and V_2 are fixed points for φ, this becomes V_1–$\varphi(X)$–V_2. By the trichotomy property for betweenness (Definition IB.1 Property B.2), this is a contradiction to V_1–X–V_2–$\varphi(X)$.

Therefore every point of $\mathbb{L}$ is a fixed point for φ, and by Definition CAP.25 φ is an axial affinity with axis $\mathbb{L}$. □

Theorem AA.9. *Let $\mathcal{P}$ be a Euclidean/LUB plane and let φ be a belineation of $\mathcal{P}$ which has three noncollinear fixed points. Then $\varphi = \iota$ (the identity mapping of $\mathcal{P}$).*

Proof. Let A, B, and C be noncollinear fixed points of φ. By Theorem AA.8 every point on $\overleftrightarrow{AB}$ is a fixed point of φ. Since by Theorem COBE.2 every belineation is a collineation, we may apply Exercise CAP.3 to get $\varphi = \iota$. □

Theorem AA.10. *Let $\mathcal{P}$ be a Euclidean/LUB plane. If α and β are belineations of $\mathcal{P}$ and if A, B, and C are noncollinear points on $\mathcal{P}$ such that $\alpha(A) = \beta(A)$, $\alpha(B) = \beta(B)$, and $\alpha(C) = \beta(C)$, then $\alpha = \beta$.*

Proof. By Theorem COBE.3 β^{-1} is a belineation; therefore $\alpha \circ \beta^{-1}$ is a belineation. Moreover $(\alpha \circ \beta^{-1})(A) = A$, $(\alpha \circ \beta^{-1})(B) = B$, and $(\alpha \circ \beta^{-1})(C) = C$. By Theorem AA.9 $\alpha \circ \beta^{-1} = \iota$, but that means $\alpha = \beta$. □

Theorems AA.9 and AA.10 are generalizations (to all belineations) of Theorems NEUT.24 and NEUT.25 (Chapter 8), which are valid for isometries in a neutral plane.

Theorem AA.11. *Let $\mathcal{P}$ be a Euclidean/LUB plane and let φ be a nonidentity collineation of $\mathcal{P}$ which has distinct fixed points A and B. Then φ is an axial affinity iff φ is a belineation.*

Proof. If φ is an axial affinity, by Theorem AX.4 φ is a belineation. It is interesting to note (but not needed for the proof) that by Theorem CAP.26, both A and B are members of the axis of φ. Conversely, if φ is a belineation, by Theorem AA.8, φ is an axial affinity having $\mathbb{M} = \overleftrightarrow{AB}$ as its axis. □

19.2 Summaries for belineations

Remark AA.12. We have shown a number of relationships between the principal types of belineations, namely isometries, axial affinities, dilations, and their subcategories, and have explored their characteristics. These results are scattered throughout the book, mainly in Chapters 3 (CAP), 8 (NEUT), 10 (ROT), 12 (ISM), and 16 (AX). To make more of this information conveniently available in one place, we display it here in a way that we hope is helpful. The following theorems are needed to justify the diagram below.

Theorem AA.13. *A nonidentity belineation α on a Euclidean/LUB plane $\mathcal{P}$ is a line reflection iff it is a stretch and an isometry.*

Proof. Suppose α is a line reflection with axis $\mathcal{M}$; by Remark NEUT.1.5 it is a belineation; if $Q \notin \mathcal{M}$, by Theorem NEUT.22 $\overleftrightarrow{Q\alpha(Q)}$ is a fixed line intersecting $\mathcal{M}$, so by Definition AX.0, α is a stretch. By Definition NEUT.3 it is an isometry.

Conversely, suppose α is a stretch and an isometry; since it is not the identity, by Theorem ISM.17 it must be either a rotation, translation, glide reflection, or a line reflection. The first three of these have either one or no fixed point; since α is a stretch, it has a whole line of fixed points, so these options are ruled out. The only choice left is a line reflection. □

Theorem AA.14. *A nonidentity belineation α on a Euclidean/LUB plane $\mathcal{P}$ is a point reflection about a point O iff it is a rotation about O and a dilation with fixed point O.*

Proof. Let α be a point reflection about O. By Definition ROT.1, α is a rotation about O. By Theorem ISM.3(C) α is a dilation with fixed point O.

Conversely, suppose α is a dilation and a rotation (both with fixed point O). Since α is a dilation, by Theorem CAP.18 every line containing O is a fixed line. By Theorem ROT.19, a rotation with a fixed line is a point reflection. Therefore α is a point reflection. □

Remark AA.15 (Justification for Figure 19.1). Throughout we assume that all mappings are nonidentity belineations of a Euclidean/LUB plane.

(A) By Theorem ISM.17, an **isometry** is one of the following types:

  (1) a **line reflection**: by Definition NEUT.1 this is an axial affinity, having a line $\mathcal{M}$ of fixed points.
  (2) a **rotation**: by Theorem ROT.2 this has exactly one fixed point.
  (3) a **translation** or a **glide reflection**: these have no fixed point, by Definition CAP.6 and Theorem ISM.13.

(B) By Definition CAP.17 and Theorem CAP.18, a **dilation** has exactly one fixed point.

(C) By Definition CAP.25, an **axial affinity** has a line $\mathcal{M}$ (its axis) of fixed points; by Theorem AX.3 this is either a **stretch** or a **shear**.

(D) By Theorem AA.13 the set of all **line reflections** is the intersection of the set of all stretches and the set of all isometries.

(E) By Theorem AA.14 the set of all **point reflections** about a point O is the intersection of the set of rotations about O and the set of dilations with fixed point O.

Figure 19.1 illustrates these relationships with a Venn diagram; the bold boxes represent the sets of isometries, axial affinities, and dilations. Definitions NEUT.1 and NEUT.3, Theorem AX.4, and Theorem DLN.8 show that all these mappings are belineations.

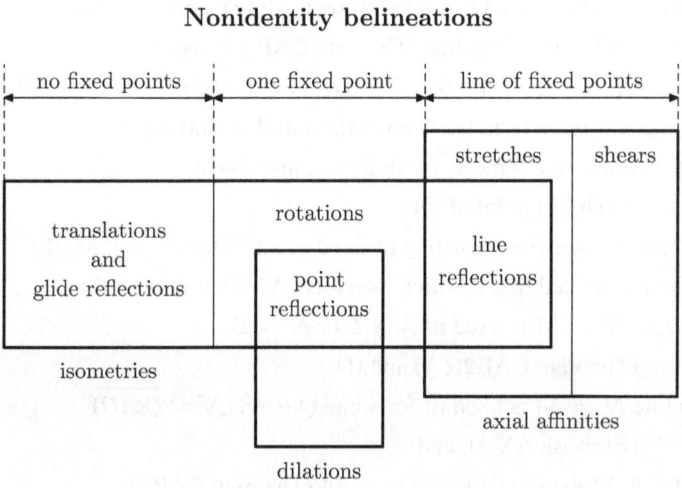

Fig. 19.1 Showing relationships between different types of nonidentity belineation.

Remark AA.16 (Summaries of actions of belineations). In this remark, $\mathcal{L}$, $\mathcal{N}$, and $\mathcal{M}$ will denote lines, and α a nonidentity belineation on a Euclidean/LUB plane $\mathcal{P}$.

If α is a **translation**:

$\mathcal{L}$ is fixed iff $\mathcal{L}$ is parallel to a fixed line (Theorem CAP.8);
$\mathcal{L}$ is fixed iff for some $Q \in \mathcal{P}$, $\mathcal{L} = \overleftrightarrow{Q\alpha(Q)}$ (Theorem CAP.8); and
if $\mathcal{L}$ is not fixed, then $\alpha(\mathcal{L}) \parallel \mathcal{L}$ (Definition CAP.6).

If α is a **glide reflection**:

$\alpha = \mathcal{R}_{\mathcal{L}} \circ \tau$ where τ is a translation and $\mathcal{L}$ is a fixed line for τ (Definition ISM.12);

$\mathcal{L}$ is the only fixed line (Theorem ISM.13);

if $\mathcal{M} \parallel \mathcal{L}$, then $\alpha(\mathcal{M}) \parallel \mathcal{L}$ (Exercise ISM.8(A)); and

if $\mathcal{M} \perp \mathcal{L}$, then $\alpha(\mathcal{M}) \perp \mathcal{L}$ (Exercise ISM.8(B)).

If α is a **rotation, not a point reflection**:

there is exactly one fixed point O (Theorem ROT.2);

thus if $O \in \mathcal{L}$ then $O \in \alpha(\mathcal{L})$; and

there are no fixed lines (contrapositive of Theorem ROT.19(A)).

If α is a **dilation**:

there is exactly one fixed point O (Theorem CAP.18(B));

if $O \notin \mathcal{L}$, then $\alpha(\mathcal{L}) \parallel \mathcal{L}$ (Definition CAP.17); and

$O \in \mathcal{L}$ iff $\mathcal{L}$ is a fixed line (Theorem CAP.20(A));

$\mathcal{L}$ is fixed iff for some $Q \in \mathcal{P} \setminus \{O\}$, $\mathcal{L} = \overleftrightarrow{Q\alpha(Q)}$ (Theorem CAP.20(B)).

If α is a **point reflection, (so is a rotation and a dilation)**:

its action is the same as for dilations, just above.

If α is a **stretch (an axial affinity)**:

there is a line $\mathcal{M}$ consisting of fixed points (Definition CAP.25);

there is a fixed line $\mathcal{L}$ which intersects $\mathcal{M}$ (Definition AX.0);

a line $\mathcal{N} \neq \mathcal{M}$ is fixed iff $\mathcal{N} \parallel \mathcal{L}$ or $\mathcal{N} = \mathcal{L}$ (Theorem CAP.26(A) and (D));

a line $\mathcal{N} \neq \mathcal{M}$ is fixed iff for some $Q \notin \mathcal{M}$, $\mathcal{N} = \overleftrightarrow{Q\alpha(Q)}$ (Exercise AX.2); and

if $\mathcal{L} \parallel \mathcal{M}$ then $\alpha(\mathcal{L}) \parallel \alpha(\mathcal{M}) = \mathcal{M}$ (Theorem CAP.3);

If α is a **line reflection (so is an isometry, an axial affinity, and a stretch)**:

its action is the same as for stretches, just above; and

a line $\mathcal{N} \neq \mathcal{M}$ is a fixed line iff $\mathcal{N} \perp \mathcal{M}$ (Theorem NEUT.44).

If α is a **shear (an axial affinity)**:

there is a line $\mathcal{M}$ consisting of fixed points (Definition CAP.25);

a line is a fixed line iff it is parallel (or equal) to $\mathcal{M}$ (Definition AX.0); and

a line $\mathcal{N} \neq \mathcal{M}$ is fixed iff for some $Q \notin \mathcal{M}$, $\mathcal{N} = \overleftrightarrow{Q\alpha(Q)}$ (Exercise AX.2).

There are no exercises for this chapter.

Chapter 20
Ratios of Sensed Segments (RS)

Acronym: *RS*

Dependencies: *all prior Chapters 1 through 19*

New Axioms: *none*

New Terms Defined: *sensed segment; initial, final point, sensed length (of a sensed segment); ratio in which X separates the points A and B*

Abstract: This chapter proves two classical theorems of geometry, due to Menelaus of Alexandria (c. 70–140) and to Giovanni Ceva (1647–1734). The proofs use the machinery of ratios of sensed segments.

20.1 Basic theorems on sensed segments

The development in this chapter is based on that of Martin, *Transformation Geometry: An Introduction to Symmetry*, Chapter 14 (Springer, 1982) [14].

Up through Remark RS.7, we assume that $\mathbb{L}$ is a line in a Euclidean/LUB plane $\mathcal{P}$, which has been built into a complete ordered field with origin O and unit U. A, B, C, and D are points of $\mathbb{L}$; by Corollary REAL.35.1 there exist real numbers a, b, c, and d such that $A = aU$, $B = bU$, $C = cU$, and $D = dU$.

For Theorems RS.1 through RS.3 we add the further assumptions (together comprising the "blanket hypotheses" for them) that $\mathbb{L}$ has also been equipped with another origin O' and another unit U'; then by Corollary REAL.35.1 there exist real

E.J. Specht et al., *Euclidean Geometry and its Subgeometries*,
DOI 10.1007/978-3-319-23775-6_20

numbers h and u such that $O' = hU$ and $U' = uU$, and real numbers a', b', c', and d' such that $A = aU = a'U'$, $B = bU = b'U'$, $C = cU = c'U'$, and $D = dU = d'U'$.

Theorem RS.1. *Under the blanket hypotheses, for any distinct points $A = aU = a'U'$ and $B = bU = b'U'$ in $\mathbb{L}$, $b' - a' = u(b - a)$.*

Proof. $B - A = bU - aU = (b - a)U$. Also

$$B - A = b'U' - a'U' = (b' - a')U' = (b' - a')uU$$

so that $(b - a)U = (b' - a')uU$ and

$$\begin{aligned} O &= (b - a)U - (b' - a')uU = ((b - a) - (b' - a')u)U \\ &= ((b - a) - (b' - a')u)U. \end{aligned}$$

By Corollary REAL.34(C), $(b - a) - (b' - a')u = 0$ and $(b - a) = u(b' - a')$. □

Theorem RS.2. *Under the blanket hypotheses, $\overrightarrow{OU} \cong \overrightarrow{O'U'}$ iff $u - h = 1$ or $u - h = -1$.*

Proof. $U' = O' + (U' - O') = \tau_{O'}(U' - O')$. Applying $\tau_{-O'}$ to both sides, we have

$$\begin{aligned} \tau_{-O'}(U') &= \tau_{-O'}(O' + (U' - O')) = \tau_{-O'}(\tau_{O'}(U' - O')) \\ &= U' - O' = uU - hU = (u - h)U = \delta_{u-h}(U) \end{aligned}$$

where δ_{u-h} is as in Definition REAL.38. Since $\tau_{-O'}$ is an isometry, by Theorem NEUT.15(5)

$$\begin{aligned} \tau_{-O'}(\overrightarrow{O'U'}) &= \overrightarrow{(\tau_{-O'}(O'))(\tau_{-O'}(U'))} = \overrightarrow{O(\tau_{-O'}(U'))} = \\ &= \overrightarrow{O(U' - O')} = \overrightarrow{O(uU - hU)} = \overrightarrow{O(\delta_{u-h}(U))}. \end{aligned}$$

Thus $\overrightarrow{O'U'} \cong \overrightarrow{O(\delta_{u-h}(U))}$ since $\tau_{-O'}$ is an isometry.

(Case 1: $\delta_{u-h}(U) \in \overleftrightarrow{OU}$.) If $\overrightarrow{O'U'} \cong \overrightarrow{OU}$ then $\overrightarrow{O(\delta_{u-h}(U))} \cong \overrightarrow{OU}$ and by Property R.4 of Definition NEUT.2, $U = \delta_{u-h}(U) = (u-h)U$ so that $u-h = 1$. Conversely, if $u - h = 1$ then $\delta_{u-h}(U) = U$ and $\overrightarrow{O'U'} \cong \overrightarrow{OU}$.

(Case 2: $\delta_{u-h}(U) \in \overleftrightarrow{O\mathcal{R}_O(U)}$.) We know that $\overrightarrow{O\mathcal{R}_O(U)} \cong \overrightarrow{OU}$ because $\mathcal{R}_O$ is an isometry with O as a fixed point. If $\overrightarrow{O'U'} \cong \overrightarrow{OU}$, then

$$\overrightarrow{O\mathcal{R}_O(U)} \cong \overrightarrow{OU} \cong \overrightarrow{O\mathcal{R}_O(U)} = \overrightarrow{O(-U)},$$

and by Property R.4 of Definition NEUT.2, $-U = (u - h)U$ so that $u - h = -1$. Conversely, if $u - h = -1$ then $\delta_{u-h}(U) = -U = \mathcal{R}_O(U)$ and hence, by our first calculation,

$$\overrightarrow{O'U'} \cong \overrightarrow{O(\delta_{u-h}(U))} = \overrightarrow{O\mathcal{R}_O(U)} \cong \overrightarrow{OU}.$$

Here we have used Corollary REAL.34(C) several times. □

Theorem RS.3. *Under the blanket hypotheses, if $A \neq B$ and $C \neq D$, $\frac{b'-a'}{d'-c'} = \frac{b-a}{d-c}$. Therefore the ratio $\frac{b-a}{d-c}$ is independent of the origin and the unit chosen for $\mathbb{L}$.*

Proof. By Theorem RS.1, $\frac{b'-a'}{d'-c'} = \frac{u(b-a)}{u(d-c)} = \frac{b-a}{d-c}$. □

Definition RS.4. (A) A **sensed segment** is an ordered pair $(\overleftrightarrow{AB}, A)$ where the first element of the pair is a closed segment $\overleftrightarrow{AB}$ and the second element is one of its endpoints, called the **initial** point of the segment. The other endpoint is called the **final** point of the segment.

(B) We will denote the sensed segment $(\overleftrightarrow{AB}, A)$ by the symbol $[AB\rangle$.

(C) If an origin O and a unit U has been chosen for the line $\mathbb{L} = \overleftrightarrow{AB}$, and if $A = aU$ and $B = bU$ for some real numbers a and b, then the **sensed length** of $[AB\rangle = [(aU)(bU)\rangle$ will be the real number $b - a$.

Remark RS.5. Suppose $A \neq B$, $C \neq D$ are collinear points on the plane, an origin O and a unit U have been chosen for the line $\mathbb{L} = \overleftrightarrow{AB}$, and $A = aU$, $B = bU$, $C = cU$, and $D = dU$ for some real numbers a, b, c, and d. Then by Theorem RS.3 the ratio of the sensed lengths $\frac{b-a}{d-c}$ is independent of the origin and the unit chosen for $\mathbb{L}$. Therefore it is legitimate to speak of *the* ratio $\frac{b-a}{d-c}$, so long as it is understood that *some* origin and *some* unit have been chosen for $\mathbb{L}$.

Moreover, it is quite legitimate, since their numerators and denominators are real numbers, to multiply two such ratios and shuffle the numerators and denominators about as we would in any other fractions; for instance,

$$\frac{b-a}{d-c} \cdot \frac{b'-a'}{d'-c'} = \frac{b-a}{d'-c'} \cdot \frac{b'-a'}{d-c}$$

and so forth.

Definition RS.6. If $X = xU$ is any point on $\mathbb{L}$ other than A or B, $\frac{x-a}{b-x}$ is called the **ratio in which X separates the points A and B.**

Remark RS.7. (A) We will indulge now in a bit of what the French call "abuse of notation." There will be times when we don't want to bother saying, before an argument involving sensed segments, that an origin O and a unit U has been chosen for the line and $A = aU$, $B = bU$, $C = cU$, and $D = dU$ for some real numbers a, b, c, and d. In such cases we may, as a notational convention, write $[AB\rangle$ when we mean $b - a$, and $\frac{[AB\rangle}{[CD\rangle}$ in place of $\frac{b-a}{d-c}$. Doing so is legitimate because this ratio is independent of our choice of origin and unit. The ratio in which X separates the points A and B may then be written as $\dfrac{[AX\rangle}{[XB\rangle}$. If there is virtue in this convention, it is that it keeps the geometric character of the argument more clearly in view. We will use it at various places in the rest of this chapter.

(B) Since $\dfrac{[AB\rangle}{[CD\rangle}$ means $\frac{b-a}{d-c}$, several facts become apparent:

(1) Since $\frac{b-a}{d-c} = \frac{a-b}{c-d} = -\frac{b-a}{c-d} = -\frac{a-b}{d-c}$, we have
$$\frac{[AB\rangle}{[CD\rangle} = \frac{[BA\rangle}{[DC\rangle} = -\frac{[AB\rangle}{[DC\rangle} = -\frac{[BA\rangle}{[CD\rangle}.$$

(2) For any $X = xU$, if A–X–B then a–x–b, so that either $a < x < b$ or $b < x < a$ and $\dfrac{[AX\rangle}{[XB\rangle} = \dfrac{x-a}{b-x} > 0$.

(3) If X–A–B or A–B–X, then x–a–b or a–b–x, that is, $x < a < b$, $a < b < x$, $x > a > b$, or $a > b > x$. In any of these cases
$$\frac{[AX\rangle}{[XB\rangle} = \frac{x-a}{b-x} < 0.$$

(4) $[AB\rangle^{-1} = {}_{\overline{1}}[AB\rangle$, $\dfrac{[AB\rangle}{[AB\rangle} = 1$, and $\dfrac{[AB\rangle}{[CD\rangle} \cdot \dfrac{[CD\rangle}{[EF\rangle} = \dfrac{[AB\rangle}{[EF\rangle}$.

The following theorem clarifies the relationship between ratios of free segments and ratios of sensed segments.

Theorem RS.8. *Let $\mathbb{L}$ and $\mathbb{M}$ be intersecting lines in a Euclidean/LUB plane $\mathcal{P}$, which have been built into complete ordered fields, where $\mathbb{L}$ has unit U and $\mathbb{M}$ has unit U'. Let A, B, C, and D be points of $\mathbb{L}$, and A', B', C', and D' be points of $\mathbb{M}$, and suppose that for some real numbers a, b, c, d, a', b', c', and d', $A = aU$, $B = bU$, $C = cU$, and $D = dU$, and $A' = a'U'$, $B' = b'U'$, $C' = c'U'$, and $D' = d'U'$.*

(A) *If $[AB\rangle = [CD\rangle$ or $[AB\rangle = [DC\rangle$, then $[\overleftrightarrow{AB}] = [\overleftrightarrow{CD}]$.*

(B) *If $\dfrac{[\overleftrightarrow{AB}]}{[\overleftrightarrow{CD}]} = \dfrac{[\overleftrightarrow{A'B'}]}{[\overleftrightarrow{C'D'}]}$, then $\dfrac{[AB\rangle}{[CD\rangle} = \dfrac{[A'B'\rangle}{[C'D'\rangle}$ or $\dfrac{[AB\rangle}{[CD\rangle} = -\dfrac{[A'B'\rangle}{[C'D'\rangle}$.*

Proof. (A) In the first instance, $\overleftrightarrow{AB} = \overleftrightarrow{CD}$ so $[\overleftrightarrow{AB}] = [\overleftrightarrow{CD}]$; in the second $\overleftrightarrow{AB} = \overleftrightarrow{DC} \cong \overleftrightarrow{CD}$ so that in either case $[\overleftrightarrow{AB}] = [\overleftrightarrow{CD}]$.

(B) Since both the hypotheses and conclusion of this part are expressed in terms of ratios of sensed segments, it follows from Theorem RS.3 that the choice of origin and unit is irrelevant. Thus we may, for convenience in the proof, assume that the point of intersection of both lines is their common origin O, and that $\overleftrightarrow{OU} \cong \overleftrightarrow{OU'}$. That is, if $\mathbb{K}$ is the line of symmetry of $\angle UOU'$, then since $\mathcal{R}_{\mathbb{K}}$

is an isometry, $\mathcal{R}_{\mathcal{K}}(\overrightarrow{OU}) = \overrightarrow{O\mathcal{R}_{\mathcal{K}}(U)}$; thus $\overrightarrow{OU'} \cong \overrightarrow{OU} \cong \overrightarrow{O\mathcal{R}_{\mathcal{K}}(U)}$ and by Property R.4 of Definition NEUT.2, $\mathcal{R}_{\mathcal{K}}(U) = U'$. By Theorem OF.15

$$\begin{aligned}
[\overrightarrow{AB}] &= [\overrightarrow{O(B-A)}] = [\overrightarrow{O(b-a)U}] = [\overrightarrow{O|b-a|U}],\\
[\overrightarrow{CD}] &= [\overrightarrow{O(D-C)}] = [\overrightarrow{O(d-c)U}] = [\overrightarrow{O|d-c|U}],\\
[\overrightarrow{A'B'}] &= [\overrightarrow{O(B'-A')}] = [\overrightarrow{O(b'-a')U'}] = [\overrightarrow{O|b'-a'|U'}], \text{ and}\\
[\overrightarrow{C'D'}] &= [\overrightarrow{O(D'-C')}] = [\overrightarrow{O(d'-c')U'}] = [\overrightarrow{O|d'-c'|U'}].
\end{aligned}$$

By Definition SIM.7, $[\overrightarrow{CD}]^{-1} = [\overrightarrow{O(D-C)^{-1}}] = [\overrightarrow{O\frac{1}{(d-c)}U}] = [\overrightarrow{O\frac{1}{|d-c|}U}]$, so that by Remark REAL.43, Theorem REAL.23, and Theorem REAL.25

$$\frac{[\overrightarrow{AB}]}{[\overrightarrow{CD}]} = \left[\overrightarrow{O\left|\frac{b-a}{d-c}\right|U}\right] \text{ and } \frac{[\overrightarrow{A'B'}]}{[\overrightarrow{C'D'}]} = \left[\overrightarrow{O\left|\frac{b'-a'}{d'-c'}\right|U'}\right].$$

Putting this with our hypothesis,

$$\left[\overrightarrow{O\left|\frac{b-a}{d-c}\right|U}\right] = \left[\overrightarrow{O\left|\frac{b'-a'}{d'-c'}\right|U'}\right]. \quad (*)$$

By Definition REAL.38 and Theorem REAL.42, define δ be the dilation with fixed point O such that for all A

$$\delta(A) = \left|\frac{b-a}{d-c}\right|A.$$

Since $\mathbb{K}$ contains the point O, $\delta(K)$ also contains O, so that Theorem DLN.17 says that δ commutes with $\mathcal{R}_{\mathcal{K}}$, which is an isometry. Then

$$\begin{aligned}
\mathcal{R}_{\mathcal{K}}\left(\left|\frac{b-a}{d-c}\right|U\right) &= \mathcal{R}_{\mathcal{K}}(\delta(U)) = \delta(\mathcal{R}_{\mathcal{K}}(U))\\
&= \delta(U') = \left|\frac{b-a}{d-c}\right|U'.
\end{aligned}$$

By Theorem DLN.8, δ is a belineation so Theorem COBE.5(5) applies, and

$$\overrightarrow{O\left|\frac{b-a}{d-c}\right|U'} \cong \overrightarrow{O\left|\frac{b-a}{d-c}\right|U} \text{ and } \left[\overrightarrow{O\left|\frac{b-a}{d-c}\right|U'}\right] = \left[\overrightarrow{O\left|\frac{b-a}{d-c}\right|U}\right].$$

Combining this with (*), we have

$$\left[\overrightarrow{O\left|\frac{b-a}{d-c}\right|U'}\right] = \left[\overrightarrow{O\left|\frac{b'-a'}{d'-c'}\right|U'}\right].$$

By Property R.4 of Definition NEUT.2

$$\left|\frac{b-a}{d-c}\right|U' = \left|\frac{b'-a'}{d'-c'}\right|U'$$

and by Corollary REAL.34(C)

$$\left|\frac{b-a}{d-c}\right| = \left|\frac{b'-a'}{d'-c'}\right| \text{ so that } \frac{b-a}{d-c} = \frac{b'-a'}{d'-c'} \text{ or } \frac{b-a}{d-c} = -\frac{b'-a'}{d'-c'},$$

which is to say $\dfrac{[AB\rangle}{[CD\rangle} = \dfrac{[A'B'\rangle}{[C'D'\rangle}$ or $\dfrac{[AB\rangle}{[CD\rangle} = -\dfrac{[A'B'\rangle}{[C'D'\rangle}$. □

Theorem RS.9. *Let $\mathbb{L}_1$ and $\mathbb{L}_2$ be distinct lines on a Euclidean/LUB plane which are built into complete ordered fields which have a common origin O and respective units U_1 and U_2.*

Let $A_1 = a_1U_1$, $B_1 = b_1U_1$, and $C_1 = c_1U_1$ be distinct points on $\mathbb{L}_1$ and $A_2 = a_2U_2$, $B_2 = b_2U_2$, and $C_2 = c_2U_2$ be distinct points on $\mathbb{L}_2$ such that $\overleftrightarrow{A_1A_2} \parallel \overleftrightarrow{B_1B_2} \parallel \overleftrightarrow{C_1C_2}$; then

(0) $\dfrac{[OA_1\rangle}{[OA_2\rangle} = \dfrac{[OB_1\rangle}{[OB_2\rangle} = \dfrac{[OC_1\rangle}{[OC_2\rangle}$, *or* $\dfrac{a_1}{a_2} = \dfrac{b_1}{b_2} = \dfrac{c_1}{c_2}$,

(1) $\dfrac{[A_1B_1\rangle}{[A_1C_1\rangle} = \dfrac{[A_2B_2\rangle}{[A_2C_2\rangle}$, *or* $\dfrac{b_1-a_1}{c_1-a_1} = \dfrac{b_2-a_2}{c_2-a_2}$,

(2) $\dfrac{[B_1A_1\rangle}{[B_1C_1\rangle} = \dfrac{[B_2A_2\rangle}{[B_2C_2\rangle}$, *or* $\dfrac{a_1-b_1}{c_1-b_1} = \dfrac{a_2-b_2}{c_2-b_2}$, *and*

(3) $\dfrac{[C_1B_1\rangle}{[C_1A_1\rangle} = \dfrac{[C_2B_2\rangle}{[C_2A_2\rangle}$, *or* $\dfrac{b_1-c_1}{a_1-c_1} = \dfrac{b_2-c_2}{a_2-c_2}$.

Proof. Let $\mathbb{M} = \text{par}(0, \overleftrightarrow{A_1A_2})$. Without loss of generality we may assume that U_1 and U_2 have been chosen so that both are on the same side of $\mathbb{M}$, and so that $\overline{OU_1} \cong \overline{OU_2}$. By Exercise PSH.14 both members of each of the pairs $\{A_1, A_2\}$, $\{B_1, B_2\}$, and $\{C_1, C_2\}$ belong to the same side of $\mathbb{M}$.

By Theorem SIM.19, $\dfrac{[\overline{OA_1}]}{[\overline{OA_2}]} = \dfrac{[\overline{OB_1}]}{[\overline{OB_2}]} = \dfrac{[\overline{OC_1}]}{[\overline{OC_2}]} = \dfrac{[\overline{Oa_1U_1}]}{[\overline{Oa_2U_2}]}$.

Let δ_{a_2} be the dilation with fixed point O such that for all points $A \in \mathcal{P}$, $\delta_{a_2}(A) = a_2A$. Again, as in the proof of Theorem RS.8 above, δ_{a_2} is a belineation. Let $\mathcal{R}_{\mathbb{K}}$ be the reflection over $\mathbb{K}$, the line of symmetry of $\angle U_1OU_2$; then since $\overline{OU_1} \cong \overline{OU_2}$, $\mathcal{R}_{\mathbb{K}}(U_2) = U_1$. From Theorem DLN.17

$$\mathcal{R}_{\mathbb{K}}(a_2U_2) = \mathcal{R}_{\mathbb{K}}(\delta_{a_2}(U_2)) = \delta_{a_2}(\mathcal{R}_{\mathbb{K}}(U_2)) = \delta_{a_2}(U_1) = a_2U_1$$

and hence

$$\overline{O(a_2U_2)} = \delta_{a_2}(\overline{OU_2}) \cong \delta_{a_2}(\overline{OU_1}) = \overline{O(a_2U_1)}$$

so that

$$\frac{[\overline{OA_1}]}{[\overline{OA_2}]} = \frac{[\overline{Oa_1U_1}]}{[\overline{Oa_2U_1}]} = [\overline{O(\tfrac{a_1}{a_2}U_1)}]$$

by Remark REAL.43. Here we have implicitly used Theorem REAL.42 by assuming that the same dilation accomplishes multiplication by a_2 on both lines. By similar reasoning,

$$\frac{[\overline{OB_1}]}{[\overline{OB_2}]} = \frac{[\overline{Ob_1U_1}]}{[\overline{Ob_2U_1}]} = [\overline{O(\tfrac{b_1}{b_2}U_1)}]$$

and

$$\frac{[\overline{OC_1}]}{[\overline{OC_2}]} = \frac{[\overline{Oc_1U_1}]}{[\overline{Oc_2U_1}]} = [\overline{O(\tfrac{c_1}{c_2}U_1)}]$$

so that

$$[\overline{O(\tfrac{a_1}{a_2}U_1)}] = [\overline{O(\tfrac{b_1}{b_2}U_1)}] = [\overline{O(\tfrac{c_1}{c_2}U_1)}]$$

and

$$\overline{O(\tfrac{a_1}{a_2}U_1)} \cong \overline{O(\tfrac{b_1}{b_2}U_1)} \cong \overline{O(\tfrac{c_1}{c_2}U_1)}.$$

We have already noted that both members of each of the pairs $\{A_1, A_2\}$, $\{B_1, B_2\}$, $\{C_1, C_2\}$, and $\{U_1, U_2\}$ belong to the same side of $\mathbb{M}$; therefore both real numbers of each of the pairs $\{a_1, a_2\}$, $\{b_1, b_2\}$, and $\{c_1, c_2\}$ have the same sign, hence all their ratios are positive, and all the points $\frac{a_1}{a_2}U_1$, $\frac{b_1}{b_2}U_1$, and $\frac{c_1}{c_2}U_1$ are points of $\overrightarrow{OU_1}$. By Property R.4 of Definition NEUT.2, $\frac{a_1}{a_2}U_1 = \frac{b_1}{b_2}U_1 = \frac{c_1}{c_2}U_1$. Therefore by Theorem REAL.34(C), $\dfrac{a_1}{a_2} = \dfrac{b_1}{b_2} = \dfrac{c_1}{c_2}$, proving conclusion (0).

By arithmetic,

$\dfrac{b_1}{a_1} = \dfrac{b_2}{a_2}$, hence $\dfrac{b_1}{a_1} - 1 = \dfrac{b_2}{a_2} - 1$ and $\dfrac{b_1 - a_1}{a_1} = \dfrac{b_2 - a_2}{a_2}$;

$\dfrac{c_1}{a_1} = \dfrac{c_2}{a_2}$, hence $\dfrac{c_1}{a_1} - 1 = \dfrac{c_2}{a_2} - 1$ and $\dfrac{c_1 - a_1}{a_1} = \dfrac{c_2 - a_2}{a_2}$;

$\dfrac{c_1}{b_1} = \dfrac{c_2}{b_2}$, hence $\dfrac{c_1}{b_1} - 1 = \dfrac{c_2}{b_2} - 1$ and $\dfrac{c_1 - b_1}{b_1} = \dfrac{c_2 - b_2}{b_2}$; and

$\dfrac{b_1}{c_1} = \dfrac{b_2}{c_2}$, hence $\dfrac{b_1}{c_1} - 1 = \dfrac{b_2}{c_2} - 1$ and $\dfrac{b_1 - c_1}{c_1} = \dfrac{b_2 - c_2}{c_2}$.

(1) $\dfrac{b_1 - a_1}{a_1} \cdot \dfrac{a_1}{c_1 - a_1} = \dfrac{b_2 - a_2}{a_2} \cdot \dfrac{a_2}{c_2 - a_2}$, and $\dfrac{b_1 - a_1}{c_1 - a_1} = \dfrac{b_2 - a_2}{c_2 - a_2}$ proving conclusion (1).

(2) $\dfrac{b_1 - a_1}{a_1} \cdot \dfrac{b_1}{c_1 - b_1} = \dfrac{b_2 - a_2}{a_2} \cdot \dfrac{b_2}{c_2 - b_2}$; since $\dfrac{b_1}{a_1} = \dfrac{b_2}{a_2}$, $\dfrac{b_1 - a_1}{c_1 - b_1} = \dfrac{b_2 - a_2}{c_2 - b_2}$, proving conclusion (2).

(3) $\dfrac{b_1 - c_1}{c_1} \cdot \dfrac{a_1}{c_1 - a_1} = \dfrac{b_2 - c_2}{c_2} \cdot \dfrac{a_2}{c_2 - a_2}$; since $\dfrac{a_1}{c_1} = \dfrac{a_2}{c_2}$, $\dfrac{b_1 - c_1}{c_1 - a_1} = \dfrac{b_2 - c_2}{c_2 - a_2}$, proving conclusion (3). □

20.2 Theorems of Menelaus and Ceva

Menelaus of Alexandria (c. 70–140) developed spherical geometry in his only extant work, *Sphaerica*. This work survived in an Arabic translation, and contains the following theorem. The lunar crater Menelaus is named after him in recognition of his contributions to astronomy.

Theorem RS.10 (Menelaus). *Let $\mathcal{P}$ be a Euclidean/LUB plane, A, B, and C be noncollinear points on $\mathcal{P}$, D, E, and F be points such that $D \in (\overleftrightarrow{BC} \setminus \{B, C\})$, $E \in (\overleftrightarrow{AC} \setminus \{A, C\})$, and $F \in (\overleftrightarrow{AB} \setminus \{A, B\})$, then D, E, and F are collinear iff* $\dfrac{[AF\rangle}{[FB\rangle} \cdot \dfrac{[BD\rangle}{[DC\rangle} \cdot \dfrac{[CE\rangle}{[EA\rangle} = -1.$

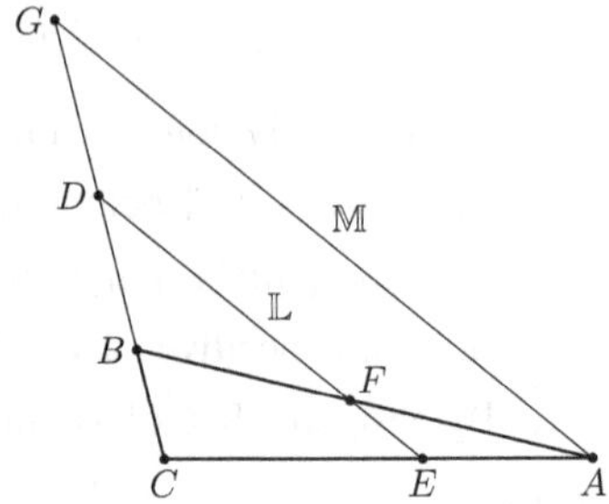

Fig. 20.1 For Menelaus' Theorem RS.10.

Proof. See Figure 20.1. We assume that the lines $\overleftrightarrow{BC}$, $\overleftrightarrow{CA}$, and $\overleftrightarrow{AB}$ have been built into ordered fields, each having an origin and a unit, so that the previous considerations of this chapter apply to them.

(I) If D, E, and F belong to the line $\mathbb{L}$, then let $\mathbb{M} = \text{par}\,(A, \mathbb{L})$. By Exercises I.1 and IP.4 $\mathbb{M}$ and $\overleftrightarrow{BC}$ intersect at a point G. Applying Theorem RS.9 to the lines $\overleftrightarrow{BC}$ and $\overleftrightarrow{CA}$ and to the points D, G, and C on the first of these lines, and to the points E, A, and B on the second line, $\dfrac{[CE\rangle}{[EA\rangle} = \dfrac{[CD\rangle}{[DG\rangle}$. Applying the same theorem to the lines $\overleftrightarrow{BC}$ and $\overleftrightarrow{AB}$ and to the points D, G, and B on the first of these lines, and to the points A, F, B on the second line, $\dfrac{[BF\rangle}{[FA\rangle} = \dfrac{[BD\rangle}{[DG\rangle}$, that is $\dfrac{[AF\rangle}{[FB\rangle} = \dfrac{[GD\rangle}{[DB\rangle}$. By Remark RS.7 and arithmetic,

$$\frac{[AF\rangle}{[FB\rangle} \cdot \frac{[BD\rangle}{[DC\rangle} \cdot \frac{[CE\rangle}{[EA\rangle} = \frac{[GD\rangle}{[DB\rangle} \cdot \frac{[BD\rangle}{[DC\rangle} \cdot \frac{[CD\rangle}{[DG\rangle} = \frac{[GD\rangle}{[DG\rangle} \cdot \frac{[BD\rangle}{[DB\rangle} \cdot \frac{[CD\rangle}{[DC\rangle}$$
$$= (-1)(-1)(-1) = -1.$$

(II) If $\dfrac{[AF\rangle}{[FB\rangle} \cdot \dfrac{[BD\rangle}{[DC\rangle} \cdot \dfrac{[CE\rangle}{[EA\rangle} = -1$, then $\overleftrightarrow{EF}$ and $\overleftrightarrow{BC}$ intersect at a point D'. For, if $\overleftrightarrow{EF}$ and $\overleftrightarrow{BC}$ were parallel, then by Theorem RS.9 we would have $\dfrac{[AF\rangle}{[FB\rangle} = \dfrac{[AE\rangle}{[EC\rangle} =$

$\dfrac{[EA\rangle}{[CE\rangle}$, thus $\dfrac{[AF\rangle}{[FB\rangle}\cdot\dfrac{[CE\rangle}{[EA\rangle}$ would be equal to 1 and by our initial assumption, $\dfrac{[BD\rangle}{[DC\rangle}$ would be equal to -1. This would contradict Exercise RS.1. By part (I)

$$\frac{[AF\rangle}{[FB\rangle}\cdot\frac{[BD'\rangle}{[D'C\rangle}\cdot\frac{[CE\rangle}{[EA\rangle}=\frac{[AF\rangle}{[FB\rangle}\cdot\frac{[BD\rangle}{[DC\rangle}\cdot\frac{[CE\rangle}{[EA\rangle}.$$

Thus $\dfrac{[BD'\rangle}{[D'C\rangle}=\dfrac{[BD\rangle}{[DC\rangle}$ and by Exercise RS.2 $D'=D$, so that D, E, and F are collinear. □

Remark RS.11. Giovanni Ceva (1647–1734) was an Italian mathematician who studied geometry for most of his life. He first published Theorem RS.13 in 1678, in his work *De lineis rectis*. According to Audun Holme (*Geometry, Our Cultural Heritage*, 2nd ed, Springer, Heidelberg, pp. 193–194 (2010) [11]) the theorem was proved much earlier by Yusuf Al-Mu'taman ibn Hūd, an eleventh-century king of Zaragoza. Ceva also rediscovered and published Menelaus' Theorem.

Definition RS.12. If A is a corner of a triangle and B and C are its other two corners and $D\in\overleftrightarrow{BC}\setminus\{B,C\}$, then the line $\overleftrightarrow{AD}$ is called a **Cevian** in Ceva's honor. We shall call the Cevian $\overleftrightarrow{AD}$ an **interior** Cevian if $D\in\overline{BC}$, that is B–D–C, and an **exterior** Cevian if $D\notin\overline{BC}$, that is D–B–C or B–C–D.

Theorem RS.13 (Ceva). *Let $\mathcal{P}$ be a Euclidean/LUB plane, A, B, and C be noncollinear points on $\mathcal{P}$, and D, E, and F be points such that $D\in(\overleftrightarrow{BC}\setminus\{B,C\})$, $E\in(\overleftrightarrow{AC}\setminus\{A,C\})$, and $F\in(\overleftrightarrow{AB}\setminus\{A,B\})$. Then these two statements are equivalent:*

(1) *$\overleftrightarrow{AD}$, $\overleftrightarrow{BE}$, and $\overleftrightarrow{CF}$ are either concurrent or are parallel*

(2) $\dfrac{[AF\rangle}{[FB\rangle}\cdot\dfrac{[BD\rangle}{[DC\rangle}\cdot\dfrac{[CE\rangle}{[EA\rangle}=1.$

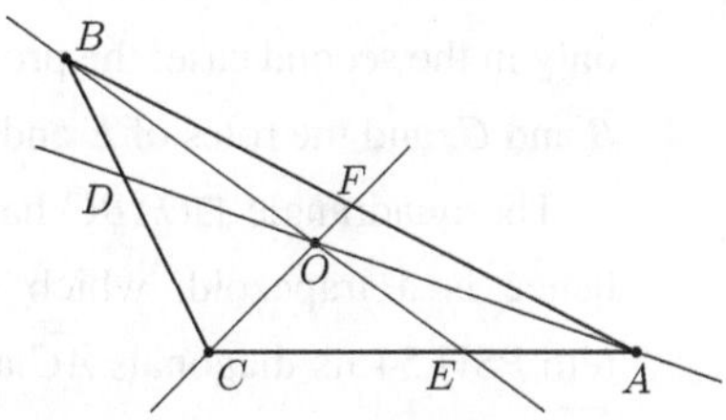

Fig. 20.2 For Ceva's Theorem RS.13.

Proof. See Figure 20.2.

(I) If $\overleftrightarrow{AD}$, $\overleftrightarrow{BE}$, and $\overleftrightarrow{CF}$ are concurrent at O, then we apply Theorem RS.10 (Menelaus) to $\triangle ABD$ and points F, O, C and to $\triangle ACD$ and points E, O, and B. Thus $\frac{[AO\rangle}{[OD\rangle}\cdot\frac{[BF\rangle}{[FA\rangle}\cdot\frac{[DC\rangle}{[CB\rangle}=-1$ and $\frac{[AO\rangle}{[OD\rangle}\cdot\frac{[CE\rangle}{[EA\rangle}\cdot\frac{[DB\rangle}{[BC\rangle}=-1$ so that $\left(\frac{[AO\rangle}{[OD\rangle}\cdot\frac{[CE\rangle}{[EA\rangle}\cdot\frac{[DB\rangle}{[BC\rangle}\right)/\left(\frac{[AO\rangle}{[OD\rangle}\cdot\frac{[BF\rangle}{[FA\rangle}\cdot\frac{[DC\rangle}{[CB\rangle}\right)=\frac{-1}{-1}=1.$

Rewriting this as a product, we have by Remark RS.7

$$\begin{aligned}1&=\frac{[AO\rangle}{[OD\rangle}\cdot\frac{[OD\rangle}{[AO\rangle}\cdot\frac{[CE\rangle}{[EA\rangle}\cdot\frac{[FA\rangle}{[BF\rangle}\cdot\frac{[DB\rangle}{[BC\rangle}\cdot\frac{[CB\rangle}{[DC\rangle}\\&=\frac{[CE\rangle}{[EA\rangle}\cdot\frac{[FA\rangle}{[BF\rangle}\cdot\frac{[BD\rangle}{[CB\rangle}\cdot\frac{[CB\rangle}{[DC\rangle}=\frac{[CE\rangle}{[EA\rangle}\cdot\frac{[FA\rangle}{[BF\rangle}\cdot\frac{[BD\rangle}{[DC\rangle},\end{aligned}$$

and therefore $\frac{[AF\rangle}{[FB\rangle}\cdot\frac{[BD\rangle}{[DC\rangle}\cdot\frac{[CE\rangle}{[EA\rangle}=1.$

(II) Assume $\overleftrightarrow{AD}$, $\overleftrightarrow{BE}$, and $\overleftrightarrow{CF}$ are parallel. $\overleftrightarrow{CF}\parallel\overleftrightarrow{BE}$, so applying Theorem RS.9 to $\overleftrightarrow{AB}$ and $\overleftrightarrow{AC}$, we have $\frac{[AF\rangle}{[FB\rangle}=\frac{[AC\rangle}{[CE\rangle}$.

Likewise since $\overleftrightarrow{AD}\parallel\overleftrightarrow{BE}$, we may apply Theorem RS.9 to $\overleftrightarrow{BC}$ and $\overleftrightarrow{AC}$ to get $\frac{[CA\rangle}{[AE\rangle}=\frac{[CD\rangle}{[DB\rangle}$, that is (*cf* Remark RS.7) $\frac{[BD\rangle}{[DC\rangle}=\frac{[AE\rangle}{[CA\rangle}$. Hence

$$\begin{aligned}\frac{[AF\rangle}{[FB\rangle}\cdot\frac{[BD\rangle}{[DC\rangle}\cdot\frac{[CE\rangle}{[EA\rangle}&=\frac{[AC\rangle}{[CE\rangle}\cdot\frac{[AE\rangle}{[CA\rangle}\cdot\frac{[CE\rangle}{[EA\rangle}=\frac{[AC\rangle}{[CA\rangle}\cdot\frac{[AE\rangle}{[EA\rangle}\cdot\frac{[CE\rangle}{[CE\rangle}\\&=(-1)(-1)(1)=1.\end{aligned}$$

(III) Conversely, assume that $\frac{[AF\rangle}{[FB\rangle}\cdot\frac{[BD\rangle}{[DC\rangle}\cdot\frac{[CE\rangle}{[EA\rangle}=1$. We want to show that if this is true, either all the Cevians are parallel or they are concurrent. If they are not all parallel, then two of them must intersect; thus it will suffice to show that if two of them intersect at a point O, then the third one must also contain this point. We choose our notation so that $\overleftrightarrow{BE}\cap\overleftrightarrow{CF}=\{O\}$.

Suppose the lines $\overleftrightarrow{AO}$ and $\overleftrightarrow{BC}$ fail to intersect, so are parallel. There are two cases: either $O\in B$-side of $\overleftrightarrow{AC}$, or in the opposite side. We shall give a proof only in the second case; the proof of the first case is similar, where the roles of B and C, and the roles of E and F are interchanged.

The quadrangle $\square OABC$ has a pair of opposite sides which are parallel, hence is a trapezoid, which by Theorem PSH.53.1 is rotund; by Theorem PSH.54 its diagonals $\overline{AC}$ and $\overline{BO}$ intersect at a point which is E, so that $\overleftrightarrow{BE}$ is an interior Cevian.

If $F \in \overline{AB}$, the point of intersection O of $\overline{CF}$ with $\overleftrightarrow{BE}$ would belong to enc $\triangle ABC$ and $\overleftrightarrow{AO}$ could not be parallel to $\overleftrightarrow{BC}$. Thus $F \notin \overline{AB}$, so that $\overleftrightarrow{CF}$ is an exterior Cevian. By Remark RS.7(B)(2) and (3), $\frac{[AE\rangle}{[EC\rangle} > 0$ and $\frac{[AF\rangle}{[FB\rangle} < 0$.

Since O is outside the triangle, we have B–E–O, as well as A–E–C. We have already shown F–A–B so that F is on the opposite side of $\overleftrightarrow{AO}$ from B and C, and hence F–O–C.

By Theorem SIM.19, $\frac{[\overline{FA}]}{[\overline{FO}]} = \frac{[\overline{FB}]}{[\overline{FC}]}$ so that $\frac{[\overline{FA}]}{[\overline{FB}]} = \frac{[\overline{FO}]}{[\overline{FC}]}$. By Theorem SIM.18(IV) $\triangle CBF \sim \triangle OAF$, since $\angle OFA$ is common to both. Then by Theorem SIM.19, $\frac{[\overline{EA}]}{[\overline{EC}]} = \frac{[\overline{EO}]}{[\overline{EB}]}$. Since $\angle CEB \cong \angle AEO$, then by Theorem SIM.18(IV) $\triangle CEB \sim \triangle AEO$ and hence $\frac{[\overline{CE}]}{[\overline{EA}]} = \frac{[\overline{CB}]}{[\overline{OA}]}$. Also, since $\angle CFB \cong \angle OFA$, then by Theorem SIM.18(IV) $\triangle CFB \sim \triangle OFA$ and hence $\frac{[\overline{CB}]}{[\overline{OA}]} = \frac{[\overline{BF}]}{[\overline{AF}]}$. Therefore $\frac{[\overline{CE}]}{[\overline{EA}]} = \frac{[\overline{CB}]}{[\overline{OA}]} = \frac{[\overline{BF}]}{[\overline{AF}]}$.

By Theorem RS.8(B), either $\frac{[CE\rangle}{[EA\rangle} = \frac{[BF\rangle}{[AF\rangle}$ or $\frac{[CE\rangle}{[EA\rangle} = -\frac{[BF\rangle}{[AF\rangle}$. As noted above, $\frac{[AE\rangle}{[EC\rangle} > 0$ and $\frac{[AF\rangle}{[FB\rangle} < 0$, so $\frac{[FB\rangle}{[FA\rangle} = \frac{[BF\rangle}{[AF\rangle} > 0$ and $\frac{[CE\rangle}{[EA\rangle} > 0$. Therefore $\frac{[CE\rangle}{[EA\rangle} = \frac{[BF\rangle}{[AF\rangle}$, so that

$$\frac{[AF\rangle}{[FB\rangle} \cdot \frac{[CE\rangle}{[EA\rangle} = \frac{[AF\rangle}{[FB\rangle} \cdot \frac{[FB\rangle}{[FA\rangle} = -1.$$

By Exercise RS.1, $\frac{[BD\rangle}{[DC\rangle} \neq -1$, so that it is impossible for

$$\frac{[AF\rangle}{[FB\rangle} \cdot \frac{[BD\rangle}{[DC\rangle} \cdot \frac{[CE\rangle}{[EA\rangle} = 1.$$

Therefore $\overleftrightarrow{AO}$ intersects $\overleftrightarrow{BC}$ at some point D'.

(IV) From part (I),

$$\frac{[AF\rangle}{[FB\rangle} \cdot \frac{[BD'\rangle}{[D'C\rangle} \cdot \frac{[CE\rangle}{[EA\rangle} = 1.$$

Since

$$\frac{[AF\rangle}{[FB\rangle} \cdot \frac{[BD\rangle}{[DC\rangle} \cdot \frac{[CE\rangle}{[EA\rangle} = 1$$

by assumption we have

$$\frac{[AF\rangle}{[FB\rangle} \cdot \frac{[BD'\rangle}{[D'C\rangle} \cdot \frac{[CE\rangle}{[EA\rangle} \cdot \frac{[FB\rangle}{[AF\rangle} \cdot \frac{[DC\rangle}{[BD\rangle} \cdot \frac{[EA\rangle}{[CE\rangle} = \frac{[BD'\rangle}{[D'C\rangle} \cdot \frac{[DC\rangle}{[BD\rangle} = 1,$$

that is $\dfrac{(d'-b)}{(c-d')} \cdot \dfrac{(c-d)}{(d-b)} = 1$. By arithmetic this becomes

$$\begin{aligned}
&(d'-b)(c-d) = (d-b)(c-d'),\\
&d'c - d'd - bc + bd = dc - dd' - bc + bd',\\
&d'c - d'd + bd = dc - dd' + bd',\\
&d'(c-d) + bd = dc - d'(d-b),\\
&d'(c-d) + d'(d-b) = dc - bd = d(c-b),\\
&d'(c-b) = d(c-b) \text{ and } d' = d.
\end{aligned}$$

Thus $D = dU = d'U = D'$ and the Cevian $\overleftrightarrow{AD}$ passes through O, proving the theorem. □

20.3 Exercises for ratios of sensed segments

Answers to starred (*) exercises may be accessed from the home page for this book at www.springer.com.

Exercise RS.1*. If $a \neq b$ are real numbers, then for any real number x,

$$\frac{x-a}{b-x} \neq -1.$$

Exercise RS.2*. If $a \neq b$ are real numbers, and x and y are any real numbers, if $\dfrac{x-a}{b-x} = \dfrac{y-a}{b-y}$ then $x = y$.

Exercise RS.3*. Let A, B, and X be points on a line $\mathbb{L}$ in the Euclidean/LUB plane $\mathcal{P}$, where $A \neq B$. Make a graph of the function $f(X) = \dfrac{[AX\rangle}{[XB\rangle}$.

Exercise RS.4*. If statement (2) of Ceva's theorem is true, that is if $\dfrac{[AF\rangle}{[FB\rangle} \cdot \dfrac{[BD\rangle}{[DC\rangle} \cdot \dfrac{[CE\rangle}{[EA\rangle} = 1$, then the number of exterior Cevians is either zero or two, the other Cevians being interior.

Chapter 21
Consistency and Independence of Axioms; Other Matters Involving Models

Acronyms: *LA, LB, LC, FM, DZI, MLT, PSM, LE, BI, MMI, RSI, DZII, DZIII*
Dependencies: *Sections 1.1 through 1.6 of Chapter 1, and other parts of the book as axioms are shown to hold in a given model*
New Axioms: *none*
New Terms Defined: *model, consistent, independent, strongly independent, sequentially independent, coordinate space; the following terms for coordinate space: line, plane, segment, ray; dot product, orthogonal, c-perpendicular, length, norm, normalize, distance; c-midpoint; between, quadratic distance, transfer mapping, induced mirror mapping*

Abstract: The first part of this lengthy chapter shows that Cartesian (coordinate) space satisfies all thirteen of the axioms of the main development of this book. This means that the axioms are consistent since there is a model, that is, an actual mathematical system, in which all are valid. The second part constructs, for each axiom, a model in which all previously listed axioms are true, but the new one is false. This shows that the newly added axiom is independent of those previously invoked. In the third part, models are exhibited showing the mutual independence of various properties of the definitions of betweenness, mirror mappings, and reflections. The fourth part consists of models showing the insufficiency of the incidence and betweenness axioms for creation of a satisfactory geometry.

E.J. Specht et al., *Euclidean Geometry and its Subgeometries*,
DOI 10.1007/978-3-319-23775-6_21

21.1 Euclid meets Descartes: synthetic *vs.* coordinate geometry

We have now traversed most of our intended theory; it has been a long and rather arduous journey. We have followed a *synthetic* approach to geometry, starting with undefined terms *point, line,* and *plane*. We then introduced, in order, axioms I.0, I.1, I.2, I.3, I.4, I.5, BET, PSA, REF, PS, and LUB.

Unlike Euclid, we have not tried to decide whether these axioms are universally "true" or "self-evident." In accordance with modern axiomatics, we only explored the ways that the various theorems in Euclidean geometry depend on these statements and on each other.

Euclid did not have number systems available to him, and he never made the connection, so clear to us moderns, between geometry and algebra. It would be some 1900 years before the French mathematician and philosopher René Descartes (1596–1650) made this connection by his invention of *coordinate* (or *analytic*) geometry, which laid the foundation for much modern mathematics, including calculus.

In Descartes' geometry, space $\mathbb{R}^3$ consists of points which are ordered triples (x_1, x_2, x_3) of numbers; these numbers are called the *Cartesian* coordinates of the point, and can be interpreted as displacements from the three base planes passing through an origin. The plane $\mathbb{R}^2$ is made of ordered pairs of numbers describing displacements from two main axes. To us this approach seems perfectly natural; we can visualize it easily and it seems "real" to us.

At the end of Chapter 18, we saw that Euclid's synthetic approach leads to the conclusion that every plane is essentially a Cartesian coordinate plane. Euclid thus meets Descartes, and we are the beneficiaries.

21.2 Our models and their implications

In this chapter we will be concerned with creating **models**, by which we mean actual "concrete" mathematical systems having certain useful characteristics. In the next two sections, 21.3 and 21.4, we will develop properties of coordinate space and the coordinate plane, in preparation for the following sections in which we develop the models themselves. These models fall into four categories, and each category will be developed in its own section, as listed below.

1. **Consistency model, Section 21.5:** This model will be based on 3-dimensional Cartesian space, together with its 2-dimensional subspaces, or planes, which together satisfy all our axioms. The existence of such a model rules out two embarrassing possibilities: one is that the whole development we have undertaken is *vacuous*, in the sense that there is no mathematical system to which it could apply. Not that our lack of knowledge of such a system would *prove* that the theory is vacuous—it might mean merely that our understanding is inadequate. But happily, that possibility is ruled out by this model.
   It also rules out the possibility that our axioms are not **consistent**—that there might be contradictions among them. If such contradictions existed, there could be no mathematical system in which they are all true. The fact that the axioms lead to useful results does not ensure that they are consistent, because the possibility exists that there might be contradictions among them that did not interfere with our development.
2. **Axiom independence models, Section 21.6:** We use several quite different models to show **independence** of the axioms; these are based on sets of numbers or sets of pairs or triples of numbers such as the natural numbers $\mathbb{N}$, the integers $\mathbb{Z}$, the rational numbers $\mathbb{Q}$, the real algebraic numbers $\mathbb{A}$, or the real numbers $\mathbb{R}$.
3. **Property independence models, Section 21.7:** We use linear models based on Cartesian space or Cartesian planes to show the independence of various properties of definitions, thus assuring that these definitions are stated with reasonable economy.
4. **Insufficiency models, Section 21.8:** Here we will show that the incidence and betweenness axioms by themselves (as set forth in Chapters 1 and 4) are insufficient to create a satisfactory geometry—that is, that Axiom PSA is necessary. More specifically, we show that in a geometry where only the incidence and betweenness axioms are invoked, there can be several circumstances which are highly offensive to our intuition; for instance there can be a segment having two different sets of endpoints.

   The models used in this section are based on Model DZI (initially developed in Subsection 21.6.3), consisting of the set $\mathbb{Z}^3$ of all ordered triples (x_1, x_2, x_3) where x_1, x_2, and x_3 are integers; that is, the set of all points of Cartesian space having integer coordinates, sometimes called (*lattice points*).

21.2.1 List of axioms for reference

For the convenience of the reader, we again list our set of axioms. For a listing of relevant definitions, see the chapters referenced.

Incidence axioms (Chapter 1).

Axiom I.0. Lines and planes exist and are subsets of space $\mathcal{S}$.

Axiom I.1. There exists exactly one line through two distinct points.

Axiom I.2. There exists exactly one plane through three noncollinear points.

Axiom I.3. If two distinct points lie in a plane, then any line through the points is contained in the plane.

Axiom I.4. If two distinct planes have a nonempty intersection, then their intersection has at least two members.

Axiom I.5. **(A)** There exist at least two distinct points on every line.

(B) There exists at least one noncollinear set of three points on every plane.

(C) There exists at least one noncoplanar set of four points in space.

Betweenness Axiom BET (Chapter 4). There exists a betweenness relation on space $\mathcal{S}$, satisfying Properties B.0 through B.3 of Definition IB.1.

Plane Separation Axiom PSA (Chapter 5). Let $\mathcal{L}$ be a line and $\mathcal{E}$ and $\mathcal{F}$ be opposite sides of $\mathcal{L}$; if $Q \in \mathcal{E}$ and $R \in \mathcal{F}$, then $\overline{QR} \cap \mathcal{L} \neq \emptyset$.

Reflection Axiom REF (Chapter 8). On the Pasch plane $\mathcal{P}$, there exists a set $\mathbb{REF}$ of reflections satisfying Properties R.1 through R.6 of Definition NEUT.2.

Parallel Axiom PS (Chapters 2 and 11). Given a line $\mathcal{L}$ and a point P not belonging to $\mathcal{L}$, there exists *exactly* one line $\mathcal{M}$ such that $P \in \mathcal{M}$ and $\mathcal{L} \parallel \mathcal{M}$.

Least Upper Bound Axiom LUB (Chapter 18). Let $\mathbb{L}$ be an ordered field with origin O and unit U. Every nonempty subset $\mathcal{E}$ of $\mathbb{L}$ which is bounded above has a least upper bound lub $\mathcal{E}$.

21.3 Coordinate space: linear Model LM3

In this chapter, the reader should temporarily put aside essentially everything from the main development, from Chapter 1 Section 1.8 through Chapter 20. Here we will use, as a starting point, the basics of coordinate geometry, as briefly outlined below.

Nomenclature and notation: In this and following sections, when we speak of an **ordered field** $\mathbb{F}$ we will mean one of the fields $\mathbb{Q}$, $\mathbb{A}$, or $\mathbb{R}$ which has been equipped with the natural ordering, all as described in Chapter 1, Section 1.5 under the title "Number systems."

If $\mathbb{F}$ should be specifically one of the fields $\mathbb{Q}$, $\mathbb{A}$, or $\mathbb{R}$, we will attach the appropriate letter suffix: that is, the 3-dimensional Model LM3 over the real algebraic numbers $\mathbb{A}$ would be called "Model LM3A"; the 2-dimensional Model LM2 over $\mathbb{Q}$, the rational numbers, would be "Model LM2Q."

We habitually write ordered pairs and triples horizontally, as (x_1, x_2) or (x_1, x_2, x_3); however, there will be occasions where we wish to write them vertically, as $\begin{pmatrix} x_1 \\ x_2 \end{pmatrix}$. The disadvantage of the vertical notation is obvious—it takes up a lot of space on the page. But when the entries of a pair or triple are long or elaborate, it sometimes makes things clearer to display them vertically.

We begin by summarizing for $n = 3$ the definition from Chapter 1, Section 1.5, under the title "Vector spaces of n-tuples."

3-dimensional coordinate space: The set $\mathbb{F}^3$ of ordered triples (3-tuples) (a_1, a_2, a_3) of elements of $\mathbb{F}$ is a vector space called 3-*dimensional coordinate space*, where we have defined $(a_1, a_2, a_3) + (b_1, b_2, b_3) = (a_1 + b_1, a_2 + b_2, a_3 + b_3)$, and $t(a_1, a_2, a_3) = (ta_1, ta_2, ta_3)$ for $t \in \mathbb{F}$. The triple $O = (0, 0, 0)$ is the origin, or zero element, of $\mathbb{F}^3$.

Let $E_1 = (1, 0, 0)$, $E_2 = (0, 1, 0)$, and $E_3 = (0, 0, 1)$. The set $\mathcal{E} = \{(1, 0, 0), (0, 1, 0), (0, 0, 1)\} = \{E_1, E_2, E_3\}$ is linearly independent and spans $\mathbb{F}^3$, so the dimension of $\mathbb{F}^3$ is 3.

In our main development (Chapters 1 Section 1.8 through Chapter 20) "line" and "plane" were undefined terms. Here, in our Models LM3 and LM2, they start out as defined objects. In the first definition we will give the meanings of space, lines, planes, segments, and rays in our linear Model LM3; these are easily reducible to Model LM2. Eventually it will be shown that these meanings coincide with those given in our main development.

Definition LA.1. **Space** for Model LM3 is the vector space $\mathbb{F}^3$, the set of ordered triples $A = (a_1, a_2, a_3)$ of members of an ordered field $\mathbb{F}$. The model also includes lines, planes, segments, and rays as follows:

(1) A subset $\mathcal{L}$ is a **line** in $\mathbb{F}^3$ iff there exist distinct points A and B of $\mathbb{F}^3$ such that $\mathcal{L} = \{A + t(B - A) \mid t \in \mathbb{F}\}$; this line is denoted by $\overleftrightarrow{AB}$.

(2) A subset $\mathcal{P}$ is a **plane** in $\mathbb{F}^3$ iff there exist noncollinear points A, B, and C of $\mathbb{F}^3$ such that $\mathcal{P} = \{A + s(B - A) + t(C - A) \mid (s, t) \in \mathbb{F}^2\}$; this plane is denoted by $\overleftrightarrow{ABC}$.

(3) Let $A = (a_1, a_2, a_3)$, $B = (b_1, b_2, b_3)$, and $X = (x_1, x_2, x_3)$ be points of $\mathbb{F}^3$, where $A \neq B$; and let t denote a real number.

(3A) The **open segment** from A to B is
$$\overset{\dashv\vdash}{AB} = \{A + t(B - A) \mid 0 < t < 1\}.$$

(3B) The **closed segment** from A to B is
$$\overset{\sqsubset\sqsupset}{AB} = \{A + t(B - A) \mid 0 \leq t \leq 1\}.$$

(3C) The **half-open segment** from A to B is either
$$\overset{\sqsubset\vdash}{AB} = \{A + t(B - A) \mid 0 \leq t < 1\} \text{ or}$$
$$\overset{\dashv\sqsupset}{AB} = \{A + t(B - A) \mid 0 < t \leq 1\}.$$

(3D) The **closed ray** with initial point A is $\overset{\sqsubset\!\rightarrow}{AB} = \{A + t(B - A) \mid t \geq 0\}$.

(3E) The **open ray** with initial point A is $\overset{\dashv\!\rightarrow}{AB} = \{A + t(B - A) \mid t > 0\}$.

In the following, many of the proofs will be relegated to exercises, the solutions of which are accessible online from the home page of this book at www.springer.com.

Remark LA.2. (1) $\overleftrightarrow{AB} = \{A + t(B - A) \mid t \in \mathbb{F}\}$
$$= \{B + (1 - t)(A - B) \mid t \in \mathbb{F}\} = \overleftrightarrow{BA}.$$

(2) $\overleftrightarrow{ABC}$ is the same plane for any permutation of the points A, B, and C; that is,
$$\overleftrightarrow{ABC} = \overleftrightarrow{ACB} = \overleftrightarrow{BAC} = \overleftrightarrow{CAB} = \overleftrightarrow{BCA} = \overleftrightarrow{CBA}$$

We will provide a proof for the first two equalities only:

(A) $\overleftrightarrow{ABC} = \{A + s(B - A) + t(C - A) \mid (s, t) \in \mathbb{F}^2\}$
$$= \{A + t(C - A) + s(B - A) \mid (t, s) \in \mathbb{F}^2\} = \overleftrightarrow{ACB}.$$

(B) $\overleftrightarrow{ABC} = \{A + s(B - A) + t(C - A) \mid (s, t) \in \mathbb{F}^2\}$
$$= \{B + (1 - s - t)(A - B) + t(C - B) \mid (s, t) \in \mathbb{F}^2\}$$
$$= \{B + u(A - B) + t(C - B) \mid (u, t) \in \mathbb{F}^2\} = \overleftrightarrow{BAC}.$$

The other proofs are similar and are left to the reader as Exercise LM.1.

Theorem LA.3. *Distinct points A, B, and C in* $\mathbb{F}^3$ *are collinear iff* $B - A$ *and* $C - A$ *are linearly dependent.*

Proof. The proof is Exercise LM.2. □

Theorem LA.4. *Distinct points A, B, C, and D in $\mathbb{F}^3$ are coplanar iff $B-A$, $C-A$, and $D-A$ are linearly dependent.*

Proof. The proof is Exercise LM.3. □

Theorem LA.5. *Let A and B be distinct points in $\mathbb{F}^3$. For each $t \in \mathbb{F}$ define $\varphi(t) = A + t(B - A)$, so that φ maps $\mathbb{F}$ into $\overleftrightarrow{AB}$. Then φ is a bijection (one-to-one mapping) of $\mathbb{F}$ onto $\overleftrightarrow{AB}$.*

Proof. The proof is Exercise LM.4. □

Theorem LA.6. *If A, B, and C are noncollinear points of $\mathbb{F}^3$ and if s and t are any numbers in $\mathbb{F}$, then the equality*

$$\varphi(s,t) = X = A + s(B - A) + t(C - A)$$

defines a bijection of $\mathbb{F}^2$ onto $\overleftrightarrow{ABC}$.

Proof. By Definition LA.1(2), for every $(s,t) \in \mathbb{F}^2$, $\varphi(s,t) = A+s(B-A)+t(C-A)$ is a member of $\overleftrightarrow{ABC}$, and every point of this plane is such a point; thus φ maps $\mathbb{F}^2$ onto $\overleftrightarrow{ABC}$.

Suppose that for some (s,t) and (u,v) in $\mathbb{F}^2$, $\varphi(s,t) = \varphi(u,v)$. Then

$$A + s(B - A) + t(C - A) = A + u(B - A) + v(C - A),$$

so that $(s-u)(B-A)+(t-v)(C-A) = O$. By the contrapositive of Theorem LA.3, $B - A$ and $C - A$ are linearly independent, so that $s = u$ and $t = v$, proving that φ is 1–1. □

Remark LA.7 (On notation). We use the notation $\mathcal{E}+X$ to mean $\{A+X \mid A \in \mathcal{E}\}$, where $\mathcal{E}$ is a subset of $\mathbb{F}^3$ and X is any member of $\mathbb{F}^3$.

For any two subsets $\mathcal{E}$ and $\mathcal{G}$ of $\mathbb{F}^3$, $\mathcal{E} + X \subseteq \mathcal{G} + X$ iff $\mathcal{E} \subseteq \mathcal{G}$. For if $\mathcal{E} + X \subseteq \mathcal{G} + X$, then for every $A \in \mathcal{E}$ there exists a $B \in \mathcal{G}$ such that $A + X = B + X$, and $A = A + X - X = B + X - X = B$, proving that $\mathcal{E} \subseteq \mathcal{G}$. Showing the converse is trivial. It follows easily from the fact that two sets are equal iff each is a subset of the other, that $\mathcal{E} + X = \mathcal{G} + X$ iff $\mathcal{E} = \mathcal{G}$.

Since $Y \in \mathcal{E}$ iff $Y + X \in \mathcal{E} + X$, it is also true that $Y \notin \mathcal{E}$ iff $Y + X \notin \mathcal{E} + X$; thus $\mathcal{E} + X$ is a proper subset of $\mathbb{F}^3$ iff $\mathcal{E}$ is a proper subset.

In the following Remarks LA.8 and LA.9 we will write each assertion in italics and follow it with its justification.

Remark LA.8 (Of lines).

(A) *If $A = O$, the line $\mathcal{L} = \{A + t(B - A) \mid t \in \mathbb{F}\}$ contains the point O.* To see this, let $t = 0$.

(B) *A line $\mathcal{L}$ in $\mathbb{F}^3$ containing the origin O is $\{sC \mid s \in \mathbb{F}\}$ for some $C \in \mathbb{F}^3$.*

A line $\mathcal{L} = \{A + t(B - A) \mid t \in \mathbb{F}\}$ contains the origin O iff there exists a number t_0 such that $A + t_0(B - A) = O$. In this case define $s = t - t_0$, and $C = B - A$. Then $A + t(B - A) = A + (s + t_0)C = (A + t_0C) + sC = O + sC$.

(C) *A line $\mathcal{L}$ in $\mathbb{F}^3$ containing the origin O is a subspace of $\mathbb{F}^3$, having dimension 1; a subspace of $\mathbb{F}^3$ of dimension 1 is a line containing O.*

If $X \in \mathcal{L}$ and $Y \in \mathcal{L}$, then from part (B), there exist numbers s and t such that $X = sC$ and $Y = tC$; then $X + Y = sC + tC = (s + t)C \in \mathcal{L}$. For any number u, $uX = usC \in \mathcal{L}$. Since every member of $\mathcal{L}$ is a scalar multiple of C, its dimension is 1. Conversely, if $\mathcal{L}$ is a subspace of $\mathbb{F}^3$ having dimension 1, there is a vector $C \neq O$ such that for every point $X \in \mathcal{L}$, $X = sC$ for some number s; thus by Definition LA.1(1), $\mathcal{L}$ is a line containing O.

(D) *If $\mathcal{L}$ is a line and Y a member of $\mathbb{F}^3$, then $\mathcal{L} + Y$ is a line. Here we are using the notation of Remark LA.7.*

Let $\mathcal{L} = \{A + t(B - A) \mid t \in \mathbb{F}\}$ according to Definition LA.1(1). Then

$$\begin{aligned}\mathcal{L} + Y &= \{A + t(B - A) \mid t \in \mathbb{F}\} + Y\\ &= \{(A + Y) + t((B + Y) - (A + Y)) \mid t \in \mathbb{F}\}\end{aligned}$$

which is a line by Definition LA.1(1).

(E) *If $Y \in \mathcal{L}$, then by part (D) $\mathcal{L} - Y$ is a line; it contains the origin O because $Y - Y = O$; and by part (C), it is a 1-dimensional subspace of $\mathbb{F}^3$.*

Remark LA.9 (Of planes).

(A) *If $A = O$, the plane $\mathcal{P} = \{A + s(B - A) + t(C - A) \mid (s, t) \in \mathbb{F}^2\}$ contains the point O.* To see this, let $s = t = 0$.

(B) *A plane $\mathcal{P}$ in $\mathbb{F}^3$ containing the origin O is $\{uE + vF \mid (u, v) \in \mathbb{F}^2\}$ for some linearly independent vectors D and E.*

To see this, note that a plane $\mathcal{P} = \{A + s(B - A) + t(C - A) \mid (s, t) \in \mathbb{F}^2\}$ contains O iff for some numbers s_0 and t_0, $A + s_0(B - A) + t_0(C - A) = O$. Let $u = s + s_0$, $v = t + t_0$, $E = B - A$, and $F = C - A$. Then

$$\begin{aligned}A + s(B - A) + t(C - A) &= A + (u - s_0)D + (v - t_0)E\\ &= (A - s_0D - t_0E) + uD + vE = O + uD + vE,\end{aligned}$$

so that $\mathcal{P} = \{uE + vF \mid (u, v) \in \mathbb{F}^2\}$. The vectors $D = B - A$ and $E = C - A$ are linearly independent by Theorem LA.3.

(C) *A plane $\mathcal{P}$ in $\mathbb{F}^3$ containing the origin O is a subspace of $\mathbb{F}^3$ having dimension 2; a subspace of $\mathbb{F}^3$ of dimension 2 is a plane containing O.*

If $X \in \mathcal{P}$ and $Y \in \mathcal{P}$, then using part (B), there exist linearly independent vectors E and E and numbers s and t such that $X = sD + tE$ and $Y = uD + vE$. Then $X + Y = sD + tE + uD + vE = (s + u)D + (t + v)E \in \mathcal{P}$, and for any number w, $wX = wsD + wtE \in \mathcal{P}$, showing that $\mathcal{P}$ is a subspace. Moreover, the dimension of $\mathcal{P}$ is 2, since its members are all the linear combinations of D and E, which are linearly independent. Conversely, if $\mathcal{P}$ is a subspace of $\mathbb{F}^3$ having dimension 2, there exist linearly independent vectors D and E such that for every point $X \in \mathcal{P}$, $X = sD + tE$ for some numbers s and t; thus by Definition LA.1(2), $\mathcal{P}$ is a plane containing O.

(D) *If $\mathcal{P}$ is a plane as defined in Definition LA.1(2), then for any $Y \in \mathbb{F}^3$, $\mathcal{P} + Y$ is a plane in $\mathbb{F}^3$.*

If $\mathcal{P} = \{A + s(B - A) + t(C - A) \mid (s, t) \in \mathbb{F}^2\}$ is a plane (as defined in Definition LA.1(2)), then for any $Y \in \mathbb{F}^3$,

$$\begin{aligned}\mathcal{P} + Y &= \{A + s(B - A) + t(C - A) + Y \mid (s, t) \in \mathbb{F}^2\}\\ &= \{(A+Y)+s((B+Y)-(A+Y))+t((C+Y)-(A+Y)) \mid (s, t) \in \mathbb{F}^2\}\end{aligned}$$

which by definition is a plane containing the points $A + Y$, $B + Y$, and $C + Y$.

(E) *If $\mathcal{P}$ is a plane in $\mathbb{F}^3$ and $Y \in \mathcal{P}$, then by part (D) $\mathcal{P}-Y$ is a plane; it contains the origin O because $Y - Y = O$; and by part (C), it is a 2-dimensional subspace of $\mathbb{F}^3$.*

Definition LA.1(1) says that a set is a line iff its points are all the points $X = A + t(B - A)$ for some distinct A and B belonging to the set; this begs the question of whether these two points completely determine the line. Likewise, a set is a plane iff its points are all the points $X = A + s(B - A) + t(C - A)$ for some noncollinear points A, B, and C of the set. Again, it is not immediately clear that these three points completely determine the plane. These questions are answered in the affirmative by the following theorem.

Theorem LA.10. (A) *Let A and B be any two points of $\mathbb{F}^3$; there is exactly one line $\overleftrightarrow{AB}$ containing these two points.*

(B) *Let A, B, and C be three noncollinear points of $\mathbb{F}^3$; there is exactly one plane $\overleftrightarrow{ABC}$ containing these three points.*

Proof. (A) Let C and D be distinct points of $\mathbb{F}^3$ such that $\overleftrightarrow{CD}$ is a line containing both A and B. By Definition LA.1(1) there exist distinct numbers t_1 and t_2

such that $A = C + t_1(D - C)$ and $B = C + t_2(D - C)$. Then for every $X = A + u(B - A) \in \overleftrightarrow{AB}$,

$$\begin{aligned} X &= A + u(B - A) \\ &= C + t_1(D - C) + u((C + t_2(D - C)) - (C + t_1(D - C))) \\ &= C + (t_1 + u(t_2 - t_1))(D - C) \in \overleftrightarrow{CD}. \end{aligned}$$

If Y is any point of $\overleftrightarrow{AB}$, it is also a point of $\overleftrightarrow{CD}$, and by Remark LA.8(D), both $\overleftrightarrow{AB} - Y$ and $\overleftrightarrow{CD} - Y$ are subspaces of $\mathbb{F}^3$; by Remark LA.8(C) the dimension of both subspaces is 1. Since $\overleftrightarrow{AB} - Y \subseteq \overleftrightarrow{CD} - Y$, it follows from the Dimension Criterion (Chapter 1 Section 1.5) that $\overleftrightarrow{AB} - Y = \overleftrightarrow{CD} - Y$; by Remark LA.7, $\overleftrightarrow{AB} = \overleftrightarrow{CD}$.

(B) Let D, E, and F be noncollinear points in $\mathbb{F}^3$ such that A, B, and C are members of $\overleftrightarrow{DEF}$; then by Definition LA.1(2) there exist pairwise distinct ordered pairs (u_1, v_1), (u_2, v_2), and (u_3, v_3) such that

$$\begin{aligned} A &= D + u_1(E - D) + v_1(F - D), &\quad (*) \\ B &= D + u_2(E - D) + v_2(F - D), &\quad (**) \\ C &= D + u_3(E - D) + v_3(F - D). &\quad (***) \end{aligned}$$

By Definition LA.1(2), if $X \in \overleftrightarrow{ABC}$, there exist numbers u and v such that $X = A + u(B - A) + v(C - A)$. This becomes, on substitution by equations (*) through (***)

$$\begin{aligned} X &= D + u_1(E - D) + v_1(F - D) \\ &\qquad + u\Big((D + u_2(E-D) + v_2(F-D))\Big) - u\Big(D + u_1(E-D) + v_1(F-D)\Big) \\ &\qquad + v\Big(D + u_3(E-D) + v_3(F-D)\Big) - v\Big(D + u_1(E-D) + v_1(F-D)\Big) \\ &= D + u_1(E - D) + v_1(F - D) \\ &\qquad + uu_2(E - D) + uv_2(F - D) - uu_1(E - D) - uv_1(F - D) \\ &\qquad + vu_3(E - D) + vv_3(F - D) - vu_1(E - D) - vv_1(F - D) \\ &= D + (u_1 + uu_2 - uu_1 + vu_3 - vu_1)(E - D) \\ &\qquad + (v_1 + uv_2 - uv_1 + vv_3 - vv_1)(F - D) \in \overleftrightarrow{DEF}. \end{aligned}$$

Hence $\overleftrightarrow{ABC} \subseteq \overleftrightarrow{DEF}$.

If Y is any point of $\overleftrightarrow{ABC}$, it is also a point of $\overleftrightarrow{DEF}$, and by Remark LA.9(D), both $\overleftrightarrow{ABC} - Y$ and $\overleftrightarrow{DEF} - Y$ are subspaces of $\mathbb{F}^3$; by Remark LA.9(C) the dimension of both subspaces is 2. Since $\overleftrightarrow{ABC} - Y \subseteq \overleftrightarrow{DEF} - Y$, it follows from the Dimension Criterion (Chapter 1 Section 1.5) that $\overleftrightarrow{ABC} - Y = \overleftrightarrow{DEF} - Y$; by Remark LA.7, $\overleftrightarrow{ABC} = \overleftrightarrow{DEF}$. This completes the proof. □

Corollary LA.10.1. (A) *For any line* $\mathcal{L} \subseteq \mathbb{F}^3$, *and any distinct points A and B of* $\mathcal{L}$, $\mathcal{L} = \{A + t(B - A) \mid t \in \mathbb{F}\}$.

(B) *For any plane* $\mathcal{P} \subseteq \mathbb{F}^3$, *and any noncollinear points A, B, and C of* $\mathcal{P}$, $\mathcal{P} = \{A + s(B - A) + t(C - A) \mid (s,t) \in \mathbb{F}^2\}$.

Remark LA.11. (A) Every line $\mathcal{L}$ is a proper subset of any plane containing it; for if a line contains all the points of a plane, that line would, by definition, contain noncollinear points, a contradiction.

(B) Every plane $\mathcal{P}$ is a proper subset of $\mathbb{F}^3$. Again by Remark LA.9(C), the dimension of $\mathcal{P} - Y$ is 2, and by the Dimension Criterion of Chapter 1 Section 1.5, $\mathcal{P} - Y$ is a proper subset of $\mathbb{F}^3$. Pick $X \in \mathbb{F}^3$ such that $X \notin \mathcal{P} - Y$; then $X + Y \notin \mathcal{P}$.

This can also be seen as follows: let $A = (0,0,0)$, $B = (1,0,0)$, $C = (0,1,0)$, and $D = (0,0,1)$. If some plane should contain the entire space, it would contain all these points, including A, B, and C, which are noncollinear. By Definition LA.1(2) a point X belongs to this plane iff for some s and t, $X = A + s(B - A) + t(C - A)$, and since $A = O$, this is just $X = sB + tC$. The point D does not belong to this plane, because there are no scalars s and t such that $s(1,0,0) + t(0,1,0) = (0,0,1)$. By Theorem LA.10, this is the only plane containing A, B, and C; hence no plane contains $\mathbb{F}^3$.

Definition/Remark LA.12. (1) If $\mathbb{F}$ is one of the fields $\mathbb{Q}$, $\mathbb{A}$, or $\mathbb{R}$ the **dot product**[1] $A \bullet B$ of $A = (a_1, a_2, a_3)$ and $B = (b_1, b_2, b_3)$ in $\mathbb{F}^3$ is the scalar $a_1b_1 + a_2b_2 + a_3b_3$. It is a straightforward computation to show that for any vectors A, B, and C of $\mathbb{F}^3$ and any scalar s, $(A + B) \bullet C = A \bullet C + B \bullet C$, $A \bullet (B + C) = A \bullet B + A \bullet C$, and $s(A \bullet B) = sA \bullet B = A \bullet sB$.

(2) Two vectors A and B in $\mathbb{F}^n$ are said to be **orthogonal** iff $A \bullet B = 0$. We denote this by $A \perp B$.

(3) If $\mathcal{L}$ and $\mathcal{M}$ are two lines in a plane, then $\mathcal{L}$ is **c-perpendicular** to $\mathcal{M}$ iff for some distinct points Q and P in $\mathcal{L}$ and some distinct points R and S in $\mathcal{M}$, $(Q - P) \bullet (R - S) = 0$. In this case we write $\mathcal{L} \perp \mathcal{M}$, just as we write $(Q - P) \perp (R - S)$ to indicate that these vectors are orthogonal. The "c-" added to "perpendicularity" is to remind the reader that this definition is only for coordinate space, and is different from Definition NEUT.31 (*cf* Theorem LC.46).

---

[1] Sometimes called the **inner product**. In vector space theory, the inner product of two vectors A and B is sometimes denoted (A, B), but we will adhere to the notation $A \bullet B$.

Definition/Remark LA.13. (1) If $\mathbb{F}$ is one of the fields $\mathbb{A}$ or $\mathbb{R}$, so that non-negative numbers have square roots, define the **length** or **norm** of a vector $A = (a_1, a_2, a_3)$ as $\|A\| = \sqrt{A \bullet A} = \sqrt{a_1^2 + a_2^2 + a_3^2}$. Clearly $\|A\| = 0$ iff $A = O = (0, 0, 0)$. For any vector $A = (a_1, a_2, a_3)$, $\|\frac{A}{\|A\|}\| = 1$, since

$$\|\tfrac{A}{\|A\|}\| = \sqrt{(\tfrac{a_1}{\|A\|})^2 + (\tfrac{a_2}{\|A\|})^2 + (\tfrac{a_n}{\|A\|})^2} = \sqrt{\tfrac{a_1^2}{\|A\|^2} + \tfrac{a_2^2}{\|A\|^2} + \tfrac{a_3^2}{\|A\|^2}}$$
$$= \sqrt{\tfrac{a_1^2+a_2^2+a_3^2}{\|A\|^2}} = \sqrt{\tfrac{\|A\|^2}{\|A\|^2}} = 1.$$

We say that the vector A is **normalized** when it is divided by its norm, thus acquiring length 1.

(2) If $\mathbb{F}$ is one of the fields $\mathbb{A}$ or $\mathbb{R}$, and $A = (a_1, a_2, a_3)$ and $B = (b_1, b_2, b_3)$ are two points of $\mathbb{F}^3$, the **distance** $\operatorname{dis}(A, B)$ between A and B is the length of the difference vector, that is,

$$\|A - B\| = \sqrt{(A - B) \bullet (A - B)} = \sqrt{(a_1 - b_1)^2 + (a_2 - b_2)^2 + (a_3 - b_3)^2}.$$

(3) Let $\mathbb{F}$ be any ordered field; if $A = (a_1, a_2, a_3)$ and $B = (b_1, b_2, b_3)$ are two points (members) of $\mathbb{F}^3$, the **c-midpoint** of $\overleftrightarrow{AB}$ is the point

$$\tfrac{A+B}{2} = \left(\tfrac{a_1+b_1}{2}, \tfrac{a_2+b_2}{2}, \tfrac{a_3+b_3}{2}\right) = A + \tfrac{B-A}{2} = B + \tfrac{A-B}{2}.$$

The "c-" added to "midpoint" is intended to remind the reader that this definition is different from that of Definition NEUT.3(C) (*cf* Theorem LC.47).

If $\mathbb{F}$ is one of the fields $\mathbb{A}$ or $\mathbb{R}$, and $M = \frac{A+B}{2}$ is the c-midpoint of the line segment connecting A and B,

$$\begin{aligned}
\operatorname{dis}(A, M) &= \|A - M\| \\
&= \sqrt{(a_1 - (\tfrac{a_1+b_1}{2}))^2 + (a_2 - (\tfrac{a_2+b_2}{2}))^2 + (a_3 - (\tfrac{a_3+b_3}{2}))^2} \\
&= \sqrt{(\tfrac{a_1-b_1}{2})^2 + (\tfrac{a_2-b_2}{2})^2 + (\tfrac{a_3-b_3}{2})^2} = \sqrt{(\tfrac{b_1-a_1}{2})^2 + (\tfrac{b_2-a_2}{2})^2 + (\tfrac{b_3-a_3}{2})^2} \\
&= \sqrt{(b_1 - (\tfrac{a_1+b_1}{2}))^2 + (b_2 - (\tfrac{a_2+b_2}{2}))^2 + (b_3 - (\tfrac{a_3+b_3}{2}))^2} \\
&= \|B - M\| = \operatorname{dis}(B, M).
\end{aligned}$$

We now derive two well-known properties for norms.

Theorem LA.13.1. *Let $\mathbb{F}$ be one of the fields $\mathbb{A}$ or $\mathbb{R}$. If $A = (a_1, a_2, a_3)$ and $B = (b_1, b_2, b_3)$ are members of $\mathbb{F}^3$ and $x \in \mathbb{F}$, then*

(A) $\|A + B\| \le \|A\| + \|B\|$ *(triangle inequality);*

(B) $\|xA\| = |x|\|A\|$*; and*

(C) $A = O$ *iff* $\|A\| = 0$.

This is also true for points of $\mathbb{F}^2$.

Proof. We give a proof for $\mathbb{F}^3$, which is easily reducible to $\mathbb{F}^2$. The proof of (A) depends on the *Cauchy-Schwarz-Bunyakovski* inequality[2] which we derive first.

For any numbers a_1, a_2, a_3, b_1, b_2, and b_3, the following expression is greater or equal to 0 because it is the sum of squares:

$$
\begin{aligned}
&(a_1b_1 - a_1b_1)^2 + (a_1b_2 - a_2b_1)^2 + (a_1b_3 - a_3b_1)^2 \\
&\qquad (a_2b_1 - a_1b_2)^2 + (a_2b_2 - a_2b_2)^2 + (a_2b_3 - a_3b_2)^2 \\
&\qquad\qquad (a_3b_1 - a_1b_3)^2 + (a_3b_2 - a_2b_3)^2 + (a_3b_3 - a_3b_3)^2 \\
&= (a_1^2b_1^2 - 2a_1^2b_1^2 + a_1^2b_1^2) + (a_1^2b_2^2 - 2a_1a_2b_1b_2 + a_2^2b_1^2) + (a_1^2b_3^2 - 2a_1a_3b_1b_3 + a_3^2b_1^2) \\
&\quad + (a_2^2b_1^2 - 2a_1a_2b_1b_2 + a_1^2b_2^2) + (a_2^2b_2^2 - 2a_2^2b_2^2 + a_2^2b_2^2) + (a_2^2b_3^2 - 2a_2a_3b_2b_3 + a_3^2b_2^2) \\
&\quad + (a_3^2b_1^2 - 2a_1a_3b_1b_3 + a_1^2b_3^2) + (a_3^2b_2^2 - 2a_2a_3b_2b_3 + a_2^2b_3^2) + (a_3^2b_3^2 - 2a_3^2b_3^2 + a_3^2b_3^2) \\
&\geq 0.
\end{aligned}
$$

In each of the nine groupings above, the first and last term occurs again as a first or last term in some grouping; in the next equality we combine these into a single grouping. Likewise, each of the middle terms of these nine groupings that does not include squared numbers is repeated; in the next equality we combine these in a second grouping, so that

$$
\begin{aligned}
&2(a_1^2b_1^2 + a_1^2b_2^2 + a_1^2b_3^2 + a_2^2b_1^2 + a_2^2b_2^2 + a_2^2b_3^2 + a_3^2b_1^2 + a_3^2b_2^2 + a_3^2b_3^2) \\
&\qquad - 2(a_1^2b_1^2 + a_2^2b_2^2 + a_3^2b_3^2 + 2a_1a_2b_1b_2 + 2a_1a_3b_1b_3 + 2a_2a_3b_2b_3) \geq 0.
\end{aligned}
$$

Dividing by 2 this becomes

$$(a_1^2 + a_2^2 + a_3^2)(b_1^2 + b_2^2 + b_3^2) - (a_1b_1 + a_2b_2 + a_3b_3)^2 \geq 0$$

as can be seen by multiplying out the last expression. This is

$$\|A\|^2\|B\|^2 - (A \bullet B)^2 \geq 0, \text{ or } \|A\|^2\|B\|^2 \geq (A \bullet B)^2.$$

Therefore $\|A\|\,\|B\| \geq A \bullet B$, the Cauchy-Schwarz-Bunyakovski inequality.

(A) Using this inequality,

$$
\begin{aligned}
\|A + B\|^2 &= (A + B) \bullet (A + B) = \|A\|^2 + 2A \bullet B + \|B\|^2 \\
&\leq \|A\|^2 + 2\|A\|\,\|B\| + \|B\|^2 = (\|A\| + \|B\|)^2
\end{aligned}
$$

so that $\|A + B\| \leq (\|A\| + \|B\|$, as required.

(B) If x is any number, then

$$
\begin{aligned}
\|xA\| &= \sqrt{x^2a_1^2 + x^2a_2^2 + x^2a_3^2} = \sqrt{x^2(a_1^2 + a_2^2 + a_3^2)} \\
&= |x|\sqrt{a_1^2 + a_2^2 + a_3^2} = |x|\|A\|.
\end{aligned}
$$

(C) $\|A\| = \sqrt{a_1^2 + a_2^2 + a_3^2} = 0$ iff $a_1 = a_2 = a_3 = 0$. □

[2] Some sources indicate that Bunyakovski's contribution was to the integral or infinite-dimensional form of this inequality. We include his name here out of deference to our late beloved co-author Harold T. Jones, whom we remember as being quite insistent on its inclusion.

Theorem LA.14 (Orthogonality and linear independence). *If $A_1, A_2, \ldots, A_n$ are non-zero members of $\mathbb{F}^3$ and are pairwise orthogonal (each is orthogonal to the others), they are linearly independent.*

Proof. Suppose that for some numbers $x_1, x_2, \ldots, x_n$, $x_1A_1 + x_2A_2 + \ldots + x_nA_n = O$. For each $i \in [1; n]$

$$\begin{aligned} 0 = A_i \bullet O &= A_i \bullet (x_1A_1 + x_2A_2 + \ldots + x_nA_n) \\ &= A_i \bullet x_1A_1 + A_i \bullet x_2A_2 + \ldots + A_i \bullet x_nA_n) \\ &= x_1(A_i \bullet A_1) + x_2(A_i \bullet A_2) + \ldots + x_n(A_i \bullet A_n). \end{aligned}$$

Since $A_j \bullet A_k = 0$ whenever $j \neq k$, this becomes $0 + 0 + A_i \bullet x_iA_i = x_i\|A_i\|^2$. Since $A_i \neq O$, $\|A_i\| \neq 0$ so that $x_i = 0$. Therefore $x_1 = x_2 = \ldots = x_n = 0$, showing that the vectors $A_1, A_2, \ldots, A_n$ are linearly independent. □

The following theorem is a standard result from elementary linear algebra.

Theorem LA.15.

(A) *Let $A = (a_1, a_2)$ and $B = (b_1, b_2)$ be members of $\mathbb{F}^2$. Then A and B are linearly dependent iff* $\begin{vmatrix} a_1 & b_1 \\ a_2 & b_2 \end{vmatrix} = 0.$

(B) *Let $A = (a_1, a_2, a_3)$, $B = (b_1, b_2, b_3)$, and $C = (c_1, c_2, c_3)$ be members of $\mathbb{F}^3$. Then A, B, and C are linearly dependent iff* $\begin{vmatrix} a_1 & b_1 & c_1 \\ a_2 & b_2 & c_2 \\ a_3 & b_3 & c_3 \end{vmatrix} = 0.$

Proof. The proof is Exercise LM.5. □

Remark LA.16. In coordinate geometry a plane is usually defined as a set

$$\mathcal{E} = \{(x_1, x_2, x_3) \mid ax_1 + bx_2 + cx_3 + d = 0\} \ (*)$$

where a, b, c, and d are numbers in $\mathbb{F}$, and at least one of a, b, or c is nonzero. The equation $ax_1 + bx_2 + cx_3 + d = 0$ is an *equation of the plane*.

Note that if it were true that $a = b = c = 0$, then $d = 0$ and every point in $\mathbb{F}^3$ would satisfy $ax_1 + bx_2 + cx_3 + d = 0$. We rule out this case by requiring one of a, b, or c to be nonzero.

Most readers, being familiar with coordinate geometry, will not find it difficult to accept equation (*) as describing a plane; but our Definition LA.1(2) (defining a plane) is not the same as this, and a somewhat cumbersome proof is required to show these are equivalent. In the next three Theorems LA.17 through LA.19

we lay out the argument that this is so: a set $\mathcal{E}$ satisfies equation (*) iff it satisfies Definition LA.1(2). For those who wish to see the details worked out, the proofs of these theorems are provided (on line) as solutions to Exercises LM.6 through LM.8.

Theorem LA.17. *Let a, b, c, and d be members of $\mathbb{F}$, where at least one of a, b, c is nonzero; let $\mathcal{E}$ be the set of all points $(x_1, x_2, x_3) \in \mathbb{F}^3$ such that $ax_1 + bx_2 + cx_3 + d = 0$, as defined in Remark LA.16.*

(A) *$\mathcal{E}$ is a proper subset of $\mathbb{F}^3$.*
(B) *If $X = (x_1, x_2, x_3) \in \mathcal{E}$, there exist two other points $Y = (y_1, y_2, y_3)$ and $Z = (z_1, z_2, z_3)$ in $\mathcal{E}$ such that X, Y, and Z are noncollinear, which is to say (by Theorem LA.3) that the vectors $Y - X$ and $Z - X$ are linearly independent.*

Proof. The proof is Exercise LM.6. □

Theorem LA.18. *Let $X = (x_1, x_2, x_3)$, $Y = (y_1, y_2, y_3)$, and $Z = (z_1, z_2, z_3)$ be noncollinear points in $\mathbb{F}^3$, so that $\overleftrightarrow{XYZ}$ is a plane as in Definition LA.1(2). Then there exist numbers a, b, c, and d in $\mathbb{F}$, where not all of a, b, or c are zero, such that*

$$\overleftrightarrow{XYZ} = \{(w_1, w_2, w_3) \mid aw_1 + bw_2 + cw_3 + d = 0\}$$

Proof. The proof is Exercise LM.7. □

Theorem LA.19. *Let a, b, c, and d be numbers in $\mathbb{F}$, where not all of a, b, or c are zero. Then the set*

$$\mathcal{E} = \{(w_1, w_2, w_3) \mid aw_1 + bw_2 + cw_3 + d = 0\}$$

is a plane in $\mathbb{F}^3$ as defined by Definition LA.1(2). It follows immediately from Theorem LA.18 that our two definitions of a plane are equivalent.

Proof. The proof is Exercise LM.8. □

We now turn to a series of theorems which yield, in Theorem LA.22, a criterion for two lines in $\mathbb{F}^3$ to be parallel. In the next theorem we use the notation for determinants and matrices set forth in Chapter 1 Section 1.5.

Theorem LA.20. *Let the entries in the matrix $\begin{pmatrix} a_1 & b_1 & c_1 \\ a_2 & b_2 & c_2 \\ a_3 & b_3 & c_3 \end{pmatrix}$ be numbers (in $\mathbb{F}$) such that*

$$\begin{vmatrix} a_1 & b_1 & c_1 \\ a_2 & b_2 & c_2 \\ a_3 & b_3 & c_3 \end{vmatrix} = 0. \quad (*)$$

If there exist no ordered pairs (s,t) *belonging to* $\mathbb{F}^2$ *such that all the equations*

$$\begin{pmatrix} a_1 s + b_1 t = c_1, \\ a_2 s + b_2 t = c_2, \\ a_3 s + b_3 t = c_3 \end{pmatrix} \quad (**)$$

are true, then $\begin{vmatrix} a_1 & b_1 \\ a_2 & b_2 \end{vmatrix} = 0$, $\begin{vmatrix} a_1 & b_1 \\ a_3 & b_3 \end{vmatrix} = 0$, *and* $\begin{vmatrix} a_2 & b_2 \\ a_3 & b_3 \end{vmatrix} = 0$.

Proof. By Theorem LA.15(B), equation (*) says the vectors $A = (a_1, a_2, a_3)$, $B = (b_1, b_2, b_3)$, and $C = (c_1, c_2, c_3)$ are linearly dependent, so there exist numbers s, t, and u, not all zero, such that $sA + tB + uC = 0$.

If $u \neq 0$, then $\frac{s}{u}A + \frac{t}{u}B + C = 0$, that is to say, $-\frac{s}{u}A - \frac{t}{u}B = C$ so all of the equations (**) are true, contradicting our hypothesis. Therefore $u = 0$ so that $sA + tB = O$. All the pairs of vectors (in $\mathbb{F}^2$) (a_1, a_2) and (b_1, b_2), (a_1, a_3) and (b_1, b_3), (a_2, a_3) and (b_2, b_3) are linearly dependent; thus by Theorem LA.15(A), $\begin{vmatrix} a_1 & b_1 \\ a_2 & b_2 \end{vmatrix} = 0$, $\begin{vmatrix} a_1 & b_1 \\ a_3 & b_3 \end{vmatrix} = 0$, *and* $\begin{vmatrix} a_2 & b_2 \\ a_3 & b_3 \end{vmatrix} = 0$. □

Theorem LA.21. *If* $A = (a_1, a_2, a_3)$ *and* $B = (b_1, b_2, b_3)$ *are members of* $\mathbb{F}^3$ *such that* $A \neq O$, $a_1b_2 - a_2b_1 = 0$, $a_1b_3 - a_3b_1 = 0$, *and* $a_2b_3 - a_3b_2 = 0$, *then there exists a member* k *of* $\mathbb{F}$ *such that* $B = kA$.

Proof. Since $A \neq O$, at least one of a_1, a_2, and a_3 must be nonzero.

If $a_1 \neq 0$, then $b_3 = \frac{b_1}{a_1}a_3$, $b_2 = \frac{b_1}{a_1}a_2$, and $b_1 = \frac{b_1}{a_1}a_1$, so in this case $k = \frac{b_1}{a_1}$.
If $a_2 \neq 0$, then $b_3 = \frac{b_2}{a_2}a_3$, $b_2 = \frac{b_2}{a_2}a_2$, and $b_1 = \frac{b_2}{a_2}a_1$, so in this case $k = \frac{b_2}{a_2}$.
If $a_3 \neq 0$, then $b_3 = \frac{b_3}{a_3}a_3$, $b_2 = \frac{b_3}{a_3}a_2$, and $b_1 = \frac{b_3}{a_3}a_1$, so in this case $k = \frac{b_3}{a_3}$. □

Theorem LA.22. *Let* $A = (a_1, a_2, a_3)$, $B = (b_1, b_2, b_3)$, *and* $C = (c_1, c_2, c_3)$ *be noncollinear points of* $\mathbb{F}^3$ *and let* $D = (d_1, d_2, d_3)$ *be a point distinct from* C; *then* $\overleftrightarrow{CD} \parallel \overleftrightarrow{AB}$ *iff there exists a number* k *different from* 0 *such that* $D - C = k(B - A)$.

Proof. (I) Assume there exists a number $k \neq 0$ such that $D - C = k(B - A)$; then

$$D = C + (D - C) = C + k(B - A) = A + k(B - A) + (C - A),$$

so that by Definition LA.1(2) $D \in \overleftrightarrow{ABC}$, and A, B, C, and D are coplanar.

By Definition LA.1(1), a point $X \in \overleftrightarrow{CD}$ iff for some t, $X = C + t(D - C)$. By our assumption that $D - C = k(B - A)$, this becomes $X = C + tk(B - A)$. Also,

$X \in \overleftrightarrow{BA}$ iff for some number s, $X = A + s(B - A)$. If the two lines $\overleftrightarrow{CD}$ and $\overleftrightarrow{BA}$ have a point in common, there exist s and t such that both these are true, so that

$$C + tk(B - A) - (A + s(B - A)) = (tk - s)(B - A) + (C - A) = O, \qquad (*)$$

By Theorem LA.3, $C - A$ and $B - A$ are linearly independent, so if this is true, both the coefficient of $B - A$ and the coefficient of $C - A$ must be zero, which is impossible, since the coefficient of $C - A$ is 1. Therefore, for every choice of s and t, $C + t(D - C) - (A + s(B - A)) \neq O$, and the lines $\overleftrightarrow{CD}$ and $\overleftrightarrow{AB}$ are parallel.

Note that in the case where $s = t = 0$, the left-hand side of (*) reduces to $C - A$ which is not equal to O because C and A are distinct.

(II) If $\overleftrightarrow{AB} \parallel \overleftrightarrow{CD}$, then (*cf* Chapter 2 Definition IP.0) A, B, C, D are coplanar.
By Theorem LA.4, $B - A$, $C - A$, and $D - A$ are linearly dependent. By Theorem LA.15(B)

$$\begin{vmatrix} b_1 - a_1 & c_1 - a_1 & d_1 - a_1 \\ b_2 - a_2 & c_2 - a_2 & d_2 - a_2 \\ b_3 - a_3 & c_3 - a_3 & d_3 - a_3 \end{vmatrix} = 0.$$

Since $\overleftrightarrow{AB} \cap \overleftrightarrow{CD} = \emptyset$, there exist no members (s, t) of $\mathbb{F}^2$ such that

$$A + s(B - A) - (C + t(D - C)) = O, \text{ or}$$

$$s(B - A) - t(D - C) - (C - A) = O.$$

In other words, there is no member (s, t) of $\mathbb{F}^2$ such that these three equations hold:

$$s(b_1 - a_1) - t(d_1 - c_1) - (c_1 - a_1) = 0,$$
$$s(b_2 - a_2) - t(d_2 - c_2) - (c_2 - a_2) = 0,$$
$$s(b_3 - a_3) - t(d_3 - c_3) - (c_3 - a_3) = 0.$$

By Theorem LA.20,

$$\begin{vmatrix} b_1 - a_1 & d_1 - c_1 \\ b_2 - a_2 & d_2 - c_2 \end{vmatrix} = 0, \begin{vmatrix} b_1 - a_1 & d_1 - c_1 \\ b_3 - a_3 & d_3 - c_3 \end{vmatrix} = 0, \text{ and}$$

$$\begin{vmatrix} b_2 - a_2 & d_2 - c_2 \\ b_3 - a_3 & d_3 - c_3 \end{vmatrix} = 0.$$

Hence by Theorem LA.21 there exists a number k different from 0 such that $D - C = k(B - A)$. □

Theorem LA.23. *Let $\mathcal{L}$ and $\mathcal{M}$ be lines in a plane $\mathcal{P} \subseteq \mathbb{F}^3$; then if $\mathcal{L}$ and $\mathcal{M}$ are c-perpendicular, they intersect.*

Proof. Let A and B be distinct points of $\mathcal{L}$, and let C and D be distinct points of $\mathcal{M}$, and suppose these two lines are c-perpendicular. By Definition/Remark LA.12(3), $(B - A) \bullet (D - C) = 0$; if these lines do not intersect, they are parallel, since both lie in plane $\mathcal{P}$. By Theorem LA.22 there exists a number $k \neq 0$ such that $D - C = k(B - A) = 0$; then

$$\begin{aligned} 0 &= (B - A) \bullet (D - C) = (B - A) \bullet k(B - A) \\ &= k(B - A) \bullet (B - A) = k\|B - A\|^2. \end{aligned}$$

Since $B - A \neq O$, by Theorem LA.13.1(C) $\|B - A\| \neq 0$, so that $k = 0$, a contradiction. □

21.4 Coordinate plane: linear Model LM2

Reflections must be done on planes, so we now explore some of the basic structure on the plane $\mathbb{F}^2$ that will be needed to define them.

Throughout this section $\mathbb{F}$ is one of the ordered fields $\mathbb{Q}$, $\mathbb{A}$, or $\mathbb{R}$; we will routinely refer to a member of $\mathbb{F}$ as a number, and to the additive and multiplicative identities as 0 and 1, respectively. The dot product and norm are defined for any vector space over one of these fields.

Definition LB.1. Model LM2 is the vector space $\mathbb{F}^2$, the set of ordered pairs $A = (a_1, a_2) \in \mathbb{F}^2$. The model also includes lines, segments, and rays as defined above in Definition LA.1, with the substitution of $\mathbb{F}^2$ for $\mathbb{F}^3$, and the substitution of ordered pairs for ordered triples.

Remark LB.2. (A) The plane $\mathbb{F}^2$ can be regarded as a subset of $\mathbb{F}^3$ by identifying each point $A = (a_1, a_2) \in \mathbb{F}^2$ with the point $A = (a_1, a_2, 0) \in \mathbb{F}^3$. We will sometimes call this plane a "base plane" of $\mathbb{F}^3$. Lines, segments, and rays in the plane $\mathbb{F}^2$ are defined as in parts (1) and (3) of Definition LA.1, where it is understood that points A, B, X, *etc.* are ordered pairs rather than ordered triples.

(B) Let $E_1 = (1, 0)$ and $E_2 = (0, 1)$. The set $\mathcal{E} = \{(1, 0), (0, 1)\} = \{E_1, E_2\}$ is linearly independent; every vector $A = (a_1, a_2) = a_1(1, 0) + a_2(0, 1)$ is a linear combination of the vectors E_1 and E_2, which therefore constitute a *basis* for $\mathbb{F}^2$. Hence the *dimension* of $\mathbb{F}^2$ is 2.

(C) According to Definition LA.1(1) $\mathcal{L}$ is a *line* in $\mathbb{F}^2$ iff there exist distinct points $A = (a_1, a_2)$ and $B = (b_1, b_2)$ such that $\mathcal{L} = \{A + t(B - A) \mid t \in \mathbb{F}\}$. An equivalent formulation which is sometimes useful is this: $\mathcal{L}$ is a line in $\mathbb{F}^2$ iff there exist points $A = (a_1, a_2)$ and $C = (c_1, c_2) \neq (0, 0)$ such that $\mathcal{L} = \{A + tC) \mid t \in \mathbb{F}\}$.

Remark LB.3. In coordinate geometry a line on a plane is usually defined as a set

$$\mathcal{L} = \{(x_1, x_2) \mid ax_1 + bx_2 + c = 0\}$$

where a, b, and c are numbers in $\mathbb{F}$, and at least one of a or b is nonzero; equation $ax_1 + bx_2 + c = 0$ is called an *equation of the line*.

Note that if it were true that $a = b = 0$, then $c = 0$ and every point in $\mathbb{F}^2$ would satisfy $ax_1 + bx_2 + c = 0$. We rule out this case by requiring one of a or b to be nonzero.

Again, (*cf* Remark LA.16) most readers will readily accept the equation $ax_1 + bx_2 + c = 0$ as describing a line, and as in the previous section, a somewhat cumbersome proof is required to show that this equation and our Definition LA.1(1) are equivalent. In the next three Theorems LB.4 through LB.6 we give the argument that this is so. For those who wish to see the details worked out, the proofs of these theorems are provided (online at the home page for this book at www.springer.com) as solutions to Exercises LM.9 through LM.11.

Theorem LB.4. *Let a, b, c, a′, b′, and c be numbers in* $\mathbb{F}$*, and suppose at least one of a or b, and at least one of a′ or b′ is nonzero. Then*

(A) $\mathcal{L} = \{(x_1, x_2) \mid ax_1 + bx_2 + c = 0\} \neq \mathbb{F}^2$*;*

(B) *there exist at least two distinct points in* $\mathcal{L}$*; and*

(C) *both* $ax_1 + bx_2 + c = 0$ *and* $a'x_1 + b'x_2 + c' = 0$ *are equations for* $\mathcal{L}$ *iff there exists a number* $k \neq 0$ *such that* $a' = ka$, $b' = kb$, *and* $c' = kc$.

Proof. The proof is Exercise LM.9. □

Theorem LB.5. *Let* $X = (x_1, x_2)$ *and* $Y = (y_1, y_2)$ *be distinct points in* $\mathbb{F}^2$*, and let* $\overleftrightarrow{XY}$ *be the line containing both X and Y according to Definition LA.1(1). Then* $\overleftrightarrow{XY} = \{(w_1, w_2) \mid aw_1 + bw_2 + c = 0\}$*, where* $a = y_2 - x_2$*,* $b = x_1 - y_1$*, and* $c = x_2(y_1 - x_1) - x_1(y_2 - x_2)$.

Proof. The proof is Exercise LM.10. □

Theorem LB.6. *Let a, b, and c be numbers in* $\mathbb{F}$*, where at least one of a or b is nonzero. Then the set*

$$\mathcal{L} = \{(w_1, w_2) \mid aw_1 + bw_2 + c = 0\}$$

is a line in $\mathbb{F}^2$ as defined by Definition LA.1(1). It follows immediately from Theorem LB.5 that our two definitions of a line (in a plane) are equivalent.

Proof. The proof is Exercise LM.11. □

Remark LB.7. In the following we will frequently refer to two lines $\mathcal{L}$ and $\mathcal{M}$ as being *c-perpendicular*. The reader may wish to refer again to Definition/Remark LA.12(3) where this notion is defined. Again we emphasize that this is not (yet) the same notion of perpendicularity as was developed in neutral geometry (*cf* Definition NEUT.31). In Theorem LC.46, after we have defined the "correct" reflections over lines on $\mathbb{F}^2$ we shall see that the two notions coincide. We may, however, use the same symbol $\mathcal{L} \perp \mathcal{M}$ to indicate c-perpendicularity of the lines $\mathcal{L}$ and $\mathcal{M}$ as we do to indicate that they are perpendicular in the sense of neutral geometry.

Theorem LB.8. *Let $\overleftrightarrow{XY}$ and $\overleftrightarrow{ZW}$ be two lines on $\mathbb{F}^2$ according to Definition LA.1(1), where $X = (x_1, x_2) \neq Y = (y_1, y_2)$ and $W = (w_1, w_2) \neq Z = (z_1, z_2)$. Then according to Theorem LB.5*

$$\overleftrightarrow{XY} = \{(t_1, t_2) \mid a_1t_1 + b_1t_2 + c_1 = 0\}$$

where $a_1 = y_2 - x_2$, $b_1 = x_1 - y_1$, and $c_1 = x_2(y_1 - x_1) - x_1(y_2 - x_2)$; also

$$\overleftrightarrow{ZW} = \{(t_1, t_2) \mid a_2t_1 + b_2t_2 + c_2 = 0\},$$

where $a_2 = w_2 - z_2$, $b_2 = z_1 - w_1$, and $c_2 = z_2(w_1 - z_1) - z_1(w_2 - z_2)$.

Then $\overleftrightarrow{XY}$ is c-perpendicular to $\overleftrightarrow{ZW}$ iff $(X - Y) \bullet (Z - W) = 0$, which is true iff $a_1a_2 + b_1b_2 = 0$.

Proof. By Theorem LB.5

$$\begin{aligned}(X - Y) \bullet (Z - W) &= (x_1 - y_1)(z_1 - w_1) + (x_2 - y_2)(z_2 - w_2)\\ &= b_1b_2 + (-a_1)(-a_2) = b_1b_2 + a_1a_2.\end{aligned}$$

By Definition/Remark LA.12(3), $\overleftrightarrow{XY} \perp \overleftrightarrow{ZW}$ iff $(X - Y) \bullet (Z - W) = 0$ which, by the calculation, is true iff $b_1b_2 + a_1a_2 = 0$.

Note that if the coefficients a_1 and b_1 are both multiplied by the same nonzero number, or if a_2 and b_2 are multiplied by the same nonzero number, the condition will remain true. □

Remark LB.9. In the equation $ax_1 + bx_2 + c = 0$, when $a = 0$, $x_2 = -c/b$. Then all the second coordinates of points on the line $\mathcal{L}$ defined by the equation are the same. This situation is usually described by saying that the line $\mathcal{L}$ is "horizontal."

Likewise if $b = 0$ the line is "vertical." In the proof of Theorem LB.8, those cases where $a_1 = 0$ are exactly those where $b_2 = 0$, and those cases where $a_2 = 0$ are exactly those where $b_1 = 0$. That is, one of the lines is horizontal and the other (being c-perpendicular to it) is vertical.

Theorem LB.10. *Let*

$$\mathcal{L} = \{(x_1, x_2) \mid a_1x_1 + b_1x_2 + c_1 = 0\} \text{ and}$$
$$\mathcal{M} = \{(x_1, x_2) \mid a_2x_1 + b_2x_2 + c_2 = 0\}$$

be two lines in the plane $\mathbb{F}^2$. If they are c-perpendicular, they must intersect.

Proof. This follows directly from Theorem LA.23. A direct proof is possible using the equations for lines as given here; the proof is Exercise LM.12. □

Theorem LB.11. *Let $\mathbb{F}$ be either $\mathbb{A}$ or $\mathbb{R}$; let A and B be distinct members of $\mathbb{F}^2$, and let $r > 0$ be any number. Then there exists a unique point C on $\overrightarrow{AB}$ such that* $\text{dis}(A, C) = r$.

Proof. In this proof we use the fact that if $\mathbb{F}$ is $\mathbb{A}$ or $\mathbb{R}$, norms exist in $\mathbb{F}^2$. By Definition LA.1(3D) every point C of the ray $\overrightarrow{AB}$ is of the form $C = A + t(B - A)$ for some number $t \geq 0$. Let $t = \frac{r}{\|B-A\|}$, which is ≥ 0; then

$$\text{dis}(A, C) = \|C - A\| = \|A + \tfrac{r}{\|B-A\|}(B - A) - A\| = \|\tfrac{r}{\|B-A\|}(B - A)\|.$$

By Theorem LA.13.1(B) this is $|\frac{r}{\|B-A\|}|\|(B - A)\| = r$. □

Theorem LB.12. *Let $\mathcal{L}$ be any line on $\mathbb{F}^2$; by Theorem LB.5 there exist numbers a, b, and c in $\mathbb{F}$ such that $(a, b) \neq (0, 0)$ and*

$$\mathcal{L} = \{(x_1, x_2) \mid (x_1, x_2) \in \mathbb{F}^2 \text{ and } ax_1 + bx_2 + c = 0\}$$

Let (u_1, u_2) be a point on $\mathbb{F}^2$. Then there is one and only one line $\mathcal{M}$ through (u_1, u_2) which is c-perpendicular to $\mathcal{L}$, namely

$$\mathcal{M} = \{(x_1, x_2) \mid (x_1, x_2) \in \mathbb{F}^2 \text{ and } bx_1 - ax_2 - bu_1 + au_2 = 0\}.$$

Proof. By Theorem LB.8, for every number d,

$$\mathcal{M} = \{(x_1, x_2) \mid (x_1, x_2) \in \mathbb{F}^2 \text{ and } bx_1 - ax_2 + d = 0\}$$

is a line which is c-perpendicular to $\mathcal{L}$, since $a(b) + b(-a) = 0$. $\mathcal{M}$ is the line through (u_1, u_2) which is c-perpendicular to $\mathcal{L}$ iff $bu_1 - au_2 + d = 0$, which is true iff $d = -bu_1 + au_2$. Hence

$$\mathcal{M} = \{(x_1, x_2) \mid (x_1, x_2) \in \mathbb{F}^2 \text{ and } bx_1 - ax_2 - bu_1 + au_2 = 0\}.$$ □

Theorem LB.13. (A) *Let a, b, c_1, and c_2 be numbers in the ordered field $\mathbb{F}$ such that $(a, b) \neq (0, 0)$, and let*

$$\mathcal{L} = \{(x_1, x_2) \mid (x_1, x_2) \in \mathbb{F}^2 \text{ and } ax_1 + bx_2 + c_1 = 0\},$$

and

$$\mathcal{M} = \{(x_1, x_2) \mid (x_1, x_2) \in \mathbb{F}^2 \text{ and } ax_1 + bx_2 + c_2 = 0\}$$

Then $\mathcal{L}$ *and* $\mathcal{M}$ *are parallel iff* $c_1 \neq c_2$.

(B) *Let* $\mathcal{L}$, $\mathcal{M}$, *and* $\mathcal{N}$ *be distinct lines on* $\mathbb{F}^2$ *such that* $\mathcal{L}$ *and* $\mathcal{M}$ *are c-perpendicular and* $\mathcal{L}$ *and* $\mathcal{N}$ *are c-perpendicular. Then* $\mathcal{M}$ *and* $\mathcal{N}$ *are parallel.*

Proof. (A) $\mathcal{L} \nparallel \mathcal{M}$ iff there exists a member (x_1, x_2) of $\mathbb{F}^2$ such that $ax_1 + bx_2 + c_1 = ax_1 + bx_2 + c_2$, which is true iff $c_1 = c_2$.

(B) Let $\mathcal{L} = \{(x_1, x_2) \mid (x_1, x_2) \in \mathbb{F}^2, \text{ and } ax_1 + bx_2 + c = 0\}$. By Theorem LB.12 there exist numbers d and e such that

$$\mathcal{M} = \{(x_1, x_2) \mid (x_1, x_2) \in \mathbb{F}^2 \text{ and } bx_1 - ax_2 + d = 0\}$$

and

$$\mathcal{N} = \{(x_1, x_2) \mid (x_1, x_2) \in \mathbb{F}^2 \text{ and } bx_1 - ax_2 + e = 0\}.$$

Now $\mathcal{M} = \mathcal{N}$ iff $d = e$; since these are distinct lines, $d \neq e$. If $\mathcal{M}$ and $\mathcal{N}$ are not parallel, there exists a member (x_1, x_2) of $\mathbb{F}^2$ such that $bx_1 - ax_2 + d = bx_1 - ax_2 + e$ so that $d = e$, a contradiction. Thus $\mathcal{M} \parallel \mathcal{N}$, completing the proof. □

Theorem LB.14. *Let*

$$\mathcal{L} = \{(x_1, x_2) \mid (x_1, x_2) \in \mathbb{F}^2 \text{ and } ax_1 + bx_2 + c = 0\}.$$

Let (u_1, u_2) *be a point on* $\mathbb{F}^2$ *and*

$$\mathcal{M} = \{(x_1, x_2) \mid (x_1, x_2) \in \mathbb{F}^2 \text{ and } bx_1 - ax_2 - bu_1 + au_2 = 0\}.$$

be the line through (u_1, u_2) *which is c-perpendicular to* $\mathcal{L}$, *according to Theorem LB.12. Then*

$$\mathcal{L} \cap \mathcal{M} = (y_1, y_2) = \left(\tfrac{b^2u_1 - abu_2 - ac}{a^2+b^2}, \tfrac{-abu_1 + a^2u_2 - bc}{a^2+b^2}\right).$$

Proof. By Cramer's rule the solution to $\left\{\begin{array}{rcl} ay_1 + by_2 &=& -c \\ -by_1 + ay_2 &=& -bu_1 + au_2 \end{array}\right\}$ is

$$y_1 = \frac{\begin{vmatrix} -c & b \\ -bu_1 + au_2 & a \end{vmatrix}}{\begin{vmatrix} a & b \\ -b & a \end{vmatrix}} = \frac{-ac + b(bu_1 - au_2)}{a^2+b^2} = \frac{b^2u_1 - abu_2 - ac}{a^2+b^2}$$

and

$$y_2 = \frac{\begin{vmatrix} a & -c \\ -b & -bu_1 + au_2 \end{vmatrix}}{\begin{vmatrix} a & b \\ -b & a \end{vmatrix}} = \frac{a(-bu_1 + au_2) - bc}{a^2+b^2} = \frac{-abu_1 + a^2u_2 - bc}{a^2+b^2}.$$

It is a straightforward calculation to verify that the point (y_1, y_2) found above belongs to both the lines $\mathcal{L}$ and $\mathcal{M}$. □

Remark LB.15. We may become more comfortable with the slightly complicated result of Theorem LB.14 by noting the following:

(A) If $a = 0$ ($\mathcal{L}$ is a "horizontal" line), then $y_1 = \frac{b^2 u_1}{b^2} = u_1$, and $y_2 = \frac{-bc}{b^2} = \frac{-c}{b}$. Thus the line through (u_1, u_2) and (y_1, y_2) is "vertical."

(B) If $b = 0$ ($\mathcal{L}$ is a "vertical" line), then $y_1 = \frac{-ac}{a^2} = \frac{-c}{a}$, and $y_2 = \frac{a^2 u_2}{a^2} = u_2$, Thus the line through (u_1, u_2) and (y_1, y_2) is "horizontal."

(C) In the case that $(u_1, u_2) \in \mathcal{L}$, if $a = 0$, $u_2 = \frac{-c}{b} = y_2$, and if $b = 0$, $u_1 = \frac{-c}{a} = y_1$.

(D) If $a \neq 0$ and $b \neq 0$, and $(u_1, u_2) \in \mathcal{L}$, $u_2 = \frac{-au_1}{b} - \frac{c}{b}$ and
$$y_1 = \frac{b^2u_1 - abu_2 - ac}{a^2+b^2} = \frac{b^2u_1 - a(-au_1 - c) - ac}{a^2+b^2} = \frac{a^2+b^2}{a^2+b^2}u_1 = u_1.$$
Also, if $(u_1, u_2) \in \mathcal{L}$, $u_1 = -\frac{bu_2}{a} - \frac{c}{a}$ so that
$$y_2 = \frac{-b(-bu_2 - c) + a^2u_2 - bc}{a^2+b^2} = \frac{b^2u_2 + cb + a^2u_2 - bc}{a^2+b^2} = \frac{a^2+b^2}{a^2+b^2}u_2 = u_2.$$
These calculations assure us that the formulas for the intersection point (y_1, y_2) are valid, as they should be, when $(u_1, u_2) \in \mathcal{L}$.

Definition LB.16 (Mirror mapping over a line).

(A) (Coordinate-free form) Let $\mathcal{L}$ be a line on $\mathbb{F}^2$ and let U be any point of $\mathbb{F}^2$. Let $\mathcal{M}$ be the line (whose existence is guaranteed by Theorem LB.12) such that $U \in \mathcal{M}$ and $\mathcal{M}$ is c-perpendicular to $\mathcal{L}$. By Theorem LB.10, $\mathcal{L}$ and $\mathcal{M}$ must intersect at some point Y. Define $\Phi(U) = Y + (Y - U) = 2Y - U$. For a visualization see Figure 21.1 below.

(B) (Coordinate form) Let a, b, and c be numbers in $\mathbb{F}$ such that
$$\mathcal{L} = \{(x_1, x_2) \mid (x_1, x_2) \in \mathbb{F}^2 \textit{ and } ax_1 + bx_2 + c = 0\}.$$
By Theorem LB.12, the unique line $\mathcal{M}$ which is c-perpendicular to $\mathcal{L}$ and contains $U = (u_1, u_2)$ is
$$\mathcal{M} = \{(x_1, x_2) \mid (x_1, x_2) \in \mathbb{F}^2 \textit{ and } bx_1 - ax_2 - bu_1 + au_2 = 0\}.$$
By Theorem LB.14, the point of intersection between $\mathcal{L}$ and $\mathcal{M}$ is
$$Y = (y_1, y_2) = \begin{pmatrix} \frac{b^2u_1 - abu_2 - ac}{a^2+b^2} \\ \frac{-abu_1 + a^2u_2 - bc}{a^2+b^2} \end{pmatrix}.$$

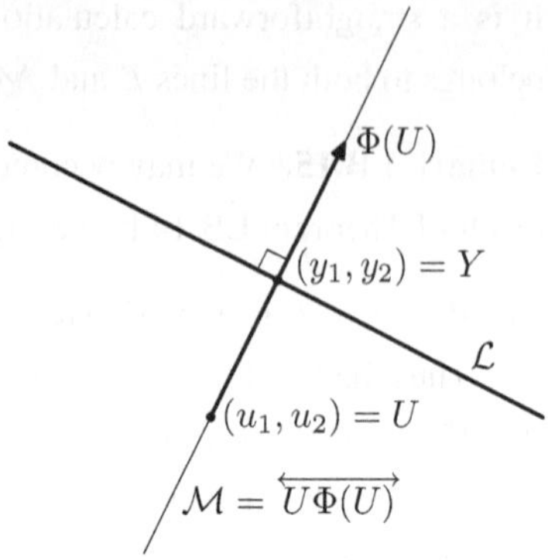

Fig. 21.1 Showing action of the mapping Φ (to be named $\mathcal{R}_\mathcal{L}$ in Definition LC.24).

Define

$$\begin{aligned}\Phi(U) &= \Phi(u_1, u_2) = Y + (Y - U) = 2Y - U\\ &= 2\begin{pmatrix}\frac{b^2u_1 - abu_2 - ac}{a^2+b^2}\\ \frac{-abu_1 + a^2u_2 - bc}{a^2+b^2}\end{pmatrix} - \begin{pmatrix}u_1\\ u_2\end{pmatrix} = \begin{pmatrix}\frac{(b^2-a^2)u_1 - 2abu_2 - 2ac}{a^2+b^2}\\ \frac{-2abu_1 + (a^2-b^2)u_2 - 2bc}{a^2+b^2}\end{pmatrix}.\end{aligned}$$

It is clear that for each line $\mathcal{L}$ in $\mathbb{F}^2$, the mapping Φ in Definition LB.16 is defined on all points $X \in \mathbb{F}^2$.

Theorem LB.17. *In this theorem we use the notation of Definition LB.16.*

(A) *Every point of $\mathcal{L}$ is a fixed point for Φ.*
(B) *For every $U \in \mathbb{F}^2 \setminus \mathcal{L}$, the line $\mathcal{M} = \overleftrightarrow{U\Phi(U)}$; this line is c-perpendicular to $\mathcal{L}$.*
(C) *The point Y of intersection of $\mathcal{L}$ and $\mathcal{M}$ is the c-midpoint of the segment $[U\Phi(U)]$.*
(D) *The line $\mathcal{M} = \overleftrightarrow{U\Phi(U)}$ named in Definition LB.16 is a fixed line for Φ.*
(E) *A line $\mathcal{N}$ is c-perpendicular to $\mathcal{L}$ iff it is a fixed line for Φ.*

Proof. (A) If $U = Y$, $\Phi(U) = Y + (Y - U) = U$, so points of $\mathcal{L}$ are fixed points for Φ. This may also be calculated numerically: if $U = (u_1, u_2) \in \mathcal{L}$, by Theorem LB.14 $c = -au_1 - bu_2$; substituting this value into the coordinate expression for $\Phi(U)$ shows that $\Phi(u_1, u_2) = (u_1, u_2)$.

(B) Since $\Phi(U) = Y + (Y - U)$, then $\Phi(U) = Y + (-1)(U - Y)$ so that by Definition LA.1(1) $\Phi(U)$ belongs to the line $\mathcal{M} = \overleftrightarrow{UY}$. Then both $\mathcal{M}$ and $\overleftrightarrow{U\Phi(U)}$ contain both U and $\Phi(U)$; by Theorem LA.10(A) $\overleftrightarrow{U\Phi(U)} = \mathcal{M}$; this line was defined to be c-perpendicular to $\mathcal{L}$.

(C) The point Y of intersection of $\mathcal{L}$ and $\mathcal{M}$ is the c-midpoint of the segment $\overline{U\Phi(U)}$, since

$$\frac{U+\Phi(U)}{2} = \frac{U+Y+(Y-U)}{2} = \frac{Y+Y}{2} = Y.$$

(D) Let $X \in \mathcal{M}$ but $X \notin \mathcal{L}$; by Definition LB.16, $\Phi(X)$ is on a line $\mathcal{M}'$ which also contains X and is c-perpendicular to $\mathcal{L}$. By Theorem LB.12 there is only one line through X which is c-perpendicular to $\mathcal{L}$, so that $\mathcal{M}' = \mathcal{M}$, so that $\Phi(X) \in \mathcal{M}$. Therefore $\mathcal{M}$ is a fixed line for Φ.

(E) If $\mathcal{N}$ is c-perpendicular to $\mathcal{L}$, choose $U \in \mathcal{N}$ so that $U \notin \mathcal{L}$; then Definition LB.16 defines $\Phi(U)$ to be a point on $\mathcal{N}$, and by part (D), $\mathcal{N}$ is a fixed line. Conversely, if $\mathcal{N}$ is a fixed line and $U \in \mathcal{N} \backslash \mathcal{L}$, $\Phi(U) \in \mathcal{N} = \overleftrightarrow{U\Phi(U)}$ which by part (B) is c-perpendicular to $\mathcal{L}$. □

21.5 Axiom consistency: a linear model

In this section we show that the incidence, betweenness, parallel, and Plane Separation axioms hold in Model LM3, that every mapping Φ (defined by Definition LB.16) over a line $\mathcal{L}$ in $\mathbb{F}^2$ is a mirror mapping, and that the collection of all such mappings is a reflection set. Using this result we will show that there is a reflection set on every plane $\mathcal{P}$ in $\mathbb{F}^3$. A proof that Axiom LUB holds on Model LM3R completes the demonstration that our axioms are consistent.

21.5.1 *Incidence Axioms I.0–I.5 are valid in a linear model*

Remark LC.1. The acronym LC is meant to suggest "consistency using linear models."

In this subsection, we will show that the Axioms I.0, I.1, I.2, I.3, I.4, and I.5 are all true for Model LM3R, and thus consistent; this proof is essential for showing the consistency of our complete axiom set.[3] Most of our theorems do not require that the underlying field $\mathbb{F}$ contain square roots of its non-negative members (as do $\mathbb{A}$

[3]The proof that the incidence axioms are consistent is actually redundant, since we have already exhibited a discrete model in Chapter 1, Section 1.9 for which all they are all true.

and $\mathbb{R}$), so will be stated for the more general context of Model LM3 or Model LM2 and a generic ordered field $\mathbb{F}$, which may be any of $\mathbb{Q}$, $\mathbb{A}$, or $\mathbb{R}$.

Theorem LC.2. *For Model LM3, Axiom I.0 and Axiom I.1 are both true.*

Proof. Axiom I.0 is true because lines and planes are subsets of space. To see that Axiom I.1 is true, let A and B be any two points of $\mathbb{F}^3$; by Definition LA.1(1) there exists at least one line containing both, *i.e.* $\overleftrightarrow{AB}$. That there is only one such line is immediate from Theorem LA.10(A). □

Theorem LC.3. *Axiom I.2 is true for Model LM3.*

Proof. If A, B, and C are any noncollinear points of $\mathbb{F}^3$, by Definition LA.1(2) there exists at least one plane containing all three points, *i.e.* $\overleftrightarrow{ABC}$. That there is only one such plane is immediate from Theorem LA.10(B). □

Theorem LC.4. *Axiom I.3 is true for Model LM3.*

Proof. We show that if two points are on a plane, then a line through those points is a subset of the plane. Let A, B, and C be any noncollinear points in $\mathbb{F}^3$ and let D and E be any distinct points on the plane $\overleftrightarrow{ABC}$. By Definition LA.1(1) $\overleftrightarrow{DE} = \{D + t(E - D) \mid t \in \mathbb{F}\}$. By Definition LA.1(2) there exist numbers u_1, u_2, v_1, and v_2 such that

$$D = A + u_1(B - A) + u_2(C - A) \text{ and}$$
$$E = A + v_1(B - A) + v_2(C - A).$$

If X is any point of $\overleftrightarrow{DE}$, there exists a number t such that $X = D + t(E - D)$. Then

$$\begin{aligned} X &= D + t(E - D) \\ &= A + u_1(B - A) + u_2(C - A) \\ &\qquad + t\Big(v_1(B - A) + v_2(C - A) - u_1(B - A) - u_2(C - A)\Big) \\ &= A + (u_1 + t(v_1 - u_1))(B - A) + (u_2 + t(v_2 - u_2))(C - A)). \end{aligned}$$

Let $s_1 = u_1 + t(v_1 - u_1)$ and $s_2 = u_2 + t(v_2 - u_2)$; then

$$X = A + s_1(B - A) + s_2(C - A)$$

which is a member of $\overleftrightarrow{ABC}$. Therefore $\overleftrightarrow{DE} \subseteq \overleftrightarrow{ABC}$. □

We next prove that Axiom I.5 is true, because it is needed to prove that Axiom I.4 is true.

Theorem LC.5. *Axiom I.5 is true for Model LM3.*

Proof. (A) By Definition LA.1(1), a line $\mathcal{L}$ contains two distinct points A and B. This shows that Axiom I.5(A) is satisfied.

(B) By Definition LA.1(2), a plane $\mathcal{P}$ contains three noncollinear points A, B, and C. This shows that Axiom I.5(B) is satisfied.

(C) By Remark LA.11(B), no plane contains all of $\mathbb{F}^3$; in particular, the four points $(0,0,0)$, $(1,0,0)$, $(0,1,0)$, and $(0,0,1)$ are noncoplanar. Therefore $\mathbb{F}^3$ contains four noncoplanar points and Axiom I.5(C) is satisfied. □

In the next series of results we will habitually refer to "points" (A, B, G_1, G_2) of a plane $\mathcal{P}$ which is a subset of space $\mathbb{F}^3$. In the same discussion we will speak of the differences between such "points" ($B - A$, $G_1 - A$, *etc.*) as "vectors" because this seems more natural when emphasizing orthogonality and direction.

Lemma LC.6. *Let $\mathcal{P}$ be a plane in $\mathbb{F}^3$, and let $A = (a_1, a_2, a_3)$ and $B = (b_1, b_2, b_3)$ be distinct points of $\mathcal{P}$.*

(A) *There exists a point $C = (c_1, c_2, c_3) \in$ of $\mathcal{P}$, distinct from both A and B, such that $(B - A) \perp (C - A)$.*

(B) *There exists a point $C = (c_1, c_2, c_3) \notin \mathcal{P}$ such that $(B - A) \perp (C - A)$.*

Proof. (A) If all points $X \in \mathcal{P}$ were collinear with A and B, then by Definition LA.1(2) $\mathcal{P}$ would not be a plane; hence there exists a member X of $\mathcal{P}$ which is not collinear with A and B.

(B) By Remark LA.11 $\mathcal{P}$ is a proper subset of $\mathbb{F}^3$, so there exists a point $X \notin \mathcal{P}$, and A, B, and X are noncollinear because $\overleftrightarrow{AB} \subseteq \mathcal{P}$ does not contain X.

The following calculation is the same for both (A) and (B). If $(X - A) \perp (B - A)$ let $C = X$, and we are done. If $(X - A) \not\perp (B - A)$, then let $C = (c_1, c_2, c_3)$ be the point $C = A + s(X - A) + t(B - A)$ where

$$s = -\frac{1}{(B - A) \bullet (X - A)} \text{ and } t = \frac{1}{(B - A) \bullet (B - A)}.$$

Note that both s and t are nonzero. Then

$$\begin{aligned}(B - A) \bullet (C - A) &= (B - A) \bullet (s(X - A) + t(B - A)) \\ &= -\frac{1}{(B - A) \bullet (X - A)}(B - A) \bullet (X - A) \\ &\quad + \frac{1}{(B - A) \bullet (B - A)}(B - A) \bullet (B - A) = 0\end{aligned}$$

so that $(B - A) \perp (C - A)$. Again, we consider the two parts separately.

(A) We have seen that $C = A + s(X - A) + t(B - A)$; by Definition LA.1(2), since A, B, and X belong to $\mathcal{P}$, so does C. If $C = A$, $s(X - A) + t(B - A) = O$ and $X - A$ and $B - A$ would be linearly dependent; by Theorem LA.3, X, A, and B would be collinear; this was ruled out at the beginning of the proof. Therefore C is a member of $\mathcal{P}$ that is distinct from A.

(B) Again, $C = A + s(X - A) + t(B - A)$, where $s \neq 0$ and $t \neq 0$. If $C \in \mathcal{P}$, we see that $C - A = s(X - A) + t(B - A)$ and

$$X = A + \tfrac{1}{s}(C - A) - \tfrac{t}{s}(B - A).$$

By Definition LA.1(2) $X \in \mathcal{P}$, contradicting our original choice for X. This shows that $C \notin \mathcal{P}$. □

Theorem LC.6.1. *Let $\mathcal{P}$ be a plane in $\mathbb{F}^3$, and let $A = (a_1, a_2, a_3)$ and $B = (b_1, b_2, b_3)$ be distinct points of $\mathcal{P}$. Then there exist three points G_1, G_2, and G_3, of $\mathbb{F}^3$, all distinct from A, such that*

(A) *G_1 and G_2 belong to $\mathcal{P}$, and for every point $X \in \mathcal{P}$, there exist scalars s and t in $\mathbb{F}$ such that $X - A = s(G_1 - A) + t(G_2 - A)$;*
(B) *the vectors $G_1 - A$, $G_2 - A$, and $G_3 - A$ are pairwise orthogonal;,*
(C) *for every point $X \in \mathbb{F}^3$, there exist scalars x_1, x_2, and x_3 in $\mathbb{F}$ such that $X - A = x_1(G_1 - A) + x_2(G_2 - A) + x_3(G_3 - A)$; and*
(D) *for every point $X \in \mathbb{F}^3$ such that $(X - A) \perp (G_3 - A)$, $X \in \mathcal{P}$.*

Proof. In this proof we shall frequently use the various statements of Remark LA.7 without further reference.

Let $G_1 = B$. From Lemma LC.6(A) there exists a point $G_2 \in \mathcal{P}$ such that $(G_1 - A) \perp (G_2 - A)$ and $G_2 \neq A$. By Remark LA.9(E) $\mathcal{P} - A$ is a subspace of $\mathbb{F}^3$ having dimension 2, and the vectors $G_1 - A$ and $G_2 - A$ are linearly independent members of this subspace (because they are orthogonal), thus forming a basis for it. Then $X \in \mathcal{P}$ iff $X - A \in \mathcal{P} - A$ iff for some scalars s and t, $X - A = s(G_1 - A) + t(G_2 - A)$, that is, $X = A + s(G_1 - A) + t(G_2 - A)$.

Moreover, by the Dimension Criterion of Chapter 1 Section 1.5, $\mathcal{P} - A$ and $\mathcal{P}$ are proper subsets of $\mathbb{F}^3$. By Lemma LC.6(B) there exists a point $D \in \mathbb{F}^3$ such that $D \notin \mathcal{P}$, $D - A \notin \mathcal{P} - A$, and $(D - A) \perp (G_1 - A)$.

Since $D \notin \mathcal{P}$, and both G_2 and A belong to $\mathcal{P}$, D, A, and G_2 are noncollinear; let $\mathcal{Q} = \overleftrightarrow{DAG_2}$. By Lemma LC.6(A) there exists a point $G_3 \in \mathcal{Q}$ such that $(G_3 - A) \perp (G_2 - A)$. Since $G_3 - A = s(G_2 - A) + t(D - A)$,

$$(G_3 - A) \bullet (G_1 - A) = s(G_2 - A) \bullet (G_1 - A) + t(D - A) \bullet (G_1 - A) = 0 + 0 = 0,$$

so that $G_3 - A$ is orthogonal to both $G_2 - A$ and $G_1 - A$. Since $(G_1 - A) \perp (G_2 - A)$, $G_1 - A$, $G_2 - A$, and $G_3 - A$ are pairwise orthogonal and thus by Theorem LA.14 are linearly independent; thus they span $\mathbb{F}^3$. This completes the proof of (C) and (B); the first paragraph proves assertion (A).

(D) Suppose $(X - A) \perp (G_3 - A)$; by part (C) there exist numbers x_1, x_2, and x_3 such that $X - A = x_1(G_1 - A) + x_2(G_2 - A) + x_3(G_3 - A)$. Then

$$\begin{aligned} 0 &= (X - A) \bullet (G_3 - A) = \big(x_1(G_1 - A) + x_2(G_2 - A) + x_3(G_3 - A)\big) \bullet (G_3 - A) \\ &= x_1(G_1 - A) \bullet (G_3 - A) + x_2(G_2 - A) \bullet (G_3 - A) + x_3(G_3 - A) \bullet (G_3 - A) \\ &= x_1 \cdot 0 + x_2 \cdot 0 + x_3(G_3 - A) \bullet (G_3 - A) = x_3\|G_3 - A\|^2. \end{aligned}$$

Since $\|G_3 - A\| \neq 0$, $x_3 = 0$, so that $X - A = x_1(G_1 - A) + x_2(G_2 - A)$. By Definition LA.1(2), $X \in \mathcal{P}$. □

Theorem LC.7. *Axiom I.4 is true for Model LM3.*

Proof. We show that if two planes intersect, their intersection must contain at least two points. Let $\mathcal{P}$ and $\mathcal{Q}$ be any two planes in $\mathbb{F}^3$, and suppose that they intersect at some point A. By Theorem LC.6.1, there exist points P_1, P_2, and P_3 in $\mathbb{F}^3$, all distinct from A, such that both P_1 and P_2 are members of $\mathcal{P}$ and the vectors $P_1 - A$, $P_2 - A$, and $P_3 - A$ are pairwise orthogonal.

Similarly, there exist points Q_1, Q_2, and Q_3 in $\mathbb{F}^3$, all distinct from A, such that both Q_1 and Q_2 are members of $\mathcal{Q}$, and the vectors $Q_1 - A$, $Q_2 - A$, and $Q_3 - A$ are pairwise orthogonal.

Now if $P_3 - A$ and $Q_3 - A$ are linearly dependent, one is a scalar multiple of the other and the two planes $\mathcal{P}$ and $\mathcal{Q}$ are the same, because for every point $X \in \mathcal{Q}$, $(X-A)\bullet(P_3-A) = 0$ and hence by Theorem LC.6.1(D), $X \in \mathcal{P}$; likewise every point of $\mathcal{P}$ belongs to $\mathcal{Q}$. In this case, there are infinitely many points in the intersection and the theorem is proved.

If, on the other hand, $P_3 - A$ and $Q_3 - A$ are linearly independent, by Theorem LA.3, A, P_3 and Q_3 are noncollinear and by Definition LA.1(2) there exists a plane $\overleftrightarrow{AP_3Q_3}$ containing the points A, P_3, and Q_3.

Applying Theorem LC.6.1(B) to this plane, there exists a point $Y \in \mathbb{F}^3$ such that $Y - A \neq O$ and is orthogonal to both $P_3 - A$ and $Q_3 - A$. By two applications of Theorem LC.6.1(D), $Y \in \mathcal{P}$ and $Y \in \mathcal{Q}$ so that Y is a point in $\mathcal{P} \cap \mathcal{Q}$ distinct from A. By Theorem LC.4 (Axiom I.3) the line containing A and Y is a subset of both $\mathcal{P}$ and $\mathcal{Q}$, proving the theorem. □

21.5.2 Betweenness Axiom BET is valid on a linear model

Definition LC.8 (A betweenness relation on $\mathbb{F}^3$). Let X, Y, and Z be collinear points in $\mathbb{F}^3$, and let A and B be distinct points collinear with these points. Then

(X, Y, Z) belongs to the betweenness relation, that is to say Y is **between** X and Z (notation: X–Y–Z) iff there exist distinct numbers t_1, t_2, and t_3 such that

$$X = A + t_1(B - A),\ Y = A + t_2(B - A), \text{ and } Z = A + t_3(B - A)$$

and t_1–t_2–t_3; that is, either $t_1 < t_2 < t_3$ or $t_1 > t_2 > t_3$.

Remark LC.9. By Theorem LA.5, X, Y, and Z are distinct since t_1, t_2, and t_3 are distinct.

Theorem LC.10 (Betweenness is well defined). *Betweenness does not depend on the choice of A and B in Definition LC.8. That is, if $A \neq B$, and $C \neq D$ are points collinear with X, Y, and Z, then statement (A) below is true iff statement (B) is true:*

(A) *there exist distinct numbers t_1, t_2, and t_3 such that*

$$X = A + t_1(B - A),\ Y = A + t_2(B - A), \text{ and } Z = A + t_3(B - A)$$

and t_1–t_2–t_3;

(B) *there exist distinct numbers s_1, s_2, and s_3 such that*

$$X = C + s_1(D - C),\ Y = C + s_2(D - C), \text{ and } Z = C + s_3(D - C)$$

and s_1–s_2–s_3.

Proof. By Definition LA.1(2) there exist numbers t_1, t_2, t_3, s_1, s_2, and s_3 such that

$$X = A + t_1(B - A) = C + s_1(D - C), \qquad (1)$$

$$Y = A + t_2(B - A) = C + s_2(D - C), \qquad (2)$$

$$\text{and } Z = A + t_3(B - A) = C + s_3(D - C). \qquad (3)$$

Suppose X–Y–Z according to Definition LC.8, using the points A and B; then t_1–t_2–t_3, that is to say, either $t_1 < t_2 < t_3$ or $t_1 > t_2 > t_3$. Subtracting (2) from (1) we have

$$(t_1 - t_2)(B - A) = (s_1 - s_2)(D - C) \qquad (4)$$

and subtracting (3) from (2) we have

$$(t_2 - t_3)(B - A) = (s_2 - s_3)(D - C). \qquad (5)$$

Since the coefficients $t_1 - t_2$ and $t_2 - t_3$ are nonzero, there exists a number a such that $a(t_1 - t_2) = (t_2 - t_3)$; multiplying through equation (4) by a we get

$$a(t_1 - t_2)(B - A) = a(s_1 - s_2)(D - C).$$

The left-hand side is $(t_2 - t_3)(B - A)$, so by (5) we get

$$a(s_1 - s_2)(D - C) = (s_2 - s_3)(D - C).$$

Now $t_1 - t_2$ and $t_2 - t_3$ have the same sign because either $t_1 < t_2 < t_3$ or $t_1 > t_2 > t_3$, so $a > 0$ and hence $s_1 - s_2$ and $s_2 - s_3$ have the same sign, that is, either $s_1 < s_2 < s_3$ or $s_1 > s_2 > s_3$. Thus X–Y–Z is true using C and D in Definition LC.8. □

Theorem LC.11 (First alternative definition of betweenness). *Let X, Y, and Z be distinct points in $\mathbb{F}^3$. Then X–Y–Z iff there exist distinct points A and B in $\mathbb{F}^3$ and numbers t_1, t_2, and t_3 with $t_1 < t_2 < t_3$, such that $X = A + t_1(B - A)$, $Y = A + t_2(B - A)$, and $Z = A + t_3(B - A)$ are all true.*

Proof. If the alternative condition is true, then $t_1 < t_2 < t_3$; this implies that X–Y–Z according to Definition LC.8.

If t_1–t_2–t_3 according to Definition LC.8, either $t_1 < t_2 < t_3$ or $t_1 > t_2 > t_3$; if $t_1 < t_2 < t_3$, the alternative holds. If $t_1 > t_2 > t_3$, $X = B + (1 - t_1)(A - B)$, $Y = B + (1 - t_2)(A - B)$, and $Z = B + (1 - t_3)(A - B)$, and $1 - t_1 < 1 - t_2 < 1 - t_3$ so that X–Y–Z according to the alternative. So in either case, the alternative definition holds. □

Theorem LC.12 (Second alternative definition of betweenness). *Let X, Y, and Z be distinct points in $\mathbb{F}^3$. Then X–Y–Z iff for some number s with* 0–s–1 *(that is, $0 < s < 1$), $Y = X + s(Z - X)$.*

Proof. In Definition LC.8 we may choose A and B to be any points collinear with X, Y, and Z. Let $X = A$ and $Z = B$; then $X = X + 0(Z - X)$ and $Z = X + 1(Z - X)$. By that definition, X–Y–Z iff there exists a number s such that $Y = X + s(Z - X)$ and for some number s, $Y = X + s(Z - X)$ and 0–s–1 which is the criterion of this theorem. □

Theorem LC.13 (Segments and rays). *The definitions of segments and rays given in Definition LA.1(3) are equivalent to those given in Definitions IB.3 and IB.4 of Chapter 4. Specifically, if we let A and B be distinct points in $\mathbb{F}^3$ and give $\overline{AB}$, $\overrightarrow{AB}$, etc., their meanings in Definitions IB.3 and IB.4, and if s and t denote numbers in $\mathbb{F}$, the following statements are true:*

(1) $\overline{AB} = \{A + t(B - A) \mid 0 < t < 1\}$
$= \{B + (1 - t)(A - B) \mid 0 < t < 1\}$
$= \{B + s(A - B) \mid 0 < s < 1\} = \overline{BA}$.

(2) $\overline{AB} = \{A + t(B - A) \mid 0 \le t \le 1\}$
$= \{B + (1 - t)(A - B) \mid 0 \le t \le 1\}$
$= \{B + s(A - B) \mid 0 \le s \le 1\} = \overline{BA}$.

(3) $\overline{AB} = \{A + t(B - A) \mid 0 \le t < 1\}$
$= \{B + (1 - t)(A - B) \mid 0 \le t < 1\}$
$= \{B + s(A - B) \mid 0 < s \le 1\} = \overline{BA}$.

(4) $\overset{\rceil\!-\!\rceil}{AB} = \{A + t(B - A) \mid 0 < t \leq 1\}$
$= \{B + (1 - t)(A - B) \mid 0 < t \leq 1\}$
$= \{B + s(A - B) \mid 0 \leq s < 1\} = \overset{\sqsubset\!-\!\sqsubset}{BA}$.
(5) $\overset{\rceil\!\rightarrow}{AB} = \{A + t(B - A) \mid t > 0\}$.
(6) $\overset{\sqsubset\!\rightarrow}{AB} = \{A + t(B - A) \mid t \geq 0\}$.

Proof. (1) Definition IB.3 states that $X \in \overset{\rceil\!-\!\lceil}{AB}$ iff A–X–B; by the second alternative definition of betweenness (Theorem LC.12) this means that for some t with $0 < t < 1, X = A + t(B - A)$.

The proofs of (2), (3), and (4) are obvious from (1) and simple set theory observations about the inclusion of the end points A and B.

(5) Definition IB.4 states that $X \in \overset{\rceil\!\rightarrow}{AB}$ iff A–X–B or A–B–X or $X = B$. By Theorem LC.12, A–X–B means that for some t with $0 < t < 1, X = A + t(B - A)$.

Also, A–B–X means that for some s with $0 < s < 1$, $B = A + s(X - A)$, that is, $B - A - sX + sA = B - A + sA - sX = 0$, or $sX = B - A + sA$. Since $0 < s < 1$, we may let $t = \frac{1}{s}$, so that $t > 1$, and $X = A + t(B - A)$. Therefore $X \in \overset{\rceil\!\rightarrow}{AB}$ iff for some t with $X = A + t(B - A)$, where either $0 < t < 1, t > 1$, or $t = 1$, that is, where $t > 0$.

(6) $X \in \overset{\sqsubset\!\rightarrow}{AB}$ iff $X \in \overset{\rceil\!\rightarrow}{AB}$ or $X = A$, that is, for some t with $t \geq 0, X = A + t(B - A)$. □

Theorem LC.14. *The betweenness relation of Definition LC.8 satisfies Properties B.0, B.1, B.2, and B.3 of Definition IB.1. This shows that Axiom BET is valid for* $\mathbb{F}^3$ *(and hence for any plane of* $\mathbb{F}^3$*).*

Proof. The properties of the betweenness relation follow immediately from Definition LC.8 and the corresponding properties of betweenness for members of the field $\mathbb{F}$, as listed in Chapter 1 Section 1.5 under the title "Number systems." □

Remark LC.15. Theorems LC.2 through LC.14 show that the incidence Axioms I.0, I.1, I.2, I.3, I.4, and I.5, and the betweenness Axiom BET are consistent, since they are all true for Model LM3.

21.5.3 Parallel Axiom PS is valid on a linear model

Theorem LC.16. *The parallel axiom PS is true for Model LM3, and therefore for Model LM2.*

Proof. Let $\mathcal{L}$ be any line in $\mathbb{F}^3$ and let A and B be distinct points on $\mathcal{L}$. Let C be any member of $\mathbb{F}^3 \setminus \overleftrightarrow{AB}$, let t be any nonzero number, and let $D = C + t(B - A)$. Then by Theorem LA.22 $\overleftrightarrow{CD}$ and $\overleftrightarrow{AB}$ are parallel to each other.

If E is a point of $\mathbb{F}^3$ distinct from C such that $\overleftrightarrow{CE}$ and $\overleftrightarrow{AB}$ are parallel to each other, then by Theorem LA.22 there exists a nonzero number s such that $E-C = s(B-A)$. Since $B - A = \frac{1}{t}(D - C)$, $E = C + \frac{s}{t}(D - C)$. By Definition LA.1(1) $E \in \overleftrightarrow{CD}$. Hence there is a unique line through C which is parallel to $\overleftrightarrow{AB}$. □

Remark LC.17. We have inserted the above theorem out of the order of presentation in the development, since we need Axiom PS to prove the next theorem, Theorem LC.18.

21.5.4 Plane Separation Axiom PSA is valid on a linear model

The following three theorems show that Axiom PSA holds for every plane in $\mathbb{F}^3$.

Theorem LC.18. *Let A, B, and C be noncollinear points in $\mathcal{P}$, a plane in $\mathbb{F}^3$. The C-side of $\overleftrightarrow{AB}$ is equal to*

$$\mathcal{E} = \{A + s(B - A) + t(C - A) \mid (s, t) \in \mathbb{F}^2 \text{ and } t > 0\}.$$

Proof. (I) ($\mathcal{E} \subseteq$ C-side of $\overleftrightarrow{AB}$.) Let X be any member of $\mathcal{E}$; then there exists a number s and a positive number t such that $X = A + s(B - A) + t(C - A)$. By Definition LA.1(1), $A + s(B - A)$ is a point on $\overleftrightarrow{AB}$. By Theorem LC.13(5), X is a member of $\overrightarrow{(A + s(B - A))C}$. Since we know that the incidence axioms hold, and there exists a betweenness relation on this model, we can apply Theorem IB.14 to conclude that $X = A + s(B - A) + t(C - A)$ is a member of the C-side of $\overleftrightarrow{AB}$.

(II) (C-side of $\overleftrightarrow{AB} \subseteq \mathcal{E}$.) Let X be any member of the C-side of $\overleftrightarrow{AB}$. By Theorem LC.16 we may use Axiom PS, the strong form of the Parallel Axiom. Let $\mathcal{L}$ be the line through X which is parallel to $\overleftrightarrow{BC}$. By Exercise IP.4, $\mathcal{L}$ and $\overleftrightarrow{AB}$ intersect at a point V. By Definition LA.1(1) there exists a number s such that $V = A + s(B - A)$. Using Axiom PS and Exercise IP.4, let $\mathcal{M}$ be the line through C which is parallel to $\overleftrightarrow{AB}$ and let W be the point of intersection of $\overleftrightarrow{VX}$ and $\mathcal{M}$. Again using Definition LA.1(1) and Theorem LC.13(5) there exists a (unique) positive number t such that $V + t(W - V) = X$. That is,

$$X = A + s(B - A) + t(W - V). \quad (*)$$

Since $\overleftrightarrow{CW} \parallel \overleftrightarrow{AB}$, by Theorem LA.22 there exists a number $k \neq 0$ such that $W - C = k(B - A)$, so that $W = C + k(B - A)$. Also, since $V \in \overleftrightarrow{AB}$, by Definition LA.1(1) there exists a number v such that $V = A + v(B - A)$. Substituting into equation (*), we have

$$\begin{aligned} X &= A + s(B - A) + t(W - V) \\ &= A + s(B - A) + t(C + k(B - A) - (A + v(B - A))) \\ &= A + (s + t(k - v))(B - A) + t(C - A) \end{aligned}$$

so that $X \in \mathcal{E}$. Thus the C-side of $\overleftrightarrow{AB}$ is a subset of $\mathcal{E}$. □

Theorem LC.19. *Let A, B, and C be noncollinear points in $\mathcal{P}$, a plane in $\mathbb{F}^3$. Let C' be a point such that C'–B–C, then* the C'-side of $\overleftrightarrow{AB}$ *(which we denote by $\mathcal{E}'$) is equal to $\{A + s(B - A) + t(C - A) \mid (s, t) \in \mathbb{F}^2$ and $t < 0\}$.*

Proof. This is word-for-word the proof of LC.18 except that C is replaced by C' and the inequality $>$ is replaced by $<$. □

Theorem LC.20. *Axiom PSA (The Plane Separation Axiom) is true for Model LM3 and Model LM2.*

Proof. Let A, B, and C be noncollinear points in $\mathbb{F}^3$, let D be a point on $\overleftrightarrow{AB}$, and let E be a point such that C–D–E. By Definition LA.1(2) the points A, B, and C define the plane $\overleftrightarrow{ABC} \subseteq \mathbb{F}^3$.

Then the C-side of $\overleftrightarrow{AB}$ and the E-side of $\overleftrightarrow{AB}$ are opposite sides of $\overleftrightarrow{AB}$ in the plane $\overleftrightarrow{ABC}$ (cf Definition IB.11). If X is any member of the C-side of $\overleftrightarrow{AB}$ and Y is any member of the E-side of $\overleftrightarrow{AB}$, then we show that $\overleftrightarrow{AB} \cap \overline{XY}$ is a singleton (cf the note following Axiom PSA).

By Theorems LC.18 and LC.19 there exist numbers s_1, s_2, $t_1 > 0$, and $t_2 < 0$ such that

$$X = A + s_1(B - A) + t_1(C - A) \text{ and } Y = A + s_2(B - A) + t_2(C - A).$$

By Remark LC.13, $\overline{XY} = \{X + u(Y - X) \mid u \in \mathbb{F}$ and $0 < u < 1\}$. By Definition LA.1(1), a point $Z \in \overleftrightarrow{AB}$ iff there exists a number v such that $Z = A + v(B - A)$. We want to find a point U that is in both $\overline{XY}$ and in $\overleftrightarrow{AB}$, that is to say, we seek numbers u and v such that $0 < u < 1$ and

$$\begin{aligned} U &= X + u(Y - X) \\ &= A + s_1(B - A) + t_1(C - A) \\ &\quad + u\big(A + s_2(B - A) + t_2(C - A) - (A + s_1(B - A) + t_1(C - A))\big) \\ &= A + s_1(B - A) + u(s_2 - s_1)(B - A) + u(t_2 - t_1)(C - A) + t_1(C - A) \\ &= A + v(B - A). \end{aligned}$$

This is true iff

$$\big(s_1 + u(s_2 - s_1) - v\big)(B - A) + \big(t_1 + u(t_2 - t_1)\big)(C - A) = O. \quad (*)$$

Since A, B, and C are noncollinear, by Theorem LA.3 $B - A$ and $C - A$ are linearly independent. Therefore equation (*) is true iff $s_1 + u(s_2 - s_1) - v = 0$ and $t_1 + u(t_2 - t_1) = 0$, that is, iff $u = \frac{t_1}{t_1 - t_2}$ and $v = s_1 + \frac{t_1(s_2 - s_1)}{t_1 - t_2}$. Then

$$U = X + u(Y - X) = X + \tfrac{t_1}{t_1 - t_2}(Y - X) \in \overleftrightarrow{XY}$$

and also

$$U = A + v(B - A) = A + \big(s_1 + \tfrac{t_1(s_2 - s_1)}{t_1 - t_2}\big)(B - A) \in \overleftrightarrow{AB}.$$

Since $t_1 > 0$ and $t_2 < 0$, it follows that $0 < u = \frac{t_1}{t_1 - t_2} < 1$; by Remark LC.13, $X + u(Y - X) \in \overline{XY}$, and $X–U–Y$. Therefore, the segment $\overline{XY}$ intersects $\overleftrightarrow{AB}$ at the point U; by Exercise I.1, since $\overleftrightarrow{XY}$ and $\overleftrightarrow{AB}$ are distinct, the point of intersection is a singleton.

This shows that Axiom PSA holds for $\mathbb{F}^3$. □

Remark LC.21. Theorems LC.2 through LC.20 show that the incidence Axioms I.0, I.1, I.2, I.3, I.4, and I.5, the betweenness Axiom BET, Axiom PS, and Axiom PSA are consistent, since they are all true for Model LM3.

21.5.5 *Reflection Axiom REF is valid on Model LM2A and Model LM2R*

Remark LC.22. We return to the development of the last section, culminating in Definition LB.16, which defined a single mapping Φ over every line $\mathcal{L}$ as follows: for any $U \in \mathbb{F}^2$, $\Phi(U) = Y + (Y - U)$, where Y is the point of intersection of $\mathcal{L}$ and $\mathcal{M}$, and $\mathcal{M}$ is the line containing U that is c-perpendicular to $\mathcal{L}$.

It will sometimes be useful to have the coordinate-wise definition available: suppose the equation for $\mathcal{L}$ is $ax_1 + bx_2 + c = 0$, where not both a and b are zero, and $U = (u_1, u_2)$; then

$$\Phi(u_1, u_2) = \big(\tfrac{(b^2 - a^2)u_1 - 2abu_2 - 2ac}{a^2 + b^2}, \tfrac{-2abu_1 + (a^2 - b^2)u_2 - 2bc}{a^2 + b^2}\big),$$

and the point of intersection

$$Y = (y_1, y_2) = \big(\tfrac{b^2u_1 - abu_2 - ac}{a^2 + b^2}, \tfrac{-abu_1 + a^2u_2 - bc}{a^2 + b^2}\big).$$

Theorem LC.23. *Definition LB.16 defines a single mirror mapping over a given line $\mathcal{L}$ in $\mathbb{F}^2$.*

Proof. Clearly Definition LB.16 defines only one mapping over the line $\mathcal{L}$. We show that the mapping Φ of Definition LB.16 satisfies Properties (A), (B), (C), and (D) of Definition NEUT.1, and therefore is a mirror mapping over $\mathcal{L}$. We use the notation from Definition LB.16, which is cited in Remark LC.22 above.

(A: Every point of $\mathcal{L}$ is a fixed point for Φ.) This is Theorem LB.17(A). See Exercise LM.14 for a coordinate-wise proof.

(B: If $U \notin \mathcal{L}$, then $\Phi(U)$ is on the opposite side of $\mathcal{L}$ from U.) Since $Y \in \mathcal{L}$ and $Y \in [U\Phi(U)]$, and $Y \neq U$ and $Y \neq \Phi(U)$, then $Y \in \,]U\Phi(U)[$. Hence U and $\Phi(U)$ are on opposite sides of $\mathcal{L}$, by Definition IB.11.

(C: For every $U \in \mathbb{F}^2$, $\Phi(\Phi(U)) = U$.) By Theorem LB.17(D) $\mathcal{M} = \overleftrightarrow{U\Phi(U)}$ is a fixed line for Φ, so that $\Phi(\Phi(U)) \in \mathcal{M}$; Y is the point of intersection of $[U\Phi(U)]$ with $\mathcal{L}$; substituting $\Phi(U)$ for U in the definition of Φ, we have

$$\begin{aligned}\Phi(\Phi(U)) &= Y + (Y - \Phi(U)) = Y + (Y - (Y + (Y - U)))\\ &= Y - (Y - U) = U,\end{aligned}$$

so that $\Phi \circ \Phi = \iota$, the identity map.

(D: Φ preserves betweenness.) For this case, we use the coordinate form of Definition LB.16 and by direct calculation show that for every triple (u_1, u_2), (x_1, x_2), (v_1, v_2) of points on $\mathbb{F}^2$, if (u_1, u_2)–(x_1, x_2)–(v_1, v_2), then $\Phi(u_1, u_2)$–$\Phi(x_1, x_2)$–$\Phi(v_1, v_2)$.

If (x_1, x_2) is between (u_1, u_2) and (v_1, v_2), then by the second alternative definition of betweenness (Theorem LC.12) there exists a number t such that $0 < t < 1$ and $\begin{pmatrix} x_1 \\ x_2 \end{pmatrix} = \begin{pmatrix} u_1 + t(v_1 - u_1) \\ u_2 + t(v_2 - u_2) \end{pmatrix}$. Then

$$\begin{aligned}
\Phi(x_1, x_2) &= \Phi\begin{pmatrix} x_1 \\ x_2 \end{pmatrix} = \Phi\begin{pmatrix} u_1 + t(v_1 - u_1) \\ u_2 + t(v_2 - u_2) \end{pmatrix}\\
&= \begin{pmatrix} \frac{b^2-a^2}{a^2+b^2}(u_1 + t(v_1 - u_1)) - \frac{2ab}{a^2+b^2}(u_2 + t(v_2 - u_2)) - \frac{2ac}{a^2+b^2} \\ \frac{-2ab}{a^2+b^2}(u_1 + t(v_1 - u_1) + \frac{a^2-b^2}{a^2+b^2}(u_2 + t(v_2 - u_2)) - \frac{2bc}{a^2+b^2} \end{pmatrix}\\
&= \begin{pmatrix} \frac{b^2-a^2}{a^2+b^2}u_1 - \frac{2ab}{a^2+b^2}u_2 - \frac{2ac}{a^2+b^2} \\ \frac{-2ab}{a^2+b^2}u_1 + \frac{a^2-b^2}{a^2+b^2}u_2 - \frac{2bc}{a^2+b^2} \end{pmatrix}\\
&\quad + t\begin{pmatrix} \frac{b^2-a^2}{a^2+b^2}v_1 - \frac{2ab}{a^2+b^2}v_2 - \frac{2ac}{a^2+b^2} - \left(\frac{b^2-a^2}{a^2+b^2}u_1 - \frac{2ab}{a^2+b^2}u_2 - \frac{2ac}{a^2+b^2}\right) \\ \frac{-2ab}{a^2+b^2}v_1 + \frac{a^2-b^2}{a^2+b^2}v_2 - \frac{2bc}{a^2+b^2} - \left(\frac{-2ab}{a^2+b^2}u_1 + \frac{a^2-b^2}{a^2+b^2}u_2 - \frac{2bc}{a^2+b^2}\right) \end{pmatrix}\\
&= \Phi(u_1, u_2) + t(\Phi(v_1, v_2) - \Phi(u_1, u_2)).
\end{aligned}$$

By Theorem LC.12, $\Phi(x_1, x_2)$ is between $\Phi(u_1, u_2)$ and $\Phi(v_1, v_2)$. □

Definition LC.24. From now on we denote the mirror mapping Φ (defined over a line $\mathcal{L}$ by Definition LB.16) by $\mathcal{R}_{\mathcal{L}}$, the symbol we used in our main development for a mirror mapping or reflection. We will sometimes refer to the mapping $\mathcal{R}_{\mathcal{L}}$ as the **LB.16 mapping** over the line $\mathcal{L}$.

This notation anticipates the proof in the following sequence of theorems that the set of all such mappings $\mathcal{R}_{\mathcal{L}}$ is indeed a reflection set (as in Definition NEUT.2) on $\mathbb{A}^2$ or $\mathbb{R}^2$. However, most of these theorems are valid for an arbitrary ordered field $\mathbb{F}$.

Theorem LC.25 (Existence and uniqueness: Properties R.1 and R.2). *The set of all mappings $\mathcal{R}_{\mathcal{L}}$ on $\mathbb{F}^2$ satisfies Properties R.1 and R.2 of Definition NEUT.2.*

Proof. The proof is obvious from the fact that Definition LB.16 defines exactly one mapping $\Phi = \mathcal{R}_{\mathcal{L}}$ over each line $\mathcal{L}$. □

The following definition provides a way around the lack of a notion of distance in the case that $\mathbb{F} = \mathbb{Q}$, and makes it possible to show that the set of all mappings $\mathcal{R}_{\mathcal{L}}$ on $\mathbb{Q}^2$ satisfies Properties R.3 and R.4 of Definition NEUT.2.

Definition LC.26. Let $\mathbb{F}$ be any ordered field, and let X and Y be any points of $\mathbb{F}^3$. Define $\text{qdi}(X, Y) = (X - Y) \bullet (X - Y)$ to be the **quadratic distance** between X and Y.

Remark LC.26.1. (A) If $X = (x_1, x_2)$ and $Y = (y_1, y_2)$ are in $\mathbb{F}^2$, the quadratic distance between X and Y is $\text{qdi}(X, Y) = (x_1 - y_1)^2 + (x_2 - y_2)^2$.

(B) If $\mathbb{F} = \mathbb{A}$ or $\mathbb{F} = \mathbb{R}$, so that $\mathbb{F}$ contains square roots of its non-negative members, $\text{dis}^2(X, Y) = \text{qdi}(X, Y)$; that is, $\text{dis}(X, Y) = \sqrt{\text{qdi}(X, Y)} = \sqrt{(X - Y) \bullet (X - Y)} = \|X - Y\|$. (Also, $\|X\| = \sqrt{\text{dis}(X, O)}$.) In this environment a mapping φ preserves quadratic distance iff it preserves distance.

Theorem LC.27 ($\mathcal{R}_{\mathcal{L}}$ preserves qdi and dis). *Let $\mathbb{F}$ be an ordered field and let $\mathcal{L}$ be any line in $\mathbb{F}^2$ where*

$$\mathcal{L} = \{(x_1, x_2) \mid (x_1, x_2) \in \mathbb{F}^2 \text{ and } ax_1 + bx_2 + c = 0\}$$

and $(a, b) \neq (0, 0)$. Let $X = (x_1, x_2)$ and $Y = (y_1, y_2)$ be any points of $\mathbb{F}^2$, and let $\mathcal{R}_{\mathcal{L}}$ be the mapping Φ over $\mathcal{L}$ as in Definition LB.16.

(A) *Then $\text{qdi}(\mathcal{R}_{\mathcal{L}}(X), \mathcal{R}_{\mathcal{L}}(Y)) = \text{qdi}(X, Y)$, that is, $\mathcal{R}_{\mathcal{L}}$ preserves quadratic distance.*

(B) *If $\mathbb{F} = \mathbb{A}$ or $\mathbb{F} = \mathbb{R}$ ($\mathbb{F}$ contains square roots of its non-negative members), then*

$$\text{dis}(\mathcal{R}_{\mathcal{L}}(X), \mathcal{R}_{\mathcal{L}}(Y)) = \|\mathcal{R}_{\mathcal{L}}(X) - \mathcal{R}_{\mathcal{L}}(Y)\| = \|X - Y\| = \text{dis}(X, Y),$$

that is, $\mathcal{R}_{\mathcal{L}}$ preserves distance.

Proof. Let Γ_1 and Γ_2 be the mappings such that for every member (x_1, x_2) of $\mathbb{F}^2$

$$\Gamma_1(x_1, x_2) = \tfrac{b^2-a^2}{a^2+b^2}x_1 - \tfrac{2ab}{a^2+b^2}x_2 - \tfrac{2ac}{a^2+b^2} \text{ and}$$
$$\Gamma_2(x_1, x_2) = \tfrac{-2ab}{a^2+b^2}x_1 - \tfrac{a^2-b^2}{a^2+b^2}x_2 - \tfrac{2bc}{a^2+b^2}.$$

Then, by Definition LB.16 $\mathcal{R}_{\mathcal{L}}(x_1, x_2) = (\Gamma_1(x_1, x_2), \Gamma_2(x_1, x_2))$. If $X = (x_1, x_2)$ and $Y = (y_1, y_2)$ are any two points of $\mathbb{F}^2$, by Exercise LM.16,

$$\begin{aligned}
&\mathrm{qdi}(\mathcal{R}_{\mathcal{L}}(X), \mathcal{R}_{\mathcal{L}}(Y)) = \mathrm{qdi}(\mathcal{R}_{\mathcal{L}}(x_1, x_2), \mathcal{R}_{\mathcal{L}}(y_1, y_2))\\
&= (\Gamma_1(x_1, x_2) - \Gamma_1(y_1, y_2))^2 + (\Gamma_2(x_1, x_2) - \Gamma_2(y_1, y_2))^2\\
&= (x_1 - y_1)^2 + (x_2 - y_2)^2 = \mathrm{qdi}((x_1, x_2), (y_1, y_2)) = \mathrm{qdi}(X, Y).
\end{aligned}$$

This proves part (A). Part (B) follows immediately by taking square roots of both sides. □

Corollary LC.27.1. *Let $\mathbb{F}$ be an ordered field and let $\mathcal{L}$ be any line in $\mathbb{F}^2$.*

(A) *Every composition of mirror mappings $\mathcal{R}_{\mathcal{M}}$ (as in Definition LB.16) over lines $\mathcal{M}$ in $\mathbb{F}^2$ preserves quadratic distance.*

(B) *If $\mathbb{F} = \mathbb{A}$ or $\mathbb{F} = \mathbb{R}$ ($\mathbb{F}$ contains square roots of its non-negative members), every composition of mirror mappings $\mathcal{R}_{\mathcal{M}}$ (as in Definition LB.16) over lines $\mathcal{M}$ in $\mathbb{F}^2$ preserves distance.*

Exercise NEUT.0 shows that on a coordinate plane there can be a mirror mapping $\Psi \neq \mathcal{R}_{\mathcal{L}}$. The following theorem shows that no such mapping can preserve quadratic distance.

Theorem LC.28. *For any ordered field $\mathbb{F}$, if $A = (a_1, a_2, a_3)$, $B = (b_1, b_2, b_3)$, and $C = (c_1, c_2, c_3)$ are points in $\mathbb{F}^3$ such that $C \in \overrightarrow{AB}$ and* $\mathrm{qdi}(A, B) = \mathrm{qdi}(A, C)$, *then $C = B$.*

Proof. By Definition LA.1(3E), since $C \in \overrightarrow{AB}$ there exists a number $t > 0$ such that $C = A + t(B - A)$, that is, $c_1 = a_1 + t(b_1 - a_1)$, $c_2 = a_2 + t(b_2 - a_2)$, and $c_3 = a_3 + t(b_3 - a_3)$. If $\mathrm{qdi}(A, B) = \mathrm{qdi}(A, C)$, then

$$\begin{aligned}
&(a_1 - b_1)^2 + (a_2 - b_2)^2 + (a_3 - b_3)^2 = (a_1 - c_1)^2 + (a_2 - c_2)^2 + (a_3 - c_3)^2\\
&= (a_1 - a_1 - t(b_1 - a_1))^2 + (a_2 - a_2 - t(b_2 - a_2))^2\\
&\quad + (a_3 - a_3 - t(b_3 - a_3))^2\\
&= t^2(b_1 - a_1)^2 + t^2(b_2 - a_2)^2 + t^2(b_3 - a_3)^2\\
&= t^2((a_1 - b_1)^2 + (a_2 - b_2)^2) + (a_3 - b_3)^2).
\end{aligned}$$

Hence $t^2 = 1$, so $t = 1$ and $C = A + (B - A) = B$. □

Theorem LC.29 (Only $\mathcal{R}_\mathcal{L}$ preserves qdi **or** dis**).** *Let $\mathbb{F}$ be an ordered field and let $\mathcal{L}$ be any line in $\mathbb{F}^2$; let Ψ be a mirror mapping over $\mathcal{L}$ such that $\Psi \neq \mathcal{R}_\mathcal{L}$, the mirror mapping of Definition LB.16. Then*

(A) *Ψ does not preserve quadratic distance, and*

(B) *if $\mathbb{F}$ contains square roots of its non-negative members (is either $\mathbb{A}$ or $\mathbb{R}$), Ψ does not preserve distance.*

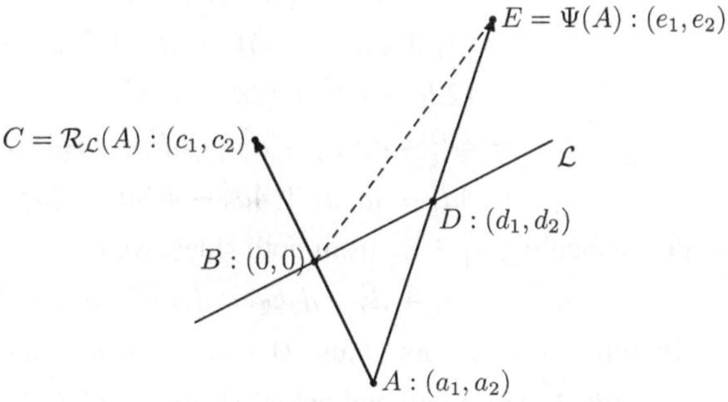

Fig. 21.2 For Theorem LC.29

Proof. Let $A = (a_1, a_2)$ be a point not on $\mathcal{L}$ such that $\Psi(A) \neq \mathcal{R}_\mathcal{L}(A)$. Let $C = (c_1, c_2) = \mathcal{R}_\mathcal{L}(A)$ and $E = (e_1, e_2) = \Psi(A)$. The line $\overleftrightarrow{A\mathcal{R}_\mathcal{L}(A)}$ is a fixed line for $\mathcal{R}_\mathcal{L}$, is perpendicular to $\mathcal{L}$, and intersects $\mathcal{L}$ at some point B which is a fixed point for $\mathcal{R}_\mathcal{L}$ (and Ψ). Choose a coordinate system so that $B = (0, 0)$. For a visualization, see Figure 21.2.

The line $\overleftrightarrow{A\Psi(A)}$ is a fixed line for Ψ, intersects $\mathcal{L}$ at some point $D = (d_1, d_2)$, which is a fixed point for Ψ (and $\mathcal{R}_\mathcal{L}$). All points of $\mathcal{L}$ are fixed points for both $\mathcal{R}_\mathcal{L}$ and Ψ.

We know that the mapping $\mathcal{R}_\mathcal{L}$ preserves quadratic distance, so that

$$a_1^2 + a_2^2 = \text{qdi}(A, B) = \text{qdi}(\mathcal{R}_\mathcal{L}(A), B) = c_1^2 + c_2^2.$$

Assume that Ψ also preserves quadratic distance. Then

$$\begin{aligned}(a_1 - d_1)^2 + (a_2 - d_2)^2 &= \text{qdi}(A, D) = \text{qdi}(E, D) \\ &= (e_1 - d_1)^2 + (e_2 - d_2)^2. \; (*)\end{aligned}$$

By Definition LA.1(1), since A, D, and E are collinear, there exists a number t such that $E = A + t(D - A)$, that is, $e_1 = a_1 + t(d_1 - a_1)$ and $e_2 = a_2 + t(d_2 - a_2)$. Substituting into the right-hand side of equation (*),

$$\begin{aligned}&(a_1 - d_1)^2 + (a_2 - d_2)^2 = (e_1 - d_1)^2 + (e_2 - d_2)^2 \\ &\quad = (a_1 + t(d_1 - a_1) - d_1)^2 + (a_2 + t(d_2 - a_2) - d_2)^2\end{aligned}$$

$$= ((1-t)a_1 - (1-t)d_1)^2 + ((1-t)a_2 - (1-t)d_2)^2$$
$$= (1-t)^2((a_1 - d_1)^2 + (a_2 - d_2)^2)$$

so that $(1-t)^2 = 1$, which is true iff $t = 0$ or $t = 2$. If $t = 0$, then $A = E$ which is impossible because A and E are on opposite sides of $\mathcal{L}$. Thus $E = A + 2(D - A)$, that is, $e_1 = a_1 + 2(d_1 - a_1)$ and $e_1 = a_2 + 2(d_2 - a_2)$.

Now Ψ maps A to E, and $B = (0,0)$ is a fixed point for both $\mathcal{R}_\mathcal{L}$ and Ψ, and Ψ preserves quadratic distance qdi. Therefore

$$\begin{aligned} a_1^2 + a_2^2 = \text{qdi}(A,B) &= \text{qdi}(E,B) = e_1^2 + e_2^2 \\ &= (a_1 + 2(d_1 - a_1))^2 + (a_2 + 2(d_2 - a_2))^2 \\ &= (2d_1 - a_1)^2 + (2d_2 - a_2)^2 \\ &= 4d_1^2 - 4d_1a_1 + a_1^2 + 4d_2^2 - 4d_2a_2 + a_2^2 \\ &= 4d_1^2 - 4d_1a_1 + 4d_2^2 - 4d_2a_2 + (a_1^2 + a_2^2) \end{aligned}$$

so that, canceling $a_1^2 + a_2^2$ from both sides, we have

$$d_1^2 - d_1a_1 + d_2^2 - d_2a_2 = d_1(d_1 - a_1) + d_2(d_2 - a_2) = 0.$$

Therefore the vectors D and $D - A$ are orthogonal and the lines $\overleftrightarrow{BD} = \mathcal{L}$ and $\overleftrightarrow{AD} = \overleftrightarrow{A\Psi(A)}$ are c-perpendicular. Since $\mathcal{L}$ and $\overleftrightarrow{A\mathcal{R}_\mathcal{L}(A)}$ are c-perpendicular, and there is only one c-perpendicular to a line at a point, $\overleftrightarrow{A\mathcal{R}_\mathcal{L}(A)} = \overleftrightarrow{A\Psi(A)}$; and A, B, $\mathcal{R}_\mathcal{L}(A)$, and $\Psi(A)$ are collinear. Both $\mathcal{R}_\mathcal{L}$ and Ψ preserve quadratic distance, so $\text{qdi}(B, \mathcal{R}_\mathcal{L}(A)) = \text{qdi}(B, A) = \text{qdi}(B, \Psi(A))$; $\mathcal{R}_\mathcal{L}(A)$ and $\Psi(A)$ are on the same side of $\mathcal{L}$ and hence $\Psi(A) \in \overrightarrow{B\mathcal{R}_\mathcal{L}(A)}$. By Theorem LC.28, $\mathcal{R}_\mathcal{L}(A) = \Psi(A)$, contradicting the assumption that $\mathcal{R}_\mathcal{L}(A) \neq \Psi(A)$.

Thus, Ψ does not preserve quadratic distance, and in the case where $\mathbb{F}$ contains square roots of its non-negative members (is either $\mathbb{A}$ or $\mathbb{R}$), does not preserve distance. □

Corollary LA.29.1 (Closure Property R.3). *Let $\mathbb{F}$ be an ordered field and let $\mathcal{L}$ be any line in $\mathbb{F}^2$. If Ψ is any mirror mapping over $\mathcal{L}$ which is a finite composition of mirror mappings $\mathcal{R}_{\mathcal{M}_k}$ as in Definition LB.16, then $\Psi = \mathcal{R}_\mathcal{L}$.*

Proof. If $\Psi \neq \mathcal{R}_\mathcal{L}$, by Theorem LC.29(A) Ψ does not preserve quadratic distance; by Corollary LC.27.1(A), Ψ does preserve quadratic distance, a contradiction. □

Theorem LC.30 (Linear scaling Property R.4). *Let $\mathbb{F}$ be an ordered field and let $\mathcal{L}$ be any line in $\mathbb{F}^2$. If α is any finite composition of mirror mappings $\mathcal{R}_{\mathcal{M}_k}$ over lines $\mathcal{M}_k$ as in Definition LB.16, and if $\alpha(\overline{AB}) = \overline{AC}$, where A, B, and C are collinear points such that $B \in \overrightarrow{AC}$, then $B = C$.*

Proof. By Theorem LC.27(A), each of the mappings $\mathcal{R}_{\mathcal{M}_k}$ preserves quadratic distance; therefore (by Corollary LC.27.1) the mapping α preserves quadratic distance; that is, $\text{qdi}(A, B) = \text{qdi}(A, C)$. Since $B \in \overrightarrow{AC}$ it follows from Theorem LC.28 that $B = C$. □

Theorem LC.31 (Angle reflection Property R.5). *Let $\mathbb{F} = \mathbb{A}$ or $\mathbb{F} = \mathbb{R}$ so that square roots of non-negative numbers exist and distance in $\mathbb{F}^2$ is defined. Then there exists a line $\mathcal{M}$ such that the mirror mapping $\mathcal{R}_{\mathcal{M}}$ (as defined by Definition LC.24) over $\mathcal{M}$ maps $\overrightarrow{CA}$ to $\overrightarrow{CB}$, and $\mathcal{R}_{\mathcal{M}}(\mathcal{M}) = \mathcal{M}$; $\mathcal{R}_{\mathcal{M}}$ is an angle reflection for $\angle ACB$, and* $\text{dis}(A, C) = \text{dis}(\mathcal{R}_{\mathcal{M}}(A), C)$.

Proof. Let $A = (a_1, a_2)$, $B = (b_1, b_2)$, and $C = (c_1, c_2)$ be noncollinear points on $\mathbb{F}^2$ so that $\angle ACB$ is defined. Since in this environment the distance between two points is defined (as in Definition/Remark LA.13(2)) we may divide $A-C$ and $B-C$ by their lengths $\|A - C\|$ and $\|B - C\|$ respectively to locate points on the two rays of the angle which are a distance 1 from C. Thus, without loss of generality, we may assume that $\text{dis}(C, A) = 1$ and $\text{dis}(C, B) = 1$.

Let $\mathcal{L} = \overleftrightarrow{AB}$, and let $M = (\frac{a_1+b_1}{2}, \frac{a_2+b_2}{2})$ be the c-midpoint of $\overline{AB}$. By Exercise LM.13 $\overleftrightarrow{AB} = \mathcal{L}$ is the set of all pairs $(x_1, x_2) \in \mathbb{F}^2$ such that

$$(b_2 - a_2)x_1 - (b_1 - a_1)x_2 - a_1(b_2 - a_2) + a_2(b_1 - a_1) = 0.$$

Let $\mathcal{M}$ be the set of all pairs $(x_1, x_2) \in \mathbb{F}^2$ such that

$$(b_1 - a_1)x_1 + (b_2 - a_2)x_2 + (a_1 - b_1)c_1 + (a_2 - b_2)c_2 = 0.$$

Claim: $\mathcal{M}$ is c-perpendicular to $\mathcal{L}$ and contains both C and M.

First, note that $(b_1 - a_1)(b_2 - a_2) + (-(b_1 - a_1)(b_2 - a_2)) = 0$, so that by the criterion of Theorem LB.8, $\mathcal{M} \perp \mathcal{L}$.

To show that C is a point of $\mathcal{M}$, we substitute its coordinates (c_1, c_2) into the formula for $\mathcal{M}$ to get

$$(b_1 - a_1)c_1 + (b_2 - a_2)c_2 + (a_1 - b_1)c_1 + (a_2 - b_2)c_2 = 0.$$

Finally, we verify by direct calculation that $M = (\frac{a_1+b_1}{2}, \frac{a_2+b_2}{2}) \in \mathcal{M}$. From $\text{dis}(C, A) = 1$ and $\text{dis}(C, B) = 1$, it follows that

$$(a_1 - c_1)^2 + (a_2 - c_2)^2 = 1 \text{ and } (b_1 - c_1)^2 + (b_2 - c_2)^2 = 1,$$

so that

$$2a_1c_1 + 2a_2c_2 = a_1^2 + c_1^2 + a_2^2 + c_2^2 - 1 \quad (*)$$

and

$$2b_1c_1 + 2b_2c_2 = b_1^2 + b_2^2 + c_1^2 + c_2^2 - 1. \quad (**)$$

Substituting $\frac{a_1+b_1}{2}$ and $\frac{a_2+b_2}{2}$ for x_1 and x_2, respectively, into the formula for $\mathcal{M}$, we have

$$\begin{aligned}(b_1 - a_1)\left(\tfrac{b_1+a_1}{2}\right) + (b_2 - a_2)\left(\tfrac{b_2+a_2}{2}\right) + (a_1 - b_1)c_1 + (a_2 - b_2)c_2 \\ = \tfrac{1}{2}(b_1^2 - a_1^2 + b_2^2 - a_2^2 + 2a_1c_1 - 2b_1c_1 + 2a_2c_2 - 2b_2c_2),\end{aligned}$$

and by (*) and (**) this last expression becomes

$$\tfrac{1}{2}(b_1^2 - a_1^2 + b_2^2 - a_2^2 + a_1^2 + c_1^2 + a_2^2 + c_2^2 - 1 - b_1^2 - b_2^2 - c_1^2 - c_2^2 + 1) = 0,$$

so that $M = \left(\frac{a_1+b_1}{2}, \frac{a_2+b_2}{2}\right) \in \mathcal{M}$. This proves the Claim.

Define $\mathcal{R}_\mathcal{M} = \Phi$ to be the mirror mapping over the line $\mathcal{M}$ as in Definition LB.16. We have already defined $M = \frac{A+B}{2}$ to be the c-midpoint of $\overline{AB}$, and we have defined $\mathcal{M}$ so that M is the point of intersection of $\mathcal{M}$ and $\overleftrightarrow{AB}$ and is thus a fixed point for $\mathcal{R}_\mathcal{M}$. By Theorem LB.17(C) M is the c-midpoint of $\overline{A\mathcal{R}_\mathcal{M}(A)}$, that is, $M = \frac{A+\mathcal{R}_\mathcal{M}(A)}{2} = \frac{A+B}{2}$. A simple computation shows that $\mathcal{R}_\mathcal{M}(A) = B$.

Since $\mathcal{R}_\mathcal{M}(C) = C$, by Theorem NEUT.15(3) $\mathcal{R}_\mathcal{M}(\overrightarrow{CA}) = \overrightarrow{CB}$. (Here we are entitled to use Theorem NEUT.15 because we have already shown that $\mathcal{R}_\mathcal{M}$ is a mirror mapping.) Since LB.16 mirror mappings preserve distance, $\text{dis}(A, C) = \text{dis}(\mathcal{R}_\mathcal{M}(A), C)$. □

Remark LC.31.1. In Subsection 21.7.3, Theorem RSI.3 will prove that the set of LB.16 reflections on Model LM2Q ($\mathbb{Q}^2$) fails to satisfy Property R.5; the current set of remarks shows that this set satisfies all the other reflection properties. We will discuss the significance of this at the beginning of Subsection 21.6.4.

Theorem LC.32 (Midpoint existence Property R.6). *Let $\mathbb{F}$ be an ordered field, and let A and B be any distinct points of $\mathbb{F}^2$. Let $M = \frac{A+B}{2}$ be the c-midpoint of $\overline{AB}$. Then there exists a line $\mathcal{M}$ containing M which is c-perpendicular to $\overleftrightarrow{AB}$, and a mirror mapping $\mathcal{R}_\mathcal{M}$ over $\mathcal{M}$ such that $\mathcal{R}_\mathcal{M}(A) = B$, $\mathcal{R}_\mathcal{M}(M) = M$, $\mathcal{R}_\mathcal{M}(\overline{AM}) = \overline{BM}$, and M is a midpoint of $\overline{AB}$.*

Proof. By Theorem LB.12, the line $\mathcal{M}$ exists and is c-perpendicular to $\overleftrightarrow{AB}$. Let $\mathcal{R}_\mathcal{M}$ be the mirror mapping Φ of Definition LB.16 over $\mathcal{M}$; by Theorem LB.17(D) $\overleftrightarrow{A\mathcal{R}_\mathcal{M}(A)}$ is a fixed line for $\mathcal{R}_\mathcal{M}$, and by definition is c-perpendicular to $\mathcal{M}$, as is $\overleftrightarrow{AB}$. By Theorem LB.12 $\overleftrightarrow{A\mathcal{R}_\mathcal{M}(A)} = \overleftrightarrow{AB}$ since they both contain the point A. By definition $\overleftrightarrow{A\mathcal{R}_\mathcal{M}(A)}$ intersects $\mathcal{M}$ at the c-midpoint $\frac{A+\mathcal{R}_\mathcal{M}(A)}{2}$ and we already know that $\overleftrightarrow{AB}$ intersects $\mathcal{M}$ at the c-midpoint $\frac{A+B}{2}$. Since these lines are the same, their intersections with $\mathcal{M}$ are the same point M, and $\frac{A+\mathcal{R}_\mathcal{M}(A)}{2} = \frac{A+B}{2} = M$; by a simple computation, $\mathcal{R}_\mathcal{M}(A) = B$.

Since $M \in \mathcal{M}$, $\mathcal{R}_{\mathcal{M}}(M) = M$; by Theorem NEUT.15 (which we can use because we have shown $\mathcal{R}_{\mathcal{M}}$ to be a mirror mapping), $\mathcal{R}_{\mathcal{M}}(\overline{AM}) = \overline{BM}$. Thus by Definition NEUT.3(C), M is a midpoint of $\overline{AB}$. □

Theorem LC.33 (Summary: Axiom REF is valid for Model LM2A and Model LM2R). *Suppose $\mathbb{F}$ is either $\mathbb{A}$ or $\mathbb{R}$. Then $\{\mathcal{R}_{\mathcal{L}} \mid \mathcal{L}$ is a line in $\mathbb{F}^2\}$ is a reflection set on $\mathbb{F}^2$, where each mapping $\mathcal{R}_{\mathcal{L}}$ is as in Definition LC.24. Thus Axiom REF holds on Models LM2A and LM2R, and every mirror mapping $\mathcal{R}_{\mathcal{L}}$ may legitimately be called a reflection.*

Proof. By Theorem LC.25, Properties R.1 and R.2 of Definition NEUT.2 (existence and uniqueness) hold. By Corollary LA.29.1 Property R.3 (closure) is true. By Theorem LC.30 Property R.4 (linear scaling) is true. By Theorem LC.31, Property R.5 (angle reflection) is true provided $\mathbb{F}$ is either $\mathbb{A}$ or $\mathbb{R}$. Finally, Theorem LC.32 shows that Property R.6 (existence of midpoint) is true. □

21.5.6 *On an arbitrary plane in $\mathbb{F}^3$*

Remark LC.34. In this subsection, $\mathbb{F}$ is either the field $\mathbb{A}$ of real algebraic numbers or the field $\mathbb{R}$ of real numbers, so that norms of vectors exist. This opens the possibility of adding to part (B) of Theorem LC.6.1 the provision that the norms of the vectors $G_1 - A$, $G_2 - A$, and $G_3 - A$ are all equal to 1. For if we let $H_i = A + \frac{G_i - A}{\|G_i - A\|}$, so that $\|H_i - A\| = \|\frac{G_i - A}{\|G_i - A\|}\| = 1$, $H_i - A$ is a scalar multiple of $G_i - A$ and orthogonality is not disturbed. For reference we repeat the statement of Theorem LC.6.1 with this modification:

Modified Theorem LC 6.1. *Let $\mathcal{P}$ be a plane in $\mathbb{F}^3$, where $\mathbb{F}$ is either $\mathbb{A}$ or $\mathbb{R}$; let $A = (a_1, a_2, a_3)$ and $B = (b_1, b_2, b_3)$ be distinct points of $\mathcal{P}$. Then there exist three points G_1, G_2, and G_3, of $\mathbb{F}^3$, all distinct from A, such that*

(A) *G_1 and G_2 belong to $\mathcal{P}$, and for every point $X \in \mathcal{P}$, there exist scalars s and t in $\mathbb{F}$ such that $X - A = s(G_1 - A) + t(G_2 - A)$;*
(B) *the vectors $G_1 - A$, $G_2 - A$, and $G_3 - A$ are pairwise orthogonal and $\|G_1 - A\| = \|G_2 - A\| = \|G_3 - A\| = 1$;*
(C) *for every point $X \in \mathbb{F}^3$, there exist scalars x_1, x_2, and x_3 in $\mathbb{F}$ such that $X - A = x_1(G_1 - A) + x_2(G_2 - A) + x_3(G_3 - A)$; and*
(D) *for every point $X \in \mathbb{F}^3$ such that $(X - A) \perp (G_3 - A)$, $X \in \mathcal{P}$.*

Definition LC.35. Let $\mathcal{P}$ be a plane in $\mathbb{F}^3$, where $\mathbb{F}$ is either $\mathbb{A}$ or $\mathbb{R}$. Let A be a point of $\mathcal{P}$, and let G_1 and G_2 be the points of $\mathbb{F}^3$ as defined in the Modified Theorem LC.6.1. That is, $G_1 - A$ and $G_2 - A$ are orthogonal with norm 1, and for every $X \in \mathcal{P}$ there exist scalars s and t such that $X - A = s(G_1 - A) + t(G_2 - A)$. (By Theorem LA.3, A, G_1, and G_2 are noncollinear.)

For each point $X = (s, t)$ of $\mathbb{F}^2$ define $\varphi(s, t) = A + s(G_1 - A) + t(G_2 - A)$. We shall refer to this mapping as a **transfer** mapping from $\mathbb{F}^2$ to $\mathcal{P}$.

Remark LC.35.1. (A) By Theorem LA.6, φ is a bijection of $\mathbb{F}^2$ onto $\mathcal{P}$. Here G_1 takes the place of B in that theorem, G_2 takes the place of C, and $\mathcal{P}$ takes the place of $\overleftrightarrow{ABC}$.

(B) The main application of transfer mappings will be to facilitate the definition of reflections on an arbitrary plane, using the reflections already defined on $\mathbb{F}^2$, determine the properties of these reflections, and show that they satisfy the properties of Definition NEUT.2, thus confirming that Axiom REF holds on every plane.

(C) Notice that we speak of *a* transfer mapping, since each choice of $\mathcal{P}$, and of points A, G_1, and G_2 on $\mathcal{P}$, defines a different transfer mapping. This will not cause difficulty, because most of our applications of transfer mappings will be in a given environment with the plane and the points of the plane already specified.

(D) For the record, we make no use in the following of G_3, which just "goes along for the ride."

Theorem LC.36 (φ and φ^{-1} preserve lines). *Let $\mathbb{F}$ be either $\mathbb{A}$ or $\mathbb{R}$, and let φ be the transfer mapping which maps $\mathbb{F}^2$ onto a plane $\mathcal{P}$, as defined in Definition LC.35.*

(A) *If $\mathcal{L}$ is a line in $\mathbb{F}^2$, then $\varphi(\mathcal{L})$ is a line in $\mathcal{P}$.*
(B) *If $\mathcal{M}$ is a line in $\mathcal{P}$, then $\varphi^{-1}(\mathcal{M})$ is a line in $\mathbb{F}^2$.*
(C) *Let C and D be distinct points on $\mathbb{F}^2$; for any point $X \in \mathbb{F}^2$, $X = C + t(D - C)$ iff*

$$\varphi(X) = \varphi(C + t(D - C)) = \varphi(C) + t(\varphi(D) - \varphi(C)).$$

Proof. We use the meanings of A, G_1, and G_2 as given in Definition LC.35.

(A) Let $\mathcal{L}$ be a line in $\mathbb{F}^2$, and let $C = (c_1, c_2)$ and $D = (d_1, d_2)$ be distinct points of $\mathcal{L}$. By Definition LA.1(1) and Corollary LA.10.1, $X \in \mathcal{L}$ iff for some t, $X = C + t(D - C)$.

Observe first that $\varphi(C) = A + c_1(G_1 - A) + c_2(G_2 - A)$ and $\varphi(D) = A + d_1(G_1 - A) + d_2(G_2 - A)$. Then for every $X \in \mathcal{L}$,

$$\begin{aligned}\varphi(X) &= \varphi(C + t(D - C)) = \varphi((c_1, c_2) + t((d_1, d_2) - (c_1, c_2)))\\ &= \varphi(c_1 + t(d_1 - c_1), c_2 + t(d_2 - c_2))\\ &= A + (c_1 + t(d_1 - c_1))(G_1 - A) + (c_2 + t(d_2 - c_2))(G_2 - A)\\ &= (A + c_1(G_1 - A) + c_2(G_2 - A))\\ &\quad + t\big((d_1 - c_1)(G_1 - A) + (d_2 - c_2)(G_2 - A)\big)\\ &= (A + c_1(G_1 - A) + c_2(G_2 - A))\\ &\quad + t\big((A + d_1(G_1 - A) + d_2(G_2 - A)) - (A + c_1(G_1 - A) + c_2(G_2 - A))\big)\\ &= \varphi(C) + t(\varphi(D) - \varphi(C)).\end{aligned}$$

Thus φ maps the line $\overleftrightarrow{CD}$ into the line $\overleftrightarrow{\varphi(C)\varphi(D)}$; φ also maps onto this line because every point in it is a point $\varphi(C) + t(\varphi(D) - \varphi(C))$ for some t, and so is the image under φ of $C + t(D - C)$.

(B) With appropriate adjustments for the fact that φ maps $\mathbb{F}^2$ onto $\mathcal{P}$ rather than onto $\mathbb{F}^2$, this is the proof of Theorem CAP.1(D') from Chapter 3.

(C) By the calculation in part (A), if $X = C + t(D - C)$, then $\varphi(X) = \varphi(C + t(D - C)) = \varphi(C) + t(\varphi(D) - \varphi(C))$. Conversely, suppose that $\varphi(X) = \varphi(C) + t(\varphi(D) - \varphi(C))$; this is $\varphi(C + t(D - C))$, and since φ is one-to-one, $X = C + t(D - C)$. □

Theorem LC.37. *Let $\mathbb{F}$ be either $\mathbb{A}$ or $\mathbb{R}$, and let φ be the transfer mapping which maps $\mathbb{F}^2$ onto a plane $\mathcal{P}$, as defined in Definition LC.35.*

(A: φ and φ^{-1} preserve intersections of lines.) *Two lines $\mathcal{L}$ and $\mathcal{M}$ in $\mathbb{F}^2$ intersect at a point X iff the lines $\varphi(\mathcal{L})$ and $\varphi(\mathcal{M})$ intersect at $\varphi(X)$.*

(B: φ and φ^{-1} preserve betweenness.) *For every X, Y, and Z in $\mathbb{F}^2$, X–Y–Z iff $\varphi(X)$–$\varphi(Y)$–$\varphi(Z)$.*

(C: φ and φ^{-1} preserve segments and rays.)

(1) *For any two points X and Z in $\mathbb{F}^2$, $\varphi(\overline{]XZ[}) = \overline{]\varphi(X)\varphi(Z)[}$, $\varphi(\overline{[XZ]}) = \overline{[\varphi(X)\varphi(Z)]}$, $\varphi(\overline{[XZ[}) = \overline{[\varphi(X)\varphi(Z)[}$, and $\varphi(\overline{]XZ]}) = \overline{]\varphi(X)\varphi(Z)]}$.*

(2) *For any two points X and Z in $\mathbb{F}^2$, $\varphi(\overrightarrow{]XZ}) = \overrightarrow{]\varphi(X)\varphi(Z)}$ and $\varphi(\overrightarrow{[XZ}) = \overrightarrow{[\varphi(X)\varphi(Z)}$.*

(3) *Similar results hold where φ is replaced by ψ^{-1}.*

(D: φ and φ^{-1} preserve c-midpoints.) *M is the c-midpoint of* $\overline{XZ} \subseteq \mathbb{F}^2$ *iff* $\varphi(M)$ *is the c-midpoint of* $\varphi(\overline{XZ}) = \overline{\varphi(X)\varphi(Z)}$.

Proof. (A) A point $P \in \mathcal{L} \cap \mathcal{M}$ iff $P \in \mathcal{L}$ and $P \in \mathcal{M}$, which by Theorem LC.36 is true iff $\varphi(P) \in \varphi(\mathcal{L})$ and $\varphi(P) \in \varphi(\mathcal{M})$, which is true iff $\varphi(P) \in \varphi(\mathcal{L}) \cap \varphi(\mathcal{M})$.

(B) Using Theorem LC.12 (the second alternate definition of betweenness), X–Y–Z iff for some number s such that $0 < s < 1$, $Y = X + s(Z - X)$. By Theorem LC.36(C) $Y = X + s(Z - X)$ iff $\varphi(Y) = \varphi(X) + s(\varphi(Z) - \varphi(X))$; again using Theorem LC.12, this is true iff $\varphi(X)$–$\varphi(Y)$–$\varphi(Z)$.

(C) By Theorem LC.13, the definitions of Definition LA.1 for segments and rays are equivalent to those of Definitions IB.3 and IB.4 given in Chapter 4, which use betweenness to define segments and rays.

(1) By Definition IB.3, for any two points X and Z in $\mathbb{F}^3$, $Y \in \overline{XZ}$ iff X–Y–Z, which by part (B) is true iff $\varphi(X)$–$\varphi(Y)$–$\varphi(Z)$, which is true iff $\varphi(Y) \in \overline{\varphi(X)\varphi(Z)}$. The proofs for closed and half-closed segments follow from the observation that in those case, endpoints are included.

(2) By Definition IB.4, for any two points X and Z in $\mathbb{F}^2$, $Y \in \overrightarrow{XZ}$ iff X–Y–Z or X–Z–Y or $Y = Z$. Since φ is a bijection that preserves betweenness, this is true iff $\varphi(X)$–$\varphi(Y)$–$\varphi(Z)$ or $\varphi(X)$–$\varphi(Z)$–$\varphi(Y)$ or $\varphi(Y) = \varphi(Z)$, that is, iff $\varphi(Y) \in \overrightarrow{\varphi(X)\varphi(Z)}$. For a closed ray the proof follows from the fact that the endpoint is included.

(3) The arguments for φ^{-1} are similar to those just above.

(D) According to Definition/Remark LA.13(3) the c-midpoint of $\overline{XZ}$ is the point $M = X + \frac{1}{2}(Z - X)$; also by Theorem LC.36(C), $M = X + \frac{1}{2}(Z - X)$ iff $\varphi(M) = \varphi(X) + \frac{1}{2}(\varphi(Z) - \varphi(X))$, which is true iff $\varphi(M)$ is the c-midpoint of $\overline{\varphi(X)\varphi(Z)}$. □

Theorem LC.38. *Let* $\mathbb{F}$ *be either* $\mathbb{A}$ *or* $\mathbb{R}$, *and let* φ *be the transfer mapping which maps* $\mathbb{F}^2$ *onto a plane* $\mathcal{P}$, *as defined in Definition LC.35.*

(A: φ and dot products.) *For every* $X = (x_1, x_2)$, $Y = (y_1, y_2)$, *and* $Z = (z_1, z_2)$ *in* $\mathbb{F}^2$, $(X - Z) \bullet (Y - Z) = (\varphi(X) - \varphi(Z)) \bullet (\varphi(Y) - \varphi(Z))$.

(B: φ and φ^{-1} preserve c-perpendicularity of lines.) *Two lines* $\mathcal{L}$ *and* $\mathcal{M}$ *in* $\mathbb{F}^2$ *are c-perpendicular iff* $\varphi(\mathcal{L})$ *and* $\varphi(\mathcal{M})$ *are c-perpendicular.*

(C: φ and norms.) *For every* $X \in \mathbb{F}^2$, $\|X\| = \|\varphi(X) - A\|$.

(D: φ and φ^{-1} preserve distance.) *For every X and Z in* $\mathbb{F}^2$, $\mathrm{dis}(X, Z) = \|X - Z\| = \|\varphi(X) - \varphi(Z)\| = (\mathrm{dis}(\varphi(X), \varphi(Z)))$.

Proof. (A) Let $X = (x_1, x_2)$, $Y = (y_1, y_2)$, and $Z = (z_1, z_2)$ be pairwise distinct points of $\mathbb{F}^2$. Then

$$\begin{aligned}(X - Z) \bullet (Y - Z) &= (x_1 - z_1, x_2 - z_2) \bullet (y_1 - z_1, y_2 - z_2) \\ &= (x_1 - z_1)(y_1 - z_1) + (x_2 - z_2)(y_2 - z_2). \ (*)\end{aligned}$$

On the other hand,

$$\begin{aligned}\varphi(X) &= A + x_1(G_1 - A) + x_2(G_2 - A), \\ \varphi(Y) &= A + y_1(G_1 - A) + y_2(G_2 - A), \text{ and} \\ \varphi(Z) &= A + z_1(G_1 - A) + z_2(G_2 - A),\end{aligned}$$

so that

$$\begin{aligned}\varphi(X) - \varphi(Z) &= (x_1 - z_1)(G_1 - A) + (x_2 - z_2)(G_2 - A) \text{ and} \\ \varphi(Y) - \varphi(Z) &= (y_1 - z_1)(G_1 - A) + (y_2 - z_2)(G_2 - A).\end{aligned}$$

Then

$$\begin{aligned}&(\varphi(X) - \varphi(Z)) \bullet (\varphi(Y) - \varphi(Z)) \\ &\quad = ((x_1 - z_1)(G_1 - A) + (x_2 - z_2)(G_2 - A)) \\ &\qquad \bullet (y_1 - z_1)(G_1 - A) + (y_2 - z_2)(G_2 - A) \\ &\quad = (x_1 - z_1)(G_1 - A) \bullet (y_1 - z_1)(G_1 - A) \\ &\qquad + (x_1 - z_1)(G_1 - A) \bullet (y_2 - z_2)(G_2 - A) \\ &\qquad\quad + (x_2 - z_2)(G_2 - A) \bullet (y_1 - z_1)(G_1 - A) \\ &\qquad\qquad + (x_2 - z_2)(G_2 - A) \bullet (y_2 - z_2)(G_2 - A).\end{aligned}$$

The second and third terms of this sum are 0, since $G_1 - A$ and $G_2 - A$ are orthogonal; since

$$(G_1 - A) \bullet (G_1 - A) = 1 = (G_1 - A) \bullet (G_2 - A),$$

this reduces to

$$(x_1 - z_1)(y_1 - z_1) + 0 + 0 + (x_2 - z_2)(y_2 - z_2)$$

which by equation (*) is $(X - Z) \bullet (Y - Z)$.

(B) Suppose $\mathcal{L}$ and $\mathcal{M}$ are lines in $\mathbb{F}^2$; by Theorem LA.23 (or LB.10) they intersect at a point Z. By Theorem LC.37(A) $\varphi(\mathcal{L})$ and $\varphi(\mathcal{M})$ intersect at the point $\varphi(Z)$. Choose $X \in \mathcal{L}$ and $Y \in \mathcal{M}$ to be distinct from Z; then by Definition/Remark LA.12(3) these lines are c-perpendicular iff $(X - Z) \bullet (Y - Z) = 0$. By part (A), this is true iff $(\varphi(X) - \varphi(Z)) \bullet (\varphi(Y) - \varphi(Z)) = 0$, which is true iff $\varphi(\mathcal{L})$ and $\varphi(\mathcal{M})$ are c-perpendicular.

By Theorem LC.36 every line in $\mathcal{P}$ is the image under φ of a line in $\mathbb{F}^2$; thus any two lines in $\mathcal{P}$ are images under φ of two intersecting lines in $\mathbb{F}^2$. If $\varphi(\mathcal{L})$ and $\varphi(\mathcal{M})$ are c-perpendicular lines in $\mathcal{P}$ by Theorem LA.23 they intersect at some point $\varphi(Z)$; and by Theorem LC.37(A) $\mathcal{L}$ and $\mathcal{M}$ intersect at the point Z. Then by the equivalence shown in the preceding paragraph, they are c-perpendicular, showing that φ^{-1} preserves c-perpendicularity.

(C) We continue to use the notation of Definition LC.35. Let $Z = O$ so that $\varphi(Z) = A$, and let $X = Y$. Then by part (A) $X \bullet X = (\varphi(X) - A)) \bullet (\varphi(X) - A)$, and taking square roots, $\|X\| = \|\varphi(X) - A\|$.

(D) In part (A) let X and Z be any points of $\mathbb{F}^2$, and let $Y = X$. Then for every X and Z in $\mathbb{F}^2$,

$$\begin{aligned}(\text{dis}(X, Z))^2 &= \|X - Z\|^2 = (X - Z) \bullet (X - Z) \\ &= (\varphi(X) - \varphi(Z)) \bullet (\varphi(X) - \varphi(Z)) \\ &= \|\varphi(X) - \varphi(Z)\|^2 = (\text{dis}(\varphi(X), \varphi(Z)))^2;\end{aligned}$$

taking square roots completes the proof. □

Definition LC.39. Let $\mathbb{F}$ be an ordered field $\mathbb{A}$ or $\mathbb{R}$, let $\mathcal{P}$ be any plane in $\mathbb{F}^3$, and let $\mathcal{L}$ be any line on $\mathcal{P}$. Define $\mathcal{S}_{\mathcal{L}} = \varphi \circ \mathcal{R}_{\varphi^{-1}(\mathcal{L})} \circ \varphi^{-1}$. We shall refer to this mapping as the **induced mirror mapping** for the line $\mathcal{L}$ on $\mathcal{P}$. That is, it is induced by a mirror mapping on $\mathbb{F}^2$ and the mapping φ.

We may at times wish to describe an induced mapping in terms of a line on $\mathbb{F}^2$ instead of on the plane $\mathcal{P}$; in such a case we may let $\mathcal{N} = \varphi^{-1}(\mathcal{L})$ or $\mathcal{L} = \varphi(\mathcal{N})$; then

$$\mathcal{S}_{\varphi(\mathcal{N})} = \varphi \circ \mathcal{R}_{\varphi(\varphi^{-1}(\mathcal{L}))} \circ \varphi^{-1} = \varphi \circ \mathcal{R}_{\mathcal{L}} \circ \varphi^{-1}.$$

Theorem LC.40. *Let $\mathbb{F}$ be either $\mathbb{A}$ or $\mathbb{R}$, and let $\mathcal{L}$ be a line on the plane $\mathcal{P}$. Then the induced mirror mapping $\mathcal{S}_{\mathcal{L}}$ is a mirror mapping over $\mathcal{L}$ on the plane $\mathcal{P}$.*

Proof. From Theorem LC.23 we know that the mapping $\mathcal{R}_{\varphi^{-1}(\mathcal{L})}$ is a mirror mapping over the line $\varphi^{-1}(\mathcal{L}) \subseteq \mathbb{F}^2$, as it satisfies Properties (A), (B), (C), and (D) of Definition NEUT.1.

(A: All points of $\mathcal{L}$ are fixed under $\mathcal{S}_{\mathcal{L}}$.) Let X be any point of $\mathcal{L}$. Then $\varphi^{-1}(X)$ is a point of $\varphi^{-1}(\mathcal{L})$, which consists of fixed points of $\mathcal{R}_{\varphi^{-1}(\mathcal{L})}$, and

$$\mathcal{S}_{\mathcal{L}}(X) = \varphi(\mathcal{R}_{\varphi^{-1}(\mathcal{L})}(\varphi^{-1}(X))) = \varphi(\varphi^{-1}(X)) = X.$$

(B: If $X \notin \mathcal{L}$, $\mathcal{S}_{\mathcal{L}}(X)$ is on the opposite side of $\mathcal{L}$ from X.) Since φ is a bijection, $\varphi^{-1}(X) \notin \varphi^{-1}(\mathcal{L})$. Since $\mathcal{R}_{\varphi^{-1}(\mathcal{L})}$ is a mirror mapping, $\mathcal{R}_{\varphi^{-1}(\mathcal{L})}(\varphi^{-1}(X))$ is on the opposite side of $\varphi^{-1}(\mathcal{L})$ from $\varphi^{-1}(X)$ and the segment

$$\overline{(\varphi^{-1}(X))(\mathcal{R}_{\varphi^{-1}(\mathcal{L})}(\varphi^{-1}(X)))}$$

intersects $\varphi^{-1}(\mathcal{L})$ at some point P. By Theorem LC.37(C)(1),

$$\varphi\left(\overline{(\varphi^{-1}(X))(\mathcal{R}_{\varphi^{-1}(\mathcal{L})}(\varphi^{-1}(X)))}\right)$$
$$= \overline{\varphi(\varphi^{-1}(X))\varphi(\mathcal{R}_{\varphi^{-1}(\mathcal{L})}(\varphi^{-1}(X)))} = \overline{X\mathcal{S}_{\mathcal{L}}(X)}.$$

Then $\varphi(P)$, a member of $\mathcal{L}$, belongs to this segment, so X and $\mathcal{S}_{\mathcal{L}}(X)$ are on opposite sides of $\mathcal{L}$.

(C: $\mathcal{S}_{\mathcal{L}}$ is its own inverse.) By the definition of $\mathcal{S}_{\mathcal{L}}$, and the fact that $\mathcal{R}_{\varphi^{-1}(\mathcal{L})}$ is its own inverse, being a mirror mapping,

$$\mathcal{S}_{\mathcal{L}} \circ \mathcal{S}_{\mathcal{L}} = \varphi \circ (\mathcal{R}_{\varphi^{-1}(\mathcal{L})} \circ (\varphi^{-1} \circ \varphi) \circ \mathcal{R}_{\varphi^{-1}(\mathcal{L})}) \circ \varphi^{-1} = \iota.$$

(D: $\mathcal{S}_{\mathcal{L}}$ preserves betweenness.) $\mathcal{S}_{\mathcal{L}}$ is the composition of three mappings, one of which ($\mathcal{R}_{\varphi^{-1}(\mathcal{L})}$) is known to preserve betweenness because it is a mirror mapping; the other two mappings are φ and φ^{-1} which preserve betweenness by Theorem LC.37(B). Therefore $\mathcal{S}_{\mathcal{L}}$ preserves betweenness. □

Theorem LC.41. *Let $\mathbb{F}$ be either $\mathbb{A}$ or $\mathbb{R}$; then every induced mirror mapping $\mathcal{S}_{\mathcal{L}}$ on a plane $\mathcal{P}$ preserves distance.*

Proof. By Theorem LC.38(D) both φ and φ^{-1} preserve distance. By Theorem LC.27(B), $\mathcal{R}_{\varphi^{-1}(\mathcal{L})}$ preserves distance. Therefore $\mathcal{S}_{\mathcal{L}} = \varphi \circ \mathcal{R}_{\varphi^{-1}(\mathcal{L})} \circ \varphi^{-1}$, the composition of these three mappings, preserves distance. □

Theorem LC.42. *If $\mathbb{F} = \mathbb{A}$ or $\mathbb{F} = \mathbb{R}$, the set of all induced mirror mappings $\mathcal{S}_{\mathcal{L}}$ on $\mathcal{P}$ is a reflection set as in Definition NEUT.2.*

Proof. We show that the set of all induced mirror mappings $\mathcal{S}_{\mathcal{L}}$ on $\mathcal{P}$ satisfy Properties R.1 through R.6 of Definition NEUT.2.

R.1 and R.2 (Existence and uniqueness) For every line $\mathcal{L}$ in the plane $\mathcal{P}$, Definition LC.39 defines $\mathcal{S}_{\mathcal{L}}$ to be a single mapping over $\mathcal{L}$; this is shown to be a mirror mapping by Theorem LC.40.

R.3 (Closure) Every induced mirror mapping $\mathcal{S}_{\mathcal{L}}$ over a line $\mathcal{L}$ belongs to the set of all such mappings, whether or not it is a composition of other mappings.

R.4 (Linear scaling) By Theorem LC.41 every induced mirror mapping $\mathcal{S}_{\mathcal{L}}$, and thus every composition of such mappings, preserves distance. By the same argument as in Theorem LC.30, the linear scaling property holds.

R.5 (Angle reflection) Let $\angle ACB$ be any angle in the plane $\mathcal{P}$. By Theorem LC.37(C)(2) φ^{-1} is a bijection of $\mathcal{P}$ onto $\mathbb{F}^2$ which preserves rays, so

$$\varphi^{-1}(\angle ACB) = \angle(\varphi^{-1}(A))(\varphi^{-1}(C))(\varphi^{-1}(B))$$

is an angle in $\mathbb{F}^2$.

By Theorem LC.31, there exists a line $\mathcal{M}$ in $\mathbb{F}^2$ such that the mirror mapping $\mathcal{R}_{\mathcal{M}}$ is an angle reflection for $\angle(\varphi^{-1}(A))(\varphi^{-1}(C))(\varphi^{-1}(B))$. Then $\mathcal{R}_{\mathcal{M}}(\varphi^{-1}(A)) \in \overrightarrow{\varphi^{-1}(C)\varphi^{-1}(B)}$. By Theorem LC.37(C)(2),

$$\begin{aligned} \mathcal{S}_{\varphi(M)}(A) &= \varphi(\mathcal{R}_{\mathcal{M}}(\varphi^{-1}(A))) \\ &\in \varphi(\overrightarrow{\varphi^{-1}(C)\varphi^{-1}(B)}) = \overrightarrow{\varphi(\varphi^{-1}(C))\varphi(\varphi^{-1}(B))} = \overrightarrow{CB}. \end{aligned}$$

Therefore $\mathcal{S}_{\varphi(M)}$ is an angle reflection for $\angle ACB$.

R.6 (Existence of a midpoint) Let $\overline{AB}$ be any closed segment in $\mathcal{P}$. Then by Theorem LC.37(C), $\varphi^{-1}(\overline{AB}) = \overline{\varphi^{-1}(A)\varphi^{-1}(B)}$. By Theorem LC.32 there exists a midpoint, that is, a point $M \in \overline{\varphi^{-1}(A)\varphi^{-1}(B)}$ and a line $\mathcal{M}$ containing M such that the mirror mapping $\mathcal{R}_{\mathcal{M}}$ satisfies $\mathcal{R}_{\mathcal{M}}(\varphi^{-1}(A)) = \varphi^{-1}(B)$ and $\mathcal{R}_{\mathcal{M}}(M) = M$. We show that $\varphi(M)$ is a midpoint for $\overline{AB}$.

Since $\mathcal{S}_{\varphi(\mathcal{M})} = \varphi \circ \mathcal{R}_{\varphi^{-1}(\varphi(\mathcal{M}))} \circ \varphi^{-1} = \varphi \circ \mathcal{R}_{\mathcal{M}} \circ \varphi^{-1}$,

$$\mathcal{S}_{\varphi(\mathcal{M})}(A) = \varphi(\mathcal{R}_{\mathcal{M}}(\varphi^{-1}(A))) = \varphi(\varphi^{-1}(B)) = B.$$

Also,

$$\mathcal{S}_{\varphi(\mathcal{M})}(\varphi(M)) = \varphi(\mathcal{R}_{\mathcal{M}}(\varphi^{-1}(\varphi(M)))) = \varphi(\mathcal{R}_{\mathcal{M}}(M)) = \varphi(M).$$

By Theorem NEUT.15 (which we may use because $\mathcal{S}_{\varphi(\mathcal{M})}$ is a mirror mapping),

$$\mathcal{S}_{\varphi(\mathcal{M})}(\overline{\varphi(M)A}) = \overline{\mathcal{S}_{\varphi(\mathcal{M})}(\varphi(M))\mathcal{S}_{\varphi(\mathcal{M})}(A)} = \overline{\varphi(M)B}$$

showing that $\varphi(M)$ is a midpoint of $\overline{AB}$. □

21.5.7 *Least upper bound Axiom LUB is valid on a linear model*

The following theorem is the only one in this part that requires the field $\mathbb{F}$ to be the real numbers $\mathbb{R}$, not merely one that contains square roots of its non-negative members.

Theorem LC.43 (Axiom LUB in Model LM3R). *Axiom LUB holds for any plane in Model LM3R (based on $\mathbb{R}^3$). In other words, for any line $\mathcal{L} \subseteq \mathbb{R}^3$ which has been built into an ordered field, with origin O and unit U, every nonempty subset $\mathcal{E}$ of $\mathcal{L}$ which is bounded above has a least upper bound* lub $\mathcal{E}$.

Proof. By Definition LA.1(1), $\mathcal{L}$ is a line in $\mathbb{R}^3$ iff there exist distinct points A and B of $\mathbb{R}^3$ such that $\mathcal{L} = \{A+s(B-A) \mid s \in \mathbb{R}\}$. If we let $s = 0$, $X = A+0(B-A) = A$; if we let $s = 1$, $X = A+1(B-A) = B$. For any real numbers s and t, if $X = A+s(B-A)$ and $Y = A+t(B-A)$ define $X+Y = A+(s+t)(B-A)$ and $X \cdot Y = A+(st)(B-A)$.

With these definitions, $\mathcal{L}$ is a field, where A is the origin, and B is the unit. Moreover, the mapping $\varphi(s) = A + s(B - A)$ is an isomorphism from the set $\mathbb{R}$ of real numbers onto $\mathcal{L}$. Define $X < Y$ iff $s < t$. Then $\mathcal{L}$ is an ordered field, and the mapping φ is order-preserving.

Let $\mathcal{E}$ be a nonempty subset of $\mathcal{L}$ which is bounded above. That is to say, $\mathcal{E}$ has an upper bound $Y \in \mathcal{L}$. Equivalently, for some real number t, and every $X = A + s(B - A) \in \mathcal{E}, X = A + s(B - A) \leq Y = A + t(B - A)$, which is true iff $s \leq t$. Define

$$\mathcal{E}' = \{s \mid s \in \mathbb{R} \textit{ and } A + s(B - A) \in \mathcal{E}\}.$$

Then for every $s \in \mathcal{E}'$, $s \leq t$, so that $\mathcal{E}'$ is bounded above; we know that the LUB property holds for real numbers, so there exists a real number t_0 which is the least upper bound of $\mathcal{E}'$.

Claim. $A + t_0(B - A)$ is the least upper bound of $\mathcal{E}$.

(I) t_0 is an upper bound for $\mathcal{E}'$ means that for every $s \in \mathcal{E}'$, $s \leq t_0$, so $A + s(B - A) \leq A + t_0(B - A)$. Therefore $A + t_0(B - A)$ is an upper bound for $\mathcal{E}$.

(II) If $A + u(B - A)$ is any upper bound for $\mathcal{E}$, then for every $s \in \mathcal{E}'$, $A + s(B - A) \leq A + u(B - A)$ so that $s \leq u$. Now t_0 is the least upper bound of $\mathcal{E}'$, so that $t_0 \leq u$, that is $A + t_0(B - A) \leq A + u(B - A)$. Thus $A + t_0(B - A)$ is less than or equal to every upper bound for $\mathcal{E}$, so that $A + t_0(B - A)$ is the least upper bound of $\mathcal{E}$.

□

21.5.8 Axioms I.0–I.5, BET, PSA, REF, PS, and LUB are consistent

Theorem LC.44 (Summary showing consistency).

(A) *Axioms I.0, I.1, I.2, I.3, I.4, I.5, BET, PSA, REF, PS, and LUB are all true for Model LM3R, where space is* $\mathbb{R}^3$*; hence these axioms are consistent.*

(B) *Model LM3R is a Euclidean/LUB space.*

Proof. Theorems LC.2 through LC.7 show that the incidence Axioms I.0, I.1, I.2, I.3, I.4, and I.5 are all true on Model LM3R.

Theorem LC.14 shows that there exists a betweenness relation on $\mathbb{F}^3$ satisfying Properties B.0, B.1, B.2, and B.3 of Definition IB.1, so that Axiom BET is valid for Model LM3.

Theorem LC.20 shows that the Plane Separation Axiom PSA is true for Model LM3.

Theorem LC.33 shows that Axiom REF is true on the plane of Model LM2A and Model LM2R.

Theorem LC.42 shows that Axiom REF is true on any plane in $\mathbb{A}^3$ or in $\mathbb{R}^3$, and hence for Model LM3A and for Model LM3R.

Theorem LC.16 shows that the parallel Axiom PS is true for Model LM3.

Theorem LC.43 shows that Axiom LUB holds for any line $\mathcal{L}$ in $\mathbb{R}^3$ which has been built into an ordered field; thus Axiom LUB holds for Model LM3R.

Therefore, all the axioms are true in Model LM3R; they are consistent and Model LM3R is a Euclidean/LUB space. □

Remark LC.45. (A) In our main development (Chapters 1–20), planes and lines were initially undefined objects, whose properties were specified entirely by the axioms they obey. Since all our axioms hold for Model LM3R, every line and plane defined by Definition LA.1 is a line or plane as specified in the original development.

(B) Theorem LC.44 shows that all the axioms of our main development are true for Model LM3R (based on $\mathbb{R}^3$). Thus we can invoke any of the theorems from Chapters 1 through 20 for our space $\mathbb{R}^3$ and plane $\mathbb{R}^2$.

(C) In Theorem LC.13, we showed that the definitions of segments and rays given in Definition LA.1(3) are equivalent to their definitions as given in Definitions IB.3 and IB.4 of Chapter 4. The next theorems will show that the definitions of *c-perpendicular* and *c-midpoint* are equivalent, respectively, to those of *perpendicular* and *midpoint* from the main development.

Theorem LC.46 (C-perpendicular = perpendicular). *Suppose $\mathbb{F} = \mathbb{A}$ or $\mathbb{F} = \mathbb{R}$, and let $\mathcal{P}$ be any plane in $\mathbb{F}^3$. Then two lines in $\mathcal{P}$ are c-perpendicular iff they are perpendicular.*

Proof. By Remark LC.45 just above, we can use any results from the main development, including the results of Chapter 8 (NEUT). By Theorem LA.23 (or Theorem LB.10), if two lines $\mathcal{L}$ and $\mathcal{M}$ in the plane $\mathbb{F}^2$ are c-perpendicular they must intersect.

(A) By Theorem LB.17(E), a line $\mathcal{M}$ in $\mathbb{F}^2$ is c-perpendicular to $\mathcal{L}$ iff $\mathcal{M}$ is a fixed line for $\mathcal{R}_\mathcal{L}$. By Theorem NEUT.32 this is true iff $\mathcal{M}$ is perpendicular to $\mathcal{L}$.

(B) The general case: by Theorem LC.38(B) two lines $\mathcal{L}$ and $\mathcal{M}$ in a plane $\mathcal{P}$ are c-perpendicular iff $\varphi^{-1}(\mathcal{M})$ is c-perpendicular to $\varphi^{-1}(\mathcal{L})$. By part (A) this is true iff $\varphi^{-1}(\mathcal{M})$ is a fixed line for $\mathcal{R}_{\varphi^{-1}(\mathcal{L})}$, that is,

$$\mathcal{R}_{\varphi^{-1}(\mathcal{L})}(\varphi^{-1}(\mathcal{M})) \subseteq \varphi^{-1}(\mathcal{M}).$$

This is true iff

$$\mathcal{S}_\mathcal{L}(\mathcal{M}) = \varphi(\mathcal{R}_{\varphi^{-1}(\mathcal{L})}(\varphi^{-1}(\mathcal{M}))) \subseteq \varphi(\varphi^{-1}(\mathcal{M})) = \mathcal{M},$$

that is, $\mathcal{M}$ is a fixed line for $\mathcal{S}_\mathcal{L}$. By Theorem NEUT.32 this is equivalent to saying that $\mathcal{M}$ is perpendicular to $\mathcal{L}$. We can use this theorem since Theorems LC.40 and LC.42 show that $\mathcal{S}_\mathcal{L}$ is a reflection. □

Theorem LC.47 (C-midpoint = midpoint of a segment in $\mathbb{F}^3$). *Suppose $\mathbb{F} = \mathbb{A}$ or $\mathbb{F} = \mathbb{R}$, and let $\mathcal{P}$ be any plane in $\mathbb{F}^3$. If X and Y are distinct points in $\mathbb{F}^3$, then a point M is the c-midpoint of $\overline{XY}$ iff M is the midpoint of this segment.*

Proof. Again we can use the results of Chapter 8 (NEUT). Also, we know that the set of all mirror mappings $\mathcal{R}_\mathcal{L}$ over lines $\mathcal{L}$ in $\mathbb{F}^2$ is a reflection set, as is the set of all induced mappings $\mathcal{S}_\mathcal{L}$ over lines $\mathcal{L}$ in a plane $\mathcal{P}$.

(A) We first prove the theorem for two points of $\mathbb{F}^2$. If X and Y are distinct points on $\mathbb{F}^2$, then $M = \frac{X+Y}{2}$ is the c-midpoint of $\overline{XY}$, so by Theorem LC.32 M is a midpoint for this segment.

Conversely, if M is a midpoint of $\overline{XY}$, by Theorem NEUT.52 there exists a line $\mathcal{M}$ containing M such that $\mathcal{R}_\mathcal{M}(X) = Y$ and $\mathcal{R}_\mathcal{M}(M) = M$. Since all our reflections are defined by Definition LB.16, $\mathcal{M}$ is c-perpendicular to $\overleftrightarrow{XY}$, and by the same definition, M is the c-midpoint of $\overline{XY}$.

(B) In the general case, let X and Y be two points in $\mathbb{F}^3$, $\mathcal{L}$ a line containing these points, and let $\mathcal{P}$ be a plane containing $\mathcal{L}$. Let φ be a transfer mapping from $\mathbb{F}^2$ onto $\mathcal{P}$ defined by Definition LC.35.

We first show that if M is the c-midpoint of $\overline{XY}$, it is a midpoint. By Theorem LC.37(D), M is the c-midpoint of $\overline{XY}$ iff $\varphi^{-1}(M)$ is the c-midpoint of $\overline{\varphi^{-1}(X)\varphi^{-1}(Y)}$. By part (A) above, this is so iff $\varphi^{-1}(M)$ is a midpoint of $\overline{\varphi^{-1}(X)\varphi^{-1}(Y)}$. By Definition NEUT.3(C) and Theorem NEUT.52 this is true iff there exists a reflection mapping $\mathcal{R}_\mathcal{N}$ such that $\mathcal{R}_\mathcal{N}(\varphi^{-1}(M)) = \varphi^{-1}(M)$ and $\mathcal{R}_\mathcal{N}(\varphi^{-1}(X)) = \varphi^{-1}(Y)$. This is true iff

$$\mathcal{S}_{\varphi(\mathcal{N})}(M) = \varphi(\mathcal{R}_\mathcal{N}(\varphi^{-1}(M))) = \varphi(\varphi^{-1}(M)) = M$$

and

$$\mathcal{S}_{\varphi(\mathcal{N})}(X) = \varphi(\mathcal{R}_{\mathcal{N}}(\varphi^{-1}(X))) = \varphi(\varphi^{-1}(Y)) = Y,$$

so that by Theorem NEUT.15,

$$\mathcal{S}_{\varphi(\mathcal{N})}(\overleftrightarrow{XM}) = \overleftrightarrow{\mathcal{S}_{\varphi(\mathcal{N})}(X)\mathcal{S}_{\varphi(\mathcal{N})}(M)} = \overleftrightarrow{YM},$$

and M is a midpoint of $\overleftrightarrow{XY}$.

Conversely, suppose M is a midpoint of $\overleftrightarrow{XY}$; by Theorem LC.42 the set of all induced mappings $\mathcal{S}_{\mathcal{L}}$ on lines $\mathcal{L}$ of $\mathcal{P}$ is a reflection set, so by Theorem NEUT.52, there exists $\mathcal{S}_{\mathcal{L}}$, one of these reflections, such that $\mathcal{S}_{\mathcal{L}}(X) = Y$ and $\mathcal{S}_{\mathcal{L}}(M) = M$. By Definition LC.39,

$$\mathcal{S}_{\mathcal{L}} = \varphi \circ \mathcal{R}_{\varphi^{-1}(\mathcal{L})} \circ \varphi^{-1},$$

where $\mathcal{R}_{\varphi^{-1}(\mathcal{L})}$ is a reflection over the line $\varphi^{-1}(\mathcal{L})$ in $\mathbb{F}^2$. Then $\mathcal{R}_{\varphi^{-1}(\mathcal{L})} = \varphi^{-1} \circ \mathcal{S}_{\mathcal{L}} \circ \varphi$. Since $\mathcal{S}_{\mathcal{L}}(X) = Y$,

$$\mathcal{R}_{\varphi^{-1}(\mathcal{L})}(\varphi^{-1}(X)) = \varphi^{-1}(\mathcal{S}_{\mathcal{L}}(\varphi(\varphi^{-1}(X)))) = \varphi^{-1}(\mathcal{S}_{\mathcal{L}}(X)) = \varphi^{-1}(Y).$$

Also from $\mathcal{S}_{\mathcal{L}}(M) = M$,

$$\mathcal{R}_{\varphi^{-1}(\mathcal{L})}(\varphi^{-1}(M)) = \varphi^{-1}(\mathcal{S}_{\mathcal{L}}(\varphi(\varphi^{-1}(M)))) = \varphi^{-1}(\mathcal{S}_{\mathcal{L}}(M)) = \varphi^{-1}(M);$$

so that $\varphi^{-1}(M)$ is a midpoint of $\overleftrightarrow{\varphi^{-1}(X)\varphi^{-1}(Y)}$.

By part (A), $\varphi^{-1}(M)$ is the c-midpoint of this interval, and by Theorem LC.37(D) M is the c-midpoint of $\overleftrightarrow{XY}$. □

21.6 Independence of Axioms

In contrast to some other treatments of geometry, this one has taken the hard road of axiom *independence*. Much of the detailed and tedious work in our earlier chapters came about because of our pursuit of this goal.

We say that Axiom A is **independent** of a set $\mathcal{B}$ of axioms iff Axiom A cannot be logically deduced from any of the axioms in $\mathcal{B}$, or from any combination of them.

If Axiom A were a logical consequence of the set $\mathcal{B}$, it would be impossible to exhibit a model in which Axiom A is false, but all the axioms in $\mathcal{B}$ are true. Therefore, exhibiting such a model shows that Axiom A cannot be proved from the axioms in $\mathcal{B}$, that is, Axiom A is *independent* of the axioms in $\mathcal{B}$.

Ideally, we would like each axiom in our system to be independent of all the other axioms in the system. We might call this *strong* independence. But strong independence is too ambitious a goal.

It is probably not possible to construct a set of axioms equivalent to those on our list in which no axiom can be proved from any combination of the others.

This should not surprise us, for axioms that are added to the list often depend on previously given axioms and may contain language from which a previous axiom can be inferred. For example, we can't even *state* the Plane Separation Axiom (PSA) without the existence of a betweenness relation, because the definition of PSA uses segments and their definition depends on betweenness. The same goes for Axiom REF.

We settle, instead, for something less, called *sequential* independence, meaning that each axiom in a list is independent of all those preceding it on the list. To show this for any particular axiom, we need to find a model in which that axiom is false, but all its predecessors on the list are true.[4] In our case, we show that each of the Axioms I.0, I.1, I.2, I.3, I.4, I.5, BET, PSA, and LUB is independent of those preceding it on the list. Axiom REF is independent of Axioms I.0, I.1, I.5(A), BET, PSA, and LUB. Axiom PS is independent of all other axioms except possibly for Axiom REF; the resolution of that issue appears to belong to hyperbolic geometry, which is beyond the scope of this book.

Table of independence models

for Axioms I.0–I.5(C), BET, PSA, REF, PS, and LUB.

Subsection	Theorem(s)	Model	True	False
21.6.1	FM.2–.8	Various discrete	Each of I.0–I.5(C) is independent of the others	
21.6.2	FM.10	FM.1	I.0–I.5	BET
21.6.3	DZI.5–.8	DZI ($\mathbb{Z}^3$)	I.0–I.5,BET	PSA
21.6.4	MLT.3–.9	MLT	I.0,.1,.5(A)(B),BET,PSA,PS,LUB	REF
21.6.5	PSM.3–.5	PSM	I.0–.5,BET,PSA	PS
21.6.7	LE.1	LM3A	I.0–.5,BET,PSA,REF,PS	LUB

In a later section of the chapter, we give attention to the independence of the various *properties* within the definitions for betweenness, mirror mappings, and reflections. These results are perhaps less important than those showing the independence of the axioms.

[4]Sometimes a model constructed for such a purpose may strike the reader as quite strange, even bizarre; this should not be too surprising, given that we are asking it to have non-standard properties.

21.6.1 Incidence Axioms I.0– I.5 are independent (Model FM)

This subsection deals with axiom independence using finite models. We also include, as a by-product, an "extra" proof of the consistency of the incidence axioms, which has already been shown using linear models in Theorems LC.2 through LC.7. The models and theorems in this subsection will be named FM.*n* (suggesting "finite model").

Theorem FM.1. *The incidence axioms are consistent.*

Proof. For Model FM.1, let space $\mathcal{S}$ be the set of points $\{1, 2, 3, 4\}$; let the lines be the six doubletons, namely $\{1, 2\}$, $\{1, 3\}$, $\{1, 4\}$, $\{2, 3\}$, $\{2, 4\}$, $\{3, 4\}$, all of which are subsets of $\mathcal{S}$; and let the planes be the four triples, $\{1, 2, 3\}$, $\{1, 2, 4\}$, $\{1, 3, 4\}$, $\{2, 3, 4\}$, which are subsets of $\mathcal{S}$. Clearly Axiom I.0 is true.

Simple checking verifies that Axioms I.1, I.2, I.3, and I.5 are true. Since

$$\{1, 2, 3\} \cap \{1, 2, 4\} = \{1, 2\}, \{1, 2, 3\} \cap \{1, 3, 4\} = \{1, 3\},$$
$$\{1, 2, 3\} \cap \{2, 3, 4\} = \{2, 3\}, \{1, 2, 4\} \cap \{1, 3, 4\} = \{1, 4\},$$
$$\{1, 2, 4\} \cap \{2, 3, 4\} = \{2, 4\}, \text{ and } \{1, 3, 4\} \cap \{2, 3, 4\} = \{3, 4\},$$

then I.4 is true. □

Theorem FM.2. *Axiom I.0 is independent of the other incidence axioms.*

Proof. For Model FM.2, let space $\mathcal{S}$ consist of the set of points $\{1, 2, 3, 4\}$; then space does not contain 0; Definition I.0 says that space is the set of all points, so that 0 is not a point. Let the lines be $\{0, 1, 2\}, \{1, 3\}, \{1, 4\}, \{2, 3\}$, $\{2, 4\}$, and $\{3, 4\}$, and let the planes be $\{0, 1, 2, 3\}$, $\{0, 1, 2, 4\}$, $\{1, 3, 4\}$, and $\{2, 3, 4\}$. The first assertion of Axiom I.0 is false because there is a line $\{0, 1, 2\}$ which is not a subset of space. The second assertion of Axiom I.0 is false because there is a plane $\{0, 1, 2, 3\}$ that is not a subset of space.

It is easy to check that Axioms I.1, I.2, I.3, and I.5 are true. Since

$$\{0, 1, 2, 3\} \cap \{0, 1, 2, 4\} = \{0, 1, 2\}, \{0, 1, 2, 3\} \cap \{1, 3, 4\} = \{1, 3\},$$
$$\{0, 1, 2, 3\} \cap \{2, 3, 4\} = \{2, 3\}, \{0, 1, 2, 4\} \cap \{1, 3, 4\} = \{1, 4\},$$
$$\{0, 1, 2, 4\} \cap \{2, 3, 4\} = \{2, 4\}, \text{ and } \{1, 3, 4\} \cap \{2, 3, 4\} = \{3, 4\},$$

I.4 is true. □

Without Axiom I.0, there is nothing in Axioms I.1 through I.5 requiring that all the members of a line or a plane must be points. The proof just above puts a "non-point," namely 0, into a line and a plane but not into $\mathcal{S}$. Axioms I.1 through I.5 then apply to the points that are in the lines and planes of the model, but not to any non-points.

Theorem FM.3. *Axiom I.1 is independent of the other incidence axioms.*

Proof. (A) The following Model FM.3(A) shows that the uniqueness part of Axiom I.1 is independent of the other axioms and of the existence part of Axiom I.1. Let space $\mathcal{S}$ be the set of points $\{1, 2, 3, 4, 5, 6\}$; let the lines be $\{1, 2, 3\}$, $\{1, 2, 4\}$, $\{1, 3, 4\}$, $\{2, 3, 4\}$, $\{1, 5\}$, $\{1, 6\}$, $\{2, 5\}$, $\{2, 6\}$, $\{3, 5\}$, $\{3, 6\}$, $\{4, 5\}$, $\{4, 6\}$, and $\{5, 6\}$; and let the planes be $\{1, 2, 3, 4, 5\}$, $\{1, 2, 3, 4, 6\}$, $\{1, 5, 6\}$, $\{2, 5, 6\}$, $\{3, 5, 6\}$, and $\{4, 5, 6\}$. Then I.0 is clearly true.

Since $\{1, 2, 3\} \cap \{1, 2, 4\} = \{1, 2\}$, the points 1 and 2 belong to two different lines, so the uniqueness of I.1 fails; however, every pair of points is contained in a line, so the existence is true.

The triples of noncollinear points are: $\{1, 2, 5\}$, $\{1, 2, 6\}$, $\{1, 3, 5\}$, $\{1, 3, 6\}$, $\{1, 4, 5\}$, $\{1, 4, 6\}$, $\{1, 5, 6\}$, $\{2, 3, 5\}$, $\{2, 3, 6\}$, $\{2, 4, 5\}$, $\{2, 4, 6\}$, $\{2, 5, 6\}$, $\{3, 4, 5\}$, $\{3, 4, 6\}$, $\{3, 5, 6\}$, and $\{4, 5, 6\}$. Since each of these triples is a subset of one and only one plane, Axiom I.2 is true.

Each of the four lines $\{1, 2, 3\}$, $\{1, 2, 4\}$, $\{1, 3, 4\}$, and $\{2, 3, 4\}$ having three members has the following property: for each pair of points contained in the line, if the pair is contained in a plane, then the line is contained in that plane. Hence Axiom I.3 is true.

Since $\{1, 2, 3, 4, 5\} \cap \{1, 2, 3, 4, 6\} = \{1, 2, 3, 4\}$,
$\{1, 2, 3, 4, 5\} \cap \{1, 5, 6\} = \{1, 5\}$, $\{1, 2, 3, 4, 5\} \cap \{2, 5, 6\} = \{2, 5\}$,
$\{1, 2, 3, 4, 5\} \cap \{3, 5, 6\} = \{3, 5\}$, $\{1, 2, 3, 4, 5\} \cap \{4, 5, 6\} = \{4, 5\}$,
$\{1, 2, 3, 4, 6\} \cap \{1, 5, 6\} = \{1, 6\}$, $\{1, 2, 3, 4, 6\} \cap \{2, 5, 6\} = \{2, 6\}$,
$\{1, 2, 3, 4, 6\} \cap \{3, 5, 6\} = \{3, 6\}$, $\{1, 2, 3, 4, 6\} \cap \{4, 5, 6\} = \{4, 6\}$,
$\{1, 5, 6\} \cap \{2, 5, 6\} = \{5, 6\}$, $\{1, 5, 6\} \cap \{3, 5, 6\} = \{5, 6\}$,
$\{1, 5, 6\} \cap \{4, 5, 6\} = \{5, 6\}$, $\{2, 5, 6\} \cap \{3, 5, 6\} = \{5, 6\}$,
$\{2, 5, 6\} \cap \{4, 5, 6\} = \{5, 6\}$, and $\{3, 5, 6\} \cap \{4, 5, 6\} = \{5, 6\}$,
I.4 is true. Finally, I.5 is clearly true.

(B) The following Model FM.3(B) shows that the existence part of Axiom I.1 is independent of the other axioms. Let space $\mathcal{S}$ be the set of points $\{1, 2, 3, 4\}$; let the lines be $\{1, 2\}$ and $\{2, 3\}$; let the planes be the triples $\{1, 2, 3\}$, $\{1, 2, 4\}$, $\{1, 3, 4\}$, and $\{2, 3, 4\}$. Then I.0 is clearly true.

The existence part of Axiom I.1 is false, since there is no line containing both the points 1 and 3.

Every set of three noncollinear points is a plane, so Axiom I.2 is true.

Axiom I.3 is true because each line is a doubleton. In other words, the only pairs of points that belong to lines are $\{1, 2\}$ and $\{2, 3\}$; each of these pairs *is* the line to which it belongs; so if it is a subset of a plane, the line containing it is a subset of that plane.

Since there are only four points in space, and every plane contains three points, the intersection of any two planes must contain at least two points, so that Axiom I.4 is true. Alternatively, we can verify that the intersection of any two planes contains at least two points. Simple checks verify that Axiom I.5 is true. □

Theorem FM.4. *Axiom I.2 is independent of the other incidence axioms.*

Proof. (A) The following Model FM.4(A) shows that the uniqueness part of Axiom I.2 is independent of the other axioms and of the existence part of Axiom I.2. Let space $\mathcal{S}$ consist of the set of points $\{1, 2, 3, 4, 5\}$; let the lines be $\{1, 2\}$, $\{1, 3\}$, $\{1, 4\}$, $\{1, 5\}$, $\{2, 3\}$, $\{2, 4\}$, $\{2, 5\}$, $\{3, 4\}$, $\{3, 5\}$, $\{4, 5\}$; and let the planes be $\{1, 2, 3, 4\}$, $\{1, 2, 3, 5\}$, $\{1, 4, 5\}$, $\{2, 4, 5\}$, and $\{3, 4, 5\}$. Then Axiom I.0 is true because every point is in $\mathcal{S}$. Axiom I.1, Axiom I.3, and Axiom I.5(A) are true because each line is a doubleton. Uniqueness of Axiom I.2 is false since $\{1, 2, 3, 4\} \cap \{1, 2, 3, 5\} = \{1, 2, 3\}$, but existence is true because every triple is contained in some plane.

Since

$$\{1, 2, 3, 4\} \cap \{1, 2, 3, 5\} = \{1, 2, 3\}, \{1, 2, 3, 4\} \cap \{1, 4, 5\} = \{1, 4\},$$
$$\{1, 2, 3, 4\} \cap \{2, 4, 5\} = \{2, 4\}, \{1, 2, 3, 4\} \cap \{3, 4, 5\} = \{3, 4\},$$
$$\{1, 2, 3, 5\} \cap \{1, 4, 5\} = \{1, 5\}, \{1, 2, 3, 5\} \cap \{2, 4, 5\} = \{2, 5\}$$
$$\{1, 2, 3, 5\} \cap \{3, 4, 5\} = \{3, 5\}, \{1, 4, 5\} \cap \{2, 4, 5\} = \{4, 5\},$$
$$\{1, 4, 5\} \cap \{3, 4, 5\} = \{4, 5\}, \text{ and } \{2, 4, 5\} \cap \{3, 4, 5\} = \{4, 5\},$$

I.4 is true. Since every line has two points, I.5(A) is true; since every plane contains at least three points, I.5(B) is true; the points 1, 2, 4, and 5 are noncoplanar, so I.5(C) is true.

(B) The following Model FM.4(B) shows that the existence part of Axiom I.2 is independent of the other axioms. Let space $\mathcal{S}$ consist of the set of points $\{1, 2, 3, 4\}$; let the lines be $\{1, 2\}$, $\{1, 3\}$, $\{1, 4\}$, $\{2, 3\}$, $\{2, 4\}$, and $\{3, 4\}$; and let the planes be $\{1, 2, 3\}$, $\{1, 2, 4\}$, and $\{1, 3, 4\}$.

Then $\{2, 3, 4\}$ is a noncollinear set of three points which is not contained in a plane, so the existence part of Axiom I.2 is false.

Axiom I.0 is true because every point is in $\mathcal{S}$. Axioms I.1 and I.3 are true because each line is a doubleton. Axiom I.4 is true by verifying that the intersection of any two planes contains at least two points. Simple checks verify that Axiom I.5 is true. □

Theorem FM.5. *Axiom I.3 is independent of the other incidence axioms.*

Proof. For Model FM.5, let space $\mathcal{S}$ consist of the set of points $\{1,2,3,4,5\}$; let the lines be $\{1,2,3\}$, $\{1,4\}$, $\{1,5\}$, $\{2,4,5\}$, $\{3,4\}$, and $\{3,5\}$; and let the planes be $\{1,2,4,5\}$, $\{1,3,4\}$, $\{1,3,5\}$, and $\{2,3,4,5\}$. Then Axiom I.0 is trivially true.

The pairs in $\mathcal{S}$ are $\{1,2\}$, $\{1,3\}$, $\{1,4\}$, $\{1,5\}$, $\{2,3\}$, $\{2,4\}$, $\{2,5\}$, $\{3,4\}$, $\{3,5\}$, and $\{4,5\}$. Since each of these pairs is a subset of one and only one line, Axiom I.1 is true.

The noncollinear triples in $\mathcal{S}$ are $\{1,2,4\}$, $\{1,2,5\}$, $\{1,3,4\}$, $\{1,3,5\}$, $\{1,4,5\}$, $\{2,3,4\}$, $\{2,3,5\}$, and $\{3,4,5\}$. Since each of these triples is a subset of one and only one plane, Axiom I.2 is true, and every plane contains one of them, so Axiom I.5(B) is true.

Axiom I.3 is false since $\{1,2\} \subseteq \{1,2,4,5\}$, but $\{1,2,3\} \not\subseteq \{1,2,4,5\}$. Since

$$\{1,2,4,5\} \cap \{1,3,4\} = \{1,4\}, \{1,2,4,5\} \cap \{1,3,5\} = \{1,5\},$$
$$\{1,2,4,5\} \cap \{2,3,4,5\} = \{2,4,5\}, \{1,3,4\} \cap \{1,3,5\} = \{1,3\},$$
$$\{1,3,4\} \cap \{2,3,4,5\} = \{3,4\}, \text{ and } \{1,3,5\} \cap \{2,3,4,5\} = \{3,5\},$$

Axiom I.4 is true.

Axiom I.5(A) is true since all lines have at least two points; direct verification shows that every plane has at least three points that are not collinear, so Axiom I.5(B) is true; Axiom I.5(C) is true since $\{1,3,4,5\}$ is not a subset of any plane. □

Theorem FM.6. *Axiom I.4 is independent of the other incidence axioms.*

Proof. For Model FM.6 let space $\mathcal{S}$ consist of the set of points $\{1,2,3,4,5\}$; let the lines contained in $\mathcal{S}$ be $\{1,2\}$, $\{1,3\}$, $\{1,4\}$, $\{1,5\}$, $\{2,3\}$, $\{2,4\}$, $\{2,5\}$, $\{3,4\}$, $\{3,5\}$, and $\{4,5\}$, *i.e.* the doubletons in $\mathcal{S}$; and let the planes contained in $\mathcal{S}$ be $\{1,2,3\}$, $\{1,2,4\}$, $\{1,2,5\}$, $\{1,3,4\}$, $\{1,3,5\}$, $\{1,4,5\}$, and $\{2,3,4,5\}$. Then Axiom I.0 is trivially true. Axioms I.1 and I.3 are both true since each line is a pair of points.

The noncollinear triples in $\mathcal{S}$ are $\{1,2,3\}$, $\{1,2,4\}$, $\{1,2,5\}$, $\{1,3,4\}$, $\{1,3,5\}$, $\{1,4,5\}$, $\{2,3,4\}$, $\{2,3,5\}$, $\{2,4,5\}$, and $\{3,4,5\}$. Since each of these triples is a subset of one and only one plane, Axiom I.2 is true.

Axiom I.4 is false, however, since $\{1,2,5\} \cap \{1,3,4\} = \{1\}$.

Axiom I.5(A) is true because every line has two points; Axiom I.5(B) is true because every plane has at least three points which are thus noncollinear; Axiom I.5(C) is true since $\{1, 2, 3, 4\}$ is not a subset of any plane. □

Theorem FM.7. *Each of the Axioms I.5(A), I.5(B), and I.5(C) is independent of the other incidence axioms.*

Proof. (A) For Model FM.7(A), let space $\mathcal{S}$ be the set of points $\{1, 2, 3, 4\}$; let the lines be the sets $\{1\}$, $\{1, 2\}$, $\{1, 3\}$, $\{1, 4\}$, $\{2, 3\}$, $\{2, 4\}$, and $\{3, 4\}$; and let the planes be the four triples $\{1, 2, 3\}$, $\{1, 2, 4\}$, $\{1, 3, 4\}$, and $\{2, 3, 4\}$.

A simple check shows that Axioms I.0, I.1, I.2, I.3, and I.4 are true. Axiom I.5(A) is false because the line $\{1\}$ contains only one point; Axiom I.5(B) is true because the only sets of three noncollinear points are planes; Axiom I.5(C) is true because there is no plane containing all points of $\{1, 2, 3, 4\}$, so this set is noncoplanar.

(B) For Model FM.7(B), let space $\mathcal{S}$ be the set of points $\{1, 2, 3, 4\}$; let the lines be the sets $\{1, 2\}$, $\{1, 3\}$, $\{1, 4\}$, and $\{2, 3, 4\}$; and let the planes be the four triples $\{1, 2, 3\}$, $\{1, 2, 4\}$, $\{1, 3, 4\}$, and $\{2, 3, 4\}$.

A simple check shows that Axioms I.0, I.1, and I.2 are true. If a pair of points is one of $\{1, 2\}$, $\{1, 3\}$, or $\{1, 4\}$, the line containing the pair is the same set and hence is a subset of any plane containing the pair. If a pair of points is one of $\{2, 3\}$, $\{2, 4\}$, or $\{3, 4\}$, the only line containing it is $\{2, 3, 4\}$, and the only plane containing it is $\{2, 3, 4\}$. Therefore Axiom I.3 is true.

Axiom I.4 is true since the intersection of any two planes contains two points. Axiom I.5(A) is true since every line is a pair or a triple; Axiom I.5(B) is false, since the plane $\{2, 3, 4\}$ is also a line, hence all its points are collinear; and Axiom I.5(C) is true since every plane is a triple, so that the points in $\mathcal{S}$ are noncoplanar.

(C) For Model FM.7(C), let space $\mathcal{S}$ consist of the set of points $\{1, 2, 3\}$, let lines be $\{1, 2\}$, $\{1, 3\}$, and $\{2, 3\}$, and let the one and only one plane be $\{1, 2, 3\}$ or all space; then Axioms I.0, I.1, I.2, and I.3 are easily seen to be true.

Axiom I.4 is vacuously true. Axiom I.5(A) is true; Axiom I.5(B) is true since there is only one plane and it consists of a noncollinear set of three points. Axiom I.5(C) is false because $\mathcal{S}$ is a plane. □

Remark FM.8. If we let space $\mathcal{S}$ consist of a single point 1, and let lines and planes be the same as $\mathcal{S}$ (so there is a single line and a single plane), then Axiom I.0 is obviously true, I.1, I.2, I.3, and I.4 are vacuously true, and all the assertions of Axiom I.5 are false.

Remark FM.9. Theorems FM.2 through FM.7 show that the incidence axioms are strongly independent, hence sequentially independent.

21.6.2 Betweenness Axiom BET is independent of Axioms I.0–I.5 (Model FM)

Theorem FM.10. *Axiom BET is independent of the incidence axioms, so that by Remark FM.9(B), Axioms I.0, I.1, I.2, I.3, I.4, I.5, and BET are sequentially independent.*

Proof. Here we use Model FM.1, where $\mathcal{S} = \{1, 2, 3, 4\}$ (as in Theorem FM.1), with the same definitions of lines and planes. That is, the lines are the sets $\{1, 2\}$, $\{1, 3\}$, $\{1, 4\}$, $\{2, 3\}$, $\{2, 4\}$, and $\{3, 4\}$, planes are the sets $\{1, 2, 3\}$, $\{1, 2, 4\}$, $\{1, 3, 4\}$, and $\{2, 3, 4\}$. By Theorem FM.1, Axioms I.0 through I.5 are true. We show that there cannot exist a betweenness relation (which has Properties B.0 through B.3) on this model.

Let us designate the members of $\mathcal{S}$ by letters, so that $\mathcal{S} = \{A, B, C, D\} = \{1, 2, 3, 4\}$. Assume there is a betweenness relation on $\mathcal{S}$ which contains at least one triple (A, B, C), that is A–B–C. By Axiom B.0, A, B, and C are collinear and distinct. By definition of the model, the lines are sets having two points, that is, doubletons, so that $\{A, B, C\}$ cannot be collinear; this contradicts the existence of a betweenness relation. □

21.6.3 Plane Separation Axiom PSA is independent of Axioms I.0–I.5 and BET (Model DZI)

To establish the independence of the Plane Separation Axiom (PSA) from each of the incidence, parallel, and betweenness axioms, we develop a discrete Model DZI. We shall use the acronym DZI in this subsection, as well as for the single theorem to be proved later in Subsection 21.8.1.

Whenever we refer to Model LM3Q in this subsection, it will be understood that space is $\mathbb{Q}^3$, where $\mathbb{Q}$ is the field of rational numbers.

Definition DZI.1. (1) Space $\mathcal{S}$ for Model DZI is $\mathbb{Z}^3$, the set of ordered triples of integers.[5] We denote members of $\mathbb{Z}^3$ by capital letters, and their coordinates by subscripted lowercase letters. For example, $X = (x_1, x_2, x_3)$, where x_1, x_2, and x_3 are integers.

(2) A nonempty subset $\mathcal{P}$ of $\mathbb{Z}^3$ is a **plane** for Model DZI iff for some plane $\mathcal{V}$ in $\mathbb{Q}^3$, $\mathcal{P} = \mathcal{V} \cap \mathbb{Z}^3$. (Here $\mathcal{V}$ is as in Definition LA.1(2).) Since the definition requires that $\mathcal{P}$ be nonempty, there must exist at least one point $A \in \mathcal{V}$ such that all the coordinates a_1, a_2, and a_3 are integers.

(3) A nonempty subset $\mathcal{L}$ of $\mathbb{Z}^3$ is a **line** for Model DZI iff for some line $\mathcal{M}$ in $\mathbb{Q}^3$, $\mathcal{L} = \mathcal{M} \cap \mathbb{Z}^3$. (Here $\mathcal{M}$ is as in Definition LA.1(1).) Since the definition requires that $\mathcal{L}$ be nonempty, there must exist at least one point $A \in \mathcal{M}$ such that all the coordinates a_1, a_2, and a_3 are integers, so that $A \in \mathcal{L}$.

Theorem DZI.2. *Let A be a member of $\mathbb{Z}^3$, and let B be any member of $\mathbb{Q}^3$, and let $\mathcal{M} = \overleftrightarrow{AB}$, a line in $\mathbb{Q}^3$. Then*

(A) *at least one member C of $\mathcal{M} \setminus \{A\}$ belongs to $\mathbb{Z}^3$, and hence to $\mathcal{L} = \mathcal{M} \cap \mathbb{Z}^3$;*
(B) *infinitely many members of $\mathcal{M}$ belong to $\mathcal{L} = \mathcal{M} \cap \mathbb{Z}^3$.*

Proof. By Definition LA.1(1),
$$\mathcal{M} \setminus \{A\} = \{A + t(B - A) \mid t \in \mathbb{Q} \text{ and } t \neq 0\}.$$
Since $B-A$ is a member of $\mathbb{Q}^3 \setminus \{(0,0,0)\}$, each of its coordinates $b_1 - a_1$, $b_2 - a_2$, and $b_3 - a_3$ is a rational number which can be expressed as the quotient of two integers whose greatest common divisor is 1. If we let t be the least common multiple of the denominators of the coordinates of $B - A$, then each of the coordinates of $t(B - A)$ is an integer and $t(B - A) \in \mathbb{Z}^3$. Then since $A \in \mathbb{Z}^3$, the point $C = A + t(B - A) \in \mathbb{Z}^3$, and since $t \neq 0$, $C \neq A$, proving part (A). Part (B) follows immediately from the observation that for any integer k, the point $A + kt(B - A) \in \mathbb{Z}^3$. □

Thus, if $\mathcal{M}$ is any line in $\mathbb{Q}^3$, $\mathcal{M} \cap \mathbb{Q}^3$ is a line in $\mathbb{Z}^3$ iff it contains at least one point of $\mathbb{Z}^3$.

Definition DZI.3. Let A, B, and C be distinct members of $\mathbb{Z}^3$. Then B is **between** A and C (that is, A–B–C) iff there exist distinct members P and Q of $\mathbb{Z}$ and integers a, b, and c such that $A = P + a(Q - P)$, $B = P + b(Q - P)$, $C = P + c(Q - P)$, and either $a < b < c$, or $c < b < a$.

---

[5] "Z" is the first letter of the German word "Zahl" for number.

Theorem DZI.4. *Let A be a member of $\mathbb{Z}^3$, and let B and C be any members of $\mathbb{Q}^3$ such that A, B, and C are noncollinear, so that $\mathcal{V} = \overleftrightarrow{ABC}$ is a plane in $\mathbb{Q}^3$. Let $\mathcal{P} = \mathcal{V} \cap \mathbb{Z}^3$. Then for any point $X \in \mathcal{V}$ there exists some integer $u \neq 0$ such that $A + u(X - A) \in \mathcal{P} \setminus \{A\}$; moreover, for every integer k, $A + ku(X - A) \in \mathcal{P}$.*

Proof. Since $X \in \mathcal{V}$, by Definition LA.1(2), there exist rational numbers s and t such that $X = A + s(B - A) + t(C - A)$; then by the proof of Theorem DZI.2, there exists an integer $u \neq 0$ such that $A + u(X - A) \neq A$ has integer coordinates, and for any integer k, $A + ku(X - A)$ has integer coordinates and is a member of $\mathbb{Z}^3$. Then

$$\begin{aligned} A + ku(X - A) &= A + ku(A + s(B - A) + t(C - A) - A) \\ &= A + kus(B - A) + kut(C - A) \in \mathcal{P}, \end{aligned}$$

since by Definition LA.1(2), $A + ku(X - A) \in \mathcal{V}$. □

Theorem DZI.5. *Each of the Axioms I.0, I.1, I.2, I.3, I.4, I.5, and BET is true for Model DZI.*

Proof. (1) Since lines and planes for Model DZI are subsets of $\mathbb{Z}^3$, Axiom I.0 is true for DZI.

(2) If A and B are distinct members of $\mathbb{Z}^3$, by Theorem LC.2 (or Theorem LA.10(A)) there is exactly one line $\overleftrightarrow{AB}$ in $\mathbb{Q}^3$ containing both A and B. Since every line in $\mathbb{Z}^3$ is the intersection of a line in $\mathbb{Q}^3$ with $\mathbb{Z}^3$, there is exactly one line in $\mathbb{Z}^3$ containing both A and B. So Axiom I.1 is true for Model DZI.

(3) Let A, B, and C be noncollinear members of $\mathbb{Z}^3$; by Theorem LC.3 (or Theorem LA.10(B)) there exists a unique plane $\overleftrightarrow{ABC}$ in $\mathbb{Q}^3$ through these points. Since every plane in $\mathbb{Z}^3$ is the intersection of a plane in $\mathbb{Q}^3$ with $\mathbb{Z}^3$, there is exactly one plane in $\mathbb{Z}^3$ containing A, B, and C. So Axiom I.2 is true for Model DZI.

(4) Let A and B be any points of $\mathbb{Z}^3$; let $\mathcal{P}$ be any plane for Model DZI containing A and B. Then by Definition DZI.1(2) there exists a plane $\mathcal{V}$ in $\mathbb{Q}^3$ such that both A and B are members of $\mathcal{P} = \mathcal{V} \cap \mathbb{Z}^3$. By Theorem LC.3, $\overleftrightarrow{AB} \subseteq \mathcal{V}$, so that $\overleftrightarrow{AB} \cap \mathbb{Z}^3 \subseteq \mathcal{V} \cap \mathbb{Z}^3$. By Definition DZI.1(3), $\overleftrightarrow{AB} \cap \mathbb{Z}^3$ is the unique line in $\mathbb{Z}^3$ containing both A and B, and this is a subset of $\mathcal{P}$. Thus Axiom I.3 is true for Model DZI.

(5) Let $\mathcal{P}_1$ and $\mathcal{P}_2$ be planes for Model DZI, which have the point A in common. Let $\mathcal{V}_1$ and $\mathcal{V}_2$ be the planes for Model LM3Q such that $\mathbb{Z}^3 \cap \mathcal{V}_1 = \mathcal{P}_1$ and $\mathbb{Z}^3 \cap \mathcal{V}_2 = \mathcal{P}_2$. By Theorem LC.7, Axiom I.4 holds for Model LM3Q, so there exists a member B of $\mathcal{V}_1 \cap \mathcal{V}_2$ such that $B \neq A$. Then by part (4) above, $\overleftrightarrow{AB} \subseteq \mathcal{V}_1 \cap \mathcal{V}_2$.

By Theorem DZI.2 there exists a member C of $\overleftrightarrow{AB}$ such that $C \in \mathbb{Z}^3 \setminus \{A\}$; therefore

$$C \in (\mathcal{V}_1 \cap \mathcal{V}_2) \cap \mathbb{Z}^3 = (\mathcal{V}_1 \cap \mathbb{Z}^3) \cap (\mathcal{V}_2 \cap \mathbb{Z}^3) = \mathcal{P}_1 \cap \mathcal{P}_2,$$

and $C \neq A$. Therefore Axiom I.4 is true for Model DZI.

(6) By Definition DZI.1(3) $\mathcal{L}$ is a line in $\mathbb{Z}^3$ iff for some line $\mathcal{M}$ in $\mathbb{Q}^3$, $\mathcal{L} = \mathcal{M} \cap \mathbb{Z}^3$; $\mathcal{L}$ is nonempty so it must contain at least one point A; by Theorem DZI.2, there is at least one other point in $\mathcal{L}$, so that Axiom I.5(A) is true.

By Definition DZI.1(2) $\mathcal{P}$ is a plane in $\mathbb{Z}^3$ iff for some plane $\mathcal{V}$ in $\mathbb{Q}^3$, $\mathcal{P} = \mathcal{V} \cap \mathbb{Z}^3$; $\mathcal{P}$ is nonempty so it must contain at least one point A; by Definition LA.1(2) there exist two points B and C in $\mathcal{V}$ such that A, B, and C are noncollinear. By Theorem DZI.4 there exist integers $u \neq 0$ and $v \neq 0$ such that $D = A + u(B - A)$ and $E = A + v(C - A)$ are members of $\mathcal{P}$, and $D \neq A$ and $E \neq A$.

Claim: A, D, and E are noncollinear. If $E \in \overleftrightarrow{AD}$, then for some integer t,

$$E = A + t(D - A) = A + t(A + u(B - A) - A) = A + tu(B - A)$$

so that $E \in \overleftrightarrow{AB}$; but then $E = A + v(C - A) = A + tu(B - A)$ and hence $v(C - A) = tu(B - A)$ which is true only if $v = tu = 0$, since $B - A$ and $C - A$ are linearly independent by Theorem LA.3. This implies that $E = A$ which contradicts Theorem DZI.4 which says that $E \neq A$. Thus A, D, and E are noncollinear and Axiom I.5(B) is true.

Finally, the points $(0, 0, 0)$, $(1, 0, 0)$, $(0, 1, 0)$, and $(0, 0, 1)$ are noncoplanar showing that Axiom I.5(C) is true.

(7) The proof of Theorem LC.14 holds for Model DZI, where the properties of betweenness for integers as listed in Chapter 1 Section 1.5 under the title "Number systems" are substituted for the same properties for the field $\mathbb{F}$. □

Theorem DZI.6. *The parallel axiom PS is true for Model DZI.*

Proof. Let $\mathcal{L}$ be a line in $\mathbb{Z}^3$ for Model DZI and let H be a member of $\mathbb{Z}^3 \setminus \mathcal{L}$. By Definition DZI.1(3) there exists a line $\mathcal{M}$ for Model LM3Q in $\mathbb{Q}^3$ such that $\mathcal{L} = \mathcal{M} \cap \mathbb{Z}^3$. Since Axiom PS holds for Model LM3Q, there exists a unique line $\mathcal{N} \subseteq \mathbb{Q}^3$ through H which is parallel to $\mathcal{M}$. Let $\mathcal{J} = \mathbb{Z}^3 \cap \mathcal{N}$. By Definition IP.0(B), $\mathcal{N} \cap \mathcal{M} = \emptyset$ and $\mathcal{N}$ and $\mathcal{M}$ are coplanar. By elementary set theory

$$\mathcal{L} \cap \mathcal{J} = (\mathcal{M} \cap \mathbb{Z}^3) \cap (\mathcal{N} \cap \mathbb{Z}^3) = \emptyset$$

and $\mathcal{L}$ and $\mathcal{J}$ are coplanar. So $\mathcal{L} \parallel \mathcal{J}$. Since $\mathcal{N}$ is unique, $\mathcal{J}$ is unique. □

Definition DZI.7. Let $\mathcal{P}$ be a plane in $\mathbb{Z}^3$, that is, for Model DZI, and let $\mathcal{L}$ be a line in $\mathcal{P}$, and let $\mathcal{M}$ be the line for Model LM3Q such that $\mathcal{L} = \mathcal{M} \cap \mathbb{Z}^3$. Then

(A) a set $\mathcal{E}$ is a **side** of $\mathcal{L}$ (contained in $\mathbb{Z}^3$) iff $\mathcal{E} = \mathcal{F} \cap \mathbb{Z}^3$ and $\mathcal{F}$ is a side of $\mathcal{M}$ (in $\mathbb{Q}^3$);

(B) two sets $\mathcal{E}_1$ and $\mathcal{E}_2$ are **opposite sides** of $\mathcal{L}$ iff $\mathcal{F}_1$ and $\mathcal{F}_2$ are opposite sides of $\mathcal{M}$ (in $\mathbb{Q}^3$), $\mathcal{E}_1 = \mathcal{F}_1 \cap \mathbb{Z}^3$, and $\mathcal{E}_2 = \mathcal{F}_2 \cap \mathbb{Z}^3$.

Theorem DZI.8. *The Plane Separation Axiom PSA is false for Model DZI.*

Proof. Let $\mathcal{L}$ be the line for Model DZI through $(0,0,0)$ and $(1,0,0)$. The points $(0,1,0)$ and $(0,-1,0)$ are on opposite sides of $\mathcal{L}$ since

$$\overline{(0,1,0)(0,-1,0)} \cap \mathcal{L} = \{(0,0,0)\}.$$

The point $(1,-1,0)$ is on the $(0,-1,0)$-side of $\mathcal{L}$ since

$$\overline{(1,-1,0)(0,-1,0)} \cap \mathcal{L} = \emptyset.$$

Therefore $(0,1,0)$ and $(1,-1,0)$ are on opposite sides of $\mathcal{L}$. Let $\mathcal{M}$ be the line for Model LM3Q such that $\mathcal{L} = \mathcal{M} \cap \mathbb{Z}^3$; let $\mathcal{N} = \overleftrightarrow{(0,1,0)(1,-1,0)}$, a line in Model LM3Q, and let $\mathcal{J} = \mathcal{N} \cap \mathbb{Z}^3$. Then $\mathcal{M} \cap \mathcal{N} = \{(\frac{1}{2},0,0)\} \notin \mathbb{Z}^3$, so that $\mathcal{L} \cap \mathcal{J} = \emptyset$. Thus the segment (in Model DZI) $\overline{(0,1,0)(1,-1,0)}$ is empty, and does not intersect $\mathcal{L}$; therefore Axiom PSA is false for Model DZI. □

Remark DZI.9. (A) Theorems DZI.5, DZI.6, and DZI.8 show that Axiom PSA is independent of each of the following axioms: I.0, I.1, I.2, I.3, I.4, I.5, BET, and PS. This is a stronger result than is needed to show that Axiom PSA is independent of all preceding axioms.

(B) One additional result based on Model DZI, Theorem DZI.10, is included later in Subsection 21.8.1; we put it there because it is relevant to the larger discussion in that section showing that the incidence and betweenness axioms are not adequate for developing a satisfactory geometry.

21.6.4 Axiom REF is independent of Axioms I.0, I.1, I.5(A), BET, PSA, PS, and LUB (Model MLT)

In Theorem RSI.3 of Subsection 21.7.3, we will prove that the set of LB.16 reflections on Model LM2Q ($\mathbb{Q}^2$) fails to satisfy Property R.5; that this set satisfies all the other reflection properties was established in Subsection 21.5.5 Theorems LC.25 through LC.33. One might be tempted to think that this establishes the independence of Axiom REF, but that is not so; it does not rule out the possibility that there could exist another set of mirror mappings on $\mathbb{Q}^2$ which would satisfy all of Properties R.1 through R.6, in which case Axiom REF would hold.

The result we offer with respect to the independence of Axiom REF is the following, in which we display the **Moulton plane**, invented in 1902 by the American astronomer Forest Ray Moulton.[6] Points in the Moulton plane are the points in the coordinate plane $\mathbb{R}^2$, and lines are ordinary lines of this plane, with the exception that lines with a negative slope "break" where they pass from the left side to the right side of the y-axis, their slope on the right side being double that on the left.

We will show that in such a plane, Axioms I.1, I.5(A), I.5(B), BET, PSA, PS, and LUB are all true, but Axiom REF is false, thus proving that Axiom REF is independent of this set of axioms. Axioms I.0, I.2, I.3, I.4, and I.5(C) deal with space; since we are dealing with a model for a plane we do not consider them. It could be an interesting challenge to construct a space $\mathcal{S}$ in which the Moulton plane could be embedded, and in which all axioms other than Axiom REF hold. We have not pursued this possibility.

Definition MLT.1. For Model MLT (the **Moulton plane**), the **plane** is the Euclidean plane $\mathbb{R}^2$, that is, the vector space consisting of all ordered pairs (x, y) of real numbers.

(A) A **line** for Model MLT is a set of one of the following types:

Type V (vertical line): $\{(c, y) \mid y \in \mathbb{R}\}$, where c is some real number;

Type H (horizontal line): $\{(x, d) \mid x \in \mathbb{R}\}$ where d is some real number;

Type P (positive slope line): $\{(x, y) \mid y = ax + b \textit{ and } x \in \mathbb{R}\}$, where $a > 0$ and b are real numbers; or

Type N (negative slope broken[7] line):

$$\{(x, y) \mid y = ax + b \text{ and } x \leq 0\} \cup \{(x, y) \mid y = 2ax + b \text{ and } x \geq 0\};$$

where a and b are real numbers and $a < 0$.

---

[6]Moulton, Forest Ray (1902), *A Simple Non-Desarguesian Plane Geometry* [16], Transactions of the American Mathematical Society (Providence, R.I.: American Mathematical Society) 3 (2): 192-Ű195, ISSN 0002-9947, JSTOR 1986419. The "non-Desarguesian" property is related to, but different from the nonplanar "Proposition of Desargues" which we proved in Chapter 1 as Theorem I.10.

[7]Here we take "broken" to mean that the line is continuous but its slope is discontinuous at the point where it crosses the y-axis

(B) We will need, on occasion, to distinguish lines and segments in the the coordinate plane Model LM2R from those in Model MLT, and we will do this by the following notational conventions: lines and segments in Model LM2R will be designated as $\mathcal{L}_c$, $\overleftrightarrow{AB}_c$, $\overset{[\,]}{AB}_c$, $\overset{]\,[}{AB}_c$, and $\overset{[\,[}{AB}_c$, *etc.* Lines and segments in Model MLT will be designated $\mathcal{L}_m$, $\overleftrightarrow{AB}_m$, $\overset{[\,]}{AB}_m$, $\overset{]\,[}{AB}_m$, and $\overset{[\,[}{AB}_m$, *etc.*

(C) If $\mathcal{L}$ is a line in either model which intersects the y-axis, then $\{(x,y) \in \mathcal{L} \mid x \leq 0\}$ will be called its **left-hand ray** or simply the **left ray** (in symbols, $lr(\mathcal{L})$), and $\{(x,y) \in \mathcal{L} \mid x \geq 0\}$ its **right-hand ray** or simply the **right ray** (in symbols, $rr(\mathcal{L})$). Note that both rays include the point of intersection with the y-axis. For a type N line, which is the union of two LM2R rays, the **slope of the left (right) ray** will mean the slope of the Model LM2R line containing that ray. The slope of a line $\mathcal{L}$ or ray $\overrightarrow{XY}$ will be denoted by the symbol $sl(\mathcal{L})$ or $sl(\overrightarrow{XY})$.

The terminology of right and left ray will often be used for lines of type N, but will also apply to lines of type P and type H, where the two rays have the same slope. It should also be observed that for any line $\mathcal{L}$ with slope, that is, of type P, H, or N, $sl(lr(\mathcal{L})) \geq sl(rr(\mathcal{L}))$—equality occurs if $\mathcal{L}$ is of type P or type H.

(D) If a line $\mathcal{L}$ in $\mathbb{R}^2$ is not vertical, for every point $C = (c_1, c_2) \in \mathbb{R}^2$ there exists a point $A = (a_1, a_2) \in \mathcal{L}$ with $a_1 = c_1$; if $a_2 < c_2$, we say that C lies **above** $\mathcal{L}$ (and **above** A); if $a_2 > c_2$, C lies **below** $\mathcal{L}$ (and **below** A).

If a line $\mathcal{L}$ in $\mathbb{R}^2$ is vertical, then for every point $C = (c_1, c_2) \in \mathbb{R}^2$ there exists a point $A = (a_1, a_2) \in \mathcal{L}$ with $a_2 = c_2$; if $c_1 > a_1$ we say that C lies to the **right** of $\mathcal{L}$ (to the **right** of A); if $c_1 < a_1$, C lies to the **left** of $\mathcal{L}$ (to the **left** of A). In either case, the line $\overleftrightarrow{AC}_c$ is a horizontal line.

(E) We shall refer to $\{(0,y) \mid y \in \mathbb{R}\}$ as the **y-axis** of the plane, and to $\{(x,0) \mid x \in \mathbb{R}\}$ as the **x-axis** of the plane.

(F) We define an order relation on any line in Model MLT as follows:

(1) If $\mathcal{L}$ is a vertical line, and $P = (a,b)$ and $Q = (a,c)$ are points on $\mathcal{L}$, $P < Q$ iff $b < c$.

(2) If $\mathcal{L}$ is of type H or type P, it is a line in Model LM2R; it intersects the y-axis at some point O, which we take as the origin. Let U be the point a distance 1 (along the line) to the right of O; this will be the unit. For each real number r define $\alpha(r) = rU$. For any two points P and Q on $\mathcal{L}$, define $P < Q$ iff $\alpha^{-1}(P) < \alpha^{-1}(Q)$. This definition yields an ordering "$<$" on $\mathcal{L}$ according to Definition ORD.1.

(3) If $\mathcal{L}_m$ is a line of type N in Model MLT, let $\mathcal{M}_c$ be the line (in Model LM2R) containing the left-hand ray of $\mathcal{L}_m$ and $\mathcal{M}'_c$ the Model LM2R line containing the right-hand ray. Then $\mathcal{M}_c \cap \mathcal{M}'_c = \{O\}$ where O is a point on the y-axis. Let O be the origin for both $\mathcal{M}_c$ and $\mathcal{M}'_c$, and let U be the unit for $\mathcal{M}_c$, U' the unit for $\mathcal{M}'_c$, and $V = -U$. The left ray of $\mathcal{L}_m$ is $\overrightarrow{OV}$, and the right ray of $\mathcal{L}_m$ is $\overrightarrow{OU'}$.

Let α be the mapping (as defined in (2)) from $\mathbb{R}$ onto $\mathcal{M}_c$ and α' the corresponding mapping from $\mathbb{R}$ onto $\mathcal{M}'_c$. Define β as follows: for any real number r, if $r < 0$ define $\beta(r) = \alpha(r)$; if $r > 0$ define $\beta(r) = \alpha'(r)$; and define $\beta(0) = O$. Define $P < Q$ iff $\beta^{-1}(P) < \beta^{-1}(Q)$. Then "<" orders the line $\mathcal{M}$ according to Definition ORD.1. (The proof of this is Exercise MLT.6.)

Note that if P is on the left ray and Q is on the right, then $P < Q$.

Remark MLT.1.1. (A) By Exercise MLT.4, every line in coordinate space Model LM2R or in Model MLT that is not vertical intersects every vertical line.

(B) From Subsections 21.5.1, 21.5.2, 21.5.3, and 21.5.4 we know that in the coordinate plane the incidence, betweenness, parallel, plane separation, and LUB axioms hold. We now show that all these axioms (which can be stated for a plane) hold in the Moulton plane.

Theorem MLT.2. *In the Moulton plane, let $\mathcal{L}$ and $\mathcal{M}$ be distinct lines having negative slope. Then the left ray of $\mathcal{L}$ has the same slope as the left ray of $\mathcal{M}$ iff both right rays have the same slope iff $\mathcal{L} \parallel \mathcal{M}$.*

Proof. The first equivalence is obvious, since the slope of the right ray is twice that of the left ray.

Suppose now the left rays have the same slope and the right rays have the same slope. Distinct lines on the coordinate plane are parallel iff they have the same slope, so that neither the left rays nor the right rays of $\mathcal{L}$ and $\mathcal{M}$ can intersect, and hence $\mathcal{L} \parallel \mathcal{M}$.

To prove the converses, assume that $\mathcal{L} \parallel \mathcal{M}$; let $(0, b)$ and $(0, c)$ be the center points (that is, the points of intersection of the line with the y-axis) of $\mathcal{L}$ and $\mathcal{M}$, respectively, and choose the notation so that $b > c$.

If the slope of the left ray of $\mathcal{L}$ is less negative than the slope of the left ray of $\mathcal{M}$, these rays will intersect, contradicting our assumption that the lines are parallel. If the slope of the left ray of $\mathcal{L}$ is more negative than the slope of the left ray of $\mathcal{M}$, the rays do not intersect; however, their right rays will intersect, and this contradicts the parallelism of the two lines. Thus the left rays must have the same slope. A similar argument shows that the right rays must have the same slope. □

Theorem MLT.3 (Incidence axioms). *The incidence Axioms I.1, I.5(A), and I.5(B) are true on Model MLT.*

Proof. Axiom I.5 (A) is obviously true since every line in the model has more than one point. Axiom I.5(B) is true because given a line, there exist points not on that line. (Again, we do not deal with Axioms I.0, I.2, I.3, I.4, or I.5(C), because these involve space.)

We show that Axiom I.1 holds: there exists exactly one line $\overleftrightarrow{AB}_m$ through two distinct points A and B. We know that in the coordinate plane, there is a line $\overleftrightarrow{AB}_c$ through two such points. If $\overleftrightarrow{AB}_c$ is vertical or horizontal or has positive slope, let $\overleftrightarrow{AB}_m = \overleftrightarrow{AB}_c$.

If the slope of $\overleftrightarrow{AB}_c$ is negative, it is not a line in Model MLT. In this case, if A and B are on the same side of (or possibly on) the y-axis, let P be the point of intersection of $\overleftrightarrow{AB}_c$ and the y-axis.

If A and B are on the left side of (or possibly on) the y-axis, define $\overleftrightarrow{AB}_m$ to be the union of $\overrightarrow{PA}$ and $\overrightarrow{PC}$, where C is on the right side and the slope of $\overrightarrow{PC}$ is twice the slope of $\overrightarrow{PA}$. Then $\overleftrightarrow{AB}_m$ is the unique Model MLT line through both A and B. A similar argument will produce the same result if A and B are on the right side of (or possibly on) the y-axis.

Now suppose $A = (a_1, a_2)$ is on the left side of the y-axis, and $B = (b_1, b_2)$ is on the right side, still assuming that the coordinate line connecting them has negative slope. Then $a_1 < 0$ and $b_1 > 0$.

Let $c = \dfrac{-a_1b_2 + 2b_1a_2}{2b_1 - a_1}$; then the point $P = (0, c)$ lies on the Model MLT line through both A and B. For the slope of the left-hand ray $\overrightarrow{PA}$ is $-\dfrac{c - a_2}{a_1} = \dfrac{b_2 - a_2}{2b_1 - a_1}$, and the slope of the right-hand ray $\overrightarrow{PB}$ is $\dfrac{b_2 - c}{b_1} = 2\dfrac{b_2 - a_2}{2b_1 - a_1}$, which is twice the left-hand slope.

The proof of uniqueness, that any Model MLT line passing through A and B must also contain the point $P = (0, c)$, is Exercise MLT.1. □

Theorem MLT.4 (Axiom BET). *Axiom BET is true on Model MLT; that is, there exists a betweenness relation on Model MLT.*

Proof. Let A, B, and C be any three collinear points on Model MLT. If these three points lie on a line of type V, H, or P, or if they all lie on either the left or right ray of a line of type N, define A–B–C iff A–B–C as points in Model LM2R. Since our definition of ray includes the endpoint with first coordinate 0, this includes the possibility that one of these points lies on the y-axis.

If A, B, and C lie on a line $\mathcal{L}$ of type N, and two of them, say A and B, lie on one of its rays, whilst C lies in the other ray, there are two cases:

(Case 1: one of A or B is on the y-axis.)

(Subcase 1A: B is on the y-axis.) Define A–B–C (equivalently, C–B–A);

(Subcase 1B: A lies on the y-axis.) Define B–A–C (C–A–B).

These subcases are mutually exclusive because A and B are distinct points, and cannot both belong to the y-axis.

(Case 2: neither A nor B lies on the y-axis.) Let W be the point of intersection of the ray containing A and B with the y-axis;

(Subcase 2A A–B–W as points in Model LM2R.) Define A–B–C (C–B–A);

(Subcase 2B) B–A–W as points in Model LM2R.) Define B–A–C, (C–A–B).

These subcases are mutually exclusive because by trichotomy for Model LM2R, we cannot have both A–B–W and B–A–W. Therefore, in either case, A–B–C and B–A–C are mutually exclusive.

By definition, the betweenness relation described just above satisfies **Property B.0** and **Property B.1** of Definition IB.1. To show **Property B.2, trichotomy**, we need only consider sets $\{A, B, C\}$ of points which are not all on a Model LM2R line, and are not all on one or the other of the rays of a line of type N—for in these cases, trichotomy already holds from Model LM2R.

That is, we consider only sets $\{A, B, C\}$ where two points are on one ray (possibly on the y-axis) and the remaining point is on the other ray of a line of type N (but not on the y-axis). The possibilities are:

1) A and B are on one ray (possibly on the y-axis) and C is on the other. In this case, the possibilities are A–B–C and B–A–C.
2) B and C are on one ray (possibly on the y-axis) and A is on the other. In this case, the possibilities are B–C–A and C–B–A.
3) A and C are on one ray (possibly on the y-axis) and B is on the other. In this case, the possibilities are A–C–B and C–A–B.

As pointed out above, in each possibility the two alternatives are mutually exclusive. Thus, if A–B–C, B–A–C is ruled out by 1) and A–C–B is ruled out by 2); if B–A–C, A–B–C is ruled out by 1) and A–C–B is ruled out by 3); if A–C–B, C–A–B is ruled out by 3) and C–B–A is ruled out by 2). It follows that for any collinear points A, B, and C exactly one of A–B–C, B–A–C, and A–C–B can be true; this establishes trichotomy, Property B.2.

Finally, we show that **Property B.3, extension**, holds. If A and B are distinct points and the slope of the line $\overleftrightarrow{AB}$ is positive, or zero, or if $\overleftrightarrow{AB}$ is vertical, then by the extension property for the coordinate plane Model LM2R there exists a point C such that A–B–C in Model MLT. If A and B are distinct points on a line $\mathcal{L}_m$ of type N, there are three cases:

(Case 1: both A and B lie on the same side of the y-axis, but neither lies on the y-axis.) Let X be the point of intersection of $\mathcal{L}_m$ with the y-axis. Then either A–B–X or B–A–X; in the first instance, choose $C = X$; then A–B–C in Model MLT. In the second instance, by the extension property for betweenness in Model LM2R choose C such that A–B–C in Model LM2R; then $C \in \overrightarrow{XB}$ so A–B–C in Model MLT.

(Case 2: both A and B lie on the same side of the y-axis, and one of them lies on the y-axis.) If A is on the y-axis, by extension for betweenness in Model LM2R choose C such that A–B–C in Model LM2R; then $C \in \overrightarrow{AB}$ so A–B–C in Model MLT.

If B is on the y-axis, since A and B lie on the same side, $\overrightarrow{BA}$ is a ray (either left or right) of $\mathcal{L}_m$. Let C be any point (other than B) of the other ray of $\mathcal{L}_m$. Then A–B–C.

(Case 3: A and B lie on opposite sides of the y-axis.) Let X be the point of intersection of $\mathcal{L}_m$ and the y-axis. By extension for betweenness in Model LM2R choose C such that X–B–C in Model LM2R. Then A–B–C in Model MLT.

It follows that the betweenness relation defined above satisfies all of Properties B.0 through B.3 of Definition IB.1. □

Theorem MLT.5 (Axiom PSA). *The Plane Separation Axiom is true in Model MLT. That is, if we let $\mathcal{L}_m$ be a line in Model MLT, and let $\mathcal{E}$ and $\mathcal{F}$ be opposite sides of $\mathcal{L}_m$, then if $P \in \mathcal{E}$, and $Q \in \mathcal{F}$, $\overline{PQ}_m \cap \mathcal{L}_m \neq \emptyset$.*

The proof is by a series of claims. The first of these is intuitively quite evident, and pertains to Model LM2R. We will summarize the results and finish the proof in the Summary for PSA.

Claim 1. *Let $\mathcal{L}_c$ be a line in Model LM2R ($\mathbb{R}^2$).*

(A) *If $\mathcal{L}_c$ is nonvertical, and if $\mathcal{E}$ is the set of all points above $\mathcal{L}_c$, and $\mathcal{F}$ is the set of all points below the line, conclusions (1)–(3) follow.*

(B) *If $\mathcal{L}_c$ is vertical, and if $\mathcal{E}$ is the set of all points to the right of $\mathcal{L}_c$, and $\mathcal{F}$ is the set of all points to the left of the line, conclusions (1)–(3) follow.*

(1) *If A and B are any distinct points of $\mathcal{E}$, or any distinct points of $\mathcal{F}$, $\overleftrightarrow{AB} \cap \mathcal{L} = \emptyset$.*

(2) *For any $A \in \mathcal{E}$ and $B \in \mathcal{F}$, $\overline{AB} \cap \mathcal{L} \neq \emptyset$.*

(3) *$\mathcal{E}$ and $\mathcal{F}$ are the sides of the line $\mathcal{L}$, are opposite, and $\mathcal{E} \cup \mathcal{F} \cup \mathcal{L}_c = \mathbb{R}^2$.*

The proof of Claim 1 depends on Theorem LC.18 and is Exercise MLT.2.

Claim 2. *Let $\mathcal{L}_m$ be a nonvertical line in Model MLT, and let O be the point of intersection of $\mathcal{L}_m$ with the y-axis. Let $X = (x_1, x_2)$ and $Y = (y_1, y_2)$ be distinct points. If both X and Y are above, or both are below $\mathcal{L}_m$, then $\overleftrightarrow{XY}_m \cap \mathcal{L}_m = \emptyset$.*

Proof. By Axiom I.1 (which is true in Model MLT) there exists a line $\overleftrightarrow{XY}_m$ containing both X and Y, which intersects the y-axis at some point P.

(Case 1: Both $x_1 \le 0$ and $y_1 \le 0$, or both $x_1 \ge 0$ and $y_1 \ge 0$.) That is, both X and Y lie in the same ray (left or right) of $\overleftrightarrow{XY}_m$, that is, lying either above (or below) $lr(\mathcal{L}_m)$ or $rr(\mathcal{L}_m)$ as the case may be. Then X and Y both lie above (or below) the Model LM2R line containing that ray, Then $\overleftrightarrow{XY}_c = \overleftrightarrow{XY}_m$, so that by Claim 1, $\overleftrightarrow{XY}_m \cap \mathcal{L}_m = \overleftrightarrow{XY}_c \cap \mathcal{L}_m = \emptyset$.

(Case 2: Both X and Y lie above $\mathcal{L}_m$, $x_1 < 0$, $y_1 > 0$, and P lies above O.) Then P, X, and Y all lie above $\mathcal{L}_m$. By Case 1, $\overleftrightarrow{XP}_m \cap \mathcal{L}_m = \emptyset = \overleftrightarrow{PY}_m \cap \mathcal{L}_m$, so that $\overleftrightarrow{XY}_m \cap \mathcal{L}_m = \emptyset$. If X, Y, and P all lie below $\overleftrightarrow{XY}_m$ a similar argument gives the same result.

(Case 3: Both X and Y lie above $\mathcal{L}_m$, $x_1 < 0$, $y_1 > 0$, either $P = O$ or P lies below O, and the slope $sl(\overrightarrow{PX}) < sl(lr(\mathcal{L}_m))$.) Thus if $\mathcal{L}_m$ is of type N, so is $\overleftrightarrow{XY}_m$.

If $\mathcal{L}_m$ is of type H or type P, $sl(\overrightarrow{PY}) \le sl(\overrightarrow{PX}) < sl(lr(\mathcal{L}_m))$. If $\mathcal{L}_m$ is of type N, then $\overleftrightarrow{XY}_m$ is of type N and $sl(\overrightarrow{PY}) = 2sl(\overrightarrow{PX}) < 2sl(lr(\mathcal{L}_m)) = sl(rr(\mathcal{L}_m))$. In either case, since P lies below O, all points of $\overrightarrow{PY}$ lie below $\mathcal{L}_m$, which is impossible, since Y is above that line.

(Case 4: Both X and Y lie below $\mathcal{L}_m$, $x_1 < 0$, and $y_1 > 0$, and either $P = O$ or P lies above O.) This too leads to a contradiction. The proof is Exercise MLT.3.

Thus both cases 3 and 4 are ruled out. □

Claim 3. *If $\mathcal{L}$ is a vertical line in Model MLT, and if X and Y both lie to the right, or both lie to the left of $\mathcal{L}$, then $\overleftrightarrow{XY}_m \cap \mathcal{L} = \emptyset$.*

Proof. $\mathcal{L}$ is a line in Model LM2R as well as in Model MLT.

(Case 1: X and Y have the same first coordinate.) In this case, $\overleftrightarrow{XY}_m$ is a vertical line which does not intersect $\mathcal{L}$.

(Case 2: $\mathcal{L}$ is the y-axis.) If both X and Y lie on the same side of the y-axis, and the slope of $\overleftrightarrow{XY}_c$ is non-negative, $\overline{\overline{XY}}_m = \overline{\overline{XY}}_c$; this is also true if the slope is negative, since the change in slope for a type N line in Model MLT occurs at the y-axis. By Claim 1, both X and Y belong to the same Model LM2R side of $\mathcal{L}$, so that $\overline{\overline{XY}}_m \cap \mathcal{L} = \overline{\overline{XY}}_c \cap \mathcal{L} = \emptyset$.

(Case 3: $\mathcal{L}$ is not the y-axis, and both X and Y lie on the same side of $\mathcal{L}$.) Let P be the point of intersection of $\overleftrightarrow{XY}_m$ and the y-axis.

(Subcase 3a: Both X and Y lie on the same side of the y-axis, or one of them lies on the y-axis.) Then $\overline{\overline{XY}}_m = \overline{\overline{XY}}_c$ which by Claim 1 is disjoint from $\mathcal{L}$.

(Subcase 3b: X and Y lie on opposite sides of the y-axis.) Then both $\overline{\overline{XP}}_m = \overline{\overline{XP}}_c$ and $\overline{\overline{PY}}_m = \overline{\overline{PY}}_c$ both of which are disjoint from $\mathcal{L}$ by Subcase 3a above; hence $\overline{\overline{XY}}_m = \overline{\overline{XP}}_c \cup \overline{\overline{PY}}_c$ is disjoint from $\mathcal{L}$. □

Claim 4. *If $\mathcal{L}_m$ is a nonvertical line in Model MLT, and one of X or Y lies above $\mathcal{L}_m$ and the other lies below, then $\overline{XY}_m \cap \mathcal{L}_m \neq \emptyset$.*

Proof. Let O be the point of intersection of $\mathcal{L}_m$ and the y-axis.

(Case 1: X and Y have the same first coordinate.) It is a simple matter to calculate the point between X and Y which belongs to $\mathcal{L}$. In the remaining cases we assume that $\overleftrightarrow{XY}$ is nonvertical.

(Case 2: X and Y both lie on the same side of the y-axis, or on the y-axis.) Then $\overline{XY}_m = \overline{XY}_c$. Since Axiom PSA holds for Model LM2R, by Claim 1 $\overline{XY}_m \cap \mathcal{L}_m = \overline{XY}_c \cap \mathcal{L}_m \neq \emptyset$.

(Case 3: X lies to the left of the y-axis and above $\mathcal{L}_m$ and Y lies to the right of the y-axis and below $\mathcal{L}_m$.) Suppose $\overleftrightarrow{XY}_m$ intersects the y-axis at the point P.

If P is above O, then by Case 2, $\overline{PY}_m \cap \mathcal{L}_m \neq \emptyset$. If $P = O$, $\overline{XY}_m \cap \mathcal{L}_m \neq \emptyset$. If P is below O, then by Case 2, $\overline{XP}_m \cap \mathcal{L} \neq \emptyset$. It follows that since $\overline{XP}_m$ and $\overline{PY}_m$ are subsets of $\overline{XY}_m$, $\overline{XY}_m \cap \mathcal{L}_m \neq \emptyset$.

The proof for the case where X lies to the left of the y-axis and below $\mathcal{L}_m$, and Y lies to the right of the y-axis and above $\mathcal{L}_m$, is similar to Case 3. □

Claim 5. *If $\mathcal{L}$ is a vertical line in Model MLT, and let $X = (x_1, x_2)$ and $Y = (y_1, y_2)$ be distinct points; if X lies to the left of $\mathcal{L}$ and Y lies to the right, then $\overline{XY}_m \cap \mathcal{L}_m \neq \emptyset$.*

Proof. The proof is Exercise MLT.4, which also proves the observation in Remark MLT.1.1 that every nonvertical line intersects every vertical line. □

Summary for Theorem MLT.5 (PSA) Let $\mathcal{L}_m$ be any nonvertical line in Model MLT; every point P not on $\mathcal{L}_m$ is either above or below this line. Let P be any point above, and Q be any point below $\mathcal{L}_m$. By Claim 2 of Theorem MLT.5, every point above the line is in the P-side of $\mathcal{L}_m$ and every point below the line is in the Q-side of $\mathcal{L}_m$. These two sides cannot intersect. Therefore they are the only possible sides for $\mathcal{L}_m$.

By similar reasoning, if $\mathcal{L}_m$ is a vertical line in Model MLT, P is a point to the right of, and Q is a point to the left of $\mathcal{L}_m$, by Claim 3 of Theorem MLT.5, the P-side of $\mathcal{L}_m$ and the Q-side of $\mathcal{L}_m$ are the only possible sides for $\mathcal{L}_m$.

In either case, $\mathbb{R}^2$ is the union of these two sides and $\mathcal{L}_m$. Then by Claims 4 and 5 of Theorem MLT.5, for every $X \in$ the P-side of $\mathcal{L}_m$ and every $Y \in$ the Q-side of $\mathcal{L}_m$, $\overline{XY} \cap \mathcal{L}_m \neq \emptyset$, showing that these two sides are opposite, and also that Axiom PSA holds.

Theorem MLT.6 (Axiom PS). *The Strong Parallel Axiom PS holds in Model MLT. That is, given a line $\mathcal{L}$ and a point P not belonging to $\mathcal{L}$, there exists exactly one line $\mathcal{M}$ such that $P \in \mathcal{M}$ and $\mathcal{L} \parallel \mathcal{M}$.*

Proof. For any line $\mathcal{L}$ other than a type N line, there exists exactly one line not of type N through P, since Axiom PS holds for lines in the coordinate plane. Moreover, the only lines parallel to a line of type N are those of type N, so there can be no line of type N parallel to $\mathcal{L}$. If $\mathcal{L}$ is of type N, consider three cases:

(Case 1: P is on the y-axis.) Using Axiom PS on the coordinate plane, choose $\mathcal{L}_1$ to be the unique line containing P parallel to the left ray of $\mathcal{L}$, and $\mathcal{L}_2$ as the unique line containing P parallel to the right ray of $\mathcal{L}$; let $\mathcal{M}$ be the union of the left ray of $\mathcal{L}_1$ and the right ray of $\mathcal{L}_2$; both these rays contain P, so that $\mathcal{M}$ is the unique line of type N which contains P and is parallel to $\mathcal{L}$.

(Case 2: P lies on the negative (left) side of the y-axis.) Using Axiom PS on the coordinate plane, let $\mathcal{L}_1$ be the unique line containing P parallel to the left ray of $\mathcal{L}$; this line intersects the y-axis at some point Q; let $\mathcal{L}_2$ be the (unique) line containing Q with slope twice that of $\mathcal{L}_1$. Let $\mathcal{M}$ be the union of the left ray of $\mathcal{L}_1$ and the right ray of $\mathcal{L}_2$; both these rays contain Q, so that $\mathcal{M}$ is the unique line of type N which contains P and is parallel to $\mathcal{L}$.

(Case 3: P lies on the positive (right) side of the y-axis.) The proof is similar to that of Case 2. □

Theorem MLT.7 (Axiom LUB). *Axiom LUB is true in Model MLT. Let $\mathcal{L}$ be a line which is equipped with the order relation of Definition MLT.1(F). Then every nonempty subset $\mathcal{E}$ of $\mathcal{L}$ which is bounded above has a least upper bound lub $\mathcal{E}$.*

Proof. Let $\mathcal{L}$ be a line of type V, type H, or type P. The ordering of Definition MLT.1(F) on any lines of these types is just the standard ordering of lines in Model LM2R, and since Axiom LUB is true on this model, it is true for these lines.

Let $\mathcal{L}_m$ be a line of type N in Model MLT. As in Definition MLT.1(F), O is the origin and U' the unit; $\mathcal{M}_c$ is the Model LM2R line containing the left ray of $\mathcal{L}_m$, and $\mathcal{M}'_c$ is the Model LM2R line containing the right ray. Let $\mathcal{E}$ be a nonempty subset of $\mathcal{L}_m$, which is bounded above. If $\mathcal{E} \cap \overrightarrow{OU'} \neq \emptyset$, the upper bound for $\mathcal{E}$ is an upper bound for this intersection, and since Axiom LUB holds on $\mathcal{M}'_c$, there is a least upper bound for $\mathcal{E} \cap \overrightarrow{OU'}$, which belongs to $\overrightarrow{OU'} \subseteq \mathcal{M}'_c$. This is the least upper bound for $\mathcal{E}$.

If $\mathcal{E} \cap \overrightarrow{OU'} = \emptyset$, then $\mathcal{E} \subseteq \overrightarrow{OV}$ which is a subset of $\mathcal{M}_c$; then $\mathcal{E}$ is bounded above by O, and since LUB holds for $\mathcal{M}_c$, $\mathcal{E}$ has a least upper bound in $\mathcal{M}_c$, and this upper bound belongs to $\overrightarrow{OV} \subseteq \mathcal{L}_m$.

It follows that every bounded nonempty subset of a line in Model MLT has a least upper bound, so that Axiom LUB holds. □

Theorem MLT.8. *Suppose Axiom REF holds on Model MLT, and let k and $b > 0$ be real numbers; suppose further that there exists a reflection φ over the horizontal line $\mathcal{K} = \{(x, k) \mid x$ is any real number $\}$.*

(A) *All the fixed lines for φ are vertical.*

(B) *φ maps every point $(x, k-b)$ to the point $(x, k+\dfrac{b}{\sqrt{2}})$, so that the line $y = k-b$ maps to the line $y = k + \dfrac{b}{\sqrt{2}}$.*

Proof. Preamble: since Axiom REF holds, we may apply all the theorems of neutral geometry. Let $\mathcal{L} = \{(x, k-b) \mid x$ is any real number$\}$ be the horizontal line b units below and parallel to $\mathcal{K}$. By Exercise NEUT.1, $\mathcal{M} = \varphi(\mathcal{L})$ is a line parallel to $\mathcal{K}$; by Definition NEUT.1(B) this line is on the opposite side of $\mathcal{K}$ from $\mathcal{L}$. Therefore for some real number $c > 0$,

$$\mathcal{M} = \{(x, k+c) \mid x \text{ is any real number}\}.$$

By Theorem NEUT.22(E), all fixed lines for a reflection are parallel. Therefore (given the numbers b and c as above) there exists a real number d such that for any real number a,

$$\varphi(a, k-b) = (a+d, k+c). \qquad (1)$$

So if $P = (a, k-b)$ is an arbitrary point of $\mathcal{L}$, the number d measures the horizontal offset of $\varphi(P)$ from P; if $d > 0$, $\varphi(P)$ lies to the right of the vertical line through P; if $d < 0$, it lies to the left.

We complete our preamble by defining $O = (0, k)$, the intersection of $\mathcal{K}$ with the y-axis.

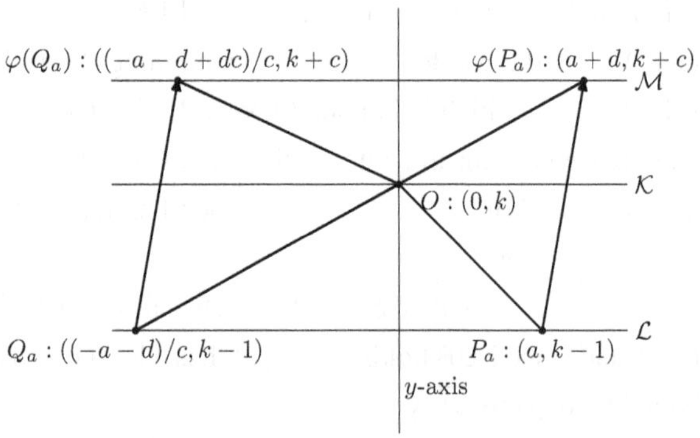

Fig. 21.3 For part (A) of Theorem MLT.8.

(A) We now prove that $d = 0$, showing that the fixed lines of φ are vertical. For this part, let $b = 1$, so that $\mathcal{L}$ is the horizontal line 1 unit below $\mathcal{K}$. $\varphi(\mathcal{L})$ is a horizontal line c units above $\mathcal{K}$, but we don't yet know the value of c—we just know it is a fixed number determined by $b = 1$. Again, d is the offset defined in the preamble, and it is determined by $b = 1$ and c.

Choose $a_0 > 0$ so that for any $a \geq a_0$, $a + d > 0$; for any such a, define $P_a = (a, k-1)$; then $\varphi(P_a) = \varphi(a, k-1) = (a+d, k+c)$ lies to the right of the y-axis.

The slope of the line $\overleftrightarrow{O\varphi(P_a)}$ is positive, so this line (in Model MLT) is a line in the coordinate plane. Let $Q_a = (x, k-1)$ be the point of intersection of $\overleftrightarrow{O\varphi(P_a)}$ with $\mathcal{L}$. To determine x we calculate the slope of this line in two different ways: first, using O and $\varphi(P_a)$, the slope is $\dfrac{c}{a+d} > 0$. Using O and Q_a the slope is $\dfrac{1}{-x}$, so that $\dfrac{c}{a+d} = \dfrac{1}{-x}$ and hence $x = \dfrac{-a-d}{c}$; thus $Q_a = (\dfrac{-a-d}{c}, k-1)$.

By Theorem NEUT.15(1), $\varphi(\overleftrightarrow{Q_a\varphi(P_a)}) = \overleftrightarrow{\varphi(Q_a)P_a}$, and this line has negative slope, so it "breaks" at the y-axis. The slope of the ray $\overrightarrow{OP_a}$ (as a line in coordinate space) is $-1/a$ so the slope of the ray $\overrightarrow{O\varphi(Q_a)}$ must be $-\frac{1}{2a}$.

Since all fixed lines are parallel (that is, by equation (1) above), the first coordinate of $\varphi(Q_a)$ is $\frac{-a-d}{c} + d$; since $\varphi(Q_a) \in \mathcal{M}$, $Q = (\frac{-a-d}{c} + d, k + c)$. Then the slope of the ray $\overrightarrow{O\varphi(Q_a)}$ is

$$\frac{-c}{\frac{a+d}{c} - d} = \frac{-c^2}{a + d - dc} = -\frac{1}{2a},$$

so that for all $a \geq a_0$,

$$a + d - dc = 2ac^2 \text{ or } a(2c^2 - 1) = d(1 - c). \qquad (2)$$

If $c = 1$, equation (2) becomes $a(2 - 1) = a = 0$; but $a > 0$ by hypothesis, so we have a contradiction. Therefore $c \neq 1$.

If $2c^2 - 1 \neq 0$, then $a = \frac{d(1-c)}{2c^2-1}$, and a is completely determined by the values of c and d, which are fixed numbers for this argument; thus there can be only one number a for which this can be true, contradicting the fact that equation (2) is true for all $a \geq a_0$. Therefore $2c^2 - 1 = 0$, so that $d(1 - c) = 0$; since $1 - c \neq 0$, $d = 0$. Thus all fixed lines are vertical.

Fig. 21.4 For part (B) of Theorem MLT.8.

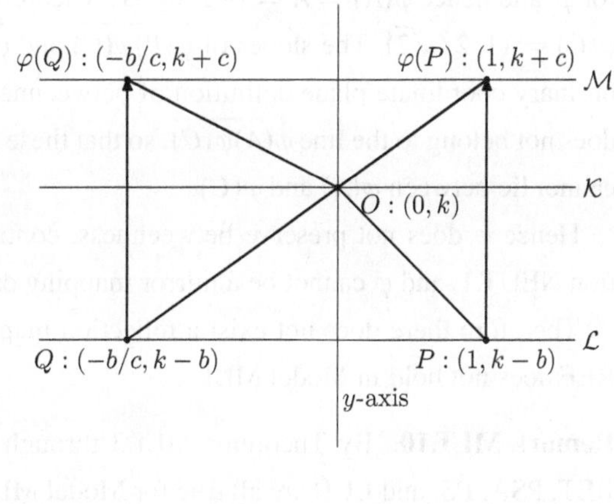

(B) Referring again to the preamble, let $b > 0$ be any real number, so that $\mathcal{L}$ is the horizontal line b units below $\mathcal{K}$. Let $a = 1 > 0$ (for this argument, a can be fixed) and let $P = (1, k - b)$. As in the preamble, there exists a number $c > 0$ such that $\varphi(P)$ lies on

$$\mathcal{M} = \{(x, k + c) \mid x \text{ is any real number}\}.$$

By part (A), $\varphi(P) = (1, k + c)$.

As before, let $O = (0, k)$ be the point of intersection of the y-axis and $\mathcal{K}$; then $\overleftrightarrow{O\varphi(P)}$ intersects $\mathcal{L}$ at some point Q. The slope from O to $\varphi(P)$ is c; the first coordinate of Q is easily calculated to be $-b/c$ so that $Q = (-b/c, k-b)$. Thus $\varphi(Q) = (-b/c, k+c)$.

Now the slope of the (right-hand) ray $\overrightarrow{OP}$ is $-b < 0$ and the slope of the left-hand ray will be $-b/2$. The latter is also $-\dfrac{c}{b/c} = -\dfrac{c^2}{b}$ so that $\dfrac{c^2}{b} = \dfrac{b}{2}$, or $2c^2 = b^2$, that is, $c = \dfrac{b}{\sqrt{2}}$. This completes the proof. □

Theorem MLT.9. *Axiom REF does not hold on Model MLT; that is, there is no reflection set on this model.*

Proof. It is possible to show this by invoking properties of non-Desarguesian planes, but we will give a proof using Theorem MLT.8 above, showing that if Axiom REF holds on Model MLT, we get a contradiction.

Let $A = (-2, 0)$, $B = (0, -1)$, and $C = (1, -2)$. By our definition of betweenness (*cf* Theorem MLT.4) A–B–C, as the coordinate line $\overleftrightarrow{BC}$ has slope -1 and the coordinate line $\overleftrightarrow{AB}$ has slope $-1/2$.

Let φ be the reflection over the x-axis. Since A lies on the x-axis, it is a fixed point for φ, and hence $\varphi(A) = A = (-2, 0)$. By Theorem MLT.8, $\varphi(B) = (0, 1/\sqrt{2})$, and $\varphi(C) = (1, 2/\sqrt{2})$. The slopes of $\overleftrightarrow{\varphi(B)\varphi(C)}$ and $\overleftrightarrow{\varphi(A)\varphi(B)}$ are positive, so that the ordinary coordinate plane definition of betweenness holds. But $\varphi(B) = (0, 1/\sqrt{2})$ does not belong to the line $\overleftrightarrow{\varphi(A)\varphi(C)}$, so that these points are not collinear, and $\varphi(B)$ cannot lie between $\varphi(A)$ and $\varphi(C)$.

Hence φ does not preserve betweenness, contradicting Property (D) of Definition NEUT.1, and φ cannot be a mirror mapping or a reflection.

Therefore there does not exist a reflection mapping over the x-axis, and Axiom REF does not hold in Model MLT. □

Remark MLT.10. By Theorems MLT.3 through MLT.7, Axioms I.0, I.1, I.5(A), BET, PSA, PS, and LUB are all true for Model MLT. By Theorem MLT.9, there can be no reflection set on Model MLT. This shows that Axiom REF is independent of this set of axioms.

21.6.5 Parallel Axiom PS is independent of Axioms I.0–I.5, BET, PSA (Model PSM)

We now construct a model which establishes the independence of the parallel axiom PS from the incidence and betweenness axioms and the Plane Separation Axiom PSA. This model we designate Model PSM.

Definition PSM.1. For Model PSM, space $\mathcal{S}$ is an "open" unit cube—that is, the unit cube which does not include any of its "surface" points. More precisely,

$$\mathcal{S} = \{X = (x_1, x_2, x_1) \mid X \in \mathbb{F}^3, 0 < x_1 < 1, 0 < x_2 < 1, \text{ and } 0 < x_3 < 1\},$$

where $\mathbb{F}$ is an ordered field.

We define $\mathcal{L}$ to be a **line** for Model PSM iff for some distinct points X and Y of $\mathcal{S}$, $\mathcal{L} = \overleftrightarrow{XY} \cap \mathcal{S} \neq \emptyset$, where $\overleftrightarrow{XY}$ is the line through these two points as in Definition LA.1(1).

$\mathcal{P}$ is a **plane** for Model PSM iff for some noncollinear members X, Y, and Z of $\mathcal{S}$, $\mathcal{P} = \overleftrightarrow{XYZ} \cap \mathcal{S} \neq \emptyset$, where $\overleftrightarrow{XYZ}$ is the plane through these points as in Definition LA.1(2).

If X, Y, and Z are points of $\mathcal{S}$, then Y is **between** X and Z iff Y is between X and Z in Model LM3.

Let $\mathcal{P} = \mathcal{Q} \cap \mathcal{S}$ be a plane for Model PSM, where $\mathcal{Q}$ is a plane in $\mathbb{F}^3$; let $\mathcal{L} = \mathcal{M} \cap \mathcal{S}$ be a line in $\mathcal{P}$, where $\mathcal{M}$ is a line in $\mathbb{F}^3$. Then for any point $A \in \mathcal{P} \setminus \mathcal{L}$, a set $\mathcal{E}$ is the A-**side** of $\mathcal{L}$ for Model PSM iff $\mathcal{E} = \mathcal{F} \cap \mathcal{S}$, where $\mathcal{F}$ is the A-side for the line $\mathcal{M}$ in $\mathbb{F}^3$, according to Definition IB.11.

Theorem PSM.2. *Let $X = (x_1, x_2, x_3)$ be any point of $\mathcal{S}$, and let $Y = (y_1, y_2, y_3)$ be any other point of $\mathbb{F}^3$. Then*

(A) *there exists a number $t > 0$ such that $Z = X + t(Y - X) \in \mathcal{S}$ (so that $Z \neq X$);*
(B) *if $Y \in \mathcal{S}$, there exists a number $t > 1$ such that $Z = X + t(Y - X) \in \mathcal{S}$.*

Proof. (A) Since $X \in \mathcal{S}$, for each $i \in \{1, 2, 3\}$, $0 < x_i < 1$. For each such i there are three possible cases:

(Case 1: $y_i - x_i = 0$.) Let $t_i = 2$. Then $0 < x_i + t_i(y_i - x_i) = x_i < 1$.

(Case 2: $y_i - x_i > 0$.) Then $\frac{1-x_i}{y_i-x_i} > 0$ since $x_i < 1$. Choose $t_i > 0$ so that $0 < t_i < \frac{1-x_i}{y_i-x_i}$. Then

$$t_i(y_i - x_i) < \frac{1-x_i}{y_i-x_i}(y_i - x_i) = 1 - x_i$$

so that $x_i+t_i(y_i-x_i) < 1$; since $y_i-x_i > 0$, $t_i > 0$ and $x_i > 0$, $0 < x_i+t_i(y_i-x_i) < 1$.

(Case 3: $y_i - x_i < 0$.) Then $0 < -\frac{x_i}{y_i-x_i}$; choose t_i so that $0 < t_i < -\frac{x_i}{y_i-x_i}$. Then

$$t_i(y_i - x_i) > -\frac{x_i}{y_i-x_i}(y_i - x_i) = -x_i$$

so that $x_i + t_i(y_i - x_i) > 0$; since $t_i(y_i - x_i) < 0$ and $x_1 < 1$ it follows that $0 < x_i + t_i(y_i - x_i) < 1$.

Let $t = \min\{t_1, t_2, t_3\}$. Then for every $i \in \{1, 2, 3\}$ $t_i > 0$ so that $t > 0$, and $0 < x_i + t(y_i - x_i) < 1$, so that $X + t(Y - X) \in \mathcal{S}$. This completes the proof of part (A).

(B) Now we assume that $Y \in \mathcal{S}$, so that for all $i \in \{1, 2, 3\}$, $0 < y_i < 1$. Again we have three cases:

(Case 1: $y_i - x_i = 0$.) Let $t_i = 2$. Then $0 < x_i + t_i(y_i - x_i) = x_i < 1$.

(Case 2: $y_i - x_i > 0$.) Then $\frac{1-x_i}{y_i-x_i} > 1$ because $1 - x_i > y_i - x_i > 0$, and we may choose t_i so that $0 < 1 < t_i < \frac{1-x_i}{y_i-x_i}$; as in part (A), $0 < x_i + t_i(y_i - x_i) < 1$.

(Case 3: $y_i - x_i < 0$.) Then $-\frac{x_i}{y_i-x_i} = \frac{x_i}{x_i-y_i} > 1$ because $x_i > x_i - y_i > 0$. We may choose t_i so that $0 < 1 < t_i < -\frac{x_i}{y_i-x_i}$. Then by the same reasoning as in Case 3 of part (A), $0 < x_i + t_i(y_i - x_i) < 1$.

Let $t = \min\{t_1, t_2, t_3\}$; $t > 1$ since for every i, $t_i > 1$. Then for every $i \in \{1, 2, 3\}$, $0 < x_i + t(y_i - x_i) < 1$, so that $X + t(Y - X) \in \mathcal{S}$. This completes the proof of part (B). □

Theorem PSM.3. *$\mathcal{S}$ is convex; each of Axioms I.0, I.1, I.2, I.3, I.4, I.5, and BET is true for Model PSM.*

Proof. (A) Convexity: we show that if $X = (x_1, x_2, x_1)$ and $Y = (y_1, y_2, y_3)$ are any two points of $\mathcal{S}$, for any number t such that $0 < t < 1$, $X + t(Y - X) \in \mathcal{S}$. Since X and Y belong to $\mathcal{S}$, for every $i \in \{1, 2, 3\}$, $0 < x_i < 1$ and $0 < y_i < 1$; then $x_i + t(y_i - x_i) = (1 - t)x_i + ty_i > 0$, since $1 - t > 0$, $x_i > 0$, $t > 0$ and $y_i > 0$. Also, since $x_i < 1$ and $y_i < 1$, $(1 - t)x_i + ty_i < (1 - t) + t = 1$. Therefore $X + t(Y - X) \in \mathcal{S}$ proving its convexity.

(B) It is trivial to show that Axiom I.0 holds, since all planes and lines are subsets of $\mathcal{S}$. Two points in $\mathcal{S}$ are also points in $\mathbb{F}^3$ and since Axiom I.1 holds for Model LM3, there is exactly one line in LM3 containing these points, and the intersection of this line with $\mathcal{S}$ is the unique line containing both points. Therefore Axiom I.1 holds for Model PSM. Similar reasoning shows that Axiom I.2 holds.

(C) If two points X and Y lie in a plane $\mathcal{P} = \mathcal{Q} \cap \mathcal{S}$, where $\mathcal{Q}$ is a plane in $\mathbb{F}^3$, then the line $\overleftrightarrow{XY} \subseteq \mathcal{Q}$ since Axiom I.3 holds for Model LM3. Then $\mathcal{L} = \overleftrightarrow{XY} \cap \mathcal{S} \subseteq \mathcal{Q} \cap \mathcal{S} = \mathcal{P}$ so Axiom I.3 holds for Model PSM.

(D) Let $\mathcal{P}_1 = \mathcal{Q}_1 \cap \mathcal{S}$ and $\mathcal{P}_2 = \mathcal{Q}_2 \cap \mathcal{S}$ be two planes in $\mathcal{S}$, where $\mathcal{Q}_1$ and $\mathcal{Q}_2$ are planes of $\mathbb{F}^3$. Suppose $X = (x_1, x_2, x_3) \in (\mathcal{P}_1 \cap \mathcal{P}_2)$. Then $X \in (\mathcal{Q}_1 \cap \mathcal{Q}_2)$ and since Axiom I.4 holds for Model LM3, there exists at least one other point $Y \in (\mathcal{Q}_1 \cap \mathcal{Q}_2)$. Since Axiom I.3 holds for Model LM3, $\overleftrightarrow{XY} \subseteq (\mathcal{Q}_1 \cap \mathcal{Q}_2)$. By Theorem PSM.2(A) there exists a point $Z = X + t(Y - X) \in (\overleftrightarrow{XY} \cap \mathcal{S})$ where $Z \neq X$. Hence $Z \in (\mathcal{P}_1 \cap \mathcal{P}_2)$. This shows that Axiom I.4 holds for Model PSM.

(E) To show that Axiom I.5(A) holds for Model PSM, let $\mathcal{L}$ be any line in $\mathcal{U}$. Then for some line $\mathcal{M}$ in $\mathbb{F}^3$, $\mathcal{L} = \mathcal{M} \cap \mathcal{S}$, and by definition there exists a point $X \in \mathcal{L}$. By Theorem LC.5, Axiom I.5(A) holds for Model LM3, so there exists a point $Y \neq X$ such that $Y \in \mathcal{M}$. By Theorem PSM.2(A), there exists a point $Z = X + t(Y - X) \in \mathcal{S}$ such that $Z \in \mathcal{L} = \overleftrightarrow{XY} \cap \mathcal{S}$, and $Z \neq X$. Thus Axiom I.5(A) holds for Model PSM.

(F) To show that Axiom I.5(B) holds for Model PSM, let $\mathcal{P}$ be any plane in $\mathcal{U}$. Then for some plane $\mathcal{Q}$ in $\mathbb{F}^3$, $\mathcal{P} = \mathcal{Q} \cap \mathcal{S}$. By definition there exists a point $X \in \mathcal{L}$; by Theorem LC.5, Axiom I.5(B) holds for Model LM3, so there exist points $Y \neq X$ and $Z \neq X$ such that both Y and Z are members of $\mathcal{Q}$, and X, Y, and Z are noncollinear. By Theorem LC.4, Axiom I.3 holds for Model LM3, so both $\overleftrightarrow{XY}$ and $\overleftrightarrow{XZ}$ are subsets of $\mathcal{Q}$. By Theorem PSM.2(A) there exist points Y' and Z' and nonzero numbers s and t such that $Y' = X + s(Y - X) \in \mathcal{S}$ and $Z' = X + t(Z - X) \in \mathcal{S}$. If $Z' \in \overleftrightarrow{XY'}$, that is, X, Y', and Z' are collinear, then for some number $u \neq 0$, $Z' = X + u(Y' - X)$, so that

$$\begin{aligned}(X + s(Z - X) - X) - u(X + s(Y - X) - X) \\ = t(Z - X) - us(Y - X) = 0.\end{aligned}$$

By Theorem LA.3, $Y - X$ and $Z - X$ are linearly independent, hence both t and us are 0; but by construction, $t \neq 0$ so we have a contradiction. Therefore X, Y', and Z' are noncollinear members of $\mathcal{P}$, and Axiom I.5(B) holds.

(G) Axiom I.5(C) holds, since the points $(\frac{1}{2}, \frac{1}{2}, \frac{1}{2})$, $(\frac{3}{4}, \frac{1}{2}, \frac{1}{2})$, $(\frac{1}{2}, \frac{3}{4}, \frac{1}{2})$, and $(\frac{1}{2}, \frac{1}{2}, \frac{3}{4})$ are noncoplanar points in $\mathcal{S}$.

(H) Axiom BET holds for Model PSM. It is easy to see that Properties B.0, B.1, and B.2 of Definition IB.1 hold in Model PSM, since they hold in Model LM3. To see that Property B.3 holds, let $X = (x_1, x_2, x_1)$ and $Y = (y_1, y_2, y_1)$ be any two points of $\mathcal{S}$.

By Theorem PSM.2(B), there exists a point $Z = (z_1, z_2, z_3) \in \mathcal{S}$ and a number $t > 1$ such that $Z = X+t(Y-X)$, so that X, Y, and Z are collinear points. Then by Definition LC.8, X–Y–Z since $X = X+0(Y-X)$, $Y = X+1(Y-X)$, and $Z = X+t(Y-X)$ and 0–1–t. Thus Property B.3 of Definition IB.1 holds in Model PSM. □

Theorem PSM.4. *The Plane Separation Axiom PSA is true for Model PSM.*

Proof. Since $\mathcal{S}$ is convex, the proof of Theorem LC.20 is valid for Model PSM, so the Plane Separation Axiom PSA is true for Model PSM. □

Theorem PSM.5. *Neither the strong nor the weak form of the parallel axiom (PS or PW) is true for Model PSM.*

Proof. Let $\mathcal{M}_1 = \overleftrightarrow{(\frac{1}{2}, 0, \frac{1}{2})(\frac{1}{2}, 1, \frac{1}{2})}$, and let the point $P = (\frac{1}{2}, \frac{1}{2}, \frac{3}{4})$. Then $P \notin \mathcal{M}_1$. Let $\mathcal{M}_2 = \overleftrightarrow{(\frac{1}{2}, \frac{1}{2}, \frac{3}{4})(\frac{1}{2}, 1, \frac{3}{4})}$ and let $\mathcal{M}_3 = \overleftrightarrow{(\frac{1}{2}, \frac{1}{2}, \frac{3}{4})(\frac{1}{2}, 1, \frac{7}{8})}$.

Let $\mathcal{L}_1 = \mathcal{M}_1 \cap \mathcal{S}$, $\mathcal{L}_2 = \mathcal{M}_2 \cap \mathcal{S}$, and $\mathcal{L}_3 = \mathcal{M}_3 \cap \mathcal{S}$. Then $\mathcal{L}_2 \cap \mathcal{L}_3 = \{P\}$ and both $\mathcal{L}_2$ and $\mathcal{L}_3$ are parallel to $\mathcal{L}_1$ (because the parts of the lines $\mathcal{M}_2$ and $\mathcal{M}_3$ which lie within $\mathcal{S}$ do not intersect $\mathcal{M}_1$). This shows that there are distinct lines through the point P which are parallel to the line $\mathcal{L}_1$, and the weak form of the parallel axiom does not hold. □

Remark PSM.6. Since the weak form of the parallel axiom is false for Model PSM, the parallel axioms are independent of each of the incidence axioms, betweenness axioms, and the Plane Separation Axiom PSA.

21.6.6 Independence of parallel Axiom PS from Axioms I.0–I.5, BET, PSA, and REF

In Chapter 2 of this book we discussed the parallel axioms and the classification of geometries into elliptic geometries (no parallel lines), Euclidean geometry, and hyperbolic geometries. Since parallel lines exist in neutral geometry (Property PE, Theorem NEUT.48(B)), neutral geometry is incompatible with elliptic geometry. However, hyperbolic geometry is neutral geometry combined with a denial of Axiom PW. That is, given a line $\mathcal{L}$, and a point P not on $\mathcal{L}$, there may be multiple lines through P parallel to $\mathcal{L}$.

Thus, to show that Axiom PS is independent of Axioms I.0–I.5, BET, and PSA as well as Axiom REF, it would be sufficient to exhibit a model for a hyperbolic plane, where Axiom PW is false, in which a reflection set exists over every line. Hyperbolic geometry is beyond the purview of this book, so we leave this issue to others.

For a readable account of the history of the parallel axiom and non-Euclidean geometries, including hyperbolic geometry, we suggest Marvin J. Greenberg's *Euclidean and non-Euclidean geometries, development and history*, 4th ed., W. H. Freeman, 2008 [8]. Chapter 7 of this volume is devoted to the issue of the independence of the parallel axiom.

21.6.7 Axiom LUB is independent of Axioms I.0–I.5, BET, PSA, REF, and PS (Model LM3A)

Our goal in this subsection is to prove that the LUB axiom is independent of all the other axioms we have introduced. To do this we build Model LM3A (meaning Model LM3 built on the field $\mathbb{A}$ of real algebraic numbers) for which Axiom LUB is false but each of the other axioms is true. Here we use the acronym LE.

Theorem LE.1. *Axiom LUB is independent of all previous axioms.*

Proof. Let space for Model LM3A be $\mathbb{A}^3$, where $\mathbb{A}$ is the field of real algebraic numbers, those real numbers that can be generated by adding, subtracting, multiplying, and dividing rational numbers, and finding roots of polynomial equations with rational coefficients. As we have pointed out previously, $\mathbb{A}$ contains the square roots of any of its non-negative numbers, so that norms and distances are defined in $\mathbb{A}^3$.

In Model LM3A we assume that reflections have been defined on each plane according to Definition LB.16 and Definition LC.24. Indeed, the entire development of Model LM3, up through Theorem LC.42, holds for $\mathbb{A}^3$. In particular, by Remark LC.21, Axioms I.0–I.5, BET, PS, and PSA hold; by Theorem LC.33 Axiom REF holds on $\mathbb{A}^2$; by Theorem LC.42, it holds on any plane in $\mathbb{A}^3$, and hence in $\mathbb{A}^3$.

Since π is a transcendental real number (*i.e.*, is nonalgebraic), it is not a member of the field $\mathbb{A}$. Let $X \in \mathbb{A}^2$ and let $\mathcal{E} = \{tX \mid t \in \mathbb{A} \textit{ and } t < \pi\}$; then $\operatorname{lub} \mathcal{E} = \pi X$. Hence $\mathcal{E}$ is a nonempty subset of $\mathbb{A}$ which is bounded above but $\operatorname{lub} \mathcal{E}$ does not belong to $\mathbb{A}$, and Axiom LUB is false for Model LM3A. Since all axioms prior to this one are true for this model, Axiom LUB is independent. □

Remark LE.2 (Summary of axiom independence). In earlier parts of this section we proved that each of the Axioms I.0–I.5, BET, PSA, and PS is independent of its predecessors on this list; now, in Theorem LE.1, we have shown that Axiom LUB is independent of all other axioms. In Subsection 21.6.4 we showed that Axiom REF is independent of Axioms I.0, I.1, I.5(A), BET, PSA, PS, and LUB.

As we stated above in Subsection 21.6.6, proof of the independence of Axiom PS from Axiom REF depends on considerations from hyperbolic geometry, which are beyond the scope of this book.

21.7 Independence of definition properties

In this section we show independence, not of axioms, but of various properties of definitions. This is done for the definitions of betweenness, mirror mapping, and reflection by constructing models and equipping them, respectively, with "pseudo" relations, mappings, or sets of mappings that satisfy some but not all of the required properties.

Table of independence models
for definition properties.

Subsection	Theorem	Model	Relations/maps/sets	True	False
21.7.1	BI.0	LM3	various	B.1–.3	B.0
	BI.1,1.1	LM3	pseudo-	B.0	B.1–.3
	BI.2,2.1	LM3	betweenness	B.0,.1	B.2,.3
	BI.3	$LM3\backslash x+$	relations	B.1–.3	B.0
21.7.2	MMI.1	LM2	various	(A),(C),(D)	(B)
	MMI.2	LM2	pseudo-	(A),(B)	(C)
	MMI.3	LM2	mirror mappings	(A),(B),(C)	(D)
21.7.3	RSI.1	LM2R	various	R.1,.3,.5,.6	R.2,.4
	RSI.2	LM2R	pseudo-	R.1,.2,.6	R.3–.5
	RSI.3	LM2Q	reflection sets	R.1–.4,.6	R.5

21.7.1 Independence of betweenness properties

This subsection deals with independence of the various properties of Definition IB.1, which defines a betweenness relation. We name our theorems BI.*n*, suggesting "betweenness independence." We will use Model LM3, based on $\mathbb{F}^3$ as our model, except in the last case, where we use a variant thereof.

Theorem BI.0. *Property B.0 is independent of the other betweenness properties of Definition IB.1.*

Proof. (A) Let r, s, and t be numbers, and let $\mathcal{L} = \overleftrightarrow{(0,0,0)(1,0,0)}$. We define a "pseudo-betweenness" relation PB on $\mathbb{F}^3$ as follows: first, we include in PB all the triples of the betweenness relation defined in Definition LC.8. In the proof below, we will freely use the fact that Properties B.1, B.2, and B.3 hold for this definition, as shown in Theorem LC.14. We also include in the relation PB the two triples

$$((1,0,0),(1,0,0),(2,0,0)) \text{ and } ((2,0,0),(1,0,0),(1,0,0)).$$

Then, according to PB,

(1) point $(1,0,0)$ is between $(1,0,0)$ and $(2,0,0)$,

(2) point $(1,0,0)$ is between $(2,0,0)$ and $(1,0,0)$, but

(3) point $(2,0,0)$ is not between $(1,0,0)$ and $(2,0,0)$.

(B) Property B.0 is false for relation PB since the entries in the ordered triple $((1,0,0),(1,0,0),(2,0,0))$ are not distinct.

(C) Property B.1 (symmetry) is true for relation PB, because it holds for all triples in the relation of Definition LC.8, as well as for the triples

$$(1,0,0)\text{–}(1,0,0)\text{–}(2,0,0) \text{ and } (2,0,0)\text{–}(1,0,0)\text{–}(1,0,0).$$

(D) Property B.2 (trichotomy) is true for Model BI.0. It holds for all collinear triples of distinct points, and holds vacuously for any triple comprised only of the points $(1,0,0)$ and $(2,0,0))$, since these points are not distinct.

(E) Property B.3 is true for relation PB. If $A = (1,0,0)$ and $B = (2,0,0)$, we may let $C = (3,0,0)$; similarly, if $A = (2,0,0)$ and $B = (1,0,0)$, we may let $C = (0,0,0)$; in either case, by the definition of betweenness, A–B–C.

Now suppose no more than one of A or B belongs to $\{(1,0,0),(2,0,0)\}$. Then since B.3 is true for the betweenness relation of Definition LC.8, there exists a C such that A–B–C. □

Theorem BI.1. *Property B.1 is independent of Property B.0.*

Proof. (A) On Model LM3 let $A = (0,0,0)$ and $B = (1,0,0)$ and suppose that the points Y_1, Y_2, and Y_3 belong to $\mathcal{L} = \overleftrightarrow{AB} = \overleftrightarrow{(0,0,0)(1,0,0)}$. Define a pseudo-betweenness relation PC as follows: let

$$Y_1 = A + t_1(B - A),\ Y_2 = A + t_2(B - A), \text{ and } Y_3 = A + t_3(B - A).$$

Then define $Y_1–Y_2–Y_3$ iff $t_1 < t_2 < t_3$.

If no more than one of the points Y_1, Y_2, and Y_3 belongs to $\overleftrightarrow{AB}$, then define $Y_1–Y_2–Y_3$ as in Definition LC.8. Relation PC is then well-defined.

(B) If $Y_1–Y_2–Y_3$ and the points Y_1, Y_2, and Y_3 belong to $\mathcal{L}$, then $t_1 < t_2 < t_3$ and Y_1, Y_2, and Y_3 are distinct and collinear. If no more than one point of Y_1, Y_2, or Y_3 belong to $\mathcal{L}$, then by Definition LC.8 and Remark LC.9 these points are distinct and collinear. Therefore Property B.0 holds for relation PC.

(C) Let $Y_1 = A + 0(B - A)$, $Y_2 = A + 1(B - A)$, and $Y_3 = A + 2(B - A)$. Since $0 < 1 < 2$, $Y_1–Y_2–Y_3$. But it is not the case that $2 < 1 < 0$ so it is false that $Y_3–Y_2–Y_1$. Therefore Property B.1 is false for relation PC. □

Remark BI.1.1. It may be of some slight interest that for relation PC, Properties B.2 and B.3 are false.

(A) Property B.2 says that for any collinear points Y_1, Y_2, and Y_3, exactly one of $Y_1–Y_2–Y_3$, $Y_2–Y_1–Y_3$ or $Y_1–Y_3–Y_2$ is true. Let

$$Y_1 = A + 2(B - A),\ Y_2 = A + 1(B - A), \text{ and } Y_3 = A + 0(B - A).$$

Then neither $2 < 1 < 0$, $1 < 2 < 0$, nor $2 < 0 < 1$ is true, so that none of the conditions in the trichotomy property is true, and Property B.2 does not hold.

(B) Property B.3 says that for any two points Y_1 and Y_2, there exists a third point Y_3 such that $Y_1–Y_2–Y_3$. If we let Y_1 and Y_2 be as defined just above, we see that since $2 < 1$ is false, there can be no number t such that $2 < 1 < t$, and there can be no point Y_3 satisfying $Y_1–Y_2–Y_3$.

Theorem BI.2. *Property B.2 is independent of Properties B.0 and B.1 of Definition IB.1.*

Proof. (A) On Model LM3, let $A = (0, 0, 0)$ and $B = (1, 0, 0)$. Define a pseudo-betweenness relation PD as follows: if Y_1, Y_2, and Y_3 belong to $\mathcal{L} = \overleftrightarrow{AB} = \overleftrightarrow{(0,0,0)(1,0,0)}$, let

$$Y_1 = A + t_1(B - A),\ Y_2 = A + t_2(B - A), \text{ and } Y_3 = A + t_3(B - A),$$

and define $Y_1–Y_2–Y_3$ iff $|t_1| < |t_2| < |t_3|$ or $|t_3| < |t_2| < |t_1|$.

If no more than one of Y_1, Y_2, or Y_3 belongs to $\overleftrightarrow{AB}$, then define $Y_1–Y_2–Y_3$ as in Definition LC.8. Then relation PD is well-defined.

(B) Property B.2 is false. Let

$$Y_1 = A + (-2)(B - A),\ Y_2 = A + 2(B - A), \text{ and } Y_3 = A + 3(B - A).$$

Then Y_1, Y_2, and Y_3 are distinct points on $\mathcal{L}$, but $|-2| = |2|$. Then

$$|-2| < |2| < |3| \text{ and } |3| < |-2| < |2|$$

are false, so that Y_1–Y_2–Y_3 is false;

$$|2| < |-2| < |3| \text{ and } |3| < |-2| < |2|,$$

are false, so that Y_2–Y_1–Y_3 is false; and

$$|-2| < |3| < |2| \text{ and } |2| < |3| < |-2|$$

are false, so that Y_1–Y_3–Y_2 is false. Thus trichotomy does not hold for this particular choice of Y_1, Y_2, and Y_3.

(C) Property B.0 is true for relation PD. If Y_1, Y_2, and Y_3 are members of $\mathcal{L}$ and Y_1–Y_2–Y_3, either $|t_1| < |t_2| < |t_3|$ or $|t_3| < |t_2| < |t_1|$. Then the numbers t_1, t_2, and t_3 are distinct and so Y_1, Y_2, and Y_3 are distinct. If not all of Y_1, Y_2, and Y_3 belong to $\mathcal{L}$ and Y_1–Y_2–Y_3, the points are distinct because Property B.0 is true for the betweenness relation of Definition LC.8.

(D) Property B.1 is true for relation PD. If Y_1, Y_2, and Y_3 are members of $\mathcal{L}$ and Y_1–Y_2–Y_3, either $|t_1| < |t_2| < |t_3|$ or $|t_3| < |t_2| < |t_1|$, and therefore Y_3–Y_2–Y_1. If not all of Y_1, Y_2, and Y_3 belong to $\mathcal{L}$ and Y_1–Y_2–Y_3, then Y_3–Y_2–Y_1 because Property B.1 holds for the betweenness relation of Definition LC.8. □

Remark BI.2.1. Property B.3 is false for relation PD. To see this, let

$$Y_1 = A + (-2)(B - A) \text{ and } Y_2 = A + 2(B - A).$$

Since $|-2| = |2|$, there is no point $Y_3 = A + t(B - A)$ such that $|-2| < |2| < |t|$ or $|2| < |-2| < |t|$, hence no point Y_3 such that Y_1–Y_2–Y_3.

Theorem BI.3 (Model BI). *Property B.3 is independent of the other betweenness properties of Definition IB.1.*

Proof. For Model BI, space is $\mathcal{S} = \mathbb{F}^3 \setminus \overrightarrow{(0,0,0)(1,0,0)}$. $\mathcal{L}$ is a line in $\mathcal{S}$ iff there exists a line $\mathcal{M}$ in $\mathbb{F}^3$ (as in Definition LA.1(1)) such that $\mathcal{L} = \mathcal{M} \cap \mathcal{S}$. $\mathcal{E}$ is a plane in $\mathcal{S}$ iff there exists a plane $\mathcal{P}$ in $\mathbb{F}^3$ (as in Definition LA.1(2)) such that $\mathcal{E} = \mathcal{P} \cap \mathcal{S}$. Betweenness is defined as in Definition LC.8.

Properties B.0, B.1, and B.2 were proved in Theorem LC.14 to be true for betweenness as in Definition LC.8. These proofs also hold for space $\mathcal{S}$. The only property that does not carry over to $\mathcal{S}$ is the extensibility Property B.3; this property does not hold since there is no point W such that $(0,0,0)$ is between $(-1,0,0)$ and W. □

Theorem BI.4 (Summary). (A) *Each of the betweenness Properties B.1, B.2, and B.3 is independent of those preceding it on the list.*

(B) *Property B.0 is independent of Properties B.1, B.2, and B.3.*

Proof. (A) follows immediately from Theorems BI.1, BI.2, and BI.3, and (B) from Theorem BI.0. □

21.7.2 Independence of mirror mapping properties

This subsection deals with independence of the various properties of Definition NEUT.1, which defines *mirror mapping*. We name our theorems MMI.*n*, suggesting "mirror mapping independence." We will use Model LM2, based on $\mathbb{F}^2$, as our model.

Theorem MMI.1. *Property (B) of Definition NEUT.1 is independent of Properties (A), (C), and (D).*

Proof. We show that there exists a mapping $\varphi_{\mathcal{L}}$ on $\mathbb{F}^2$ which satisfies Properties (A), (C), and (D) of Definition NEUT.1, but not Property (B). Let $\mathcal{L}$ be any line in $\mathbb{F}^2$, and define $\varphi_{\mathcal{L}} = \iota$, the identity mapping. Then all points of $\mathcal{L}$ are fixed points for ι, so that Property (A) is true; $\iota \circ \iota = \iota$ so Property (C) is true; and since A–B–C is the same as $\iota(A)$–$\iota(B)$–$\iota(C)$, Property (D) is true. Finally, since all points are fixed points for ι, Property (B) is false. □

Theorem MMI.2. *Property (C) of Definition NEUT.1 is independent of Properties (A) and (B).*

Proof. We show that there exists a mapping $\varphi_{\mathcal{L}}$ on $\mathbb{F}^2$ satisfying Properties (A) and (B) of Definition NEUT.1, but not Property (C). Let $\mathcal{L} = \{(x_1, x_2) \mid x_2 = 0$.

For all (x_1, x_2) such that $x_2 \geq 0$, define $\varphi_{\mathcal{L}}(x_1, x_2)$ to be $\Phi(x_1, x_2)$ as in Definition LB.16; for all (x_1, x_2) such that $x_2 < 0$, define $\varphi_{\mathcal{L}}(x_1, x_2)$ to be $\Psi(x_1, x_2)$ as defined in Exercise NEUT.0.

If $(x_1, x_2) \in \mathcal{L}$, $\varphi_{\mathcal{L}}(x_1, x_2) = \Phi(x_1, x_2) = (x_1, x_2)$ so that Property (A) holds. If $x_2 \geq 0$, then $\varphi_{\mathcal{L}}(x_1, x_2) = \Phi(x_1, x_2) = (x_1, -x_2)$ which is on the side of $\mathcal{L}$ opposite to (x_1, x_2); if $x_2 < 0$, then $\varphi_{\mathcal{L}}(x_1, x_2) = \Psi(x_1, x_2) = (x_1 - x_2, -x_2)$ is on the side of $\mathcal{L}$ opposite to (x_1, x_2). Therefore Property (B) holds.

Finally,

$$\begin{aligned}\varphi_{\mathcal{L}}(\varphi_{\mathcal{L}}(0,1)) &= \Psi(\Phi(0,1)) = \Psi(0,-1) \\ &= (0-(-1), -(-1)) = (1,1) \neq (0,1)\end{aligned}$$

so that Property (C) does not hold. □

Theorem MMI.3. *Property (D) of Definition NEUT.1 is independent of Properties (A), (B), and (C).*

Proof. We show that there exists a mapping $\varphi_{\mathcal{L}}$ on $\mathbb{F}^2$ which satisfies Properties (A), (B), and (C) of Definition NEUT.1, but does not satisfy Property (D). Let $\mathcal{L} = \{(x_1, x_2) \mid x_2 = 0\}$. For all $(x_1, x_2) \in \mathbb{Q}^2 \setminus \mathcal{E}$, where $\mathcal{E} = \{(1,1),(2,1),(1,-1),(2,-1)\}$, define $\varphi_{\mathcal{L}}$ to be Φ as in Definition LB.16; define $\varphi_{\mathcal{L}}(1,1) = (2,-1)$, $\varphi_{\mathcal{L}}(2,1) = (1,-1)$, $\varphi_{\mathcal{L}}(1,-1) = (2,1)$, $\varphi_{\mathcal{L}}(2,-1) = (1,1)$.

If $(x_1, x_2) \in \mathcal{L}$, $\varphi_{\mathcal{L}}(x_1, x_2) = (x_1, x_2)$ so $\varphi_{\mathcal{L}}$ satisfies Property (A) of Definition NEUT.1. If $(x_1, x_2) \notin \mathcal{E}$, (x_1, x_2) and $\varphi_{\mathcal{L}}(x_1, x_2)$ are on opposite sides of $\mathcal{L}$ by Theorem LC.23; a brief inspection shows that points $(x_1, x_2) \in \mathcal{E}$ map to the opposite side of $\mathcal{L}$; therefore Property (B) is satisfied. If $(x_1, x_2) \notin \mathcal{E}$, $\varphi_{\mathcal{L}}(\varphi_{\mathcal{L}}(x_1, x_2)) = (x_1, x_2)$ by Theorem LC.23; if $(x_1, x_2) \in \mathcal{E}$, this follows immediately from the definition of $\varphi_{\mathcal{L}}$ for points of $\mathcal{E}$. Thus $\varphi_{\mathcal{L}}$ satisfies Property (C).

However, Property (D) is not satisfied. For $(0,1)$–$(1,1)$–$(2,1)$, and also $\varphi_{\mathcal{L}}(0,1) = (0,-1)$, $\varphi_{\mathcal{L}}(1,1) = (2,-1)$ and $\varphi_{\mathcal{L}}(2,1) = (1,-1)$; it follows, then, that $(0,-1)$–$(2,-1)$–$(1,-1)$ is false. □

21.7.3 Independence of reflection properties

This subsection deals with independence of the various properties of Definition NEUT.2, which defines *reflection set*. We name our theorems RSI.*n*, suggesting "reflection set independence." In the statements of the theorems "Property R.*n*" will refer to a property of Definition NEUT.2. We will use Model LM2R, based on $\mathbb{R}^2$ as our model, except in the last case, where we use Model LM2Q, based on $\mathbb{Q}^2$.

For the convenience of the reader, we repeat the statement of **Exercise NEUT.0**, which will be used in this subsection: If for each pair (u_1, u_2) of real numbers on the plane $\mathbb{R}^2$, we define $\Phi(u_1, u_2) = (u_1, -u_2)$ and $\Psi(u_1, u_2) = (u_1 - u_2, -u_2)$, then both Φ and Ψ are mirror mappings over the x-axis.

Theorem RSI.1. *Properties R.2 and R.4 are independent of Properties R.1, R.3, R.5, and R.6. That is, there is a set $\mathcal{E}$ of mirror mappings for which R.2 and R.4 are false, but R.1, R.3, R.5 and R.6 are true.*

Proof. Let $\mathcal{E}$ be that set of mirror mappings on $\mathbb{R}^2$ consisting of (A) all LB.16 mirror mappings over lines in $\mathbb{R}^2$, (B) the mapping Ψ defined as follows: for any point

(u_1, u_2), $\Psi(u_1, u_2) = (u_1 - u_2, -u_2)$ (by Exercise NEUT.0 this is a mirror mapping over the line $\mathcal{L} = \overleftrightarrow{(0,0)(1,0)}$ and is different from the LB.16 mapping $\mathcal{R}_{\mathcal{L}}$ over the same line), together with (C) all mirror mappings formed by composition of these mappings.

Then Property R.1 is true and Property R.2 is false, since we have more than one mirror mapping over $\mathcal{L}$. Property R.3 is true because by definition all mirror mappings formed by composition from the members of $\mathcal{E}$ are included in it. Since $\mathcal{E}$ contains all mappings $\mathcal{R}_{\mathcal{L}}$ defined in Definition LC.24, Property R.5 is true by Theorem LC.31. Property R.6 is true by Theorem LC.32.

Property R.4 is false. To see this, let $\mathcal{R}_{\mathcal{L}}$ and Ψ be as defined above. By Property R.5 let $\mathcal{R}_{\mathcal{M}}$ be the angle reflection for the angle $\angle(0,1)(0,0)(1,1)$; this is an LB.16 mapping so it preserves distance.

Then $\mathcal{R}_{\mathcal{M}}(\Psi(\mathcal{R}_{\mathcal{L}}(0,1))) = (0, \sqrt{2})$ and $\mathcal{R}_{\mathcal{M}}(\Psi(\mathcal{R}_{\mathcal{L}}(0,0))) = (0,0)$; The mapping $\mathcal{R}_{\mathcal{M}} \circ \Psi \circ \mathcal{R}_{\mathcal{L}}$ is an isometry which carries $(0,0)$ to $(0,0)$ and $(0,1)$ to $(0, \sqrt{2})$, and both the latter belong to the same ray from $(0,0)$. Therefore Property R.4 does not hold, since $1 \neq \sqrt{2}$. □

Theorem RSI.2. *Properties R.3, R.4, and R.5 are independent of Properties R.1, R.2, and R.6.*

Proof. Let $\mathcal{L} = \overleftrightarrow{(0,0)(1,0)}$. Let the set $\mathcal{E}$ of mirror mappings over lines in $\mathbb{R}^2$ consist of all LB.16 mirror mappings $\mathcal{R}_{\mathcal{M}}$ over lines $\mathcal{M}$ other than $\mathcal{L} = \overleftrightarrow{(0,0)(1,0)}$, together with the map Ψ defined in Exercise NEUT.0, which is a mirror map over $\mathcal{L}$. Then for each point $(u_1, u_2) \in \mathbb{R}^2$, $\Psi(u_1, u_2) = (u_1 - u_2, -u_2))$. In particular, $\Psi(1,-1) = (2,1)$.

I. **Property R.1 and Property R.2 are true** since exactly one mapping is defined for each line on the plane.

II. **Property R.3 is false.** Let $\mathcal{K} = \overleftrightarrow{(0,0)(1,1)}$; then $\mathcal{K}$ is the line of symmetry for $\angle(1,0)(0,0)(0,1)$ since the LB.16 mapping $\mathcal{R}_{\mathcal{K}}$ takes $(1,0)$ to $(0,1)$ and $(0,0)$ to itself. By Theorem NEUT.27, $\varphi = \mathcal{R}_{\mathcal{K}} \circ \Psi \circ \mathcal{R}_{\mathcal{K}}$ is a mirror mapping over $\mathcal{M} = \mathcal{R}_{\mathcal{K}}(\mathcal{L}) = \overleftrightarrow{(0,0)(0,1)}$.

    Then $\mathcal{R}_{\mathcal{K}}(1,-1) = (-1,1)$ so that
$$\begin{aligned}\varphi(1,-1) &= \mathcal{R}_{\mathcal{K}}(\Psi(\mathcal{R}_{\mathcal{K}}(1,-1))) = \mathcal{R}_{\mathcal{K}}(\Psi(-1,1)) \\ &= \mathcal{R}_{\mathcal{K}}(-2,-1) = (-1,-2).\end{aligned}$$
    But $\mathcal{R}_{\mathcal{M}}(1,-1) = (-1,-1) \neq \varphi(1,-1)$, so that $\varphi \neq \mathcal{R}_{\mathcal{M}}$. This shows that Property R.3 is false.

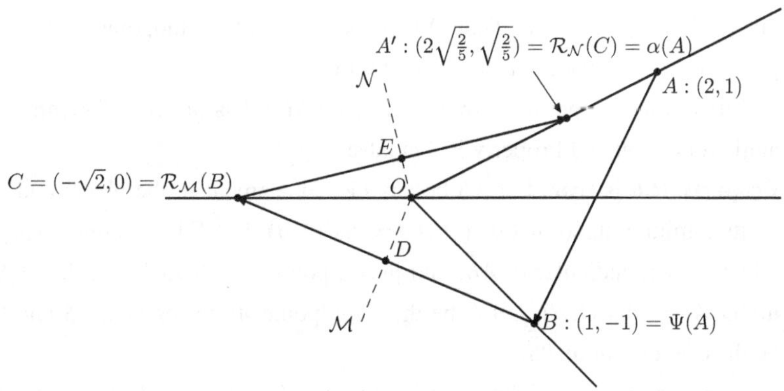

Fig. 21.5 Showing action of α, where Property R.4 fails.

III. **Property R.4 is false.** See figure 21.5. Let $O = (0, 0)$, $A = (2, 1)$, $B = (1, -1) = \Psi(A)$, let $C = (-\sqrt{2}, 0)$, and let $A' = (2\frac{\sqrt{2}}{\sqrt{5}}, \frac{\sqrt{2}}{\sqrt{5}})$ so that
$$\text{dis}(O, B) = \text{dis}(O, C) = \text{dis}(O, A') = \sqrt{2}.$$
Let D be the c-midpoint of $\overline{BC}$ and E be the c-midpoint of $\overline{CA'}$, and let $\mathcal{M} = \overleftrightarrow{OD}$ and $\mathcal{N} = \overleftrightarrow{OE}$. By an argument similar to that in Theorem LC.31, $\mathcal{M}$ is a line of symmetry and $\mathcal{R}_{\mathcal{M}}$ is an angle reflection for $\angle BOC$ and $\mathcal{N}$ is a line of symmetry and $\mathcal{R}_{\mathcal{N}}$ is an angle reflection for $\angle COA = \angle COA'$. Moreover, $\mathcal{R}_{\mathcal{M}}(B) = C$ and $\mathcal{R}_{\mathcal{N}}(C) = A'$.

Now $\mathcal{R}_{\mathcal{N}}(\mathcal{R}_{\mathcal{M}}(B))$ is a point on $\overrightarrow{OA}$. Both $\mathcal{R}_{\mathcal{M}}$ and $\mathcal{R}_{\mathcal{N}}$ preserve distance, so $\text{dis}(\mathcal{R}_{\mathcal{N}}(\mathcal{R}_{\mathcal{M}}(B)), O) = \text{dis}(B, O) = \sqrt{2}$, and therefore $\mathcal{R}_{\mathcal{N}}(\mathcal{R}_{\mathcal{M}}(B)) = A'$. Let $\alpha = \mathcal{R}_{\mathcal{N}} \circ \mathcal{R}_{\mathcal{M}} \circ \Psi$. Then $\alpha(A) = \mathcal{R}_{\mathcal{N}}(\mathcal{R}_{\mathcal{M}}(\Psi(A))) = \mathcal{R}_{\mathcal{N}}(\mathcal{R}_{\mathcal{M}}(B)) = \mathcal{R}_{\mathcal{N}}(C) = A'$ which is a member of $\overrightarrow{OA}$. By Theorem NEUT.15(5), $\alpha(\overline{OA}) = \overline{OA'}$. But $\text{dis}(O, A') = \sqrt{2}$ and $\text{dis}(O, A) = \sqrt{5}$, contradicting Property R.4.

IV. **Property R.5 is false.** Let $X = (1, -1)$ and $Y = (1, 1)$, and $O = (0, 0)$; we show that $\angle XOY$ has no line of symmetry.

If $\angle XOY$ has a line of symmetry, it must be either the line $\mathcal{L} = \overleftrightarrow{(0, 0)(1, 0)}$ or some other line. If $\mathcal{L}$ is a line of symmetry for $\angle XOY$, since Ψ is the mirror mapping over $\mathcal{L}$, Ψ is an angle reflection for this angle. Then $\Psi(X) = \Psi(1, -1) = (2, 1)$ which is not on $\overrightarrow{OX}$, so Ψ is not an angle reflection for $\angle AOB$ contradicting our assumption that $\mathcal{L}$ is a line of symmetry for $\angle XOY$.

If some line $\mathcal{M} \neq \mathcal{L}$ is a line of symmetry for $\angle XOY$, then $\mathcal{R}_{\mathcal{M}}$ is an angle reflection for this angle, and since $\mathcal{R}_{\mathcal{M}}$ is an LB.16 mapping, and $\text{dis}(X, O) =$

$\mathrm{dis}(Y, O)$, $\mathcal{R}_{\mathcal{M}}(X) = Y$. Then $\mathcal{M}$ intersects $\overline{XY}$ at its midpoint, which is the point $(1, 0)$ and hence $\mathcal{M} = \mathcal{L}$, a contradiction.

Either way we get a contradiction, so $\angle XOY$ has no line of symmetry or angle reflection and Property R.5 is false.

V. **Property R.6 is true.** Let $\overline{AB}$ be any closed segment in the plane, and let O be its c-midpoint, so that $\mathrm{dis}(O, A) = \mathrm{dis}(O, B)$. Let $\mathcal{S}$ be the line through O which is perpendicular to $\overleftrightarrow{AB}$, and pick a point S on $\mathcal{S}$ such that $\mathrm{dis}(O, S) = \mathrm{dis}(O, A) = \mathrm{dis}(O, B)$. Let C be the c-midpoint of the segment $\overline{AS}$ and let D be the c-midpoint of $\overline{BS}$.

In the arguments to follow, we must ensure that all our reflections are LB.16 mappings, so we can use preservation of distance. The cases below are needed to eliminate the possibility that $\mathcal{L}$ might appear as a line of symmetry.

Fig. 21.6 For part V showing existence of midpoint O, Case 1A.

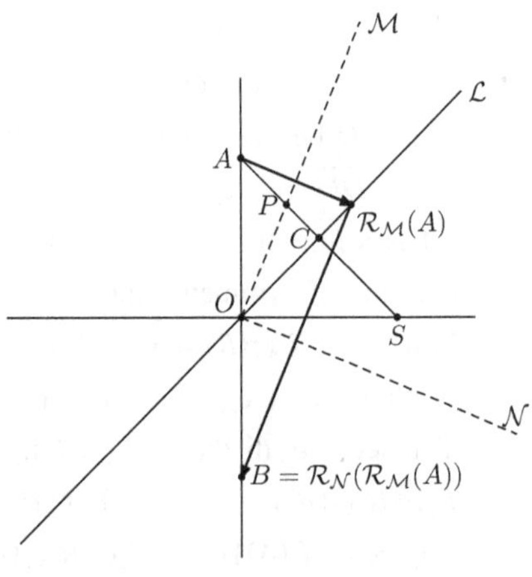

(Case 1A: $O \in \mathcal{L}$ and $C \in \mathcal{L}$.) Let $\mathcal{M}$ be the line of symmetry for $\angle AOC$ and let P be the point of intersection of $\overline{AC}$ with $\mathcal{M}$, which intersection is guaranteed by Theorem PSH.39 (crossbar).

Next let $\mathcal{N}$ be the line of symmetry for $\angle COB$. Since neither $\mathcal{N}$ nor $\mathcal{M}$ is $\mathcal{L}$, the mappings $\mathcal{R}_{\mathcal{N}}$ and $\mathcal{R}_{\mathcal{M}}$ are LB.16 mirror mappings and hence by Theorem LC.27 both of them, and hence their composition, preserve distance. Hence $\mathrm{dis}(A, O) = \mathrm{dis}(\mathcal{R}_{\mathcal{N}}(\mathcal{R}_{\mathcal{M}}(A)), O)$.

$\mathcal{R}_{\mathcal{M}}$ maps A to a point $\mathcal{R}_{\mathcal{M}}(A)$ on $\overrightarrow{OC} \subseteq \mathcal{L}$; $\mathcal{R}_{\mathcal{N}}$ maps $\mathcal{R}_{\mathcal{M}}(A)$ to a point $\mathcal{R}_{\mathcal{N}}(\mathcal{R}_{\mathcal{M}}(A))$ on $\overrightarrow{OB}$. Therefore $\mathcal{R}_{\mathcal{N}}(\mathcal{R}_{\mathcal{M}}(A)) = B$, and since $\mathcal{R}_{\mathcal{N}}(\mathcal{R}_{\mathcal{M}}(O)) = O$, by Theorem NEUT.15(5) $\mathcal{R}_{\mathcal{N}}(\mathcal{R}_{\mathcal{M}}(\overline{OA})) = \overline{OB}$. Thus

O is a midpoint of $\overline{AB}$ and Property R.6 of Definition NEUT.2 is satisfied. See figure 21.6 below.

(Case 1B: $O \in \mathcal{L}$ and $D \in \mathcal{L}$.) Let $\mathcal{M}$ be the line of symmetry for $\angle BOD$, and let P be the point of intersection of $\overline{BD}$ with $\mathcal{M}$.

Then let $\mathcal{N}$ be the line of symmetry for $\angle COB$. Since neither $\mathcal{N}$ nor $\mathcal{M}$ is $\mathcal{L}$, the mappings $\mathcal{R}_{\mathcal{N}}$ and $\mathcal{R}_{\mathcal{M}}$ are LB.16 mirror mappings and hence by Theorem LC.27 both of them, and hence their composition, preserve distance. Hence $\text{dis}(B, O) = \text{dis}(\mathcal{R}_{\mathcal{M}}(\mathcal{R}_{\mathcal{N}}(B)), O)$.

$\mathcal{R}_{\mathcal{M}}$ maps B to a point $\mathcal{R}_{\mathcal{M}}(B)$ on $\overrightarrow{OD} \subseteq \mathcal{L}$; $\mathcal{R}_{\mathcal{N}}$ maps $\mathcal{R}_{\mathcal{M}}(B)$ to a point $\mathcal{R}_{\mathcal{N}}(\mathcal{R}_{\mathcal{M}}(B))$ on $\overrightarrow{OA}$. Therefore $\mathcal{R}_{\mathcal{N}}(\mathcal{R}_{\mathcal{M}}(B)) = A$, and since $\mathcal{R}_{\mathcal{N}}(\mathcal{R}_{\mathcal{M}}(O)) = O$, by Theorem NEUT.15(5) $\mathcal{R}_{\mathcal{N}}(\mathcal{R}_{\mathcal{M}}(\overline{OB})) = \overline{OA}$. Thus O is a midpoint of $\overline{AB}$ and Property R.6 of Definition NEUT.2 is satisfied.

(Case 1C: $O \in \mathcal{L}$ and neither the c-midpoint C of $\overline{AS}$ or the c-midpoint D of $\overline{BS}$ lies on $\mathcal{L}$.) The proof in this case is similar to that of Case 2 below.

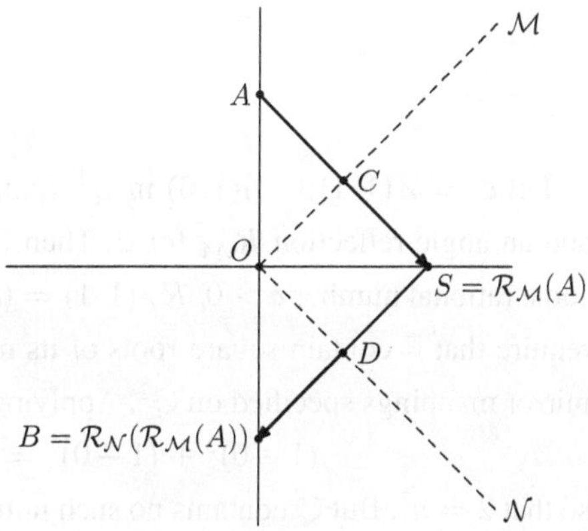

Fig. 21.7 For part V showing existence of midpoint O, Case 2.

(Case 2: $O \notin \mathcal{L}$.) In this case, no line containing O can be $\mathcal{L}$. See figure 21.7 below.

Let $\mathcal{M} = \overleftrightarrow{OC}$ and $\mathcal{N} = \overleftrightarrow{OD}$; then $\mathcal{M}$ is the line of symmetry for $\angle AOS$ and $\mathcal{N}$ is the line of symmetry for $\angle SOB$. Since both $\mathcal{R}_{\mathcal{M}}$ and $\mathcal{R}_{\mathcal{N}}$ are LB.16 mirror mappings, by Theorem LC.27 both of these mappings, and hence their composition, preserve distance. Hence $\text{dis}(A, O) = \text{dis}(\mathcal{R}_{\mathcal{N}}(\mathcal{R}_{\mathcal{M}}(A)), O)$ so that $\mathcal{R}_{\mathcal{N}}(\mathcal{R}_{\mathcal{M}}(A)) = B$, and since $\mathcal{R}_{\mathcal{N}}(\mathcal{R}_{\mathcal{M}}(O)) = O$, by Theorem NEUT.15(5) $\mathcal{R}_{\mathcal{N}}(\mathcal{R}_{\mathcal{M}}(\overline{OA})) = \overline{OB}$, Thus O is a midpoint of $\overline{AB}$ and Property R.6 of Definition NEUT.2 is satisfied. □

Theorem RSI.3. *Property R.5 is independent of all other properties of Definition NEUT.2.*

Proof. Model LM2Q is the 2-dimensional linear model based on $\mathbb{Q}^2$, where $\mathbb{Q}$ is the ordered field of rational numbers. Let $\mathcal{E} = \{\mathcal{R}_{\mathcal{L}} \mid \mathcal{L} \text{ is a line in } \mathbb{Q}^2\}$, that is, the set of all LB.16 mirror mappings over lines in $\mathbb{Q}^2$. The development of the linear model through Theorem LC.32 is valid, with the single exception of Theorem LC.31, which deals with Property R.5. Thus all the properties of Definition NEUT.2 hold for this model, except for Property R.5.

We now show that Property R.5 is false, by showing that there is at least one angle in the plane which has no angle reflection. See Figure 21.8.

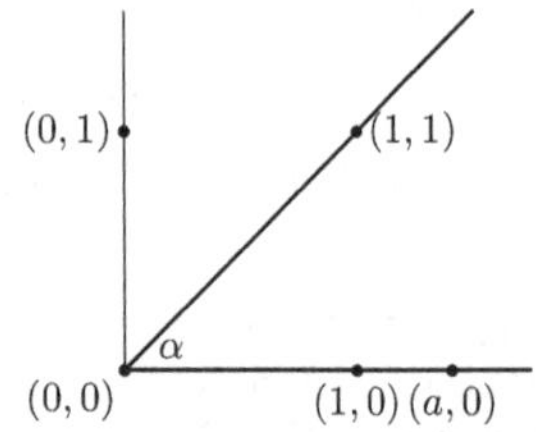

Fig. 21.8 For Theorem RSI.3.

Let $\alpha = \angle(1,1)(0,0)(1,0)$ in $\mathbb{Q}^2$. Suppose there exists a line of symmetry $\mathcal{M}$ and an angle reflection $\mathcal{R}_{\mathcal{M}}$ for α. Then $\mathcal{R}_{\mathcal{M}}(\overrightarrow{(0,0)(1,1)}) = \overrightarrow{(0,0)(1,0)}$, and for some rational number $a > 0$, $\mathcal{R}_{\mathcal{L}}(1,1) = (a,0)$. Note that Exercise LM.16 does not require that $\mathbb{F}$ contain square roots of its non-negative numbers, so is valid for the mirror mappings specified on $\mathbb{Q}^2$. Applying it to $\mathcal{R}_{\mathcal{M}}$, we have

$$(1-0)^2 + (1-0)^2 = (a-0)^2 + (0-0)^2,$$

so that $2 = a^2$. But $\mathbb{Q}$ contains no such number a, so this is impossible. Hence there is no line of symmetry and there is no angle reflection for α. □

Remark RSI.4. (A) It might be tempting to think that the result of Theorem RSI.3 above, together with the proofs in Subsection 21.5.5 that the set of LB.16 mirror mappings on Model LM2Q ($\mathbb{Q}^2$) satisfies all the other reflection properties, shows the independence of Axiom REF. But that is not so; it does not rule out the possibility that there could exist another set of mirror mappings on $\mathbb{Q}^2$ which would satisfy all Properties R.1 through R.6, in which case Axiom REF would hold (*cf* Subsection 21.6.4).

(B) We leave the reader with a challenge to find other independence relationships among the properties of Definition NEUT.2; the ones listed above in Theorems RSI.1–RSI.3 are the ones we have found convenient to prove.

In particular, we would be delighted if it could be shown that Property R.6 is independent of Properties R.1–R.5. We would be even more delighted if the contrary were shown—that is, if it were proved that in the presence of a reflection set having Properties R.1–R.5, every segment on a neutral plane has a midpoint. Our attempts to prove this have not been successful. One of these attempts eventuated in the result showing that midpoints exist for segments in the neutral plane, provided Axiom PW also holds. This result is part of the Supplementary materials, which may be accessed from the home page for this book at www.springer.com.

21.8 Insufficiency of Incidence and Betweenness axioms

As we stated earlier, in the introduction to Section 21.2, this section will show that the incidence and betweenness axioms by themselves (as set forth in Chapter 4) are insufficient to create a satisfactory geometry. To put it more bluntly, IB geometry is not very useful. Most of this section will use two models (DZII and DZIII) based on Model DZI, which was developed in Subsection 21.6.3 above.

Much of our work will have to do with the line $\mathcal{L}$ in Model DZI which contains $(0, 0, 0)$ and $(1, 0, 0)$. We define betweenness on $\mathcal{L}$ by letting $P = (0, 0, 0)$ and $Q = (1, 0, 0)$ in Definition DZI.3. Then for points $A = (a, 0, 0)$, $B = (b, 0, 0)$, and $C = (c, 0, 0)$ on $\mathcal{L}$ (where a, b, and c are integers), A–B–C iff $a < b < c$ or $c < b < a$. If we assign an ordering to $\mathcal{L}$ by specifying that $A < B$ iff $a < b$ in the normal ordering of integers, it becomes apparent that A–B–C iff $a < b < c$ or $c < b < a$ iff a–b–c iff $A < B < C$ or $C < B < A$.

21.8.1 "Property B.4" does not replace Axiom PSA (DZI)

In this subsection we show that in Model DZI, "Property B.4" is true; but denseness does not hold for this model since there are no points of the model between $(0, 0, 0)$ and $(1, 0, 0)$.

This helps complete the argument that even if "Property B.4" had been included in Definition IB.1, the incidence and betweenness axioms would still be inadequate for developing a satisfactory geometry, and its presence would not remove the need for the Plane Separation Axiom (PSA). (*cf* Remark IB.4.2)

We number this theorem as we do to place it in sequence with items DZI.1 through DZI.9 from Subsection 21.6.3.

Theorem DZI.10. *The incidence and betweenness axioms hold for Model DZI, as does "Property B.4," which was proposed but not adopted for Definition IB.1.*

Proof. Theorem DZI.5 (Subsection 21.6.3) shows that the incidence and betweenness axioms hold for Model DZI. Let $A = (a, 0, 0)$, $B = (b, 0, 0)$ and $C = (c, 0, 0)$; then *A–B–C* iff *a–b–c*, *etc.*

Assume that *A–B–C* and *A–C–D*; by the introduction to this section this is true iff ($a < b < c$ or $c < b < a$) and ($a < c < d$ or $d < c < a$). This statement is equivalent to

$$(a < b < c \text{ and } a < c < d) \text{ or } (c < b < a \text{ and } d < c < a),$$

that is to say, ($a < b < c < d$ or $d < c < b < a$); thus *B–C–D*, showing that B.4(a) is true.

A similar proof shows that if *A–B–C* and *B–C–D*, then *A–B–D*; so B.4(b) is true.

□

21.8.2 *Strange results without Axiom PSA (DZII)*

The next model DZII is identical to Model DZI except that the definition of betweenness is altered. We shall refer to the point $(2, 0, 0)$ simply as 2, the point $(3, 0, 0)$ as 3, *etc.*

Definition DZII.1. Let $\mathbb{B}_1$ denote the betweenness relation defined for Model DZI in Definition DZI.3. Define the betweenness relation for Model DZII as

$$\mathbb{B}_2 = (\mathbb{B}_1 \setminus \{(2, 3, 4), (4, 3, 2)\}) \cup \{(3, 2, 4), (4, 2, 3)\}.$$

That is, substitute the triples $(3, 2, 4)$ and $(4, 2, 3)$ for the triples $(2, 3, 4)$ and $(4, 3, 2)$. In yet other words, 3–2–4 and 4–2–3 are true but the other possibilities forbidden by trichotomy are false.

Theorem DZII.2. *The incidence and betweenness axioms hold for Model DZII.*

Proof. The incidence axioms are valid for Model DZII because they are valid for Model DZI. By Theorem DZI.5 (Subsection 21.6.3) the relation $\mathbb{B}_1$ satisfies Properties B.0, B.1, B.2, and B.3 of Definition IB.1. We must show that $\mathbb{B}_2$ satisfies these properties.

(B.0) If $\{A, B, C\}$ is any set of points other than $\{2, 3, 4\}$, and one of the ordered triples (A, B, C), (B, A, C), or (A, C, B) belongs to $\mathbb{B}_2$, then that triple belongs to $\mathbb{B}_1$ so that the points A, B, and C are distinct and collinear. The points 2, 3, and 4 are distinct and collinear, so that any ordered triple belonging to $\mathbb{B}_2$ consists of distinct and collinear points.

(B.1) For every set $\{A, B, C\}$ of collinear points other than $\{2, 3, 4\}$, *A*–*B*–*C* iff *C*–*B*–*A*, since this is true for Model DZI. If $\{A, B, C\} = \{2, 3, 4\}$, and *A*–*B*–*C*, then either $A = 3$, $B = 2$ and $C = 4$, or $A = 4$, $B = 2$ and $C = 3$; whichever it is, the other is true by Definition DZII.1.

(B.2) For every set $\{A, B, C\}$ of collinear points other than $\{2, 3, 4\}$, exactly one of *A*–*B*–*C*, *B*–*A*–*C* and *A*–*C*–*B* is true, because this is the case for Model DZI. For the set $\{2, 3, 4\}$, the statement 3–2–4 is true, while 2–3–4 and 3–4–2 are false; also 4–2–3 is true, while 2–4–3 and 4–3–2 are false. Thus, trichotomy holds for the set $\{2, 3, 4\}$, also.

(B.3) For every set $\{A, B\}$ of distinct points such that not both belong to the line $\mathcal{L}$, Property B.3 holds since it holds for Model DZI. Now suppose both A and B belong to $\mathcal{L}$; if $A < B$, let C be a point on $\mathcal{L}$ such that $C > \max(B, 6)$. The set $\{A, B, C\}$ contains the point C which is not in $\{2, 3, 4\}$ so the points A, B, and C are ordered numerically, that is, $A < B < C$; therefore *A*–*B*–*C*. Similarly, if $A > B$, let C be a point on $\mathcal{L}$ such that $C < \min(B, 1)$; again the set $\{A, B, C\}$ contains the point C which is not in $\{2, 3, 4\}$ so the points A, B, and C are ordered numerically, that is, $A > B > C$; therefore *A*–*B*–*C*. Therefore Property B.3 holds for Model DZII. □

The following theorem shows that "Property B.4" is not a consequence of the other properties of Definition IB.1.

Theorem DZII.3. *Both Property B.4(a) and Property B.4(b) of Definition IB.1 are false for Model DZII.*

Proof. In Model DZII, 1–2–3 and 1–3–4 are both true; if Property B.4(a) were true, then 2–3–4; by Definition DZII.1 this is false.

We know from the definition of Model DZII that 4–2–3; since $\{2,3,5\}$ contains one point not in $\{2,3,4\}$, this triple is ordered numerically, and 2–3–5 is true. If Property B.4(b) were true, 4–2–5, which is false by trichotomy. Therefore Property B.4(b) is false. □

Next we show decisively that things don't "work right" without Axiom PSA.

Theorem DZII.4. *Let $\mathcal{L}$ be the line in Model DZII which contains both* $(0,0,0)$ *and* $(1,0,0)$.

(A) *There exist points A, B, and C on $\mathcal{L}$ such that A–B–C and $\overrightarrow{BA} \cup \overrightarrow{BC} \neq \overleftrightarrow{AC}$. That is, the union of "opposing" rays is not the whole line.*
(B) *There exist points A, B, and C on $\mathcal{L}$ such that A–B–C and $\overline{AB} \cup \overline{BC} \neq \overline{AC}$.*
(C) *There exist points A, B, C, and D on $\mathcal{L}$ such that $A \neq B$ and $\overrightarrow{AC} = \overrightarrow{BD}$.*
(D) *There exist points A, B, and C on $\mathcal{L}$ such that $A \neq B$, $C \in \overleftrightarrow{AB}$, but $\overrightarrow{AC} \neq \overrightarrow{AB}$.*

Proof. (A) By Definition IB.4

$$\overrightarrow{24} = \{2,4\} \cup \{x \mid x \in \mathbb{Z} \text{ and } (2\text{–}x\text{–}4 \text{ or } 2\text{–}4\text{–}x)\}$$
$$= \{2\} \cup \{x \mid x \in \mathbb{Z} \textit{ and } x \geq 4\}.$$

Here we have used the fact that there is no point x such that 2–x–4 (2–3–4 is false by the definition of Model DZII). On the other hand,

$$\overrightarrow{21} = \{1,2\} \cup \{x \mid x \in \mathbb{Z} \text{ and } (1\text{–}x\text{–}2 \text{ or } 2\text{–}1\text{–}x)\};$$
$$= \{1,2\} \cup \{x \mid x \in \mathbb{Z} \textit{ and } x < 1\}.$$

Again, we have used the fact that there is no point x such that 1–x–2. It follows that $\overrightarrow{24} \cup \overrightarrow{21} = \mathbb{Z} \setminus \{3\}$, and the union of these two "opposing" rays is not the whole line.

(B) By Definition IB.3, $\overline{24} = \{2,4\}$, since in Model DZII, 2–3–4 is false. Also, $\overline{45} = \{4,5\}$, because there is no point between 4 and 5. Therefore

$$\overline{24} \cup \overline{45} = \{2,4,5\}.$$

But 2–3–5 and 2–4–5, so that $\overline{25} = \{2,3,4,5\}$, and therefore

$$\overline{24} \cup \overline{45} \neq \overline{25}.$$

Thus if A–B–C, we cannot conclude that $\overline{AC} = \overline{AB} \cup \overline{BC}$.

(C) By Definition IB.4,

$$\overrightarrow{34} = \{3,4\} \cup \{x \mid x \in \mathbb{Z} \text{ and } (3\text{–}x\text{–}4 \text{ or } 3\text{–}4\text{–}x)\};$$
$$= \{2,3,4\} \cup \{x \mid x \in \mathbb{Z} \textit{ and } x > 4\} = \{x \mid x \in \mathbb{Z} \textit{ and } x \geq 2\}.$$

Here we have used the fact that by definition of Model DZII, 3–2–4. On the other hand,

$$\overrightarrow{25} = \{2, 5\} \cup \{x \mid x \in \mathbb{Z} \text{ and } (2\text{–}x\text{–}5 \text{ or } 2\text{–}5\text{–}x)\}.$$

By definition of Model DZII, both 2–3–5 and 2–4–5, so this is

$$= \{2, 3, 4, 5\} \cup \{x \mid x \in \mathbb{Z} \text{ and } 2\text{–}5\text{–}x\} = \{x \mid x \in \mathbb{Z} \textit{ and } x \geq 2\}.$$

Therefore $\overrightarrow{34} = \overrightarrow{25}$, and this ray has two different initial points, 2 and 3.

(D) By Definition IB.4,

$$\overrightarrow{24} = \{2, 4\} \cup \{x \mid x \in \mathbb{Z} \text{ and } (2\text{–}x\text{–}4 \text{ or } 2\text{–}4\text{–}x)\};$$
$$= \{2, 4\} \cup \{x \mid x \in \mathbb{Z} \text{ and } 2\text{–}4\text{–}x\} = \{2\} \cup \{x \mid x \in \mathbb{Z} \textit{ and } x \geq 4\}.$$

On the other hand,

$$\overrightarrow{25} = \{2, 5\} \cup \{x \mid x \in \mathbb{Z} \text{ and } (2\text{–}x\text{–}5 \text{ or } 2\text{–}5\text{–}x)\}.$$

By definition of Model DZII, both 2–3–5 and 2–4–5, so this is

$$= \{2, 5\} \cup \{3, 4\} \cup \{x \mid x \in \mathbb{Z} \text{ and } 2\text{–}5\text{–}x\}$$
$$= \{2, 3, 4, 5\} \cup \{x \mid x \in \mathbb{Z} \textit{ and } x > 5\} = \{x \mid x \in \mathbb{Z} \textit{ and } x \geq 2\}.$$

Since $3 \notin \overrightarrow{24}$, $\overrightarrow{24} \neq \overrightarrow{25}$, even though $4 \in \overrightarrow{25}$ and $5 \in \overrightarrow{24}$. □

21.8.3 Segment and triangle strangeness without Axiom PSA (DZIII)

The next Model DZIII is identical to Model DZI except that the definition of betweenness is altered. As before, we let $\mathbb{B}_1$ denote the betweenness relation defined for Model DZI, as in Definition DZI.3. We shall refer to the point $(2, 0, 0)$ simply as 2, $(3, 0, 0)$ as 3, *etc.*

Definition DZIII.1. Define the betweenness relation for Model DZIII as

$$\mathbb{B}_3 = (\mathbb{B}_1 \setminus \{(2, 3, 4), (4, 3, 2), (3, 4, 5), (5, 4, 3)\})$$
$$\cup\{(3, 2, 4), (4, 2, 3), (3, 5, 4), (4, 5, 3)\}.$$

That is, substitute the triples $(3, 2, 4)$ and $(4, 2, 3)$ for the triples $(2, 3, 4)$ and $(4, 3, 2)$, and substitute the triples $(3, 5, 4)$ and $(4, 5, 3)$ for the triples $(3, 4, 5)$ and $(5, 4, 3)$. In yet other words, 3–2–4, 4–2–3, 3–5–4, and 4–5–3 are true, and the other possibilities forbidden by trichotomy are false.

Theorem DZIII.2. *The incidence and betweenness axioms hold for Model DZIII.*

Proof. The incidence axioms are valid for Model DZIII because they are valid for Model DZI. By Theorem DZI.5 (Subsection 21.6.3) the relation $\mathbb{B}_1$ satisfies Properties B.0, B.1, B.2, and B.3 of Definition IB.1. We must show that $\mathbb{B}_3$ satisfies these properties.

(B.0) If $\{A, B, C\}$ is any set of points other than $\{2, 3, 4\}$ or $\{3, 4, 5\}$, and one of the ordered triples (A, B, C), (B, A, C), or (A, C, B) belongs to $\mathbb{B}_3$, then that triple belongs to $\mathbb{B}_1$ so that the points A, B, and C are distinct and collinear. The points 2, 3, and 4 are distinct and collinear, and the points 3, 4, and 5 are distinct and collinear, so that any ordered triple belonging to $\mathbb{B}_3$ consists of distinct and collinear points.

(B.1) For every set $\{A, B, C\}$ of collinear points other than $\{2, 3, 4\}$ or $\{3, 4, 5\}$, *A–B–C* iff *C–B–A*, since this is true for Model DZI. The same argument as given for Property B.1 in Theorem DZII.2 shows that 3–2–4 iff 4–2–3, and a similar argument shows that 3–5–4 iff 4–5–3.

(B.2) For every set $\{A, B, C\}$ of collinear points other than $(2, 3, 4)$ or $(3, 4, 5)$, exactly one of *A–B–C*, *B–A–C* and *A–C–B* is true, because this is the case for Model DZI. For the set $\{2, 3, 4\}$, the statement 3–2–4 is true, while 2–3–4 and 2–4–3 are false; also 4–2–3 is true, while 4–3–2 and 3–4–2 are false. For the set $\{3, 4, 5\}$, the statement 3–5–4 is true, while 5–3–4 and 3–4–5 are false; also 4–5–3 is true, while 5–4–3 and 4–3–5 are false. Thus, trichotomy holds for these triples, also.

(B.3) For every set $\{A, B\}$ of distinct points such that not both belong to the line $\mathcal{L}$, Property B.3 holds since it holds for Model DZI. Now suppose both A and B belong to $\mathcal{L}$; if $A < B$, let C be a point on $\mathcal{L}$ such that $C > \max(B, 6)$. Since $C \notin \{2, 3, 4, 5\}$, the points A, B, and C are ordered numerically, that is, $A < B < C$; thus *A–B–C*. Similarly, if $A > B$, let C be a point on $\mathcal{L}$ such that $C < \min(B, 1)$. Again, since $1 \notin \{2, 3, 4, 5\}$, the points A, B, and C are ordered numerically, that is, $A > B > C$; thus *A–B–C*. Therefore Property B.3 holds for Model DZIII. □

Remark DZIII.3. In the next theorem we show that in Model DZIII there is a segment with two different sets of end points. Note that if this is true, the two sets of end points must be pairwise disjoint. For suppose the contrary, that A, B, and C are distinct collinear points and $\overline{AC} = \overline{BC}$. Then since $B \in \overline{AC}$, *A–B–C*; also, since $A \in \overline{BC}$, *B–A–C*. This is a contradiction to the trichotomy Property B.2 of Definition IB.1.

Theorem DZIII.4. *Let $\mathcal{L}$ be the line in Model DZIII which contains both* $(0,0,0)$ *and* $(1,0,0)$.

(A) *There exist distinct points A, B, C, and D on $\mathcal{L}$ such that $\overline{AB} = \overline{CD}$. That is, there exists a segment having two different sets of end points.*

(B) *There exist points A, B, C, U, V, and W such that $\{A, B, C\}$ and $\{U, V, W\}$ are noncollinear sets and $\{A, B, C\} \neq \{U, V, W\}$, such that $\triangle ABC = \triangle UVW$. That is, there exists a triangle that has two different sets of corners.*

Proof. (A) $\overline{34} = \{2, 3, 4, 5\}$ because by definition of Model DZIII, 3–2–4 and 3–5–4.

In this model, 2–3–5 and 2–4–5 are as in Model DZI. Therefore $\overline{25} = \{2, 3, 4, 5\} = \overline{34}$. Hence both the sets $\{2, 5\}$ and $\{3, 4\}$ are endpoints for the same segment: a segment does not completely determine its endpoints.

(B) Note first that the point $A = (2, 1, 0)$ is in the plane containing $\mathcal{L}$, and that there is no point of Model DZIII between A and any of the points of the segment defined in part (A) above—that is, between A and any of the points $B = (2, 0, 0)$, $C = (3, 0, 0)$, $D = (4, 0, 0)$, or $E = (5, 0, 0)$. Therefore $\overline{AB} = \{A, B\}$, $\overline{AC} = \{A, C\}$, $\overline{AD} = \{A, D\}$, $\overline{AE} = \{A, E\}$. Also, from part (A), $\overline{BE} = \{B, C, D, E\} = \overline{CD}$.

It follows that

$$\begin{aligned}\triangle ABE &= \overline{AB} \cup \overline{BE} \cup \overline{EA} = \{A, B, C, D, E\}\\ &= \overline{AC} \cup \overline{CD} \cup \overline{DA} = \triangle ABE.\end{aligned}$$

Thus by Definition IB.7, both the sets $\{A, B, E\}$ and $\{A, C, D\}$ are sets of corners for this triangle. Thus, in Model DZIII, specifying a triangle does not completely determine its corners. □

21.9 Exercises for models

Answers to starred (*) exercises may be accessed from the home page for this book at www.springer.com.

Exercise LM.1. Using Definition LA.1(2) complete the proof of Remark LA.2; that is, prove that $\overleftrightarrow{ABC} = \overleftrightarrow{CAB} = \overleftrightarrow{BCA} = \overleftrightarrow{CBA}$.

Exercise LM.2*. Prove Theorem LA.3: distinct points A, B, and C are collinear iff $B-A$ and $C-A$ are linearly dependent.

Exercise LM.3*. Prove Theorem LA.4: distinct points A, B, C, and D in $\mathbb{F}^3$ are coplanar iff $B-A$, $C-A$, and $D-A$ are linearly dependent.

Exercise LM.4*. Prove Theorem LA.5: if A and B are distinct points in $\mathbb{F}^3$, define, for each real number t, $\varphi(t) = A + t(B-A)$. Then φ is a one-to-one mapping of $\mathbb{F}$ onto $\overleftrightarrow{AB}$.

Exercise LM.5.

(A) Prove Theorem LA.15: (A) Two points $A = (a_1, a_2)$ and $B = (b_1, b_2)$ of $\mathbb{F}^2$ are linearly dependent iff $\begin{vmatrix} a_1 & b_1 \\ a_2 & b_2 \end{vmatrix} = 0$. A solution is provided for this part.

(B) Three points $A = (a_1, a_2, a_3)$, $B = (b_1, b_2, b_3)$, and $C = (c_1, c_2, c_3)$ of $\mathbb{F}^3$ are linearly dependent iff $\begin{vmatrix} a_1 & b_1 & c_1 \\ a_2 & b_2 & c_2 \\ a_3 & b_3 & c_3 \end{vmatrix} = 0$.

Exercise LM.6*. Prove Theorem LA.17: Let a, b, c and d be members of $\mathcal{F}$, where at least one of a, b, c is nonzero; let $\mathcal{E}$ be the set of all points $(x_1, x_2, x_3) \in \mathcal{F}^3$ such that $ax_1 + bx_2 + cx_3 + d = 0$, as defined in Remark LA.16.

(A) $\mathcal{E}$ is a proper subset of $\mathcal{F}^3$.

(B) If $X = (x_1, x_2, x_3) \in \mathcal{E}$, there exist two other points $Y = (y_1, y_2, y_3)$ and $Z = (z_1, z_2, z_3)$ in $\mathcal{E}$ such that X, Y, and Z are noncollinear, which is to say (by Theorem LA.3) that the vectors $Y - X$ and $Z - X$ are linearly independent.

Exercise LM.7*. Prove Theorem LA.18: Let $X = (x_1, x_2, x_3)$, $Y = (y_1, y_2, y_3)$, and $Z = (z_1, z_2, z_3)$ be noncollinear points in $\mathbb{F}^3$, so that $\overleftrightarrow{XYZ}$ is a plane as in Definition LA.1(2). Then there exist numbers a, b, c, and d in $\mathbb{F}$, where not all of a, b, or c are zero, such that

$$\overleftrightarrow{XYZ} = \{(w_1, w_2, w_3) \mid aw_1 + bw_2 + cw_3 + d = 0\};$$

Exercise LM.8*. Prove Theorem LA.19: Let a, b, c, and d be numbers in $\mathbb{F}$, where not all of a, b, or c are zero. Then the set

$$\mathcal{E} = \{(w_1, w_2, w_3) \mid aw_1 + bw_2 + cw_3 + d = 0\}$$

is a plane in $\mathbb{F}^3$ as defined by Definition LA.1(2).

Exercise LM.9*. Prove Theorem LB.4: For any numbers a, b, c, a', b', and c in $\mathbb{F}$, where at least one of a or b, and at least one of a' or b' is nonzero, then

(A) $\mathcal{L} = \{(x_1, x_2) \mid ax_1 + bx_2 + c = 0\} \neq \mathbb{F}^2$;
(B) there exist at least two distinct points in $\mathcal{L}$; and
(C) both $ax_1 + bx_2 + c = 0$ and $a'x_1 + b'x_2 + c' = 0$ are equations for $\mathcal{L}$ iff there exists a number $k \neq 0$ such that $a' = ka$, $b' = kb$, and $c' = kc$.

Exercise LM.10*. Prove Theorem LB.5: Let $X = (x_1, x_2)$ and $Y = (y_1, y_2)$ be distinct points in $\mathbb{F}^2$, and let $\overleftrightarrow{XY}$ be the line containing both X and Y according to Definition LA.1(1). Then $\overleftrightarrow{XY} = \{(w_1, w_2) \mid aw_1 + bw_2 + c = 0\}$, where $a = y_2 - x_2$, $b = x_1 - y_1$, and $c = x_2(y_1 - x_1) - x_1(y_2 - x_2)$.

Exercise LM.11. Prove Theorem LB.6: Let a, b, and c be numbers in $\mathbb{F}$, where at least one of a or b is nonzero. Then the set

$$\mathcal{L} = \{(w_1, w_2) \mid aw_1 + bw_2 + c = 0\}$$

is a line in $\mathbb{F}^2$ as defined by Definition LA.1(1).

Exercise LM.12. Let

$$\mathcal{L} = \{(x_1, x_2) \mid a_1x_1 + b_1x_2 + c_1 = 0\} \text{ and}$$
$$\mathcal{M} = \{(x_1, x_2) \mid a_2x_1 + b_2x_2 + c_2 = 0\}$$

be two lines in $\mathbb{F}^2$. Using the equations above, show that if they are c-perpendicular, they must intersect.

Exercise LM.13*. Show that the line $\mathcal{L}$ on $\mathbb{R}^2$ through the distinct points (u_1, u_2) and (v_1, v_2) is

$$\{(x_1, x_2) \mid (x_1, x_2) \in \mathbb{R}^2 \textit{ and } (v_2 - u_2)(x_1 - u_1) - (v_1 - u_1)(x_2 - u_2) = 0\}.$$

Exercise LM.14*. Show that for every member (x_1, x_2) on the line

$$\mathcal{L} = (x_1, x_2) \mid (x_1, x_2) \in \mathbb{R}^2 \textit{ and } ax_1 + bx_2 + c = 0,$$

the formula for $\Phi(x_1, x_2)$ given in Definition LB.16 yields $\Phi(x_1, x_2) = (x_1, x_2)$. For a coordinate-free proof, see Theorem LC.23(A).

Exercise LM.15*. In the plane $\mathbb{F}$, if a line $\mathcal{L}$ is c-perpendicular to a line $\mathcal{M}$ and if line $\mathcal{M}$ and line $\mathcal{N}$ are parallel, then $\mathcal{L}$ is c-perpendicular to line $\mathcal{N}$.

Exercise LM.16*. Let $\mathbb{F}$ be an ordered field, and let $\mathcal{R}_{\mathcal{L}} = \Phi$ be the mapping defined by Definition LB.16 and Definition LC.24 over the line

$$\mathcal{L} - \{(x_1, x_2) \mid (x_1, x_2) \in \mathbb{F}^2 \textit{ and } ax_1 + bx_2 + c = 0\}.$$

where $(a, b) \neq (0, 0)$. Define Γ_1 and Γ_2 to be the mappings such that $\mathcal{R}_{\mathcal{L}}(x_1, x_2) = (\Gamma_1(x_1, x_2), \Gamma_2(x_1, x_2))$. Then if $X = (x_1, x_2)$ and $Y = (y_1, y_2)$ are any points of $\mathbb{F}^2$,

$$\begin{aligned}&\big(\Gamma_1(x_1, x_2) - \Gamma_1(y_1, y_2)\big)^2 + \big(\Gamma_2(x_1, x_2) - \Gamma_2(y_1, y_2)\big)^2 \\ &= (x_1 - y_1)^2 + (x_2 - y_2)^2.\end{aligned}$$

In case $\mathbb{F}$ contains square roots of its non-negative numbers, so that distance is defined, this says that $\text{dis}^2(\mathcal{R}_{\mathcal{L}}(X), \mathcal{R}_{\mathcal{L}}(Y)) = \text{dis}^2(X, Y)$.

Exercise MLT.1*. Prove the uniqueness of the line found in Theorem MLT.3, which passes through both points A and B.

Exercise MLT.2*. Prove Claim 1 of the proof of Theorem MLT.5.

Exercise MLT.3*. Prove that Case 4 of Claim 2 of the proof of Theorem MLT.5 leads to a contradiction.

Exercise MLT.4*. Let $X = (x_1, x_2)$ and $Y = (y_1, y_2)$ be two points in Model MLT, where $x_1 < y_1$, and let d be any real number such that $x_1 < d < y_1$. Then there exists a real number e such that the point $Z = (d, e)$ is the point of intersection of $\mathcal{L}$ and $\overleftrightarrow{XY}_m$; also $Z \in \overline{XY}_m$. This proves that every nonvertical line intersects every vertical line.

Exercise MLT.5*. Prove that in Model MLT, every line parallel to a line of type N is a line of type N.

Exercise MLT.6. Prove that the relation "<" defined (for lines of type N) in part (3) of Definition MLT.1(F) is an order relation according to Definition ORD.1. Note that this proof will involve Model MLT rays, which may lie partly in one side and partly on the other side of the y-axis (and hence don't look like the Model LM2R rays we considered in the text).

References

1. Bachmann, F.: Aufbau der Geometrie aus dem Spiegelungsbegriff, 2nd edn. Grundlehren der mathematischen Wissenschaften. Springer, Berlin (1973). ISBN 978-3642655388
2. Behnke, H., Bachmann, F., Fladt, K., Kunle, H. (eds.): Fundamentals of Mathematics, vol. 2. MIT, Cambridge (1974). ISBN 978-0262020695. Chapter 3, Affine and Projective Planes, R. Lingenberg and A. Bauer; Chapter 4, Euclidean Planes, J. Diller and J. Boczec; also Chapter 5, Absolute Geometry, F. Bachmann, W. Pejas, H. Wolff, and A. Bauer; tr. by S. Gould
3. Birkhoff, G.D., Beatley, R.: Basic Geometry, 2nd edn. AMS Chelsea Publishing Co., New York (2000, Reprint). ISBN 978-0821821015
4. Burton, D.M.: The History of Mathematics: An Introduction, 7th edn. McGraw Hill, New York (2010). ISBN 978-0073383156
5. Doneddu, A.: Étude de géométries planes ordonnées. Rend. Circ. Mat. Palermo Ser. II **19**(1), 27–67 (1970). ISSN 0009-725X
6. Euclid of Alexandria: Elements. The Thirteen Books of Euclid's Elements (3 vols., tr. by Thomas L. Heath, Cambridge University Press, 1925 edn.) (facsimile). Dover, New York (1956). ISBN 978-0486600888 (vol. 1), ISBN 978-0486600895 (vol. 2), ISBN 978-0486600901 (vol. 3); also available in a single volume from Digireads.com. ISBN 978-1420934762
7. Ewald, G.: Geometry: An Introduction. Ishi Press (2013). ISBN 978-4871877183
8. Greenberg, M.J.: Euclidean and Non-Euclidean Geometries, Development and History, 4th edn. W.H. Freeman and Co., New York (2008). ISBN 978-0716799481
9. Halmos, P.R.: Finite Dimensional Vector Spaces, 2nd edn. Springer, New York (1974, 1987). ISBN 978-0387900933
10. Hilbert, D.: Foundations of Geometry, 2nd edn. (tr. by P. Bernays). Open Court, Chicago (1971). Also available via Project Gutenberg at http://www.gutenberg.org/ebooks/17384 in a translation by E.J. Townsend, Open Court (1950). ISBN 978-0875481647
11. Holme, A.: Geometry, Our Cultural Heritage, 2nd edn. Springer, Heidelberg (2010). ISBN 978-3642144400
12. Karzel, H., Sperner, E.: Begründer einer neuen Ordnungstheorie. Mitteilungen der Mathematischen Gesellschaft in Hamburg, Hamburg **25**, 33–44 (2006)
13. Lumiste, U.: Foundations of Geometry. Estonian Mathematical Society, Tartu (2009)
14. Martin, G.E.: Transformation Geometry: An Introduction to Symmetry, Chapter 14. Springer, New York (1982). ISBN 978-0387906362
15. Millman, R.S., Parker, G.D.: Geometry, a Metric Approach with Models, 2nd edn, Chapter 10. Springer, New York (1993). ISBN 978-0387974125

E.J. Specht et al., *Euclidean Geometry and its Subgeometries*,
DOI 10.1007/978-3-319-23775-6

16. Moulton, F.R.: A simple non-desarguesian plane geometry. Trans. Am. Math. Soc. **3**(2), 192Ű-195 (1902). American Mathematical Society, Providence. ISSN 0002-9947, JSTOR 1986419
17. Pasch, M.: Vorlesungen Über Neuere Geometrie (Lectures Over More Recent Geometry), Ulan Press (2012, English reprint) ASIN: B009ZJXQ3A
18. Rosen, K.: Discrete Mathematics and Its Applications, 7th edn. McGraw Hill, Boston (2011). ISBN 978-0073383095
19. Sperner, E.: Die Ordnungsfunktionen einer Geometrie. Mathematische Annalen (Springer) **121**, 107–130 (1949)
20. Tarski, A.: What is elementary geometry? In: Henkin, L., Suppes, P., Tarski, A. (eds.) The Axiomatic Method, with Special Reference to Geometry and Physics. Proceedings of an International Symposium at the Univ. of Calif., Berkeley, 26 December 1957–4 January 1958. Studies in Logic and the Foundations of Mathematics, pp. 16Ű-29. North-Holland, Amsterdam. Brouwer Press (2007, Reprint). ISBN 978-1443728126. MR0106185
21. Tarski, A.: In: Tarski, J. (ed.) Introduction to Logic: And to the Methodology of Deductive Sciences, 4th edn. Dover, New York (1995). ISBN 978-0486284620
22. van der Waerden, B.L.: De logische grondslagen der Euklidische meetkunde (Dutch). Chr. Huygens **13**, 65–84, 257–274 (1934)

Index

E.J. Specht et al., *Euclidean Geometry and its Subgeometries*,
DOI 10.1007/978-3-319-23775-6